COMPUTATIONAL METHODS FOR COMPLEX LIQUID–FLUID INTERFACES

PROGRESS IN COLLOID AND INTERFACE SCIENCE

Series Editors
Reinhardt Miller and Libero Liggieri

Physical-Chemical Mechanics of Disperse Systems and Materials
Eugene D. Shchukin and Andrei S. Zelenev

Computational Methods for Complex Liquid–Fluid Interfaces
Mohammad Taeibi Rahni, Mohsen Karbaschi, and Reinhard Miller

Colloid and Interface Chemistry for Nanotechnology
Peter Kralchevsky, Reinhard Miller, and Francesca Ravera

Drops and Bubbles in Contact with Solid Surfaces
Michele Ferrari, Libero Liggieri, and Reinhardt Miller

Bubble and Drop Interfaces
Reinhardt Miller and Libero Liggieri

Interfacial Rheology
Reinhardt Miller and Libero Liggieri

COMPUTATIONAL METHODS FOR COMPLEX LIQUID–FLUID INTERFACES

Edited by

Mohammad Taeibi Rahni

Mohsen Karbaschi

Reinhard Miller

CRC Press
Taylor & Francis Group
Boca Raton London New York

CRC Press is an imprint of the
Taylor & Francis Group, an **informa** business

CRC Press
Taylor & Francis Group
6000 Broken Sound Parkway NW, Suite 300
Boca Raton, FL 33487-2742

First issued in paperback 2019

© 2016 by Taylor & Francis Group, LLC
CRC Press is an imprint of Taylor & Francis Group, an Informa business

No claim to original U.S. Government works

ISBN-13: 978-1-4987-2208-7 (hbk)
ISBN-13: 978-0-367-37713-7 (pbk)

Visit the Taylor & Francis Web site at
http://www.taylorandfrancis.com

and the CRC Press Web site at
http://www.crcpress.com

Contents

SECTION I Introduction to Interfacial Phenomena

SECTION II Modern Computational Approaches for Analyzing Interfacial Problems

SECTION III   Applied Multiscale Computational Methodologies

SECTION IV Specific Interfacial Physics Simulations

Series Foreword

Liquid interfaces are omnipresent in our daily life and represent essential elements in many technologies and final products. Emulsions and foams, for example, consist, to a large extent, of interfaces, and their behavior as material is controlled by the corresponding interfacial properties. Such practical systems are often dealt with under highly dynamic conditions so that, in particular, fluid dynamics comes into play.

To master dynamic technologies, such as high speed coatings, spraying, and emulsification, a detailed knowledge on the specific behavior of interfaces is required. Often, experiments for their quantitative determination are extremely complex and a clear analysis of data gained by various methods appears impossible without the corresponding computational simulations on molecular, mesoscopic, and macroscopic levels. Many experimental methods in surface science have very strong limitations with respect to dynamic conditions, except when adequate computational simulations are available to complement them.

The present book is the fifth volume in the series Progress in Colloid and Interface Science published by CRC Press. Its title, *Computational Methods for Complex Liquid–Fluid Interfaces*, indicates that it deals with a very broad spectrum of works at liquid interfaces studied experimentally and described by complementary simulations.

The book is edited by the team of editors from both the Max Planck Institute of Colloids and Interfaces in Potsdam, Germany, and the Sharif University of Technology in Tehran, Iran; The editors, Mohammad Taeibi Rahni, Mohsen Karbaschi, and Reinhard Miller, have been involved in the fields of fluid dynamics simulations and experiments for a very long period of time. In addition, they invited many experts from highly credential institutions from all over the world to write about different aspects of the given topics.

First, introductions to surface science dynamics and various types of simulations are presented, spanning over molecular dynamics to macroscale simulations, based on many different conventional and novel computational techniques. The instructive examples are given to the three different levels of simulations, that is, molecular, mesoscale, and macroscale.

The book is meant for a broad spectrum of readers, starting partly with fundamentals of surface science of interfacial layers and their experimental characterization, passing through more tutorial chapters on various levels of simulations to specific applications, which will be useful to specialists in respective fields.

Reinhard Miller
Potsdam, Germany

Libero Liggieri
Genoa, Italy
The Series Editors

Preface

Many important phenomena in nature, science, and technology contain multiphase flows. Some examples of such flows include emulsification, foaming, coating, spray painting, spray combustion, boiling, coal slurry transport, cavitation, sedimentation, fluidized bed, raining, snowing, and volcanic rock motion. Even though there has been a great deal of research conducted in this important field, the complete dynamics of such flows are not yet fully understood due to their complex interphase coupling, whereby different phases strongly affect one another.

Liquid–fluid interfaces are omnipresent in most modern technologies and their quantitative characterization is essential for the optimum use of such technologies. Although an increasing number of specialized related experimental methods exist and many new methods are being developed and improved, the unambiguous data analysis was and still is a bottleneck for further progress in the quantitative understanding of many related phenomena.

In addition, nonequilibrium properties have shown to be very essential in interfacial flows and the interactions between physicochemical properties and hydrodynamics make interfacial dynamics considerably complicated. In particular, processes like emulsification, foaming, or coating require the understanding of highly dynamic properties. Another type of complexity in liquid–fluid interfaces is related to the interactions between the interface and the flow fields of adjacent liquid phases. Thus, any change in the properties of the interface can directly affect the flow fields in both phases. This includes interrelations between the physicochemical properties and hydrodynamics, which usually make interfacial dynamics considerably complicated. Such two-way interactions are of course more complicated in systems containing surface-active molecules, as the mechanisms of transport and adsorption/desorption of surfactant molecules to and from the interface are involved. On the other hand, the quantitative understanding of the mutual dynamical bulk–interface interactions is the main challenge in studying the stability of many systems, such as foams and emulsions. Thus, the properties of the adsorbed layers depend on rather complex mutual bulk–interface interrelations. Especially, the complicated dynamical behaviors existing in both phases (e.g., inside and outside bubbles/drops) have attracted much interest in fundamental and applied research works. This, for example, has led to studying the mechanisms of drop formation and detachment from a capillary tip, which, under certain conditions, are of great interest.

On the other hand, real-world numerical simulations have often been computationally too intensive to predict many flows with large interfaces. However, they have provided necessary insight needed for modeling actual flow problems. In about the past three decades, several numerical works have been performed in which computational fluid dynamics (CFD) has been used. The enormous progress made by various types of such simulations (in molecular, mesoscopic, and macroscopic levels) has provided opportunities for a much better foundation of particularly highly dynamic experimental methods. However, most of these works have not included much details on interfacial thermodynamics, especially variations of molecular properties along the interface, wherein the use of integrated effects can lead to nonphysical solutions.

Of course, if we are dealing with interfacial flows, our mathematical modeling is quite different from multiphase flows in which small interfaces are present, for example, spray flow problems. In interfacial flows, the main concern is to model the transport of liquid–fluid interfaces, which in general are moving and deforming at the same time. One main difference in the mathematical modeling of such flows is having (1) one-fluid (one-field) or (2) two-fluid (two-field) models. In the first type of model, one set of governing equations are used for both phases, while in the latter different sets of equations are used for different phases. Note, the first model type is more common, while in the second one the interface represents a complex boundary condition for both sets of equations used. In conventional CFD techniques of handling large interfaces, the most popular approaches are front tracking, VOF (volume of fluid), and level set methods.

Fortunately, over about the past one and a half decades new CFD ideas, namely, *meshless* methods, such as, smoothed particle hydrodynamics (SPH) method and lattice Boltzmann method (LBM), have been developed and widely used to study interfacial flows. In such methods, the fluid itself is discretized (instead of the media containing it). Then, the mathematical modeling is somehow similar to the one used in molecular dynamics (MD), wherein particle collisions are also taken into account. Due to its numerous advantages, it seems that LBM has been found to be more useful and thus has become more popular in simulations of many interfacial flows. Even though it is not yet fully developed and has some drawbacks for certain flow physics, it seems at the present time that for the simulation of complex liquid–fluid interfaces, it is a rather accurate and relatively simple tool.

The main objective of this book is to highlight the most important computational challenges involved in the two-way coupling of complex liquid–fluid interfaces. In other words, the book is dedicated to present the state of the art of computational methodologies and numerous simulation techniques for the quantification of many interfacial quantities.

The subjects covered in the book are divided into three main aspects of investigations at liquid–fluid interfaces. Section I of the book (Chapters 1 through 5) is dedicated to the fundamentals of liquid–fluid interfaces, that is, the definition of the most important quantities and their experimental investigations. Besides the description of the state of the art of various characterization methods, the theoretical background is described and solutions are presented and discussed.

Section II presents the most important numerical techniques used in interfacial flow problems. It starts with microscale techniques for quantum chemical calculations, treating small systems of molecules by MD computational approaches (Chapter 6). Then, Chapters 7 and 8 cover more modern meshless numerical techniques, that is, SPH and LB methods, which are categorized as state-of-the-art mesoscale CFD techniques. From a macroscale point of view, however, Chapter 9 deals with the most important conventional CFD techniques, which have mostly been used over about the last three decades. These techniques have been tremendously useful in understanding many aspects of interfacial flows.

On the other hand, Sections III and IV deal with different applications of the techniques described in Section II. However, in Section III (Chapters 10 through 18) emphasis is on the technicalities of correctly using the computational techniques developed for interfacial flows, while Section IV (Chapters 19 through 24) emphasizes on the simulation of certain interesting interfacial flow physics, for example, acoustic cavitation-bubble dynamics.

Many people have helped us tremendously during the preparation of this manuscript and of course without their support this work would not have been finalized on time. First, we thank our coauthors, who accepted to work with us and prepared and sent their manuscripts on time. Second, the continuous support of our group members at MPI in Potsdam was a great help in the preparation of the manuscript.

Mohammad Taeibi Rahni

Mohsen Karbaschi

Reinhard Miller

Editors

MATLAB® is a registered trademark of The MathWorks, Inc. For product information, please contact:

The MathWorks, Inc.
3 Apple Hill Drive
Natick, MA 01760-2098 USA
Tel: 508-647-7000
Fax: 508-647-7001
E-mail: info@mathworks.com
Web: www.mathworks.com

Editors

Mohammad Taeibi Rahni received both his BSc and MSc from the University of Texas at Austin and his PhD from the University of Illinois, Urbana-Champaign. He is a guest researcher at the Max Planck Institute of Colloids and Interfaces in Potsdam, Germany, and has a full time faculty member position at Sharif University of Technology, Tehran, Iran. His main area of interest is computational fluid dynamics (CFD), with particular emphasis on the application of various computational techniques to interfacial phenomena.

Mohsen Karbaschi earned his PhD in chemical engineering from Sharif University of Technology, Tehran, Iran, linked to the Max Planck Institute of Colloids and Interfaces in Potsdam, Germany. He has worked as a researcher at the Max Planck Institute since 2011. His primary research interests lie in the area of surface science and multiphase flows, particularly with respect to different applications of computational fluid dynamics simulations and experimental analysis.

Reinhard Miller studied mathematics at the University of Rostock, Rostock, Germany, and did his PhD and habilitation at the Berlin Academy of Sciences, Berlin, Germany. He works at the Max Planck Institute of Colloids and Interfaces in Potsdam, Germany. His scientific interests are experimental investigations of adsorption layers at liquid interfaces under dynamic conditions, interfacial rheology, stability of foams, and emulsions.

Contributors

Eugene V. Aksenenko
Institute of Colloid Chemistry and Chemistry
  of Water
Ukrainian National Academy of Sciences
Kyiv (Kiev), Ukraine

Alidad Amirfazli
Department of Mechanical Engineering
York University
Toronto, Ontario, Canada

Shelley L. Anna
Departments of Chemical Engineering and
  Mechanical Engineering
Center for Complex Fluids Engineering
Carnegie Mellon University
Pittsburgh, Pennsylvania

Masoumeh Azadegan
Electrical Engineering Group
School of Engineering
Tarbiat Modarres University
Tehran, Iran

Dariush Bastani
Chemical and Petroleum Engineering
  Department
Sharif University of Technology
Tehran, Iran

Elena A. Belyaeva
Sechenov Institute of Evolutionary Physiology
  and Biochemistry
Donetsk National Technical University
Donetsk, Ukraine

Thomas Boeck
Institute of Thermodynamics and Fluid
  Mechanics
Technical University of Ilmenau
Ilmenau, Germany

Michael R. Booty
Department of Mathematical Sciences
and
Center for Applied Mathematics and Statistics
New Jersey Institute of Technology
Newark, New Jersey

Dieter Bothe
Center of Smart Interfaces, Mathematical
  Modeling and Analysis
Technical University of Darmstadt
Darmstadt, Germany

Richard A. Campbell
Laue-Langevin Institute
Grenoble, France

Huanchen Chen
Department of Mechanical Engineering
University of Alberta
Edmonton, Alberta, Canada

Mehdi Daemi
Aerospace Engineering Department
Sharif University of Technology
Tehran, Iran

Joël De Coninck
Department of Physics
University of Mons
Mons, Belgium

Kathrin Dieter-Kissling
Center of Smart Interfaces, Mathematical
  Modeling and Analysis
Technical University of Darmstadt
Darmstadt, Germany

Trang Nhu Do
Department of Chemistry
and
Waterloo Institute for Nanotechnology
University of Waterloo
Waterloo, Ontario, Canada

Kerstin Eckert
Institute of Fluid Mechanics
Technical University of Dresden
Dresden, Germany

Valentin B. Fainerman
Medical Physiochemical Center
Donetsk Medical University
Donetsk, Ukraine

Zhi-Gang Feng
Department of Mechanical Engineering
University of Texas at San Antonio
San Antonio, Texas

Sashikumaar Ganesan
Supercomputer Education and Research Centre
Indian Institute of Science
Bangalore, India

Majid Haghshenas
Mechanical Engineering Department
University of Central Florida
Orlando, Florida

Andreas Hahn
Institute for Analysis and Computational
    Mathematics
Otto von Guericke University Magdeburg
Magdeburg, Germany

Mohammad Taghi Hamidi Beheshti
Electrical Engineering Group
School of Engineering
Tarbiat Modarres University
Tehran, Iran

Jari Jalkanen
Forschungszentrum Jülich GmbH
Institute for Advanced Simulation
Jülich Supercomputing Centre
Jülich, Germany

Aliyar Javadi
Max-Planck-Institute for Colloids and
    Interfacial Research
Potsdam, Germany
and
Chemical Engineering Department
University of Tehran
Tehran, Iran

Mohsen Karbaschi
Max-Planck-Institute for Colloids and
    Interfacial Research
Potsdam, Germany

Elena S. Kartashynska
Physical and Organical Chemistry Department
Donetsk National Technical University
Donetsk, Ukraine

Mikko Karttunen
Department of Chemistry
and
Waterloo Institute for Nanotechnology
University of Waterloo
Waterloo, Ontario, Canada
and
Department of Mathematics and Computer
    Science
Institute for Complex Molecular Systems
MetaForum, Eindhoven, the Netherlands

Mehran Kiani
Aerospace Engineering Department
Sharif University of Technology
Tehran, Iran

Thomas Köllner
Institute of Thermodynamics and Fluid
    Mechanics
Technical University of Ilmenau
Ilmenau, Germany

Nina M. Kovalchuk
Department of Chemical Engineering
University of Loughborough
Loughborough, United Kingdom
and
Department of Micro Kinetics of Native
    Disperse Systems
Institute of Biocolloidal Chemistry
National Academy of Sciences of Ukraine
Kyiv (Kiev), Ukraine

Volodja I. Kovalchuk
Department of Micro Kinetics of Native
    Disperse Systems
Institute of Biocolloidal Chemistry
National Academy of Sciences of Ukraine
Kyiv (Kiev), Ukraine

Jürgen Krägel
Max-Planck-Institute for Colloids and
    Interfacial Research
Potsdam, Germany

Christoph Lehrenfeld
Institute for Geometry and Practical
    Mathematics
RWTH Aachen University
Aachen, Germany

Libero Liggieri
CNR—Institute for Energetics and Interphases
Genoa, Italy

Bin Liu
Department of Chemistry
and
Waterloo Institute for Nanotechnology
University of Waterloo
Waterloo, Ontario, Canada

Giuseppe Loglio
Department of Organic Chemistry
University of Florance
Florence, Italy

Marzieh Lotfi
Max-Planck-Institute for Colloids and
    Interfacial Research
Potsdam, Germany
and
Chemical and Petroleum Engineering
    Department
Sharif University of Technology
Tehran, Iran

Holger Marschall
Center of Smart Interfaces, Mathematical
    Modeling and Analysis
Technical University of Darmstadt
Darmstadt, Germany

Hamid Reza Massah
Acoustical Engineering Society
Tehran, Iran

Iman Mazaheri
Mechanical and Aerospace Engineering
    Department
Azad University
Tehran, Iran

Marcel B.J. Meinders
Wageningen University and Research Centre
Top Institute of Food and Nutrition
Wageningen, the Netherlands

Efstathios E. Michaelides
Department of Engineering
Texas Christian University
Fort Worth, Texas

Reinhard Miller
Max-Planck-Institute for Colloids and
    Interfacial Research
Potsdam, Germany

Zahra Mokhtari-Wernosfaderani
Acoustical Engineering Society
Tehran, Iran

Mahmoud Najafi
Mathematics Department
Kent State University
Kent, Ohio

Stefan Odenbach
Institute of Fluid Mechanics
Technical University of Dresden
Dresden, Germany

Chiara Pesci
Center of Smart Interfaces, Mathematical
    Modeling and Analysis
Technical University of Darmstadt
Darmstadt, Germany

Mohammad Taeibi Rahni
Aerospace Engineering Department
Sharif University of Technology
Tehran, Iran
and
Max-Planck-Institut für Kolloid- und
    Grenzflächenforschung
Potsdam, Germany

Arnold Reusken
Institute for Geometry and Practical
    Mathematics
RWTH Aachen University
Aachen, Germany

Massoud Rezavand
Mechanical and Aerospace Engineering
    Department
Azad University
Tehran, Iran

Leonard M.C. Sagis
Laboratory of Physics and Physical Chemistry
    of Food
Department AFSG
Wageningen University
Wageningen, the Netherlands
and
Department of Materials, Polymer Physics
ETH Zurich
Zurich, Switzerland

Karin Schwarzenberger
Institute of Fluid Mechanics
Technical University of Dresden
Dresden, Germany

Michael Siegel
Department of Mathematical Sciences
and
Center for Applied Mathematics and Statistics
New Jersey Institute of Technology
Newark, New Jersey

Kristin Simon
Institute for Analysis and Computational
    Mathematics
Otto von Guericke University Magdeburg
Magdeburg, Germany

Tian Tang
Department of Mechanical Engineering
University of Alberta
Edmonton, Alberta, Canada

Lutz Tobiska
Institute for Analysis and Computational
    Mathematics
Otto von Guericke University Magdeburg
Magdeburg, Germany

Ruud G.M. Van der Sman
Wageningen University and Research Centre
Top Institute Food and Nutrition
Wageningen, the Netherlands

Dieter Vollhardt
Max-Planck-Institute for Colloids and
    Interfacial Research
Potsdam, Germany

Yuri B. Vysotsky
Physical and Organical Chemistry Department
Donetsk National Technical University
Donetsk, Ukraine

Jirasak Wong-ekkabut
Faculty of Science
Department of Physics
Kasetsart University
Bangkok, Thailand

Section I

Introduction to Interfacial Phenomena

1 Thermodynamics of Adsorption at Liquid Interfaces

*Valentin B. Fainerman, Reinhard Miller,
and Eugene V. Aksenenko*

CONTENTS

1.1 INTRODUCTION

The thermodynamic description of adsorption layers is important for the quantitative understandings of liquid/fluid interfaces in equilibrium. A thermodynamic analysis of adsorption layers provides the equation of state that expresses the surface tension (surface pressure) as a function of surface layer composition. Moreover, we get the adsorption isotherm, which determines the dependence of the adsorption of each component as a function of the bulk concentrations. The values of surface tension, adsorption, and dilational viscoelasticity calculated on the basis of these equations can be directly compared with experimental data. Various theoretical models for the description of adsorption layers at liquid interfaces have been proposed so far [1–30], using approaches based on surface thermodynamics. The analysis of chemical potentials of the surface layer components based on the Butler equation [31] turned out to be extremely productive in studies of interfacial layers for various systems [1–6,17–23]. It is very important to note here that the Butler equation for the chemical potential was derived in the framework of a rigorous Gibbs thermodynamics, for example, by Defay and Prigogine [2] and Rusanov [5].

The adsorption isotherm and the equation of state for adsorption layers proposed by Frumkin [14] better describe the adsorption of low molecular weight surfactants for systems that deviate from an ideal (Langmuir) behavior. Differences in the molecular area are of obvious relevance in mixed monolayers. In such a situation, the smaller molecules are increasingly adsorbed with increasing surface pressure. The analytical expression for this general physicochemical principle of Braun–Le Châtelier was first applied by Joos [3] to adsorption layers of surfactants and proteins. It shows that the surface pressure acts as a self-regulation mechanism in adsorption layers.

Various aspects of the Gibbs thermodynamics of interface, including the equations of state for soluble and insoluble surfactants, based on statistical mechanics and assuming the compressibility of adsorbed molecules, the excluded area and phase transitions in the interfacial layer were developed by Rusanov [6,19,20]. The adsorption behavior of proteins at a solution/air or solution/oil interface is rather different from that of low molecular weight surfactants. In particular, proteins can unfold at the interface and their partial molar surface area is larger and decreases with increasing surface pressure.

The concept of chemical potentials of a surface layer was applied to develop new models for the adsorption layer of surfactants. This relates to models assuming molecular reorientations [32], interfacial aggregation [33], and a 2D surface compressibility [34]. This concept was also applied to develop models for mixed surfactants solutions [35], for solutions of proteins [36], and for protein/surfactant mixtures [37,38]. The same method was employed to describe the simultaneous adsorption of surfactants from an aqueous solution and oil molecules from the gaseous oil vapor phase [39] and also the coadsorption of surfactant and oil molecules at the solution/oil interface [40]. The chemical potential method, while assuming another choice of the Gibbs' dividing surface (zero adsorption of solvent), was also used to derive the equations of state for insoluble 2D monolayers (single-component compressible monolayers, which exhibit a first-order or second-order phase transition, mixed monolayer) [41–45].

The aim of this introduction chapter is to derive the equations of state and adsorption isotherms for different surfactant and protein systems using equations for the chemical potential of the components. Some theoretical models will be compared with selected experimental surface tension isotherms.

1.2   THERMODYNAMIC AND CHEMICAL POTENTIALS OF THE SURFACE LAYER

The main results of this paragraph were presented in [33]. At constant pressure P and temperature T, the variation of the free energy G is given by

$$dG = -Ad\gamma + \sum \mu_i^s dm_i^s, \tag{1.1}$$

where
    A is the interfacial area
    γ is the surface tension
    μ_i^s is the chemical potentials
    m_i^s is the excess of moles of component I
    the superscript "s" refers to the surface (interface)

dG is a total differential, and therefore, at constant γ, P and T, for each component j at the interface, the following Maxwell relationship holds:

$$\left(\frac{\partial \mu_j^s}{\partial \gamma} \right)_{m_j^s} = -\left(\frac{\partial A}{\partial m_j^s} \right)_\gamma. \tag{1.2}$$

The values of m_j^s coincide with real numbers of molecules in the monolayer. By definition, the derivative on the right-hand side of Equation 1.2 is the partial surface molar area ω of the jth component:

$$\left(\frac{\partial A}{\partial m_j^s} \right)_\gamma = \omega_j \text{ for constant } \gamma, P, T, \text{ and numbers of molecules } i \neq j. \tag{1.3}$$

The partial surface molar area is constant for the right choice of the position of the dividing surface. It is the parameter of state and possesses all properties of partial values.

The chemical potential μ_i^s of component i in the surface layer depends on the composition of the layer x_i^s and its surface tension. This dependence is given by the following equation:

$$\mu_i^s = \mu_i^{0s}(T,P,\gamma) + RT \ln f_i^s x_i^s. \tag{1.4}$$

The standard chemical potential of component i $\mu_i^{0s}(T,P,\gamma)$ depends on temperature, pressure, surface tension, and the activity coefficients f_i. By introducing an explicit dependence of $\mu_i^{0s}(T,P,\gamma)$ on the surface tension into Equation 1.4, it can be expressed as a function of pressure and temperature only. Together with Equations 1.2 and 1.3, Equation 1.4 transforms into

$$\mu_i^s = \mu_i^{0s}(T,P) - \int_0^\gamma \omega_i d\gamma + RT \ln f_i^s x_i^s. \tag{1.5}$$

In contrast to Equation 1.4, this equation contains a standard chemical potential $\mu_i^{0s}(T,P) = \mu_i^{0s}$ that is independent of surface tension γ. It was shown in [2,5] that assuming ω_i to be also independent of γ and integrating Equation 1.5, we obtain

$$\mu_i^s = \mu_i^{0s} + RT \ln f_i^s x_i^s - \gamma\omega_i. \tag{1.6}$$

This equation is called the Butler equation [31] and can be applied to the concept of Gibbs' dividing surface with positive constant adsorptions values for all components, including the solvent. This convention was proposed by Lucassen-Reynders [21,22] and Joos [18].

However, there is a deficiency in Equation 1.6, pointed out by Rusanov [6,19,20]: in this equation, the values ω_i are constant, while all partial molar areas depend in fact on interfacial tension or adsorption. In a liquid-condensed (LC) insoluble monolayer, the molar area ω can be approximated by a linear dependence on the surface pressure $\Pi = \gamma_0 - \gamma$ [46–48]:

$$\omega = \omega_{01}(1 - \varepsilon\Pi) \tag{1.7}$$

where
γ_0 is the surface tension of the solvent
ω_{01} is the molar area at $\Pi = 0$

The parameter ε is called the 2D relative surface layer compressibility coefficient and characterizes the intrinsic compressibility of molecules in the adsorption layer. The same idea of compressibility of adsorbed molecules has been used in [49–51] for adsorption layers of soluble surfactants. While for an insoluble monolayer in the LC state, the surface coverage θ is close to unity, for surfactant adsorption layer, we must use a better expression in Equation 1.7 [52]:

$$\omega = \omega_{01}(1 - \varepsilon\Pi\theta). \tag{1.8}$$

Equations for the chemical potentials of surface layers can be derived also when taking into account the 2D compressibility of molecules, and with Equation 1.7, one obtains [53]

$$\mu_i^s = \mu_i^{0s} + RT \ln f_i^s x_i^s - \gamma\omega_i^0[1 - \varepsilon\Pi/2]. \tag{1.9}$$

This equation is a modified Butler and can be used for deriving a Frumkin-type equation of state and adsorption isotherms in the framework of the 2D solution theory. In [53], the theoretical isotherms of surface tension $\gamma(c)$ and dilational surface elasticity ($E_0 = d\Pi/d\ln\Gamma$) were calculated using this rigorous model of Equation 1.9 or the approximate model of Equation 1.6. The obtained results were compared with experimental data for different surfactants, and both models agree well with the experimental $\Pi(c)$ and $E_0(c)$ data [54–56].

It is possible to disregard the surface layer compressibility in Equation 1.6, if the inaccuracy thus introduced is "compensated" by small adjustments of other parameters (adsorption equilibrium constant and intermolecular interaction parameter). However, calculations using the Frumkin model and neglecting the surface layer compressibility lead to strong deviations of the calculated dilation viscoelasticity from the experimental data. If the more rigorous Equation 1.8 is used instead of Equation 1.7 (in this case, it is convenient to express θ via the surface pressure Π), the obtained results are very similar. Thus, the thermodynamic model based on the Butler equation can be applied to derive the equations of state and adsorption isotherms. This was successfully demonstrated for a number of systems, for example, in [33,57] and in the subsequent sections of this chapter.

By setting equal expressions of the type of Equation 1.6 for the chemical potentials at the surface to those in the solution bulk, we can derive equations of state and adsorption isotherms [33]:

$$\mu_i^\alpha = \mu_i^{0\alpha} + RT \ln f_i^\alpha x_i^\alpha, \tag{1.10}$$

where

The superscript "α" refers to the bulk solution

$\mu_i^{0\alpha}$ are the standard chemical potentials as functions of pressure and temperature

At equilibrium, we get [33,57]

$$\mu_i^{0s} + RT \ln f_i^s x_i^s - \gamma\omega_i = \mu_i^{0\alpha} + RT \ln f_i^\alpha x_i^\alpha. \tag{1.11}$$

By introducing a standard state (pure liquid), $x_0^s = 1, f_0^s = 1, x_0^\alpha = 1, f_0^\alpha = 1, \gamma = \gamma_0$, and $\mu_0^{0s} - \gamma_0\omega_0 = \mu_0^{0\alpha}$; for infinite dilution $x_i^\alpha \to 0$, we get an additional normalization of the potentials for all components [33,58]:

$$\mu_{0i}^s + RT \ln x_i^s\Big|_{x_i^\alpha \to 0} - \gamma_0\omega_i = \mu_{0i}^\alpha + RT \ln x_i^\alpha\Big|_{x_i^\alpha \to 0} \tag{1.12}$$

and

$$\mu_{0i}^\alpha - \mu_{0i}^s = -\gamma_0\omega_i + RT \ln K_i. \tag{1.13}$$

The parameters $K_i = (x_i^s/x_i^\alpha)_{x_i^\alpha \to 0}$ are the distribution coefficients at infinite dilution. From we can derive [33,57,59]

$$\ln \frac{f_0^s x_0^s}{f_0^\alpha x_0^\alpha} = -\frac{(\gamma_0 - \gamma)}{RT}, \tag{1.14}$$

$$\ln \frac{f_i^s x_i^s/f_{(0)i}^s}{K_i f_i^\alpha x_i^\alpha/f_{(0)i}^\alpha} = -\frac{(\gamma_0 - \gamma)\omega_i}{RT}, \tag{1.15}$$

which are the most general relationships to derive isotherms for nonionic surfactants. After Lucassen-Reynders, the dividing surface can be localized for a two-component system such that the total adsorption of the solvent and surfactant is equal to $1/\omega_1$ [21–23,33], that is,

$$\Gamma_0 + \Gamma_1 = \frac{1}{\omega_1} = \Gamma_1^\infty. \tag{1.16}$$

The definition of the dividing surface for single molecules adsorbing in several adsorption states at the interface of for surfactant mixtures can be given in a more general way [33,60]:

$$\sum_{i=0}^{n} \Gamma_i = \frac{1}{\omega}. \tag{1.17}$$

For this choice of the dividing surface, the part of the surface occupied by a surfactant molecule is equal to its mole fraction in the surfactant mixture at the surface ($x_i^s = \Gamma_i / \sum_{j \geq 0} \Gamma_j \equiv \Gamma_i \omega$). In [5,57,59,61], equations for averaging the molar area ω_i of all adsorbing components have been proposed, such as

$$\omega = \frac{\left(\sum_{i \geq 1} \Gamma_i \omega_i\right)}{\left(\sum_{i \geq 1} \Gamma_i\right)}. \tag{1.18}$$

1.3 ADSORPTION OF SURFACTANTS FROM SOLUTIONS

1.3.1 NONIONIC SURFACTANTS, IONIC, AND MIXTURES

The main equations discussed in this section were presented in [33]. As mentioned earlier, the equation of state for a nonideal surface layer of an ideal bulk solution can be derived from Equation 1.14:

$$\Pi = \frac{RT}{\omega_0} \left(\ln x_0^s + \ln f_0^s \right). \tag{1.19}$$

Combining Equations 1.14 and 1.15, a corresponding adsorption isotherm is obtained:

$$\ln = \frac{f_i^s x_i^s}{K_i x_i^\alpha} = \frac{\omega_i}{\omega_0} \left(\ln x_0^s + \ln f_0^s \right), \tag{1.20}$$

which takes a simple form when for the partial molar areas in Equation 1.19 similar values are considered

$$\Pi = \frac{RT}{\omega_0} \left[\ln \left(1 - \sum_{i>1} \theta_i \right) + \ln f_0^s \right], \tag{1.21}$$

$$K_i x_i = \frac{\theta_i f_i^s}{\left(1 - \sum_{i \geq 1} \theta_i \right)^{n_i} \left(f_0^s \right)^{n_i}}. \tag{1.22}$$

Here, it is assumed that the degree of surface coverage $\theta_j = \Gamma_j \omega_j$ is equal to the molar fraction x_j^s and $n_i = \omega_i / \omega_0$. In the framework of the regular solution theory for nonideal enthalpy, we can calculate the activity coefficients f_i^{sH} in terms of intermolecular interactions (see [61–63]):

$$RT \ln f_k^{sH} = \sum_i \sum_j \left(A_{ik}^s - \frac{1}{2} A_{ij}^s \right) \theta_i \theta_j. \tag{1.23}$$

The coefficients $A_{ij}^s = U_{ii}^s + U_{jj}^s - 2U_{ij}^s, U_{ii}^s,$ and U_{ij}^s are the energies of interaction between the species. For a nonideal entropy of mixing, Lucassen-Reynders [64] derived the following equations:

$$\ln f_j^{sE} = 1 - n_j \sum_i \left(\frac{\theta_i}{n_i} \right). \tag{1.24}$$

Assuming the additivity of enthalpy and entropy in the free energy, we get

$$f_i^s = f_i^{sH} \cdot f_i^{sE} \quad \text{or} \quad \ln f_i^{sH} + \ln f_i^{sE}. \tag{1.25}$$

Then, from Equations 1.21 and 1.22, we can derive a set of equations similar to that proposed in [65]:

$$\Pi = -\frac{RT}{\omega_0} \left[\ln \left(1 - \theta_1 - \theta_2 \right) + \theta_1 \left(1 - \frac{1}{n_1} \right) + \theta_2 \left(1 - \frac{1}{n_2} \right) + a_1 \theta_1^2 + a_2 \theta_2^2 + 2a_{12} \theta_1 \theta_2 \right], \tag{1.26}$$

$$b_i c_i = \frac{\theta_i}{\left(1 - \theta_1 - \theta_2 \right)^{n_i}} \exp \left(-2a_i \theta_i - 2a_{12} \theta_j \right) \cdot \exp \left[\left(1 - n_i \right) \left(a_1 \theta_1^2 + a_2 \theta_2^2 + 2a_{12} \theta_1 \theta_2 \right) \right], \tag{1.27}$$

with $a_1 = A_{01}$, $a_2 = A_{02}$, and $a_{12} = (A_{01} + A_{02} - A_{12})/2$. c_i are here the bulk concentrations and b_i the adsorption equilibrium constants. For an ideal enthalpy of mixing ($a_1 = a_2 = a_{12} = 0$), the given set of equations simplifies to [33]

$$\Pi = \frac{RT}{\omega_0} \left[\ln \left(1 - \theta_1 - \theta_2 \right) + \theta_1 \left(1 - \frac{1}{n_1} \right) + \theta_2 \left(1 - \frac{1}{n_2} \right) \right], \tag{1.28}$$

$$b_i c_i = \frac{\theta_i}{\left(1 - \theta_i - \theta_2 \right)^{n_i}}. \tag{1.29}$$

And for an ideal entropy of mixing of all components in the surface layer, $n_1 = n_2 = 1$, equations are obtained from Equations 1.26 and 1.27, which represent a generalized Frumkin model [33,65,66]:

$$\Pi = -\frac{RT}{\omega_0} \left[\ln \left(1 - \theta_1 - \theta_2 \right) + a_1 \theta_1^2 + a_2 \theta_2^2 + 2a_{12} \theta_1 \theta_2 \right], \tag{1.30}$$

$$b_i c_i = \frac{\theta_i}{\left(1 - \theta_1 - \theta_2 \right)} \exp \left(-2a_i \theta_i - 2a_{12} \theta_j \right). \tag{1.31}$$

When we reduce this set of equation to the case of a single component ($\theta_2 = 0$ and $c_2 = 0$), we get the classical Frumkin model [14]:

$$\Pi = \frac{RT}{\omega_0}\left[\ln(1-\theta) + a\theta^2\right], \qquad (1.32)$$

$$bc = \frac{\theta}{(1-\theta)}\exp(-2a\theta). \qquad (1.33)$$

For ideal surface layers of a solution containing n-components, Equations 1.26 and 1.27 yield the generalized von Szyszkowski–Langmuir equation of state

$$\Pi = -\frac{RT}{\omega_0}\ln\left(1 - \sum_{i\geq 1}\theta_i\right) = \frac{RT}{\omega_0}\ln\left(1 + \sum_{i\geq 1}b_i c_i\right) \qquad (1.34)$$

and the generalized Langmuir adsorption isotherm

$$b_i c_i = \frac{\theta_i}{\left(1 - \sum_{i\geq 1}\theta_i\right)}. \qquad (1.35)$$

So far, we assumed that the partial molar areas of the components or states do not differ essentially. If this does not hold, we obtain [36,64]

$$x_k^s = \frac{\theta_k}{n_k\sum_{i\geq 0}(\theta_i/n_i)}. \qquad (1.36)$$

Then instead of Equation 1.24, the entropic contribution to the activity coefficients reads [36]

$$\ln f_j^{sE} = 1 - n_j\sum_{i\geq 0}\left(\frac{\theta_i}{n_i}\right) + \ln\left[n_j\sum_{i\geq 0}\left(\frac{\theta_i}{n_i}\right)\right]. \qquad (1.37)$$

With Equation 1.37 inserted into Equations 1.19 and 1.20, we can obtain the equations of state and adsorption isotherm for a binary surfacing mixture [36] for a protein adsorbing in a single adsorption state [67].

In Equations 1.32 and 1.33 of the Frumkin type, the constant a reflects the intermolecular (van der Waals) attraction as well as the electrostatic repulsion between adsorption ionic surfactants. In [68], it was shown for the adsorption of sodium dodecyl sulfate (SDS) at the solution/air interface that the degree of counterion binding is up to 94%–96%, while at the solution/oil interface, it is only 88%–90%. Thus, when analyzing the adsorption of ionic surfactants at different interfaces, we can assume that the surface layer is electroneutral [35,69]. In [21,23], Lucassen-Reynders also assumed that the surface is a 2D solution described by Equation 1.6 and applied an electroneutral dividing surface. Therefore, the chemical potentials, μ_{jk}, of electroneutral combinations of ions j and k must not be replaced by electrochemical potentials as they refer only to single ions, and a generalized Butler equation for 1:1 ionic surfactants applies [23,35]:

$$\mu_{jk}^s = \mu_{jk}^{0s} + RT \ln f_j^s f_k^s x_j^s x_k^s - \gamma \omega_{jk}. \tag{1.38}$$

Here, the mole fractions in Equation 1.6 are replaced by the mean ionic product $(x_j^s x_k^s)^{1/2}$ and a convention of the location of the dividing surface results: $\omega_{jk} = 2\omega_0$. The factor 2 is caused by the dissociation of a molecule into two ions, and the constant total amount of surface excess is now $\Gamma_0 + \Gamma_j + \Gamma_k = 1/\omega_0 = 2/\omega_{jk}$.

While the surface equation of state is yet given by Equation 1.26, the distribution of surfactant between the surface and the solution bulk has to be determined for electroneutral combinations of ions, say R and X. When the molar concentration c_i has to be replaced by the average ionic product $(c_R c_X)^{1/2}$, we get an adsorption isotherm identical to the one for the surfactant RX. For an anionic surfactant with R^- being the surface-active ion and X^+ the counterion, plus the inorganic electrolyte XY with the same counterion X^+, we get instead of Equation 1.15 the following relationship [35,59]:

$$\ln \frac{f_R^s f_X^s x_R^s x_X^s / f_{(0)RX}^s}{K_{RX} f_R^\alpha f_X^\alpha x_R^\alpha x_X^\alpha / f_{(0)RX}^\alpha} = -\frac{(\gamma_0 - \gamma)\omega_{RX}}{RT}. \tag{1.39}$$

Equations 1.40 and 1.41 are the equations of state and adsorption isotherm obtained for an electroneutral ($x_R^s = x_X^s$) nonideal surface layer and a nonideal bulk solution of ionic surfactant in the presence or absence of an added electrolyte [35,59]:

$$\frac{\Pi \omega_0}{RT} = \frac{\Pi \omega_{RX}}{2RT} = -\ln(1 - \theta) - a\theta^2 \tag{1.40}$$

$$b\left(c_{R^-} c_{X^+}\right)^{1/2} f_\pm = \frac{\theta}{1 - \theta} \exp\left(-2a\theta\right) \tag{1.41}$$

Here, the surface mole fraction of the surfactant RX is given by $\theta (= 2x_R^s = \Gamma_{RX}/\Gamma_{RX}^\infty)$ with $\Gamma_{RX}^\infty = 1/2\omega_0 = 1/\omega_{RX}$, and the adsorption constant b is defined as $2K_{RX}^{1/2}$. Note that $f_\pm$ are the average activity coefficients in the solution bulk for $c_{X^+} = c_{RX} + c_{XY}$ and $c_{R^+} = c_{RX}$.

For an ideal surface layer, that is, $a = 0$, Equation 1.41 is simplified to

$$\theta = \frac{bf_\pm \left(c_R c_X\right)^{1/2}}{1 + bf_\pm \left(c_R c_X\right)^{1/2}}. \tag{1.42}$$

This leads to the following equation of state [21–23]:

$$\Pi = 2RT\Gamma_{RX}^\infty \ln\left[bf_\pm \left(c_{RX} c_{X^-}\right)^{1/2} + 1\right]. \tag{1.43}$$

In the review article [35], the approach for an electroneutral 2D solution was applied to ionic surfactant mixtures, in particular for mixtures of surfactants of the same charge and for mixtures of anionic and cationic surfactants, where the electric double-layer effects were also negligible.

In mixtures of anionic R^-X^+ and cationic R^+Y^- surfactants [35], extra effects are expected, because the pairs R^-R^+ have a very high surface activity, while the contribution of R^-X^+ and R^+Y^- to the adsorption layer properties is more or less negligible [70]. Neglecting any effects of free surfactant molecules, the adsorption isotherm and equation of state are given by

$$b_{R^+R^-}\left(c_{R^+}c_{R^-}\right)^{1/2} = \frac{\theta_{R^+R^-}}{1-\theta_{R^+R^-}}\exp\left(-2a_{R^+R^-}\theta_{R^+R^-}\right), \tag{1.44}$$

$$\Pi = -\frac{2RT}{\omega_{R^+R^-}}\left[\ln\left(1-\theta_{R^+R^-}\right)+a_{R^+R^-}\theta^2_{R^+R^-}\right]. \tag{1.45}$$

The adsorption constant for the ion pair R^-R^+ can be approximated by the respective adsorption constants for R^-X^+ and R^+Y^- [35]: $b_{R^-R^+}=\left(b_{RX}b_{RY}\right)/V$ with V being the average molar volume of the surfactants at the surface.

The experimental dependencies for individual solutions and mixtures of nonionic surfactants have been extensively presented in [33,57,71]. The results of theoretical calculations exhibit good agreement with the experimental data. One example for solutions of the cationic surfactant C_{14}TAB (tetradecyl trimethyl ammonium bromide) is shown in Figure 1.1. The experimental surface tension dependencies on activity taken from [56,72,73] are compared with values calculated using Equations 1.40 and 1.41. It is seen that the Frumkin model satisfactorily describes these experimental data. If the compressibility of C_{14}TAB molecules is taken into account via Equation 1.8, then the surface tension isotherm obtained in this way coincides with that shown in Figure 1.1 ($\varepsilon=0$). However, when the nonideality of entropy caused by the compressibility is taken into account in the equations of state and adsorption isotherms, then the surface tension isotherms become slightly different [51]. At the same time, if the compressibility coefficient is taken to be $\varepsilon=0.008$ m/mN for the same sets of parameters as in Figure 1.1 ($\omega_{RX}=3.7\cdot10^5$ m²/mol, $b=1.0\cdot10^3$ L/mol, $a=1.4$), then the agreement with the

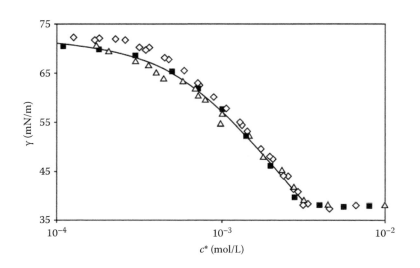

FIGURE 1.1 Dependence of equilibrium surface tension for the C_{14}TAB solutions on the activity c^*: ($\lozenge$), data [73]; ($\triangle$), data [74]; ($\blacksquare$), data [56]; theoretical curve calculated for the Frumkin model; the parameter values are discussed in the text. (Data from Monroy, F. et al., *Colloids Surf. A*, 143, 251, 1998; Schelero, N. et al., *Colloids Surf. A: Physicochem. Eng. Aspects*, 413, 115, 2012; Stubenrauch, C. et al., *J. Phys. Chem. B*, 109, 1505, 2005.)

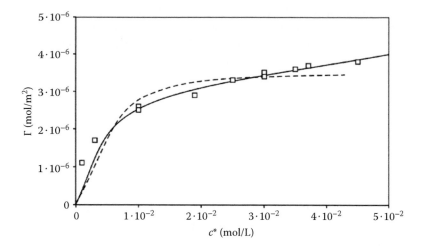

FIGURE 1.2 Dependence of the adsorption of $C_{14}TAB$ on the activity c^* ($\square$) [55]; theoretical dependencies were calculated for the Frumkin model (dotted line) and 2D compressibility model (black line).

experimental dilational elasticity [56] and adsorption [55] on the $C_{14}TAB$ activity becomes much better, which is shown in Figure 1.2. The calculated surface tension isotherms for additional members of the same homologous series, $C_{10}TAB$, $C_{12}TAB$, and $C_{16}TAB$, are presented elsewhere [69].

In Figure 1.3, the equilibrium surface tension isotherms are shown as a function of the mean ion activity $\left(c_{R^+}c_{R^-}\right)^{1/2}$ for 1:1 mixtures of the anionic surfactant Na alkanoate ($CH_3(CH_2)_n COONa$) and the cationic surfactant $C_{14}TAB$ (filled symbols) and for a fixed bulk concentration $C_{14}TAB$ of 10^{-4} mol/L (open symbols) [74]. As one can see, in these coordinates the two dependencies for each alkanoate merge into one curve, in agreement with the experimental results for various ratios of anionic and cationic surfactants in the mixture, as shown in [35]. The theoretical curves were calculated in [74] from Equations 1.44 and 1.45 using the parameters summarized in Table 1.1.

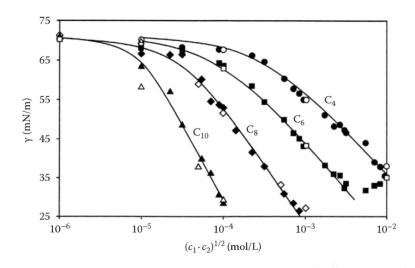

FIGURE 1.3 Surface tension isotherms for mixed solutions of $C_{14}TAB$ and Na alkanoates of different alkyl chain lengths as a function of the mean ion concentration $(c_1 \times c_2)^{1/2}$; C_4—($\blacktriangle$), C_6—($\blacklozenge$), C_8—($\blacksquare$), and C_{10} ($\bullet$); open symbols, 1:1 mixtures; closed symbols, mixtures at a fixed $C_{14}TAB$ concentration 10^{-4} mol/L; black solid line, theoretical curves calculated from Equations 1.44 and 1.45.

TABLE 1.1

Isotherm Parameters of C_{14}TAB Mixed with Na Alkanoates

Alkanoate Chain Length	ω (10^5 m²/mol)	b (10^3 L/mol)	a
C_4	4.6	2.97	0
C_6	4.4	11.1	0
C_8	3.5	23.7	0.2
C_{10}	3.3	40.4	1.2

Source: According to Fainerman, V.B. and Lucassen-Reynders, E.H., *Adv. Colloid Interface Sci.*, 96, 295, 2002.

The formed complexes have a significantly higher adsorption constant b, about 1000 times larger than that for the single alkanoate in absence of C_{14}TAB. For solutions of pure C_{14}TAB, the isotherm parameters were the same as those used for the calculations in Figure 1.1. One can see that at a C_{14}TAB concentration of 10^{-4} mol/L, the surface tension decrease does not exceed 1 mN/m. Only for mixtures with Na butanoate the adsorption constant b is close to the one for pure C_{14}TAB.

1.3.2 COMPRESSIBILITY, REORIENTATION, AND AGGREGATION OF ADSORBED MOLECULES

Equations 1.7 through 1.9 presented in Section 1.2 describe the surfactant molecules' compressibility and provide new options for the understanding and interpretation of the interfacial behavior at higher surface coverage. The high-frequency limit of the elasticity $E_0 = d\Pi/d\ln\Gamma$ can be calculated from the equation of state taking into account the dependence of Γ on Π. For the Frumkin model, Equations 1.32 and 1.33 with $\omega = \omega_0 = $const. we obtain

$$E_0 = \frac{RT}{\omega_0}\left(\frac{\theta}{1-\theta} - 2a\theta^2\right). \tag{1.46}$$

If we assume an intrinsic compressibility, as given by Equation 1.7, we get

$$E_0 = \frac{\dfrac{\theta}{1-\theta} - 2a\theta^2}{\dfrac{\omega_0}{RT} + \dfrac{\varepsilon\theta}{1-\varepsilon\Pi}\left(\dfrac{1}{1-\theta} - 2a\theta\right)}. \tag{1.47}$$

It is interesting to compare the values of the limiting elasticity for $\theta \rightarrow 1$ with each other calculated from different equations. For $\theta \rightarrow 1$, Equation 1.46 yields $E_0 \rightarrow \infty$, which is in contradiction to all experimental results. In contrast, from Equation 1.47, we get

$$E_0 = \frac{1-\varepsilon\Pi}{\varepsilon}. \tag{1.48}$$

For a compressibility of $\varepsilon = 0.005$ m/mN (a usual for common surfactants) and $\Pi = 40$ mN/m, we get from Equation 1.48 a value of about $E_0 = 160$ mN/m, which is in qualitative agreement with experimental data. The results obtained assuming the influence of compressibility of adsorbed molecules for various surfactants and proteins on the surface tension, adsorption, and dilational rheology are summarized in a recent review [75].

The equations describing the reorientation of surfactant molecules in the adsorption layer can be derived from Equations 1.19 and 1.20. Any change in the molecular orientation leads to a variation of the partial molar area ω_i. Taking into account the nonideality of both, enthalpy and entropy, in the surface layer, we obtain for the solvent [76]

$$\ln f_0^s x_0^s = \ln\left(1 - \sum_{i_{min}}^{i_{max}} \Gamma_i \omega_i\right) + 1 - \omega_0 \sum_{i_{min}}^{i_{max}} \Gamma_i + a\left(\sum_{i_{min}}^{i_{max}} \Gamma_i \omega_i\right)^2 \tag{1.49}$$

and for each state j of the molecule in the adsorption layer [33,76]

$$\ln(f_j^s x_j^s / f_{j0}^s) = \ln(\omega_0 \Gamma_j) - \frac{\omega_j}{\omega_0}\left(1 + \sum_{i_{min}}^{i_{max}} (\omega_0 \Gamma_i)\right) + a\frac{\omega_j}{\omega_0}\left[\left(\sum_{i_{min}}^{i_{max}} (\Gamma_i \omega_i)\right)^2 - 1\right]. \tag{1.50}$$

From Equations 1.19, 1.20, and 1.50, we get the respective surface equation of state

$$-\frac{\Pi\omega_0}{RT} = \ln(1 - \Gamma\omega) + \Gamma(\omega - \omega_0) + a(\Gamma\omega)^2 \tag{1.51}$$

and the adsorption isotherm

$$b_i c = \frac{\Gamma_i \omega_0}{(1 - \Gamma\omega)^{\omega_i/\omega_0}} \exp\left(-\frac{\omega_i}{\omega_0}(2a\Gamma\omega)\right). \tag{1.52}$$

Here, $\Gamma = \Sigma_{i\geq1}\Gamma_i$ is the total adsorption, and ω is the molar area averaged over all n states. In [32,33], it was assumed that the adsorption constant increases with increasing molar area ω_i, according to a power law with a constant exponent α. With this, we can relate all b_i to state 1 of the minimum partial molar area:

$$b_i = b_1\left(\frac{\omega_i}{\omega_1}\right)^\alpha. \tag{1.53}$$

When we allow only two coexisting orientations of an adsorbed surfactant molecule at the surface with the molar areas ω_1 and ω_2 (with $\omega_2 > \omega_1$), from Equation 1.52, we get the adsorption isotherms for both adsorption states:

$$bc = \frac{\Gamma_1 \omega_0}{(1 - \Gamma\omega)^{\omega_1/\omega_0}} \exp\left(-\frac{\omega_1}{\omega_0}(2a\Gamma\omega)\right), \tag{1.54}$$

$$bc = \frac{\Gamma_2 \omega_0}{(\omega_2/\omega_1)^\alpha (1 - \Gamma\omega)^{\omega_2/\omega_0}} \exp\left(-\frac{\omega_2}{\omega_0}(2a\Gamma\omega)\right). \tag{1.55}$$

The ratio of these two adsorbed states is

$$\frac{\Gamma_1}{\Gamma_2} = \frac{(\omega_1/\omega_2)^\alpha}{(1-\Gamma\omega)^{(\omega_2-\omega_1)/\omega_0}} \exp\left(-\frac{(\omega_2-\omega_1)}{\omega_0}(2a\Gamma\omega)\right). \tag{1.56}$$

A simplified model for surfactant molecules showing reorientation effects was proposed in [32,33] without taking a nonideality of entropy and enthalpy into account. The ratio of adsorptions for $\alpha = 0$ reads

$$\frac{\Gamma_2}{\Gamma_1} = \exp\left(\frac{\omega_2-\omega_1}{\omega}\right)\exp\left[-\frac{\Pi(\omega_2-\omega_1)}{RT}\right]. \tag{1.57}$$

This expression represents the general physicochemical principle of Braun–Le Châtelier, applied for the first time by Joos [3,77] to adsorption layers. Combined models of reorientation and compression were discussed in [34] and successfully applied to several other surfactant systems [78–80].

The adsorption of binary mixtures of surfactants able to change orientation was described in [81,82] by the following equation of state, similar to Equation 1.26 when assuming $a_i = 0$:

$$-\frac{\Pi\omega_0^*}{RT} = \ln(1-\theta_1-\theta_2) + \theta_1(1-\omega_{10}/\omega_1) + \theta_2(1-\omega_{20}/\omega_2) + 2a\theta_1\theta_2 \tag{1.58}$$

with

$$\omega_0^* = \frac{\omega_{10}\theta_1 + \omega_{20}\theta_2}{\theta_1 + \theta_2}, \tag{1.59}$$

where
   the parameters $\theta_i = \omega_i\Gamma_i$ are the surface coverages
   Γ_i is the adsorptions of component i
   ω_{i0} is the molar area in the state of minimum area at zero surface pressure
   a is the constant of interaction between the two surfactants 1 and 2

This equation is an approximation and implies that the values ω_{i0} are constant and close to each other. The total adsorption of each of the two surfactants Γ_i consists of the adsorptions of both surfactants each two states: $\Gamma_i = \Gamma_{i1} + \Gamma_{i2}$ (the first subscript refers to the surfactant, the second to the adsorption state). Taking ω_{i1} and ω_{i2} as the molar areas of surfactants adsorbed in states 1 and 2 (with $\omega_{i2} > \omega_{i1}$), the surface coverage $\theta_i = \omega_i\Gamma_i = \omega_{i1}\Gamma_{i1} + \omega_{i2}\Gamma_{i2}$ is obtained. The molar areas of the surfactants as functions of surface pressure Π and total surface coverage $\theta = \theta_1 + \theta_2$ read [81,82]

$$\omega_{i1} = \omega_{i0}(1-\varepsilon_i\Pi\theta), \quad (i=1,2) \tag{1.60}$$

(ε_i are the 2D relative surface layer compressibility coefficients). The adsorption isotherms for the states i_1 and i_2 of each of the surfactants i read

$$b_ic_i = \frac{\Gamma_{i1}\omega_{i0}}{(1-\theta)^{\omega_{i1}/\omega_{i0}}} \exp(-2a\theta_j), \quad (i,j=1,2;i \neq j), \tag{1.61}$$

$$b_i c_i = \frac{\Gamma_{i2}\omega_{i0}}{(\omega_{i2}/\omega_{i1})^{\alpha_i}(1-\theta)^{\omega_{i2}/\omega_{i0}}}\exp(-2a\theta_j), \quad (i,j=1,2; i\neq j). \tag{1.62}$$

The equations of state and adsorption isotherms for surfactant mixtures assuming intermolecular interactions (described by a_i) were presented in [83].

For a binary surfactant mixture, assuming a diffusion-controlled matter exchange, the dynamic dilational surface elasticity E is given by [52]

$$E = \frac{1}{B}\left(\frac{\partial\Pi}{\partial\ln\Gamma_1}\right)_{\Gamma_2}\left[\sqrt{\frac{i\Omega}{D_1}}q_{11}+\sqrt{\frac{i\Omega}{D_2}}q_{12}\frac{\Gamma_2}{\Gamma_1}+\frac{i\Omega}{\sqrt{D_1 D_2}}(q_{11}q_{22}-q_{12}q_{21})\right]$$

$$+\frac{1}{B}\left(\frac{\partial\Pi}{\partial\ln\Gamma_2}\right)_{\Gamma_1}\left[\sqrt{\frac{i\Omega}{D_1}}q_{21}\frac{\Gamma_1}{\Gamma_2}+\sqrt{\frac{i\Omega}{D_2}}q_{22}+\frac{i\Omega}{\sqrt{D_1 D_2}}(q_{11}q_{22}-q_{12}q_{21})\right], \tag{1.63}$$

with Ω being the angular frequency of oscillation, D_i the diffusion coefficients of the surfactants, and $B = 1 + \sqrt{i\Omega/D_1}\,q_{11} + \sqrt{i\Omega/D_2}\,q_{22} + (i\Omega/\sqrt{D_1 D_2})\cdot(q_{11}q_{22}-q_{12}q_{21})$. The coefficients $q_{ij}=(\partial\Gamma_i/\partial c_j)|_{c_{k\neq j}}$ are the partial derivatives of the adsorptions $\Gamma_j=\Gamma_j(c_1,c_2)$ of the jth component as functions of c_1 and c_2. The partial derivatives of the surface pressure $[\partial\Pi/\partial(\ln\Gamma_1)]_{\Gamma_2}$ and $[\partial\Pi/\partial(\ln\Gamma_2)]_{\Gamma_1}$ are functions of the adsorptions. The real and imaginary parts of the viscoelasticity $E=E_r+iE_i$ can be obtained from Equation 1.63 as described in [52].

Figure 1.4 shows the experimental equilibrium surface tension isotherm as a function of the total concentration for mixed $C_{12}EO_5/C_{14}EO_8$ solutions at a mixing ratio of 3:1 [81], while the data for the individual solutions of $C_{12}EO_5$ and $C_{14}EO_8$ were taken from [83]. The theoretical isotherms (solid lines) were calculated with the reorientation model using Equations 1.58 through 1.62 and the model parameters summarized in Table 1.2.

One can see that the used model is in a perfect agreement with the experimental data. Note, using the same parameters also the adsorption values $\Gamma(c)$ for the individual solutions agree well with data measured directly by neutron reflection [84].

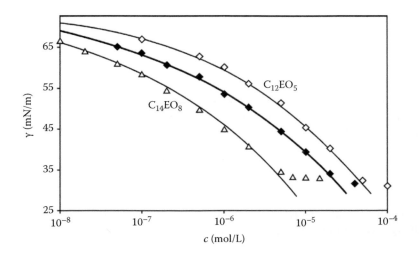

FIGURE 1.4 Equilibrium surface tension isotherms for solutions of $C_{12}EO_5$ ($\diamondsuit$) and $C_{14}EO_8$ ($\triangle$) and their mixtures at a molar mixing ratio of 3:1 ($\blacklozenge$); the theoretical curves were calculated with Equations 1.58 through 1.62 using the parameters given in Table 1.2. (According to Fainerman, V.B. et al., *Langmuir*, 26, 284, 2010.)

TABLE 1.2

Model Parameters for the Two Surfactants Presented in Figure 1.4

	ω_{i1} (m²/mol)	ω_{i2} (m²/mol)	α_i	ε_i (m/mN)	b_i (m³/mol)
$C_{12}EO_5$	$4.2 \cdot 10^5$	$1.0 \cdot 10^6$	2.5	0.006	$5.1 \cdot 10^3$
$C_{14}EO_8$	$5.7 \cdot 10^5$	$1.0 \cdot 10^6$	2.8	0.007	$1.0 \cdot 10^5$

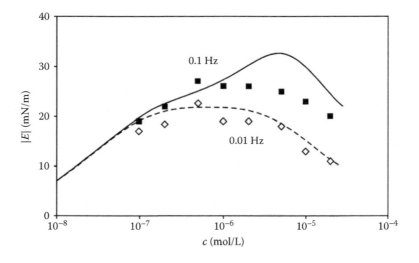

FIGURE 1.5 Dependence of the viscoelasticity modulus $|E|$ on concentration for a 3:1 $C_{12}EO_5/C_{14}EO_8$ mixture at an oscillation frequency of 0.01 Hz ($\Diamond$) and 0.1 Hz ($\blacksquare$); curves calculated using Equations 1.58 through 1.63. (Data taken from Fainerman, V.B. et al., *J. Phys. Chem. B*, 114, 4503, 2010.)

The measured viscoelasticity modulus $|E|$ as a function of concentrations of $C_{12}EO_5$ and $C_{14}EO_8$ in mixed solutions at frequencies 0.01 and 0.1 Hz are shown in Figure 1.5. The theoretical curves were calculated using Equation 1.63 and $D_i = 10^{-9}$ m²/s for both surfactants. To determine the six partial derivatives in Equation 1.63, Equations 1.58 through 1.62 were used. The approximate theoretical model for multicomponent surfactant solutions proposed in [82] provide a good description of the results in Figures 1.4 and 1.5. This is true also for results obtained for $C_{14}EO_8$, $C_{12}EO_5$, and $C_{10}EO_5$ in a wide range of total concentration and oscillation frequencies.

When 2D aggregates are formed at the surface due to interaction between adsorbed surfactant molecules, there exists an equilibrium between monomers and n-mers, which can be described by the following simple relation [85]:

$$\mu_n^s = n\mu_1^s. \tag{1.64}$$

The adsorption of such 2D aggregates (n-mers) is described by the relationship

$$\Gamma_n = K_n \Gamma_1^n \omega^{n-1} \exp\left(\Pi\Delta\omega/RT\right), \tag{1.65}$$

where $K_n = \exp\left\{\left[\left(n\mu_1^o - \mu_n^o\right) - \gamma_0\Delta\omega\right]/RT\right\}$ is the aggregation constant and $\Delta\omega = n\omega_1 - \omega_n$. With $\Delta\omega = 0$, $\omega_0 = \omega_1$, assuming ideal enthalpy of mixing and introducing a critical aggregation adsorption Γ_c, the equation of state for surface layers containing small 2D aggregates takes the form

$$\Pi = -\frac{RT}{\omega_1}\left[\ln\left[1 - \omega_1\Gamma_1\left(1 + n\left(\frac{\Gamma_1}{\Gamma_c}\right)^{n-1}\right)\right] + \omega_1 n\Gamma_1\left(\frac{\Gamma_1}{\Gamma_c}\right)^{n-1}\left(1 - \frac{1}{n}\right)\right], \qquad (1.66)$$

where $n = n_2 = \omega_n/\omega_1$ and $n_1 = \omega_1/\omega_0 = 1$. The corresponding adsorption isotherm reads

$$bc = \frac{\Gamma_1\omega_1}{\left[1 - \Gamma_1\omega_1\left(1 + n\left(\Gamma_1/\Gamma_c\right)^{n-1}\right)\right]}. \qquad (1.67)$$

When large aggregates, so-called clusters ($n \gg 1$), are formed, the approximate equations $\Gamma_1 \cong \Gamma_c$ and $1 + \left(\Gamma_1/\Gamma_c\right)^{n-1} \cong 1$ are valid [46]. The equation of state and adsorption isotherm for such large aggregates were discussed in [33], and extensive theoretical calculations of surface tension dependencies for various aggregation numbers are compared with experimental data obtained for surfactant solutions, which exhibit 2D aggregation and cluster formation at interfaces.

1.3.3 Adsorption of Oil Molecules from the Vapor Phase

The water/oil vapor interface is a situation in between water/air and water/alkane systems. It turned out that the presence of oil molecules in the gas phase can influence the adsorption of surfactants significantly, although oil molecules have no amphiphilic structure. The adsorption of surfactants at the water/oil vapor interfaces is governed by the mutual interaction between surfactant and oil molecules. The oil molecules enhance the surfactant's adsorption from the water bulk. In turn, the adsorption layer of surfactants boosts the oil molecules to assemble from the vapor phase at the interface [39,86–92].

In [91], the simultaneous adsorption of C_nTAB from an aqueous solution and hexane from an air phase saturated with hexane vapor was described by a model originally proposed for the competitive adsorption of surface-active components from liquid bulk phases [40]. In this model, the equation of state for the mixed adsorption layer is described, similarly to the mixture of two surfactants, by Equation 1.30, and the average molar area of the oil (component 1) and the surfactant (component 2) molecules at zero surface pressure is expressed by Equation 1.59. Also a compressibility of surfactant molecules is assumed, which is expressed, similarly to Equations 1.8 or 1.60, by a linear dependence on surface pressure Π and the total surface coverage $\theta = \theta_1 + \theta_2$. For alkane as oil saturated in the vapor phase, the adsorption isotherm reads

$$d_1 P_1 = \frac{\theta_1}{\left(1 - \theta_1 - \theta_2\right)}\exp\left[-2a_1\theta_1 - 2a_{12}\theta_2\right]. \qquad (1.68)$$

The adsorption isotherm for the water-soluble surfactant, similarly to Equation 1.31, is

$$b_2 c_2 = \frac{\theta_2}{\left(1 - \theta_1 - \theta_2\right)}\exp\left[-2a_2\theta_2 - 2a_{12}\theta_1\right], \qquad (1.69)$$

where

The parameters d_1 and b_2 are the surface activity coefficients of the two adsorbing components
P_1 is the partial pressure of alkane vapors
c_2 is the surfactant concentration

The equilibrium surface tension isotherms of $C_{10}EO_8$ solutions in a wide concentrations range in presence of hexane vapor were studied in [89]. A good agreement between the experimental data and the model given earlier was achieved when the constant b_2 was several times larger as compared to its value for the same $C_{10}EO_8$ solutions in the absence of hexane vapor. This fact was ascribed to the coadsorption of hexane from vapor phase at the $C_{10}EO_8$ solution surface leading to an increase in the surface activity of $C_{10}EO_8$. In [91], a systematic study of the adsorption behavior of members of the homologous series C_nTAB ($n = 10$, 12, 14, and 16) at the solution/hexane vapor interface was presented. The quasi-equilibrium surface tensions were compared with results obtained for the solution/air and solution/liquid hexane interfaces. The good agreement with the experimental data confirms a significant increase of C_nTAB adsorption caused by the presence of hexane. This, in turn, leads to a remarkable increase of the adsorption equilibrium constant and to high values of the intermolecular interaction constant.

The dependence of surface tensions on the partial hexane vapor pressure was discussed in [93]. The surface tensions were taken after 1000 s, that is, at adsorption equilibrium but yet before any measurable effect of a condensed hexane film that forms on the aqueous drop surface. For example, the surface tension at maximum vapor pressure of 20,000 Pa was 68.5 mN/m. The model data were calculated with Equations 1.32 and 1.33; the hexane adsorption parameters were estimated to be $d_1 = 1.4 \cdot 10^{-5}$ 1/Pa, $\omega_{10} = 2.0 \cdot 10^5$ m²/mol, $a_1 = 0.7$, $\varepsilon_1 = 0$. Note that the d_1 value is lower than the one obtained in [89]. This difference can be ascribable to the much longer adsorption time of 8000 s in [89]. Therefore, first effects of alkane condensation on the drop surface and formation of liquid films and lenses could have occurred. The calculations performed in [93] were dedicated to the water/hexane vapor interface for the following surfactants: SDS, $C_{12}TAB$ (dodecyl trimethyl ammonium bromide) and $C_{12}DMPO$ (dodecyl dimethyl phosphine oxide). The experiments were done in the same way as for the pure water/hexane vapor interface, that is, 300 s after the formation of a surfactant solution drop during 300 s, the respective hexane/squalene mixture was added to the cell. Mixtures of hexane with squalene allow to tune the partial hexane vapor pressure between 20,000 and 0 Pa. The theoretical dependencies were calculated using the adsorption model described earlier. The experimental data and theoretical results obtained for $C_{12}TAB$ solutions are shown in Figure 1.6 as an example. The calculated data are obtained for the following model parameters for $C_{12}TAB$: $\omega_{20} = 2.5 \cdot 10^5$ m²/mol, $a_2 = 0$, and $\varepsilon_2 = 0.005$ m/mN, while the values for pure hexane were

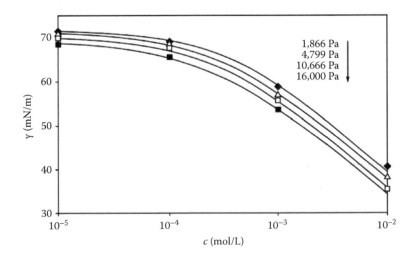

FIGURE 1.6 Interfacial tension of $C_{12}TAB$ solutions as a function of bulk concentrations at different hexane partial vapor pressures; symbols, experimental data; solid curves, calculated with the model using the parameters in Table 1.3.

TABLE 1.3

Parameters b_2 and a_{12} for C_{12}TAB Solutions at Different P_1 Values

P_1 (Pa)	a_{12}	b_2 (m³/mol)
1,866	0.6	2.55
3,466	0.9	2.69
4,799	1.0	3.02
7,066	1.0	3.27
10,666	0.7	3.85
13,333	1.0	4.17
16,000	0.8	4.25

already given in the preceding text. The values for b_2 and a_{12} at different partial pressure P_1 are summarized in Table 1.3.

The value of the hexane vapor pressure influences the C_{12}TAB adsorption. This influence is expressed by the a_{12} value that changes in the range between 0.6 and 1.0. Moreover, the value of b_2 increases by a factor of 1.8 in the studied partial hexane vapor pressure range. Somewhat different results were obtained for other surfactants in [93]. For SDS, we obtain only $a_{12} = 0.2$, but b_2 increased by a factor of 3 in the same studied range of partial hexane vapor pressures. For C_{12}DMPO, $a_{12} = 2.5$, while the b_2 values were virtually the same for all partial vapor pressures of hexane and by a factor of 25–40 larger than those obtained for C_{12}TAB [93].

Figure 1.7 illustrates the dependence of the coverage of the adsorption layer by C_{12}TAB and hexane molecules as a function of the C_{12}TAB concentration. One can see that with increasing partial hexane vapor pressure, its adsorptions become higher, and also the C_{12}TAB adsorption increases insignificantly. Also shown in Figure 1.7 for the sake of comparison are the coverage values for C_{12}DMPO solutions at the maximum partial pressure of hexane vapors. Strong intermolecular attractions lead to the remarkable increase in the adsorption of both the hexane and C_{12}DMPO molecules at low surfactant bulk concentrations.

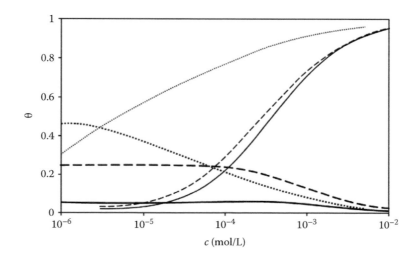

FIGURE 1.7 Dependence of the adsorption layer coverage θ by surfactants molecules (thin lines) and hexane molecules (bold lines) on the surfactant concentrations. In the solutions of C_{12}TAB: solid lines, hexane vapor pressure 3,466 Pa; dashed lines, hexane vapor pressure 16,000 Pa. In the solutions of C_{12}DMPO: dotted lines, hexane vapor pressure 16,000 Pa.

1.3.4   COMPETITIVE ADSORPTION OF ALKANE AND SURFACTANT AT THE WATER/OIL INTERFACE

The thermodynamics and kinetics of adsorption and the dilational rheology of surfactants at the aqueous solution/alkane (oil) interface have been studied in many papers [28,30,94–106]. Traditional theories derived for the water/air interface were used and specific interactions between the surfactant and oil molecules were assumed not to affect the surfactant adsorption from the aqueous solution. The approach proposed in [40] represents a new thermodynamic picture for describing the adsorption of surfactants at liquid/liquid interfaces. This new approach is based on the physical picture that the oil molecules are part of the adsorption layers and compete with the adsorbing surfactant molecules.

We want to consider first the results obtained for the adsorption of C_nTAB at the aqueous solution/alkane interface. The equilibrium interfacial tension isotherms for aqueous solutions of various C_nTABs ($n = 10$, 12, 14, and 16) in phosphate buffer at the solution/hexane interfaces are shown in Figure 1.8.

The symbols represent the experimental data, while the dashed curves are theoretical calculations of the adsorption isotherms of C_nTABs without considering the coadsorption of the alkane (so-called Frumkin ionic compressibility [FIC] model). The FIC model assumes, however, the intrinsic compressibility of surfactant molecules at the interface, the dissociation of C_nTAB into ions, and the presence of added inorganic electrolyte [98]. This means that the calculations were performed using Equations 1.7, 1.40, and 1.41, while the activity coefficients were calculated as proposed in [56]. The interfacial tension of the pure water/hexane interface was 50 mN/m. The model parameters used in the calculations are given in [98].

A second theoretical model used to describe the data in Figure 1.8 consists of the set of Equations 1.32, 1.33, 1.59, and 1.60 assumes that the oil phase does provide only a hydrophobic environment for the alkyl chains of the adsorbing surfactant molecules at the interface, but the alkane molecules themselves compete for the space at the interface. In this model, the compressibility given by Equation 1.60 applies only to the adsorbed surfactant molecules. The molar concentration of hexane c_1 can be calculated from its molecular mass and density, which for 100% hexane is $c_1 = 7.58$ mol/L. The value of b_1 can then be calculated via the surface tension decrease from the value for water/air down to the value for the pure water/hexane interface. Assuming $\omega_1 = 3.3 \cdot 10^5$ m^2/mol and $\varepsilon_1 = 0$,

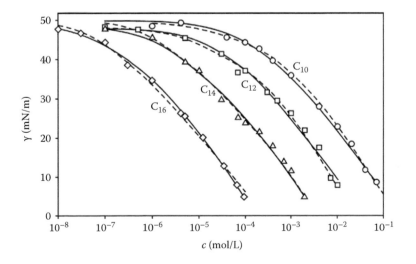

Figure 1.8   Interfacial tension isotherms of c_nTAB at the solution/hexane interface in the presence of phosphate buffer (10 mM, pH 7); symbols are experimental data taken from Pradines, V. et al., *Colloids surf.* A, 371, 22, 2010; dashed lines were calculated from Equations 1.7, 1.40, and 1.41; solid lines were calculated from Equations 1.32, 1.33, 1.60, and 1.61 using the parameters listed in Table 1.4.

TABLE 1.4

Summary of Parameters Involved in Equations 1.32, 1.33, 1.58, and 1.60 for Aqueous Solutions of C_nTAB at the Water/Hexane Interface with a Compressibility Coefficient for All Surfactants of $\varepsilon_2 = 0.005$ m/mN

C_nTAB	b_2 (10^5 L/mol)	ω_{20} (10^5 m²/mol)	a_2	a_{12}
C_{10}TAB	0.43	3.45	0.8	1.1
C_{12}TAB	2.0	3.5	1.3	1.4
C_{14}TAB	15.0	3.35	1.0	1.7
C_{16}TAB	120	3.2	1.6	1.7

we obtain $b_1 = 2.6$ L/mol. The equilibrium interfacial tension isotherms for aqueous solutions of some C_nTABs at the solution/hexane interfaces given in Figure 1.8 are complemented by curves calculated with the coadsorption model (solid curves), using the model parameters summarized in Table 1.4. It is seen from Figure 1.8 that a good agreement exists between the calculated data and experimental points. Note that due to the very high molar concentration of alkane, their adsorption process is very fast. Moreover, the adsorption of alkane molecules leads only to a minor variation of the interfacial structure at very short times, as it was mentioned in [2]. This time is probably of the order of 10^{-4} s or less and experimentally not accessible.

The chain length of the oil phase can have an influence on the adsorption properties of the surfactants. The effect of the oil phase on the adsorption behavior of surfactants at liquid/liquid interfaces was studied for aqueous solutions of C_nTAB at the interfaces to alkanes of different chain lengths [95,104–106]. In [106], interfacial tension measurements have been performed for C_{10}TAB and C_{12}TAB at the interface between water (pH 7) and hexane, heptane, octane, nonane, decane, dodecane, and tetradecane. Figure 1.9 illustrates these experimental data and the results of theoretical calculations (black lines) using Equations 1.7, 1.40, and 1.41, assuming the presence of inorganic electrolyte for the case of the solution/octane and solution/tetradecane interfaces. The model parameters for the coadsorption of C_{12}TAB and the studied alkanes are listed in Table 1.5 as example.

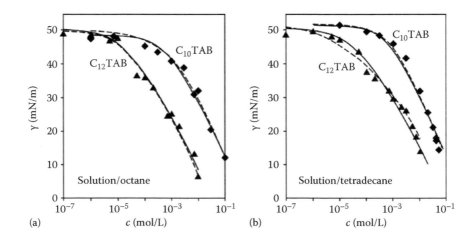

FIGURE 1.9 Interfacial tension isotherms of C_{10}TAB (◆) and C_{12}TAB (▲) at the solution/alkane interface in the presence of phosphate buffer (10 mM, pH 7): (a) solution/octane and (b) solution/tetradecane. Dashed lines are calculated from Equations 1.7 and 1.40 и 1.41 in Mucic, N. et al., *J. Colloid Interface Sci.*, 410, 181, 2013; solid lines are calculated from Equations 1.32, 1.33, 1.60, and 1.61 with the parameters listed in Table 1.5.

TABLE 1.5

Values of Parameters Involved in Equations 1.32, 1.33, 1.58 and 1.60 for Aqueous Solutions of C_{12}TAB at the Interface with Different Alkanes; the Compressibility C_{12}TAB Was Taken to Be $\varepsilon_2 = 0.005$ m/mN, $\omega_1 = 3.3 \cdot 10^5$ m²/mol, and $\omega_{20} = 3.5 \cdot 10^5$ m²/mol for All Alkanes

Alkane	c_1 (mol/L)	b_1 (L/mol)	b_2 (10^5 L/mol)	a_2	a_{12}
C_7	6.84	2.7	2.15	0.9	1.2
C_8	6.16	3.0	3.15	0.9	1.1
C_9	5.61	3.3	2.0	0.9	0.8
C_{10}	5.14	3.5	3.2	0.9	1.4
C_{12}	4.41	4.3	1.7	0.6	1.5
C_{14}	3.93	4.6	2.5	0.2	0.8

It is seen from Figure 1.9 that both models provide good agreement with the experiments, and for the solution/tetradecane interface, the results obtained from the coadsorption model is even significantly better. It was noted in [69] that the adsorption of surfactant molecules leads to a squeezing out of alkane molecules from the interfacial layer. However, these alkane molecules squeezed into the sublayer can also affect the surface tension value, that is, contributes additively to the surface pressure. A corresponding correction of the equation of state for such a mixed layer was proposed in [69].

As an example for a mixed surfactant/alkane layer the coverage of the adsorption layer by octane and C_nTAB molecules at the octane/solution interface is shown in Figure 1.10. In agreement with Figure 1.9a, at very low C_nTAB concentrations, the adsorption layer contains almost only of octane and water molecules, and with increasing the surfactant concentrations, more and more C_nTAB molecules are present in the adsorption layer. For all systems, there is a surfactant bulk concentration at which the monolayer coverages by surfactant amounts to 50%, corresponding to the intersection points in Figure 1.10. For C_{12}TAB, this concentration is by one order of magnitude lower than for C_{10}TAB, in agreement with the Traube rule.

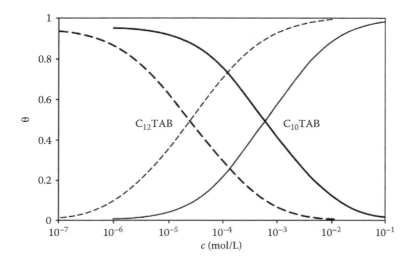

FIGURE 1.10 Dependencies of partial surface coverages θ on the surfactant concentration in aqueous solution: bold lines, octane; thin lines, surfactant; solid and dashed lines refer to the systems octane (C_{10}TAB) and octane (C_{12}TAB), respectively.

1.4 ADSORPTION OF PROTEINS AND POLYMERS

1.4.1 INDIVIDUAL PROTEIN SOLUTIONS

Due to the enormous importance of polyelectrolytes and, in particular, proteins adsorbed at fluid interfaces, there is significant progress in recent years in the development of theoretical models for the equilibrium and dynamic behavior of their adsorption layers [8,9,11–13,26,35,36,49,77,107–130]. The state-of-the-art theories applicable to equilibrium conditions were discussed in a review published in 2003 [36]. It is important to note that the adsorption behavior of proteins at solution/fluid interface is remarkably different from that of low molecular weight surfactants. Processes of denaturation of proteins at the interface can lead to their unfolding. Singer estimated the number of different conformations at the surface for a flexible-chain polymer with a 2D quasicrystal model [107], which considers the contribution of configuration entropy to the free energy. A more complicated model was proposed in [8,108] by considering the formation of loops of the macromolecules protruding into the solution bulk. Many studies tried to estimate the conformation of adsorbed polymer molecules [110–112]. This requires the configuration entropy due to the formation of loops and tails, and the enthalpy of polymer adsorption depending on the interaction between the trains and the surface. Applying a 2D quasicrystalline model for the surface layer, Silberberg [9] assumed that macromolecules can adsorb in trains and loops of various size. Later, a very detailed statistical theory was presented by Scheutjens and Fleer [12,114,115], and models for the adsorption of charged flexible polymers were developed in [11,13].

For proteins, models on the basis of the scaling theory [116,117] and thermodynamic models [119–122] are comparatively simple when compared with statistical models. They can be used for formulating the equation of state and the adsorption isotherm. A thermodynamic model for the adsorption of protein was proposed in [36]. It was further developed and allows to determine the values of interfacial tension, adsorbed amount and adsorption layer thickness, and also the dilational characteristics in a wide concentration range. A comparison with experimental data shows very good agreement.

The equations of state and adsorption isotherm developed in [26] used Equations 1.19 and 1.20 as starting point and involve also Equations 1.23, 1.36, and 1.37, which take into account the nonideality of enthalpy and entropy of the surface layer. In the model, it is assumed that protein molecules can adsorb in n states of different molar area, which varies between a maximum ω_{max} and minimum value ω_{min}. In this way, the following equation of state was obtained [36]:

$$-\frac{\Pi\omega_0}{RT} = \ln(1-\theta_p) + \theta_p(1-\omega_0/\omega_p) + a_p\theta_p^{\,2}, \tag{1.70}$$

where
   a_P is the intermolecular interaction parameter
   ω_0 is the molar area of the solvent

It was assumed that ω_0 is at the same time the area occupied by one segment of the protein molecule (the area increment). Moreover, $\Gamma_P = \sum_{i=1}^{n} \Gamma_{Pi}$ is the total adsorption of proteins in all n states ($1 \le i \le n$), and the total surface coverage by protein molecules is given by $\theta_P = \omega_P\Gamma_P = \sum_{i=1}^{n} \omega_i\Gamma_{Pi}$. Here ω_P is the average molar area of the adsorbed protein, and $\omega_i = \omega_1 + (i-1)\omega_0$ is the molar area in state i, assuming an incremental change in the molar area: $\omega_1 = \omega_{min}$, $\omega_{max} = \omega_1 + (n-1)\omega_0$. The corresponding adsorption isotherm for each adsorbed state (j) of the protein is then given:

$$b_{Pj}c_P = \frac{\omega_P\Gamma_{Pj}}{(1-\theta_P)^{\omega_j/\omega_P}}\exp\left[-2a_P\left(\frac{\omega_j}{\omega_P}\right)\theta_P\right], \tag{1.71}$$

where

c_P is the protein bulk concentration

the parameters b_{Pj} are the equilibrium adsorption constants for the protein in the jth adsorbed state

Assuming that all b_{Pj} are identical, that is, $b_{Pj}=b_P$, the adsorption constant for the whole protein molecule is given by $b=\Sigma b_P=n b_P$, which leads to the distribution over all protein adsorption states:

$$\Gamma_{Pj} = \Gamma_P \frac{(1-\theta_P)^{\frac{\omega_j-\omega_1}{\omega_P}} \exp\left[2a_P\theta_P \dfrac{\omega_j-\omega_1}{\omega_P}\right]}{\sum_{i=1}^{n}(1-\theta_P)^{\frac{\omega_i-\omega_1}{\omega_P}} \exp\left[2a_P\theta_P \dfrac{\omega_i-\omega_1}{\omega_P}\right]}, \tag{1.72}$$

The set of Equations 1.70 through 1.72 describes the evolution of the states of adsorbed protein molecules as a function of bulk concentration, which agrees in many details with experimental results.

In order to improve the agreement between experimental data and theoretical results, the dependence of the adsorption activity coefficient on the protein molecule area can be introduced. This dependence was described further by Equation 1.53, which implies that the equilibrium adsorption constant in the jth state (b_j) as compared to that in the first state (b_1) is determined by the relation

$$b_j = \left(\frac{\omega_j}{\omega_1}\right)^\alpha b_1. \tag{1.73}$$

Together with Equation 1.71, we get [123]

$$b_{Pj}c_P = \frac{\omega_P \Gamma_{Pj}}{(\omega_j/\omega_1)^\alpha (1-\theta_P)^{\omega_j/\omega_P}} \exp\left[-2a_P\left(\frac{\omega_j}{\omega_P}\right)\theta_P\right]. \tag{1.74}$$

When Equation 1.72 is modified correspondingly, the adsorption equilibrium constant for the protein molecule as a whole results to

$$b = \sum_{j=1}^{n} b_j = b_1 \sum_{j=1}^{n}\left(\frac{\omega_j}{\omega_1}\right)^\alpha. $$

With increasing protein concentration, many proteins form bilayers or multilayers at interfaces. The adsorption isotherm for such multilayer adsorption can be derived by assuming that the coverages of any subsequent layer is proportional to the adsorption constant b_{P2} and the coverage of the previous layers. This leads to an approximate value for the total adsorption Γ in all layers L:

$$\Gamma \approx \Gamma_P \sum_{i=1}^{L}\left(\frac{b_{P2}c_P}{1+b_{P2}c_P}\right)^{i-1}. \tag{1.75}$$

It was shown by many experiments that there is a critical protein concentration c_P^*, above which the surface pressure increases only insignificantly, while at the same time the adsorption exhibits a strong increase. There is a critical value of the adsorption Γ^* and surface pressure Π^* corresponding to such a critical bulk concentration, which refers to a condensation or aggregation of the protein

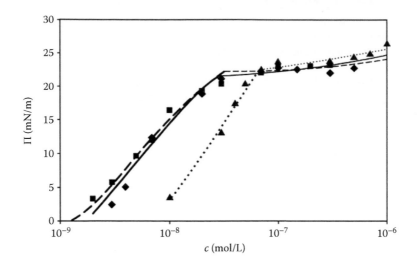

FIGURE 1.11   Adsorption isotherms of BCS in phosphate buffer at pH 5 (▲, dotted lines), pH 7 (◆, solid lines), and pH 9 (■, dashed lines). Symbols, experimental values; lines, calculated from Equations 1.70 through 1.76 using the parameters given in Wüstneck, R. et al., *Colloids Surfaces A*, 404, 17, 2002; bold lines, precritical range; thin lines, postcritical range.

molecules at the surface. The process of surface aggregation leads to changes in the average molar area of adsorbed molecules. For the postcritical surface concentration $\Gamma > \Gamma^*$, the following approximation for the surface pressure can be made (n_a is the aggregation number, see [123]):

$$\Pi = \Pi^* \left(1 + \frac{1}{n_a} \frac{\Gamma - \Gamma^*}{\Gamma^*} \right). \tag{1.76}$$

As an example, the dependence of the surface pressure Π on the β-casein (BCS) concentration for three pH values is shown in Figure 1.11 (taken from [123]). The points refer to experimental data and the lines to the theory using the parameters obtained in [123] for a surface age of 72,000 s. The effect of the solution pH on the adsorption properties and the BCS structure was also studied in [124–130] using the bubble profile tensiometry method. When a sufficiently large solution volume is used, a high accuracy of the experimental data is guaranteed because there will be almost no losses of protein caused by the adsorption at the interface [130]. A remarkable shift of the isotherm for pH 5 to higher concentrations is observed caused by the lower adsorption activity of BCS as compared to pH 7 and 9. For pH 7 and 9, the isotherms are almost identical. A rather good description of the experimental data by the theoretical isotherms is possible when the postcritical range is correctly included into the model calculations. The obtained aggregation number is about 20–40 BCS molecules per aggregate, while the number of adsorption layers estimated from the best fit of the theoretical model to the experimental rheological characteristics was 3. The experimental data for various proteins reported in [131–134] are well described with the proposed model (see also [36,47]).

1.4.2   Mixture of Proteins with Surfactants

A theory for the adsorption from mixed solutions was developed in [37] and [38] for mixtures of a protein with nonionic and ionic surfactants, respectively. The equation of state for protein/nonionic surfactant mixtures is similar to Equation 1.26 as it takes into account the nonideality of entropy only for the protein:

$$-\frac{\Pi\omega_0^*}{RT} = \ln(1-\theta_P-\theta_S)+\theta_P(1-\omega_0/\omega_P)+a_P\theta_P^2+a_S\theta_S^2+2a_{PS}\theta_P\theta_S, \qquad (1.77)$$

where
   The parameters with the subscripts S and P refer to the surfactant and protein, respectively
   Thus, $\theta_S = \omega_S \cdot \Gamma_S$ is the coverage of the surface by surfactant molecules
   Γ_S is the surfactant adsorption
   b_S is the adsorption equilibrium constant
   a_S is the corresponding interaction constant

The additional parameter a_{PS} describes the interaction between protein and surfactant molecules. Small differences between ω_0 and ω_S are accounted for by using the averaged molar area similar to Equation 1.59:

$$\omega_0^* = \frac{\omega_0\theta_P + \omega_{S0}\theta_S}{\theta_P+\theta_S}. \qquad (1.78)$$

This leads to the adsorption isotherms for the protein (in state $j=1$) and the surfactant, respectively (according to [37]):

$$b_{P1}c_P = \frac{\omega_P\Gamma_{P1}}{\left(1-\theta_P-\theta_S\right)^{\omega_1/\omega_P}}\exp\left[-2a_P\left(\frac{\omega_1}{\omega_P}\right)\theta_P - 2a_{PS}\theta_S\right], \qquad (1.79)$$

$$b_Sc_S = \frac{\theta_S}{\left(1-\theta_P-\theta_S\right)}\exp\left[-2a_S\theta_S - 2a_{PS}\theta_P\right]. \qquad (1.80)$$

The distribution over the j states of protein conformations at the interface is given by

$$\Gamma_{Pj} = \Gamma_P\frac{\left(1-\theta_P-\theta_S\right)^{\frac{\omega_j-\omega_1}{\omega_P}}\exp\left[2a_P\theta_P(\omega_j-\omega_1)/\omega_P\right]}{\sum_{i=1}^{n}\left(1-\theta_P-\theta_S\right)^{\frac{\omega_i-\omega_1}{\omega_P}}\exp\left[2a_P\theta_P(\omega_i-\omega_1)/\omega_P\right]}. \qquad (1.81)$$

According to Equation 1.60, the molar area ω_S of the surfactant molecules depends on the surface pressure Π and the total surface coverage $\theta=\theta_P+\theta_S$ (ε is the intrinsic compressibility of the surfactant molecules in the surface layer).

   The surface and bulk behavior of mixed protein/ionic surfactant solutions is essentially different from that of protein/nonionic surfactant mixtures [38]. Let us assume a protein molecule with m ionized groups at a concentration of c_P interacts with a countercharged ionic surfactant molecules of concentration c_S. Then, the ionic interaction leads to the formation of complexes. The interfacial properties of these complexes are given by the average activity of ions $\left(c_P^m c_S\right)^{1/(1+m)}$ participating in the reaction. The corresponding surface equation of state is similar to that for mixed layer containing nonionic surfactants (1.77); however, the subscript P has to be replaced by PS referring to the protein/surfactant complex.

The adsorption isotherm referring to the protein/surfactant complexes in state $j=1$ (for $\alpha=0$) and for the free surfactant molecules read [38]

$$b_{PS}c_P^{m/(1+m)}c_S^{1/(1+m)} = \frac{\omega_{PS}\Gamma_{P1}}{\left(1-\theta_{PS}-\theta_S\right)^{\omega_1/\omega_{PS}}}\exp\left[-2a_{PS}\left(\frac{\omega_1}{\omega_{PS}}\right)\theta_{PS}-2a_{SPS}\theta_S\right],\tag{1.82}$$

$$b_S(c_Sc_C)^{1/2} = \frac{\theta_S}{\left(1-\theta_{PS}-\theta_S\right)}\exp\left[-2a_S\theta_S-2a_{SPS}\theta_{PS}\right].\tag{1.83}$$

Similar isotherms can be obtained for the other possible states. In these equations, $\theta_{PS}=\omega_{PS}\Gamma_{PS}$ is the coverage of the interface by adsorbed complexes, and c_C is the surfactant counterion concentration. The parameter a_{SPS} describes the interaction of the free surfactant molecules with the protein/surfactant complexes. The distribution of adsorptions of the complex over the possible states can be obtained from Equation 1.82:

$$\Gamma_{Pj} = \Gamma_P \frac{\left(1-\theta_{PS}-\theta_S\right)^{\frac{\omega_j-\omega_1}{\omega_{PS}}}\exp\left[2a_{PS}\theta_{PS}(\omega_j-\omega_1)/\omega_{PS}\right]}{\sum_{i=1}^{n}\left(1-\theta_{PS}-\theta_S\right)^{\frac{\omega_i-\omega_1}{\omega_{PS}}}\exp\left[2a_{PS}\theta_{PS}(\omega_i-\omega_1)/\omega_{PS}\right]},\tag{1.84}$$

where j is one of the possible adsorption state. This theoretical model describes the adsorption behavior of mixed of protein/surfactant solutions sufficient well, which was demonstrated in literature for various experimental data [135–139].

The example considered here is taken from [140], where the surface tension and dilational rheology for mixtures of the biopolymer deoxyribonucleic acid (DNA) with the cationic surfactant azobenzene trimethyl ammonium bromide (AzoTAB) was discussed. In Figure 1.12, the dependence of

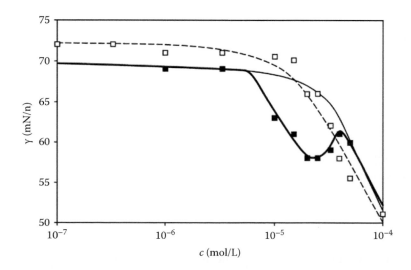

FIGURE 1.12 Equilibrium surface tension isotherms of AzoTAB solutions alone ($\square$) and mixed with DNA ($\blacklozenge$) at a fixed DNA monomer concentration of $5 \cdot 10^{-5}$ M as a function of the AzoTAB bulk concentration c; dashed curve, calculated for the individual AzoTAB using the Frumkin model; thin curve, calculated for the mixtures with $a_X=0$; bold curve, values calculated with $a_X \neq 0$ (see text).

surface tension on the concentration of the cationic AzoTAB for individual surfactant and mixtures with a fixed amount of DNA are shown.

Using the Frumkin model including an intrinsic compressibility, the theoretical isotherm of AzoTAB was calculated in [140]. DNA, due to its composition of hydrophilic groups, is almost not surface active, and the surface tension of DNA solution is 71 mN/m. A mixture of 10^{-6} M AzoTAB with DNA, however, has a surface tension of 69 mN/m. At this AzoTAB concentration, there are more than 600 molecules of the cationic surfactant per one DNA molecule in the solution. Therefore, the association number m in Equation 1.82 is very large, while the term on left-hand side is independent of m and is approximately equal to $b_{PS}c_P$. In [141], the dependence of the complex surface activity b_{PS} on c_S was introduced into Equation 1.82. For the present case of AzoTAB/DNA mixtures, this dependence is discussed here in more detail. The parameter b_{PS} on the left-hand side of Equation 1.82 can be expressed by

$$b_{PS} = b_P \quad \text{at} \quad c_s < c_0, \tag{1.85}$$

$$b_{PS} = b_P \left[1 + a_x \left(c_S - c_0 \right) \right] \quad \text{at} \quad c_0 < c_s < c_m, \tag{1.86}$$

$$b_{PS} = b_P \left[1 + a_x \left(c_m - c_0 \right) \right] \quad \text{at} \quad c_s > c_m. \tag{1.87}$$

The parameter a_x is adjustable and accounts for the influence of the surfactant concentration on the activity of DNA. At low surfactant concentrations, the parameter b_{PS} is equal to the adsorption equilibrium constant of DNA alone. Condition (1.86) corresponds to the variation of b_{PS} with the surfactant concentration c_s. Condition (1.87) implies a constant adsorption activity of the complexes at surfactant concentrations above c_m. The value of a_x can be positive and negative, depending on the surfactant aggregation in the solution bulk.

The calculated surface tension isotherm for mixtures ($a_X = 0$ and $b_{PS} = b_P$, the values of the other model parameters are given in [140]) shows that in the concentration range of 3×10^{-6} to 5×10^{-5} M AzoTAB, the theoretical model for the coadsorption of DNA/surfactant complexes is significantly different from the experimental data. We can assume that the aggregation of the AzoTAB with DNA leads not only to an increase of the adsorption activity of the complexes due to an increased effective concentration $c_P^{m/(1+m)} c_S^{1/(1+m)}$ but affects also the adsorption activity of the complex due to a change in their hydrophobicity. The influence of the AzoTAB concentration on b_{PS} can be taken into account via the coefficient a_X. This is illustrated by the bold curve in Figure 1.12, which was calculated for $c_0 = 6 \cdot 10^{-6}$ M, $c_m = 2 \times 10^{-5}$ M, and $a_X = 6.5 \times 10^5$ 1/M. The increasing part of the curve occurs at AzoTAB concentrations above 2.5×10^{-5} 10 pt, as calculated with the values $c_0 = 2.5 \times 10^{-5}$ M, $c_m = 4.0 \times 10^{-5}$ M, and $a_X = -5.9 \times 10^4$ 1/M. It is worth mentioning that the negative value of a_X corresponds to a decrease in the surface activity of the adsorbed complexes. The complicated nature of the experimental surface tension dependence for the DNA+AzoTAB mixture exhibits two extremes, which can be well described by model proposed in [38].

The experimental viscoelasticity of the DNA+AzoTAB mixed adsorption layers at surface area oscillation frequencies 0.01 and 0.2 Hz as a function of the AzoTAB concentration (measured after the adsorption equilibrium was reached) is presented in Figure 1.13. It is worth to note that the minimum of the surface tension at the AzoTAB concentration 2×10^{-5} M (see Figure 1.12) corresponds to a maximum in the elasticity (see Figure 1.13), while the surface tension maximum (at 4×10^{-5} M) coincides with the minimum in elasticity. This interrelation sounds physically reasonable [33,36,52]. The theoretical curves calculated with Equation 1.63, as modified in [140] for the protein/surfactant system, agree well with the experimental data. The parameter values (including a_x) used for the calculations were the same as those used to calculate the isotherms in Figure 1.12, while the diffusion coefficients for DNA and AzoTAB were taken as 10^{-11} and 10^{-10} m^2/s, respectively.

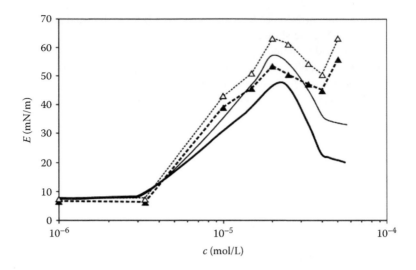

FIGURE 1.13 Dilational elasticity of the surface layer of mixed DNA+AzoTAB solutions with the monomer concentration $5 \cdot 10^{-5}$ M versus the AzoTAB concentration at the surface oscillation frequencies 0.01 (▲) and 0.2 Hz (△). The theoretical values (solid curves) were calculated from Equation 1.63; dotted lines that connect experimental points are guides for eyes.

1.4.3 Competitive Adsorption of Proteins and Oil

Protein-stabilized emulsions and foams are the most important class of food colloids [102,142]. In many papers, the adsorption behavior and interfacial rheological properties of proteins at fluid interfaces have been investigated. For example, the properties of BLG and BCS adsorption layers at water/oil interfaces were discussed extensively in [49,98,121,122,132,143–146]. The publication [147] generalized results obtained for equilibrium interface pressure for aqueous BCS solutions at pH 7 at the interfaces with tetradecane and hexane using various methods [132,143,144]. It was demonstrated that the results obtained by different methods can show important differences and the reasons were explained. The dependence of Π on the BLG concentration in a solution drop formed in hexane [146] agrees well with the data from experiments with tetradecane drop formed in aqueous protein solutions [145], as shown in Figure 1.14. Comparing the interfacial pressure isotherms of the two proteins, one can see that the concentration range of the BCS isotherms (Figure 1.11) is almost two orders of magnitude lower than that for BLG for the same Π. The results obtained for the solution drop/tetradecane interface are close to those for the hexane drop/solution interface; this should be ascribed to the fact that the adsorption activity of BLG is low, and therefore the adsorption-related losses are negligibly small. It was shown in [105] that for ordinary surfactants (sorbitan monoesters) at the aqueous surfactant solution/alkane (pentane to dodecane) interfaces, the adsorption activity of the surfactant becomes higher with decreasing molecular weight of the alkane. A similar dependence could possibly also exist for proteins.

The theoretical curve for BLG was calculated from Equations 1.70 through 1.76 with the following model parameters: $\omega_0 = 3.5 \times 10^5$ m²/mol, $\omega_1 = 5.8 \times 10^6$ m²/mol, $\omega_{max} = 1.4 \times 10^7$ m²/mol, $a = 0$, $\alpha = 1.8$, $n_a = 15$, and $L = 2$. The values of ω_{max} and ω_0 are very similar to those used in [36] for the solution/air interface. This fact points at the globular structure of BLG molecules. The theoretical results are also rather close to those published in [146]. In [147], it was shown that the adsorption of BCS can also be well described by the given theoretical model. The calculated results provide satisfactory fitting for all corresponding experimental data, and the values of the molar areas ω_1 and ω_{max} agree with those published elsewhere. Note, the values of ω_{max} for BCS at the aqueous solution/alkane interface are by a factor of 2–2.5 higher than those at the solution/air surface, due

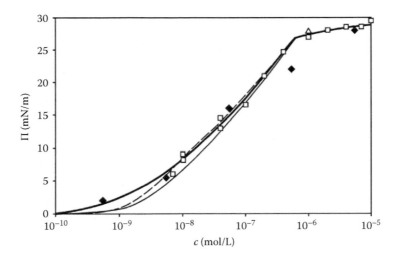

FIGURE 1.14 Dependence of Π on the BLG concentration c: ($\square$) (data from Pradines, V. et al., *J. Phys. Chem. B*, 113, 745, 2009) for a protein solution drop in hexane; ($\blacklozenge$) (data from Maldonado-Valderrama, J. et al., *Langmuir*, 26, 15901, 2010) for a tetradecane drop in protein solution; ($\triangle$) (data from Pradines, V. et al., *Colloids Surf. A*, 371, 22, 2010) for the protein solution drop/hexane interface; the dashed line corresponds to the theoretical calculations with the parameters given in the text; the solid curves were calculated using the protein, alkane coadsorption model; bold line, $a_{PS} = 0.7$; thin line, $a_{PS} = 0$.

to the larger degree of unfolding of the BCS molecules at the surface [143]. The given ω_{max} values correspond to a layer thickness of ca. 0.5 nm.

The results obtained for the proteins solutions at water/alkane interfaces can be analyzed in the framework of the model, which describes the coadsorption of protein and alkane molecules. This model was developed in [40] for solutions of surfactants. When applied to protein/alkane mixtures, this model is similar to the model for mixed protein/nonionic surfactant, described by Equations 1.77 through 1.81. In this case, the alkanes play the role of the surfactant molecules.

The calculated interfacial pressure isotherms for aqueous BLG solutions at the water/hexane interface are shown in Figure 1.14. The solid line was calculated via the coadsorption model with $c_S = 7.58$ mol/L as the concentration of 100% hexane. The model parameters for hexane were exactly the same as in Figure 1.8: $b_S = 2.6$ L/mol, $\omega_S = 3.3 \times 10^5$ m²/mol, and $\varepsilon = 0$. And also the model parameters for BLG in the calculations were the same as used for the calculation of the black curve. Only the value of $b_{P,1}$ in Equation 1.79 was increased by a factor of 45 as compared with that used in Equation 1.74, that is, from 10^4 to 4.5×10^5 m³/mol. This increase was required to properly describe the displacement of hexane molecules from the surface layer [40]. The influence of displaced alkane on the interfacial tension caused by the contribution of a secondary adsorption layer (into which the alkane is displaced) was considered in [69].

The dependence of the viscoelasticity modulus $|E|$ for BLG adsorption layers as a function of surface pressure Π at a surface area oscillation frequency of 0.1 Hz was discussed in [147]. The experiments were performed with oscillating tetradecane drops in aqueous protein solutions [145,148], and the theoretical dependencies were calculated for a diffusion coefficient of $D = 10^{-11}$ m²/s with a set of equation derived by Joos [3]. The equations also take the geometry of system into account. The dependences of the dilational viscoelasticity on Π with a maximum at 20–23 mN/m are almost identical in both papers [145,148]. A similar review of the dilational viscoelasticity modulus $|E|$ on the interfacial pressure Π for the BCS solutions at an oscillation frequency of 0.1 Hz was published in [147]. It was shown that at certain conditions, this dependence exhibits two maximums and the horizontal sections, and the calculations using the equations proposed in [3] are in satisfactory agreement with the experimental data.

1.5 INSOLUBLE MONOLAYERS

1.5.1 EQUATION OF STATE WITH CONSIDERATION OF COMPRESSIBILITY

For quite a time, insoluble monolayers are the object of interfacial science because they serve as suitable models for quite a number of systems, for example, biological model membranes [149]. One of the most important characteristics of insoluble monolayers is the dependence of surface pressure Π on the area per one surfactant molecule A. The Π-A isotherms of Langmuir monolayers show a sharp break at the main phase transition from a gaseous (G) or liquid-expanded (LE) to an LC state. There are several attempts to explain the nonhorizontal shape of Π-A isotherms in the region of the 2D phase transition [150–154]. In the studies [41–45,155], the LE-LC transitions were explained in the framework of a quasi-chemical approach. This means that the mass action law is applied to describe the monomer/aggregate equilibrium. Using the Butler equation (1.6) for the chemical potential of components within the surface layer, respective equations of state for the G or LE states of Langmuir monolayers were derived in [42]. However, as the surfactant molecules are insoluble, the relation (1.10) for the chemical potential in the solution bulk is irrelevant. To obtain the corresponding equation of state, the Butler equation was coupled with the Gibbs fundamental equation:

$$d\Pi = \sum_{i=1}^{n} \Gamma_i d\mu_i. \tag{1.88}$$

It should also be noted that the location of the dividing surface in Equation 1.88 differs from that employed by Lucassen-Reynders, see Equations 1.16 and 1.17, on which all equations of state and adsorption isotherms for soluble surfactants and polymers essentially rely. In the case considered here, the dividing surface corresponds to zero adsorption of the solvent ($\Gamma_0 = 0$). The general equation derived in this way allowed to obtain the Frumkin, the van der Waals, and the Volmer equations of state [156]. In general, the equation of state for monolayers in the fluid (G or LE) state is represented by the Volmer-type equation [42,154]:

$$\Pi = \frac{kT}{n_a(A - \omega)} - \Pi_{coh} \tag{1.89}$$

where
   k is the Boltzmann constant
   ω is the partial molecular area for monomers (or the limiting area of molecule in the gaseous state)
   Π_{coh} is the cohesion pressure
   n_a is the aggregation number of small aggregates [155] or dissociation number ($n_a < 1$)

In [42], also a generalized Volmer equation was derived for multicomponent monolayers:

$$\Pi = RT \frac{\sum_i \Gamma_i}{1 - \sum_i \Gamma_i \omega_i} - B \tag{1.90}$$

where B is the integral Volmer constant. Also the equation of state for the main phase transition using Equation 1.90 and the quasi-chemical monomer/aggregate equilibrium model was derived in [42], and the following equation of state is valid for a bimodal distribution of large clusters and monomers [155]:

$$\Pi = \frac{kT\alpha\beta/n_a}{A - \omega\left[1 + \varepsilon(\alpha\beta - 1)\right]} - \Pi_{coh}. \tag{1.91}$$

The parameter α expresses here the dependence of the aggregation constant on the surface pressure Π:

$$\alpha = \frac{A}{A_c}\exp\left[-\varepsilon\frac{\Pi - \Pi_c}{kT}\omega\right], \tag{1.92}$$

where
  β stands for the fraction of the monolayer free of aggregates: $\beta = 1 + \omega(1-\varepsilon)(\alpha-1)/A$
  Further, $\varepsilon = 1 - \omega_{cl}/\omega$ is the compressibility coefficient
  ω_{cl} is the area per monomer in a cluster
  A_c is the molecular area corresponding to the onset of the phase transition (i.e., at $\Pi = \Pi_c$)

For $A_c < 2\omega$, we get $\beta = \alpha$. The area per molecule in a cluster can be different from the area per molecule in the gaseous state of a monolayer. This can be accounted for by the compressibility coefficient ε. To reproduce the experimental findings of $\omega_{cl}(\Pi)$ dependencies for a number of monolayers [46–48], it has to be assumed that the parameter ε consists of two terms [157]: $\varepsilon = \varepsilon_0 + \eta\Pi$. With this, $\omega_{cl} = \omega(1-\varepsilon) = \omega(1-\varepsilon_0-\eta\Pi)$ was obtained, where ε_0 is the relative jump of the area per molecule during a phase transition and η is a 2D compressibility jump (or intrinsic compressibility) coefficient of the condensed monolayer. Comparing the relations described earlier with Equations 1.7 and 1.8 applicable to the soluble surfactants, one can see that the essential difference between these two types of interfacial layers is the presence of the compressibility jump in the phase transition point of insoluble monolayers. The equation $\varepsilon = \varepsilon_0 + \eta\Pi$ was generalized to take into account any further phase transitions between condensed phases [157]. This was successfully applied to experimental surface pressure isotherms for various amphiphilic monolayers [158].

Equation of state for surface layers for which the nonideality of entropy of mixing of monomers and clusters has been taken into account has the form [45]

$$\Pi = \frac{kT\alpha\beta/n_a}{A - \omega\left[1 + \varepsilon(\alpha\beta - 1)\right]} - \frac{kT}{A}\left(1 - \alpha\beta\right) - \Pi_{coh}. \tag{1.93}$$

As an example, one can compare the experimental data reported in [45] with numerical estimates obtained from Equation 1.89 in the precritical range $(A > A_c)$ and Equation 1.93 in the phase transition region $(A < A_c)$. In Figure 1.15, six experimental Π-A isotherms of $2C_{11}H_{23}$-melamine monolayers measured between 12.5°C and 29.2°C are shown. The thin lines represent the theoretical results obtained on the basis of the equations of state (1.89) and (1.93); the intersection of the dotted line (which is the tangent to the Π-A isotherms in the LC state) with the abscissa can be used to estimate the ω value [44]. It is seen that the theoretical calculations agree well with the experimental data. Note that the use of the equation of state Equation 1.91 yields results that differ insignificantly from those shown in Figure 1.15.

The experimental results obtained for a system with two phase transitions were compared with theoretical calculations in [157] as shown in Figure 1.16 for TDHPA (N-tetradecyl-β-hydroxy-propionic acid amide). The two phase transitions are denoted in the plot as I and II. A satisfactory agreement between the theory and the experiment is observed, although no arbitrary parameters are involved in the theoretical model. All required model parameters for

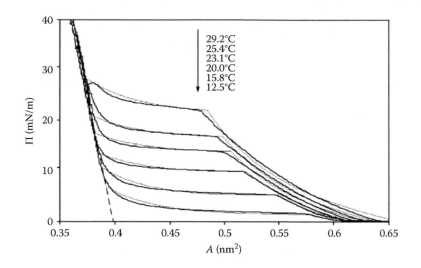

FIGURE 1.15 Experimental (bold lines) and theoretical (thin lines) Π-A isotherms of $2C_{11}H_{23}$-melamine monolayers spread on water measured in the temperature range between 12.5°C and 29.2°C; the values of the model parameters for Equation 1.93 are listed. (From Fainerman, V.B. and Vollhardt, D., *J. Phys. Chem. B*, 112, 1477, 2008.)

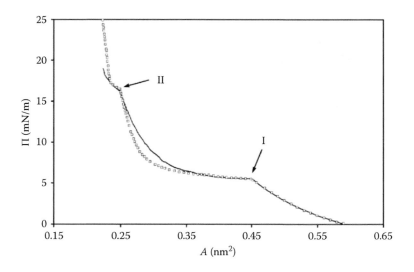

FIGURE 1.16 Comparison between experimental (points) and theoretical (line) Π-A isotherms for TDHPA monolayer; I, first phase transition point; II, second phase transition point. (From Fainerman, V.B. and Vollhardt, D., *J. Phys. Chem. B*, 107, 3098, 2003.)

Equations 1.89 and 1.93 are either calculated from the gaseous state of the monolayer or were evaluated from independent grazing incidence x-ray diffraction (GIXD) experiments. Hence, it is possible to satisfactory describe Π-A isotherms with two or more phase transitions in insoluble amphiphile monolayers.

The surface compressional modulus $K = -A(d\Pi/dA)_T$ or high-frequency dilational modulus $E_0 = \Gamma(d\Pi/d\Gamma)_T$ can be calculated either from experimental $\Pi(A)$ dependences or theoretically by differentiation of the corresponding equation of state. From Equation 1.91, we obtain

$$K = \frac{A\alpha\beta - \frac{\alpha}{A}\left[A - \omega(1-\varepsilon)\right]\left[A + \omega(1-\varepsilon)\alpha\right]}{\frac{n_a\left\{A - \omega\left[1+\varepsilon(\alpha\beta-1)\right]\right\}^2}{kT} - \omega\eta\alpha\beta(\alpha\beta-1) + \frac{\alpha}{A}\left[A - \omega(1-\varepsilon)\right]B} \tag{1.94}$$

with $B = \left\{\omega\eta(\alpha-1) + \left[A + \omega(1-\varepsilon)(2\alpha-1)\right]\left[\frac{\omega\eta(\Pi - \Pi_C)}{kT} + \frac{\omega\varepsilon}{kT}\right]\right\}.$

For $\alpha = \beta$, Equation 1.94 simplifies, and for a highly compressed monolayer, that is $A \to \omega_{cl} = \omega(1-\varepsilon)$, we obtain

$$K = \frac{(1-\varepsilon)}{\left(\frac{n_a\omega(\varepsilon\alpha)^2}{kT}\right) - \eta(\alpha^2 - 1)}. \tag{1.95}$$

We typically have $\varepsilon^2 \ll 1$, and for $A \to \omega_{cl} = \omega(1 - \varepsilon)$, we obtain (for $A_c \geq 2\omega$) $\alpha^2 \ll 1$. Then, Equation 1.95 can be approximated by $K = 1/C \cong (1 - \varepsilon)/\eta$, that is, for a highly compressed monolayer, the compressibility C is equivalent to the 2D compressibility of the condensed monolayer η. Thus, in the limiting case K coincides with E_0 for a soluble surfactant given by Equation 1.48.

In Figure 1.17, the experimental and theoretical data of the surface compressional modulus K are shown as a function of the area per molecule of TDHPA using the results given in Figure 1.16. As we can see, the surface compressional modulus K has a maximum in the two phase transition points. This means that there is a maximum compressibility of the monolayer in the region right after the beginning of the phase transition. The experimental values for K are in good agreement with estimations made with Equation 1.94. Note, however, that the theoretical K values are evaluated from independent GIXD experiments, that is, from the parameters ε_0 and μ.

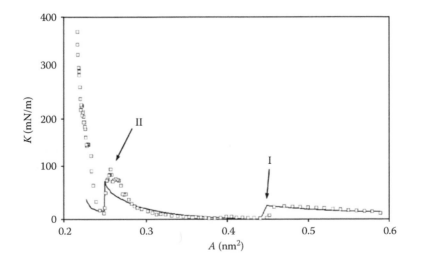

FIGURE 1.17 Experimental (symbols) and theoretical (lines) dependencies of the surface compressional modulus K on the area per TDHPA molecule; I, first phase transition point; II, second phase transition point. (From Fainerman, V.B. and Vollhardt, D., *J. Phys. Chem. B*, 107, 3098, 2003.)

1.5.2 PENETRATION OF LANGMUIR MONOLAYERS BY SOLUBLE AMPHIPHILIC MOLECULES

The main problem in the thermodynamics of penetration layers is to describe the adsorption of soluble surfactant as a function of its bulk concentration for any surface coverage by a second insoluble surfactant [159]. There are several theoretical approaches for deriving the penetration thermodynamics. One is based on the Gibbs' fundamental equation for multicomponent interfacial layers [160,161]. A second approach originally proposed in 1955 [162] starts either from Pethica's equation [163] or from more general equivalent expressions [164]. A third way proposed in [163] applied the Butler equation [31] in order to express the chemical potentials of the surface layer components. Further, in [165], the Gibbs equation was solved simultaneously with the adsorption isotherm of the soluble component and led to an equation of state for a mixed monolayer. A critical review of the existing theoretical approaches to describe the thermodynamics of soluble surfactant's penetration into an insoluble monolayer was published in [159]. Due to the complexity, so far no rigorous thermodynamic analysis of penetration systems exists; however, some simplifications, such as the assumption of constant activity coefficients for the insoluble component in the monolayer during the penetration of the soluble component, result in reasonable theories. For example, in [163], penetration layer theories were presented, first for mixtures comprised of molecules with equal partial molar area (mixture of homologues) and with an extension to include also protein penetrating into the monolayer of 2D aggregating phospholipids. This extension was based on the concept of independent segments of the protein molecules, occupying an area at the interface equal to that of the phospholipid molecules.

The theoretical models considered various mechanisms for the effect of the soluble surfactant on the aggregation of the insoluble component:

1. No effect on the aggregate formation process
2. Formation of mixed aggregates
3. Change in the aggregation constant

These mechanisms lead to different equations of state of the equation of state for the mixed monolayer, and the shape of experimental Π-A isotherms allows to draw conclusions based on the actual mechanism. The first experimental evidence has been provided for the penetration of BLG, BCS, and lysozyme into fluid-like dipalmitoyl phosphatidylcholine (DPPC) monolayers, which leads to a first-order main phase transition of DPPC even at an area per DPPC molecule two times larger than the area required for the main phase transition in a pure DPPC monolayer [159,163]. The domain formation in a condensed monolayer initiated by the penetration of proteins was visualized by Brewster angle microscopy (BAM). The phase transition is indicated by a break point in the $\Pi(t)$ curves of the penetration kinetics. The experimental results show that mixed protein/phospholipid aggregates are not formed; however, the penetration of the proteins into the monolayer generates DPPC aggregates due to the increased total surface coverage and possible changes in the lipid aggregation constant. By coupling experiments on the $\Pi(t)$ penetration kinetics with BAM and GIXD measurements, new information on phase transitions and the phase behavior of Langmuir monolayers penetrated by soluble surfactants can be obtained. Studies and the penetration of BLG at different bulk concentrations into spread DPPC monolayers have shown that the penetration of proteins happens without any specific interactions with the lipid molecules and the condensed phase consists exclusively of DPPC molecules. For mixed monolayers, the cases of no phase transitions (aggregations), and the total adsorption (DPPC + proteins) required for the onset of aggregation during protein penetration, agree very well with data predicted by the theoretical models presented in [159,163].

REFERENCES

1. J.T. Davies and E.K. Rideal, *Interfacial Phenomena*, Academic Press, New York (1963).
2. R. Defay and I. Prigogine, *Surface Tension and Adsorption*, Longmans, Green, London, U.K. (1966).

3. P. Joos, *Dynamic Surface Phenomena,* VSP, Dordrecht, the Netherlands (1999).
4. M.J. Rosen, *Surfactants and Interfacial Phenomena*, John Wiley & Sons, New York (2004).
5. A.I. Rusanov, *Phasengleichgewichte und Grenzflächenerscheinungen*, Akademie-Verlag, Berlin, Germany (1978).
6. A.I. Rusanov, *Surf. Sci. Rep.*, 58 (2005) 111.
7. L. Ter-Minassian-Saraga and I. Prigogine, *Mem. Serv. Chim. Etat*, 38 (1953) 109.
8. H.L. Frisch and R. Simha, *J. Chem. Phys.*, 27 (1957) 702.
9. A. Silberberg, *J. Chem. Phys.*, 48 (1968) 2835.
10. C.A.J. Hoeve, *J. Polym. Sci. Part C*, 34 (1971) 1.
11. F.T. Hesselink, *J. Colloid Interface Sci.*, 60 (1977) 448.
12. G.J. Fleer and J.M.H.M. Scheutjens, *Adv. Colloid Interface Sci.*, 16 (1982) 341.
13. F.A.M. Leermakers, P.L. Atkinson, E. Dickinson, and D.S. Horne, *J. Colloid Interface Sci.*, 178 (1996) 681.
14. A.N. Frumkin, *Z. Phys. Chem.* (Leipzig), 116 (1925) 466.
15. R. Douillard and J. Lefebvre, *J. Colloid Interface Sci.*, 139 (1990) 488.
16. R. Douillard, M. Daoud, J. Lefebvre, C. Minier, G. Lecannu, and J. Coutret, *J. Colloid Interface Sci.*, 163 (1994) 277.
17. S. D. Stoyanov, H. Rehage, and V. N. Paunov, *Phys. Chem. Chem. Phys.*, 6 (2004) 596.
18. P. Joos, *Bull. Soc. Chim. Belg.*, 76 (1967) 591.
19. A.I. Rusanov, *J. Chem. Phys.*, 120 (2004) 10736.
20. A.I. Rusanov, *Colloid J.*, 69 (2007) 131–143.
21. E.H. Lucassen-Reynders, *J. Colloid Sci.*, 19 (1964) 584.
22. E.H. Lucassen-Reynders, *J. Phys. Chem.*, 70 (1966) 1777.
23. E.H. Lucassen-Reynders Surface elasticity and viscosity in compression/dilation. In: *Anionic Surfactants (Physical Chemistry of Surfactant Action)*, E.H. Lukassen-Reynders (Ed.), Marcel Dekker, Inc., New York (1981) p. 1.
24. C.C. Sanchez, M.C. Fernandez, M.R. Rodriguez Nino, and J.M. Rodriguez Patino, *Langmuir*, 22 (2006) 4215.
25. T.F. Svitova, M.J. Wetherbee, and C.J. Radke, *J. Colloid Interface Sci.*, 261 (2003) 170.
26. E.M. Freer, K.S. Yim, G.C. Fuller, and C.J. Radke, *J. Phys. Chem. B*, 108 (2004) 3835.
27. P. Kralchevsky, K. Danov, C. Pishmanova, S. Kralchevska, N. Christov, K. Ananthapadmanabhan, and A. Lips, *Langmuir,* 23 (2007) 3538.
28. K.D. Danov and P.A. Kralchevsky, *Colloid J.,* 74 (2012) 172.
29. O.-S. Kwon, H. Jing, K. Shin, X. Wang, and S.K. Satija, *Langmuir*, 23 (2007) 12249.
30. S. Azizian, K. Kashimoto, T. Matsuda, H. Matsubara, T. Takiue, and M. Aratono, *J. Colloid Interface Sci.*, 316 (2007) 25.
31. J.A.V. Butler, *Proc. R. Soc. Ser. A*, 138 (1932) 348.
32. V.B. Fainerman, R. Miller, and R. Wüstneck, *J. Phys. Chem.*, 101 (1997) 6479.
33. V.B. Fainerman, E.H. Lucassen-Reynders, and R. Miller, *Colloids Surf. A*, 143 (1998) 141.
34. V.B. Fainerman, V.I. Kovalchuk, E.V. Aksenenko, M. Michel, M.E. Leser, and R. Miller, *J. Phys. Chem. B.*, 108 (2004) 13700.
35. V.B. Fainerman and E.H. Lucassen-Reynders, *Adv. Colloid Interface Sci.*, 96 (2002) 295.
36. V.B. Fainerman, E.H. Lucassen-Reynders, and R. Miller, *Adv. Colloid Interface Sci.*, 106 (2003) 237.
37. V.B. Fainerman, S.A. Zholob, M. Leser, M. Michel, and R. Miller, *J. Colloid Interface Sci.*, 274 (2004) 496.
38. V.B. Fainerman, S.A. Zholob, M.E. Leser, M. Michel, and R. Miller, *J. Phys. Chem.*, 108 (2004) 16780.
39. A. Javadi, N. Moradi, V.B. Fainerman, H. Möhwald, and R. Miller, *Colloids Surf. A*, 391 (2011) 19–24.
40. V.B. Fainerman, N. Mucic, V. Pradines, E.V. Aksenenko, and R. Miller, *Langmuir*, 29 (2013) 13783.
41. V.B. Fainerman, D. Vollhardt, and V. Melzer, *J. Phys. Chem.*, 100 (1996) 15478.
42. V.B. Fainerman and D. Vollhardt, *J. Phys. Chem. B*, 103 (1999) 145.
43. V.B. Fainerman and D. Vollhardt, *J. Phys. Chem. B*, 110 (2006) 3448.
44. V.B. Fainerman and D. Vollhardt, *J. Phys. Chem. B*, 110 (2006) 10436.
45. V.B. Fainerman and D. Vollhardt, *J. Phys. Chem. B*, 112 (2008) 1477.
46. U. Gehlert, D. Vollhardt, G. Brezesinski, and H. Möhwald, *Langmuir*, 12 (1996) 4892.
47. U. Gehlert, G. Weidemann, D. Vollhardt, G. Brezesinski, R. Wagner, and H. Möhwald, *Langmuir*, 14 (1998) 2112.
48. U. Gehlert and D. Vollhardt, *Langmuir*, 18 (2002) 688.
49. J. de Feijter and J. Benjamins, *J. Colloid Interface Sci.*, 90 (1982) 289.
50. V.B. Fainerman, R. Miller, and V.I. Kovalchuk, *J. Phys. Chem. B*, 107 (2003) 6119.

51. V.I. Kovalchuk, G. Loglio, V.B. Fainerman, and R. Miller, *J. Colloid Interface Sci.*, 270 (2004) 475.
52. E.V. Aksenenko, V.I. Kovalchuk, V.B. Fainerman, and R. Miller, *J. Phys. Chem. C*, 111 (2007) 14713.
53. V.B. Fainerman and R. Miller, *Colloids Surf. A*, 319 (2008) 8.
54. K.-D. Wantke and H. Fruhner, *J. Colloid Interface Sci.*, 237 (2001) 185.
55. E.A. Simister, E.M. Lee, R.K. Thomas, and J. Penfold, *J. Phys. Chem.*, 96 (1992) 1373.
56. C. Stubenrauch, V.B. Fainerman, E.V. Aksenenko, and R. Miller, *J. Phys. Chem. B*, 109 (2005) 1505.
57. V.B. Fainerman and R. Miller, Adsorption isotherms at liquid interfaces, in: *Encyclopedia of Surface and Colloid Science*, 2nd edn., P. Somasundaran and A. Hubbard (Eds.), Taylor & Francis, London, Vol. 1 (2009) pp. 1–15.
58. J.M. Prausnitz, *Molecular Thermodynamics of Fluid Phase Equilibria*, Prentice-Hall Englewood Cliffs, New York (1969).
59. E.H. Lucassen-Reynders, *J. Colloid Interface Sci.*, 41 (1972) 156.
60. E.H. Lucassen-Reynders, *J. Colloid Interface Sci.*, 85 (1982) 178.
61. I. Prigogine, *The Molecular Theory of Solutions*, North-Holland, Amsterdam, the Netherlands (1968).
62. E.A Guggenheim, *Mixtures*, Clarendon Press, Oxford, U.K. (1952).
63. R.C. Read, J.M. Prausnitz, and T.K. Sherwood, *The Properties of Gases and Liquids*, 3rd edn., McGraw-Hill, Inc., New York (1977).
64. E.H. Lucassen-Reynders, *Colloids Surfaces A*, 91 (1994) 79.
65. B.B. Damaskin, *Elektrochimija*, 5 (1969) 249.
66. B.B. Damaskin, *Izv. AN SSSR, Ser. Chim.*, 2 (1969) 346.
67. E.H. Lucassen-Reynders, A. Cagna, and J. Lucassen, *Colloids Surf. A*, 186 (2001) 63.
68. G.S. Aleiner and O.G. Us'yarov, *Colloid J.*, 72 (2010) 731.
69. V.B. Fainerman, E.V. Aksenenko, N. Mucic, A. Javadi, and R. Miller, *Soft Matter*, 10 (2014) 6873.
70. E.H. Lucassen-Reynders, J. Lucassen, and D. Giles, *J. Colloid Interface Sci.*, 81 (1981) 150.
71. V.B. Fainerman, R. Miller, E.V. Aksenenko, and A.V. Makievski, Equilibrium adsorption properties of surfactants at the solution-fluid interface, in: *Surfactants—Chemistry, Interfacial Properties and Application*, Studies in Interface Science, V.B. Fainerman, D. Möbius, and R. Miller (Eds.), Vol. 13, Elsevier, Amsterdam, 2001, pp. 189.
72. V. Bergeron, *Langmuir*, 13 (1997) 3474.
73. F. Monroy, J. Giermanska Khan, and D. Langevin, *Colloids Surf. A*, 143 (1998) 251.
74. N. Schelero, R. von Klitzing, V.B. Fainerman, and R. Miller, *Colloids Surf. A Physicochem. Eng. Aspects*, 413 (2012) 115.
75. V.B. Fainerman, V.I. Kovalchuk, M.E. Leser, and R. Miller, Effect of the intrinsic compressibility on the dilational rheology of adsorption layers of surfactants, proteins and their mixtures, in: *Colloid and Interface Science Series*, T. Tadros (Ed.), Vol. 1, Wiley-VCH, Weinheim (2007) pp. 307.
76. V.B. Fainerman, S.A. Zholob, E.H. Lucassen-Reynders, and R. Miller, *J. Colloid Interface Sci.*, 261 (2003) 180.
77. P. Joos and G. Serrien, *J. Colloid Interface Sci.*, 145 (1991) 291.
78. Y.-C. Lee, H.-S. Liu, and S.-Y. Lin, *Colloids Surf. A*, 212 (2003) 123.
79. A. Ritacco and J. Busch, *Langmuir*, 20 (2004) 3648.
80. M.A. Valenzuela, M.P. Garate, and A.F. Olea, *Colloids Surf. A*, 307 (2007) 28.
81. V.B. Fainerman, E.V. Aksenenko, J.T. Petkov, and R. Miller, *J. Phys. Chem. B*, 114 (2010) 4503.
82. V.B. Fainerman, E.V. Aksenenko, J.T. Petkov, and R. Miller, *Soft Matter*, 7 (2011) 6873.
83. V.B. Fainerman, E.V. Aksenenko, S.V. Lylyk, J.T. Petkov, J. Yorke, and R. Miller, *Langmuir*, 26 (2010) 284.
84. J.R. Lu, R.K. Thomas, and J. Penford, *Adv. Colloid Interface Sci.*, 84 (2000) 143.
85. V.B. Fainerman and R. Miller, *Langmuir*, 12 (1996) 6011.
86. R.H. Müller and C.M. Keck, *J. Biotechnol.*, 113 (2004) 151.
87. L.L. Schramm, E.N. Stasiuk, and D.G. Marangoni, *Annu. Rep. Prog. Chem.*, 99 (2003) 3.
88. A.J. Prosser and E.I. Franses, *Colloids Surf. A*, 178 (2001) 1.
89. V.B. Fainerman, E.V. Aksenenko, V.I. Kovalchuk, A. Javadi, and R. Miller, *Soft Matter*, 7 (2011) 7860.
90. A. Javadi, N. Moradi, H. Möhwald, and R. Miller, *Soft Matter*, 6 (2010) 4710.
91. N. Mucic, N. Moradi, A. Javadi, E.V. Aksenenko, V.B. Fainerman, and R. Miller, *Colloids Surf. A.*, 442 (2014) 50.
92. A. Lou and B.A. Pethica, *Langmuir*, 13 (1997) 4933.
93. N. Mucic, N. Moradi, A. Javadi, E.V. Aksenenko, V.B. Fainerman, and R. Miller, Effect of partial vapor pressure on the co-adsorption of surfactants and hexane at the water/hexane vapor interface, *Colloids Sur. A*, 480 (2015) 79.
94. M.M. Knock, G.R. Bell, E.K. Hill, H.J. Turner, and C.D. Bain, *J. Phys. Chem. B*, 107 (2003) 10801.

95. M.L. Schlossmann and A.M. Tikhonov, *Annu. Rev. Phys. Chem.*, 59 (2008) 153.
96. H. Fang and D.O. Shah, *J. Colloid Interface Sci.*, 205 (1998) 531.
97. T.G. Gurkov, T.S. Horozov, I.B. Ivanov, and R.P. Borwankar, *Colloids Surf. A*, 87 (1994) 81.
98. V. Pradines, V.B. Fainerman, E.V. Aksenenko, J. Krägel, N. Mucic, and R. Miller, *Colloids Surf. A*, 371 (2010) 22.
99. T. Takiue, T. Tottori, K. Tatsuta, H. Matsubara, H. Tanida, K. Nitta, T. Uruga, and M. Aratono, *J. Phys. Chem. B*, 116 (2012) 13739.
100. A. Bonfillon and D. Langevin, *Langmuir*, 9 (1993) 2172.
101. A. Goebel and K. Lunkenheimer, *Langmuir*, 13 (1997) 369.
102. E. Dickinson, *Colloids Surf. B*, 15 (1999) 161.
103. H. Dominguez and M. L. Berkowitz, *J. Phys. Chem. B*, 104 (2000) 5302.
104. K. Medrzycka and W. Zwierzykowski, *J. Colloid Interface Sci.*, 230 (2000) 67.
105. L. Peltonen, J. Hirvonen, and J. Yliruusi, *J. Colloid Interface Sci.*, 240 (2001) 272.
106. N. Mucic, N.M. Kovalchuk, E.V. Aksenenko, V.B. Fainerman, and R. Miller, *J. Colloid Interface Sci.*, 410 (2013) 181.
107. S.J. Singer, *J. Chem. Phys.*, 16 (1948) 872.
108. H.L. Frisch and R. Simha, *J. Chem. Phys.*, 24 (1956) 652.
109. J.T. Davies, *Biochim. Biophys. Acta*, 11 (1953) 165.
110. A. Silberberg, *J. Phys. Chem.*, 66 (1962) 1872, 1884.
111. F.L. McCrackin, *J. Chem. Phys.*, 47 (1967) 1980.
112. R.I. Feigin and D.H. Napper, *J. Colloid Interface Sci.*, 71 (1979) 11.
113. C.A.J. Hoeve, *J. Chem. Phys.*, 44 (1966) 1505.
114. J.M.H.M. Scheutjens and G.J. Fleer, *J. Phys. Chem.*, 83 (1979) 1619.
115. J.M.H.M. Scheutjens and G.J. Fleer, *J. Phys. Chem.*, 84 (1980) 178.
116. P.G. de Gennes, *Scaling Concepts in Polymer Physics*, Cornell University Press, Ithaca, NY (1979)
117. P.G. de Gennes, *Adv. Colloid Interface Sci.*, 27 (1987) 189.
118. H.B. Bull, *J. Biol. Chem.*, 185 (1950) 27.
119. P. Joos, *Biochim. Biophys. Acta*, 375 (1975) 1.
120. L. Ter-Minassian-Saraga, *J. Colloid Interface Sci.*, 80 (1981) 393.
121. J. Benjamins, A. Cagna, and E.H. Lucassen-Reynders, *Colloids Surf. A*, 114 (1996) 245.
122. E.H. Lucassen-Reynders and J. Benjamins: in *Food Emulsions and Foams: Interfaces, Interactions and Stability*, E. Dickinson and J.M. Rodriguez Patino (Eds.), Special Publication No. 227, Royal Society of Chemistry, Cambridge, U.K. (1999), p. 195.
123. R. Wüstneck, V.B. Fainerman, E. Aksenenko, C.S. Kotsmar, V. Pradines, J. Krägel, and R. Miller, *Colloids Sur. A*, 404 (2012) 17–24.
124. C. Moitzi, I. Portnaya, O. Glatter, O. Ramon, and D. Danino, *Langmuir*, 24 (2008) 3020.
125. C. Holt and L. Sawyer, *J. Chem. Soc.-Faraday Trans.*, 89 (1993) 2683.
126. P.X. Qi, E.D. Wickham, and H.M. Farrell, *Protein J.*, 23 (2004) 389.
127. H. Sjögren and S. Ulvenlund, *Biophys. Chem.*, 116 (2005) 11.
128. A. Chakraborty and S. Basak, *J. Photochem. Photobiol. B-Biol.*, 87 (2007) 191.
129. E. Dickinson, *Int. Dairy J.*, 9 (1999) 305.
130. V.B. Fainerman, S.V. Lylyk, A.V. Makievski, and R. Miller, *J. Colloid Interface Sci.*, 275 (2004) 305.
131. J. Benjamins, J.A. de Feijter, M.T.A. Evans, D.E. Graham, and M.C. Phillips, *Disc. Faraday Soc.*, 59 (1978) 218.
132. D.E. Graham and M.C. Phillips, *J. Colloid Interface Sci.*, 70 (1979) 415.
133. D.E. Graham and M.C. Phillips, *J. Colloid Interface Sci.*, 70 (1979) 427.
134. J.A. Feijter, J. Benjamins, and F.A. Veer, *Biopolymers*, 17 (1978) 1760.
135. J. Maldonado-Valderrama, A. Martin-Molina, A. Martin-Rodriguez, M.A. Cabrerizo-Vilchez, M.J. Galvez-Ruiz, and D. Langevin, *J. Phys. Chem. C*, 111 (2007) 2715.
136. R. Miller, V.S. Alahverdjieva, and V.B. Fainerman, *Soft Matter*, 4 (2008) 1141.
137. V.S. Alahverdjieva, V.B. Fainerman, E.V. Aksenenko, M.E. Leser, and R. Miller, *Colloids Surf. A*, 317 (2008) 610.
138. V.S. Alahverdjieva, D.O. Grigoriev, V.B. Fainerman, E.V. Aksenenko, R. Miller, and H. Möhwald, *J. Phys. Chem. C*, 112 (2008) 2136.
139. P. Reis, R. Miller, M.E. Leser, H.J. Watzke, V.B. Fainerman, and K. Holmberg, *Langmuir*, 24 (2008) 5781.
140. N. Moradi, Y. Zakrevskyy, S. Santer, A. Javadi, E.V. Aksenenko, V.B. Fainerman, and R. Miller, Surface tension and dilation rheology of DNA solutions in mixtures with cationic surfactant Azo-TAB, submitted to Colloids Surfaces A.

141. A.A. Sharipova, S.B. Aidarova, V.B. Fainerman, E.V. Aksenenko, N.E. Bekturganova, Y.I. Tarasevich, and R. Miller, *Colloids Surf. A*, 2014 (460) 11.
142. E. Dickinson, and R. Miller (Eds.), *Food Colloids—Fundamentals of Formulation*. Special publication No. 258, Royal Society of Chemistry, Cambridge, U.K. (2001).
143. J. Maldonado-Valderrama, V.B. Fainerman, M.J. Gálvez-Ruiz, A. Marín-Rodriguez, M.A. Cabrerizo-Vílchez, and R. Miller, *J. Phys. Chem. B*, 109 (2005) 17608.
144. A. Dan, R. Wüstneck, J. Krägel, E.V. Aksenenko, V.B. Fainerman, and R. Miller, *Food Hydrocolloids*, 34 (2014) 193.
145. J. Maldonado-Valderrama, P.J. Wilde, V.J. Morris, V.B. Fainerman, and R. Miller, *Langmuir*, 26 (2010) 15901.
146. V. Pradines, J. Krägel, V.B. Fainerman, and R. Miller, *J. Phys. Chem. B*, 113 (2009) 745.
147. R. Miller, E.V. Aksenenko, I.I. Zinkovych, and V.B. Fainerman, Adsorption of proteins at the aqueous solution/alkane interface: Co-adsorption of protein and alkane, *Adv. Colloid Interface Sci.*, accepted
148. E.H. Lucassen-Reynders, J. Benjamins, and V.B. Fainerman, *Curr. Opin. Colloid Interface Sci.*, 15 (2010) 264.
149. C. Stefaniu, G. Brezesinski, and H. Möhwald, *Adv. Colloid Interface Sci.*, 208 (2014) 197.
150. T. Smith, *Adv. Colloid Interface Sci.*, 3 (1972) 161.
151. E. Ruckenstein and A. Bhakta, *Langmuir*, 10 (1994) 2694.
152. J.N. Israelachvili, *Langmuir*, 10 (1994) 3774.
153. E. Ruckenstein and B. Li, *J. Phys. Chem.*, 100 (1996) 3108.
154. E. Ruckenstein and B. Li, *J. Phys. Chem.*, 102 (1998) 981.
155. V.B. Fainerman, D. Vollhardt, and G. Emrich, *J. Phys. Chem. B*, 104 (2000) 8536.
156. M. Volmer, *Z. Phys. Chem.* (Leipzig), 115 (1925) 253.
157. V.B. Fainerman and D. Vollhardt, *J. Phys. Chem. B*, 107 (2003) 3098.
158. D. Vollhardt and V.B. Fainerman, *Adv. Coll. Interface Sci.*, 127 (2006) 83.
159. D. Vollhardt and V.B. Fainerman, *Adv. Colloid Interface Sci.*, 86(2000) 103.
160. K. Motomura, I. Hayami, M. Aratono, and R. Matuura, *J. Colloid Interface Sci.*, 87 (1982) 333.
161. I. Panaiotov, L. Ter-Minassian-Saraga, and G. Albrecht, *Langmuir*, 1 (1985) 395.
162. B.A. Pethica, *Trans. Faraday Soc.*, 1955, 51, 1402.
163. V.B. Fainerman and D. Vollhardt, *Langmuir*, 15 (1999) 1784.
164. D.M. Alexander and G.T. Barnes, *J. Chem. Soc. Faraday Trans. 1*, 76 (1980) 118.
165. S. Sundaran and K.J. Stebe, *Langmuir*, 12 (1996) 2028; 13 (1997) 1729.

2 Nonequilibrium Thermodynamics of Interfaces

Leonard M.C. Sagis

CONTENTS

2.1 INTRODUCTION

Chapter 1 of this book discussed the thermodynamics of interfaces, in equilibrium with their adjoining bulk phases. Multiphase systems such as foams and emulsions are often produced or processed under conditions far from equilibrium, where they are subjected to high deformation rates and significant temperature gradients. Their response to these gradients can be highly nonlinear as a result of an intricate coupling of dynamic interfacial processes and subphase processes. For example, inhomogeneities in the flow along an interface can cause compositional gradients, where regions that are compressed coexist with expanded regions (Figure 2.1). The difference in concentration along the interface drives diffusion of mass along the interface (surface diffusion) and diffusive exchange between the interface and the subphase. But the concentration gradients also induce in-plane gradients in the surface rheological properties (such as the surface shear viscosity, or surface dilatational modulus), leading to a nonlinear rheological response, which affects the deformation of the interface.

Similar couplings of transfer modes can also be observed when temperature gradients are present in the system. These drive energy transfer along and across the interface but, in view of the temperature dependence of surface rheological properties, also affect the deformation of the interface. Moreover, the temperature may also induce mass diffusion along and across the interface (thermophoresis), an effect that is often negligible in single-phase bulk systems but can be highly important in describing mass transfer *across* interfaces [1–3]. Subphase mass transfer can also induce *transitions* in behavior of an interface. For example, when an interface is stabilized by negatively charged rodlike particles (e.g., protein fibrils), adsorption of divalent ions like calcium or magnesium to the interface can cross-link the fibrils [4], inducing a transition from liquid-like to solid-like behavior (Figure 2.2). Rodlike particle stabilized interfaces can also undergo transitions by an applied deformation (Figure 2.3): by compressing the interface its state can change from isotropic to nematic [5]. Whether and when this occurs depends on the degree of mass transfer of the particles between bulk

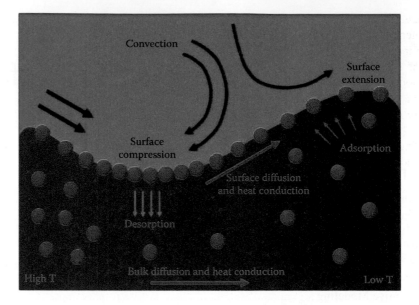

FIGURE 2.1   Coupled mass, momentum, and energy transfer in multiphase systems, in the bulk, along the interface, and between bulk and interface.

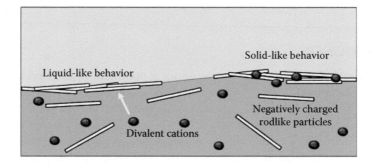

FIGURE 2.2   Effect of mass transfer between bulk and interface on interfacial behavior: adsorption of positively charged divalent ions cross-links adsorbed protein fibrils, inducing a change from liquid-like to solid-like (2D gel) behavior.

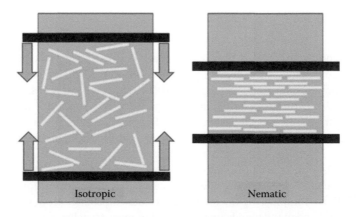

FIGURE 2.3   Example of a deformation-induced phase transition in an interface stabilized by rodlike particles: upon compression the state of the interface changes from isotropic to 2D nematic.

and interface. If they are soluble in the subphase, the compression may cause the particles to desorb from the interface, prior to reaching the surface concentration at which the transition occurs.

To describe the behavior and time evolution of multiphase systems under dynamic conditions and account for the coupling of mass, momentum, and energy transfer, we need to resort to a generalization of classical equilibrium thermodynamics, generally referred to as nonequilibrium thermodynamics (NET) [6].

There are several NET frameworks currently available to describe the dynamics of multiphase systems [7]. A first classification of these frameworks can be made based on the description of the interface. The two principal methods to describe interfaces in multiphase flows are the diffuse interface model [8] and the Gibbs dividing surface model [9]. In the former, the interface is modeled as a thin 3D region, in which densities and material properties change rapidly but continuously from their value in one bulk phase to their value in the adjoining bulk phase [8]. In the Gibbs dividing surface model, the interface is modeled as a 2D surface, placed sensibly within the interfacial region [9]. Bulk properties are extrapolated up to the surface, and the difference between actual and extrapolated fields is accounted for by associating excess properties (e.g., a surface excess density or surface excess energy) with this dividing surface. The Gibbs dividing surface model finds application mainly in macroscopic flow problems, where the interfacial thickness $d \ll L$, where L is a characteristic dimension of the flow problem. The diffuse interface model is typically used to describe problems where $d \simeq L$, such as droplet coalescence, phase separation in immiscible blends, or nano- and microfluidic flows.

In this chapter, we will focus only on NET frameworks employing the Gibbs dividing surface model. At the core of all Gibbs-based NET frameworks is the derivation of a set of partial differential equations for the (coupled) time rate of change of mass, momentum, and energy of a multiphase system, and the derivation of expressions for the fluxes that appear in these balances, which together with an equation of state and a set of appropriate boundary conditions, provide a complete description of the dynamics of a multiphase system.

One of the first NET frameworks to be generalized to multiphase systems was classical irreversible thermodynamics (CIT) [10–15]. This framework is particularly suited for describing multiphase systems near equilibrium, with interfaces with a viscous behavior (Section 2.3.1). By introducing internal (or structural) variables [16], the CIT framework can also be used to derive descriptions for (nonlinear) viscoelastic interfaces (Section 2.3.2). A second framework we will review here is extended irreversible thermodynamics (EIT) [17–19]. In this framework, the surface fluxes for mass, momentum, and energy are treated as additional independent variables. EIT is also suitable for constructing constitutive models for viscoelastic interfaces (Section 2.3.2). Other frameworks capable of describing multiphase systems exist, such as rational thermodynamics [7,20], extended rational thermodynamics [7,21], or GENERIC [7,22–27], but these are outside the scope of this chapter. See the recent articles [21–27] for discussions of these frameworks.

The first, and one of the most important steps in all NET frameworks, is to select a comprehensive set of system variables, capable of describing the dynamics of the system we wish to study. For a single-phase system without a complex microstructure, the classical choice of system variables is the bulk density, ρ; the momentum density, $\boldsymbol{m} = \rho \boldsymbol{v}$ (where $\boldsymbol{v}$ is the bulk velocity field); the internal energy density, $\bar{U} = \rho \hat{U}$ (where $\hat{U}$ is the internal energy per unit mass); and, in case we are dealing with an N-component system, the densities of the individual components, $\rho_{(J)}$ ($J = 1,...,N-1$). For a multiphase system with excesses mass, momentum, and energy, associated with the dividing surface, the most straightforward choice for the surface variables would be to introduce an excess variable for each of the aforementioned bulk variables. This means we assume that our relevant surface variables are the surface mass density, ρ^s; the surface momentum density, $\boldsymbol{m}^s = \rho^s \boldsymbol{v}^s$ (where $\boldsymbol{v}^s$ is the surface velocity field); the surface internal energy per unit area, $\bar{U}^s = \rho^s \hat{U}^s$ (where $\hat{U}^s$ is the surface internal energy per unit mass); and the surface densities of the individual components, $\rho^s_{(J)}$ ($J = 1,...,N - 1$).

In Section 2.2, we will show that the inclusion of these surface excess variables in the conservation principles for a multiphase system allows us to derive balance equations for these variables [7,10–15,19–21,25–27]. These balance equations contain contributions from fluxes, and in Section 2.3.1, we will show how CIT can be used to derive constitutive equations for these fluxes, which are consistent with the second law of thermodynamics, for simple viscous interfaces without a complex microstructure. In Section 2.3.2, we show how, with the introduction of internal variables [16], the CIT framework can also be used to derive constitutive equations for interfaces with a complex microstructure, with (nonlinear) viscoelastic behavior. We also discuss how such equations can be derived using the EIT framework.

2.2 CONSERVATION PRINCIPLES

In this section, we will discuss the conservation of mass, momentum, moment of momentum, energy, and entropy, for multiphase systems with excess variables associated with their dividing surfaces. We will show that the inclusion of excess variables in the conservation principles leads to a set of partial differential equations for these excess variables, describing their time evolution as a result of in-plane convective and diffusion processes, and convective and diffusive exchange between the dividing surface and the adjoining bulk phases [7,10–15,19–21,25–27]. These balances are often referred to as *jump* balances [7,20], and together with a set of boundary conditions, they couple the motion and deformation of the dividing surfaces to the time evolution of the bulk phases.

2.2.1 CONSERVATION OF MASS

The principle of conservation of mass states that the mass of a multiphase system should be constant in time. For a multiphase system with mass associated with its dividing surfaces, we can express this principle as [7,11,15,20]

$$\frac{\mathrm{d}}{\mathrm{d}t}\left[\int_R \rho\,\mathrm{d}V + \int_\Sigma \rho^s\,\mathrm{d}\Omega\right] = 0, \tag{2.1}$$

where
  R denotes the domain of the bulk phases
  Σ denotes the domain of the dividing surfaces
  $\mathrm{d}V$ denotes a volume integration
  $\mathrm{d}\Omega$ denotes an area integration

Evaluating the time derivative on the left-hand side of Equation 2.1, we find that the principle of conservation of mass requires that at any point in the bulk phase R

$$\frac{\mathrm{d_b}\rho}{\mathrm{d}t} = -\rho\nabla\cdot\boldsymbol{v}, \tag{2.2}$$

which is generally referred to as the equation of continuity [7,11,15,20]. Here, ∇ is the 3D gradient operator. In addition to this expression, we find that conservation of mass also implies that at any point on the dividing surface Σ

$$\frac{\mathrm{d_s}\rho^s}{\mathrm{d}t} = -\rho^s\nabla_s\cdot\boldsymbol{v}^s - \left[\rho\left(\boldsymbol{v}-\boldsymbol{v}^s\right)\cdot\boldsymbol{n}\right]. \tag{2.3}$$

This equation is referred to as the jump overall mass balance [7,20]. At every point on the dividing surface, it describes the time rate of change of the overall surface mass density, as a result of in-plane convective processes (first term on the right-hand side), and exchange of mass with the adjoining bulk phases (second term on the right-hand side). The operator ∇_s is the surface gradient operator [2,15]. The material derivatives appearing in Equations 2.2 and 2.3 are defined as

$$\frac{d_b\psi}{dt} = \frac{\partial\psi}{\partial t} + (\nabla\psi)\cdot v \qquad \frac{d_s\psi^s}{dt} = \frac{\partial\psi^s}{\partial t} + (\nabla_s\psi^s)\cdot \dot{y}. \tag{2.4}$$

The velocity $\dot{y}$ is the intrinsic surface velocity, defined as $\dot{y} \equiv v^s - u$, where u is the speed of displacement of the interface [20]. The double square brackets in Equation 2.3, describing the exchange of mass between the interface and the adjoining bulk phases, are defined as [20]

$$\left[\psi n\right] = \psi^I n^I + \psi^{II} n^{II}, \tag{2.5}$$

where
   ψ^I and ψ^{II} are, respectively, the value of ψ in bulk phases I and II, evaluated at the dividing surface
   The vector n^I is the unit vector normal to the dividing surface, pointing in the direction of phase I (and hence, $n^I = -n^{II}$)

When our system is a multicomponent system, we can also explore what the principle of conservation of mass implies for each individual component in the mixture. For each component in the mixture, mass conservation implies that its time rate of change is equal to the rate at which the component is converted by chemical reactions in the bulk phase and at the dividing surfaces. Mathematically, we can express this as [7,11,15,20]

$$\frac{d}{dt}\left[\int_R \rho\omega_{(J)}\, dV + \int_\Sigma \rho^s\omega^s_{(J)}\, d\Omega\right] = \int_R r_{(J)}\, dV + \int_\Sigma r^s_{(J)}\, d\Omega, \tag{2.6}$$

where
   $\omega_{(J)}$ and $\omega^s_{(J)}$ are, respectively, the bulk and surface mass fraction of component J ($J = 1,\dots,N$)
   $r_{(J)}$ is the rate at which mass of component J is converted per unit volume, by reactions in the bulk phase
   $r^s_{(J)}$ is the rate at which J is converted per unit area, by reactions in the dividing surface

Evaluating the time derivative on the left-hand side of Equation 2.6, we find that in the bulk phases,

$$\rho\frac{d_b\omega_{(J)}}{dt} = -\nabla\cdot j_{(J)} + r_{(J)}, \tag{2.7}$$

which is the familiar differential mass balance for component J in the mixture. At any point on the dividing surface, Equation 2.6 requires

$$\rho^s\frac{d_s\omega^s_{(J)}}{dt} = -\nabla_s\cdot j^s_{(J)} + r^s_{(J)} - \left[\rho\left(\omega_{(J)} - \omega^s_{(J)}\right)\left(v - v^s\right)\cdot n + j_{(J)}\cdot n\right]. \tag{2.8}$$

This balance is referred to as the jump species mass balance [7,20] and describes the time rate of change of the surface excess concentration of component J, as a function of (in order of appearance, on the right-hand side of the equation) in-plane diffusion, chemical reactions, convective transport between the interface and bulk phases, and diffusive transfer between the interface and bulk phases. The mass flux vectors appearing in Equations 2.7 and 2.8 are defined as

$$\boldsymbol{j}_{(J)} = \rho_{(J)}(\boldsymbol{v}_{(J)} - \boldsymbol{v}), \quad \boldsymbol{j}^{\mathrm{s}}_{(J)} = \rho^{\mathrm{s}}_{(J)}(\boldsymbol{v}^{\mathrm{s}}_{(J)} - \boldsymbol{v}^{\mathrm{s}}). \tag{2.9}$$

The jump species mass balance is an important balance. When interfaces are deformed, surface concentrations typically change, and in-plane gradients may develop, which drive mass transfer along the interface (Gibbs–Marangoni effect) and exchange of mass with the adjoining bulk phases. These mass-transfer processes affect the composition of the interface and hence its local material properties, such as surface tension or surface viscosities. The jump mass balance then needs to be solved simultaneously with the jump momentum balance, which we will discuss next.

2.2.2 CONSERVATION OF MOMENTUM AND MOMENT OF MOMENTUM

The principle of conservation of momentum states that the time rate of change of momentum of a system is equal to the sum of the body forces acting on the material in the bulk phases and dividing surfaces and the stresses (or contact forces) acting on the system through its outer boundaries. For a multiphase system with excess properties associated with its dividing surface, we can formulate this principle as [7,11,15,20]

$$\frac{\mathrm{d}}{\mathrm{d}t}\left[\int_R \rho \boldsymbol{v}\,\mathrm{d}V + \int_\Sigma \rho^{\mathrm{s}} \boldsymbol{v}^{\mathrm{s}}\,\mathrm{d}\Omega\right] = \int_R \rho \boldsymbol{b}\,\mathrm{d}V + \int_\Sigma \rho^{\mathrm{s}} \boldsymbol{b}^{\mathrm{s}}\,\mathrm{d}\Omega + \int_S \boldsymbol{T} \cdot \boldsymbol{n}\,\mathrm{d}\Omega + \int_C \boldsymbol{T}^{\mathrm{s}} \cdot \boldsymbol{\mu}\,\mathrm{d}L, \tag{2.10}$$

where
   $\boldsymbol{b}$ and $\boldsymbol{b}^{\mathrm{s}}$ are the body forces per unit mass acting on, respectively, material in the bulk phase and material in the dividing surfaces
   $\boldsymbol{T}$ is the stress tensor in the bulk phase
   $\boldsymbol{T}^{\mathrm{s}}$ is the surface stress tensor
   S and C domains are the outer bounding surface of the system and the curve formed by the intersection of this surface with the dividing surface Σ
   $\boldsymbol{\mu}$ is the unit vector, perpendicular to C, tangent to Σ, and pointing outward from the system
   $\mathrm{d}L$ denotes a line integration

Evaluating the time derivative on the left-hand side of Equation 2.10 and using the divergence and surface divergence theorems, we find that the principle of conservation of momentum implies that at each point in the bulk phase, we must satisfy the differential momentum balance:

$$\rho \frac{\mathrm{d}_{\mathrm{b}}\boldsymbol{v}}{\mathrm{d}t} = \nabla \cdot \boldsymbol{T} + \rho \boldsymbol{b}. \tag{2.11}$$

In addition, on each point on the dividing surface Σ, we must satisfy

$$\rho^{\mathrm{s}} \frac{\mathrm{d}_{\mathrm{s}}\boldsymbol{v}^{\mathrm{s}}}{\mathrm{d}t} = \nabla_{\mathrm{s}} \cdot \boldsymbol{T}^{\mathrm{s}} + \rho^{\mathrm{s}} \boldsymbol{b}^{\mathrm{s}} - \left[\!\left[\rho\left(\boldsymbol{v} - \boldsymbol{v}^{\mathrm{s}}\right)\left(\boldsymbol{v} - \boldsymbol{v}^{\mathrm{s}}\right) \cdot \boldsymbol{n} - \boldsymbol{T} \cdot \boldsymbol{n}\right]\!\right]. \tag{2.12}$$

The latter equation is often referred to as the jump momentum balance [7,20]. Often, we will find it convenient to split the stress tensors in these two equations in a hydrostatic and deviatoric contribution, according to

$$T = -p\boldsymbol{I} + \sigma, \quad \boldsymbol{T}^s = \gamma\boldsymbol{P} + \sigma^s, \tag{2.13}$$

where

p is the thermodynamic pressure

σ is the extra stress tensor in the bulk phase

γ is the surface tension

σ^s is the surface extra stress tensor

$\boldsymbol{I}$ and $\boldsymbol{P}$ are, respectively, the 3D and 2D unit tensors [7,20]

Substituting Equations 2.13 in 2.12, we obtain

$$\rho^s \frac{d_s \boldsymbol{v}^s}{dt} = \nabla_s \gamma + 2\gamma H \boldsymbol{n} + \nabla_s \cdot \sigma^s + \rho^s \boldsymbol{b}^s - \left[\rho\left(\boldsymbol{v} - \boldsymbol{v}^s\right)\left(\boldsymbol{v} - \boldsymbol{v}^s\right) \cdot \boldsymbol{n} + P\boldsymbol{n} - \sigma \cdot \boldsymbol{n}\right]. \tag{2.14}$$

Here, H is the mean curvature of the dividing surface. This expression allows us to calculate the time rate of change of momentum associated with the dividing surface, as a result of (in order of appearance on the right-hand side of the equation) surface tension gradients (Marangoni stresses), curvature-induced stresses, deviatoric stresses, body forces, and stresses exerted on the interface by the adjoining bulk phases (respectively, inertial, hydrostatic, and deviatoric stresses). Equation 2.14 is a generalized form of the Young–Laplace equation, which is often used in the analysis of droplet-based tensiometry methods, such as profile analysis tensiometry or bubble pressure tensiometry [7,28]. For slow surface deformations, the term on the left-hand side of the equation will be negligibly small. If in addition the deformation is uniform and deviatoric stresses and external force fields are negligible, as well as the inertial and deviatoric stresses exerted on the interface by the bulk phases, then Equation 2.14 reduces to

$$2\gamma H\boldsymbol{n} - \left[p\boldsymbol{n}\right] = 0. \tag{2.15}$$

And for a spherical bubble with radius R, this can be written as

$$\frac{\gamma}{R} = p^{II} - p^{I}, \tag{2.16}$$

where p^{II} is the pressure in the interior of the bubble. This is the familiar Laplace equation for the pressure difference across the interface of a spherical bubble [20]. In tensiometry experiments on interfaces with a complex microstructure, the deviatoric contributions to (2.14) are often not negligible, which may lead to a strain and droplet size dependence of the dilatational moduli [29].

The principle of conservation of moment of momentum states that the time rate of change of the total moment of momentum of a system is equal to the total torque applied on the system. If we assume that the body forces and stresses we previously introduced are the only contributions to the applied torque, we can formulate this principle as [20]

$$\frac{d}{dt}\left[\int_R \rho\boldsymbol{r} \times \boldsymbol{v} dV + \int_\Sigma \rho^s \boldsymbol{r} \times \boldsymbol{v}^s d\Omega\right] = \int_R \rho\boldsymbol{r} \times \boldsymbol{b} dV + \int_\Sigma \rho^s \boldsymbol{r} \times \boldsymbol{b}^s d\Omega$$

$$+ \int_S \boldsymbol{r} \times (\boldsymbol{T} \cdot \boldsymbol{n}) d\Omega + \int_C \boldsymbol{r} \times (\boldsymbol{T}^s \cdot \mu) dL, \tag{2.17}$$

where $\boldsymbol{r}$ is the position vector. Evaluating the time derivative on the left-hand side of (2.17), using the divergence and surface divergence theorems, and invoking the differential and jump momentum balance, we find that [20]

$$\varepsilon : \boldsymbol{T} = 0, \quad \varepsilon : \boldsymbol{T}^s = 0, \tag{2.18}$$

which implies that both stress tensors are symmetric. The tensor ε is the skew-symmetric Levi-Civita tensor [20]. Note that when there are additional sources of moment of momentum, beyond those resulting from the body and contact forces (such as those induced by a suitable rotating electric field), the stress tensors may no longer be symmetric. In the remainder of this chapter, we will assume that both the bulk and surface stress tensors *are* symmetric.

2.2.3 Conservation of Energy

The principle of conservation of energy states that the time rate of change of the sum of the internal and kinetic energy of a system is equal to the work done on the system by the sum of all body and contact forces, plus the energy transmitted to the interior of the system by radiation, plus the energy transmitted to the system through its outer boundaries. For a multiphase system with mass, momentum, and energy associated with its dividing surface, we can formulate this principle as [7,11,15,20]

$$\frac{d}{dt}\left[\int_R \rho \left(\hat{U} + \frac{1}{2}v^2 \right) dV + \int_\Sigma \rho^s \left(\hat{U}^s + \frac{1}{2}[v^s]^2 \right) d\Omega \right] = \int_R \sum_J \rho_{(J)} \boldsymbol{b}_{(J)} \cdot \boldsymbol{v}_{(J)} dV$$

$$+ \int_\Sigma \sum_J \rho^s_{(J)} \boldsymbol{b}^s_{(J)} \cdot \boldsymbol{v}^s_{(J)} d\Omega + \int_S \boldsymbol{v} \cdot (\boldsymbol{T} \cdot \boldsymbol{n}) d\Omega + \int_C \boldsymbol{v}^s \cdot (\boldsymbol{T}^s \cdot \boldsymbol{\mu}) dL$$

$$+ \int_R \rho \hat{Q} \, dV + \int_\Sigma \rho^s \hat{Q}^s \, d\Omega + \int_S \boldsymbol{q} \cdot \boldsymbol{n} d\Omega + \int_C \boldsymbol{q}^s \cdot \boldsymbol{\mu} dL, \tag{2.19}$$

where

$\hat{Q}$ is the rate of energy transfer per unit mass to the material in the bulk phase by radiation
$\hat{Q}^s$ is the rate of radiative energy transfer per unit mass to the material in the dividing surface
$\boldsymbol{q}$ is the energy flux vector
$\boldsymbol{q}^s$ is the surface energy flux vector and $v^2 = \boldsymbol{v} \cdot \boldsymbol{v}$

In the first two terms on the right-hand side of this equation, we have allowed for the fact that the body forces may vary for different components. Evaluating the time derivative on the left-hand side of the equation, using the divergence and surface divergence theorems, the differential and jump mass balances, and the differential and jump momentum balances, we find that the principle of conservation of energy requires that at each point in the bulk phase

$$\rho \frac{d_b \hat{U}}{dt} = \boldsymbol{\sigma} : \nabla \boldsymbol{v} - p\nabla \cdot \boldsymbol{v} + \sum_J \boldsymbol{j}_{(J)} \cdot \boldsymbol{b}_{(J)} - \nabla \cdot \boldsymbol{q} + \rho \hat{Q}. \tag{2.20}$$

In addition, we must require at each point on the dividing surface

$$\rho^s \frac{d_s \hat{U}^s}{dt} = \boldsymbol{\sigma} : \nabla_s \boldsymbol{v}^s + \gamma \nabla_s \cdot \boldsymbol{v}^s + \sum_{J=1}^N \boldsymbol{j}^s_{(J)} \cdot \boldsymbol{b}^s_{(J)} - \nabla_s \cdot \boldsymbol{q}^s + \rho^s \hat{Q}^s$$

$$- \left[\rho \left(\hat{U} + p\hat{V} - \hat{U}^s + \frac{1}{2}|\boldsymbol{v} - \boldsymbol{v}^s|^2 \right)(\boldsymbol{v} - \boldsymbol{v}^s) \cdot \boldsymbol{n} + \boldsymbol{q} \cdot \boldsymbol{n} - (\boldsymbol{v} - \boldsymbol{v}^s) \cdot \boldsymbol{\sigma} \cdot \boldsymbol{n} \right]. \tag{2.21}$$

These balances are generally referred to as, respectively, the differential energy balance and the jump energy balance [7,20]. In Equation 2.21, $\hat{V} = 1/\rho$ denotes the volume per unit mass, and the colon in the first term on the right-hand side denotes a double contraction between the two tensors. The jump energy balance allows us to calculate the time evolution of the surface internal energy per unit mass, as a result of in-plane viscous dissipation, the work done by surface tension and body forces, in-plane conduction, radiative heat transfer, convective and conductive exchange with the bulk phases, and viscous friction between the interface and the bulk phases.

2.2.4 ENTROPY BALANCE

The final principle we will consider in this section is the entropy balance. This principle states that for any system the time rate of change of its total entropy must be equal to the entropy production in its interior, plus the entropy transmitted to the system through its outer boundaries. Mathematically, we can formulate this as [7,11,15]

$$\frac{d}{dt}\left[\int_R \rho\hat{S}dV + \int_\Sigma \rho^s\hat{S}^s d\Omega\right] = \int_R \rho\hat{E}dV + \int_\Sigma \rho^s\hat{E}^s d\Omega - \int_S j_S \cdot n d\Omega - \int_C j_S^s \cdot \mu dL, \qquad (2.22)$$

where

$\hat{E}$ is the rate of entropy production per unit mass in the bulk phase
$\hat{E}^s$ is the rate of entropy production per unit mass in the dividing surface
j_S is the bulk entropy flux vector
j_S^s is the surface entropy flux vector

Evaluating the time derivative on the left-hand side of this equation and using the divergence and surface divergence theorems, we find that at any point in the bulk phase, we must satisfy

$$\rho\frac{d_b\hat{S}}{dt} = -\nabla \cdot j_S + \rho\hat{E}. \qquad (2.23)$$

In addition, at every point on the dividing surface, we must require

$$\rho^s\frac{d_s\hat{S}^s}{dt} = -\nabla_s \cdot j_S^s + \rho^s\hat{E}^s - \left[\rho\left(\hat{S} - \hat{S}^s\right)(v - v^s) \cdot n + j_S \cdot n\right]. \qquad (2.24)$$

These equations are, respectively, the differential and jump entropy balances [7,20]. To satisfy the second law of thermodynamics, we must impose

$$\hat{E} \geq 0, \quad \hat{E}^s \geq 0, \qquad (2.25)$$

where the entropy production rates are zero for reversible processes, and greater than zero for irreversible processes. The entropy balances provide an important tool in the construction of admissible constitutive equations, which do not violate the second law of thermodynamics. We will illustrate this in the next section.

2.3 CONSTITUTIVE MODELING

When we examine the balance equations we derived in the previous section more closely, we observe that we still need to supply some additional information before we can solve them: (1) we need to choose appropriate constitutive equations for the fluxes that appear in these balances, and (2) we need to construct the boundary conditions that couple the bulk dynamics to the time evolution of the dividing surface. The various NET frameworks differ in how they construct these equations. Here, we will primarily follow the procedures used in the CIT framework [6]. In Section 2.3.1, we will first discuss constitutive modeling for simple interfaces, and in Section 2.3.2, we briefly discuss modeling the behavior of interfaces with a complex microstructure.

2.3.1 SIMPLE INTERFACES

In the CIT framework, the entropy balance is used to guide us in the construction of constitutive equations for the fluxes and the boundary conditions that couple the bulk and jump equations [6,7,11,15]. We will illustrate this procedure here for the surface fluxes, using the jump entropy balance. The derivation of constitutive equations for bulk fluxes is described in detail elsewhere [6] and is very similar to the development we present here.

To construct equations for the surface fluxes, we first need to choose a functional dependence for the surface entropy per unit mass [6,7,11,15]. In the CIT framework, it is typical to assume local equilibrium, that is, although the system as a whole is not in a global state of equilibrium, locally, at any point in the system, the material is assumed to be in equilibrium with the material in its immediate neighborhood [6]. This means that locally, the surface entropy per unit mass depends on the same set of variables, as the entropy of an interface in global equilibrium:

$$\hat{S}^s = \hat{S}^s(\hat{U}^s, \hat{A}, \omega_{(J)}^s, \dots, \omega_{(N-1)}^s). \tag{2.26}$$

Here, $\hat{A} = 1/\rho^s$ is the surface area per unit mass. If we now take the surface material time derivative of $\hat{S}^s$ and invoke (2.26), we obtain using the chain rule of differentiation

$$\rho^s \frac{d_s \hat{S}^s}{dt} = \frac{\rho^s}{T^s} \frac{d_s \hat{U}^s}{dt} - \frac{\gamma \rho^s}{T^s} \frac{d_s \hat{A}}{dt} - \frac{\rho^s}{T^s} \sum_{J=1}^{N} \mu_{(J)}^s \frac{d_s \omega_{(J)}^s}{dt}, \tag{2.27}$$

where

μ_J^s is the chemical potential of component J
T^s is the surface temperature

Next, we substitute the jump energy balance (2.21), the jump mass balance (2.3), and the jump component mass balance (2.8) in this expression and subsequently introduce the result into the jump entropy balance (2.24), to find an expression for the surface rate of entropy production per unit mass:

$$\rho^s \hat{E}^s = \frac{1}{T^s} \bar{\sigma}^s : \bar{D}^s + \frac{\mathrm{tr}\sigma^s}{T^s} \mathrm{tr} D^s - \frac{1}{T^s} \sum_{J=1}^{N} j_{(J)}^s \cdot d_{(J)}^s - \frac{1}{(T^s)^2} \left(q^s - \sum_{J=1}^{N} \mu_{(J)}^s j_{(J)}^s \right) \cdot \nabla_s T^s$$

$$- \frac{1}{T^s} \left\| \begin{array}{c} \rho \left((\hat{U} + p\hat{V}) \left[\dfrac{T - T^s}{T} \right] + \sum_{J=1}^{N} T^s \left(\dfrac{\tilde{\mu}_{(J)}}{T} - \dfrac{\tilde{\mu}_{(J)}^s}{T^s} \right) \omega_{(J)} \right) (v - v^s) \cdot n \\[2ex] + \frac{1}{2} \rho \left((v - v^s)^2 + \dfrac{v^2 T^s - (v^s)^2 T}{T} \right) (v - v^s) \cdot n \\[2ex] + q \cdot n \left[\dfrac{T - T^s}{T} \right] - (v - v^s) \cdot \sigma \cdot n + T^s \sum_{J=1}^{N} j_{(J)} \cdot n \left(\dfrac{\tilde{\mu}_{(J)}}{T} - \dfrac{\tilde{\mu}_{(J)}^s}{T^s} \right) \end{array} \right\| \tag{2.28}$$

$$\geq 0.$$

The bar over the tensors $\boldsymbol{\sigma}^s$ and $\boldsymbol{D}^s$ denotes the symmetric traceless part of these tensors, so, $\overline{\boldsymbol{\sigma}}^s = \boldsymbol{\sigma}^s - 1/2\,\boldsymbol{P}(\mathrm{tr}\boldsymbol{\sigma}^s)$ and $\overline{\boldsymbol{D}}^s = \boldsymbol{D}^s - 1/2\,\boldsymbol{P}(\mathrm{tr}\boldsymbol{D}^s)$.

The tensor $\boldsymbol{D}^s$ is the surface rate of deformation tensor, equal to $1/2\left(\boldsymbol{P}\cdot\nabla_s\boldsymbol{v}^s + [\nabla_s\boldsymbol{v}^s]^{\mathrm{T}}\cdot\boldsymbol{P}\right)$.

The chemical potentials $\tilde{\mu}_{(J)}$ and $\tilde{\mu}^s_{(J)}$ are, respectively, the velocity modified bulk and surface chemical potentials of component J, equal to $\tilde{\mu}_{(J)} = \mu_{(J)} - 1/2v^2$ and $\tilde{\mu}^s_{(J)} = \mu^s_{(J)} - 1/2(v^s)^2$.

T is the temperature in the bulk phase, and the vector $\boldsymbol{d}^s_{(J)} \equiv \nabla_s\mu^s_{(J)} - \boldsymbol{b}^s_{(J)}$.

In arriving at this result, we have assumed that the reaction rates $r^s_{(J)}$ and the rate of radiative energy transmission per unit mass $\hat{Q}^s$ are both zero.

Equation 2.28 is a bilinear form consisting of products of fluxes and driving forces. To ensure Equation 2.28 is satisfied, we assume that the fluxes that appear in this expression depend linearly on all driving forces of equal tensorial order [6,7,11,15]. For the traceless part of the surface extra stress tensor, and its trace, this implies, we choose

$$\overline{\boldsymbol{\sigma}}^s = \overline{\boldsymbol{\sigma}}^s(\overline{\boldsymbol{D}}^s) = 2\varepsilon_s\overline{\boldsymbol{D}}^s, \tag{2.29}$$

$$\mathrm{tr}\boldsymbol{\sigma}^s = \mathrm{tr}\boldsymbol{\sigma}^s(\mathrm{tr}\boldsymbol{D}^s) = \varepsilon_d\mathrm{tr}\boldsymbol{D}^s, \tag{2.30}$$

where

ε_s is the surface shear viscosity

ε_d is the surface dilatational viscosity

Both these viscosities may depend on temperature and composition of the interface, but not on the rate of deformation. Combining these two expressions, we obtain

$$\boldsymbol{\sigma}^s = \left(\varepsilon_d - \varepsilon_s\right)\left(\mathrm{tr}\boldsymbol{D}^s\right)\boldsymbol{P} + 2\varepsilon_s\boldsymbol{D}^s, \tag{2.31}$$

which is the linear Boussinesq model [30–32], the surface equivalent of the Newtonian fluid model.

For the mass and heat flux vectors, Equation 2.28 suggests the following functional dependence:

$$\boldsymbol{j}^s_{(J)} = \boldsymbol{j}^s_{(J)}(\boldsymbol{d}^s_{(J)}, \nabla_s T^s) = -\sum_K D^s_{(JK)}\boldsymbol{d}^s_{(K)} - \alpha^s_{(J)}\nabla_s\ln T^s, \tag{2.32}$$

$$\boldsymbol{q}^s - \sum_{J=1}^{N}\mu^s_{(J)}\boldsymbol{j}^s_{(J)} = T^s\boldsymbol{j}^s_S = T^s\boldsymbol{j}^s_S(\boldsymbol{d}^s_{(J)}, \nabla_s T^s) = -\sum_J\alpha^s_{(J)}\boldsymbol{d}^s_{(J)} - \lambda^s\nabla_s\ln T^s, \tag{2.33}$$

where

the coefficients $D^s_{(JK)}$ denote the components of the $N \times N$ diffusion matrix

$\alpha^s_{(J)}$ is the surface thermal diffusion coefficient for component J

λ^s is the surface thermal conductivity

Using the fact that $\mu^s_{(J)} = H^s_{(J)} - TS^s_{(J)}$, where $H^s_{(J)}$ and $S^s_{(J)}$ are the partial surface enthalpy and entropy of component J, the left-hand side of Equation 2.33 can be written as

$$\boldsymbol{q}^s - \sum_{J=1}^{N}\mu^s_{(J)}\boldsymbol{j}^s_{(J)} \equiv \boldsymbol{\epsilon}^s + \sum_{J=1}^{N}T^s S^s_{(J)}\boldsymbol{j}^s_{(J)}, \tag{2.34}$$

where $\boldsymbol{\epsilon}^s = \boldsymbol{q}^s - \sum_{J=1}^{N}H^s_{(J)}\boldsymbol{j}^s_{(J)}$ is often referred to as the *measurable* heat flux.

Note that Equations 2.32 and 2.33 contain couplings between mass and energy transfer, which are the surface equivalents of the Soret effect (mass transfer driven by gradients in temperature, also known as thermodiffusion or thermophoresis) and Dufour effect (the energy flux driven by concentration gradients). These effects are typically negligible in the bulk phase, and for in-plane mass and energy transfer, but are often highly relevant in the description of mass and energy transfer *across* interfaces [1–3]. When the Soret effect and contributions from forced diffusion are negligible, Equation 2.32 reduces to

$$\boldsymbol{j}_{(J)}^{\mathrm{s}} = -\sum_{K} D_{(JK)}^{\mathrm{s}} \nabla_{\mathrm{s}} \mu_{(K)}^{\mathrm{s}}, \tag{2.35}$$

which is the surface equivalent of Fick's law. For an interface with uniform surface composition, the Dufour effect is negligible, and Equation 2.33 reduces to

$$\boldsymbol{q}^{\mathrm{s}} = -\frac{\lambda^{\mathrm{s}}}{T^{\mathrm{s}}} \nabla_{\mathrm{s}} T^{\mathrm{s}}. \tag{2.36}$$

This expression is the surface equivalent of Fourier's law.

Apart from the constitutive expressions for the in-plane fluxes, Equation 2.28 allows us also to construct constitutive expressions for the fluxes describing exchange between the interface and the adjoining bulk phases. For this, we have to focus on the double bracket term in this expression. Following along the lines we used for deriving the constitutive equations for the surface fluxes, we obtain ($M,N = \mathrm{I,II}$)

$$\boldsymbol{\sigma}^{M} \cdot \boldsymbol{n}^{M} - \rho^{M} \boldsymbol{v}^{M} \left(\boldsymbol{v}^{M} - \boldsymbol{v}^{\mathrm{s}} \right) \cdot \boldsymbol{n}^{M} = \sum_{N=I}^{II} \zeta^{M,N} T^{\mathrm{s}} \cdot \left(\frac{\boldsymbol{v}^{N}}{T^{N}} - \frac{\boldsymbol{v}^{\mathrm{s}}}{T^{\mathrm{s}}} \right), \tag{2.37}$$

$$\boldsymbol{q}^{M} \cdot \boldsymbol{n}^{M} + \rho^{M} \left[\hat{U}^{M} + p^{M} \hat{V}^{M} + \frac{1}{2} (\mathrm{v}^{M})^{2} \right] \left(\boldsymbol{v}^{M} - \boldsymbol{v}^{\mathrm{s}} \right) \cdot \boldsymbol{n}^{M} - \boldsymbol{v}^{M} \cdot \boldsymbol{\sigma}^{M} \cdot \boldsymbol{n}^{M}$$
$$= -\frac{T^{M} - T^{\mathrm{s}}}{R_{K}^{M}} - \sum_{J} \Lambda_{(J)}^{TM} T^{M} T^{\mathrm{s}} \left(\frac{\tilde{\mu}_{(J)}^{M}}{T^{M}} - \frac{\tilde{\mu}_{(J)}^{\mathrm{s}}}{T^{\mathrm{s}}} \right), \tag{2.38}$$

$$\boldsymbol{j}_{(J)}^{M} \cdot \boldsymbol{n}^{M} + \rho_{(J)}^{M} \left(\boldsymbol{v}^{M} - \boldsymbol{v}^{\mathrm{s}} \right) \cdot \boldsymbol{n}^{M} = -\Lambda_{(J)}^{M} \left(\frac{\tilde{\mu}_{(J)}^{M}}{T^{M}} - \frac{\tilde{\mu}_{(J)}^{\mathrm{s}}}{T^{\mathrm{s}}} \right) - \Lambda_{(J)}^{TM} \left(T^{M} - T^{\mathrm{s}} \right), \tag{2.39}$$

where
  $\zeta^{M,N}$ are the friction tensors, quantifying the exchange of momentum between the bulk phases and the dividing surface
  R_{K}^{M} is the Kapitza coefficient, which is the resistance against energy transfer between bulk and interface
  $\Lambda_{(J)}^{M}$ is the mass-transfer coefficient for exchange of component J between bulk and interface, driven by differences in chemical potential
  $\Lambda_{(J)}^{TM}$ is the mass-transfer coefficient for the transfer of component J between bulk and interface, driven by temperature differences

These equations are not only constitutive equations for the exchange of mass, momentum, and energy between the bulk phases and the dividing surface. They also act as boundary conditions, coupling the differential equations for the bulk density, velocity, and energy fields, with their respective jump balances.

The entropy balance allows us also to make a statement about the sign of the coefficients we introduced in the constitutive equations [6]. For a system without exchange between the bulk phase and the dividing surface, Equation 2.28 reduces to (substituting Equations 2.29, 2.30, 2.32, and 2.33)

$$\varepsilon_s \overline{\boldsymbol{D}}^s : \overline{\boldsymbol{D}}^s + \varepsilon_d (\mathrm{tr} \boldsymbol{D}^s)^2 + \sum_J \sum_K D^s_{(JK)} \boldsymbol{d}^s_{(J)} \cdot \boldsymbol{d}^s_{(K)}$$

$$+ 2 \sum_J \alpha^s_{(J)} \boldsymbol{d}^s_{(J)} \cdot \nabla_s \ln T^s + \lambda^s (\nabla_s \ln T^s)^2 \geq 0. \tag{2.40}$$

For this quadratic expression to be satisfied for any arbitrary rate of deformation, concentration gradient, or temperature gradient, we must require

$$\varepsilon_s \geq 0, \quad \varepsilon_d \geq 0, \quad D^s_{(JK)} \geq 0, \quad \lambda^s \geq 0. \tag{2.41}$$

In a similar manner, we find that the coefficients quantifying the transfer of mass and energy between the bulk phase and the dividing surface must satisfy

$$R^M_K \geq 0, \quad \Lambda^M_{(J)} \geq 0, \quad \Lambda^{TM}_{(J)} \geq 0, \tag{2.42}$$

and that the tensors $\zeta^{M,N}$ are positive semidefinite tensors.

The set of bulk and jump balances, coupled by the boundary conditions Equations 2.37 through 2.39, is not yet complete. This is perhaps most obvious when we substitute (2.31) in the jump momentum balance to find

$$\rho^s \frac{d_s \boldsymbol{v}^s}{dt} = \nabla_s \gamma + 2\gamma H \boldsymbol{n} + 2H(\varepsilon_d - \varepsilon_s)(\mathrm{tr} \boldsymbol{D}^s)\boldsymbol{n} + (\varepsilon_d - \varepsilon_s)\nabla_s(\mathrm{tr} \boldsymbol{D}^s)$$

$$+ \mathrm{tr} \boldsymbol{D}^s \nabla_s (\varepsilon_d - \varepsilon_s) + 2\varepsilon_s \nabla_s \cdot \boldsymbol{D}^s + 2(\nabla_s \varepsilon_s) \cdot \boldsymbol{D}^s$$

$$+ \rho^s \boldsymbol{b}^s - \left[\rho(\boldsymbol{v} - \boldsymbol{v}^s)(\boldsymbol{v} - \boldsymbol{v}^s) \cdot \boldsymbol{n} + p\boldsymbol{n} - \sigma \cdot \boldsymbol{n} \right]. \tag{2.43}$$

To solve this equation, we need to supply an equation of state that links the surface tension to surface composition and temperature of the interface (see Chapter 1). We also need to know the concentration and temperature dependence of the surface viscosities. The dependence of material properties on surface composition and temperature cannot be obtained from macroscopic NET frameworks. For this we have to resort either to microscopic theories (and simulations), or alternatively, we can determine them experimentally.

2.3.2 COMPLEX INTERFACES

The constitutive equations we have derived in the previous section are meaningful only for small departures from equilibrium and interfaces with a purely viscous behavior. However, many interfaces actually display viscoelastic behavior and may have a highly nonlinear response to applied deformations or temperature and concentration gradients. This is particularly true for interfaces in which the surface-active components tend to self-organize into complex microstructures, after adsorption into the interface. A wide range of interfacial microstructures can be observed in multiphase systems. For example, proteins [33,34], protein aggregates, colloidal particles [35,36], surface-active polymers [37], and some low-molecular-weight surfactants [38] can all form (quasi) 2D gels. Proteins and low-molecular-weight surfactants can also form 2D glasses [39,40]. Protein fibrils [5,41,42], semiflexible and rodlike polymers, or anisotropic colloidal particles can all form 2D liquid crystalline phases. Mixtures of immiscible surface-active components can phase-separate after adsorption, forming interfaces with a heterogeneous structure, such as 2D dispersions or 2D emulsions [43]. Mixtures of proteins and charged polysaccharides can form interfaces with a composite structure [44–46]. Whereas for simple interfaces, the surface tension is often the only interfacial parameter with a significant effect on the macroscopic behavior of a multiphase systems, for complex interfaces, rheological properties like the shear and dilatational loss and storage moduli or the interfacial bending rigidity often have a nonnegligible impact. For example, the latter has a significant effect on the dynamics of dispersions of vesicles, cells, and droplets in phase-separated biopolymer solutions [47–54], whereas the shear and dilatational properties appear to have a significant effect on stability and dynamics of foam and emulsions [55,56].

Complex interfaces have a nonlinear response to deformations because the applied deformation induces changes in the microstructure of the interface. The most common way to capture these changes in the microstructure, and their effect on overall macroscopic behavior, is to include additional variables, referred to as structural or internal variables, in the set of independent system variables [16]. The set of surface excess variables may, for example, be extended to

$$\{\rho^s, \boldsymbol{m}^s, \overline{U}^s, \rho^s_{(1)}, \ldots, \rho^s_{(N-1)}, c^s_1, \ldots, c^s_m, \boldsymbol{c}^s_1, \ldots, \boldsymbol{c}^s_n, \boldsymbol{C}^s_1, \ldots, \boldsymbol{C}^s_p\}, \tag{2.44}$$

where

c^s_m are the scalar structural variables
$\boldsymbol{c}^s_n$ are the vectorial structural variables
$\boldsymbol{C}^s_p$ denote the structural variables of a tensorial nature

Examples of variables of a scalar nature are the segment density of an adsorbed polymer or the surface fraction of particles in a Pickering stabilized interface. An example of a vectorial structural variable would be the director field of a 2D liquid crystalline interface. Examples of tensorial variables are the second moment of a particle orientation or polymer segment orientation distribution function, which describes deformation-induced changes in orientation (and in the case of polymers, also degree of stretching) of adsorbed components.

The incorporation of such variables in the set of independent surface variables leads to an additional set of partial differential equations, describing the time evolution of these variables as a result of in-plane convective, diffusive, and relaxation processes and through exchange with the adjoining bulk phases [58,59]. This time evolution is combined with an expression for the surface stress tensor in terms of the structural variables and is closed with a constitutive model for the surface configurational Helmholtz energy, in terms of these same variables. For example, for an interface with a microstructure that can be described by a single unconstrained tensor, $\boldsymbol{C}^s$, the constitutive model for the surface extra stress tensor could take the form

$$\boldsymbol{\sigma}^s = \nu^s \boldsymbol{C}^s, \tag{2.45}$$

$$\frac{\mathrm{d}_s \boldsymbol{C}^s}{\mathrm{d}t} - \boldsymbol{C}^s \cdot \left(\nabla_s \boldsymbol{v}^s\right)^{\mathrm{T}} - \left(\nabla_s \boldsymbol{v}^s\right) \cdot \boldsymbol{C}^s = 2\beta^s \boldsymbol{D}^s - \frac{1}{\tau_1} \boldsymbol{C}^s - \frac{1}{\tau_2} \boldsymbol{C}^s \cdot \boldsymbol{C}^s. \tag{2.46}$$

Here, ν^s, β^s, τ_1, and τ_2 are the scalar coefficients, which may depend on composition and temperature of the interface. The derivative on the left-hand side of (2.46) is the upper-convected surface derivative. The first term on the right-hand side of this expression, proportional to the rate of deformation tensor, drives the microstructure of the interface out of equilibrium. The last two terms describe linear and nonlinear relaxation processes, which drive the structure back to equilibrium. A model such as the one in Equation 2.46 can predict strain-thinning behavior [58].

To construct structural models capable of describing surface rheological data, we need to determine the time evolution of the microstructure experimentally (or computationally). Such information is not always available, and when this is the case, we can resort to the extended thermodynamics framework. In this framework, the fluxes are included in the set of system variables, and hence (2.44) reduces to [7,19]

$$\{\rho^s, \boldsymbol{m}^s, \overline{U}^s, \rho^s_{(1)}, \ldots, \rho^s_{(N-1)}, \text{tr}\boldsymbol{\sigma}^s, \boldsymbol{q}^s, \boldsymbol{j}^s_{(1)}, \ldots, \boldsymbol{j}^s_{(N-1)}, \overline{\boldsymbol{\sigma}}^s\}. \tag{2.47}$$

The procedure to construct constitutive equations is similar to that used in the CIT framework. We choose

$$\hat{S}^s = \hat{S}^s\left(\hat{U}^s, \hat{A}, \omega^s_{(1)}, \ldots, \omega^s_{(N-1)}, \text{tr}\boldsymbol{\sigma}^s, \boldsymbol{q}^s, \boldsymbol{j}^s_{(1)}, \ldots, \boldsymbol{j}^s_{(N-1)}, \overline{\boldsymbol{\sigma}}^s\right) \tag{2.48}$$

and use this functional to construct an expression for the surface rate of entropy production. This bilinear form is then again used to guide us in the construction of the constitutive equations. When we limit ourselves to linear models, the EIT approach leads to the following expressions for the surface extra stress tensor [7,19]:

$$\frac{d_s\overline{\boldsymbol{\sigma}}^s}{dt} + \frac{1}{\tau_s}\overline{\boldsymbol{\sigma}}^s = \frac{2\varepsilon_s}{\tau_s}\overline{\boldsymbol{D}}^s, \tag{2.49}$$

$$\frac{d_s\text{tr}\boldsymbol{\sigma}^s}{dt} + \frac{1}{\tau_d}\text{tr}\boldsymbol{\sigma}^s = \frac{2\varepsilon_d}{\tau_d}\text{tr}\boldsymbol{D}^s. \tag{2.50}$$

These expressions are the surface equivalent of the linear single-mode Maxwell model. The parameter τ_s is the surface shear relaxation time, and τ_d is the surface dilatational relaxation time. Multimode variations of this model can simply be derived by replacing (2.48) by

$$\hat{S}^s = \hat{S}^s\left(\hat{U}^s, \hat{A}, \omega^s_{(1)}, \ldots, \omega^s_{(N-1)}, \text{tr}\boldsymbol{\sigma}^s_1, \ldots, \text{tr}\boldsymbol{\sigma}^s_k, \boldsymbol{q}^s, \boldsymbol{j}^s_{(1)}, \ldots, \boldsymbol{j}^s_{(N-1)}, \overline{\boldsymbol{\sigma}}^s_1, \ldots, \overline{\boldsymbol{\sigma}}^s_k\right), \tag{2.51}$$

where $\text{tr}\boldsymbol{\sigma}^s_k$ and $\overline{\boldsymbol{\sigma}}^s_k$ represent the k-th mode of the surface extra stress tensor. For each of these modes, expressions of the form of Equations 2.49 and 2.50 are then obtained. It is also straightforward to generate nonlinear models in this framework. By switching to convective derivatives and incorporating higher order terms in the stress tensor, we obtain

$$\frac{d_s\overline{\boldsymbol{\sigma}}^s}{dt} - \overline{\boldsymbol{\sigma}}^s \cdot \left(\nabla_s\boldsymbol{v}^s\right)^T - \left(\nabla_s\boldsymbol{v}^s\right)\cdot\overline{\boldsymbol{\sigma}}^s + \frac{1}{\tau_s}\overline{\boldsymbol{\sigma}}^s + \frac{\alpha_s}{\varepsilon_s}\overline{\boldsymbol{\sigma}}^s \cdot \overline{\boldsymbol{\sigma}}^s = 2\frac{\varepsilon_s}{\tau_s}\overline{\boldsymbol{D}}^s, \tag{2.52}$$

$$\frac{d_s\text{tr}\boldsymbol{\sigma}^s}{dt} + \frac{1}{\tau_d}\text{tr}\boldsymbol{\sigma}^s + \frac{\alpha_d}{\varepsilon_d}\left(\text{tr}\boldsymbol{\sigma}^s\right)^2 = \frac{2\varepsilon_d}{\tau_d}\text{tr}\boldsymbol{D}^s. \tag{2.53}$$

These two equations are the surface equivalent of the single-mode Giesekus model [60]. The coefficients α_s and α_d are the surface shear and dilatational mobility, and to satisfy the entropy balance, we must require these coefficients to be nonnegative. The Giesekus model is just one example of a constitutive model for the surface extra stress tensor we can construct using EIT. We can also derive surface equivalents of the upper- and lower-convected Maxwell model, the Jeffreys model, or the Oldroyd-B model [61,62]. However, for modeling bulk rheological behavior, such models have all shown their limitations. Whenever possible, preference should be given to the development of structural models, in view of the direct coupling between stress signals and the time evolution of the interfacial microstructure.

Maxwell-type equations such as those in Equations 2.49 and 2.50 can also be obtained for the other fluxes. For example, for the surface energy flux vector, we find (again limiting ourselves to linear expressions)

$$\frac{\mathrm{d}_s \boldsymbol{q}^{\mathrm{s}}}{\mathrm{d}t} + \frac{1}{\tau_\lambda} \boldsymbol{q}^{\mathrm{s}} = -\frac{\lambda^{\mathrm{s}}}{\tau_\lambda T^{\mathrm{s}}} \nabla_s T^{\mathrm{s}}. \tag{2.54}$$

This is the surface equivalent of the Maxwell–Cattaneo equation for the energy flux vector [63]. The coefficient τ_λ is the relaxation time associated with surface energy transfer. Note that in the limit $\tau_\lambda \rightarrow 0$, this expression reduces to Equation 2.36. An attractive property of constitutive equations of the form of Equation 2.54 is that upon substitution in the jump energy balance, they yield a hyperbolic partial differential equation, which predicts a finite speed of propagation of thermal signals along the interface [17]. This is in contrast to Fourier's law, Equation 2.36, which yields a parabolic equation, predicting infinite speeds of propagation.

2.4   CONCLUSIONS AND OUTLOOK

In this chapter, we have shown how NET (and in particular CIT and EIT) can be used to derive descriptions for multiphase systems away from equilibrium. In the Gibbs dividing surface model, we can assign excess variables to an interface, and we have discussed how the inclusion of these surface variables in the conservation principles for mass, momentum, and energy leads to a set of time-evolution equations for these variables. We have also discussed the derivation of thermodynamically consistent constitutive equations for the fluxes appearing in these time-evolution equations and the derivation of boundary conditions that couple the time-evolution equations to the balances for the bulk variables. When this set of balance and constitutive equations is combined with an equation of state for the surface tension and the bulk pressure and expressions for the concentration and temperature dependence of the bulk and surface material properties, we have a complete set of coupled equations for the dynamics of a multiphase system.

The balance and constitutive equations we have presented here are suited to describe the behavior of interfaces with a complex microstructure, with (nonlinear) viscoelastic responses to a perturbation. So far, the nonlinear structural models that have been discussed here, but also the phenomenological ones, such as the single-mode Giesekus model, have not been used widely to analyze experimental surface rheology data or in computational studies of multiphase flows. Linear and quasilinear models have been incorporated in simulations of multiphase flows [64–73], but more work is needed to find efficient solvers that can handle nonlinear structural models. Structural models have been very successful in modeling the behavior of complex bulk phases, such as particle dispersions, polymer blends, liquid crystalline phases, and even food biopolymer gels [61,74]. In view of these successes, structural models for interfacial behavior are preferred over phenomenological ones, and more effort should be invested in deriving such models.

REFERENCES

1. J.M. Simon, D. Bedeaux, S. Kjelstrup, and E. Johannessen, *J. Phys. Chem. B*, 110 (2006) 18528.
2. K.S. Glavatskiy and D. Bedeaux, *J. Chem. Phys.*, 133 (2010) 234501.
3. O. Wilhelmse, D. Bedeaux, S. Kjelstrup, and D. Reguera, *J. Chem. Phys.*, 140 (2014) 024704.
4. S.G. Bolder, H. Hendrickx, L.M.C. Sagis, and E. van der Linden, *Appl. Rheol.*, 16 (2006) 258.
5. S. Jordens, L. Isa, I. Usov, and R. Mezzenga, *Nature Commun.*, 4 (2013) 1917.
6. S.R. de Groot and P. Mazur, *Nonequilibrium Thermodynamics*, North-Holland, Amsterdam, the Netherlands (1962).
7. L.M.C. Sagis, *Rev. Mod. Phys.*, 83 (2011) 1367.
8. J. van der Waals, The thermodynamic theory of capillarity under the hypothesis of a continuous variation of density (1893); reprinted in *J. Stat. Phys.*, 20 (1979) 200.
9. J.W. Gibbs, *The Collected Works of J. Willard Gibbs*, Vol. 1, Yale University Press, New Haven, CT (1928).
10. D. Bedeaux, A.M. Albano, and P. Mazur, *Phys. A*, 82 (1975) 438.
11. B.J.A. Zielinska and D. Bedeaux, *Phys. A*, 112 (1982) 265.
12. D. Bedeaux, *Adv. Chem. Phys.*, 64 (1986) 47.
13. A.M. Albano and D. Bedeaux, *Phys. A*, 147 (1987) 407.
14. D. Bedeaux and J. Vlieger, *Optical Properties of Surfaces*, 1 edn., Imperial College University Press, London, U.K. (2002).
15. S. Kjelstrup and D. Bedeaux, *Non-Equilibrium Thermodynamics of Heterogeneous Systems*, World Scientific Publishing, Singapore (2008).
16. I. Prigogine and P. Mazur, *Physica*, 19 (1953) 241.
17. D. Jou, J. Casas-Vásquez, and G. Lebon, *Extended Irreversible Thermodynamics*, 3rd edn., Springer, Berlin, Germany (2001).
18. G. Lebon, D. Jou, and J. Casas-Vásquez, *Understanding Non-Equilibrium Thermodynamics: Foundations, Applications, Frontiers*, Springer, Berlin, Germany (2008).
19. L.M.C. Sagis, *Phys. A*, 389 (2010) 1993.
20. J.C. Slattery, L.M.C. Sagis, and E.S. Oh, *Interfacial Transport Phenomena*, 2nd edn., Springer, New York (2007).
21. L.M.C. Sagis, *Phys. A*, 391 (2012) 979.
22. M. Grmela and H.C. Öttinger, *Phys. Rev. E*, 56 (1997) 6620.
23. H.C. Öttinger and M. Grmela, *Phys. Rev. E*, 56 (1997) 6633.
24. H.C. Öttinger, *Beyond Equilibrium Thermodynamics*, Wiley-Interscience, Hoboken, NJ (2005).
25. H.C. Öttinger, D. Bedeaux, and D.C. Venerus, *Phys. Rev. E*, 80 (2009) 021606.
26. L.M.C. Sagis, *Adv. Colloid Interface Sci.*, 153 (2010) 58.
27. L.M.C. Sagis and H.C. Öttinger, *Phys. Rev. E*, 88 (2013) 022149.
28. A. Javadi, N. Mucic, M. Karbaschi, J. Won, M. Lotfi, A. Dan, V. Ulaganathan et al., *Eur. Phys. J. ST*, 222 (2013) 7.
29. L.M.C. Sagis, K.N.P. Humblet-Hua, and S. van Kempen, *J. Phys. Condens. Matter*, 26 (2014), 464105.
30. J. Boussinesq and C.R. Hebd, *Seances Acad. Sci.*, 156 (1913) 983.
31. J. Boussinesq and C.R. Hebd, *Seances Acad. Sci.*, 156 (1913) 1035.
32. J. Boussinesq and C.R. Hebd, *Seances Acad. Sci.*, 156 (1913) 1124.
33. P. Erni, E. Windhab, and P. Fischer, *Macromol. Mater. Eng.*, 296 (2011) 249.
34. A. Torcello-Gómez, J. Maldonado-Valderrama, M.J. Gálvez-Ruiz, A. Martín-Rodríguez, M.A. Cabrerizo-Vílchez, and J. de Vicente, *J. Non-Newt. Fluid Mech.*, 166 (2011) 713.
35. D. Zang, E. Rio, D. Langevin, B. Wei, and B. Binks, *Eur. Phys. J. E*, 31 (2010) 125.
36. B. Brugger, J. Vermant, and W. Richtering, *Phys. Chem. Chem. Phys.*, 12 (2010) 14573.
37. P. Erni, P. Fischer, P. Heyer, E.J. Windhab, V. Kusnezov, and L. Läuger, *Prog. Colloid Polym. Sci.*, 129 (2004) 16.
38. H. Rehage, B. Achenbach, M. Geest, and H.W. Siesler, *Colloid Polym. Sci.*, 279 (2001) 597.
39. P. Cicuta, E. J. Stancik, and G. G. Fuller, *Phys. Rev. Lett.*, 90 (2003) 236101.
40. S.E.H.J. van Kempen, H. A. Schols, E. van der Linden, and L.M.C. Sagis, *Soft Matter*, 9 (2013) 9579.
41. L. Isa, J. Jung, and R. Mezzenga, *Soft Matter*, 7 (2011) 8127.
42. R. Mezzenga and P. Fischer, *Rep. Prog. Phys.*, 76 (2013) 046601.
43. C. Bernardini, S.D. Stoyanov, L.N. Arnaudov, and M.A. Cohen Stuart, *Chem. Soc. Rev.*, 42 (2013) 2100.

44. L.M.C. Sagis, R. de Ruiter, F.J. Rossier Miranda, J. de Ruiter, K. Schroën, A.C. van Aelst, H. Kieft, R.M. Boom, and E. van der Linden, *Langmuir*, 24 (2008) 1608.
45. K.N.P. Humblet-Hua, G. Scheltens, E. van der Linden, and L.M.C. Sagis, *Food Hydrocolloids*, 25 (2011) 569.
46. R. Fischer, *Eur. Phys. J. ST*, 222 (2013) 73.
47. J.B.A.F. Smeulders, C. Blom, and J. Mellema, *Phys. Rev. A*, 42 (1990) 3483.
48. J.B.A.F. Smeulders, C. Blom, and J. Mellema, *Phys. Rev. A*, 46 (1992) 7708.
49. K.H. de Haas, G.J. Ruiter, and J. Mellema, *Phys. Rev. E*, 52 (1995) 1891.
50. E. Scholten, L.M.C. Sagis, and E. van der Linden, *Macromolecules*, 38 (2005) 3515.
51. E. Scholten, J. Sprakel, L.M.C. Sagis, and E. van der Linden, *Biomacromolecules*, 7 (2006) 339.
52. E. Scholten, L.M.C. Sagis, and E. van der Linden, *J. Phys. Chem. B*, 110 (2006) 3250.
53. E. Scholten, L.M.C. Sagis, and E. van der Linden, *J. Phys. Chem. B*, 108 (2004) 12164.
54. L.M.C. Sagis, *J. Control. Release*, 131 (2008) 5.
55. M.A. Bos and T. van Vliet, *Adv. Colloid Interface Sci.*, 91 (2001) 437.
56. B.S. Murray, *Curr. Opin. Colloid Interface Sci.*, 7 (2002) 426.
57. P. Fischer and P. Erni, *Curr. Opin. Colloid Interface Sci.*, 12 (2007) 196.
58. L.M.C. Sagis, *Soft Matter*, 7 (2011) 7727.
59. L.M.C. Sagis, *Adv. Colloid Interface Sci.*, 206 (2014) 328.
60. H. Giesekus, *J. Non-Newt. Fluid Mech.*, 11 (1982) 69.
61. R.G. Larson, *The Structure and Rheology of Complex Fluids*, Oxford University Press, Oxford, U.K. (1999).
62. R.B. Bird, R.C. Armstrong, and O. Hassager, *Dynamics of Polymeric Liquids*, 2nd edn., Vol. 1, Wiley-Interscience, New York (1987).
63. C. Cattaneo, *Atti. Sem. Mat. Fis. Univ. Modena*, 3 (1948) 3.
64. D. Barthes-Biesel and J.M. Rallison, *J. Fluid Mech.*, 113 (1981) 251.
65. D. Barthes-Biesel, *Phys. A*, 172 (1991) 103.
66. D. Barthes-Biesel, *Prog. Colloid Polym. Sci.*, 111 (1998) 58.
67. D. Barthes-Biesel, A. Diaz, and E. Dhenin, *J. Fluid Mech.*, 460 (2002) 211.
68. F. Lac, A. Morel, and D. Barthes-Biesel, *J. Fluid Mech.*, 573 (2007) 149.
69. B.U. Felderhof, *J. Chem. Phys.*, 125 (2006) 124904.
70. T.F. Swean and A.N. Beris, *Appl. Mech. Rev.*, 47 (1994) S173.
71. M.D. Giavedoni and S. Ubal, *Ind. Eng. Chem. Res.*, 46 (2007) 5228.
72. Q. Zhang, R.A. Handler, and S.T. Fredriksson, *Int. J. Heat Mass Transf.*, 61 (2013) 82.
73. H. Liu, H. Liu, and J. Yang, *Phys. E*, 49 (2013) 13.
74. L.M.C. Sagis, M. Ramaekers, and E. van der Linden, *Phys. Rev. E*, 63 (2001) 051504.

3 Experimental Approaches and Related Theories

Marzieh Lotfi, Aliyar Javadi, Mohsen Karbaschi,
Richard A. Campbell, Volodja I. Kovalchuk, Jürgen Krägel,
Valentin B. Fainerman, Dariush Bastani, and Reinhard Miller

CONTENTS

3.1 CHARACTERIZATION OF LIQUID INTERFACES

The characterization of liquid interfaces is vital for a large number of modern technologies. In particular, the dynamic behavior of liquid interfacial layers under external perturbations is an important prerequisite for the development of new technologies and products. While Chapter 1 described in detail the situation of interfaces at equilibrium, this chapter gives a general approach to interfacial layers under nonequilibrium conditions.

In this chapter, we want to briefly describe the most frequently used experimental methodologies developed for the quantitative characterization of liquid interfacial layers, essentially under dynamic conditions. This material is then followed by Chapter 4, which is dedicated to the mechanisms controlling the process of interfacial layer formation. Chapter 5, which closes this introductory section, will show how existing theories can be applied for the interpretation of experimental data.

As very fast methods, we will describe the capillary pressure tensiometry (CPT), which can be used for both liquid/gas and liquid/liquid interfaces. At the liquid/gas interface, it is known as maximum bubble pressure tensiometry and provides data at adsorption times as short as 10^{-3} s and even less. The workhorse drop profile analysis tensiometry is for sure the most efficient technique in interfacial laboratories for studies of the adsorption process of surfactants and polymers, but much slower and can measure only from about one second surface age. It can, however, be used for very long adsorption times, such as hours and even days. For measuring the process of adsorption on a timescale of minutes and hours, the optical methods ellipsometry and neutron scattering are applicable. We will briefly present the principles of these techniques that, in contrast to the tensiometry methods, provide directly information on the adsorbed amount.

This chapter also presents oscillating drop and bubble experiments performed for getting information on the mechanical properties, such as the dilational viscoelasticity of interfacial layers. The capillary wave damping is used for the same group of interfacial parameters; it can, however, be applied at much higher frequencies, up to few hundred Hertz. The methodology with a double capillary is a specialty of the drop profile tensiometry and provides the option of changing the composition of the drop liquid during the experiment.

The two final methods are based on drops and bubbles in a flow field of liquid. They cannot be counted as methods that measure something directly but present sets of data that include all the dynamics of interfacial layers, including transport to and from the interfaces as well as surface gradients along the interface due to drag effects by the surrounding liquids.

3.2   DROP AND BUBBLE PRESSURE TENSIOMETRY

The majority of methods for measuring equilibrium and dynamic interfacial tensions are based on the mechanical response of capillary systems, where the capillary pressure is acting at a curved liquid interface. Examples of such systems are drops and bubbles. By studying the mechanical behavior of capillary systems, it is possible to obtain important information about the dynamics of adsorption, interaction of molecules at the interface, and conformational change of macromolecules. Such methods are related to CPT. For measuring the interfacial tension of liquid–liquid systems, the pressure derivative method is the most suitable. The dynamics interfacial tension and rheology can be investigated by the maximum bubble pressure method (MBPM), the growing drop/bubble, the expanded drop, and the oscillating bubble/drop methods.

Detailed description of measuring principles and the scheme of the maximum bubble pressure tensiometer bubble pressure analyzer (BPA) (SINTERFACE Technologies, Germany) is presented in [1]. This tensiometer provides a surface lifetime range from 0.1 ms to about 100 s [2]. The CPT (pressure derivative method) used by Passerone et al. [3] has been developed for interfacial tension measurements. Here, the cell is made up of two main parts connected by the capillary and containing the two liquids. The drop of one liquid is formed at the tip of the capillary inside of another liquid, creating the interface to be studied. The pressure signal is measured with a transducer placed in contact with the drop forming liquid.

3.2.1   SURFACE TENSION DETERMINATION IN MAXIMUM BUBBLE PRESSURE METHOD

In MBPM, the surface tension γ is calculated from the measured maximum capillary pressure P and the known capillary radius R_0 by using the Young–Laplace equation

$$\gamma = f \frac{R_0 \cdot P}{2}, \tag{3.1}$$

where f is a correction factor that accounts for the deviation of the bubble from sphericity. f was calculated, for example, by Sugden [4] and is tabulated as a function of the ratio R_0/a, where a is the capillary constant $a = (2\gamma/\Delta\rho g)^{1/2}$ and $\Delta\rho$ is the density difference between the two phases [5]. The correction is needed for capillary radii $R_0 > 0.1$ mm but is valid only under static conditions. For a fast bubble formation, which is a very dynamic process, the error can be even higher [6].

The measured capillary pressure P can be expressed via the excess maximum pressure in the measuring system P_s, the hydrostatic pressure of the liquid $P_H = \Delta\rho gH$, and the excess pressure P_d due to hydrodynamic and aerodynamic effects [6]

$$P = P_s - P_H - P_d, \tag{3.2}$$

The excess pressure P_d is mainly caused by the bulk viscosity of the liquid and air [6,7]. A hydrodynamic theory for MBPM, accounting for liquid penetration into the capillary, was developed mainly by Dukhin and Kovalchuk [6,8]. Accordingly, the bubble lifetime t_l can be separated into main stages [6]: after a bubble separates, the meniscus itself moves into the capillary by some depth, t_{l1}; then the meniscus moves to the end of the capillary (reverse meniscus motion), t_{l2}; and finally, the bubble radius decreases and becomes equal to the capillary radius, t_{l3}. The time interval for all these stages is called the bubble lifetime $t_l = t_{l1} + t_{l2} + t_{l3}$.

3.2.2 Bubble Time and Its Constituents

The time between bubbles t_b represents the sum of two components: lifetime and dead time. The advantage of the BPA tensiometers is their ability to measure the lifetime t_l and dead time t_d directly for lifetimes larger than 10 ms. For $t_l < 10$ ms, a procedure based on the theoretical analysis of dead time is used. The dead time t_d, that is the time during which the bubble is growing rapidly and its radius increases from R_0 to the radius R_b of the separating bubble, can be expressed by [9]

$$t_d = \frac{32l\eta}{R_0(P_s - P_H)} \left[\frac{1}{3}\left(\frac{R_b}{R_0}\right)^3 + \frac{\gamma^*}{R_0(P_s - P_H)}\left(\frac{R_b}{R_0}\right)^2 \right] \tag{3.3}$$

where
  γ^* is the surface tension at the end of the dead time interval
  l is the capillary length

This equation can be approximated by [10]

$$t_d = t_b \cdot \frac{L}{k_p P}\left(1 + \frac{3}{2}\frac{R_0}{R_b}\right) \tag{3.4}$$

where
  k_p is the Poiseuille equation constant for a capillary not immersed into the liquid ($L = k_p P$)
  L is the gas flow rate
  $P = P_s - P_H$, t_b are the time interval between successive bubbles

This method was first proposed by Kloubek [11], and it assumes that for high-frequency bubble formations, the dead time interval becomes equal to the time interval between successive bubbles. For gas flow rates lower than a critical value $L < L_c$, individual bubbles are formed with $t_l > 0$. However, in the transition point ($L = L_c$), the lifetime vanishes, and we get $t_b = t_d$ [12]. Following this protocol, we have

$$t_l = t_b - t_d = t_b\left(1 - \frac{LP_c}{L_c P}\right) \tag{3.5}$$

Thus, the lifetime can be calculated from only experimental parameters t_b, $P = P_s - P_H$, and L.

As an experimental example, the dynamic surface tension for micellar solutions of $C_{14}EO_8$ (Critical Micelle Concentration [CMC] = 7 µmol/L) as functions of the effective lifetime is shown in Figure 3.1 [2].

The data were obtained with the tensiometer BPA-1S (SINTERFACE Technologies), having a dead time of 5 ms; Figure 3.2 illustrates the dynamic surface tension for aqueous solutions of Triton X-100 at different dead times [13], as it was studied by the tensiometer BPA-1P.

The different dead times were obtained by using capillaries of different length and diameter. The lower the dead time is, the higher is the measured surface tension. The difference is especially pronounced for concentrated (1.2 mmol/L) solutions for effective lifetimes of 10 ms. Decrease in surface tension is caused by significant adsorption during the bubble dead time stage.

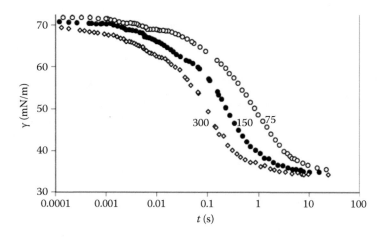

FIGURE 3.1 Dynamic surface tension for $C_{14}EO_8$ solutions as a function of the effective surface lifetime; the surfactant concentrations are given as multiples of the Critical Micelle Concentration (CMC). (Data from Fainerman, V.B. et al., *J. Colloid Interface Sci.*, 302, 40, 2006.)

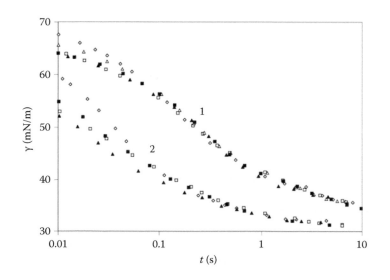

FIGURE 3.2 Dynamic surface tension for Triton X-100 solutions as a function of the effective surface lifetime t for the concentrations of (1) −0.48 mmol/L and (2) −1.2 mmol/L; open symbols, experimental results for a capillary of 0.24 mm in diameter; filled symbols, experimental results for a capillary of 0.46 mm in diameter. The dead time values were ◇, 8 ms; △, 23 ms; □ and ■, 40 ms; and ▲, 110 ms. (Data from Fainerman, V.B. et al., *J. Colloid Interface Sci.*, 304, 222, 2006.)

3.2.3 Capillary Pressure Tensiometry

Under dynamic conditions, the pressure difference ΔP across a spherical interface can be derived by the normal stress balance at the interface as it is shown in [3,14]

$$\Delta P = \frac{2\gamma}{R} + \frac{4\kappa}{R^2}\frac{dR}{dt} - \frac{4(\mu_i - \mu_e)}{R}\frac{dR}{dt} \tag{3.6}$$

where
κ is the surface dilational viscosity
μ_i and μ_e are the viscosities of the internal and external fluid phases, respectively

For an interface at mechanical equilibrium ($dR/dt = 0$), the equation reduces to Equation 3.1. The two additional terms arise from the dynamic viscosity of the interface and the bulk. The effect of the bulk viscosity is negligible for $dR/dt \ll \gamma/(\mu_i - \mu_e)$. For water/air interfaces, this means $dR/dt \ll 10^3$ cm/s. Hence, for typical experimental conditions, this contribution is significant only for highly viscous liquids [15].

The second term in Equation 3.6 can also be neglected when $(dR/dt)/R \ll \gamma/\kappa \cong 10^4$ s^{-1} holds. This leads to a quasi-equilibrium version of Equation 3.1, into which we are allowed then to include time-dependent variables. This is the basic principle of CPT, and only for very fast dynamic processes or highly viscous liquids, some specific corrections are required. For sufficiently low flow rate, dynamic effects can be neglected and we get

$$P(t) = P_{cap}(t) + P_0 = \frac{2\gamma_0}{R(t)} + P_0 \tag{3.7}$$

P_0 is a constant offset arising from the hydrostatic pressure due to the immersion depth of the capillary, which, however, can be determined via a best fit procedure of the dependence $\gamma(1/r)$ for a liquid with constant surface tension.

3.3 DROP AND BUBBLE PROFILE TENSIOMETRY

The method of drop profile analysis is superior over others for its high accuracy and very large range of surface lifetime [16,17]. It is based on fitting the Laplace equation to experimental profiles of axisymmetric bubbles and drops

$$\gamma\left(\frac{1}{R_1} + \frac{1}{R_2}\right) = \Delta P \tag{3.8}$$

where R_1 and R_2 are the principal radii of curvature. The meridional curvature can be defined as $1/R_1 = d\phi/dS$, where ϕ is the angle of the tangent to the profile and S is the arc length along the profile. The azimuthal curvature is given by $1/R_2 = \sin\phi/X$ (see Figure 3.3).

Under gravity at the drop apex, the two principal radii of curvature for axisymmetric menisci are equal $R_1 = R_2 = b$, and we obtain the following equation as starting point for a calculation algorithm [17]:

$$\frac{d\phi}{ds} = 2 + \beta z - \frac{\sin\phi}{x} \tag{3.9}$$

where
$\Delta\rho$ is the density difference between the drop and the surrounding medium
g is the gravitational acceleration constant
$\beta = \Delta\rho g b^2/\gamma$ is the dimensionless Bond number, $s = S/b$, and $x = X/b$ and $z = Z/b$ are dimensionless arc length, horizontal and vertical coordinates, respectively.

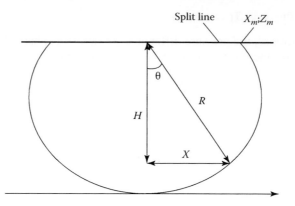

FIGURE 3.3 Definition of the spherical coordinate system. (From Dingle, N.M. et al., *J. Colloid Interface Sci.*, 286, 647, 2005.)

Together with the additional equations,

$$\frac{dx}{ds} = \cos\phi, \quad \frac{dz}{ds} = \sin\phi \tag{3.10}$$

we get a parameterized Laplace equation, which can then be solved with the initial conditions $x(0) = 0$, $z(0) = 0$, and $\phi(0) = 0$.

An analytical solution for the general case does not exist, so that numerical algorithms are required. As the first general algorithm, axisymmetric drop-shape analysis was published [19]. It acquires images of a drop or bubble, extracts the experimental profile coordinates, and fits a theoretical profile to the data points. The algorithm uses the Runge–Kutta method for solving Equations 3.9 and 3.10 and an optimization procedure for finding the best fit. For improving the methods accuracy,

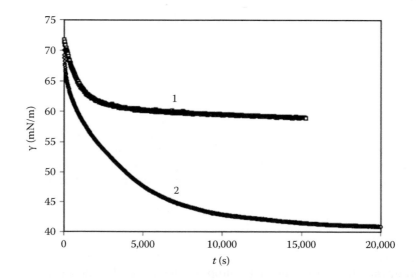

FIGURE 3.4 The dependence of dynamic surface tension of $C_{14}EO_8$ solutions at concentrations 0.002 mmol/L measured on PAT-1S by drop-shape method (1) and bubble-shape method (2). (Data from Fainerman, V.B. et al., *Colloid Surf. A*, 323, 56, 2008.)

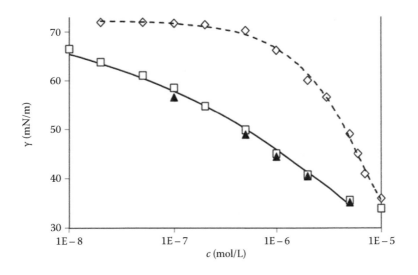

FIGURE 3.5 Dependence of equilibrium surface tension of $C_{14}EO_8$ solutions on initial concentration measured in [22] by pendant drop method (◇) and emerging bubble method (□) and in [23] by buoyant bubble method (▲); the bold curve is the theoretical isotherm. (From Fainerman, V.B. et al., *Colloid Surf. A*, 323, 56, 2008; Fainerman, V.B. et al., *Langmuir*, 29, 6964, 2013.)

advanced edge detection strategies and other correction procedures were employed [20,21]. More details of the numeric methods for solving the Laplace equation and of optimum fitting protocols can be found in [17].

The data given in Figures 3.4 and 3.5 are interesting examples showing the large potential of the technique but also some peculiarities to be considered for getting physically reasonable results. The dynamic surface tensions of a 2 μmol/L $C_{14}EO_8$ solution is shown in Figure 3.4 as measured by the drop and bubble profile methods, respectively [22]. As we can see, the results of the two methods differ from each other significantly.

Figure 3.5 illustrates the surface tension isotherms measured by the same two methods [23]. As we can see, the experimental isotherm measured by drop profile analysis, as compared to that obtained with the bubble profile methods, is essentially shifted toward higher surfactant bulk concentrations. This discrepancy is due to the adsorption of $C_{14}EO_8$ at the drop surface, which results in a depletion of the bulk concentration inside the drop [24].

3.4 ELLIPSOMETRY AND NONLINEAR OPTICAL METHODS

The methods of tensiometry and dilational rheology are indirect techniques as they measure the surface or interfacial tension as a function of surface age and surfactant bulk concentration. Then, via the fundamental Gibbs equation (see Equation 1.88), the number of adsorbed molecules can be calculated. Surface ellipsometry, neutron reflection, x-ray reflectivity, and vibrational sum frequency generation (SFG), however, are methods that give a direct access to the adsorbed amount. This and the subsequent chapter will give a very brief introduction into only the surface ellipsometry and neutron scattering techniques. For example, in [25], more recently developed methods were discussed.

In surface ellipsometry, the polarization of light is used to characterize surfaces layers. The basic idea is the fact that the polarization of light is changing upon reflection. Thus, the method measures the polarization of a laser light beam before (*P*) and after (*A*) reflection at a surface. The scheme of an ellipsometer is shown in Figure 3.6 as it was proposed by Benjamins et al. [26] who also showed how this technique is applicable to liquid/liquid interfaces by circumventing the light reflection at two interfaces via the sue of wave guide tubes.

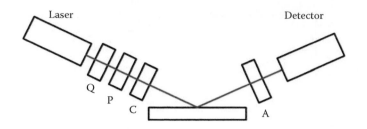

FIGURE 3.6 Scheme of a surface ellipsometer, Q, quarter plate; P, polarizer; C, compensator; A, analyzer. (According to Benjamins, J.W. et al., *Langmuir*, 18, 6437, 2002.)

The accuracy of layer thickness δ measurements is ±0.1 nm. Knowing δ, the adsorbed amount Γ can be calculated for a given bulk concentration of a surfactant C via the equation

$$\Gamma = \frac{\delta(n_1 - n_2)}{\mathrm{d}n/\mathrm{d}C} \tag{3.11}$$

where

n_1 and n_2 are the refractive indices of the adsorption layer and the bulk phase, respectively
$\mathrm{d}n/\mathrm{d}C$ is the refractive index increment as a function of the surfactant concentration

Although the data acquisition is rather fast, the ellipsometry provides kinetic changes of interfacial layer properties (δ and Γ) only on a timescale of minutes [27].

A very efficient implementation of this method is the null ellipsometry. The protocol eliminates many possible experimental errors so that this version is advantageous over the other ones [28]. Ellipsometry at liquid interfaces became a routine tool for measuring interfacial layer thickness because commercial instruments are available with high performance and software support for data handling.

In recent years, various new optical methods were developed and reached the level of routine techniques for the characterization of molecules at liquid interfaces. The vibrational SFG is one of those and is applied in many fields, for example, for studies of structures at biointerfaces [29,30]. SFG is a second-order nonlinear optical process [31]. For materials with inversion symmetry such as liquids and gases, it is interfacial specific. The main elements in SFG spectroscopy are two laser beams, of which one has a fixed wavelength and the other works with a tunable infrared wavelength. The two beams are combined at the interface, and the sum frequency of the two impinging laser fields is generated. The intensity of sum frequency beam is not only a function of the intensities of the two impinging laser beams but also on the non resonant and resonant parts of the second-order nonlinear susceptibility. Now, the resonant contribution is a function of the number density of those molecules and their orientation. Thus, SFG is sensitive to changes in the adsorbed amount at and on the molecular order in the interfacial layer. In [32], for example, adsorption layers formed from mixed solutions of sodium dodecyl sulfate (SDS) and cetyltrimethylammonium bromide (CTAB) are studied by this technique, and a long-time process of surface layer structures is observed.

3.5 NEUTRON SCATTERING AT LIQUID INTERFACES

3.5.1 SPECULAR NEUTRON REFLECTOMETRY

Neutron reflectometry (NR) is a powerful technique that can be used to measure the structure and composition of thin films at free liquid interfaces [33]. Specular NR is sensitive to structural information in the surface normal contained within a few hundred nanometers of the reflecting interface. The refractive index of a material, n, is related to the scattering lengths of the constituent atoms by

$$n^2 = 1 - \frac{\lambda}{\pi}\rho, \quad \text{where} \quad \rho = \frac{\sum_i n_{b,i}}{M_v} \tag{3.12}$$

where
    λ is the wavelength
    ρ is the scattering length density
    i is each atom in the molecule
    M_v is the molecular volume

A homogenous structural model can be applied to specular reflectivity data of simple systems to determine the interfacial composition. Additional features such as the presence of Bragg peak(s), off-specular scattering, and/or deviations in the critical edge when there is total reflection can reveal the presence of multilayers and/or lateral inhomogeneity. The technique is implemented by the reflection of a collimated beam of neutrons at grazing incident angles at a reactor or spallation source. Reflectivity, R, is measured with respect to the incident angle, θ (in monochromatic mode), or λ (in time-of-flight mode) to generate plots of reflectivity with respect to the momentum transfer, Q_z, given by

$$Q_z = \frac{4\pi \sin\theta}{\lambda} \tag{3.13}$$

The resulting plots of $R(Q_z)$ may then be fitted with an optical matrix model to extract information such as the density and thickness of individual layers.

The value of NR in the study of soft matter materials is greatly enhanced by hydrogen/deuterium substitution, which is used to boost the sensitivity of the measurement to a particular component. This technique is possible because the two isotopes have very different scattering lengths, $n_b = -3.74 \times 10^{-5}$ Å for ^{1}H and 6.67×10^{-5} Å for ^{2}H. As such, mixtures of hydrogenous and deuterated materials can be used to produce chemically identical samples with different scattering properties, the combined fitting of which increases the certainty in the structural model. For example, at the air/water interface, there is the convenience that 8.1% by volume D_2O in H_2O, which is a mixture called "air contrast matched water (ACMW)", gives a scattering contrast that is identical to air. The specular reflectivity is then dominated by the molecules in the interfacial layer, especially if they are deuterated. For a single-component monolayer, the surface excess, Γ, may be calculated from a fit to the product of ρ and layer thickness, τ, given by

$$\Gamma = \frac{\rho\tau}{N_A \sum_i n_{b,i}} \tag{3.14}$$

where N_A is Avogadro's number. The product $\rho\tau$ is largely independent of the structural model applied for measurements of the scattering excess of thin interfacial layers recorded in ACMW only at very low Q_z-values [34].

For more complex systems, the interfacial composition and structure are conventionally determined by varying the scattering contrasts of both the subphase and the components at the interface [35,36]. Recently, different approaches were carried out to determine the composition of polymer/surfactant mixtures by varying only the scattering contrast of the one component at the interface [37] and by combining the neutron data with those from ellipsometry [38]. Furthermore, a recent study compared all three approaches for the first time [39].

3.5.2 Fluid Interfaces Grazing Angles ReflectOmeter

Fluid Interfaces Grazing Angles ReflectOmeter (FIGARO) is a horizontal neutron reflectometer at the most intense neutron source in the world, the Institut Laue-Langevin in Grenoble, France [40]. A schematic of the instrument is shown in Figure 3.7. Commissioned in 2009 and already with over 50 peer-reviewed research papers, FIGARO is a world class, versatile, high-flux, time-of-flight instrument with features suitable for a range of studies in soft condensed matter, chemistry, physics, and biology at free air/liquid and liquid/liquid interfaces. The instrument exploits the time-of-flight technique where spinning discs called choppers turn the continuous stream of neutrons into pulses. As the pulses travel down the instrument, the neutrons spread out in space because those with high energy ($\lambda = 2$ Å) travel faster than those with low energy ($\lambda = 30$ Å). The detector then simply records the neutron intensity with time, and with the appropriate calibrations, the intensity with respect to the neutron wavelength is resolved.

The benefit of the reactor source is not just its high stability but that each experiment can be seamlessly tuned to the requirements of the system under study as the balance between resolution and intensity can be dialed up. This is achieved through the presence of four different chopper discs of varying separation in the range of 10–80 cm. Any pair of discs can be used to generate the neutron pulses. The uncertainty in the time of flight (and hence also the wavelength) is low if a pair of discs is used with a small separation: sharp features in the data such as interference fringes resulting from thick films can be resolved. Conversely, much higher intensity can be achieved if a pair of discs is used with a larger separation: this mode is suitable for measurements of thin films where there are not sharp features in the data. Importantly for FIGARO, and uniquely for a neutron reflectometer in time-of-flight mode, it is possible to approach the interface with neutrons from below or above the horizon. The former option is used for air/water interface studies, yet the latter option is particularly important for expanding measurements at liquid/liquid interfaces, which is a fast-growing area.

3.5.3 Applications of FIGARO at Fluid Interfaces

The instrument is equipped with a range of sample environments. Standard options include a suite of free liquid troughs for the study of adsorption as well as Langmuir troughs for the study of

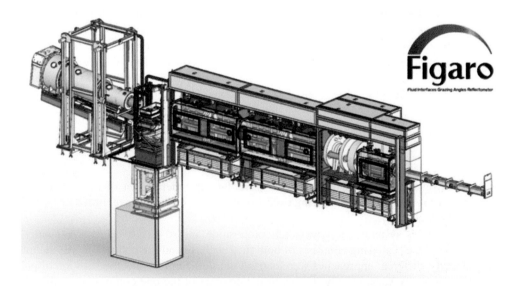

FIGURE 3.7 Schematic of the FIGARO instrument: neutrons travel from right to left. (Courtesy of Institut Laue-Langevin, Grenoble, France.)

insoluble films. More specialized options include an overflowing cylinder for the study of liquids under flow and the new development of dedicated cells for the study of free bulk liquid/liquid interfaces. A laser is used for automatic alignment of air/liquid samples by the instrument control software. This popular feature has the benefit that scientific users can carry out experiments without the need to perform any sample alignment scans. Photographs of these four options are shown in Figure 3.8.

The free liquid troughs have been used to study a range of adsorption phenomena in soft matter systems. Various types of troughs are made of different hydrophobic and hydrophilic materials, and they require liquid volumes on the order of 15–45 mL, although new ultra-low volume troughs are currently being commissioned. Sapphire windows are used for transmission of the neutrons, and tilted quartz windows are used for transmission of the alignment laser beam. All the windows are heated to avoid condensation. A range of studies have been carried out including work on, for example, polyelectrolyte/surfactant [37], protein/surfactant and protein/nanoparticle [41] mixtures.

Langmuir troughs with different areas are available for the study of insoluble films, and control of the surface area can be carried out by the instrument software. The sensor is a Wilhelmy plate made of filter paper, which is dipped in the liquid and connected to a balance allowing the measurement of the pulling force of the liquid and the modification of the surface tension. Two different types of experiments have been carried out routinely. First, there have been studies of insoluble monolayers such as phospholipids with respect to the surface pressure including their interactions with species in solution such as proteins and DNA. Second, there has been the gas-phase interaction of oxidants with insoluble monolayers from the perspective of understanding the effects of pollutants on both lung function and atmospheric processes.

The overflowing cylinder provides a continuously expanding, steady-state air/liquid interface. Liquid is supplied by gravity from a reservoir and flows up a vertical cylinder until it spills over the horizontal rim. The device is used for the study of adsorption kinetics on a subsecond timescale [42]. The slightly curved surface has been carefully calibrated yet the 1500 mL volume may be prohibitively large for some systems. The equipment has been used to good effect to determine adsorption kinetics and interfacial mechanism in polymer/surfactant mixtures.

Liquid/liquid interfaces is a rapidly growing subject area. The established approach is to deposit an oil film with a thickness of only a few micrometers while frozen on a solid crystal to limit the

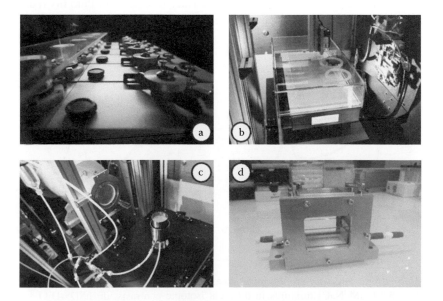

FIGURE 3.8   Free liquid sample environments available on FIGARO: (a) adsorption troughs, (b) a Langmuir trough, (c) an overflowing cylinder, and (d) a liquid/liquid interface cell.

absorption of neutrons. After the film is placed in contact with water in a confined cell, the oil is melted to provide a free liquid/liquid interface [43]. Different groups are currently investing efforts in free liquid/liquid interface measurements. This is particularly attractive on FIGARO due to its high flux and the ability to transmit through the denser phase and reflect down from the probed interface.

3.5.4 SCIENTIFIC EXAMPLE: DENDRIMER/SURFACTANT INTERACTIONS

Dendrimers have a well-defined structure formed by a core with branches that are distributed symmetrically [44]. Stepwise synthesis is carried out, and with each step, a new generation is formed that adds to the size and number of surface groups. Poly(amidoamine) (PAMAM) dendrimers comprise an ethylendiamine core and amidoamine branches with amine terminal groups, which are positively charged around neutral pH. They have been studied extensively for nanoengineering applications such as delivery vehicles of small molecules [45]. Furthermore, the combination of their cationic charge, low polydispersity, and well-defined molecular architecture makes them potential candidates for gene delivery applications [46].

Work was carried out on FIGARO to investigate the interaction of small PAMAM dendrimers of generation 2 with the anionic surfactant SDS at the air/water interface [47]. The work tied in with an ongoing project to relate the interfacial properties of strongly interacting polyelectrolyte/ surfactant mixtures to nonequilibrium features related to bulk aggregation [37,48]. Varied and rich surface structures were related systematically to variations in the bulk phase behavior. For example, whereas a thin interfacial arrangement formed in the dilute regime, a more substantial film formed in the two-phase region close to the bulk composition where there is stoichiometric binding in the bulk complexes. Interesting behavior was observed for samples in the equilibrium one-phase region close to the phase boundary. These samples contained kinetically trapped particles formed during the first moments of mixing. An example of the data is shown in Figure 3.9.

A Bragg peak in the specular reflectivity profile alluded to a multilayered arrangement at the interface. One may have been tempted to relate this to a spontaneous self-assembly of multilayers at the interface. However, an additional feature in the data indicated otherwise. There was a split critical edge of total reflection in different isotopic contrasts, which needed a patch model involving different lateral macroscopic regions across the interface (see red and blue fits required to model the different regions). This demonstrated the penetration at the interface of nanostructured particles

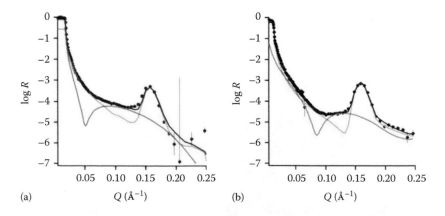

(a)   (b)

FIGURE 3.9 Neutron reflectivity profiles recorded at the air/water interface for 91 ppm poly(amidoamine)-G2/0.16 mM SDS/10 mM NaCl mixtures in different isotopic contrasts: (a) hSDS/D_2O and (b) dSDS/air contrast-matched water (ACMW). The blue and red models show fits to different patches of the interface. (Adapted from Yanez Arteta, M. et al., *J. Phys. Chem. B*, 118, 11835, 2014.)

that were retained intact as a result of charge stabilization due to the presence of excess ions. The retention of the particles at the interface is in contrast to samples in the two-phase region where homogeneous layers formed as indicated by the lack of a Bragg peak or split critical edge in the data. This difference was attributed to the dissociation and spreading of material at the interface from particles with low surface charge in the two-phase region. An involved interfacial mechanism was therefore revealed. It was concluded that small PAMAM dendrimers have high potential for applications involving the delivery of material to interfaces.

3.6 OSCILLATING DROPS AND BUBBLES

The viscoelasticity as measured by interfacial area oscillation gives additional information about adsorption layers, mechanisms of foam, and emulsion stabilization and is useful for theoretical models. The method of oscillating drops and bubbles provides viscoelasticity parameters, either via measurements of the capillary pressure (e.g., with oscillating drop and bubble analyser [ODBA] from SINTERFACE Technologies) or via drop and bubble profile analysis (e.g., with profile analysis tensiometer [PAT]-1 form SINTERFACE Technologies). The important feature of the ODBA is the option of generating high-frequency oscillations of small spherical drops/bubbles over a broad frequency range [49]. In contrast, methodology of PAT allows investigations only up to 0.1 Hz for aqueous drops in air [50].

3.6.1 HYDRODYNAMIC EFFECTS

The oscillations of drop and bubble in experiments with the ODBA are generated by a piezo drive, which can exactly control the size of small droplets with an accuracy of ± 0.0001 mm^3 in the frequency range of 0.1–300 Hz. During the generation of precise oscillations, the pressure variations are recorded as a function of time with an accurate pressure sensor. This signal contains the capillary pressure contribution along with hydrodynamic effects. The surface tension response to a harmonic oscillation of the interfacial area is measured via the pressure. The measured quantities are the phase shift ϕ between generated volume and measured surface tension oscillations and the amplitude of surface tension oscillations.

The hydrodynamic effects that are negligible for low frequencies, however, can be very significant for higher frequencies. These effects depend not only on the applied amplitude but also on the fluid's viscosities, the capillary tip size, and geometry (see, for instance, [51]).

The hydrodynamic basis for oscillation experiments in a closed measuring cell, as it is the case for the ODBA, was discussed in detail in [49]. The dynamic response of the considered measuring system is complex and includes not only the viscoelastic contribution of the studied interfacial layer but also other contributions, such as the hydrodynamics of the liquid flow through the capillary and the flow field in the bulk around the meniscus [52]. For small-amplitude harmonic oscillations, the response of the system is linear, and all contributions can be characterized by a respective complex resistance [53].

3.6.2 INTERFACIAL DILATIONAL MODULUS

The interfacial area perturbations are assumed to be small, and the interfacial area is subject to harmonic changes (oscillations):

$$A(t) = A_0 + \Delta A(t) = A_0 + \Delta A \sin(\omega t + \phi_A) \qquad (3.15)$$

where
A_0, ΔA, and ϕ_A are the mean area, the amplitude, and the phase angle of area changes, respectively
ω is the angular frequency of oscillation

For linear systems, the interfacial tension changes are also harmonic with a certain amplitude and phase shift [54]

$$\gamma(t) = \gamma_0 + \Delta\gamma \sin(\omega t + \phi_\gamma)$$
(3.16)

where γ_0, $\Delta\gamma$, and ϕ_γ are the mean surface tension value, the amplitude, and the phase shift of surface tension changes, respectively. The complex viscoelasticity carries all information of the oscillations with respect to changes in amplitude as well as in phase shift. The most general definition is given by the Fourier transform of the response $\Delta\gamma$ relative to the perturbation ΔA:

$$E(i\omega) = A_0 \frac{F[\Delta\gamma]}{F[\Delta A]}$$
(3.17)

Any complex function can be represented in a form like $E(i\omega) = E_r + iE_i$ or $E(i\omega) = |E|\exp(i\phi)$, with $|E| \equiv \dfrac{\Delta\gamma}{\Delta A/A_0}$ being the surface dilational modulus, and E and $\phi = \phi_\gamma - \phi_A$ can be considered as viscoelasticity parameters. In this way, it is possible to obtain the viscoelasticity parameters through the parameters of surface tension and interfacial area oscillations.

The dependence of the viscoelasticity on the oscillation frequency $E(i\omega)$ provides important information on the dynamic properties of a surface layer. When only one surfactant is in solution and diffusional transport controls the adsorption kinetics, for a plane surface, an analytic dependence was derived by Lucassen and van den Tempel [55]. For spherical interfaces, expressions for $E(i\omega)$ were proposed by Joos [56]: for the adsorption from the solution bulk onto a bubble surface and for the adsorption from the drop bulk onto its surface.

The dependencies of the viscoelasticity modulus for a 2 μmol/L solution of the nonionic surfactant $C_{14}EO_8$ on the surface oscillation frequency are shown in Figure 3.10, as measured by the drop and bubble profile methods [23]. To determine the equilibrium surfactant concentration

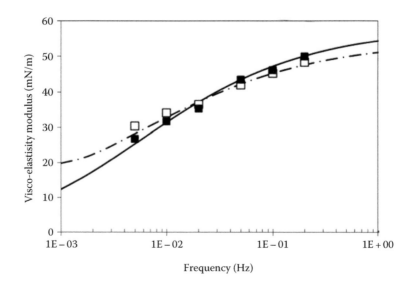

FIGURE 3.10 Dependence of viscoelasticity modulus on surface oscillation frequency at the $C_{14}EO_8$ concentration 2.0 μmol/L, measured after the equilibrium is attained in the drop (□) and by bubble profile method (■); solid lines, bubble profile method; dashed lines, drop profile method. (Data are taken from Fainerman, V.B. et al., *Langmuir*, 29, 6964, 2013.)

in the drop bulk, a correction was introduced for the surfactant losses caused by its adsorption at the drop surface. For an initial solution concentration in drops of 7 μmol/L, the equilibrium concentration in the drop was depleted down to 2 μmol/L (see Figure 3.5). We can see that the viscoelasticity modulus values obtained by the two methods are quite close to each other, however, only when a correction for the $C_{14}EO_8$ concentration was taken into account. The theoretical model proposed by Joos [56] was used and provided a good description of the dilational rheology and tensiometry data.

3.7 DAMPING OF WAVES AT INTERFACES

While experiments with oscillating drops and bubbles allow studying relaxation processes at rather low (profile analysis tensiometry) or intermediate (CPT) frequencies, the method of choice for higher frequencies is capillary wave damping. These methods, depending on the way of wave generation and wave damping analysis, span over different frequency intervals and can reach values larger than MHz, like the surface quasi-elastic light scattering technique (SQELS) [57].

Many systems have various characteristic times of relaxations so that a single method, such as the oscillating drop or bubble discussed in Section 3.6, cannot cover the requirements of a careful analysis. Wave damping is a relaxation method that, depending on the way of wave generation, can cover a quite broad frequency range. Figure 3.11 shows that mechanically generated waves overlap well with the oscillating drop/bubble methods, while the electrocapillary waves allow studies of the dilational rheology up to frequencies of 10^4 Hz [58]. The SQELS complements these relation methods toward higher frequencies up to 2 MHz [57].

The theoretical background of these wave damping methods was comprehensively presented in [59]. The surface waves propagating along a liquid–gas interface are described by the Navier–Stokes equations. If the liquid is isotropic, incompressible, and Newtonian, these equations have the following form [59]:

$$\rho\left[\frac{\partial \vec{v}}{\partial t} + (\vec{v}\cdot\vec{\nabla})\vec{v}\right] = -\vec{\nabla}p + \mu\Delta\vec{v} + \rho\vec{g} \tag{3.18}$$

and the continuity equation

$$\vec{\nabla}\cdot\vec{v} = 0 \tag{3.19}$$

where
  ρ is the liquid density
  $\vec{v}$ is the vector of liquid velocity
  p is the pressure
  μ is the liquid viscosity
  $\vec{g}$ is the gravitational acceleration

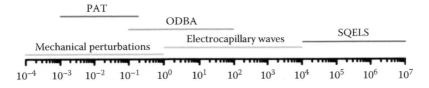

FIGURE 3.11 Schematic of the frequency range where available experimental methods can be applied.

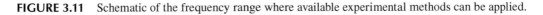

With a number of simplifications and respective boundary condition, the following dispersion relation of surface waves was derived in [60]:

$$(\rho\omega^2 - \gamma k^3 - \rho g k)(\rho\omega^2 - m k^2 E) - E k^3 (\gamma k^3 + \rho g k) + 4 i \rho \mu \omega^3 k^2 + 4 \mu^2 \omega^2 k^3 (m - k) = 0 \quad (3.20)$$

where $k = 2\pi/\lambda$ is the wave number and ω is the angular frequency and the constant $m = \sqrt{k^2 - i\omega\rho/\mu}$. This complex relation (3.20) interrelates the characteristics of the wave (k, ω) with the bulk (ρ, μ) and surface properties (γ, E).

A main drawback of surface wave techniques is that they provide information on the dilational viscoelastic modulus only through the coupling of the modes of capillary and transversal waves, which is effective only in a limited range of the E/γ ratio. Beyond this limit, only surface tension data are available.

Experimental techniques based on capillary wave damping have been critically discussed, for example, in [59,61,62]. There is a huge number of experimental data on the dilational rheology of interfacial layers covered by adsorbed surfactants, polymers, proteins, particles, and their various mixtures. As an example, Figure 3.12 shows the dilational elasticity as a function of frequency measured by Maestro et al. [63] for mixed beta-casein protein/dodecyldimethyl phosphine oxide (BCS/C_{12}DMPO) surfactant adsorption layers in a very broad frequency range between 0.01 Hz and 1 kHz. The studied C_{12}DMPO concentrations are all below the respective CMC. The measured dilational viscosity is much smaller than the elasticity over the whole frequency range so that we can conclude that the interfacial layer behaves mainly elastically. The frequency dependence $E(\omega)$ can be well described by a combination of two models, that is by the Lucassen–van den Tempel model at low frequencies and with a Maxwell model in the high-frequency range [63].

All discussions so far were based on a quasi-equilibrium system, that is not far from the adsorption equilibrium. This is, however, not true in more complex systems, for example, when concentration gradient across the bulk exists. Then, a more general approach has to be used as it is described in Chapter 2.

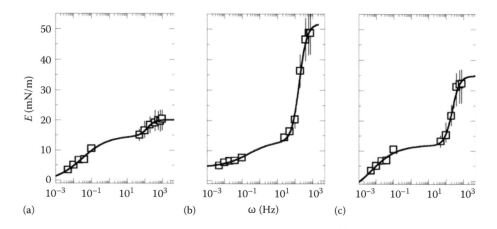

FIGURE 3.12 Frequency dependence of the dilational elastic modulus E for three mixtures of BCS (fixed at 10^{-6} mol/L) and C_{12}DMPO at 10^{-6} (a) mol/L, (b) 10^{-5} mol/L , and (c) 10^{-4} mol/L ; the lines represent the fit of data to a model. (Discussed in Maestro, A. et al., *Langmuir*, 116, 4898, 2012.)

3.8 RISING DROPS AND BUBBLES

Among the wide variety of multiphase flow situations, bubbles in flow fields are of great importance in many technological processes. The uncertainties in the flow around bubbles occur from lack of understanding the local hydrodynamics and rate processes, which govern the bubble size and thus the interfacial area between phases. In addition to the physical properties of the liquid bulk phase, also the interface characteristics can drastically affect the features of bubble rising in this liquid.

Surface active compounds can be either present in the liquid as impurities or could have been added on purpose. However, the presence of any small surfactant amounts can significantly affect the motion and shape of a rising bubble [64]. In particular, the determination of the drag force on a rising bubble with a surface contaminated by surfactants is important for understanding the basic bubble behavior in respect to industrial applications.

The physical mechanism of rising bubbles in surfactant solution was first explained by Frumkin and Levich [65]. They stated that the adsorption layer establishes a surface concentration gradient the physicochemical nature of which has been further elaborated by Levich [66]. During the movement of a bubble in presence of surfactants, an adsorption layer is formed and retards the mobility of the interface and consequently the bubble's viscous drag increases. In turn, the drag force induces a nonuniform distribution of adsorbed material over the bubble surface: the rear pole is enriched (highest surface coverage) as compared to the almost empty leading pole (Figure 3.13). The surface concentration gradient generates a Marangoni stress, which results in a retardation of the bubble motion. Changes in the behavior of a single bubble can affect, for example, significantly the structure of a whole bubble flow [67].

There is quite a number of experimental works, summarized, for example, in [68]. Most informative experimental findings are the velocity profiles of rising bubbles of various sizes in surfactant solutions. The data show that the location of the maxima in the velocity profiles and its maximum

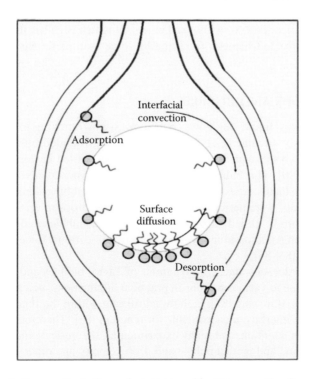

FIGURE 3.13 Concentration gradient of a surfactant caused by the convective flow around a rising bubble surface. (Redrawn from Lotfi, M. et al., *Colloids Surf. A*, 460, 369, 2014.)

value represent significant parameters for a surfactant. In a new critical review, the available experimental data and a respective analysis with existing theoretical models were given [69].

A completely new approach of stagnant cap investigations was proposed in [70], where an air bubble was fixed at the tip of a capillary and then studied in the flow field of a surfactant solution. By measurements of the capillary pressure in the bubble, the process of adsorption layer formation at the bubble surface was followed.

The majority of theoretical contributions are dedicated to the rear stagnant cap model at high Pe numbers where it can be assumed that the transport of surfactants is convection controlled. Steady hydrodynamic models for low Reynolds numbers (Stokes flow) were proposed by Sadhal and Johnson [71]. Then, Zholkovskij et al. [72] developed a nonsteady-state hydrodynamic model based on their approach and predicted a velocity profile for the case that the sorption kinetics is the slowest step in the process of the dynamic adsorption layer formation.

For high Reynolds numbers, the solution of the hydrodynamics for rising bubbles is more complex, and therefore, most of such studies were done for very low concentrations of very strong surfactants (see [73]) or for very high concentrations of very weak surfactant (see [74]). The most frequently used work for high Reynolds numbers (50–200) but intermediate surfactant concentrations was done by Fdhila and Duineveld [75]. They have obtained the angle of the stagnant cap as a function of the terminal velocity, Marangoni number, and bulk concentration. In another important work, Cuenot et al. [76] performed a numerical simulation of the surfactant adsorption at the surface of a spherical bubble, using a boundary-fitted coordinate system for having a sufficient number of grid points near the surface. This is important in order to capture the thin concentration boundary layer at the bubble surface. This model allows describing the stagnant cap around a slightly contaminated bubble surface. This model became later the basis of many other simulation studies [77].

Despite the great advances in computational fluid dynamics (CFD) for simulating deforming interfaces in two phase flows, the complexity of the boundary conditions at these deforming interfaces in presence of surfactants is still a very challenging field. There are different approaches such as front capturing [78], front tracking [79], and others, of which each has its specific advantages and weaknesses. For example, in Chapters 16, 17, and 18, some examples are given to show the potential of CFD simulations.

3.9 TAYLOR DROPS AND BUBBLES

Large gas bubbles in gas–liquid slug flows (bubble train flow) moving in a vertical pipe take bullet shapes with a nearly circular head. The bubbles are surrounded by a thin liquid film, and a liquid slug separates two subsequent bubbles. Such bubbles, called Taylor bubbles, were originally described by Taylor [80]. The bubble volume in Taylor's experiments was comparatively large (up to 34 cm^3), while in later studies by Thulasidas et al. [81], bubbles of much smaller volume, up to microfluidic experiments, were performed. Sufficiently large drops in vertical pipes also take bullet shapes and can therefore be called Taylor drops [82]. Taylor bubbles and drops flowing in narrow channels are a common situation in many microfluidic applications. Recent reviews of Taylor flow problems are given in [83–85].

During the last two decades, the hydrodynamics of Taylor bubbles and drops became an important field of research due to various reasons in practical application, such as high heat transfer rates [86], significant low axial mixing [87], high radial mixing within the liquid slugs due to recirculation [88], or adjustable interface area available for reaction [89]. Therefore, the Taylor flow regime has received enormous attention, and many experimental, theoretical, and CFD studies have been carried out to understand and predict bubble and drop velocity and other characteristic parameters [80,82,83,87–99].

Various interface capturing techniques have been used to model the Taylor flow in microchannels, such as the volume-of-fluid (VOF) approach [93], the level-set method [92], or the phase-field method [90]. Recently, Fletcher et al. [91] have shown that the main challenges in the modeling of multiphase flow in microchannels are the correct identification of the interface location and the occurrence of parasitic currents arising from modeling the surface tension. Also, there has been some confusion regarding the existence of a liquid film around the gas bubble in various CFD simulations [85]. While some researchers did not capture such a liquid film due to an insufficient grid [93], others carried out their CFD simulations assuming a direct contact of the gas bubble with the wall [90]. Moreover, Gupta et al. [85] have shown that numerical issues, such as grid refinement, require careful attention to avoid unphysical results.

Experiments have shown that the addition of surfactants changes the surface tension but has negligible effect on the bubble length [94]. Numerical simulations provide an alternative method to investigate the separate roles of surface tension and contact angle during bubble formation in microchannels [95]. Qian and Lawal [93] found that the increase in surface tension and contact angle can have an opposite influence on the bubble length and they can compensate each other as shown by numerical simulation using a VOF method.

Ratulowski and Chang [100] demonstrated that at low surfactant concentration, the interface shows slip, and at high surfactant concentration, there is no-slip condition due to Marangoni effects. According to Bretherton [96], a no-slip boundary condition at the bubble interface increases both the pressure drop over the Taylor bubble and the film thickness by a factor $4^{2/3}$. However, the real behavior of the bubble interface is in fact very complex and all experimental data are bounded by the no-shear and no-slip limits [101]. Almatroushi and Borhan [102] have observed experimentally that the presence of surfactant increases the terminal velocity of a Taylor bubble due to the reduction in surface tension and no Marangoni effect.

For the first time, Hayashi and Tomiyama have used an interface-tracking (level-set method with Eulerian approach) method to investigate the motion of a Taylor bubble contaminated by surfactants at low Morton numbers [97]. The adsorption–desorption kinetics is accounted for by using the Frumkin–Levich model, and it was found out that the surface tension reduction near the bubble nose due to the surfactant adsorption increases the terminal velocity of Taylor bubbles at low Eötvös numbers, while the terminal velocity is independent of surface tension at high Eötvös numbers. Kurimoto et al. [98] investigated effects of surfactants on the shape and terminal velocity of a Taylor drop based on experimental data and interface-tracking simulations. The result are similar to those found in [97] for Taylor bubbles, that is, the reduction in surface tension due to the adsorption of surfactant causes the increase in the terminal velocity and the elongation of a Taylor drop as shown in Figure 3.14.

3.10 SUMMARY AND OUTLOOK

Methods like bubble and drop pressure and profile analysis tensiometry provide surface and interfacial tension data, respectively. However, there are essential complications with the definition of the timescale for all those methods where during the experiment the interfacial area is changing with time. In bubble pressure tensiometry, the bubble is continuously growing so that not only the bubble surface area is increasing with time but there is also a liquid flow field directed toward the surface. Both phenomena have been considered so far essentially qualitatively, as it is shown in some more details in Chapter 4. In this respect, drop and bubble profile analysis tensiometry is a kind of static method and provides therefore clear dependencies of surface tension as a function of adsorption time. Also, ellipsometry and neutron scattering are static methods and measure adsorbed amounts as a function of time and concentration. Thus, these methodologies allow comparing experimental data directly with simple model calculations. In contrast, growing and oscillating drop and bubble pressure methods require hydrodynamic theories in order to correlate the measured pressure values with surface/interfacial parameters. As discussed earlier, at low liquid flow rates or oscillation frequencies, the hydrodynamics is more or less negligible, while at high flow rates and frequencies, the hydrodynamics dominates and can hardly be separated from the surface properties.

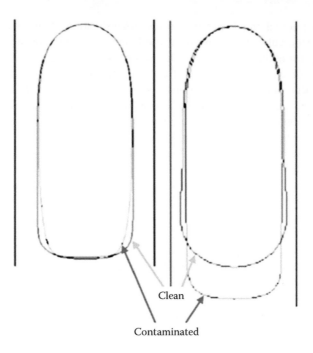

FIGURE 3.14 Comparison of shapes for drops with a clean and contaminated surface for two different case studies (different Morton numbers, Eötvös numbers, and viscosity of the continuous phase). (Details are given in Kurimoto, R. et al., *Int. J. Multiphase Flow*, 49, 8, 2013.)

In contrast to these methods, experiments with rising bubbles do not provide any direct data on the adsorption of surface active molecules at interfaces. The measured rising velocity profile, however, is a combined effect of adsorption and desorption rates, drag force of the adjacent liquid, and Marangoni flow at the bubble surface, and this allows to get access to the adsorption mechanism including adsorption/desorption rate constants. A quantitative analysis of all processes to/from and at the bubble surface is yet under consideration, as it is shown in more detail in several chapters, such as Chapters 17, 18, 22, and 23. This will demonstrate how complex situations can be handled and quantitative information obtained.

REFERENCES

1. V.B. Fainerman and R. Miller, in *Bubble and Drop Interfaces*, Vol. 2, Progress in Colloid and Interface Science, R. Miller and L. Liggieri (Eds.), Brill Publication, Leiden, the Netherlands, 2011, pp. 75–118.
2. V.B. Fainerman, V.D. Mys, A.V. Makievski, J.T. Petkov and R. Miller, Dynamic surface tension of micellar solutions in the millisecond and submillisecond time range. *J. Colloid Interface Sci.*, 302 (2006) 40–46.
3. A. Passerone, L. Liggieri, N. Rando, F. Ravera and E. Ricci, A new experimental method for the measurement of the interfacial tension between immiscible fluids at zero Bond number. *J. Colloid Interface Sci.*, 146 (1991) 152–162.
4. S. Sugden, The determination of surface tension from the maximum pressure in bubbles. Part II. *J. Chem. Soc.*, 125 (1924) 27–31.
5. R.L. Bendure, Dynamic surface tension determination with the maximum bubble pressure method. *J. Colloid Interface Sci.*, 35 (1971) 238–248.
6. V.I. Kovalchuk and S.S. Dukhin, Dynamic effects in maximum bubble pressure experiments. *Colloids Surf. A*, 192 (2001) 131–155.
7. R.L. Kao, D.A. Edwards, D.T. Wasan, and E. Chen, Measurement of interfacial dilatational viscosity at high rates of interface expansion using the maximum bubble pressure method. I. Gas—liquid surface. *J. Colloid Interface Sci.*, 148 (1992) 247–256.

8. V.I. Kovalchuk, S.S. Dukhin, V.B. Fainerman, and R. Miller, Hydrodynamic processes in dynamic bubble pressure experiments. 4. Calculation of magnitude and time of the slow oscillations. *Colloids Surf. A*, 151 (1999) 525–536.

9. V.B. Fainerman, Measurement of the dynamic surface-tension of solutions by the method of the maximal pressure in a bubble. *Kolloidn. Zh.*, 41 (1979) 111–116.

10. V.B. Fainerman, On the surface tension measurement by the bubble pressure method in the millisecond range. *Kolloidn. Zh.*, 52 (1990) 921–927.

11. J. Kloubek, Measurement of the dynamic surface tension by the maximum bubble pressure method. III. Factors influencing the measurements at high frequency of the bubble formation and an extension of the evaluation to zero age of the surface. *J. Colloid Interface Sci.*, 41 (1972) 7–16.

12. V.B. Fainerman, Adsorption kinetics from concentrated micellar solutions of ionic surfactants at the water-air interface. *Colloids Surf.*, 62 (1992) 333–347.

13. V.B. Fainerman, V.D. Mys, A.V. Makievski, and R. Miller, Application of the maximum bubble pressure technique for dynamic surface tension studies of surfactant solutions using the Sugden two-capillary method. *J. Colloid Interface Sci.*, 304 (2006) 222–225.

14. V.I. Kovalchuk, F. Ravera, L. Liggieri, G. Loglio, A. Javadi, N.M. Kovalchuk, and J. Krägel, Studies in capillary pressure tensiometry and interfacial dilational rheology, Chapter 7, in *Bubble and Drop Interfaces*, Vol. 2, Progress in Colloid and Interface Science, R. Miller and L. Liggieri (Eds.), Brill Publication, Leiden, the Netherlands, 2011, pp. 143–178.

15. N. Alexandrov, K.G. Marinova, K.D Danov, and I.B. Ivanov, Surface dilatational rheology measurements for oil/water systems with viscous oils. *J. Colloid Interface Sci.*, 339 (2009) 545–550.

16. A.W. Neumann and R.J. Good, Techniques of measuring contact angles, in *Experimental Methods in Surface and Colloid Science*, R.J. Good and R.R. Stromberg (Eds.), Vol. 11, Plenum, New York, 1979, pp. 31–91.

17. S.A. Zholob, A.V. Makievski, R. Miller, and V.B. Fainerman, Advances in calculation methods for the determination of surface tensions in drop profile analysis tensiometry, Chapter 3, in *Bubble and Drop Interfaces*, Vol. 2, Progress in Colloid and Interface Science, R. Miller and L. Liggieri (Eds.), Brill Publication, Leiden, the Netherlands, 2011, pp. 39–60.

18. N.M. Dingle, K. Tjiptowidjojo, O.A. Basaran, and M.T. Harris, A finite element based algorithm for determining interfacial tension (γ) from pendant drop profiles. *J. Colloid Interface Sci.*, 286 (2005) 647–660.

19. Y. Rotenberg, L. Boruvka, and A.W. Neumann, Determination of surface tension and contact angle from the shapes of axisymmetric fluid interfaces. *J. Colloid Interface Sci.*, 93 (1983) 169–183.

20. M. Hoorfar and A.W. Neumann, Axisymmetric drop shape snalysis (ADSA) for measuring surface tension and contact angle. *J. Adhes.*, 80 (2004) 727–743.

21. M.G. Cabezas, A. Bateni, J.M. Montanero, and A.W. Neumann, A new method of image processing in the analysis of axisymmetric drop shapes. *Colloids Surf. A*, 255 (2005) 193–200.

22. V.B. Fainerman, S.A. Zholob, J.T. Petkov, and R. Miller, $C_{14}EO_8$ adsorption characteristics studied by drop and bubble profile tensiometry. *Colloids Surf. A*, 323 (2008) 56–62.

23. V.B. Fainerman, E.V. Aksenenko, J. Krägel, and R. Miller, Viscoelasticity moduli of aqueous C14EO8 solutions as studied by drop and bubble shape methods. *Langmuir*, 29 (2013) 6964–6968.

24. A.V. Makievski, G. Loglio, J. Krägel, R. Miller, V.B. Fainerman, and A.W. Neumann, Adsorption of protein layers at the water/air interface as studied by axisymmetric drop and bubble shape analysis. *J. Phys. Chem. B*, 103 (1999) 9557.

25. D. Möbius and R. Miller (Eds.), Novel methods to study interfacial layers, in *Studies in Interface Science*, Vol. 11, Elsevier, Amsterdam, the Netherlands, 2001.

26. J.W. Benjamins, B. Jönsson, K. Thuresson, and T. Nylander, New experimental setup to use ellipsometry to study liquid-liquid and liquid-solid interfaces. *Langmuir*, 18 (2002) 6437–6444.

27. B.A. Noskov, A.V. Akentiev, A. Yu. Bilibin, D.O. Grigoriev, G. Loglio, I.M. Zorin, and R. Miller, Adsorption kinetics of non-ionic polymers. Ellipsometric study. *Mendeleev Commun.*, 15 (2005) 198–200.

28. H. Motschmann and R. Teppner, Ellipsometry in interface science, in *Novel Methods to Study Interfacial Layers*, Vol. 11, Studies in Interface, Science, D. Möbius and R. Miller (Eds.), Elsevier, Amsterdam, the Netherlands, 2001.

29. K. Engelhardt, A. Rumpel, J. Walter, J. Dombrowski, U. Kulozik, B. Braunschweig, and W. Peukert, Protein adsorption at the electrified air–water interface: Implications on foam stability. *Langmuir*, 28 (2012) 7780–7787.

30. C. Zhang, J.N. Myers, and Z. Chen, Elucidation of molecular structures at buried polymer interfaces and biological interfaces using sum frequency generation vibrational spectroscopy. *Soft Matter*, 9 (2013) 4738–4761.

31. Y.R. Shen, *The Principles of Nonlinear Optics*, John Wiley & Sons, New York, 1984.

32. A. Saha, H.P. Upadhyaya, A. Kumar, S. Choudhury, and P.D. Naik, Sum-frequency generation spectroscopy of an adsorbed monolayer of mixed surfactants at an air–water interface. *J. Phys. Chem. C*, 118 (2014) 3145–3155.

33. J.R. Lu, R.K. Thomas, and J. Penfold, Surfactant layers at the air/water interface: Structure and composition. *Adv. Colloid Interface Sci.*, 84 (2000) 143–305.

34. E.A. Simister, R.K. Thomas, J. Penfold, R. Aveyard, B.P. Binks, P. Cooper, P.D.I. Fletcher, J.R. Lu, and A. Sokolowski, Comparison of neutron reflection and surface tension measurements of the surface excess of tetradecyltrimethylammonium bromide layers at the air/water interface. *J. Phys. Chem.*, 96 (1992) 1383–1388.

35. D.J.F. Taylor, R.K. Thomas, and J. Penfold, The adsorption of oppositely charged polyelectrolyte/surfactant mixtures: Neutron reflection from dodecyl trimethylammonium bromide and sodium poly(styrene sulfonate) at the air/water interface. *Langmuir*, 18 (2002) 4748–4757.

36. M. Moglianetti, P. Li, F.L.G. Malet, S.P. Armes, R.K. Thomas, and S. Titmuss, Interaction of polymer and surfactant at the air-water interface: Poly(2-(dimethylamino)ethyl methacrylate) and sodium dodecyl sulfate. *Langmuir*, 24 (2008) 12892–12898.

37. Á. Ábraham, R. A. Campbell, and I. Varga, New method to predict the surface tension of complex synthetic and biological polyelectrolyte/surfactant mixtures. *Langmuir*, 29 (2013) 11554–11559.

38. A. Angus-Smyth, R.A. Campbell, and C. D. Bain, Dynamic adsorption of weakly interacting polymer/surfactant mixtures at the air/water interface. *Langmuir*, 28 (2012) 12479–12492.

39. H. Fauser, R. von Klitzing, and R.A. Campbell, surface adsorption of oppositely charged C14TAB-PAMPS mixtures at the air/water interface and the impact on foam film stability. *J. Phys. Chem. B*, 2014, DOI: 10.1021/jp509631b.

40. R.A. Campbell, H.P. Wacklin, I. Sutton, R. Cubitt, and G. Fragneto, FIGARO: The new horizontal neutron reflectometer at the ILL. *Eur. Phys. J. Plus*, 126 (2011) 107.

41. J.C. Ang, M. Henderson, R.A. Campbell, J.M. Lin, P.N. Yaron, A. Nelson, T.A. Faunce, and J.W. White, Human serum albumin binding to silica nanoparticles—Effect of protein fatty acid ligand. *Phys. Chem. Chem. Phys.*, 16 (2014) 10157–10168.

42. S. Manning-Benson, S.R.W. Parker, C.D. Bain, and J. Penfold, Measurement of the dynamic surface excess in an overflowing cylinder by neutron reflection. *Langmuir*, 14 (1998) 990–996.

43. A. Zarbakhsh, J. Bowers, and J.R.P. Webster, A new approach for measuring neutron reflection from a liquid/liquid interface. *Meas. Sci. Technol.*, 10 (1999) 738–743.

44. D.A. Tomalia, H. Baker, J. Dewald, M. Hall, G. Kallos, S. Martin, J. Roeck, J. Ryder, and P. Smith, A new class of polymers: Starburst-dendritic macromolecules. *Polym. J.*, 17 (1985) 117–132.

45. C.N. Moorefield and G.R. Newkome, Unimolecular micelles: Supramolecular use of dendritic constructs to create versatile molecular containers. *C. R. Chim.*, 6 (2003) 715–724.

46. J.D. Eichman, A.U. Bielinska, J.F. Kukowska-Latallo, and J.R. Baker Jr., The use of PAMAM dendrimers in the efficient transfer of genetic material into cells. *Pharm. Sci. Technol. Today*, 3 (2000) 232–245.

47. M. Yanez Arteta, R. A. Campbell, E. B. Watkins, M. Obiols-Rabasa, K. Schillen, and T. Nylander, Interactions of small dendrimers with sodium dodecyl sulfate at the air–water interface. *J. Phys. Chem. B*, 118 (2014) 11835–11848.

48. R.A. Campbell, M. Yanez Arteta, A. Angus-Smyth, T. Nylander, B. Noskov, and I. Varga, Direct impact of nonequilibrium aggregates on the structure and morphology of PDADMAC/SDS layers at the air/water interface. *Langmuir*, 30 (2014) 8664–8674.

49. V.I. Kovalchuk, J. Krägel, A.V. Makievski, G. Loglio, F. Ravera, L. Liggieri, and R. Miller, Frequency characteristics of amplitude and phase of oscillating bubble systems in a closed measuring cell. *J. Colloid Interface Sci.*, 252 (2002) 433.

50. M.E. Leser, S. Acquistapace, A. Cagna, A.V. Makievski, and R. Miller, Limits of oscillation frequencies in drop and bubble shape tensiometry. *Colloids Surf. A*, 261 (2005) 25–28.

51. F. Ravera, G. Loglio, P. Pandolfini, E. Santini, and L. Liggieri, Determination of the dilational viscoelasticity by the oscillating drop/bubble method in a capillary pressure tensiometer. *Colloids Surf. A*, 365, (2010) 2–13.

52. V.I. Kovalchuk, J. Krägel, E.V. Aksenenko, G. Loglio, and L. Liggieri, in *Novel Methods to Study Interfacial Layers*, Vol. 11, Studies in Interface Science, D. Möbius and R. Miller (Eds.), Elsevier, Amsterdam, the Netherlands, 2001, p. 485.

53. A. Javadi, J. Krägel, A.V. Makievski, V.I. Kovalchuk, N.M. Kovalchuk, N. Mucic, G. Loglio, P. Pandolfini, M. Karbaschi, and R. Miller, Fast dynamic interfacial tension measurements and dilational rheology of interfacial layers by using the capillary pressure technique. *Colloids Surf. A*, 407 (2012) 159–168.

54. E.H. Lucassen-Reynders and J. Lucassen, in *Interfacial Rheology*, Vol. 1, Progress in Colloid and Interface Science, R. Miller and L. Liggieri (Eds.), Brill Publicaiton, Leiden, the Netherlands, 2009, pp. 40–78.

55. J. Lucassen and M. van den Tempel. Longitudinal waves on visco-elastic surfaces. *J. Colloid Interface Sci.*, 41 (1972) 491.

56. P. Joos, *Dynamic Surface Phenomena*, VSP, Dordrecht, the Netherlands, 1999.

57. M. Lotfi, M. Karbaschi, A. Javadi, N. Mucic, J. Krägel, V.I. Kovalchuk, R.G. Rubio, V.B. Fainerman, and R. Miller, Dynamics of liquid interfaces under various types of external perturbations. *COCIS*, 19 (2014) 309–319.

58. F. Monroy, F. Ortega, R.G. Rubio, and M.G. Velarde, Surface rheology, equilibrium and dynamic features at interfaces, with emphasis on efficient tools for probing polymer dynamics at interfaces. *Adv. Colloid Interface Sci.*, 134–135 (2007) 175–189.

59. B. Noskov, Capillary waves in interfacial rheology, in *Interfacial Rheology*, R. Miller and L. Liggieri (Eds), Vol. 1, Progress in Colloid and Interface Science, Brill Publication, Leiden, the Netherlands, 2009, pp. 103–136.

60. R.S. Hansen and J. Ahmad, Waves at interfaces, in *Progress in Surface and Membrane Science*, J. Danielli, M. Rosenberg, and D. Cadenhead (Eds.), Vol. 4, Academic Press, New York, 1971.

61. E.H. Lucassen-Reynders and J. Lucassen, Surface dilational rheology: Past and present, in *Interfacial Rheology*, R. Miller and L. Liggieri (Eds.), Brill Publication, Leiden, the Netherlands, 2009, pp. 38–76.

62. R. Miller, J.K. Ferri, A. Javadi, J. Krägel, N. Mucic, and R. Wüstneck, Rheology of interfacial layers. *Colloid Polym. Sci.*, 288 (2010) 937–950.

63. A. Maestro, C. Kotsmar, A. Javadi, R. Miller, F. Ortega, and R.G. Rubio, Adsorption of β-casein-surfactant mixed layers at the air-water interface evaluated by interfacial rheology. *Langmuir*, 116 (2012) 4898–4907.

64. A. Prosperetti, Bubbles, *Phys. Fluids*, 16 (2004) 1852–1865.

65. A.N. Frumkin and V.G. Levich, Effect of surface-active substances on movements at the boundaries of liquid phases. *Phys. Chim.*, 21 (1947) 1183.

66. V.G. Levich, *Physicochemical Hydrodynamics*, Prentice-Hall, Englewood Cliffs, NJ, 1962.

67. M. Fukuta, S. Takagi, and Y. Matsumoto, Numerical study on the shear-induced lift force acting on a spherical bubble in aqueous surfactant solutions. *Phys. Fluids*, 20 (2008) 040704.

68. K. Małysa, J. Zawala, M. Krzan, and M. Krasowska, Bubbles rising in solutions: Local and terminal velocities, shape variations and collisions with free surface, in *Bubble and Drops Interfaces*, R. Miller and L. Liggieri (Eds.), Brill Publication, Leiden, the Netherlands, 2011, pp. 243–292.

69. S.S. Dukhin, V.I. Kovalchuk, G.G. Gochev, M. Lotfi, M. Krzan, K. Malysa, and R. Miller, Dynamics of rear stagnant cap formation at the surface of spherical bubbles rising in surfactant solutions at large Reynolds Numbers under conditions of small Marangoni number and slow sorption kinetics. *Adv. Colloid Interfaces Sci.*, doi:10.1016/j.cis.2014.10.002.

70. M. Lotfi, D. Bastani, V. Ulaganathan, R. Miller, and A. Javadi, Bubble in flow field: A new experimental protocol for investigating dynamic adsorption layers by using capillary pressure tensiometry. *Colloids Surf. A*, 460 (2014) 369–376.

71. S.S. Sadhal and R.E. Johnson, Stokes flow past bubbles and drops partially coated with thin films. Part 1. Stagnant cap of surfactant film—Exact solution. *J. Fluid Mech.*, 126 (1983) 237–250.

72. E.K. Zholkovskij, V.I. Kovalchuk, S.S. Dukhin, and R. Miller, Dynamics of rear stagnant cap formation at low Reynolds Numbers: 1. Slow sorption kinetics. *J. Colloid Interface Sci.*, 226 (2000) 51–59.

73. J.F. Harper, The rear stagnation region of a bubble rising steadily in a dilute surfactant solution. *Q. J. Mech. Appl. Math.*, 41 (1988) 203–213.

74. F.G. Andrews, R. Fike, and S. Wong, Bubble hydrodynamics and mass transfer at high Reynolds number and surfactant concentration. *Chem. Eng. Sci.*, 43 (1988) 1467.

75. R.B. Fdhila and P.C. Duineveld, The effect of surfactant on the rise of a spherical bubble at high Reynolds and Peclet numbers. *Phys. Fluids*, 8 (1996) 310.

76. B. Cuenot, J. Magnaudet, and B. Spennato, The effects of slightly soluble surfactants on the flow around a spherical bubble. *J. Fluid Mech.*, 339 (1997) 25–53.

77. F. Takemura, Adsorption of surfactants onto the surface of a spherical rising bubble and its effect on the terminal velocity of the bubble. *Phys. Fluids*, 17 (2005) 048104.

78. S. Fleckenstein and D. Bothe, Simplified modeling of the influence of surfactants on the rise of bubbles in VoF-simulations. *Chem. Eng. Sci.*, 102 (2013) 514–523.

79. M. Muradoglu and G. Tryggvason, A front-tracking method for computation of interfacial flows with soluble surfactants. *J. Comp. Phys.*, 227 (2008) 2238–2262.

80. R.M. Davies and G. Taylor, The mechanics of large bubbles rising through extended liquids and through liquids in tubes. *Proc. R. Soc. Lond. Ser. A, Math. Phys. Sci.*, 200 (1950) 375–390.

81. T. Thulasidas, M. Abraham, and R. Cerro, Bubble-train flow in capillaries of circular and square cross section. *Chem. Eng. Sci.*, 50 (1995) 183–199.

82. K. Hayashi, R. Kurimoto, and A. Tomiyama, Terminal velocity of a Taylor drop in a vertical pipe. *Int. J. Multiphase Flow*, 37 (2011) 241–251.

83. H. Marschall, S. Boden, C. Lehrenfeld, J. Carlos, D. Falconi, U. Hampel, A. Reusken, M. Wörner, and D. Bothe, Validation of interface capturing and tracking techniques with different surface tension treatments against a Taylor bubble benchmark problem. *Comput. Fluids*, 102 (2014) 336–352.

84. P. Angeli and A. Gavriilidis, Hydrodynamics of Taylor flow in small channels: A review. *P. I. Mech. Eng. C-J. Mec. Eng. Sci.*, 227 (2008) 737–751.

85. R. Gupta, D. Fletcher, and B. Haynes, Taylor flow in microchannels: A review of experimental and computational work. *J. Comput. Multiphase Flows*, 2 (2010) 1–32.

86. Y.S. Muzychka, E.J. Walsh, and P. Walsh, Heat transfer enhancement using laminar gas–liquid segmented plug flows. *J. Heat Transf. Trans. ASME*, 133 (2011) 041902–041902-9.

87. T.C. Thulasidas, M.A. Abraham, and R.L. Cerro, Dispersion during bubble-train flow in capillaries. *Chem. Eng. Sci.*, 54 (1999) 61–76.

88. G. Bercic and A. Pintar, The role of gas bubbles and liquid slug lengths on mass transport in the Taylor flow through capillaries. *Chem. Eng. Sci.*, 52 (1997) 3709–3719.

89. J.J.W. Bakker, M.M.P. Zieverink, R. Reintjens, F. Kapteijn, J.A. Moulijn, and M.T. Kreutzer, Heterogeneously catalyzed continuous-flow hydrogenation using segmented flow in capillary columns. *ChemCatChem.*, 3 (2011) 1155–1157.

90. Q. He, K. Fukagata, and N. Kasagi, Numerical simulation of gas–liquid two-phase flow and heat transfer with dry-out in a micro tube, in *Sixth International Conference on Multiphase Flow*, Leipzig, Germany, 2007 S7_Wed_C_40.

91. D.F. Fletcher, B.S. Haynes, J. Aubin, and C. Xuereb, Modelling of micro-fluidic devices, in *Handbook of Micro Reactors, Fundamentals, Operations and Catalysts*, V. Hessel, J.C. Schouten, A. Renken, and J.I. Yoshida (Eds.), WILEY-VCH Verlag GmbH & Co. KGaA, Weinheim, Germany, 2009, pp. 117–144.

92. K. Fukagata, N. Kasagi, P. Ua-arayaporn, and T. Himeno, Numerical simulation of gas–liquid two-phase flow and convective heat transfer in a micro tube. *Int. J. Heat Fluid Flow*, 28 (2007) 72–82.

93. D. Qian and A. Lawal, Numerical study on gas and liquid slugs for Taylor flow in a T-junction microchannel. *Chem. Eng. Sci.*, 61 (2006) 7609–7625.

94. M. Dang, J. Yue, G. Chen, and Q. Yuan, Formation characteristics of Taylor bubbles in a microchannel with a converging shape mixing junction. *Chem. Eng. J.*, 223 (2013) 99–109.

95. M. Dang, J. Yue, and G. Chen, Numerical simulation of Taylor bubble formation in a microchannel with a converging shape mixing junction. *Chem. Eng. J.*, 262 (2015) 616–627.

96. F.P. Bretherton, The motion of long bubbles in tubes. *J. Fluid Mech.*, 10 (1961) 166–168.

97. K. Hayashi and A. Tomiyama, Effects of surfactant on terminal velocity of a Taylor bubble in a vertical pipe. *Int. J. Multiphase Flow*, 39 (2012) 78–87.

98. R. Kurimoto, K. Hayashi, and A. Tomiyama, Terminal velocities of clean and fully-contaminated drops in vertical pipes. *Int. J. Multiphase Flow*, 49 (2013) 8–23.

99. C. Meyer, M. Hoffmann, and M. Schlüter, Micro-PIV analysis of gas–liquid Taylor flow in a vertical oriented square shaped fluidic channel. *Int. J. Multiphase Flow*, 67 (2014) 140–148.

100. J. Ratulowski and H.C. Chang, Marangoni effects of trace impurities on the motion of long gas bubbles in capillaries. *J. Fluid Mech.*, 210 (1990) 303–328.

101. M.T. Kreutzer, F. Kapteijn, J.A. Moulijn, C.R. Kleijn, and J.J. Heiszwolf, Inertial and interfacial effects on pressure drop of Taylor flow in capillaries. *AIChE J.*, 51 (2005) 2428–2440.

102. E. Almatroushi and A. Borhan, Surfactant effect on the buoyancy-driven motion of bubbles and drops in a tube. *Ann. N.Y. Acad. Sci.*, 1027 (2004) 330–341.

4 Dynamics of Interfacial Layer Formation

Aliyar Javadi, Jürgen Krägel, Volodja I. Kovalchuk,
Libero Liggieri, Giuseppe Loglio, Eugene V. Aksenenko,
Valentin B. Fainerman, and Reinhard Miller

CONTENTS

4.1   INTRODUCTION

In Chapters 1–3, the thermodynamic basis was discussed as well as the experimental opportunities to study the interfacial properties of surfactants at liquid interfaces. The description of the various experimental protocols has shown that there is no need only for one general model to describe the adsorption process at a fluid interface. On the opposite, in most cases, the measuring principles of most of the methods suitable for studies of the interfacial dynamics require quite sophisticated boundary and initial conditions in order to get a quantitative description of the experimental data. In any case, the thermodynamic baseline is the most important prerequisite for any modeling as it represents the equilibrium state for a certain adsorption layer, that is, the target for any ongoing adsorption or relaxation process. Moreover, in many cases, the thermodynamic basis is necessary to describe local equilibrium in microscopically small parts of in total nonequilibrium macroscopic systems, for example, local equilibrium between the surface and subsurface layer in the case of diffusion-controlled adsorption.

The aim of this chapter is to present the fundamentals of adsorption kinetics of surfactants and surfactant mixtures at liquid interfaces. Analytical solutions are available only for strongly simplified conditions, mostly for systems that follow a linear adsorption isotherm and at very short adsorption times. Such limitations are sometimes helpful as the obtained analytical solutions can often at least give information on the order of magnitude of certain parameters.

More complex experimental situations require respective boundary conditions, which are discussed in this chapter for the most important experiments. Also for surfactant mixtures and micellar solutions, the respective theoretical models are presented and some simple model calculations shown. All the presented models and corresponding numerical solutions shown here correspond to the actual state of the art and can be essentially used by any experimentalist because software packages have been elaborated in order to apply the models to experimental data. An overview of this software is given in Chapter 5, while we discuss here the main assumptions and results of the models.

4.2 GENERAL MECHANISMS OF THE ADSORPTION PROCESS

Our present knowledge allows describing the process of adsorption of single surfactants at liquid interfaces quantitatively. The first real model for the surfactant adsorption at the solution/air interfaces with constant surface area was derived in 1946 [1], based on a diffusional transport of the surfactant molecules to the interface. The solution of this model leads to an integral equation of the following structure:

$$\Gamma(t) = 2\sqrt{\frac{D}{\pi}}\left(c_o\sqrt{t} - \int_0^{\sqrt{t}} c(0, t-\tau)d\sqrt{\tau}\right).$$ (4.1)

The main parameters in this Volterra integral equation is the diffusion coefficient D and the surfactant bulk concentration c_o. The integral in Equation 4.1 includes the surfactant concentration in a close vicinity of the interface $c(0,t)$, called subsurface or sublayer. As result, we get the amount of adsorbed surfactants $\Gamma(t)$ as a function of time t. The application of this so-called Ward and Tordai equation (4.1) to experimental surface tension data $\gamma(t)$ requires one additional relation between $\Gamma(t)$ and the surfactant concentration $c(t)$ in the sublayer, for which typically the corresponding adsorption isotherm is used (see Chapter 1). Actually, Milner [2] discussed diffusion as responsible process for the decrease in surface tension of soap solutions with time.

The second idea to describe the adsorption process, besides the diffusion-controlled transport, assumes that the transfer of molecules from the solution close to the interface into the adsorbed state and vice versa is the key step. This mechanism is called kinetics-controlled adsorption. Several types of transfer mechanisms were proposed in the early literature, such as [3–5]. More complicated models take into account diffusion and transfer mechanisms simultaneously [6,7], leading to the so-called mixed adsorption models.

If additional molecular processes happen within the adsorption layer, such as the formation of aggregates or changes in orientation or conformation [8], these have to be considered along with the diffusional transport and the transfer step to/from the interface. An overview of existing theoretical models for the adsorption kinetics and exchange of matter and suitable experimental tools was given recently by Mucic et al. [9].

4.3 DIFFUSION-CONTROLLED ADSORPTION

While in the past, many nondiffusional models for the adsorption kinetics have been developed, at present, most experts are convinced that the transport of surfactants in the solution bulk to and from the surface is the main mechanism. The reason for speculations on adsorption barriers and other mechanisms was the low purity of surfactants used in the studies. It was demonstrated first by Mysels and Florence [10] and by Lunkenheimer and Miller [11] in a very extensive way that surface-active impurities lead to the assumption of other adsorption mechanisms as these impurities would lead to essential decreases of the diffusion coefficient of the main surfactant. However, when considering a diffusion-controlled adsorption for both, the main surfactant and the impurity, the whole process follows a diffusion-controlled mechanism. Although this situation has been discussed over a long period of time, there are still new models proposed to consider various additional mechanism acting in addition to the transport of surfactants by diffusion, such as in [12,13]. These appear to be rather artificial, while for macromolecules or proteins additional processes for changing, the conformation can be expected [14]. Also, molecular processes at the interface, such as changes in orientation [15] or 2D aggregation [16,17], can have a significant impact on the adsorption kinetics, as it is shown in Section 4.4.2.

4.3.1 BASIC MODEL OF WARD AND TORDAI

A quantitative description of adsorption kinetics processes is so far usually based on the model derived in 1946 by Ward and Tordai [1]. The various models given in literature were developed on this basis and only different boundary and initial conditions were used. The model of Ward and Tordai is based on a diffusion-controlled adsorption and assumes that the step of transfer from the subsurface to the interface is fast compared to this diffusional transport. The model is based on Fick's second diffusion equation

$$\frac{\partial c}{\partial t} + (v \ \text{grad})c = D \ \text{div grad } c, \tag{4.2}$$

which for 1D geometry simplifies to

$$\frac{\partial c}{\partial t} = D\frac{\partial^2 c}{\partial x^2} \quad \text{at} \quad x > 0, \quad t > 0, \tag{4.3}$$

when any convection or generated liquid flow is neglected. For the mathematical solution of this transport problem, additional boundary and initial conditions are required. Fick's first diffusion law serves very well as boundary condition at the surface $x = 0$:

$$\frac{\partial \Gamma}{\partial t} + (v_s \ \text{grad}_s)\Gamma = D_s \ \text{div}_s \ \text{grad}_s\Gamma - \Gamma \ \text{div}_s \ v_s + D\frac{\partial c}{\partial x}. \tag{4.4}$$

Again, when any interfacial fluxes are neglected and only diffusion between the bulk and the interface exists, this simplifies to

$$\frac{\partial \Gamma}{\partial t} = j = D\frac{\partial c}{\partial x} \quad \text{at} \quad x = 0; \quad t > 0. \tag{4.5}$$

To complete the transport problem, one additional boundary condition and initial conditions are required, such as an infinite bulk phase

$$\lim_{x \to \infty} c(x,t) = c_o \quad \text{at} \quad t > 0, \tag{4.6}$$

or a bulk phase of limited depth h with a diminishing concentration gradient at $x = h$ is assumed:

$$\frac{\partial c}{\partial x} = 0 \quad \text{at} \quad x = h; \quad t > 0. \tag{4.7}$$

A usual initial condition is a homogenous concentration in the solution bulk and a freshly formed interface without any surfactant adsorbed:

$$c(x,t) = c_o \text{ at } t = 0, \tag{4.8}$$

$$\Gamma(0) = 0. \tag{4.9}$$

The assumption of an initial load of the interface is more realistic than Equation 4.9 because the creation of a completely empty fresh interface is experimentally impossible:

$$\Gamma(0) = \Gamma_d \neq 0. \tag{4.10}$$

For this initial condition, we obtain

$$\Gamma(t) = \Gamma_d + 2\left(\frac{Dt}{\pi}\right)^{1/2} c_0 - 2\left(\frac{D}{\pi}\right)^{1/2} \int_0^{\sqrt{t}} c_s\left(t - \lambda\right) d\sqrt{\lambda}. \tag{4.11}$$

For more sophisticated experiments, other initial conditions are required; however, then analytical solutions are often impossible. New approaches for the same physical basis are published, for example, using a lattice Boltzmann simulation [18]. Such approaches allow considering many peculiarities, even solubility of the surfactants in the two liquids forming the interface and transfer from one phase into the other (see also Section 4.9).

4.3.2 Analytical and Approximate Solutions

For a linear isotherm, a simple equation was derived by Sutherland in 1952 [19]. Starting from a Frumkin isotherm Equation 1.32 and assuming negligible interaction between adsorbed molecules $a = 0$ as well as small surface coverage θ, we get the linear Henry adsorption isotherm $\Gamma_o = K c_o$. With this isotherm, the following analytical relation is obtained:

$$\Gamma(t) = \Gamma_o\left(1 - \exp\left(\frac{Dt}{K^2}\right) \text{erfc}\left(\frac{\sqrt{Dt}}{K}\right)\right), \tag{4.12}$$

the range of application of which, however, is very limited [20].

For a more general Langmuir isotherm, that is, Equation 1.32 with $a = 0$, a solution for the diffusion-controlled adsorption model via the so-called orthogonal collocation method can be derived [21]. For the dimensionless time

$$\Theta = Dt\left(\frac{c_o}{\Gamma_o}\right)^2. \tag{4.13}$$

The corresponding solution can be given in a polynomial form

$$\frac{\Gamma(\Theta)}{\Gamma_o} = \sum_{i=1}^{N} \xi_i \tau^i, \tag{4.14}$$

with ξ_i being functions of the reduced concentration bc_o and τ:

$$\tau = 1 - \frac{1}{1 + \sqrt{\Theta + \Theta^2 bc_o}}. \tag{4.15}$$

The values of the coefficients ξ_i for the range $0 \leq bc_o \leq 100$ are tabulated in [22]. Another rather complicated polynomial solution was derived in [23].

Various approximations have been derived for the comparison of the theory with experimental data [24,25]. The simplest short-time approximation reads

$$\Gamma(t) = 2c_o \sqrt{\frac{Dt}{\pi}}, \tag{4.16}$$

as obtained from Equation 4.1 by neglecting the integral term. At short times, the surface coverage is also small and a linear relation between the interfacial tension $\gamma(t)$ and $\Gamma(t)$, leading to

$$\gamma_{t \to 0} = \gamma_o - 2RTc_o \sqrt{\frac{Dt}{\pi}}, \tag{4.17}$$

where γ_o is the surface tension of the solvent. Hence, for a diffusion-controlled adsorption at constant surface area, the experimental values should give a straight line when presented in a $\gamma(\sqrt{t})$ plot and the slope provides the diffusion coefficient.

Other short-time approximations have been discussed by Fainerman et al. [26] using diffusion-controlled, barrier-controlled, and mixed kinetic models. A comparison of two long-time approximations by Hansen [27] and by Joos [28]

$$\Delta\gamma\big|_{t \to \infty} = \gamma(t) - \gamma_{eq} = \frac{RT\Gamma^2}{c_o\sqrt{\pi Dt}} \quad \text{(Hansen)} \tag{4.18}$$

and

$$\Delta\gamma\big|_{t \to \infty} = \gamma(t) - \gamma_{eq} = \frac{RT\Gamma^2}{2c_0}\left(\frac{\pi}{Dt}\right)^{1/2} \quad \text{(Joos)} \tag{4.19}$$

differ by a factor of $2/\pi$. In [24], it was shown that the Joos approximation is more accurate.

Although the model appears to be classical, historical, or even old, it is still the basic model and so-called new approaches are in most cases just small modifications, such as in [29].

4.3.3 TIME VARIABLE INTERFACIAL AREA

The adsorption kinetics models by Ward and Tordai and the approximation by Sutherland are only valid for interfaces of constant interfacial area. In experiments with Langmuir troughs, for example, different types of area deformations during the measurements can be generated by the movable barriers. Joos derived a model equation for the adsorption to or desorption from a solution surface for any possible area changes, while any liquid flow inside the solution bulk was neglected [25]. For a diffusion-controlled adsorption, he obtained an equation similar to Equation 4.1 of Ward and Tordai:

$$\Gamma f(t) = \Gamma_d + 2\left(\frac{Dt}{\pi}\right)^{1/2} c_0 - 2\left(\frac{D}{\pi}\right)^{1/2} \int_0^{\sqrt{\tau}} c_s(t-\lambda)d\sqrt{\lambda}. \tag{4.20}$$

The only difference between the two equations is the function $f(t)$, which is the relative change of the interfacial area $A(t)/A_0$ with time starting at $f = 1$:

$$f(t) = \frac{A(t)}{A_0}.$$

(4.21)

Joos showed that for this model a so-called effective surface age t_{eff} can be defined:

$$t_{eff} = \frac{\tau}{f^2(t)},$$

(4.22)

with the definition of the parameter τ via

$$\tau = \int_0^t f^2(t)\,dt = \int_0^t \left(\frac{A}{A_0}\right)^2 dt.$$

(4.23)

Another classical measuring procedure, based on continuously growing drop, is the drop volume tensiometry. In [30], the physical picture was presented for a radial flow inside a drop of radius R; however, only numerical calculations were performed. The following integral equation was derived later in [31] in analogy to the equation of Ward and Tordai Equation 4.1:

$$\frac{d\Gamma}{dt} = -\left(R(t)\right)^2 \sqrt{\frac{D}{\pi}} \int_0^t \frac{\left(\frac{dc(0,t)}{dt}\right)_{t_0}}{\int_{t_0}^t \left(R(\xi)^4\,d\xi\right)^{1/2}}\,dt_0 - \Gamma\frac{d\ln A}{dt},$$

(4.24)

and it was shown that the rate of adsorption at the surface of a radially growing drop at a linear growth of the drop volume is about 30% of that at a surface of constant area [32].

Already in 1957, Delahay and Trachtenberg [33] derived an approximation for the initial period of the adsorption process at the surface of a growing drop and obtained

$$\Gamma(t) = 2c_o\sqrt{\frac{3Dt}{7\pi}}.$$

(4.25)

Compared with Equation 4.16, the adsorption time at a growing drop surface is by a factor 7/3 longer to reach a surface concentration than at a constant surface. This factor was also discussed in [20], where a nonlinear integral equation was derived in

$$\Gamma(t) = 2c_0\sqrt{\frac{3Dt}{7\pi}} - \sqrt{\frac{D}{\pi}}t^{-2/3} \int_0^{3/7t^{7/3}} \frac{c\left(0,7/3\tau^{3/7}\right)}{\left(3/7t^{7/3} - \tau\right)}\,d\tau,$$

(4.26)

which is equivalent with the Ward and Tordai Equation (4.1). A quantitative description of growing drop processes in presence of surfactants is now possible with respective computational fluid dynamics (CFD) simulations, as, for example, shown in [34]. Also in microfluidic devises, drops of surfactant solutions are formed under definite conditions and are the object of CFD simulations [35,36]. Another particular experiment based on a coaxial capillary for a continuous bulk exchange in a single drop was used to study the desorption kinetics of surfactants [37]. These and other situations cannot be described by simple models but required CFD simulations that will be discussed further in Chapters 15 through 24 of this book.

4.4 NONDIFFUSIONAL ADSORPTION MECHANISMS

As mentioned earlier, many adsorption kinetics models were discussed in literature. These models consider specific mechanisms of the molecular transfer from the subsurface to the interface, or vice versa in the case of desorption [3–7,38].

4.4.1 KINETICS-CONTROLLED ADSORPTION MODEL

The models based only on a transfer mechanism without the consideration of the diffusional transport are called kinetic controlled. The Langmuir transfer mechanism is an often used rate equation, which reads in its general form

$$\frac{d\Gamma}{dt} = k_{ad}c_o\left[1 - \frac{\Gamma}{\Gamma_\infty}\right] - k_{des}\frac{\Gamma}{\Gamma_\infty}. \tag{4.27}$$

Under the condition $\dfrac{\Gamma_o}{\Gamma_\infty} \ll 1$, we obtain the so-called Henry rate equation

$$\frac{d\Gamma}{dt} = k_{ad}c_o - k_{des}\Gamma, \tag{4.28}$$

which in equilibrium ends up in the linear Henry isotherm with $K = k_{ad}/k_{des}$. For a Frumkin-type adsorption model, the following mechanism was proposed [31,39]:

$$\frac{d\Gamma}{dt} = k_{ad}c_o\left[1 - \frac{\Gamma}{\Gamma_\infty}\right] - k_{des}\frac{\Gamma}{\Gamma_\infty}\exp\left[a\frac{\Gamma}{\Gamma_\infty}\right]. \tag{4.29}$$

The solution of these ordinary differential equations is rather easy and leads to exponential functions.

Baret [6] was the first who proposed the coupling of a transfer mechanism with the Ward and Tordai equation (4.1) by replacing the bulk concentration c_o by the concentration $c(0,t)$ in the subsurface, right below the adsorption layer. For the Langmuir mechanism given by Equation 4.27, we obtain

$$\frac{d\Gamma(t)}{dt} = k_{ad}c(0,t)\left[1 - \frac{\Gamma(t)}{\Gamma_\infty}\right] - k_{des}\frac{\Gamma(t)}{\Gamma_\infty}. \tag{4.30}$$

Equations 4.1 and 4.30 coupled with each other yield an integrodifferential equation system that can be solved numerically [7,40].

4.4.2 REORIENTATION AND AGGREGATION AT THE INTERFACE

In a 1981 paper by Nikolov et al. [41], the possibility of different adsorption states in a surfactant adsorption layer was discussed. A series of thermodynamic models followed then as the basis for

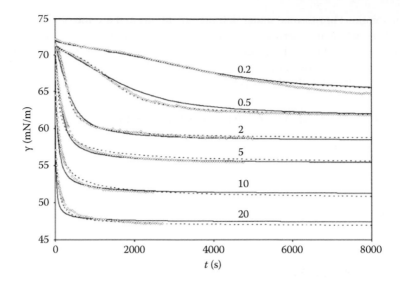

FIGURE 4.1 Dynamic surface tensions of Triton X-100 solutions measured by PAT using the emerging bubble mode; the given numbers correspond to the concentrations given in µmol/L: points, experimental data; solid curves, calculated with the Frumkin model; and dotted curves, calculated with the reorientation model.

a quantitative understanding of the coexistence of different adsorption states of soluble surfactants [15,16,42]. These new thermodynamic models, discussed in detail in Chapter 1, have an enormous impact also on understanding the adsorption dynamics at liquid interfaces. Changes in interfacial orientation or conformation and also the formation of 2D aggregates can influence the adsorption kinetics significantly [43].

The corresponding equations of state and adsorption isotherms describing an interfacial reorientation or aggregation in the adsorption layers are essentially given by Equations 1.58 and 1.66, respectively. In order to consider the processes of molecular reorientation or aggregation in a model quantitatively with particular rate constants, additional relationships have to be added to the diffusion problem as specific boundary conditions. Model calculations show impressively the large effects of these molecular processes at the interface on the overall adsorption dynamics (see, e.g., [8,44]). The type of interfacial relaxations of adsorbed molecules has also a significant influence on measurements of the dilational elasticity [45].

As an example, in Figure 4.1, we show dynamic surface tensions measured by profile analysis tensiometry for different Triton X-100 solutions [46]. Two different models have been applied to analyze the data, both based on a diffusion-controlled adsorption mechanism. The solid curves were obtained when using a Frumkin-type equation of state Equation 1.32, while the dotted curves correspond to a reorientation model Equation 1.58.

Although both models describe the data (points) adequately, the reorientation model is superior over the Frumkin model.

4.5 MIXED ADSORPTION LAYERS

For a surfactant mixture, the model looks essentially the same as for a single surfactant solution; however, instead of Equation 4.4, we need the transport equations for each of the r different compounds:

$$\frac{\partial c_i}{\partial t} + (v \text{ grad})c_i = \sum_{j=1}^{r} D_{ij} \text{ div grad } c_j. \tag{4.31}$$

At sufficiently low surfactant concentrations, the mixed diffusion coefficients D_{ij} can be replaced by the individual ones D_i (see [47]). In this way, we get an Equation 4.3 for each component r and finally the corresponding solution as a set of Equations 4.1 for each surfactant:

$$\Gamma_i(t) = 2\sqrt{\frac{D_i}{\pi}}\left(c_{oi}\sqrt{t} - \int_0^{\sqrt{t}} c_i(0, t - \tau)d\sqrt{\tau}\right), \quad i = 1, \ldots r. \tag{4.32}$$

The r integral equations are linked with each other via a multicomponent adsorption isotherm, such as Equation 1.30. For the case that all surfactants can be described by a linear adsorption isotherm, equivalent to Equation 4.12, we obtain the same equation for each surfactant [48]:

$$\Gamma_i(t) = \Gamma_{oi}\left(1 - \exp(D_i t / K_i^2)\mathrm{erfc}(\sqrt{D_i t}/K_i)\right) \quad i = 1, \ldots r. \tag{4.33}$$

Via the Gibbs fundamental equation

$$\Gamma(t) = -\frac{1}{RT}\frac{d\gamma(t)}{d\ln c} \tag{4.34}$$

and the Henry isotherm, we can calculate the changes in surface tension as a function of time from the individual adsorption values:

$$\gamma(t) = \gamma_o - RT\sum_{i=1}^{r}\Gamma_i(t) \tag{4.35}$$

For short adsorption times $t \to 0$, there is also an approximate solution for the total adsorption of all compounds, similar to Equation 4.16:

$$\Gamma_T(t) = \sum_{i=1}^{r}\Gamma_i(t) = \sum_{i=1}^{r}2c_{oi}\sqrt{\frac{D_i t}{\pi}}. \tag{4.36}$$

From this, we can learn that surface-active contaminants eventually present in the solution of a surfactant under study do not influence the dependence of $\gamma(t)$ at short adsorption times $t \to 0$. This of course is true only for the beginning of the adsorption process, while for the remaining period, impurities can take over the determining role in the adsorption layer.

The solution of the set of integral equations (4.32) coupled via a multicomponent adsorption isotherm like Equation 1.30 leads to a rather complex mathematical problem. As an example, in [49], some model calculations were performed and compared with experimental data. Also in Chapter 5, these mathematical problems will be dealt with. Dealing with mixed surfactant systems is an important matter; however, the number of papers dedicated to theoretical simulations [50–53] or experimental investigations [54,55] are rather scarce. In recent years, investigations of the dynamic interfacial properties for mixed protein/surfactant solutions have been performed more frequently. In particular, the theoretical descriptions are more sophisticated due to the more complex equations of state required for mixed protein/surfactant adsorption layers [56–60].

A new type of experiments has been performed for surfactant solutions at the interface to vapors of volatile liquids, such as alkanes [61]. It turns out that the alkane molecules, although not amphiphilic, form a mixed layer with the surfactant molecules. The presence of alkane vapor in the gas

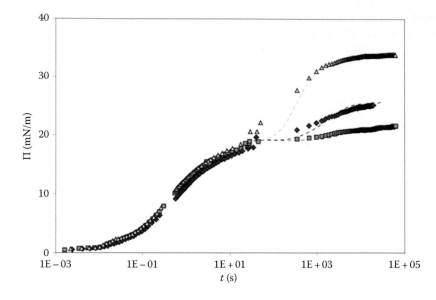

FIGURE 4.2   Dynamic surface tension of aqueous solutions of 0.3 mmol/L C_{10}DMPO mixed with C_{14}DMPO at concentrations of 0.001 (□), 0.003 (◇), and 0.01 mmol/L (△); dotted lines, calculated from Equation 1.30. (According to Miller, R. et al., *Colloids Surf. A*, 242, 123, 2004.)

phase also changes the adsorption behavior of the surfactant and a situation at the interface is reached that represents an intermediate between a solution/air and a solution/alkane interface [62]. The presence of alkane or other organic molecules in the gas phase enhances the adsorption of surfactants [63–65], an effect which is also observed for proteins at small additions of a nonionic surfactant [66].

Experimental dynamic surface pressure data for a tailored surfactant mixture are shown in Figure 4.2, together with model calculations using a set of equations (4.32) in combination with a generalized Frumkin isotherm, as given by Equations 1.30 and 1.31. The parameters of the two compounds have been taken from adsorption characteristics for the single surfactants [67]. The theoretical model described the behavior of the surfactant mixture very well.

4.6 INTERFACIAL RELAXATIONS

Methods for studies of interfacial relaxations are based on small perturbations of interfacial layers in the state of adsorption equilibrium. Small relative area changes are defined by $d\ln A \approx \Delta A/A_o \ll 1$, with A_0 being the initial area and ΔA the amplitude of the perturbation. Due to the area increase or decrease the surface tension, γ changes accordingly by $\Delta \gamma$. The ratio of the amplitudes of these two quantities yields the elasticity modulus ε, which is defined by

$$\varepsilon = -\frac{d\gamma}{d\ln\Gamma} = \frac{d\gamma}{d\ln A} = -\frac{d\gamma}{d\ln\Gamma}\frac{d\ln\Gamma}{d\ln A} \approx \frac{\Delta\gamma}{\Delta A}A_0. \qquad (4.37)$$

The two principle perturbations, transient and harmonic, were discussed by Loglio et al. [69], and it was shown that the theoretical basis of both is essentially the same. Therefore, a transient relaxation corresponds to a certain characteristic frequency. For harmonic perturbations, relaxation process leads to a phase difference between the area oscillation and the corresponding surface tension oscillation, which is a measure of the interfacial dilational viscosity.

Relaxation methods were described in Sections 3.6 and 3.7. The frequency ranges of the various existing methods overlap with each other, as shown in Figure 3.10. The most recently developed methods are the oscillating bubble or drop tensiometry. The first oscillating bubble method was described in [70] after which several other designs were published. A summary was recently given by Karbaschi et al. [71].

4.6.1 HARMONIC RELAXATIONS

For an isotropic area deformation $A(t)$, the diffusional flux at the interface can be described by the modified the first Fick's law

$$\frac{1}{A}\frac{d(\Gamma A)}{dt} = -D\frac{\partial c}{\partial t} \quad \text{at} \quad x = 0. \tag{4.38}$$

Assuming a diffusion-controlled relaxation mechanism, the solution of the transport problem can be given in the following form [72]:

$$c(x,t) = c_o + \alpha e^{\beta x} e^{i\omega t}. \tag{4.39}$$

Equation 4.38 can be changed into

$$\frac{d\ln A}{d\ln \Gamma} = -\left(1 + D\frac{\partial c/\partial x}{(d\Gamma/dc)(\partial c/\partial t)}\right). \tag{4.40}$$

Together with (4.37), we obtain the dynamic surface dilational viscosity

$$\varepsilon = -\frac{d\gamma}{d\ln \Gamma}\left(1 + D\frac{dc}{d\Gamma}\frac{\partial c/\partial x}{(\partial c/\partial t)}\right)^{-1}. \tag{4.41}$$

With the solution ansatz of Equation 4.39, we get $\varepsilon(i\omega)$ in the following form [72]:

$$\varepsilon(i\omega) = \varepsilon_0 \frac{\sqrt{i\omega}}{\sqrt{i\omega} + \sqrt{2\omega_0}}. \tag{4.42}$$

ε_0 and ω_0 are the elasticity modulus and the characteristic frequency:

$$\varepsilon_0 = -\left(\frac{d\gamma}{d\ln \Gamma}\right)_A \quad \text{and} \quad \omega_0 = \frac{D}{2}\left[\frac{dc}{d\Gamma}\right]^2. \tag{4.43}$$

For surfactant-mixed solutions, the generalized relationships were obtained in [73]:

$$\varepsilon_0 = \frac{d\gamma}{d\ln \Gamma_T}\sum_{i=1}^{r}\frac{\Gamma_i}{\Gamma_T}\frac{d\ln \Gamma_i}{d\ln A} \tag{4.44}$$

with the total adsorption $\Gamma_T = \sum_{i=1}^{r} \Gamma_i$ and the dynamic dilational viscoelasticity

$$\varepsilon(i\omega) = \varepsilon_0 \sum_{i=1}^{r} \frac{\Gamma_i}{\Gamma_T} \frac{\sqrt{i\omega}}{\sqrt{i\omega} + \sqrt{2\omega_{io}}}. \tag{4.45}$$

For a generalized linear adsorption isotherm, the adsorption Γ_i are independent and consequently the functions $\omega_{i0} = D_i/2(dc_i/d\Gamma_i)^2$.

As a peculiarity for interfaces between two immiscible liquids, the solubility of the surfactant in both adjacent phases has to be taken into consideration. Thus, the exchange of matter happens in both bulk phases. With the distribution coefficient k of the surfactant between the two liquids (see further in Section 4.9), the characteristic frequency ω_0 reads

$$\omega_o = \left(\frac{dc}{d\Gamma}\right)^2 \frac{\left(\sqrt{D_1} + \sqrt{D_2}/k\right)^2}{2}. \tag{4.46}$$

For more complex adsorption models, such as the Frumkin model or the surface reorientation and aggregation models, analytical solutions cannot be derived and numerical methods are required (see Chapter 5). A general model considering the concomitance of different relaxation processes is given in [74], which has been applied to interpret the behavior of complex amphiphilic systems, such as macromolecules or mixed nanoparticle–surfactant systems [75].

4.6.2 TRANSIENT RELAXATIONS

For a diffusion-controlled exchange of matter, the derivation of interfacial response functions were summarized in [76,77], and the following functions for a trapezoidal area change are obtained:

$$\Delta\gamma_1(t) = \frac{\Theta\varepsilon_0}{2\omega_0} \exp(2\omega_0 t)\, \text{erfc}(\sqrt{2\omega_0 t}) + \frac{2\Theta\varepsilon_0\sqrt{t}}{\sqrt{2\pi\omega_0}} - \frac{\Theta\varepsilon_0}{2\omega_0}, \quad 0 < t < t_1, \tag{4.47}$$

$$\Delta\gamma_2(t) = \Delta\gamma_1(t) - \Delta\gamma_1(t - t_1), \quad t_1 < t < t_1 + t_2, \tag{4.48}$$

$$\Delta\gamma_3(t) = \Delta\gamma_2(t) - \Delta\gamma_1(t - t_1 - t_2), \quad t_1 + t_2 < t < 2t_1 + t_2, \tag{4.49}$$

$$\Delta\gamma_4(t) = \Delta\gamma_3(t) - \Delta\gamma_1(t - 2t_1 - t_2), \quad t > 2t_1 + t_2, \tag{4.50}$$

with $\Theta = \dfrac{d\ln A}{dt} = \dfrac{1}{t_1}\ln\left(1 - \dfrac{\Delta A}{A_0}\right)$ being the relative area change. The trapezoidal area change given by Equations 4.47–4.50 can be simplified to jump, ramp, or square pulse-type perturbations. The relevant functions $\Delta\gamma(t)$ for mixed surfactant solutions as well as for relaxations at liquid/liquid interface were described in [78].

4.7 CONSIDERATION OF MICELLES IN THE BULK

For micellar solution, we have the situation of multiple equilibriums. At first, the monomers and micelles are in equilibrium in the solution bulk; however, the monomers are also in equilibrium with the adsorption layer at the interface. When we increase/decrease the area of the interfacial layer, the monomers will adsorb/desorb and disturb the equilibrium in the bulk, which in turn leads to a formation/disintegration of micelles. This leads not only to concentration gradients of monomers but also of micelles, and both are subject to diffusional transport. The presence of micelles in the solution bulk is an extra source of matter, and therefore their kinetics can be seen as an additional interfacial relaxation mechanism [79–81].

The diffusion of monomers and micelles are coupled with each other via the micellar kinetics. In literature, a fast process in the range of microseconds and a second slow process in the range of milliseconds and seconds are discussed. The fast process refers to the release or incorporation of monomers in micellar, and the slow process is a complete disintegration of micelles. Aniansson et al. [82] discussed the various mechanisms of micellar kinetics.

For a quantitative model, the diffusion of monomers and micelles as well as the micellar kinetics mechanisms have to be taken into account, as it was proposed, for example, in [25,83]. For small area jumps starting from an equilibrium state, the following relationship was derived in [84]:

$$\Delta\gamma = \frac{\Delta\gamma_0 e^{-k_m t}}{\sqrt{\omega_0 + 4k_m}} \left[\sqrt{s_2} \exp\left(s_2 t\right) \operatorname{erfc} \sqrt{s_2 t} - \sqrt{s_1} \exp\left(s_1 t\right) \operatorname{erfc} \sqrt{s_1 t} \right], \tag{4.51}$$

$$\sqrt{s_1} = \frac{-\sqrt{\omega_0} - \sqrt{\omega_0 + 4k_m}}{2}, \quad \sqrt{s_2} = \frac{-\sqrt{\omega_0} + \sqrt{\omega_0 + 4k_m}}{2}. \tag{4.52}$$

Here, $\omega_0 = D\left(\dfrac{dc}{d\Gamma}\right)^2$ is the diffusion relaxation frequency, k_1 and k_m are the rate constants for micellization and demicellization, and n is the aggregation number of the micelles. Equation 4.51 can be seen as a generalization of the Sutherland model of Equation 4.33.

For area jump of arbitrary size (expansion or compression), we get the more general equation

$$\Gamma = \sqrt{D}\left\{ \frac{c_0}{\sqrt{k_m}}\left[\left(\frac{1}{2}+k_m t\right)\operatorname{erf}\sqrt{k_m t}+\sqrt{\frac{k_m t}{\pi}}e^{-k_m t}\right] - \int_0^t c_s\left(t-\lambda\right)\left[\frac{e^{-k_m t}}{\sqrt{\pi t}}+\sqrt{k_m}\operatorname{erf}\sqrt{k_m t}\right]d\lambda \right\} \tag{4.53}$$

as an equivalent to the equation of Ward and Tordai (4.1). Also for this rather complex model, there is a short-time [25]

$$\Gamma = \sqrt{\frac{D}{k_m}}c_0\left[\left(\frac{1}{2}+k_m t\right)\operatorname{erf}\sqrt{k_m t}+\sqrt{\frac{k_m t}{\pi}}e^{-k_m t}\right] \tag{4.54}$$

and a long-time approximation

$$\Gamma = \sqrt{\frac{D}{k_m}}\left(c_0-c_s\right)\left[\left(\frac{1}{2}+k_m t\right)\operatorname{erf}\sqrt{k_m t}+\left(\frac{k_m t}{\pi}\right)^{1/2}e^{-k_m t}\right]. \tag{4.55}$$

Some new work on the theoretical description and experimental verification has been performed during the last decades showing the effect of micelles on dynamic interfacial properties [85–90].

Besides the adsorption kinetics, the frequency dependencies of the so-called dilational viscoelasticity modulus $|E|$ are very sensitive to the dynamics of micelle formation and dissolution. It was discussed in [85] that this modulus is given by

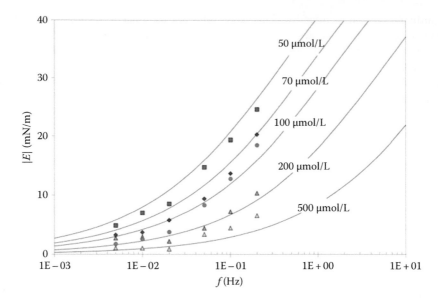

FIGURE 4.3 Viscoelasticity modulus $|E|$ as a function of frequency f for various $C_{14}EO_8$ concentrations above the CMC (7×10^{-6} mol/L); solid lines were calculated with Equation 4.56. (According to Fainerman, V.B. et al., *Langmuir*, 24, 6447, 2008.)

$$E(i\omega) = E_0 \left[1 + (1-i)\zeta \sqrt{(1+\beta)(1+\alpha\beta)} \right]^{-1}, \qquad (4.56)$$

with $\beta = (c_0 - c_k)/c_k$, $\alpha = D_m/D \approx 0.25$, and $c_k = $ CMC, c_0 is the total surfactant concentration, D and D_m are the diffusion coefficients of surfactant monomers and micelles, respectively. In Figure 4.3, data for the nonionic surfactant $C_{14}EO_8$ are summarized, measured by an oscillating bubble method, as described in Section 3.6. The calculated solid lines in Figure 4.3 show that the model described the experimental data very well.

4.8 ADSORPTION OF IONIC SURFACTANTS

Discussions of the role of the surfactant charge in the dynamics of adsorption have been made in many papers. It is shown that the presence of an electric double layer at the interface of ionic surfactant solutions influences remarkably its dynamic properties of [26,91–97]. A first theoretical model was derived by Dukhin et al. [98] and later by Borwankar and Wasan [99]. The theories were based on a quasi-equilibrium model supposing that the characteristic time of diffusion is much larger than the relaxation time of the electric double layer. In this way, the term of electrodiffusion can be omitted from the transport problem.

A numerical solution of the complete electrodiffusion problem was presented by MacLeod and Radke [100]. Although the advantage of their rigorous approach is indisputable, the numerical solution is much time consuming when applied to process experimental data, and it does not entirely elucidate the course of the underlying physical processes.

When ionic surfactants adsorb at a liquid–fluid interface, the surface charge density is changing with time, leading to the formation of an electric double layer. The surface becomes charged by the adsorbed surfactant ions and the corresponding electric field decelerates the further adsorption of surfactants. In their analysis, Danov et al. [97] described the different relaxation processes within the adsorption layer and in the adjacent solution bulk phase.

Assuming a symmetric (z:z) ionic surfactant with a corresponding indifferent symmetric (z:z) electrolyte in the solution, we get a good model to demonstrate how the electric charge of surfactant ions modifies the adsorption process as compared to nonionic surfactants. Of course, the transport of surfactant molecules in the solution bulk is still governed by diffusion, however, within an electric field. This electric field is defined by the model used. For a Stern layer model, Borwankar and Wasan [99] proposed to localize the dividing surface between the diffuse layer and the Stern layer. When the indices $i = 1$, 2, and 3 refer to the surfactant ion, the counterion, and the coion, respectively, the transport of an ion under the influence of the electric potential ψ is given by [97]

$$\frac{\partial c_i}{\partial t} = D_i \frac{\partial}{\partial x}\left(\frac{\partial c_i}{\partial x} + \frac{z_i e}{kT} c_i \frac{\partial \psi}{\partial x}\right) \quad \text{for} \quad i = 1,2,3, \tag{4.57}$$

where
  z_i is the valence
  D_i and c_i are the diffusion coefficient and bulk concentrations, respectively, of the ion i
  k and T are the Boltzmann constant and the absolute temperature, respectively

The second term in the parenthesis of Equation 4.57 is the so-called electromigration. This term accounts for the effect of the electric field on the diffusion of ions, and ψ is the electric potential given by the Poisson equation

$$\frac{\partial^2 \psi}{\partial x^2} = -\frac{4\pi e}{\varepsilon}\left[z_1 c_1 + z_2 c_2 + z_3 c_3\right], \tag{4.58}$$

where ε is the dielectric permittivity. The mass balance as boundary condition at the interface $x = 0$, that is, the change of the number of all ions Γ_i as function of time and the respective electrodiffusion from the bulk phase, takes the form

$$\frac{d\Gamma_i}{dt} = D_i\left(\frac{\partial c_i}{\partial x} + \frac{z_i e}{kT} c_i \frac{\partial \psi}{\partial x}\right) \quad \text{at} \quad x = 0 \quad \text{for} \quad i = 1, 2. \tag{4.59}$$

The electroneutrality condition for the whole system serves as an additional condition

$$\int_0^\infty \left[c_{10} - c_{20} + c_{30}\right]dx + \Gamma_{10} - \Gamma_{20} = 0. \tag{4.60}$$

As mentioned earlier, we base everything here only on symmetric z:z electrolytes for the surfactant and the salt, that is, $z_1 = -z_2 = z_3$ so that the counterions of the surfactant and the electrolyte are identical.

Danov et al. [97] derived expressions for transient relaxations of an adsorption layer induced by an external perturbation. For a small perturbation and sufficiently large times t, we get

$$\frac{\Gamma_i(t) - \Gamma_{i,eq}}{\Gamma_i(0) - \Gamma_{i,eq}} = \sqrt{\frac{\tau_i}{\pi t}}, \tag{4.61}$$

with τ_i being the adsorption relaxation times

$$\tau_i = \frac{1}{\kappa^2}\left(g_{i1} G_1 + \frac{2}{p}\frac{q}{\sqrt{D}} g_{i2} + g G_2\right)^2 \quad (i = 1,2). \tag{4.62}$$

The two relaxation times τ_1 and τ_2 depend on the coefficients g_{ij} and G_i, which in turn depend on the adsorption characteristics of the surfactant and counterions and the corresponding model for the electric double layer [97].

4.9 ADSORPTION AND MATTER TRANSFER ACROSS THE INTERFACE

To describe the adsorption of surfactants at the interface between two immiscible liquids, one additional fact has to be considered—many surfactants are soluble in both phases, that is, in water and the organic phases. This is most important for nonionic surfactants with respective distribution coefficients. The theoretical modeling for the kinetics of adsorption and matter exchange during interfacial compressions and expansions has to take the transfer of surfactant molecules across the interface into account. From a practical point of view, this phenomenon is essential, for instance, in emulsification [101].

The study of the adsorption dynamics for a nonionic surfactant at the water/oil interface presented in [102,103] is an excellent example for the situation discussed here. It allows determination of the distribution coefficient k for a surfactant in the given water/oil system. This coefficient is defined in the following way [104]:

$$k = \frac{c_1}{c_2} = \frac{v_1}{v_2} \exp\left(-\frac{\mu^{01} - \mu^{02}}{RT} \right), \tag{4.63}$$

where v_1 and v_2 and μ^{01} and μ^{02} are the surfactant's molar volumes and standard chemical potentials in phases 1 and 2, respectively. The partitioning coefficient describes the concentrations of the surfactant in the two adjacent liquid phases but also influences the adsorption process at the interface significantly [102,103]. Note the transfer of surfactant across the interface between the two liquids can lead to interfacial instabilities as it is described in later chapters of this book (see, e.g., Chapters 22 and 23).

The diffusion model for describing the adsorption dynamic at a liquid–liquid interface has to include an additional condition, that is, to assume a local equilibrium distribution of the surfactant between the two sublayers. Moreover, the model has also to consider diffusion in both liquid phases with the diffusion coefficients D_1 and D_2, respectively. Starting from the boundary condition given by Equation 4.5, we get

$$\frac{d\Gamma}{dt} = -D_1 \left. \frac{\partial c}{\partial x} \right|_{x=0^+} + D_2 \left. \frac{\partial c}{\partial x} \right|_{x=0^-}. \tag{4.64}$$

At the interface located at $x = 0$, we can use the partition equilibrium

$$c_{20}(t) = k c_{10}(t). \tag{4.65}$$

The solution of the diffusion equations in the two liquid phases with the joined boundary condition (4.64) and the classical initial conditions (4.8) in each phase plus Equation 4.9 reads then

$$\Gamma(t) = 2c_1^\infty \left[\sqrt{\frac{D_1 t}{\pi}} + k\sqrt{\frac{D_2 t}{\pi}} \right] - \frac{1}{\sqrt{\pi}} \int_0^t \left[\sqrt{D_1} + k\sqrt{D_2} \right] \frac{c_{01}(0,\tau)}{\sqrt{t-\tau}} d\tau. \tag{4.66}$$

This equation simplifies to Equation 4.1 for the case that the solubility in one phase is negligible.

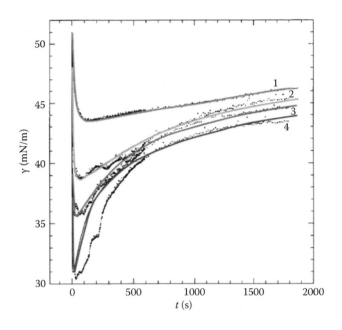

FIGURE 4.4 Experimental dynamic interfacial tension obtained for $C_{13}DMPO$ at the water/hexane interface with a drop of aqueous solution formed in hexane, points, experimental data; solid lines, calculated with the diffusion-controlled adsorption model for different $C_{13}DMPO$ concentrations such as (1) 10^{-5}, (2) 2×10^{-5}, (3) 3×10^{-5}, and (4) 5×10^{-5} mol/L. The values of the parameters used for the calculations were given in Ravera, F. et al., *Adv. Colloid Interface Sci.*, 88, 129, 2000.

As we assumed local equilibrium at the interface, the corresponding adsorption isotherm can be applied at any time t so that Equation 4.66 can be solved in the same way as done for the Ward and Tordai equation (4.1).

The surfactant transfer across an interface is particularly important when it is initially present in a single drop, as it was the case in the work of Ferrari et al. [102]. The aforementioned given diffusion-controlled model was applied in [103] to these experimental data and provided a quantitative description with the surfactant distribution coefficient as final result, as shown in Figure 4.4.

Recently, a generalization of the numerical simulation procedure was published by Gassin et al. [105] providing more accurate numerical simulations for the process of adsorption and matter transfer across the interface.

In particular cases, the solubility of surfactants in both adjacent liquid phases can be used to circumvent the disturbing effect of surfactant impurities. For example, the study of the adsorption kinetics of sodium dodecyl sulfate (SDS) at the water/air interface is typically affected by the presence of dodecanol. In contrast, at the water/oil interface, the present dodecanol transfers from the aqueous into the oil phase and does not disturb the adsorption properties of the SDS molecules [106].

4.10 SUMMARY AND OUTLOOK

In this chapter, we present a set of classical solutions of adsorption kinetic models derived so far in an analytical or semianalytical way. Also, subsequent numerical procedures have been applied to these resulting relationships in order to directly compare the theory with experiments. The available experimental data comprise essentially dynamic surface and interfacial tensions measured at a freshly formed interface or after small perturbations from the equilibrium state. Special experimental tools were developed, however, to better approach specific questions of interfacial dynamics. For example, in [12] a microtensiometer was developed to induce convection in the solution bulk, that is, minimize the diffusion effect in adsorption kinetics, in order to probe additional adsorption mechanisms at liquid interfaces. Also, the application of optical methods, as described in Chapter 3, are more and more applied to study the nonequilibrium interfacial properties of surfactant interfacial layers. Since recently, most of the neutron scattering and reflection techniques are used for adsorption studies at liquid interfaces (see Section 3.5). For example, in [107,108], neutron scattering was used to study the adsorption kinetics of surfactants. Also, fluorescence microscopy has been applied for quantifying the adsorption kinetics of special surfactants [85].

In Chapter 5, different software packages have been described. These have been elaborated over many years and allow the user now to apply comparatively complicated theoretical models for the quantitative analysis of experimental data. In later chapters of this book, such as Chapters 16, 17, 18, 22, 23, and 24, we will see how even more complex problems can be tackled with specific CFD simulations. The results given here and obtained with comparatively simple numerical methods can then serve for validation.

REFERENCES

1. A.F.H. Ward and L. Tordai, Time-dependence of boundary tensions of solutions, *J. Phys. Chem.*, 14 (1946) 453–463.
2. S.R. Milner, On surface concentration; and the formation of liquid films, *Phil. Mag.*, 13 (1907) 96–101.
3. K.S.G. Doss, Alterung der Oberflächen von Lösungen. IV. Über die Natur der Potentialschranke, welche die Anreicherung der Moleküle des gelösten, *Koll. Z.*, 86 (1939) 205–213.
4. S. Ross, The change of surface tension with time. I. Theories of diffusion to the surface, *J. Am. Chem. Soc.* 67 (1945) 990–994.
5. R.S. Hansen and T. Wallace, The kinetics of adsorption of organic acids at the water-air interface, *J. Phys. Chem.*, 63 (1959) 1085–1092.
6. J.F. Baret, Theoretical model for an interface allowing a kinetic study of adsorption, *J. Colloid Interface Sci.*, 30 (1969) 1–12.
7. R. Miller and G. Kretzschmar, Numerische Lösung für ein gemischtes Modell der diffusions-kinetik-kontrollierten Adsorption, *Colloid Polymer Sci.*, 258 (1980) 85–87.
8. F. Ravera, L. Liggieri, and R. Miller, Molecular orientation as a controlling process in adsorption dynamics, *Colloids Surf. A*, 175 (2000) 51–60.
9. N. Mucic, A. Javadi, N.M. Kovalchuk, and R. Miller, Dynamics of interfacial layers—Experimental feasibilities for adsorption kinetics and dilational rheology, *Adv. Colloid Interface Sci.*, 168 (2011) 167–178.
10. J.K. Mysels and A.T. Florence, The effect of impurities on dynamic surface tension—Basis for a valid surface purity criterion, *J. Colloid Interface Sci.*, 43 (1973) 577–582.
11. K. Lunkenheimer and R. Miller, A criterion for judging the purity of adsorbed surfactant layers, *J. Colloid Interface Sci.*, 120 (1987) 176–183.
12. N.J. Alvarez, D.R. Vogus, L.M. Walker, and S.L. Anna, Using bulk convection in a microtensiometer to approach kinetic-limited surfactant dynamics at fluid-fluid interfaces, *J. Colloid Interface Sci.*, 372 (2012) 183–191.
13. J.R. Fernandez, M.C. Muniz, and C. Nunez, A mixed kinetic-diffusion surfactant model for the Henry isotherm, *J. Math. Anal. App.*, 389 (2012) 670–684.
14. T.D. Gurkov, Adsorption kinetics under the influence of barriers at the subsurface layer, *Colloid Polymer Sci.*, 289 (2012) 1905–1915.
15. V.B. Fainerman and R. Miller, Dynamic surface tension measurements in the sub-millisecond range, *J. Colloid Interface Sci.*, 175 (1995) 118–121.
16. V.B. Fainerman, R. Miller, and R. Wüstneck, Adsorption isotherm and surface tension equation for a surfactant with changing partial molar area. 2. Non-ideal surface layer, *J. Phys. Chem. B*, 101 (1997) 6470–6483.
17. N. Anton, T.F. Vandamme, and P. Bouriat, Dilatational rheology of a gel point network formed by nonionic soluble surfactants at the oil-water interface. *Soft Matter*, 9 (2013) 1310–1318.
18. R. Skartlien, K. Furtado, E. Sollum, P. Meakin, and I. Kralova, Lattice-Boltzmann simulations of dynamic interfacial tension due to soluble amphiphilic surfactant, *Phys. A Statist. Mech. Appl.*, 390 (2011) 2291–2302.

19. K.L. Sutherland, The kinetics of adsorption at liquid surfaces, *Austr. J. Sci. Res.*, A5 (1952) 683–696.
20. R. Miller, Adsorption kinetics of surfactants at fluid interfaces—Experimental conditions and practice of application of theoretical models, *Colloids Surf.*, 46 (1990) 75–83.
21. M. Ziller and R. Miller, On the solution of diffusion controlled adsorption kinetics by means of orthogonal collocation, *Colloid Polymer Sci.*, 264 (1986) 611–615.
22. R. Miller and G. Kretzschmar, Adsorption kinetics of surfactants at fluid interfaces, *Adv. Colloid Interface Sci.*, 37 (1991) 97–121.
23. A. Yousef and B.J. McCoy, Diffusion of surfactants to an interface: Effect of a barrier of oriented water, *J. Colloid Interface Sci.*, 94 (1983) 497–501.
24. A.V. Makievski, V.B. Fainerman, R. Miller, M. Bree, L. Liggieri, and F. Ravera, Determination of equilibrium surface tension values by extrapolation via long time approximations, *Colloids Surf. A*, 122 (1997) 269–273.
25. P. Joos, *Dynamic Surface Phenomena*, VSP, Utrecht, the Netherlands, 1999.
26. V.B. Fainerman, A.V. Makievski, and R. Miller, Analysis of dynamic surface tension of Na-alkyl sulphate solutions, based on asymptotic solutions of adsorption kinetics equations, *Colloids Surf. A*, 87 (1994) 61–75.
27. R.S. Hansen, Analysis of dynamic surface tension of Na-alkyl sulphate solutions, based on asymptotic solutions of adsorption kinetics equations, *J. Phys. Chem.*, 64 (1960) 637–641.
28. I. Balbaert and P. Joos, The dynamic surface tension and the Boltzmann superposition principle, *Colloids Surf.*, 23 (1987) 259–266.
29. A.A. Ageev, V.A. Volkov, and P.R. Emel'yanov, Effect of diffusion on the rate of adsorption of surfactants, *Fibre Chem.*, 44 (2013) 381–385.
30. F.W. Pierson and S. Whittaker, Studies of the drop-weight method for surfactant solutions: I. Mathematical analysis of the adsorption of surfactants at the surface of a growing drop, *J. Colloid Interface Sci.*, 52 (1976) 203–218.
31. C.A. MacLeod and C.J. Radke, Surfactant exchange kinetics at air/water interface from the dynamic tension of growing liquid drops, *J. Colloid Interface Sci.*, 166 (1994) 73–88.
32. R. Miller, Zur Adsorptionskinetik an der Oberfläche wachsender Tropfen, *Colloid Polymer Sci.*, 258 (1980) 179–185.
33. P. Delahay and I. Trachtenberg, Adsorption kinetics and electrode processes, *J. Am. Chem. Soc.*, 79 (1957) 2355–2362.
34. K. Dieter-Kissling, M. Karbaschi, H. Marschall, A. Javadi, R. Miller, and D. Bothe, On the applicability of drop profile analysis tensiometry at high flow rates using an interface tracking method, *Colloids Surf. A*, 441 (2014) 837–845.
35. T. Glawdel and C.L. Ren, Droplet formation in microfluidic T-junction generators operating in the transitional regime. III. Dynamic surfactant effects, *Phys. Rev. E*, 86 (2012) 026308.
36. J.H. Xu, P.F. Dong, H. Zhao, C.P. Tostado, and G.S. Luo, The dynamic effects of surfactants on droplet formation in coaxial microfluidic devices, *Langmuir*, 28 (2012) 9250–9258.
37. J.K. Ferri, N. Gorevski, C.S. Kotsmar, M.E. Leser, and R. Miller, Desorption kinetics of surfactants at fluid interfaces by novel coaxial capillary pendant drop experiments, *Colloids Surf. A*, 319 (2008) 13–20.
38. J.F. Baret, Cinetique de l'adsorption aux interfaces liquide-liquide, *J. Chem. Phys.*, 65 (1968) 895–905.
39. V.B. Fainerman, Adsorption kinetics of surfactants from solutions, *Kolloid. Zh.*, 39 (1977) 106–112.
40. C.H. Chang and E.I. Franses, Modified Langmuir-Hinshelwood kinetics for dynamic adsorption of surfactants at the air/water interface, *Colloids Surf.*, 69 (1992) 189–201.
41. A. Nikolov, G. Martynov, and D. Exerowa, Associative interactions and surface tension in ionic surfactant solutions at concentrations much lower than the CMC, *J. Colloid Interface Sci.*, 81 (1981) 116–124.
42. V.B. Fainerman and R. Miller, Surface tension isotherms for surfactant adsorption layers including surface aggregation, *Langmuir*, 12 (1996) 6011–6014.
43. V.B. Fainerman, R. Miller, E.V. Aksenenko, A.V. Makievski, J. Krägel, G. Loglio, and L. Liggieri, Effect of surfactant interfacial orientation/aggregation on adsorption dynamics, *Adv. Colloid Interface Sci.*, 86 (2000) 83–101.
44. V.B. Fainerman, S.A. Zholob, E.H. Lucassen-Reynders, and R. Miller, Comparison of various models describing the adsorption of surfactant molecules capable of interfacial reorientation, *J. Colloid Interface Sci.*, 261 (2003) 180–183.
45. F. Ravera, M. Ferrari, R. Miller, and L. Liggieri, Dynamic elasticity of adsorption layers in the presence of internal reorientation processes, *J. Phys. Chem. B*, 105 (2001) 195–203.

46. V.B. Fainerman, S.V. Lylyk, E.V. Aksenenko, L. Liggieri, A.V. Makievski, J.T. Petkov, J. Yorke, and R. Miller, Adsorption layer characteristics of Triton surfactants. 2. Dynamic surface tensions and adsorption dynamics, *Colloids Surf. A*, 334 (2009) 8–15.

47. R. Haase, Grundzüge der physikalischen Chemie, Band II—Transportvorgänge, Dr. Dietrich Steinkopff Verlag, Darmstadt (1973).

48. R. Miller, K. Lunkenheimer, and G. Kretzschmar, Ein Modell für die diffusionskontrollierte Adsorption von Tensidgemischen an fluiden Phasengrenzen, *Colloid Polymer Sci.*, 257 (1979) 1118–1120.

49. Ch. Frese, S. Ruppert, M. Sugár, H. Schmidt-Lewerkühne, K.P. Wittern, V.B. Fainerman, R. Eggers, and R. Miller, Adsorption kinetics of surfactant mixtures from micellar solutions as studied by maximum bubble pressure technique, *J. Colloid Interface Sci.*, 267 (2003) 475–482.

50. E.V. Aksenenko, V.I. Kovalchuk, V.B. Fainerman, and R. Miller, Surface dilational rheology of mixed adsorption layers at liquid interfaces, *Adv. Colloid Interface Sci.*, 122 (2006) 57–66.

51. E.V. Aksenenko, V.I. Kovalchuk, V.B. Fainerman, and R. Miller, Surface dilational rheology of mixed surfactant layers at liquid interfaces, *J. Phys. Chem. C*, 111 (2007) 14713–14719.

52. V.B. Fainerman, E.V. Aksenenko, S.V. Lylyk, J.T. Petkov, J. Yorke, and R. Miller, Adsorption layer characteristics of mixed SDS/CnEOm solutions. 1. Dynamic and equilibrium surface tension, *Langmuir*, 26 (2010) 284–292.

53. D. Vitasari, P. Grassia, and P. Martin, Simulation of dynamics of adsorption of mixed protein-surfactant on a bubble surface, *Colloids Surf. A*, 438 (2013) 63–76.

54. V.B. Fainerman, E.V. Aksenenko, S.A. Zholob, J.T. Petkov, J. Yorke, and R. Miller, Adsorption layer characteristics of mixed SDS/CnEOm solutions. 2. Dilational visco-elasticity, *Langmuir*, 26 (2010) 1796–1801.

55. S. Manoli and A. Avranas, Aqueous solutions of the double chain cationic surfactants didodecyldimethylammonium bromide and ditetradecyldimethylammonium bromide with Pluronic F68: Dynamic surface tension measurements, *Colloids Surf. A*, 436 (2013) 1060–1068.

56. V.B. Fainerman, S.A. Zholob, M. Leser, M. Michel, and R. Miller, Competitive adsorption from mixed non-ionic surfactant/protein solutions, *J. Colloid Interface Sci.*, 274 (2004) 496–501.

57. Cs. Kotsmar, J. Krägel, V.I. Kovalchuk, E.V. Aksenenko, V.B. Fainerman, and R. Miller, Dilation and shear rheology of mixed BCS/surfactant adsorption layers, *J. Phys. Chem. B*, 113 (2009) 103–113.

58. Cs. Kotsmar, V. Pradines, V.S. Alahverdjieva, E.V. Aksenenko, V.B. Fainerman, V.I. Kovalchuk, J. Krägel, M.E. Leser, and R. Miller, Thermodynamics, adsorption kinetics and rheology of mixed protein-surfactant interfacial layers, *Adv. Colloid Interface Sci.*, 150 (2009) 41–54.

59. V. Ulaganathan, J. Krägel, B. Bergenstahl, and R. Miller, Adsorption and shear rheology of β-lactoglobulin/SDS mixtures at water/hexane and water/MCT interfaces, *Colloids Surf. A*, 413 (2012) 136–141.

60. A. Dan, C.S. Kotsmar, J.K. Ferri, A. Javadi, M. Karbaschi, J. Krägel, R. Wüstneck, and R. Miller, Mixed protein-surfactant adsorption layers formed in a sequential and simultaneous way at the water/air and water/oil interfaces, *Soft Matter*, 8 (2012) 6057–6065.

61. A. Javadi, N. Moradi, H. Möhwald, and R. Miller, Adsorption of alkanes from the vapour phase on water drops measured by drop profile analysis tensiometry, *Soft Matter*, 6 (2010) 4710–4714.

62. N Mucic, N. Moradi, A. Javadi, E.V. Aksenenko, V.B. Fainerman, and R. Miller, Mixed adsorption layers at the aqueous CnTAB solution/hexane vapor interface, *Colloids Surf. A*, 442 (2014) 50–55.

63. A. Javadi, N. Moradi, M. Karbaschi, V.B. Fainerman, H. Möhwald, and R. Miller, Alkane vapor and surfactants co-adsorption on aqueous solution interfaces, *Colloids Surf. A*, 391 (2011) 19–24.

64. P.N. Nguyen. T.T. Trinh Dang, G. Waton, T. Vandamme, and M.P. Krafft, A nonpolar, nonamphiphilic molecule can accelerate adsorption of phospholipids and lower their surface tension at the air/water interface, *Chem. Phys. Chem.*, 12 (2011) 2646–2652.

65. H. Martinez, E. Chacon, P. Tarazona, and F. Bresme, The intrinsic interfacial structure of ionic surfactant monolayers at water-oil and water-vapour interfaces, *Proc. R. Soc. A Math. Phys. Eng. Sci.*, 467 (2011) 1939–1958.

66. V.B. Fainerman, M. Lotfi, A. Javadi, E.V. Aksenenko, Y.I. Tarasevich, D. Bastani, and R. Miller, Adsorption of proteins at the solution/air interface influenced by added nonionic surfactants at very low concentrations for both components. 2. Effect of different surfactants and theoretical model, *Langmuir*, 30 (2014) 12812–12818.

67. K. Lunkenheimer, K. Haage, and R.Miller, On the adsorption properties of surface-chemically pure aqueous solutions of n-alkyl-dimethyl and n-alkyl-diethyl phosphine oxides, *Colloids Surf.*, 22 (1987) 215–224.

68. R. Miller, V.B. Fainerman, and E.V. Aksenenko, A simple method to estimate the dynamic surface pressure of surfactant mixtures, *Colloids Surf. A*, 242 (2004) 123–128.

69. G. Loglio, R. Miller, A. M. Stortini, U. Tesei, N. Degli Innocenti, and R. Cini, Non-equilibrium properties of fluid interfaces: Aperiodic diffusion-controlled regime, 1. Theory, *Colloids Surf. A*, 90 (1994) 251–259.

70. K. Lunkenheimer and G. Kretzschmar, Neuere Ergebnisse der Untersuchung der Elastizität von löslichen Adsorptionsschichten nach der Methode der pulsierenden Blase, *Z. Phys. Chem. (Leipzig)*, 256 (1975) 593–605.

71. M. Karbaschi, M. Taeibi Rahni, A. Javadi, Ch. Cronan, K. H. Schano, S. Faraji, J. Y. Won, J. K. Ferri, J. Krägel, and R. Miller, Dynamics of drops—Formation, growth, oscillation, detachment, and coalescence, *Adv. Colloid Interface Sci.*, doi:10.1016/j.cis.2014.10.009.

72. J. Lucassen and M. van den Tempel, Dynamic measurements of dilational properties of a liquid interface, *Chem. Eng. Sci.*, 27 (1972) 1283–1291.

73. P.R. Garrett and P. Joos, Dynamic dilational surface properties of submicellar multicomponent surfactant solutions, *J. Chem. Soc. Faraday Trans. 1*, 69 (1975) 2161–2173.

74. F. Ravera, M. Ferrari, and L. Liggieri, Modelling of dilational visco-elasticity of adsorbed layers with multiple kinetic processes, *Colloids Surf. A*, 282–283 (2006) 210–216.

75. L. Liggieri, E. Santini, E. Guzmán, A. Maestro, and F. Ravera, Wide-frequency dilational rheology investigation of mixed silica nanoparticle-CTAB interfacial layers, *Soft Matter*, 7 (2011) 7699–7709.

76. G. Loglio, U. Tesei, N. Degli-Innocenti, R. Miller, and R. Cini, Non-equilibrium surface thermodynamics. Measurement of transient dynamic surface tension for fluid/fluid interfaces by trapezoidal pulse technique, *Colloids Surf.*, 57 (1991) 335–342.

77. G. Loglio, R. Miller, U. Tesei, and R. Cini, Dilational viscoelasticity of fluid interfaces: The diffusion model for transient processes, *Colloids Surf.*, 61 (1991) 219–226.

78. R. Miller, G. Loglio, U. Tesei, and K.-H. Schano, Surface relaxations as a tool for studying dynamic interfacial behaviour, *Adv. Colloid Interface Sci.*, 37 (1991) 73–96.

79. Yu.M. Rakita and V.B. Fainerman, Diffusion kinetics of the adsorption at the interface between liquid phases in the presence of a thin structured interfacial layer, *Kolloid. Zh.*, 51 (1989) 714–720.

80. C.D. Dushkin and I.B. Ivanov, Effect of the polydispersity of diffusing micelles on the surface elasticity, *Colloids Surf.*, 60 (1991) 213–233.

81. G. Serrien, G. Geeraerts, L. Ghosh, and P. Joos, Dynamic surface properties of adsorbed protein solutions: BSA, casein and buttermilk, *Colloids Surf.*, 68 (1992) 219–233.

82. E.A.G. Aniansson, S.N. Wall, M. Almgren, H. Hoffmann, I. Kielmann, W. Ulbricht, R. Zana, J. Lang, and C. Tondre, Theory of the kinetics of micellar equilibria and quantitative interpretation of chemical relaxation studies of micellar solutions of ionic surfactant, *J. Phys. Chem.*, 80 (1976) 905–922.

83. C.D. Dushkin, Model of the quasi-monodisperse micelles with application to the kinetics of micellization, adsorption and diffusion in surfactant solutions and thin liquid films, *Colloids Surf. A*, 143 (1998) 283–299.

84. C.D. Dushkin, I.B. Ivanov, and P.A. Kralchevsky, The kinetics of the surface tension of micellar surfactant solutions, *Colloids Surf.*, 60 (1991) 235–261.

85. V.B. Fainerman, J.T. Petkov, and R. Miller, Surface dilational visco-elasticity of C14EO8 micellar solution studied by bubble profile analysis tensiometry, *Langmuir*, 24 (2008) 6447–6452.

86. V.B. Fainerman, E.V. Aksenenko, A.V. Mys, J.T. Petkov, J. Yorke, and R. Miller, Adsorption layer characteristics of mixed SDS/CnEOm solutions. 3. Dynamics of adsorption and surface dilational rheology of micellar solutions, *Langmuir*, 26 (2010) 2424–2429.

87. D. Arabadzhieva, E. Mileva, P. Tchoukov, R. Miller, F. Ravera, and L. Liggieri, Adsorption layer properties and foam film drainage of aqueous solutions of tetraethyleneglycol monododecyl ether, *Colloids Surf. A*, 392 (2011) 233–241.

88. Q. Song and M. Yuan, Visualization of an adsorption model for surfactant transport from micelle solutions to a clean air/water interface using fluorescence microscopy, *J. Colloid Interface Sci.*, 357 (2011) 179–188.

89. S. Hongxiu, Z. Lei, and L. Junji, Diffusion-controlled adsorption kinetics at the interface between air and aqueous micellar solution of heptaethylene glycol monododecyl ether, *Colloid J.*, 75 (2013) 433–436.

90. D. Arabadzhieva, B. Soklev, and E. Mileva, Amphiphilic nanostructures in aqueous solutions of triethyleneglycol monododecyl ether, *Colloids Surf. A*, 419 (2013) 194–200.

91. S.S. Dukhin, R. Miller, and G. Kretzschmar, On the theory of adsorption kinetics of ionic surfactants at fluid interfaces. 1. The effect of electric DL under quasi-equilibrium conditions, *Colloid Polymer Sci.*, 261 (1983) 335–339.

92. R.P. Borwankar and D.T. Wasan, The kinetics of adsorption of ionic surfactants at gas-liquid surfaces, *Chem. Eng. Sci.*, 41 (1986) 199–201.

93. P. van den Bogaert and P. Joos, Dynamic surface tensions of sodium myristate solutions, *J. Phys. Chem.*, 83 (1979) 2244–2248.

94. V.B. Fainerman, Kinetics of adsorption of ionic surfactants at the solution-air interface and the nature of the adsorption barrier, *Colloids Surf.*, 57 (1991) 249–266.

95. R. Miller, G. Kretzschmar, and S. S. Dukhin, On the theory of adsorption kinetics of ionic surfactants at fluid interfaces, 4. Slowing down of adsorption kinetics by the effect of a non-equilibrium diffuse layer, *Colloid Polymer Sci.*, 272 (1994) 548–553.

96. V.V. Kalinin and C.J. Radke, An ion-binding model for ionic surfactant adsorption at aqueous-fluid interfaces, *Colloids Surf. A*, 114 (1997) 337–350.

97. K.D. Danov, P.M. Vlahovska, P.A. Kralchevsky, G. Broze, and A. Mehreteab, Adsorption kinetics of ionic surfactants with detailed account for the electrostatic interactions: Effect of the added electrolyte, *Colloids Surf. A*, 156 (1999) 389–411.

98. S.S. Dukhin, R. Miller, and G. Kretzschmar, On the theory of adsorption kinetics of ionic surfactants at fluid interfaces. 2. Numerical calculations of the influence of a quasi-equilibrium electric DL, *Colloid Polymer Sci.*, 263 (1985) 420–423.

99. R.P. Borwankar and D.T. Wasan, Equilibrium and dynamics of adsorption of surfactants at fluid-fluid interface, *Chem. Eng. Sci.*, 43 (1988) 1323–1337.

100. C.A. MacLeod and C.J. Radke, Charge effects in the transient adsorption of ionic surfactants at fluid interfaces, *Langmuir*, 10 (1994) 3555–3566.

101. S.N. Maindarkar, P. Bongers, and M.A. Henson, Predicting the effects of surfactant coverage on drop size distributions of homogenized emulsions, *Chem. Eng. Sci.*, 89 (2013) 102–114.

102. M. Ferrari, L. Liggieri, F. Ravera, C. Amodio, and R. Miller, Adsorption kinetics of alkyl phosphine oxides at the water/hexane interface, 1. Pendant drop experiments, *J. Colloid Interface Sci.*, 186 (1997) 40–45.

103. F. Ravera, M. Ferrari, and L. Liggieri, Adsorption and partitioning of surfactants in liquid-liquid systems, *Adv. Colloid Interface Sci.*, 88 (2000) 129–177.

104. J. Lyklema, *Fundamentals of Interface and Colloid Science*, Vol. I, Academic Press, London, U.K. (1993).

105. P.-M. Gassin, R. Champory, G. Martin-Gassin, J.F. Dufreche, and O. Diat, Surfactant transfer across a water/oil interface: A diffusion/kinetics model for the interfacial tension evolution, *Colloids Surf. A*, 436 (2013) 1103–1110.

106. A. Javadi, N. Mucic, D. Vollhardt, V.B. Fainerman, and R. Miller, Effects of dodecanol on the adsorption kinetics of SDS at the water–hexane interface, *J. Colloid Interface Sci.*, 351 (2010) 537–541.

107. C.E. Morgan, C.J.W Breward, I.M. Griffiths, P.D. Howell, J. Penfold, R.K. Thomas, I. Tucker, J.T. Petkov, and J.R.P. Webster, Kinetics of surfactant desorption at an air-solution interface, *Langmuir*, 28 (2012) 17339–17348.

108. R.A. Campbell, H.P. Wacklin, I. Sutton, R. Cubitt, and G. Fragneto, FIGARO: The new horizontal neutron reflectometer at the ILL, *Eur. Phys. J. Plus*, 126 (2011) 107, 1–22.

5 Model-Based Computational Approach to Analyze Interfacial Problems

Eugene V. Aksenenko

CONTENTS

5.1 INTRODUCTION: THE MODEL ISOTHERM

The first attempt to provide a model approach to the theoretical description of adsorption phenomena based on a certain model of the surface layer was developed almost a hundred years ago by Irving Langmuir [1]. While the physical model on which the approach was based is, possibly, the simplest one (the adsorbed molecule occupies a certain fixed area on the surface, which prevents other molecules to adsorb therein; the interactions between the molecules are disregarded), this model remains extremely popular even at present. This popularity can easily be explained: the Langmuir model, as applied to the adsorption of a surfactant at the solution/air interface, is (in mathematical terms) expressed by a pair of well-known simple expressions:

$$\Gamma = \frac{1}{\omega}\frac{bc}{1+bc} \quad \text{the adsorption isotherm} \tag{5.1}$$

$$\gamma = \gamma_0 - \frac{RT}{\omega}\ln(1+bc) \quad \text{the equation of state} \tag{5.2}$$

Here

Γ is the adsorbed amount
γ is the surface tension
γ_0 is the surface tension of pure solvent
c is the bulk equilibrium concentration of the adsorbed species
R and T are the universal gas constant and the absolute temperature

The only parameter involved in the model is the molar area of the adsorbate molecule ω (the area covered by the surfactant molecule at the interface); the other parameter b, which is the adsorption equilibrium constant, is seen to be a scaling factor (as it enters the equations only via the product bc). As long as the physical quantities that one is interested in are only the equilibrium adsorbed amount and the equilibrium surface tension, Equations 5.1 and 5.2 can be dealt with by a suitable matrix processor (e.g., Excel) and even by a pocket calculator.

However, it soon became obvious that the behavior of adsorbed molecules at the interface is more complicated, because it is governed also by other physical phenomena, which remain unaccounted for in the Langmuir isotherm. This is explained in detail in Section 1.3, and in references therein. In mathematical terms, this leads to the fact that the equations to be solved are much more involved and could hardly be easily treated by simple means mentioned earlier. A good example here is the set of Equations 1.30 and 1.31 that describe the coadsorption of two surfactants assuming the interaction between the adsorbed molecules (the so-called generalized Frumkin model). If the intrinsic compressibility, reorientation, and aggregation of adsorbed molecules are taken into account, the equations to be solved to simulate the equilibrium behavior of the adsorbed interface layer become even more complicated. This especially concerns the mixtures of surfactants with proteins or polymers, for which the account for the conformational properties of the molecules is essential; see Sections 1.4.2 and 1.4.3.

It is important to note at this point that any further step taken in order to build the theoretical model capable to account for the physical phenomena essential to explain the behavior of the system inevitably involves the introduction of model parameter(s) *responsible* for this physical property, for example the intermolecular interaction, intrinsic compressibility coefficient, and, the areas per molecule adsorbed in various orientations at the interface. In general terms, the model-based approach to the description of properties of adsorption layers essentially relies on the calculation of ***model isotherms***. For the purpose of this chapter, this term means the combination of the following:

- *The adsorption isotherm equation(s),* that is the dependence of the adsorbed amount(s) of the ith adsorbed species Γ_i on the set $\{c\}$ of the constituents' bulk concentrations c_i, and the set $\{a\}$ of model parameters a_i involved in the model:

$$\Gamma_i = \Gamma_i(\{c\},\{a\}) \tag{5.3}$$

- *The equation of state of the adsorbed layer,* that is the dependence of the surface tension γ on the set of bulk concentrations $\{c\}$ of the constituents' bulk concentrations, and the set of model parameters $\{a\}$:

$$\gamma = \gamma(\{c\},\{a\}) \tag{5.4}$$

Some notes about the details of the calculation of model isotherms are presented further in Section 5.5. Here, it should be stressed that this set of equations is of primary importance for the model approach, because as soon as the dependencies (5.3) and (5.4) are known, that is the values of surface tension and the adsorptions of the constituents for any concentrations could be calculated, it becomes possible to determine the viscoelastic characteristics of the adsorbed layer, namely, the dependence of the dynamic complex elasticity E on the bulk concentrations and the surface oscillations frequency via the derivatives $d\Gamma_i/dc_j$ and $d\gamma/dc_i$; see Equation 1.63. Also, the kinetic curve, that is the dependence of surface tension γ as a function of time t, can be calculated (assuming a diffusion-governed adsorption kinetics) via the solution of Fick's equations with the model isotherm as the boundary condition at the interface; see, for example, [2,3].

5.2   PROBLEM-TARGETED SOFTWARE

It is obvious from the explanation in the preceding text, and taking into account the models explained in Chapter 1, that the simulation of the behavior of adsorbed layers could be a quite cumbersome computational task. Therefore, it would be very convenient for the experimentalist to also have, in addition to the software-controlled experimental equipment, which supplies *raw* experimental results of the measurements (usually in form of tabulated data), an easy-to-operate tool into which the experimental data are entered and processed by fitting these to a suitable thermodynamic model. The fitting should be controllable, allowing the user to select various options, to change the values

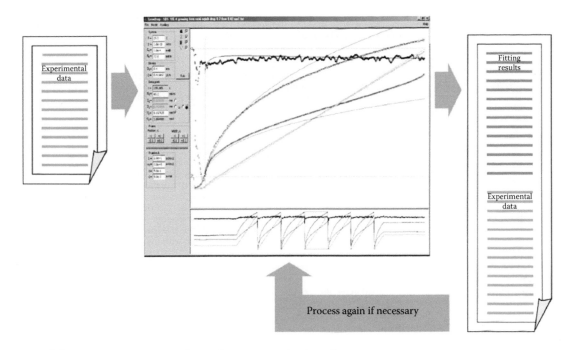

FIGURE 5.1   Sequence of experimental data processing by software.

of model parameters, etc., as schematically illustrated by Figure 5.1. The results of fitting should be shown on the screen and saved in the output file. Moreover, the results obtained from the software should be *encapsulated*, that is the output file, in addition to the fitting results, should contain also the initial experimental data, so the calculations can be continued later to refine the results if necessary.

Following these lines, a set of computer programs was developed (and is continuously under development), which implements various models of surface layers. These are outlined in Section 5.4.

5.3   GUIDELINES TO ANALYSE EXPERIMENTAL DATA USING THE SOFTWARE

During the last two decades, a number of thermodynamic models aimed at the quantitative description of the adsorption of molecules at liquid interface were developed. Some of them are described in Chapter 1, which also contains an extensive list of references. These models were implemented in the software packages, which were successfully used to explain various peculiarities of the studied systems. During this period, a kind of methodology for such studies was elaborated, which we believe to be helpful for the colleagues engaged in the investigations of adsorption phenomena at liquid interfaces. This methodology is schematically presented in the *flow chart* shown in Figure 5.2.

Quite obviously, each of such studies should be started from the comparison of the system under consideration with those already studied earlier by other colleagues. This could help to determine which properties of the system are of essential importance and which physical factors are mostly responsible for the behavior of the studied object.

Proper attention should be paid to the reliability and accuracy of the experimental data intended to be used for the fitting. For example, in a few cases dealt by us using the software, the behavior of the system was found to disobey any reasonable model based on the physicochemical properties of the constituents. After a great waste of time and effort, it was concluded that the substance in question contained a small admixture of an undefined very surface active species, which completely distorted the expected behavior of the surface layer. This was extensively analyzed, for example, in [4].

When the studied system is a mixture of surfactants, the first step should be the fitting of the experimental data (usually the concentration dependence of the surface tension, and, if available,

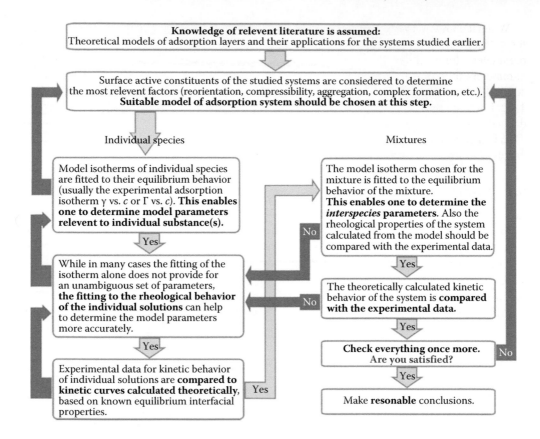

FIGURE 5.2   Fitting the models of adsorption layer—to the experimental data.

of the adsorbed amount) for the solutions of each individual component. This will provide the values of parameters involved in the surface layer model. Here, an extreme attention is to be paid to the solutions of high molecular weight compounds (e.g., proteins or polymers). While for ordinary surfactants (those with low molecular weight) the measurements present little problems, the equilibration of surface layers of high molecular weight surfactants usually requires a very long time. A suitable extrapolation seems to be of some help; however, it should be kept in mind that during the equilibration, other processes can occur in the surface layer (e.g., denaturation), which essentially affect the measured quantities. This problem was explained in more detail in [5].

However, it was shown by many examples that the fitting of equilibrium quantities does not always allow providing an unambiguous description of the system, because several models, if fitted to the measured data could exhibit a quite fair agreement (to within the experimental accuracy) with the observed values. Some examples of such situation are presented in Section 1.3. On the contrast, the viscoelasticity values have proved to be very sensitive to the choice of the model and to the values of model parameters, especially the intrinsic compressibility coefficient. Therefore, it is mandatory to verify these values by comparison of the theoretically predicted behavior of the system to the rheological data; in the case of evident discrepancies, it may be necessary to reconsider the model initially chosen.

The next step is to compare the theoretical predictions for the kinetic behavior of adsorbed layers obtained from the model with experimentally observed values. Here, in addition to other model parameters, the diffusion coefficient becomes significant.

When it comes to the analysis of a mixture, the next step is the fitting of the theoretical predictions for its equilibrium properties to the experimental data. This will help to estimate the inter-species parameters, for example a_{12} in Equations 1.30 and 1.31. It should be kept in mind, however, that the parameters found for individual species could also depend on the presence of another constituent. A good example here is the mixture of a protein with a cationic surfactant, considered in Section 1.4.2. It was shown that the properties of the DNA coadsorption with AzoTAB could be explained if a dependence of the DNA adsorption equilibrium constant on the AzoTAB concentration is assumed.

When all model parameters are estimated, the final step is to compare the kinetic behavior of the values obtained from the model with the experimental data.

5.4 SOFTWARE PACKAGES PRESENTLY AVAILABLE

In Table 5.1, the packages developed during almost two decades are summarized.

Other models and programs are under development and will be made available from the Adsorption Software website (http://www.thomascat.info/Scientific/AdSo/AdSo.htm).

The software is distributed free of charge (see the page on Terms and Conditions on the website). Downloadables are the packages that comprise executable, configuration and key files (no installation required), manuals, and examples of initial files.

Also posted on the website is an extensive list of, and the links to, references to relevant publications.

5.5 SOME HINTS TO THOSE WHO INTEND TO MAKE THEIR OWN SOFTWARE

To give the reader a taste of which problems should be solved when a model-based approach is used, the model isotherm that corresponds to the coadsorption of a protein and an ionic surfactant

---

TABLE 5.1

Presently Supported Packages

Isotherm and Rheology Fitting	Quasiequilibrium Kinetics Fitting
Individual surface-active species	**Individual surface-active species**
IsoFit—Is, Rh; *FrA, FrI, ReA, ReR*	SphereSQ (Single surfactant or single protein)—Ki; *FrA, FrI, ReA, ReR, PrA*
IsoPlot—Is; *FrA, FrI, ReA, ReR*	
Protein—Is, Rh; *PrA*	
Mixture of two surface-active species	**Mixture of two surface-active species**
IsoX2—Is, Rh; *FrA/FrA,FrA/ReR, ReR/ReR*	BubbleDQ (Mixture of two surface-active species)—Ki; *FrA/FrA,FrA/ReR, ReR/ReR*
ProteinX (Protein/surfactant mixture)—Is, Rh; *PrA*	
Abbreviations for models of adsorbed layer:	Package capabilities:
FrA—Frumkin;	Is—Isotherm;
FrI—Frumkin Ionic;	Rh—Rheology;
ReA—Reorientable surfactant, approximate account for entropy nonideality;	Ki—Kinetic curve.
ReR—Reorientable surfactant, rigorous account for entropy nonideality;	
PrA—Protein, aggregation in adsorbed layer.	

Note: Italics is used to distinguish between *models* and capabilities.

is considered in the following text as example. This model is discussed in Section 1.4.2; here, we concentrate on the mathematics and computation details. It is assumed in the model that a protein molecule with m ionized groups at a concentration of c_P interacts with counter charged ionic surfactant molecules of concentration c_S. Then, the ionic interaction leads to the formation of complexes. The interfacial properties of these complexes are given by the average activity of ions $\left(c_P^m c_S\right)^{1/(1+m)}$ participating in the reaction. This model isotherm is determined by the following set of equations:

$$-\frac{\Pi\omega_0}{RT} = \ln(1-\theta_P-\theta_S) + \theta_P(1-\omega_{P,0}/\omega_P) + a_P\theta_P^2 + a_S\theta_S^2 + 2a_{PS}\theta_P\theta_S \tag{5.5}$$

$$b_{PS,j}c_P^{m/(1+m)}c_S^{1/(1+m)} = \frac{\omega_P\Gamma_{P,j}}{\left(1-\theta_P-\theta_S\right)^{\omega_{P,j}/\omega_P}}\exp\left[-2a_P(\omega_{P,j}/\omega_P)\theta_P - 2a_{PS}\theta_S\right] \tag{5.6}$$

$$b_S\left[c_S\left(c_S+c_{ie}\right)\right]^{1/2} = \frac{\theta_S}{\left(1-\theta_P-\theta_S\right)}\exp\left[-2a_S\theta_S - 2a_{PS}\theta_P\right] \tag{5.7}$$

More specifically, Equation 5.5 is the equation of state of the adsorbed layer, and Equations 5.6 and 5.7 are the adsorption isotherm for the protein–surfactant complex *subsystem* and the adsorption isotherm for the unbound surfactant *subsystem*, respectively, see Equations 1.77, 1.82, and 1.83. Here, $\Pi = \gamma_0 - \gamma$ is the surface pressure; subscripts P and S refer to the protein and surfactant, respectively; θ_P and θ_S are the corresponding surface coverages; a_P, a_S, and a_{PS} are the model parameters that describe the intermolecular interactions. The model assumes that the protein molecule can exist at the surface in n states, each with its own molar area $\omega_{P,I} = \omega_{P,\min} + (i-1)\omega_{P,0}$, where $\omega_{P,0}$ is the area occupied by one segment of the molecule (the area increment, which is assumed to be approximately equal to the area per one solvent molecule), and the maximum molar area covered by the molecule is $\omega_{P,\max}$. Therefore, in this case, one has the adsorption isotherm for each state of adsorbed protein molecules, that is (5.6) is in fact a set of n equations, $n = (\omega_{P,\max} - \omega_{P,\min})/\omega_{P,0} + 1$, each one for its adsorption equilibrium constant $b_{PS,j}$ and partial adsorption of protein $\Gamma_{P,j}$. The total protein adsorption in all n states ($1 \le i \le n$) is

$$\Gamma_P = \sum_{i=1}^n \Gamma_{P,i} \tag{5.8}$$

and the total surface coverage by protein molecules reads

$$\theta_P = \omega_P\Gamma_P = \sum_{i=1}^n \omega_i\Gamma_{P,i} \tag{5.9}$$

where ω_P is the average molar area of the adsorbed protein. In Equation 5.5, ω_0 is the averaged molar area introduced to account for a small difference between the molar area of the solvent and the area occupied by one segment of the polymer molecule:

$$\omega_0 = \frac{\omega_{P,0}\theta_P + \omega_{S,0}\theta_S}{\theta_P + \theta_S} \tag{5.10}$$

Assuming the intrinsic compressibility ε of surfactant molecules in the surface layer, the surfactant molar area ω_S and the corresponding adsorption Γ_S depend on the surface pressure Π and the total surface coverage $\theta = \theta_P + \theta_S$:

$$\omega_S = \omega_{S0}\left[1 - \varepsilon\Pi\theta\right] \tag{5.11}$$

$$\theta_S = \Gamma_S\omega_S = \Gamma_S\omega_{S0}\left[1 - \varepsilon\Pi\theta\right] \tag{5.12}$$

where ω_{S0} is the surfactant molar area at infinitesimal coverage of surface layer. Finally, b_S and c_{ie} in Equation 5.7 are the adsorption equilibrium constant of the surfactant and the concentration of inorganic 1:1 salt in the solution.

It is assumed in the model that the surface activity of the protein molecules increases with increasing molar area, according to a power law with the constant exponent α_P. Therefore, the increase of the equilibrium adsorption constant in the jth state ($b_{PS,j}$) as compared to that in the first state ($b_{PS,1}$) is determined by the relation

$$b_{P,j} = \left(\frac{\omega_{P,j}}{\omega_{P,1}}\right)^{\alpha_P} b_{P,1} \tag{5.13}$$

This assumption drastically reduces the number of equations to be solved: combining Equations 5.6 and 5.12, one obtains

$$b_{PS,1}c_P^{m/(1+m)}c_S^{1/(1+m)} = \frac{\omega_P\Gamma_{P,j}}{(\omega_{P,j}/\omega_{P,1})^{\alpha_P}\left(1 - \theta_P - \theta_S\right)^{\omega_{P,j}/\omega_P}}\exp\left[-2\left(a_P\frac{\omega_{P,j}}{\omega_P}\theta_P + a_{PS}\theta_S\right)\right] \tag{5.14}$$

Note $b_{PS,1}$, that is the adsorption equilibrium constant for the protein molecules in the first state in the left-hand side. Then, noting that Equation 5.6 for $j = 1$ is

$$b_{PS,1}c_P^{m/(1+m)}c_S^{1/(1+m)} = \frac{\omega_P\Gamma_{P,1}}{\left(1 - \theta_P - \theta_S\right)^{\omega_{P,1}/\omega_P}}\exp\left[-2\left(a_P\frac{\omega_{P,1}}{\omega_P}\theta_P + a_{PS}\theta_S\right)\right] \tag{5.15}$$

and combining Equations 5.14 and 5.15, the adsorption in any jth state can be expressed via the adsorption in the first state:

$$\Gamma_{P,j} = \Gamma_{P,1}\left(\frac{\omega_{P,j}}{\omega_{P,1}}\right)^{\alpha_P}\left(1 - \theta_P - \theta_S\right)^{(\omega_{P,j}-\omega_{P,1})/\omega_P}\exp\left[2\left(a_P\frac{\omega_{P,j}-\omega_{P,1}}{\omega_P}\theta_P\right)\right]$$

$$= \Gamma_{P,1}\left(\frac{\omega_{P,j}}{\omega_{P,1}}\right)^{\alpha_P}\exp\left\{\frac{\omega_{P,j}-\omega_{P,1}}{\omega_P}\left[\ln\left(1 - \theta_P - \theta_S\right) + 2a_P\theta_P\right]\right\} \tag{5.16}$$

The total adsorption of protein molecules is

$$\Gamma_P = \sum_{j=1}^{n}\Gamma_{P,j} = \Gamma_{P,1}\sum_{j=1}^{n}\left(\frac{\omega_{P,j}}{\omega_{P,1}}\right)^{\alpha_P}\exp\left\{\frac{\omega_{P,j}-\omega_{P,1}}{\omega_P}\left[\ln\left(1 - \theta_P - \theta_S\right) + 2a_P\theta_P\right]\right\} \tag{5.17}$$

Therefore,

$$\Gamma_{P,1} = \frac{\Gamma_P}{\sum_{j=1}^{n} \left(\frac{\omega_{P,j}}{\omega_{P,1}}\right)^{\alpha_P} \exp\left\{\frac{\omega_{P,j}-\omega_{P,1}}{\omega_P}\left[\ln\left(1-\theta_P-\theta_S\right)+2a_P\theta_P\right]\right\}} \tag{5.18}$$

and the adsorption in any jth state of protein molecules is determined by the relation

$$\begin{aligned}
\Gamma_{P,j} &= \Gamma_{P,1}\left(\frac{\omega_{P,j}}{\omega_{P,1}}\right)^{\alpha_P} \exp\left\{\frac{\omega_{P,j}-\omega_{P,1}}{\omega_P}\left[\ln\left(1-\theta_P-\theta_S\right)+2a_P\theta_P\right]\right\} \\
&= \Gamma_P \frac{\left(\frac{\omega_{P,j}}{\omega_{P,1}}\right)^{\alpha_P} \exp\left\{\frac{\omega_{P,j}-\omega_{P,1}}{\omega_P}\left[\ln\left(1-\theta_P-\theta_S\right)+2a_P\theta_P\right]\right\}}{\sum_{j=1}^{n} \left(\frac{\omega_{P,j}}{\omega_{P,1}}\right)^{\alpha_P} \exp\left\{\frac{\omega_{P,j}-\omega_{P,1}}{\omega_P}\left[\ln\left(1-\theta_P-\theta_S\right)+2a_P\theta_P\right]\right\}}
\end{aligned} \tag{5.19}$$

The partial surface coverage by protein molecules is

$$\theta_P = \omega_P\Gamma_P = \sum_{j=1}^{n}\omega_{P,j}\Gamma_{P,j} = \Gamma_P \frac{\sum_{j=1}^{n}\omega_{P,j}\left(\frac{\omega_{P,j}}{\omega_{P,1}}\right)^{\alpha_P} \exp\left\{\frac{\omega_{P,j}-\omega_{P,1}}{\omega_P}\left[\ln\left(1-\theta_P-\theta_S\right)+2a_P\theta_P\right]\right\}}{\sum_{j=1}^{n} \left(\frac{\omega_{P,j}}{\omega_{P,1}}\right)^{\alpha_P} \exp\left\{\frac{\omega_{P,j}-\omega_{P,1}}{\omega_P}\left[\ln\left(1-\theta_P-\theta_S\right)+2a_P\theta_P\right]\right\}}$$

$$\tag{5.20}$$

where ω_P is the average molar area of the protein molecule, or

$$\omega_P = \frac{\sum_{j=1}^{n}\omega_{P,j}\left(\frac{\omega_{P,j}}{\omega_{P,1}}\right)^{\alpha_P} \exp\left\{\frac{\omega_{P,j}-\omega_{P,1}}{\omega_P}\left[\ln\left(1-\theta_P-\theta_S\right)+2a_P\theta_P\right]\right\}}{\sum_{j=1}^{n} \left(\frac{\omega_{P,j}}{\omega_{P,1}}\right)^{\alpha_P} \exp\left\{\frac{\omega_{P,j}-\omega_{P,1}}{\omega_P}\left[\ln\left(1-\theta_P-\theta_S\right)+2a_P\theta_P\right]\right\}} \tag{5.21}$$

The *first hint* is that the computational problem is best to formulate in terms of dimensionless variables, to avoid any machine number overflow:

$$\Omega_P = \frac{\omega_P}{\omega_{P,1}}, \quad \Omega_{P,0} = \frac{\omega_{P,0}}{\omega_{P,1}} \tag{5.22}$$

Then,

$$\frac{\omega_{P,j}}{\omega_{P,1}} = 1 + \left(j-1\right)\Omega_{P,0}, \quad \frac{\omega_{P,j}-\omega_{P,1}}{\omega_P} = \left(j-1\right)\frac{\Omega_{P,0}}{\Omega_P} \tag{5.23}$$

in this notation, Equation 5.21 becomes

$$\Omega_P = \frac{\sum_{j=1}^{n}\left[1+(j-1)\Omega_{P,0}\right]\left[1+(j-1)\Omega_{P,0}\right]^{a_P}\exp\left\{(j-1)\frac{\Omega_{P,0}}{\Omega_P}\left[\ln(1-\theta_P-\theta_S)+2a_P\theta_P\right]\right\}}{\sum_{j=1}^{n}\left[1+(j-1)\Omega_{P,0}\right]^{a_P}\exp\left\{(j-1)\frac{\Omega_{P,0}}{\Omega_P}\left[\ln(1-\theta_P-\theta_S)+2a_P\theta_P\right]\right\}}$$

$$=1+\Omega_{P,0}\frac{\sum_{j=1}^{n}(j-1)\left[1+(j-1)\Omega_{P,0}\right]^{a_P}\exp\left\{(j-1)\frac{\Omega_{P,0}}{\Omega_P}\left[\ln(1-\theta_P-\theta_S)+2a_P\theta_P\right]\right\}}{\sum_{j=1}^{n}\left[1+(j-1)\Omega_{P,0}\right]^{a_P}\exp\left\{(j-1)\frac{\Omega_{P,0}}{\Omega_P}\left[\ln(1-\theta_P-\theta_S)+2a_P\theta_P\right]\right\}}$$

(5.24)

Note that the term with $j = 1$ in the nominator vanishes.
Introducing new auxiliary function $\Xi_P = \Xi(\Omega_P, \theta_P, \theta_S)$

$$\Xi_P = \sum_{i=1}^{n}\left[1+(i-1)\Omega_{P,0}\right]^{a_P}\exp\left\{(i-1)\frac{\Omega_{P,0}}{\Omega_P}\left[\ln(1-\theta_P-\theta_S)+2a_P\theta_P\right]\right\}$$

$$=1+\sum_{i=2}^{n}\left[1+(i-1)\Omega_{P,0}\right]^{a_P}\exp\left\{(i-1)\frac{\Omega_{P,0}}{\Omega_P}\left[\ln(1-\theta_P-\theta_S)+2a_P\theta_P\right]\right\}$$

$$=1+\sum_{i=2}^{n}\xi_{Pi}\left(\Omega_P,\theta_P,\theta_S\right)$$

(5.25)

where

$$\xi_{P,i} = \xi_{P,i}\left(\Omega_P,\theta_P,\theta_S\right) = \left[1+(i-1)\Omega_{P,0}\right]^{a_P}\exp\left\{(i-1)\frac{\Omega_{P,0}}{\Omega_P}\left[\ln(1-\theta_P-\theta_S)+2a_P\theta_P\right]\right\} \quad (5.26)$$

one obtains

$$\Omega_P = 1+\Omega_{P,0}\frac{\sum_{j=2}^{n}(j-1)\xi_{P,j}\left(\Omega_P,\theta_P,\theta_S\right)}{1+\sum_{j=2}^{n}\xi_{P,j}\left(\Omega_P,\theta_P,\theta_S\right)}$$

(5.27)

This equation provides one relation between the variables Ω_P, θ_P, and θ_S.
Another equation is derived from the protein–surfactant complex isotherm, Equation 5.6 with $j = 1$

$$b_{PS,1}c_P^{m/(1+m)}c_S^{1/(1+m)} = \frac{\omega_P\Gamma_{P,1}}{(1-\theta_P-\theta_S)^{1/\Omega_P}}\exp\left[-2\left(a_P\frac{\theta_P}{\Omega_P}+a_{PS}\theta_S\right)\right]$$

(5.28)

It follows from Equation 5.18 that

$$\Gamma_{P,1} = \frac{\Gamma_P}{\Xi_P\left(\Omega_P,\theta_P,\theta_S\right)} \tag{5.29}$$

Then, as $\omega_P \cdot \Gamma_P = \theta_P$, the adsorption isotherm for the protein–surfactant complexes in state $j = 1$ (similar isotherms can be obtained for any of the possible n states) is

$$b_{PS,1}c_P^{m/(1+m)}c_S^{1/(1+m)} = \frac{\theta_P}{\Xi_P\left(\Omega_P,\theta_P,\theta_S\right)} \frac{\exp\left[-2\left(a_P\dfrac{\theta_P}{\Omega_P}+a_{PS}\theta_S\right)\right]}{\left(1-\theta_P-\theta_S\right)^{1/\Omega_P}} \tag{5.30}$$

and the adsorption isotherm for the unbound surfactant is given by Equation 5.7. Note that the left-hand sides of Equations 5.7 and 5.30 are the dimensionless quantities. Given the values of all model parameters, the set of model isotherm equations (5.7), (5.27), and (5.30) is sufficient to calculate the three variables Ω_P, θ_P, and θ_S for any values of c_P and c_S. These values being determined, all the thermodynamic quantities can be immediately calculated via the corresponding equations given earlier.

The *second hint* is that, as can be easily seen from Equations 5.7 and 5.30, a good care should be taken when the calculations are performed in the vicinity of the fully covered surface layer: the right-hand sides of these equations exhibit a singularity at $\theta_P + \theta_S \to 1^-$. Note that this is the feature of all model isotherms. It could be useful to transform the equations into the logarithmic form (which is straightforward) and then reformulate the mathematical problem to choose the quantity $\ln(1 - \theta_P - \theta_S)$ as one of the variables.

Also, as the set of model isotherm equations is a set of transcendental equations, it is a good question which procedure should be chosen to obtain the solution. Fortunately, here, the *third hint* could be given: until now in each particular case considered by us, the Newton–Raphson method for nonlinear systems of equations, as is explained in Chapter 9 of the fundamental publication [6], has proved itself to be unbeatable. We strongly recommend the use of the `newt` procedure supplied therein. The only case in which a simpler method was found to be preferable is when a single equation should be solved (e.g., the Frumkin model): here, the procedure `zeroin` proposed in [7] was successfully used.

Also, as the numerical methods mentioned in the preceding text require the knowledge of the initial approximation of the solution, which then should be *polished* to obtain the root, the *fourth hint* is to obtain an analytical approximation of the solution in the vicinity of a characteristic point (usually at very low surface coverage, e.g., at very low surfactant concentration) and use it as the starting value for the calculations in the first few points. Then one can tabulate the solution along the increasing concentration, extrapolating the values found in the previous points to obtain an approximation for the solution in the current point. Here again, the publication [6] gives a good advice to employ the rational (Padé) approximation; see Section 5.12 of [6].

Finally, when one has a reliable method to calculate the model isotherm, the next step is the calculation of the kinetic curve, that is the time dependence of surface tension, adsorption(s) of the component(s), and other values of interest. This requires the solution of a set of parabolic differential equations as explained in details in [2,3]. The initial approach proposed by Ward and Tordai in [8] employed the transformation of the problem to the equivalent set of integral equations. In [9], this procedure was also used; however, the algorithm based on the solution of Fick's equations based on a finite difference scheme seems to yield some characteristics of the system (the species concentrations distribution over the bulk) in a much more straightforward way; also, this approach enables one to account for an inhomogeneous initial distribution of the species throughout the solution bulk. The numerical procedure is essentially based on the Crank–Nicolson scheme as described in Section 20.2 of [6], which was found to be very stable in all cases studied. Here, the *fifth hint* to be recommended is the introduction of a spatial and temporal mesh with a nonuniform distribution of nodes. It was first proposed in [10] that the mesh that becomes gradually denser toward the interface (in spatial dimensions) and toward the time zero (in temporal dimension), that is in the regions where the concentration gradients are high, should be used.

REFERENCES

1. I. Langmuir, The adsorption of gases on plane surfaces of glass, mica and platinum, *J. Am. Chem. Soc.*, 40 (1918) 1361–1403.
2. V.B. Fainerman, S.V. Lylyk, E.V. Aksenenko, L. Liggieri, A.V. Makievski, J.T. Petkov, J. Yorke, R. Miller, Adsorption layer characteristics of Triton surfactants: Part 2. Dynamic surface tension and adsorption, *Colloids Surf. A*, 334 (2009) 8–15.
3. V.B. Fainerman, E.V. Aksenenko, S.V. Lylyk, J.T. Petkov, J. Yorke, R. Miller, Adsorption layer characteristics of mixed sodium dodecyl sulfate/C_nEO_m solutions 1. Dynamic and equilibrium surface tension, *Langmuir*, 26 (2010) 284–292.
4. K. Lunkenheimer, R. Miller, A criterion for judging the purity of adsorbed surfactant layers, *J. Colloid Interface Sci.*, 120 (1987) 176–183.
5. A. Dan, G. Gochev, R. Miller, Tensiometry and silational rheology of mixed β-lactoglobulin/ionic surfactant adsorption layers at water/air and water/hexane interfaces, *J. Colloid Interface Sci.*, 449 (2015) 383–391.
6. W.H. Press, S.A. Teukolsky, W.T. Vetterling, B.P. Flannery, *Numerical Recipes: The Art of Scientific Programming*, Cambridge University Press, Cambridge, U.K., 2007.
7. G.E. Forsythe, M.A. Malcolm, C.B. Moler, *Computer Methods for Mathematical Computations*, Prentice Hall, Englewood Cliffs, NJ, 1977.
8. A.F.H. Ward, L. Tordai, Time-dependence of boundary tensions of solutions. I. The role of diffusion in time-effects, *J. Chem. Phys.*, 14 (1946) 453–461.
9. E.V. Aksenenko, A.V. Makievski, R. Miller, V.B. Fainerman, Dynamic surface tension of aqueous alkyl dimethyl phosphine oxide solutions. Effect of the alkyl chain length, *Colloids Surf. A*, 143 (1998) 311–322.
10. L. Liggieri, F. Ravera, M. Ferrari, A. Passerone, R. Miller, Adsorption kinetics of alkylphosphine oxides at water/hexane interface. 2. Theory of the adsorption with transport across the interface in finite systems, *J. Colloid Interface Sci.*, 186 (1997) 46–52.

Section II

Modern Computational Approaches for Analyzing Interfacial Problems

Molecular-Scale Computational Techniques in Interfacial Science

Trang Nhu Do, Jari Jalkanen, and Mikko Karttunen

CONTENTS

6.1   INTRODUCTION

In this chapter, we discuss computational modeling of systems containing solid–liquid interfaces. We take a fairly practical approach: First, we provide a brief discussion about interfaces followed by a discussion about electrostatic interactions and the methods used to compute them. Interfaces pose a particular problem in computer simulations of interfacial systems, and hence we find a discussion dedicated to this topic justified. Then, in the following section, we discuss different modeling techniques at different length scales. The last sections are very practical: We provide detailed instructions as how to build solid surfaces and two case studies using biomineralization to illustrate the aspects described in the chapter. The final section uses the new technique of well-tempered metadynamics for the computation of free energy in a system of a peptide interacting with a mineral surface.

6.2   WHAT IS SPECIAL ABOUT INTERFACES?

An interface always represents a discontinuity of some kind. The bonding environment of molecules changes and a different set of collective degrees of freedom come into play. The neighbors on the denser side of a surface would prefer to keep their surroundings as close to bulk as possible, but on the other side, the electrons, which away from the interface would participate in binding, need to be allocated elsewhere. This reallocation can take place through chemistry. Commonly, the molecules at surface are encountered in oxidated or protonated forms. The bonding on the fully bonded, complete side also can become modified, leading to relaxation effects. Sometimes the surface layers minimize the contour length of unadvantageous bonding and become reconstructed by splitting into planar subdomains.

The break of periodicity at an interface often leads to polar or charged character, which is important for the interaction of the mineral with solvent or melt phase. In vacuum, the electrostatic potential of a charged layer would create an electrostatic potential that increases linearly with distance—an effect that clearly cannot extend indefinitely far. Biominerals are often in dynamic equilibrium with the solvent. The solvent typically contains a number of additives such as sodium and chloride ions that form charged, so-called Stern layers close to charged interfaces. The water, or some other polar solvents, typically responds to the dipoles and charges by orientation, which usually increases the viscosity locally. If this interaction is strong, the solvent can be totally immobilized in a thin boundary layer.

Most minerals that nucleate in biological fluids are dominated by ionic bonding. In some applications, the interaction of biomolecules with a metal surface, such as gold, titanium, or some alloy of mercury, is the subject of interest. In metals, the electrons are delocalized to a high degree and their wave character is pronounced. At an interface, part of the wave modes are reflected back to the solid, and a part extends smoothly beyond outside the metal: effects that are related to unique surface states and dipoles. Furthermore, if a charge is present above the metal, the electrons in the top layers are redistributed in a way that cancels the electric field inside the conductor. The dissipation from these induced currents will influence the motion of the charge.

When a molecule is incorporated into material, its electron density becomes redistributed. In an ionic system, the partial charges will adopt new values; in a metallic system, the charge distribution becomes more uniform, and consequently, the electrostatic potential will be more effectively screened. If the screening effect is strong, this means that the interaction range of an atom changes depending on the environment. Both the variation of the partial charges and the screening effects are difficult to account for in large-scale simulations.

In biologically occurring molecular crystals, the growth units are usually strongly anisotropic. When the supersaturation is high, which measures much the concentration is out of equilibrium, the resulting surface quality is poor—this is often referred to as fractal or nonspecific growth. With a moderate supersaturation, the growth units tend to attach to surface discontinuities, typically leading to the formation of a step system. In liquid-phase growth, the crystal components can attach directly to the steps and kinks, but in many cases, they diffuse along the surface before they become anchored.

Biological solvents are often supersaturated with respect to the formation of solid precipitates and the crystal growth is actively inhibited [1]. There is evidence that the inhibitors can attach to surface nonspecifically and pin the steps advancing along the surface. It is also conceivable that in some cases, the inhibitor adsorbs specifically to steps or kinks, and this way prevents the further association of growth units. If the growth is not completely stopped, the organic components become incorporated into the growing mineral and have a significant influence on its properties.

Generally, the interfaces have a rich phenomenology: they are less constrained than bulk systems without still being disordered. They try to find balance in a competition between the requirements of two different subsystems, and to achieve this, they exhibit properties that neither of the subsystems portrays alone. As a consequence of improvements in experimental techniques, their understanding has improved a lot: nevertheless, it is still often hard to get a complete and clear picture of all the processes and mechanisms, in particular when dynamical and complex biologically relevant systems are involved. In the last few decades, the use of computational modeling has become more common to give answers to these challenging questions.

6.3 ELECTROSTATIC INTERACTIONS

Electrostatic interactions are among the strongest interactions in classical molecular dynamics (MD) simulations. In a typical situation when simulating bulk systems, such as electrolyte solutions, proteins in a solution, and crystals, periodic boundary conditions are applied allowing for efficient use of the Ewald summation or rather the more advanced Ewald summation–based methods such as the particle-mesh Ewald (PME) [2] or particle–particle particle-mesh Ewald (P3M) [3] for computing electrostatic interactions. Recent reviews regarding the treatment of electrostatic interactions are provided in Refs. [4–7]. In the following, we will not discuss truncation methods as they should not, in general, be used in biomolecular (or other) simulations [8]. PME and P3M are fast, accurate, and flexible and their efficiency is comparable or even better (depending on the cutoff distance) than truncation methods [9].

When interfaces are present, and with confined systems, long-range interactions have to be treated with care. Long-range interactions are typically defined as $\sim 1/r^{\alpha}$, where $\alpha < d$ and d is the dimension of space. In principle, Ewald summation–based methods can be used also for interfacial systems. On the theoretical side, one may have to introduce extra terms that take into account the properties of the surfaces. The important practical problem is, however, that Ewald methods become prohibitively slow when periodicity is broken: with full periodic boundary conditions, the symmetries of the system can be exploited to reduce the computational cost of the plain Ewald sum from $\mathcal{O}(N^2)$ to $\mathcal{O}(N^{3/2})$ and down to $\mathcal{O}(N \log N)$ for PME and P3M [7]. The latter two are the most commonly used methods due to their speed and error control, and thus at least one of them is included in virtually every molecular simulation package; the plain vanilla Ewald summations is virtually never used in production simulations due to the efficiency and scalability of PME and P3M. When periodicity is broken, however, the Fourier sums cannot be combined, which leads to scaling as $\mathcal{O}(N^2)$.

6.3.1 Importance of Boundary Conditions

As discussed earlier, efficient Ewald summation and its variants rely on symmetry of space, that is, fully periodic boundary conditions. Under those conditions, they are very efficient, accurate, and readily available in common simulation software. In a typical interfacial system, one of the directions is not periodic. Here, we discuss of few of the possible approaches that can be used in such systems. As for mathematical details, we refer to the original articles and recent reviews.

The first and most important question is, is a standard technique enough or is something else needed? Simulations of lipid bilayer systems provide a stereotypical example. Due to the presence of the bilayer, the system is quite different from the usual solvent–solute system (e.g., aqueous solution of NaCl); direct application of full periodic boundary conditions creates a system of stacked bilayers. The following are some of the basic questions: What is the electrostatic screening length in the system? Is there enough water in the system to take care of screening between the bilayer stacks? In a typical bilayer simulation, the membranes are fully hydrated with additional water. It is then easy to ensure by adding an appropriate amount of water that the thickness of layer is much larger than the screening length (which can be easily calculated using the composition of the system). Under such conditions, the bilayer stacks do not interact. This is the most typical situation. The same approach can also be applied to other systems containing surfaces, but in each of the cases, the screening conditions have to be evaluated. The case studies in the following sections apply this approach and use periodic boundary conditions.

There are many cases when the aforementioned considerations do not apply. The most obvious ones are confined systems. They possess interfaces and the preceding arguments regarding the screening length do no apply. There are a few possible approaches to such cases: modifications of the Ewald-based methods and real-space methods. As for the Ewald-based variants, the older ones [10–13] are not particularly efficient. In 1999, however, Yeh and Berkowitz introduced a new variant that is conceptually simple, extremely easy to program and efficient [14]. Importantly, this method can be easily combined with PME and P3M, and hence it retains the same scaling behavior. The method involves trick of including a large volume of empty space in the direction that is not periodic. Although not an ideal solution from the conceptual point of view, it is provides a very good approach in many cases.

Real-space methods do not need such tricks as artificial inclusion of empty space. Interestingly, of the real-space methods, the fast multipole method and multigrid methods provide even better $\mathcal{O}(N)$ scaling than PME and P3M. They are, however, generally more complex to implement and tend to have larger computational overheads and hence have not become as popular as one would expect from the favorable scaling behavior. It is, however, expected that this will change in the future. Other real-space techniques include the Lekner summation [15–17]. It is conceptually appealing but scales as $\mathcal{O}(N^2)$. Its later developments [18–21] have improved the scaling and modified the scaling behavior to $\mathcal{O}(N \log N)$ for fully periodic systems and to $\mathcal{O}(N^{7/5})$ for surface systems. Let us finally mention the MMM2D method of Arnold and Holm [22,23]. The method scales as $\mathcal{O}(N^{5/3})$ and, importantly, provides clear methods to analyze the error generated due to parameter choices.

6.3.2 Software Packages and More Information

Of the methods for interfacial systems, currently (to the authors' knowledge), only very few have been implemented in the common software packages. GROMACS [24] has the method of Yeh and Berkowitz [14], and Espresso [25] has currently the most such methods implemented including MMM2D and ELC (combination of the method of Yeh and Berkowitz and MMM2D).

Several comparisons of the different methods for nonperiodic systems exist. A good demonstration of the inefficiency of the plain Ewald as applied to nonperiodic systems is provided in the 1997 paper by Widmann and Adolf [26]. Other new comparisons are available from Mazars [27] and Arnold and Holm [22,23], Solvason et al. [28], and Pollock and Glosli [29]. There are several reviews addressing both periodic and nonperiodic systems [4–7].

6.4 ATOMIC AND MULTISCALE MODELING OF INTERFACES

6.4.1 SIMULATION METHODS AND FORCE FIELDS

6.4.1.1 *Ab Initio* Calculations

Proteins have different adsorption affinities and specificities on different faces of a crystal. This suggests that the orientation and composition of the crystal surface essentially govern the chemical character of the interactions at the interface. These interactions can be described with a high level of accuracy by the first principles or *ab initio* methods, which are mostly based on density functional theory (DFT).

In DFT-based calculations, the choice of the exchange-correlation functional form is critically important, especially for a mixed organic–inorganic interface. There have been a few functionals introduced to describe this unique interface such as those of Becke-Lee-Yang-Parr [30], Perdew-Wang [31], and Perdew-Burke-Ernzerhof [32] with some success [33–35]. A more thorough review of these methods was presented by Harding et al. [36]. Although well known for their accurate electronic structure calculations, these methods suffer from inefficiency and even inaccuracy when reproducing interactions at larger length scales. Most importantly, they all fail at describing the dispersion part of the van der Waals interactions. Corrections and optimizations have been proposed to fix the problems [37–40], which in turn introduce more inaccuracy in other terms or simply are too costly.

6.4.1.2 Hybrid Quantum Mechanics and Molecular Mechanics Simulations

Due to the high computational cost, *ab initio* methods can only focus on calculating the properties of a few hundred atoms around the recognition site, which include part of the protein and the surface involved in the binding site and probably some solvent molecules mediating the binding. But the rest of the protein and a larger environment surrounding the binding site are also important and need to be characterized. Hybrid quantum mechanics and molecular mechanics (QM/MM) techniques appear to be a good approach to solve this sort of interest.

The QM/MM methods for investigating biomineralization generally divide the system into two subsystems: the QM subsystem localized at where the interaction happens and the MM subsystem involving the rest. The QM part is treated quantum mechanically, making use of the accuracy of the *ab initio* description, while the MM part is modeled with a molecular mechanical force field, taking advantage of the classical mechanics efficiency. The total energy of the system is expressed as a sum of energy of each subsystem (QM and MM) and their interaction energy. This interaction energy represents the tricky part of the methods. For organic–inorganic systems, it must be treated separately for each type of material due to the intrinsic differences of the two. Methods for describing the QM/MM boundary in both organic and inorganic systems are carefully reviewed in Ref. [36].

6.4.1.3 Atomistic Molecular Dynamics Simulations

Despite some success [41], a hybrid QM/MM scheme is only able to explore short time scales and is also limited to small length scales. Atomistic MD simulations, in which all atoms are modeled as point charges interacting with each other through classical empirical force fields, are expected to prolong the simulation time and enlarge the system size beyond what QM/MM methods can afford. Atomistic MD still preserves the ability to provide a detailed description of the interactions at the interface involving not only biomolecules and crystal surfaces but also solvent molecules and ions (see Figure 6.1).

Recent notable applications of classical MD simulations to study the nucleation and crystal growth include modeling the properties of self-assembled monolayers [42], nucleation in Lennard-Jones systems [43–45], nucleation and crystallization of silicone [46], and ice nucleation and growth [47,48].

Classical MD simulations have also been widely applied to study protein adsorption on surfaces, including amyloid aggregation on self-assembled monolayers [49,50], adsorption on hydroxyapatite (HA) surface of a fibronectin-related protein [51], bone sialoprotein [52] and matrix

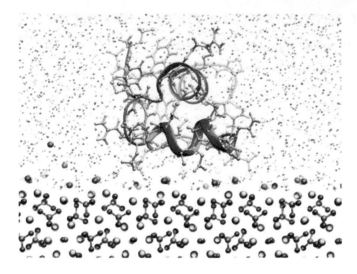

FIGURE 6.1 An example of an atomistic MD simulation that provides detailed interactions of a system containing protein (green), surface (Ca^{2+}, $PO4^{3-}$, and OH^-), Cl^- ions (gold spheres), and solvent molecules (silver spheres). The neutralizing ions and solvent molecules play important role in mediating the protein adsorption on the surface.

gamma-carboxyglutamate (γGLA) protein [53], and adsorption on calcium oxalate crystals of polyglutamic acid peptides [54] and osteopontin peptides [1,55–58].

Simulations of nucleation, crystal growth, and protein adsorption can be prohibitively expensive due to the long time scales at which the rare events happen and/or the high free-energy barriers separating the states. In these cases, classical MD simulations are still severely limited in the accessible time scales and in giving sufficient statistics of the state transitions. Enhanced sampling methods have been successfully employed to overcome this limitation of standard MD simulations and will be discussed in Section 6.4.2.

6.4.1.4   Force Fields

A force field is defined as a functional form and parameter sets describing the potential energy of a system of interacting particles, from which the force acting on a single atom is calculated at each time step of a classical molecular simulation. Finding parameter sets for a certain functional form of force fields is not a trivial task. A *good* parameterization has to ensure that calculations can produce appropriate molecular structures and interaction energies. There are no *correct* force fields with respect to both functional form and parameter set. Indeed, all force fields are empirical and only provide an estimation of the true underlying interaction energy, which controls all molecular behavior. Therefore, the accuracy of classical simulations completely relies on the validity of the chosen force fields.

Numerous force fields have been well established for organic molecules. Comprehensive discussions of these force fields can be found in Ref. [59] and many other literature reviews of biomolecular force fields. Analogously, classical force fields for several types of inorganic crystals and water models are also well developed. Harding et al. [36] provided a thorough review of these developments.

What is left as a huge challenge for simulating organic–inorganic systems is the lack of a suitable force field that can properly describe the interactions at the interface, which involves not only biomolecule-surface binding but also the effect of solvent and free ions. Force fields are always tailored to best fit the experimental or quantum mechanical data of the systems that they try to describe. With two intrinsically different systems like organic and inorganic materials, the force fields that work for one do not necessarily work for the other. This problem of force-field transferability has posed a significant challenge for the simulations at the interface. Another challenge for interface force-field development is the lack of experimental data as reference sources. Therefore, it is important to be aware of force-field limitations in the regime of organic–inorganic interactions.

6.4.2 Enhanced Sampling Simulation Methods

The classical unbiased MD simulation is the best suited technique for providing atomistic details of the interactions within the mineral crystals and at the organic–inorganic interfaces at a local equilibrium. However, crystallization from solution and protein adsorption/assembly on biomineral surfaces are transition processes whose time scales and/or free-energy barriers are well beyond the reach of brute-force MD simulations. Modeling these phenomena requires the development of enhanced sampling methods. These methods essentially come with a trade-off: losing dynamical information of the simulated systems for gaining detailed statistics of the transition processes, which in turn can be directly related to the free energy of interested states.

Enhanced sampling methods include but not limited to *tempering* approaches (simulated tempering [60,61], parallel tempering/replica exchange [62,63]) and *collective variable (CV)–based* approaches (umbrella sampling [64], steered MD [65,66], metadynamics [67,68], etc.). Here, we give a brief introduction to the most commonly used methods including replica exchange MD (REMD), umbrella sampling, and metadynamics. A more comprehensive introduction and review of these methods as well as other enhanced sampling methods can be found in a recent publication by Abrams and Bussi [69].

In REMD method, several replicas with different temperatures are simulated at the same time [62,63]. The coordinates of two replicas i and j are swapped with an acceptance probability defined as

$$\alpha = \min\{1, \exp[(1/(k_BT_i) - 1/(k_BT_j))(U(x_i) - U(x_j))]\}, \tag{6.1}$$

where

k_B is the Boltzmann constant
T_i and T_j denote the temperatures of the replicas i and j, respectively
$U(x_i)$ and $U(x_j)$ are the potential energy of the systems in replicas i and j, respectively

This method relies on the fact that the rate of barrier crossing is governed by temperature. If a system is heated and then cooled, it can undergo largely uncorrelated transitions [69]. REMD has successful applications in simulating not only protein conformational sampling but also aggregation; thus, it is suitable for studying crystallization, crystal growth, and adsorption on crystal surface.

Umbrella sampling is among the most popular nonequilibrium sampling methods [64]. It adds a biasing potential or umbrella potential, usually a simple harmonic spring, to force the system to fluctuate about a reference point in the CV space. The umbrella potential is defined as

$$V(x) = \frac{1}{2}k(s(x) - s_0)^2, \tag{6.2}$$

where

$V(x)$ denotes the biasing potential
k is a symbolic spring constant
$s(x)$ is the CV that is a function of the coordinate
s_0 is the reference point in the CV space

In the umbrella sampling scheme, the CV space is first discretized into N small *window*. Next, N sets of independent trajectories are carried out. Each set is with its own umbrella potential that *pins* the system to the corresponding reference point defining the window. A method called weighted histogram analysis method [70] is then used to combine the statistics of all the independent trajectories in each window to yield the biased probability and unbiased probability of that window. Probability distribution in the CV space eventually can be translated into free energy as a function of that CV. Umbrella sampling has solved various sampling challenges including folding and binding, especially adsorption on lipid bilayers and inorganic surfaces.

Metadynamics is a powerful CV-based method [67]. In this method, the dynamics of the system is driven not only by the potential energy of the system but also by a biasing history-dependent potential constructed as a sum of the Gaussian potentials centered along the trajectory of the CVs $s(x)$:

$$V(s(x),t) = w \sum_{t'=\tau_G, 2\tau_G, \dots < t} \exp\left(-\frac{(s(x)-s(t'))^2}{2\sigma_s^2}\right), \qquad (6.3)$$

where

$V(s(x),t)$ denotes the biasing potential in the Gaussian form
w and σ_s are the height and width of the Gaussian potentials
τ_G is the rate of their deposition
$s(x)$ is the CV that is a function of the coordinate

Constantly added biasing potentials help reduce the free-energy barrier, thus facilitating the occurrence of rare events (see Figure 6.2). Metadynamics is gaining more and more success in a vast range of sampling problems, especially nucleation/crystal growth and adsorption at organic–inorganic interfaces.

The remainder of this section discusses the applications of these methods in simulating the crystallization from solution and adsorption of organic materials (protein and peptide) on inorganic surfaces.

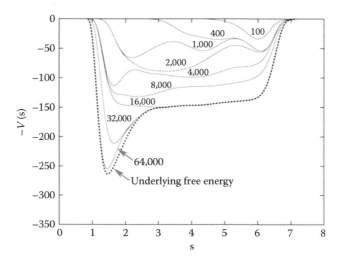

FIGURE 6.2 A schematic representation of the evolution of biasing potential in a standard metadynamics method. The numbers associated with the solid lines are the number of added Gaussian hills. The dotted line represents the underlying free energy. The negative of total biasing potential approaches the underlying free energy as the simulation converges. (From Laio, A. and Parrinello, M., *Proc. Natl. Acad. Sci. U.S.A.*, 99, 12562, 2002.)

6.4.2.1   Enhanced Sampling Simulations of Crystallization from Solution

Calcium carbonate is one of the most significant prevalent minerals on earth. It naturally occurs in limestone and other geological deposits. It forms skeleton, shells, and crusts of a vast variety of undersea organisms. It also participates in the formation of scales, resulting in vessel coating, pipe blockages, and other damaging substances in household as well as technological devices. However, a thorough insight into the growth mechanism of calcium carbonate is still lacking. Quigleya and Rodger [72] made a significant contribution to the computational studies of calcium carbonate by employing metadynamics with a combination of Steinhardt Q_l order parameters [71] and the potential energy as CVs. In this study, the free-energy of a 75 formula unit nanoparticle of calcium carbonate in water reveals that for small system sizes, the amorphous structure is clearly more favorable over a vaterite-like ordered structure. However, the practical challenge remained as the protocol, although advanced, could not be directly applied to larger systems of more nanoparticles. Tribello et al. [73] used a combination of MD and umbrella sampling to computationally examine the growth of calcium carbonate. In this study, it was found that the growth of amorphous calcium carbonate in concentrated solution had virtually no free-energy barrier and hence could be observed on affordable time scales by MD simulations, while the addition of calcium to existing calcite crystals had to overcome a large free-energy barrier, which was due to the formation of an ordered layer of water molecules above the crystal surface. Unfortunately, it was later pointed out [74] that the force field used in this study could lead to an inaccurate description of the free energy and free-energy barriers. More recently, Raiteri and Gale introduced an important force-field improvement to the model and then employed umbrella sampling with the minimum Ca-C distance as a CV [75], as previously proposed by Tribello et al. [73]. This study with the updated force field provided more insights into how and why the amorphous phase was able to grow rapidly compared to other ordered phases. Besides observing important atomistic details of the growth mechanism, this study also emphasized the key role of water in the nucleation of amorphous calcium carbonate.

In a recent study, Salvalaglio et al. [76] applied well-tempered metadynamics to investigate the molecular details of urea crystal growth. Urea, which rapidly grows from aqueous solution to form needlelike crystals, represents a valid benchmark system for examining the unwanted outcome of different growth rates of different crystal faces. The CVs to study the adsorption/desorption of both urea and additives on a given urea crystal surface include an orthogonal Cartesian position to the surface and an angle to describe molecule orientation. Accelerated atomistic simulations allowed differentiating two crystal faces by the exposure of hydrogen bonds and characterizing different adsorption free energies on the two faces.

In umbrella sampling and metadynamics simulations, the choice of CVs is critical. Giberti et al. [77] recently proposed a new CV that was able to facilitate the aggregation of ions into clusters but did not force any particular crystal pattern to occur as opposed to previously used CVs that force the ions to adopt some particular arrangements. The benchmarking system was NaCl in solution. The new CV allowed characterizing intermediate structures during the early stages of nucleation.

Metadynamics appears to be a powerful enhanced sampling method especially well suited for studying nucleation and growth. Other applications of this method includes the studies of nucleation on Lennard-Jones particles [78], ice melting [79], ice nucleation and growth [80], and calcite crystallization on self-assembled monolayers [81].

6.4.2.2   Enhanced Sampling Simulations of Adsorption on Crystal Surfaces

Interactions of organic materials such as proteins and peptides with inorganic surfaces are involved in several important biological processes, for example, hard-tissue growth and cell surface adhesion, yet the underlying physicochemical mechanisms are still not satisfactorily understood. Hoefling et al. performed extensive MD and PMF calculations based on constraint-biased simulations and obtained adsorption free energy of the amino acids as protein building blocks onto a Au (111) surface [82]. The results reasonably correlated with experimental findings and further suggested that gold surfaces may induce β-sheetlike conformations, thus shedding valuable light on future design of gold-binding peptides.

There is still a serious lack of simulation methodologies that can effectively and simultaneously tackle the two crucial challenges of simulating protein adsorption: sampling protein conformational space and overcoming strong protein–surface binding. In 2008, Wang et al. [83] presented a new enhanced sampling technique, the so-called biased REMD, which was different from the conventional REMD in the implementation of a biasing potential energy function. This biasing potential was the negative of the potential of mean force (as a function of a CV) obtained from an in-house adaptive umbrella sampling scheme. For a simple benchmarking system, biased REMD was reported to surpass both conventional REMD and umbrella sampling in simultaneously exploring the conformational space of the solute and the space of its positions over the surface [83]. In a later study, O'Brien et al. [84] applied a refined biased REMD method to examine the adsorption behavior of a peptide over a crystalline polylactide surface.

As another attempt to improve the efficiency of conventional REMD, a method called replica exchange with solute tempering was introduced [85,86]. In this method, the solute–solute and solute–solvent interactions are scaled by an *effective* temperature. In other words, the method subdivides the system into solute and solvent and scales only the parts of the Hamiltonian related to the solute. This helps saving computational expense by significantly reducing the number of required replicas compared to conventional REMD. Wright and Walsh employed this *economical* variant of REMD to study the adsorption of a peptide on a quartz surface and obtained a configurational sampling comparable to REMD at a fraction of the computational expense [87].

Deighan and Pfaendtner [88] made a significant contribution to the field by performing an exhaustive sampling of peptide adsorption using parallel tempering metadynamics in a well-tempered scheme. The method was applied to investigate the adsorption of different peptides on different self-assembled monolayer surfaces. The simulations was observed to converge quickly and allowed quantitative evaluations of binding free energies, side chain orientations, and preferred peptide conformation, which reasonably agreed with available experimental data.

Despite an immense effort in improving and refining enhanced sampling simulation methodologies, the number of successful applications of these methods to the study of organic–inorganic interfaces is still fairly limited. Inadequate sampling is the pivotal reason, but inaccurate force-field description also indirectly contributes to the discrepancy of the obtained free energies.

6.5 PRACTICALS: HOW TO BUILD SOLID SURFACES

Complex molecular crystals are usually built directly from measured crystallographic data: it is usually anything but clear what the *ideal* structure for a given set of involved molecules might be. In principle, a structure prediction would require the estimation of free energies for a large but finite number of locally optimized crystal configurations, within the range of acceptable chemical compositions. Although this kind of procedure would probably involve less uncertainties than protein structure prediction, it is seldom done in practice.

Biominerals are sometimes metastable in the thermodynamic sense: a transition to another crystal lattice would lead to a release of latent heat. In reality, the minerals can remain in the metastable state for extended periods of time when the stable configuration is hard to reach: taking the molecules from their original positions to the low-energy coordinates without steric collisions would require extremely rare events or a complete reshuffling of the crystal components. The lack of a feasible pathway is often referred to as kinetic barrier.

Kinetic barriers are common: as two examples, calcium carbonate occurs in three forms, that is, calcite, aragonite, and vaterite, and calcium oxalate appears as mono-, di-, and trihydrate, from most to least stable variant. The initial crystal formation under supersaturation tends to start from a form where the addition of new building blocks to the growing structure is easy. Later, unless a kinetic barrier prevents it, the molecules rearrange themselves into the stable form.

Biominerals are in often in dynamic equilibrium with their solvent. All the crystals exchange material with the surrounding liquid but in average, the components are more likely to remain in the stable structures. This constant exchange process will eventually lead to the dissolution of all metastable crystal nuclei, although when the solubility is low, this can take a long time. The thermodynamically stable form is usually found as the main component of larger mineral samples.

Defects and structural variation are another factor complicating the computational treatment of biominerals. Biological apatites are a well-known example of this: besides the calcium and phosphate components, they can contain hydroxide, carbonate, chlorine, and fluorine ions and also magnesium and sodium metal components. These *dopants* have a significant impact on the properties of the minerals. Their atomistic treatment is however challenging because the concentrations of the impurities are usually fairly low.

6.5.1 Mineral Coordinates

Structural information for most minerals is available in online databases. The American Mineralogist Crystal Structure Database is a comprehensive online source for mineral coordinates and several mineral information catalogs linked to it [89]. For protein crystals, there is the Marseille Protein Crystallization Database [90,91]. The protein data bank also contain crystal structures [92,93].

As an alternative, a search for the mineral from the Web of Knowledge may return the articles concerning the relevant structure determinations. If the article is not accessible, the data might nevertheless be in the Crystallography Open Database (http://www.crystallography.net/) or in the Cambridge Crystallographic Data Centre [94].

For crystal building, it would be useful be able to construct and independently manipulate several atom groups starting from the same initial data set. Each group should have its own set of arbitrary unit cell transformations and repetitions both in local and global coordinates and along with the ability to fix broken molecules. The most ubiquitous file formats cif, pdb, and xyz should also be supported.

Many of the atomic data visualization programs have options for modifying the representations and building structures, but currently none of them can conveniently handle all the useful transformations. Out of the software tested for this review, *Vesta* was found to have the largest set of desirable features [95]. Vesta can read and write a number of data formats. It allows to slice a crystalline block using an arbitrary number of lattice planes with given the Miller indices and displacements. A fairly general set of transformations can be applied to the basis. Vesta is flexible enough to lend itself even to construction of steps and kinks.

Mercury by Cambridge Crystallographic Data Centre was found to be equally powerful for construction of stacks of layers [94]. It reads cif files and it has a packing/slicing feature that allows to calculate a pair of planes with the Miller indices using adjustable distance and displacement. The inorganic builder feature of Visual Molecular Dynamics (VMD) by Theoretical and Computational Biophysics Group from University of Illinois at Urbana–Champaign could in principle be used to build the crystal from a unit cell constructed with some other program [96]. This crystal could then be sliced using the representations—features—which support an arbitrary number of logical conditions for the coordinates.

The Graphical Display Interface for Structures program and the user interface of the DL_POLY simulation suite contain options for building crystals and slicing them [97]. From the proprietary software side, Materials Studio by Accelrys, MAPS platform by Scienomics, and Crystal Maker have also options that are helpful for surface builders.

Generally, the feature sets of currently available software packages still leave room for improvement when it comes to surface, step, and kink building tasks, and no existing package really contains everything one could hope for. If the visualization is left out of consideration, MATLAB®/octave, R, python, tcl, and also moltemplate scripting environments are powerful tools for building crystals.

6.5.2 Unit Cell Construction and Orientation

Building a crystal for simulations is essentially a geometrical exercise: how to apply a sequence of transformations to reproduce a given crystallographic structure. The basis vectors of the unit cell are usually given in terms of lengths a, b, and c and angles between the vectors of lengths b/c, c/a, and a/b, in this order. We denote the cosines of these angles by C_{bc}, C_{ca}, and C_{ab} and the sine of the last angle by S_{ab}.

If the first basis vector is chosen to run along **i**-axis of a Cartesian coordinate system and the second one to be in the $\mathbf{i} \times \mathbf{j}$ plane, the basis vectors of our example would be the rows of matrix:

$$\mathbf{a} = a\mathbf{i}, \tag{6.4}$$

$$\mathbf{b} = b\left(\mathbf{i}C_{ab} + \mathbf{j}S_{ab}\right), \tag{6.5}$$

$$\mathbf{c} = c\left(\mathbf{i}C_{bc} + \mathbf{j}\frac{C_{bc} - C_{ab}C_{ca}}{S_{ab}} + \mathbf{k}\frac{\sqrt{1 - C_{bc}^2 - C_{ca}^2 - C_{ab}^2 + 2C_{ab}C_{bc}C_{ca}}}{S_{ab}}\right). \tag{6.6}$$

In a typical crystallographic data file, such as a cif file, the fractional coordinates of the atoms in the asymmetric unit are usually given as a table. In the same, or in another table, the average occupancy of the site, average vibrational amplitude, and six anisotropic displacement parameters are also given.

Sometimes, there is need to construct the crystal from data where the repeat unit is not expressed in fractional coordinates. In these cases, the fractional representation can be recovered with a multiplication by the inverse of the basis vector matrix. The unit cell is constructed from the asymmetric unit using the symmetries, usually given as a combination of rotations and reflections and a translation that is always the last operation to be applied.

Sometimes, the symmetries of a crystal are not given explicitly but with a symbol. One of the best-known and most complete sources for symmetry-related information is the Bilbao Crystallographic Server by Condensed Matter Physics Department of the University of the Basque Country (http://www.cryst.ehu.es/). For example, the symmetry of calcium oxalate monohydrate (COM) is P2$_1$. At the Bilbao Server, under "Generators/General Positions," this symmetry can be chosen from a table, which identifies it with number 14. After the symmetry is chosen, the server returns the relevant transformations as matrices, whose fourth column contains a translation vector.

6.5.3 Surfaces and Termination

Which faces are relevant depends on the crystal. Although one can often guess from the crystal structure that facets have a low surface energy, this information is usually best obtained from the experimental literature.

Especially in the construction of a large bulk systems, it is often practical to choose and rotate a single repeat unit so that the direction of the surface points up and only then build up the whole volume. The challenge with the practice of first creating a large block and then slicing it with planes is that rotations tend to become inaccurate when the distance to the pivot gets large, and this can give rise to spurious charge defects and other artifacts. It is likely that during the unit cell manipulations, some molecules become periodic: they continue across the unit cell wall to the neighboring cells (see Figure 6.3). To avoid errors in the crystal construction, it is usually best to repair the *broken* molecules, although in periodic systems, this is not always an issue.

There are often more than one way to terminate the surface, and it is usually not obvious which termination is correct. Typically, the chemical components are interlaced in the unit cell and each group of them may need to be terminated individually. The correct termination usually needs to be

FIGURE 6.3 The asymmetric unit of COM crystal, from an angle. The blue arrows identify the parts of a *periodic* or *broken* oxalate molecule. The second water has only a single hydrogen. The colors are as follows: H, white; C, grey; O, red; and Ca, green.

looked up from the experimental papers. Most biologically relevant crystals tend to be basic and therefore a good starting guess would be to terminate the surface with excess calcium. The existing surface structure databases are mainly focused on metal and semiconductor materials and contain biomineral data to a lesser degree, and currently, it is probably easiest to locate the relevant information from the Web of Knowledge.

For simulation purposes, the top and bottom faces of a planar ionic crystal need to be terminated identically; otherwise, there could be an unphysical electric field across the periodic boundaries. For the construction of the opposite facet, an incomplete unit cell is needed. Similar conclusion holds for steps and kinks, each of which needs their own incomplete unit cell.

6.5.4 BOUNDARY CONDITIONS AND CRYSTAL CONSTRUCTION

There are generally three ways to handle the periodicity: use a pair of acute and obtuse steps, use a pair of different steps, or use a pair of identical steps on opposite sides of a sloped, periodic surface. In the latter case, the surface is tilted so that together with the step, it will be continuous across the simulation cell edge.

In some cases, the first approach is not viable, for instance, if the difference between the various steps is under investigation, or the opposite step is not physically or chemically plausible, or when there are too many uncertainties associated with its structure. Simulating two steps on the same side also requires twice as large simulation cell as a single step, if the distance between the steps is large enough to avoid significant interactions between the steps.

In both cases, the size of the rectangular simulation cell and the orientation of the repeat unit must be chosen to be commensurate with each other. As a rule of thumb, strains not exceeding 2% do not usually lead to plastic deformations. The effect is further diminished in a larger simulation cell and even more if the in-plane coordinates of the crystal are relaxed before the production simulations.

If the surface is reconstructed, one way to go would be to find out the reconstructed geometry from the literature, make sure the planar lattice vectors are multiples of the ones used for the crystal, and then add them as top and bottom layers. Another possibility would be to modify the repulsive potential to increase or decrease the atomic radii and then perform a quick energy minimization.

Once the crystal looks good, it can be transformed to an appropriate format for the simulation software package using an online molecule format converter, Open Babel [98], or some of the crystal builder software packages, or just the file format documentation together with a sample file.

The further processing of the crystal, such as generation of the topology, will then be done with simulation program-specific tools. With GROMACS, the topology would be easiest to construct from the gro or pdb file using grompp, with a custom force field, and with LAMMPS, the lt file could be generated from xyz using moltemplate. As moltemplate supports array manipulation, it could be also used for parts or for the whole crystal construction.

6.5.5 SURFACE CHARGES

The partial charge assignment problem is more difficult. It is typical to assume that the partial charges within a molecule are only dependent on the overall oxidation state of the molecule. For simulation purposes, it is also a matter of practicality to have a limited selection of charge assignments for a molecule. At surfaces and interfaces, it is likely that the partial charges may vary.

A related issue is the surface charge. In simulations, systems generally need to be charge neutral. A crystalline layer however usually has an excess number of ions due to symmetry, and this charge somehow needs to be compensated for. This is not just a simulation artifact: the *in vivo* crystals cannot either have macroscopic amounts of excess charge.

The charge relaxation does not generally happen through surface defects: in direct imaging analyses, the surfaces often seem to have a highly regular Ca coverage. It is then more likely that to cancel the surface charge, there is excess negative charge in the bulk or a fraction of charge is neutralized by counterions of the solvent. It is less likely that the surface calciums assume a lowered oxidation state, although the viability of this option could in principle be assessed from the point of zero charge of the surface.

The bioapatites are reported to be often calcium deficient, while COM, which is discussed in more detail in section "Case Study 1," has a hydroxide ion instead of water in its crystal structure data. Although the handling of vacancies in a simulation is more challenging, in some cases this option could be better than the alternatives. On the other hand, the use of counterions is a standard procedure in simulations, and they are known to be relevant for *in vivo* systems as well. The magnitude of the coverage could in principle be assessed from the Langmuir–Hill equation.

The reported amount of hydroxide ions replaced with chlorine in biological apatite is somewhere between 0.01% and 0.65% [99]. The equilibrium constant between chlorapatite and HA is 845.7 [99], and since the binding events are uncommon, they can be assumed to be independent—Hill's constant should therefore be 1. Then, the concentration of Cl^- ions in the first Stern layer should be below $[Cl^-] \approx 0.01 \times 0.65 \times 847.7/(1 - 0.01 \times 0.65) \approx 5.5$ atoms per one OH^- site on the surface and above one atom per 11 such sites.

6.6 CASE STUDY 1: BUILDING SURFACE

6.6.1 UNIT CELL CONSTRUCTION AND ORIENTATION

6.6.1.1 Coordinates

We take COM as our example [1,54,100–102]. The structure is constructed with *octave* high-level numerical computation language (https://gnu.org/software/octave) [103] or its proprietary alternative MATLAB by MathWorks. These tools were chosen because their availability is good and they make the handling of the geometric operations easy. As a Cartesian reference coordinate system, we use basis vectors **i**, **j**, and **k**, whose relation to the basis vectors a, b, and c of the crystal is the same as in Equations 6.4 through 6.6. The basis vectors in this example are rows of matrix basis _ in = [a; b; c], and the fractional coordinates multiply the basis from the left.

We download the crystal data in cif format from AMCD [89], where the mineral is known as whewellite. We open the cif file in a text editor and copy the table of fractional coordinates of the atoms to another file `com.dt`, which is then loaded in octave.

The fractional coordinates of the atoms in the asymmetric unit are the first three columns of the table. In our example, the two water molecules of the primitive cell occur in two slightly different positions. We take only the molecules with the highest occupancy numbers, `r=[com(1:19,1)`, `com(1:19,2)`, `com(1:19,3)]`. Here, "1:19" refers to all integers from 1 to 19.

The position of the second hydrogen in the first water is missing from the data and it needs to be added to keep the primitive cell charge neutral. The three existing hydrogens of the waters form a column along k-axis, and we place the new hydrogen in the plane with the hydroxide ion and its closest hydrogen neighbor, on the far side:

```
rh = (r(15,:)-r(18,:))*basis_in;
rc = (r(17,:)-r(18,:))*basis_in; rc /= norm(rc);
delta = rh*cos(ang) - cross(rh, cross(rh, rc))*sin(ang)/norm(rh);
r(end+1,:) = r(18,:) + delta*inv(basis_in);
```

where the indices of O and H of the hydroxide are 18 and 15, the closest hydrogen is found at index 17, and the central angle of the water molecule is `ang = 104.5*pi/180`. The functions `cross` and `norm` take the cross product and the norm of their arguments.

The central angle of the other, complete water molecule is not equal to 104.5°, which means that unless it is treated as a new molecule type, an energy minimization or simulation treatment will induce minor changes to the structure. Of course, if the crystal is only used as a source for electrostatics, then none of the mineral coordinates will ever need to be updated.

For the generation of topology, the atoms of a molecule sometimes need to appear in a particular order. If a reordering is needed, it is easiest to do it at this stage. If, for example, the order of the atoms of oxalates in the topology was O11 C1 O12 C2 O21 O22, we could form an index vector and use it to reorganize the coordinates

```
new_indices = [1 2 7 3 8 4 9 10 11 5 12 6 13 14 15 16 17 18 19 20];
r = r(new_indices,:);
```

We do not have this applied in the rest of the example but if this were the case, the atom labels would also need to be treated in a similar way.

6.6.1.2 Symmetries

In our case, the symmetries are already given in the cif file, and they consist of identity, inversion, and a simultaneous reflection and a translation by half unit cells in the direction of the b and c basis vectors, which correspond to the second and third fractional coordinates. The complete unit cell in the fractional coordinates then reads as

```
uc = mod([r; -r; [r(:,1),0.5-r(:,2),0.5+r(:,3)];
[-r(:,1),0.5+r(:,2),0.5-r(:,3)]], 1);
```

where the `mod` function wraps the fractional coordinates back to the unit interval. One of the oxalate molecules is periodic across the unit cell boundary in a-direction. We correct this by shifting the origin by 10% in the a-direction, `uc(:,1) = mod(uc(:,1)+0.1,1)`. We also check that there are no duplicates.

Now, let us assume the crystal face under investigation is {hkl}. The set of basis vectors for the reciprocal space is proportional to the matrix

```
recip_in = [cross(b, c); cross(c, a), cross(a, b)];
```

and the normal of the face is a multiple of `sn = [h k l]*recip _ in`. We first rotate `sn` around the Cartesian **i**-axis to make the face parallel to basis vector b: the relevant rotation is

```
Rk = [sl, 0, 0; 0, sn(3), sn(2); 0, -sn(2), sn(3)]/sl;
```

where `sl = norm([sn(2), sn(3)])`. We take the transformed normal `tn = sn*Rk` and construct a rotation `Rh` that points `tn` along the Cartesian **k**-axis:

```
Rh = [tn(3), 0, tn(1); 0, tl, 0; -tn(1), 0, tn(3)]/tl;
```

where `tl = norm([tn(1), tn(3)])`. Then, the new basis where the face is oriented *up* reads as `basis _ up = basis _ in*Rk*Rh`.

6.6.2 SURFACES AND STEPS

6.6.2.1 Surface

The unit cell is now rotated but otherwise looks the same as in the beginning. We would now like to redefine the basis and the fractional coordinates so that the uppermost face of the unit cell is along the {*hkl*} surface. The *up* components of the new basis `basis _ up` are all equal to h, k, or l times the same number, respectively. To get rid of the *up* components of the first two basis vectors, an appropriate multiple of the third component needs to be subtracted, which in this case is

```
reor = eye(3); reor(1,3) = -h/l; reor(2,3) = -k/l;
basis_up = reor*basis_up;
```

The fractional coordinates needs to be compensated for this redefinition of the basis by multiplying the coordinates from the right by the inverse operation,

```
roer = eye(3); roer(1,3) = h/l; roer(2,3) = k/l;
uc = mod(uc*roer,1);
```

where the `mod` operation again wraps the coordinates back to the unit cell. The simultaneous unitary transformation of the basis from the left and the inverse transformation of the fractional coordinates from the right cancel, and therefore this redefinition leaves the atomic coordinates intact. In our example, these operations do little because the {100} had been chosen, but in a more general case they are necessary. On this stage, the unit cell looks like panel (b) of Figure 6.4.

It needs to be pointed out that in some cases, the fractions h/l and k/l may be nonintegers, while periodicity is guaranteed only for integer combinations of the basis vectors. To fix this, a supercell can be created from copies of the fractional coordinates for the unit cell. The intersections of a lattice plane are proportional to 1/h, 1/k, and 1/l. Adding n copies of the original unit cell shifted along the first basis vector and then normalized back to one multiplies the length of the first basis vector by n, and therefore the intersection point is now proportional to (1/h)/n. If n is chosen to be 1, the system is again periodic.

Another, perhaps easier solution is to construct a supercell right after the rotation of the basis by `Rk*Rh` and then select the appropriate atoms `uc _ super(B)` that are above and below suitable threshold values `min _ z` and `max _ z` by slicing,

```
B = ((uc_super(:,3) < max_z)&(uc_super(:,3) > min_z));
```

If the molecules are not in a neat plane, this may need to be combined to further selection with labels, as described in the section discussing surface termination.

6.6.2.2 Step

We use the same process to construct a step. We orient the face so that a step with the Miller indices `step = [st _ h st _ k st _ l]` will run along the **j**-axis. The direction of the step in

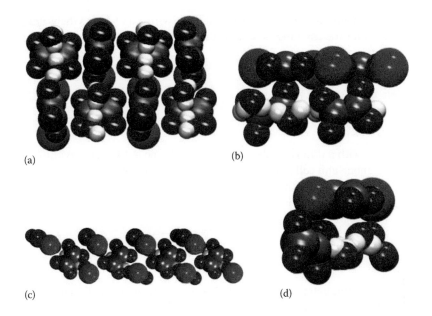

FIGURE 6.4 (a) The full unit cell of COM after repairing the broken molecules, from {100} direction. (b) The {100} face is pointing up, as seen from the negative **j**-direction. (c) This unit cell choice, in top view, has 121 and {100} planes as its faces along the **i**- and **k**-axis, corresponding to the step and surface. (d) A side view of the unit cell of panel (c) from the positive **j**-direction. The color coding is as in Figure 6.3.

the current coordinate system is st = step*recip _ in*Rk*Rh, and the transformed Miller indices using the redefined basis would be tt = st* inv(recip _ up), where recip _ up relates to basis _ up just like recip _ in relates to basis _ in. The rotation that cancels the second component of st is

Rs = [st(1), -st(2), 0; st(2), st(1), 0; 0, 0, sr]/sr;

where sr = norm([st(1), st(2)]) and the basis is now basis _ fw = basis _ up*Rs.

   Now, after this rotation, the first (**i**-) components of the basis vectors along the surface are tt(1) and tt(2) times the same number. As before, we want to reshape the unit cell so that it will have both the step and the surface as its faces. A redefinition of the basis by basis_fw = srec*basis_fw, where srec is a unit matrix except at srec(2,1) = -tt(2)/tt(1), makes the step run along the **j**-axis. Like before, to keep the coordinates intact, an opposite change in the fractional coordinates is needed. The matrix srec can be changed to its inverse by flipping the sign of its single nonzero off-diagonal element, srec(2,1) *= -1, and the fractional coordinates then become uc = mod(uc*srec,1). At this point, the unit cell looks like panels (c) and (d) of Figure 6.4.

6.6.3 SURFACE TERMINATION

6.6.3.1 Termination of Molecular Groups

Getting the termination of the surface right is crucial for interfacial simulations. The electrostatic effects from a few missing charges can easily crowd out a finer balance, which might play a role in an adsorption problem.

   The surface layers are often corrugated and parts of some molecules can be sticking out. In a complex crystal, there are usually several molecular groups that need their own, appropriately displaced bounding plane for any given surface. To do this, we need a label for each of the groups. We label all the Ca^{2+} ions with integer 1, the two oxalate ions with 2 and 3, and the water molecules with 4 and 5. The full list for the asymmetric unit reads

```
labels = [1 1 2 2 3 3 2 2 2 2 3 3 3 3 4 5 5 4 5 4];
```

and for the whole unit cell, this becomes

```
labels = reshape(labels'*ones(1, 4), 20*4, 1);
```

For each component with a different label, we can then define a different dividing line between the current and the neighbor unit cells. Initially, we set all these to zero, `shift = zeros(5,3);`. We take a look at a {121} step on a {100} surface.

In octave, for each label `i` this is accomplished with

```
B = (labels == i);
tshift = ones(ucb_size, 1)*shift(i,:);
uc(B,:) = mod(uc(B,:) + tshift,1) - tshift;
```

where the set B can be through to contain the indices of atoms with label i and the number of these atoms is denoted by `ucb _ size`.

We start with the bulky, easily distinguishable molecules. Initially, the first oxalate is periodic in **k**-direction, and the second one in **i**-direction. Setting `shift(2,3) = -0.3` and `shift(2,1) = 0.5` completes the top oxalate layer that covers the surface, and `shift(3,1) = 0.2` takes both parts of the second oxalate on the same side. Next, we bring all Ca^{2+} ions to the top layer with `shift(1,3) = -0.1`. Finally, we move all water under the top layer oxalates with `shift(4:5,1) = 0.28`.

6.6.3.2 Completion of Surface

The bottom surface and the trailing step also need to be created. The indices for this incomplete cell for the bottom surface layer can be selected with `B _ opp = (uc > f _ layer);`, where `f _ layer` is a fraction, above which the top-layer atoms are, in this case 0.8. For the trailing step, we choose the oxalates below the surface layer and the surface layer Ca^{2+} ions:

```
B_neg = (!B_opp)&((labels(:) == 2)|(labels(:) == 3));
B_ca = (labels(:) == 1)&B_opp;
```

The set B _ opp can be seen as the bottom layer in panels (c) and (d) in Figure 6.5; the set B _ neg comprises the nonhorizontal oxalates and B _ ca the calciums at the bottom step edge in the same figure.

6.6.4 Boundary Conditions and Crystal Construction

On our surface, the highly similar steps {121} and {$\overline{1}$21} would work as a pair of leading and trailing step edges, and the latter arises naturally, if the other side of the step is padded with copies of nonplanar oxalates and associated Ca^{2+} ions. Instead of this, we make the crystal periodic and use only a single step {121} on the {100} and {$\overline{12}\overline{1}$} on the {$\overline{1}$00} surface.

6.6.4.1 Continuity

Our step is running along the **j**-axis. Every time it advances along the surface by a lattice vector `basis _ bc(1,:)`, it gets shifted in the **j**-direction by `basis _ bc(1,2)` units. The crystal will be continuous if

```
n_x*basis_bc(1,2) + basis_bc(3,2) = px*basis_bc(2,2)
```

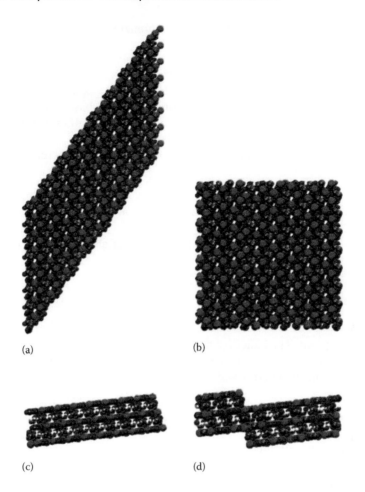

(a)   (b)

(c)   (d)

FIGURE 6.5 A 8 × 2 × 2 slab of COM crystal with a step before applying the boundary conditions, from the top and from the negative **j**-direction, panels (a) and (c). The block and the simulation cell size have been chosen to make the crystal continuous. In the panels (b) and (d), the crystal is wrapped within the rectangular simulation volume. The color coding is the same as in Figure 6.3.

where n _ x is the total number of translations in **i**-direction and px is an integer multiple of the smallest fraction of basis _ bc(2,2) that is periodically repeated in the **j**-direction. In our example, taking the fraction as 1/4 would be sufficient.

For each n _ x, we can calculate the **j**-component that would give perfect periodicity, that is, the nearest integer multiple of px*basis _ bc(2,2), and compare it to the actual **j**-component to find the number of translations with the smallest error. If we denote i _ y = basis _ bc(1,2), j _ y = basis _ bc(2,2) and k _ y = basis _ bc(3,2), then

```
shift_j = (i_y*[1:nm] - k_y)/(px*j_y);
comp = (px*j_y*round(shift_j) + k_y)./[1:nm];
errors = [abs(comp(:,1)' - i_y)/abs(i_y), [1:nm]'];
sortrows(errors)
```

where the largest acceptable number of translations nm is 20. As a result, we see that the error will be below 1% for 15, 18, 12, or 9 translations and below 2% for more than 8 translations, except when n _ x = 10 or 13. After we have chosen a value for n _ x, we set the component basis _ bc(1,2) to the corresponding value from the vector comp. Even though the error is already small, it can diminished further by energy minimizing the complete lattice at the end in the **i**- and **j**-coordinates (in-plane directions).

6.6.4.2 Orientation and Boundary Conditions

Let us now denote `tx = n _ x*basis _ bc(1,1)`, where `n _ x` is our choice. The lattice vector connecting the step's upper and lower edge is h times the third basis vector, `bh = h*basis _ fw(3,:)`. In some cases, due to symmetry, `h` might only be a one-half or one-third of the full lattice spacing; in others it might be a small integer (step bunching).

Then, the rotation that tilts the crystal while keeping it periodic and creates the step is

```
Rt = [tx + bh(1), 0, bh(3); 0, Lx, 0; -bh(3), 0, tx + bh(1)]/Lx;
```

where `Lx` is the norm if the first row. The new, transformed basis will be then `basis _ bc = basis _ fw*Rt`.

We store the incomplete unit cell of the trailing step in variable `uc _ neg` with

```
indices = reshape([1:20]'*ones(1, 4), 20*4, 1)';
us_ca = uc(B_ca,:); us_ca(:,3) -= 1.0;
uc_neg = [uc(B_neg,:); us_ca];
indices_neg = [indices(1,B_neg), indices(1,B_ca)]';
```

and denote its size by `usn`. The indices of the atoms have been stored in `indices _ neg`, and the top layer atoms have been moved to the bottom. We also need to make the labeling more granular. We store each atom name in an array with four characters

```
oxa = ['C1 C2 C1 C2 O11 O12 O21 O22 O11 O12 O21 O22 '];
at = ['CA CA ', oxa, 'HW1 HW1 HW2 OW OW HW2 '];
```

Now, the full crystal can be written in fractional coordinates as

```
for j=0:(n_y-1),
   crystal((end+1):(end+usn),:) = uc_neg + ones(usn,1)*[-1, j, 1];
   atom_labels((end+1):(end+usn),1) = indices_neg;
   group_labels((end+1):(end+usn),1) = labels(1,B_neg|B_ca);
   for i=0:(n_x-1),
      crystal((end+1):(end+usb),:) = uc_opp + ones(usb,1)*[i, j, 0];
      atom_labels((end+1):(end+usb),1) = indices(1,B_opp);
      group_labels((end+1):(end+usb),1) = labels(1,B_opp);
      for k=1:(n_z),
         crystal((end+1):(end+us),:) = uc + ones(us,1)*[i, j, k];
         atom_labels((end+1):(end+us),1) = indices(1,:)';
         group_labels((end+1):(end+us),1) = labels(1,:)';
      end
   end
end
```

This construction first adds an incomplete unit cell for the trailing edge, then the bottom surface is constructed, and finally, the rest of the atoms are made. In physical coordinates, the crystal is `uc _ phys = crystal*basis _ bc`. The horizontal dimensions of the simulation cell will be `Lx` and `n _ y*j _ y`.

We output the crystal in gro format, which is compatible with GROMACS simulations. For that, we need to define group names and numbers of atoms in each group,

```
tp = ['CA2 OXA OXA SOL SOL '];
group_sizes = [1 6 6 3 3];
```

The total number of atoms in the crystal is na = (n _ x)*(n _ y)*((n _ z)*us + usb) + n _ y*usn; with these definitions, the crystal can be saved in an external file "com.gro" with

```
outf = fopen('com.gro', 'w');
fprintf(outf,'%d\n%d\n', na, na);
prev_group_end = 0; at_id = 0; gr_id = 1;
for group_i=1:5, for i=1:ca,
    if (group_labels(i,1) == group_i),
       type_i = 4*(group_i-1);
       types = type_i+[1:4];
       at_id = at_id+1;
       atom_i = 4*(atom_labels(i,1)-1);
       atoms = atom_i+[1:4];
       format ='%5d%c%c%c%c %c%c%c%c %5d%8.3f%8.3f%8.3f\n';
       fprintf(outf, format, gr_id, tp(types), at(atoms), at_id,
       0.1*uc_phys(i,1:3));
       if (at_id - prev_group_end >= group_sizes(group_i)),
          prev_group_end = at_id;
          gr_id = gr_id+1;
          endif
    endif
end; end;


fprintf(outf,'%f %f %f\n', bl, abs(n_f*step_tilt(2,2)), 100.0);
fclose(outf);
```

The complete crystal is shown in Figure 6.5, panels (a) and (c). When the periodic boundary conditions are applied and the whole crystal is shifted by a fraction of the unit cell, the step moves in the middle. There is no noticeable break in the periodicity. We chose our crystal to be only two unit cells thick for demonstration purposes; in reality, it would be better to make it thicker.

The resulting crystal is not charge neutral: the incomplete unit cells of the bottom layer and the steps carry net positive charge. A potential source of compensating negative charge could be the hydroxide ions, as seen in the structural data. With them, each unit cell has a charge −4, which is already more than enough. By allowing a fraction of unit cells to contain hydroxide instead of water, the surface charge can be compensated without introduction of defects, lowering the oxidation state of the surface Ca atoms or some other hypothetical mechanism. Another, perhaps easier option would be to use Cl$^-$ counterions.

6.7   CASE STUDY 2: ADSORPTION OF OSTEOCALCIN ON HYDROXYAPATITE SURFACE EXPLORED BY WELL-TEMPERED METADYNAMICS

The strength of bones comes mainly from the stable hexagonal pattern of a mineral known as hydroxyapatite [104]. Proper HA pattern results in strong bones, and wrong HA pattern may cause osteoporosis. The growth of HA crystal is known to be regulated by a protein called osteocalcin (OC), which is a highly conserved and the most abundant noncollagenous protein in bones [105]. OC has been observed to adsorb to HA surface; however, it has been under debate that upon binding to HA whether OC actually helps building up HA crystal or prevents HA crystal growth. The precise mechanism through which OC influences HA growth is still unclear. The x-ray crystal structure of porcine OC revealed that five Ca^{2+} ions are coordinated by four negatively charged residues creating a spatial orientation that is complementary to the Ca^{2+} ions forming the HA crystal lattice [106]. This finding suggested that upon binding, OC is able to transfer Ca^{2+} ions to the HA crystal surface promoting in this manner the crystal growth. Other studies suggested that the protein inhibits the formation of HA. Bovine OC was found to inhibit crystallization from calcium phosphate solutions [107] and growth of HA seed crystals [108]. Chicken OC was observed to inhibit *de novo* HA formation [109]. Numerous questions about the detailed interactions and molecular mechanism of OC adsorption onto HA are left to be revealed by atomistic simulations.

Interestingly, the adsorption properties of OC was found to be dependent upon γ-carboxylation and Ca^{2+} binding. Vitamin K1 or phylloquinone, best known for its vital role in the blood coagulation cascade, is an essential cofactor in OC γ-carboxylation, which changes three OC glutamic residues glutamic acid (GLU) at the positions 17, 21, and 24 to γ-carboxyglutamyl residues (GLA). The effect of γ-carboxylation together with Ca^{2+} binding features an increased stability of α-helical portions in OC. These α-helical portions were proposed to play crucial roles in OC adsorption on HA surface [106,110]. Without the γ-carboxylation effect or the presence of Ca^{2+} ions, adsorption was not observed.

The following studies employ well-tempered metadynamics [68] to explore the binding abilities to HA surface of OC in four different conditions: decarboxylated OC (decarbox-OC hereafter), decarbox-OC with Ca^{2+} ions, γ-carboxylated OC (OC hereafter), and OC with Ca^{2+} ions.

6.7.1 COMPUTATIONAL SETUP

All simulations are atomistic with explicit water. Each system contains approximately 110,000 atoms including the 49-residue porcine OC protein, the HA crystal with 3360 PO_4^{3-} groups, 960 OH^- groups and 5760 Ca^{2+} ions, around 27,000 water molecules, and 700 counterions. The HA crystal is built using a similar method as presented in our Case Study 1 (Section 6.6). The system is simulated with the AMBER ff99SB*-ILDN force field [111–113]. All simulations are performed with GROMACS 4.6 [24] in the NVT ensemble, employing the velocity rescale algorithm for temperature coupling [114]. The PLUMED 1.3 plug-in [115] is used to perform well-tempered metadynamics.

A well-tempered metadynamics is a variant of standard metadynamics that adds an adaptive bias along the trajectory of the CVs by varying the Gaussian height:

$$w(s(t)) = w_0 \frac{\Delta T}{\Delta T + w_0 N(s,t)} \tau_G, \qquad (6.7)$$

where
w_0 is the initial Gaussian height
$N(s,t)$ is the cumulative number of times the system visits state s at time t
ΔT denotes a temperature parameter that can be tuned to increase the probability of barrier crossing

The distance between the center of mass of OC protein and the surface is used as a CV for well-tempered metadynamics in all the four cases. The binding simulation of OC to HA with Ca^{2+} ions is later refined with two additional CVs: (1) the number of contacts between aspartic acid (ASP) residues and the surface Ca^{2+} ions and (2) the number of contacts between GLU/GLA residues and the surface Ca^{2+} ions. Residues come into contact with surface when the distance between the side chain carboxylic oxygen and the surface Ca^{2+} ions is equal to or less than 2.5 Å.

6.7.2 CONVERGENCE OF SIMULATIONS AND ADSORPTION FREE-ENERGY PROFILES

In a well-tempered metadynamics scheme, a preselected maximum biasing potential is only added when a new region in the free-energy profile is explored. If the system revisits previously explored regions, depending on the number of visits, the subsequent biasing potential at these regions will decrease accordingly. This is considered one of the great advantages of well-tempered metadynamics compared to the conventional metadynamics. It allows focusing the computational effort on physically important regions, thus having more control of convergence and errors. Figure 6.6 shows the fluctuation of the added biasing potential as a function of time in four cases: decarbox-OC,

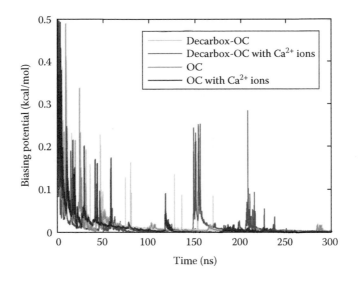

FIGURE 6.6 Decline of the external biasing potential in the course of 300 ns well-tempered metady-namics simulations of the adsorption of decarbox-OC (light blue), decarbox-OC with Ca^{2+} ions (blue), OC (light red), and OC with Ca^{2+} ions (red) onto HA surface.

decarbox-OC with Ca^{2+} ions, OC, and OC with Ca^{2+} ions. The biasing potential peaks up every time a new or less-explored region is visited and then decreases and eventually approaches zero. Toward the end of the 300 ns simulations, no more new regions are explored on the free-energy profile as a function of the OC–HA orthogonal distance.

Figure 6.7 presents the free-energy profile versus the OC–HA distance. Among the four cases, OC with Ca^{2+} ions (solid curve) exhibits the largest free-energy difference between the adsorbed and desorbed states, while for OC without Ca^{2+} ions (dashed-double-dotted curve), binding and not binding to the surface do not differ much in terms of free energy. In case of decarbox-OC, the

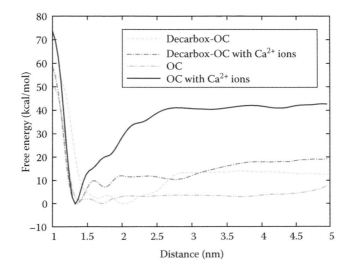

FIGURE 6.7 Free energy as a function of the orthogonal OC–HA distance in four cases: decarbox-OC (dashed line), decarbox-OC with Ca^{2+} ions (dot-dashed line), OC (dashed-double-dotted line), and OC with Ca^{2+} ions (solid line). OC with Ca^{2+} ions has the highest adsorption propensity.

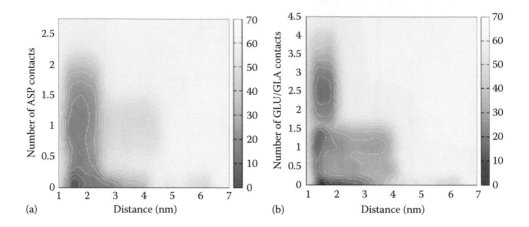

FIGURE 6.8 (a) Free energy as a function of CV1, distance, and CV2, number of ASP surface contacts. (b) Free energy as a function of CV1 and CV3—number of GLU/GLA surface contacts.

presence of Ca^{2+} ions (dot-dashed curve) helps strengthen the binding, that is, making it happen at a smaller distance to the surface with a higher free-energy barrier. These findings qualitatively support the hypotheses about the effects of γ-carboxylation and Ca^{2+} binding on the adsorption of OC to HA surface.

To have more insights into the binding of OC to HA in the presence of Ca^{2+} ions, we perform another well-tempered dynamics simulation biased not only on the aforementioned orthogonal distance (CV1) but also on two additional CVs: CV2, number of contacts between ASP residues and surface Ca^{2+} ions, and CV3, number of contacts between GLU/GLA residues and surface Ca^{2+} ions. A contact is defined when the distance between the side chain carboxylic oxygen and the surface Ca^{2+} ions is equal to or less than 2.5 Å. ASP, GLU, and GLA are the charged residues expected to play key roles in OC adsorption through electrostatic interaction with the opposite-charged surface ions. Figure 6.8 shows the 3D free-energy surface projected on the plane of CV1 and CV2 (a) and on the plane of CV1 and CV3 (b). The free energy is reconstructed after a well-tempered metadynamics simulation of 250 ns. Despite the same number of ASP and GLU/GLA residues in OC, GLU/GLA residues seem to have more contacts and a larger binding affinity to the HA surface.

6.7.3 Atomistic Details of Protein–Surface Interaction and Ion Binding Sites

A great advantage of atomistic simulation is its ability to describe molecular interactions at an atomistic detail. Figure 6.9 presents two frames of the simulation, in which the protein first approaches the surface (a) and later has more contacts with the surface (b). In Figure 6.9a, one of the two GLU residues is strongly attracted by the electrostatic field created by the surface, while other negatively charged residues are still screened by the solvent Ca^{2+} ions. In Figure 6.9b, after GLU40 strongly binds to the surface, other charged residues also start having contacts with the surface. Solvent Ca^{2+} ions still follow when OC comes down on the surface. During the 250 ns simulation, Ca^{2+} ions always bind to the negatively charged residues, especially the three GLA residues that were found to be metal binding sites in experiments [110]. Due to Ca^{2+} binding, the GLA helix containing the three GLA residues is prevented from interacting with the surface Ca^{2+} ions. The same Ca^{2+} screening effect is observed at the second helix of OC that has several solvent-exposed negatively charged residues. From our atomistic simulation, we suggest that OC strongly binds to HA surface but it does not appear to facilitate the HA crystal growth by transferring Ca^{2+} ions from the solvent to the surface as suggested before by Hoang et al. [106].

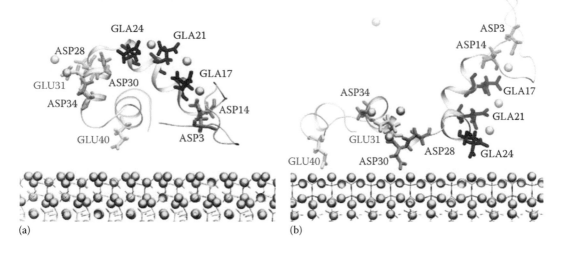

FIGURE 6.9   Snapshots of the simulation when OC first approaches the HA surface (a) and is in one of its bound conformations with the surface (b). The figures are created by VMD software. Surface and solvent Ca^{2+} ions are represented with cyan spheres. ASP, GLU, and GLA residues are in a VMD licorice representation and colored with green, light red, and red, respectively. (From Humphrey, W. et al., *J. Mol. Graph*, 14, 33, 1996.)

6.8   DISCUSSION AND CONCLUSIONS

We have provided detailed discussion and a hands-on demonstration of how systems containing solid–liquid interfaces are constructed, simulated, and analyzed including the advanced method of well-tempered metadynamics for computing free energies. We used biominerals as a particular example, but the techniques and approaches shown here are general and can be applied to any related systems.

REFERENCES

1. J. O'Young, S. Chirico, N. Al Tarhuni, B. Grohe, M. Karttunen, H.A. Goldberg and G.K. Hunter. Phosphorylation of osteopontin peptides mediates adsorption to and incorporation into calcium oxalate crystals. *Cells Tissues Organs*, 189:51–55, 2009.
2. U. Essmann, L. Perera, M. L. Berkowitz, T. Darden, H. Lee, and L. G. Pedersen. A smooth particle mesh Ewald method. *J. Chem. Phys.*, 103:8577, 1995.
3. R. W. Hockney and J. W. Eastwood. *Computer Simulation Using Particles*. Adam Hilger, Bristol, U.K., 1988.
4. M. Karttunen, J. Rottler, I. Vattulainen, and C. Sagui. Electrostatics in biomolecular simulations: Where are we now and where are we heading? *Curr. Top. Membr.*, 60:49–89, 2008.
5. M. Mazars. Long ranged interactions in computer simulations and for quasi-2D systems. *Phys. Rep.*, 500:43–116, 2011.
6. A. Arnold, K. Breitsprecher, F. Fahrenberger, S. Kesselheim, O. Lenz, and C. Holm. Efficient algorithms for electrostatic interactions including dielectric contrasts. *Entropy*, 15:4569–4588, 2013.
7. G. Andrs Cisneros, M. Karttunen, P. Ren, and C. Sagui. Classical electrostatics for biomolecular simulations. *Chem. Rev.*, 114:779–814, 2013.
8. M. Patra, M. Karttunen, M.T. Hyvönen, E. Falck, P. Lindqvist, and I. Vattulainen. Molecular dynamics simulations of lipid bilayers: Major artifacts due to truncating electrostatic interactions. *Biophys. J.*, 84:3636–3645, 2003.
9. M. Patra, M. T. Hyvönen, E. Falck, M. Sabouri-Ghomi, I. Vattulainen, and M. Karttunen. Long-range interactions and parallel scalability in molecular simulations. *Comput. Phys. Commun.*, 176:14–22, 2007.
10. D. M. Heyes, M. Barber, and J. H. R. Clarke. Molecular dynamics computer simulation of surface properties of crystalline potassium chloride. *J. Chem. Soc., Faraday Trans.*, 2:1485–1496, 1977.

11. J. Hautman and M. L. Klein. An Ewald summation method for planar surfaces and interfaces. *Mol. Phys.*, 75(2):379–395, 1992.

12. S. W. de Leeuw and J. W. Perram. Electrostatic lattice sums for semi-infinite lattices. *Mol. Sim.*, 37:1313–1322, 1979.

13. B. R. A. Nijboer and F. W. de Wette. On the calculation of lattice sums. *Physica*, 23:309–321, 1957.

14. I.-C. Yeh and M. L. Berkowitz. Ewald summation for systems with slab geometry. *J. Chem. Phys.*, 111:3155–3162, 1999.

15. J. Lekner. Summation of dipolar fields in simulated liquid-vapor interfaces. *Phys. A*, 157:826–838, 1989.

16. J. Lekner. Summation of Coulomb fields in computer-simulated disordered systems. *Phys. A*, 176:485–498, 1991.

17. J. Lekner. Coulomb forces and potentials in systems with an orthorhombic unit cell. *Mol. Sim.*, 20(6):357, 1998.

18. R. Sperb. Extension and simple proof of Lekner's summation formula for Coulomb forces. *Mol. Sim.*, 13:189–193, 1994.

19. R. Sperb. An alternative to Ewald sums, part I: Identities for sums. *Mol. Sim.*, 20:179–200, 1998.

20. R. Sperb. An alternative to Ewald sums, part 2: The Coulomb potential in a periodic system. *Mol. Sim.*, 22:199–212, 1999.

21. R. Strebel and R. Sperb. An alternative to Ewald Sums. Part 3: Implementation and results. *Mol. Sim.*, 27:61–74, 2001.

22. A. Arnold and C. Holm. A novel method for calculating electrostatic interactions in 2D periodic slab geometries. *Chem. Phys. Lett.*, 354:324–330, 2002.

23. A. Arnold and C. Holm. MMM2D: A fast and accurate summation method for electrostatic interactions in 2D slab geometries. *Comput. Phys. Commun.*, 148:327–348, 2002.

24. B. Hess, C. Kutzner, D. van der Spoel, and E. Lindahl. Gromacs 4: Algorithms for highly efficient, load-balanced, and scalable molecular simulation. *J. Chem. Theory Comput.*, 4(3):435–447, March 2008.

25. H. J. Limbach, A. Arnold, B. A. Mann, and C. Holm. ESPResSo—An extensible simulation package for research on soft matter systems. *Comput. Phys. Commun.*, 174:704–727, 2006.

26. A. H. Widmann and D. B. Adolf. A comparison of Ewald summation techniques for planar surfaces. *Comput. Phys. Commun.*, 107:167–186, 1997.

27. M. Mazars. Lekner summations and Ewald summations for quasi-two-dimensional systems. *Mol. Phys.*, 103:1241–1260, 2005.

28. D. Sølvason, H. G. Petersen, J. Kolafa, and J. W. Perram. A rigorous comparison of the Ewald method and the fast multipole method in two dimensions. *Comput. Phys. Commun.*, 87:307–318, 1995.

29. E. L. Pollock and J. Glosli. Comments on P³M, FMM, and the Ewald method for large periodic Coulomb systems. *Comput. Phys. Commun.*, 95:93–110, 1996.

30. C. Lee, W. Yang, and R. G. Parr. Development of the Colle-Salvetti correlation-energy formula into a functional of the electron density. *Phys. Rev. B*, 37:785–789, 1988.

31. J. P. Perdew. *Electronic Structure of Solids*. Akademie Verlag: Berlin, Germany, 1991.

32. J. P. Perdew, K. Burke, and M. Ernzerhof. Generalized gradient approximation made simple. *Phys. Rev. Lett.*, 77:3865–3868, 1996.

33. J. C. Sancho-Garca and J. Cornil. Assessment of recently developed exchange-correlation functionals for the description of torsion potentials in π-conjugated molecules. *J. Chem. Phys.*, 121:3096–3101, 2004.

34. J. P. Perdew, K. A. Jackson, M. R. Pederson, D. J. Singh, and C. Fiolhais. Atoms, molecules, solids, and surfaces: Applications of the generalized gradient approximation for exchange and correlation. *Phys. Rev. B*, 46:6671–6687, 1992.

35. W. Andreoni, A. Curioni, and H. Grönbeck. Density functional theory approach to thiols and disulfides on gold: Au (111) surface and clusters. *Int. J. Quant. Chem.*, 80:598–608, 2000.

36. J. H. Harding, D. M. Duffy, M. L. Sushko, P. M. Rodger, D. Quigley, and J. A. Elliott. Computational techniques at the organic-inorganic interface in biomineralization. *Chem. Rev.*, 108:4823–4854, 2008.

37. J. Nara, S. Higai, Y. Morikawa, and T. Ohno. Density functional theory investigation of benzenethiol adsorption on Au(111). *J. Chem. Phys.*, 120:6705–6711, 2004.

38. D. Fischer, A. Curioni, and W. Andreoni. Decanethiols on gold: The structure of self-assembled monolayers unraveled with computer simulations. *Langmuir*, 19:3567–3571, 2003.

39. N. Jacobi and Gy. Csanak. Dispersion forces at arbitrary distances. *Chem. Phys. Lett.*, 30:367–372, 1975.

40. F. B. van Duijnevelt, J. G. C. M. van Duijneveldt-van de Rijdt, and B. P. van Eijck. Transferable ab initio intermolecular potentials. 1. Derivation from methanol dimer and trimer calculations. *J. Phys. Chem. A*, 103:9872–9882, 1999.

41. S. Hug, G. K. Hunter, H. Goldberg, and M. Karttunen. Ab initio simulations of peptide-mineral interactions. *Phys. Proc.*, 4:51–60, 2010.
42. D. M. Duffy and J. H. Harding. Modeling the properties of self-assembled monolayers terminated by carboxylic acids. *Langmuir*, 21:3850–3857, 2005.
43. M. J. Mandell. Crystal nucleation in a three-dimensional lennard-jones system: A molecular dynamics study. *J. Chem. Phys.*, 64:3699, 1976.
44. W. Swope and H. Andersen. 106-Particle molecular-dynamics study of homogeneous nucleation of crystals in a supercooled atomic liquid. *Phys. Rev. B*, 41:7042–7054, 1990.
45. J. N. Cape, J. L. Finney, and L. V. Woodcock. An analysis of crystallization by homogeneous nucleation in a 4000-atom soft-sphere model. *J. Chem. Phys.*, 75:2366, 1981.
46. P. Beaucage and N. Mousseau. Nucleation and crystallization process of silicon using the stillinger-weber potential. *Phys. Rev. B*, 71:094102, 2005.
47. M. Matsumoto, S. Saito, and I. Ohmine. Molecular dynamics simulation of the ice nucleation and growth process leading to water freezing. *Nature*, 416:409–413, 2002.
48. L. Vrbka and P. Jungwirth. Homogeneous freezing of water starts in the subsurface. *J. Phys. Chem. B*, 110:18126–18129, 2006.
49. Q. Wang, N. Shah, J. Zhao, C. Wang, C. Zhao, L. Liu, L. Li, F. Zhou, and J. Zheng. Structural, morphological, and kinetic studies of -amyloid peptide aggregation on self-assembled monolayers. *Phys. Chem. Chem. Phys.*, 13:15200, 2011.
50. J. Zhao, Q. Wang, G. Liang, and J. Zheng. Molecular dynamics simulations of low-ordered alzheimer β-amyloid oligomers from dimer to hexamer on self-assembled monolayers. *Langmuir*, 27:14876–14887, 2011.
51. J.-W. Shen, T. Wu, Q. Wang, and H.-H. Pan. Molecular simulation of protein adsorption and desorption on hydroxyapatite surfaces. *Biomaterials*, 29:513–532, 2008.
52. G. S. Baht, J. O'Young, A. Borovina, H. Chen, C. E. Tye, M. Karttunen, G. A. Lajoie, G. K. Hunter, and H. A. Goldberg. Phosphorylation of Ser136 is critical for potent bone sialoprotein-mediated nucleation of hydroxyapatite crystals. *Biochem. J.*, 428:385–395, 2010.
53. J. O'Young, Y. Liao, Y. Xiao, J. Jalkanen, G. Lajoie, M. Karttunen, H. A. Goldberg, and G. K. Hunter. Matrix gla protein inhibits ectopic calcification by a direct interaction with hydroxyapatite crystals. *J. Am. Chem. Soc.*, 133:18406–18412, 2011.
54. B. Grohe, S. Hug, A. Langdon, J. Jalkanen, K. A. Rogers, H. A. Goldberg, M. Karttunen, and G. K. Hunter. Mimicking the biomolecular control of calcium oxalate monohydrate crystal growth: Effect of contiguous glutamic acids. *Langmuir*, 28:12182–12190, 2012.
55. B. Grohe, J. O'Young, D. Andrei Ionescu, G. Lajoie, K. A. Rogers, M. Karttunen, H. A. Goldberg, and G. K. Hunter. Control of calcium oxalate crystal growth by face-specific adsorption of an osteopontin phosphopeptide. *J. Am. Chem. Soc.*, 129:14946–14951, 2007.
56. G. K. Hunter, J. O'Young, B. Grohe, M. Karttunen, and H. A. Goldberg. The flexible polyelectrolyte hypothesis of protein-biomineral interaction. *Langmuir*, 26:18639–18646, 2010.
57. P. V. Azzopardi, J. O'Young, G. Lajoie, M. Karttunen, H. A. Goldberg, and G. K. Hunter. Roles of electrostatics and conformation in protein-crystal interactions. *PLoS One*, 5:e9330, 2010.
58. S. Hug, B. Grohe, J. Jalkanen, B. Chan, B. Galarreta, K. Vincent, F. Lagugné-Labarthet, G. Lajoie et al. Mechanism of inhibition of calcium oxalate crystal growth by an osteopontin phosphopeptide. *Soft Matter*, 8:1226, 2012.
59. A. D. Mackerell, Jr. Empirical force fields for biological macromolecules: Overview and issues. *J. Comput. Chem.*, 25:1584–1604, 2004.
60. E. Marinari and G. Parisi. Simulated tempering: A new Monte Carlo scheme. *Europhys. Lett.*, 19:451, July 1992.
61. S. Park and V. Pande. Choosing weights for simulated tempering. *Phys. Rev. E*, 76:016703, 2007.
62. U. H. E. Hansmann. Parallel tempering algorithm for conformational studies of biological molecules. *Chem. Phys. Lett.*, 281:140–150, 1997.
63. Y. Sugita and Y. Okamoto. Replica-exchange molecular dynamics method for protein folding. *Chem. Phys. Lett.*, 314:141–151, 1999.
64. G. M. Torrie and J. P. Valleau. Nonphysical sampling distributions in Monte Carlo free-energy estimation: Umbrella sampling. *J. Comput. Phys.*, 23:187–199, 1977.
65. H. Grubmüller, B. Heymann, and P. Tavan. Ligand binding: Molecular mechanics calculation of the streptavidin-biotin rupture force. *Science*, 271:997–999, 1996.
66. M. Sotomayor and K. Schulten. Single-molecule experiments in vitro and in silico. *Science*, 316:1144–1148, 2007.

67. A. Laio and M. Parrinello. Escaping free-energy minima. *Proc. Natl. Acad. Sci. U.S.A.*, 99:12562–12566, 2002.

68. A. Barducci, G. Bussi, and M. Parrinello. Well-tempered metadynamics: A smoothly converging and tunable free-energy method. *Phys. Rev. Lett.*, 100:020603, 2008.

69. C. Abrams and G. Bussi. Enhanced sampling in molecular dynamics using metadynamics, replica-exchange, and temperature-acceleration. *Entropy*, 16:163–199, December 2013.

70. S. Kumar, J. M. Rosenberg, D. Bouzida, R. H. Swendsen, and P. A. Kollman. The weighted histogram analysis method for free-energy calculations on biomolecules. I. The method. *J. Comp. Chem.*, 13:1011–1021, 1992.

71. P. J. Steinhardt, D. R. Nelson, and M. Ronchetti. Icosahedral bond orientational order in supercooled liquids. *Phys. Rev. Lett.*, 47:1297–1300, November 1981.

72. D. Quigley and P. M. Rodger. Free energy and structure of calcium carbonate nanoparticles during early stages of crystallization. *J. Chem. Phys.*, 128:221101, 2008.

73. G. A. Tribello, F. Bruneval, C. C. Liew, and M. Parrinello. A molecular dynamics study of the early stages of calcium carbonate growth. *J. Phys. Chem. B*, 113:11680–11687, 2009.

74. P. Raiteri, J. D. Gale, D. Quigley, and P. Mark Rodger. Derivation of an accurate force-field for simulating the growth of calcium carbonate from aqueous solution: A new model for the calcite-water interface. *J. Phys. Chem. C*, 114:5997–6010, 2010.

75. P. Raiteri and J. D. Gale. Water is the key to nonclassical nucleation of amorphous calcium carbonate. *J. Am. Chem. Soc.*, 132:17623–17634, 2010.

76. M. Salvalaglio, T. Vetter, F. Giberti, M. Mazzotti, and M. Parrinello. Uncovering molecular details of urea crystal growth in the presence of additives. *J. Am. Chem. Soc.*, 134:17221–17233, 2012.

77. F. Giberti, G. A. Tribello, and M. Parrinello. Transient polymorphism in NaCl. *J. Chem. Theory Comp.*, 9:2526–2530, 2013.

78. F. Trudu, D. Donadio, and M. Parrinello. Freezing of a Lennard-Jones fluid: From nucleation to spinodal regime. *Phys. Rev. Lett.*, 97:105701, 2006.

79. D. Donadio, P. Raiteri, and M. Parrinello. Topological defects and bulk melting of hexagonal ice. *J. Phys. Chem. B*, 109:5421–5424, 2005.

80. D. Quigley and P. M. Rodger. Metadynamics simulations of ice nucleation and growth. *J. Chem. Phys.*, 128:154518, 2008.

81. D. Quigley, P. M. Rodger, C. L. Freeman, J. H. Harding, and D. M. Duffy. Metadynamics simulations of calcite crystallization on self-assembled monolayers. *J. Chem. Phys.*, 131:094703, 2009.

82. M. Hoefling, F. Iori, S. Corni, and K.-E. Gottschalk. Interaction of amino acids with the Au(111) surface: Adsorption free energies from molecular dynamics simulations. *Langmuir*, 26:8347–8351, 2010.

83. F. Wang, S. J. Stuart, and R. A. Latour. Calculation of adsorption free energy for solute-surface interactions using biased replica-exchange molecular dynamics. *Biointerphases*, 3:9–18, 2008.

84. C. P. O'Brien, S. J. Stuart, D. A. Bruce, and R. A. Latour. Modeling of peptide adsorption interactions with a poly(lactic acid) surface. *Langmuir*, 24:14115–14124, 2008.

85. P. Liu, B. Kim, R. A. Friesner, and B. J. Berne. Replica exchange with solute tempering: A method for sampling biological systems in explicit water. *Proc. Natl. Acad. Sci. U.S.A.*, 102:13749–13754, 2005.

86. L. Wang, R. A. Friesner, and B. J. Berne. Replica exchange with solute scaling: A more efficient version of replica exchange with solute tempering (rest2). *J. Phys. Chem. B*, 115:9431–9438, 2011.

87. L. B. Wright and T. R. Walsh. Efficient conformational sampling of peptides adsorbed onto inorganic surfaces: Insights from a quartz binding peptide. *Phys. Chem. Chem. Phys.*, 15:4715–4726, 2013.

88. M. Deighan and J. Pfaendtner. Exhaustively sampling peptide adsorption with metadynamics. *Langmuir*, 29:7999–8009, 2013.

89. R. T. Downs and M. Hall-Wallace. The american mineralogist crystal structure database. *Am. Mineral.*, 88:247–250, 2003.

90. M. Charles, S. Veesler, and Bonneté F. Mpcd: A new interactive on-line crystallization data bank for screening strategies. *Acta Crystallograph. D*, 62:1311–1318, 2006.

91. F. Bonneté. Colloidal approach analysis of marseille protein crystallization database for protein crystallization strategies. *Cryst. Growth Des.*, 7:2176–2181, 2007.

92. H. M. Berman, J. Westbrook, Z. Feng, G. Gilliland, T. N. Bhat, H. Weissig, I. N. Shindyalov, and P. E. Bourne. The protein data bank. *Nucleic Acids Res.*, 28:235–242, 2000.

93. H. M. Berman, K. Henrick, and H. Nakamura. Announcing the worldwide protein data bank. *Nat. Struct. Biol.*, 10:98, 2003.

94. F. H. Allen. The cambridge structural database: A quarter of a million crystal structures and rising. *Acta Crystallograph.*, B58:380–388, 2002.

95. K. Momma and F. Izumi. Vesta 3 for three-dimensional visualization of crystal, volumetric and morphology data. *J. Appl. Crystallograp.*, 44:1272–1276, 2011.
96. W. Humphrey, A. Dalke, and K. Schulten. Vmd—Visual molecular dynamics. *J. Mol. Graph.*, 14:33–38, 1996.
97. I.T. Todorov and W. Smith. dl _ poly _ 3: The ccp5 national uk code for molecular-dynamics simulations. *Phil. Trans. Ser. A, Math., Phys. Eng. Sci.*, 362:1835–1852, 2004.
98. N.M. O'Boyle, M. Banck, C.A. James, C. Morley, T. Vandermeersch, and G.R. Hutchison. Open babel: An open chemical toolbox. *J. Cheminform.*, 3:33, 2011.
99. M. C. F. Magalhães, P. A. A. P. Marques, and R. N. Correia. Calcium and magnesium phosphates: Normal and pathological mineralization, in *Biomineralization: Medical Aspects of Solubility*, eds E. Konigsberger and L. C. Konigsberger. John Wiley & Sons, Weinheim, Germany, 2006.
100. S. Hug, B. Grohe, J. Jalkanen, B. Chan, B. Galarreta, K. Vincent, F. Lagugne-Labarthet, G. Lajoie, H. A. Goldberg, M. Karttunen, and G. K. Hunter. Mechanism of inhibition of calcium oxalate crystal growth by an osteopontin phosphopeptide. *Soft Matter*, 8:1226–1233, 2012.
101. B. Grohe, J. O'Young, A. Langdon, M. Karttunen, H. A. Goldberg, and G. K. Hunter. Citrate modulates calcium oxalate crystal growth by face-specific interactions. *Cells Tissues Organs*, 194:176–181, 2011.
102. B. Grohe, J. O'Young, A. Ionescu, G. Lajoie, K. A. Rogers, M. Karttunen, H. A. Goldberg, and G. K. Hunter. Control of calcium oxalate crystal growth by face–specific adsorption of an osteopontin phosphopeptide. *J. Am. Chem. Soc.*, 129:14946–14951, 2007.
103. J. W. Eaton, D. Bateman, and S. Hauberg. *GNU Octave Version 3.0.1 Manual: A High-Level Interactive Language for Numerical Computations*. CreateSpace Independent Publishing Platform, 2009.
104. M. I. Kay, R. A. Young, and A. S. Posner. Crystal structure of hydroxyapatite. *Nature*, 204:1050–1052, 1964.
105. R. K. Krishnaraju, T. C. Hart, and T. K. Schleyer. Comparative genomics and structure prediction of dental matrix proteins. *Adv. Dental Res.*, 17:100–103, 2003.
106. Q. Q. Hoang, F. Sicheri, A. J. Howard, and D. S. C. Yang. Bone recognition mechanism of porcine osteocalcin from crystal structure. *Nature*, 425:977–980, 2003.
107. P. A. Price, J. W. Poser, and N. Raman. Primary structure of the gamma-carboxyglutamic acid-containing protein from bovine bone. *Proc. Natl. Acad. Sci. U.S.A.*, 73:3374–3375, 1976.
108. R. W. Romberg, P. G. Werness, B. Lawrence Riggs, and K. G. Mann. Inhibition of hydroxyapatite-crystal growth by bone-specific and other calcium-binding proteins. *Biochemistry*, 25:1176–1180, 1986.
109. G. K. Hunter, P. V. Hauschka, A. R. Poole, L. C. Rosenberg, and H. A. Goldberg. Nucleation and inhibition of hydroxyapatite formation by mineralized tissue proteins. *Biochem. J.*, 317(Pt 1):59–64, 1996.
110. P. V. Hauschka and S. A. Carr. Calcium-dependent α-helical structure in osteocalcin. *Biochemistry*, 21:2538–2547, 1982.
111. V. Hornak, R. Abel, A. Okur, B. Strockbine, A. Roitberg, and C. Simmerling. Comparison of multiple Amber force fields and development of improved protein backbone parameters. *Proteins Struct., Funct., Bioinform.*, 65:712–725, 2006.
112. K. Lindorff-Larsen, S. Piana, K. Palmo, P. Maragakis, J. L. Klepeis, R. O. Dror, and D. E. Shaw. Improved side-chain torsion potentials for the Amber ff99SB protein force field. *Proteins Struct., Funct., Bioinform.*, 78:1950–1958, 2010.
113. R. B. Best and G. Hummer. Optimized molecular dynamics force fields applied to the helix-coil transition of polypeptides. *J. Phys. Chem. B*, 113:9004–9015, 2009.
114. G. Bussi, D. Donadio, and M. Parrinello. Canonical sampling through velocity rescaling. *J. Chem. Phys.*, 126:014101, 2007.
115. M. Bonomi, D. Branduardi, G. Bussi, C. Camilloni, D. Provasi, P. Raiteri, D. Donadio et al. PLUMED: A portable plugin for free-energy calculations with molecular dynamics. *Comput. Phys. Commun.*, 180:1961–1972, 2009.

7 Mesoscale Computational Techniques in Interfacial Science
Lattice Boltzmann Method

Mohammad Taeibi Rahni, Mohsen Karbaschi, Mehran Kiani,
Majid Haghshenas, Massoud Rezavand, and Reinhard Miller

CONTENTS

7.1 INTRODUCTION

A wide range of important phenomena in fluid dynamics contain multiphase flows, and thus such flows are of great importance to many scientists and engineers. They appear in applications, such as marine hydrodynamics and chemical, mineral, industrial, natural, and pollution control processes. Some examples of such flows include spray painting, spray combustion, boiling, coal slurry transport, emulsion, foam, cavitation, sedimentation, fluidized bed, rain, snow, and volcanic rock motion.

In general, multiphase flows are divided into two distinct categories: the ones with small interfaces (such as spray of liquid drops in a gas) and the ones with large interfaces (such as the flow of a relatively large bubble in a liquid boundary layer). Particularly, physicochemical properties and hydrodynamics make the dynamics of liquid–fluid interfaces considerably complex.

When dealing with situations where there are many small particles (solid, liquid, or gas) flowing along with a gas or liquid continuous phase, three different situations may occur: (a) void fraction is very high, that is, there are plenty of particles scattered all over the media; (b) void fraction is within its intermediate range; and (c) void fraction is high. Note, in all these three cases, one needs to somehow rely on statistical approaches (since tracking of so many particles individually is almost impossible). For case (a), there is a mathematical model called mixture, in which molecular properties of the continuous fluid are redefined and then the problem is treated very similar to a single-phase flow problem, while for cases (b) and (c), Eulerian–Eulerian and Eulerian–Lagrangian models are used, respectively. Note, Eulerian and Lagrangian refer here to different representations of the governing equations for different phases of the two-phase flow.

If we are dealing with multiphase flows in which large interfaces exist, our mathematical modeling is quite different. In such situations, the main concern is to model the liquid–fluid interface deformation and transport.

There have been numerous numerical works performed regarding complex flow problems for which analytical results are not necessarily available. Particularly, in the past three decades, most researchers have used conventional computational fluid dynamics [CFD; solving the discretized full Navier–Stokes (*FNS*) equations]. As far as multiphase flow is concerned, in conventional CFD techniques of handling large interfaces, three popular approaches are (1) front tracking, (2) volume of fluid (VOF), and (3) level set (see Chapter 9 for details of these approaches). Note besides the importance of capturing the motion and deformation of the interface correctly, one needs to be also concerned about obtaining them as sharp as possible (with least numerical dissipation).

Fortunately, in the past one and a half decades, new computational approaches, such as the lattice Boltzmann method (LBM) and the smoothed particle hydrodynamics (SPH) method, have shown to overcome most of the difficulties with conventional CFD approaches. Such methods are sometimes called meshless (or mesh-free). In such approaches, the fluid is hypothetically discretized to many small particles (instead of the media containing it). Then, the mathematical modeling is somehow similar to the ones used in molecular dynamics (MD), wherein the particle collisions are also taken into account. Note the main difference between SPH and LBM is that the first one uses Lagrangian form of FNS equations, while approximated Boltzmann equation is used in the latter.

It seems that LBM has been found to be very useful and has become very popular in simulation of many fluid flows. According to Ref. [1],

> LBM is easy to apply for complex domains, it is easy to treat multiphase and multi-component flows without a need to trace the interfaces between different phases, it can be naturally adapted to parallel processors computing, there is no need to solve Laplace equation at each time step to satisfy continuity equation of incompressible, unsteady flows, as it is in solving Navier-Stokes (N.S.) equations. Also, it can handle a problem in micro- and micro-scales with reliable accuracy. However, it needs more computer memory compared with N.S. solvers, which is not a big constraint.

Even though LBM is not yet fully developed and still has some drawbacks for certain flow physics, it seems at the present time that for simulation of complex liquid–fluid interfaces, it is a relatively accurate and simple tool.

Of course, more serious enhancements in computational techniques are still needed to achieve the level of accuracy required for many existing applications involving liquid–fluid interfaces. This is obviously because of relatively fast technological improvements. It is interesting to note that this requirement of need for better computational approaches will never stop due to the nonstop technological advancements. Also, it seems necessary for scientists and engineers to work closer and exchange their different perspectives related to interfacial problems.

7.2   BASIC LBM

As mentioned earlier, LBM is a well-known technique (based on mesoscopic ideas) in dealing with many important scientific and engineering problems and has attracted a great deal of attention, especially among fluid dynamic specialists [2,3].

Ludwig Boltzmann with his kinetic theory of gases explained the properties of dilute gases by analyzing the elementary collision processes between pairs of molecules [4]. He came up with an important general equation of motion based on the dynamics of fluid molecules.

LBM can be used to simulate single- or multiphase phenomena effectively (because of incorporation of intermolecular interactions, due to the kinetic nature of the Boltzmann equation). This for sure distinguishes LBM from other numerical methods. The kinetic nature of LBM equips it with advantages required by mesoscopic physics, such as relatively easier implementation of boundary conditions (especially for complex geometries) and more straightforward parallelization of numerical simulations [5]. These characteristics have made LBM quite suitable for simulation of interfacial phenomena [6].

As mentioned in Ref. [5], in order to reduce statistical noise in lattice gas automaton (LGA) simulations, LBM was first proposed by McNamara and Zanetti [7]. As a simplified form of MD method, the original LGA method was introduced by Frisch et al. [8]. Note particle representation is the main difference between LBM and LGA. In LBM, real numbers are used to identify the presence or the absence of particles, while in LGA only 0 and 1 are used for this purpose.

As it can be seen in any introductory LBM book, the foundations of LBM are based on an equilibrium state and that Boltzmann defined a distribution function for this state. Then, he minimized the enthalpy function to obtain a distribution function, which would be a function of flow variables. Of course, this distribution function was logarithmic and computationally too expensive. Thus, the four first terms of its Taylor series expansion are usually used as

$$f_i^{eq}(x) = w_i \rho(x) \left[1 + 3 \frac{e_i \cdot u}{c^2} + \frac{9}{2} \frac{(e_i \cdot u)^2}{c^4} - \frac{3}{2} \frac{u^2}{c^2} \right], \qquad (7.1)$$

where
    superscript eq stands for equilibrium
    ρ is the density
    u is the velocity of fluid at each point.

Note particles are only present at the grid nodes and they can get only a few distinct discrete velocities denoted by e_i.

Also note i represents the direction numbers and w_i is the weighing numbers, which are 4/9 for $i = 0$; 1/9 for 1, 2, 3, and 4; and 1/36 for 5, 6, 7, and 8 (for a 2D grid with nine distinct velocities; D2Q9).

Figure 7.1 illustrates the related discrete space and velocities for a D2Q9 case. In other words, we have nine values for the equilibrium distribution function, corresponding to the nine discrete velocities.

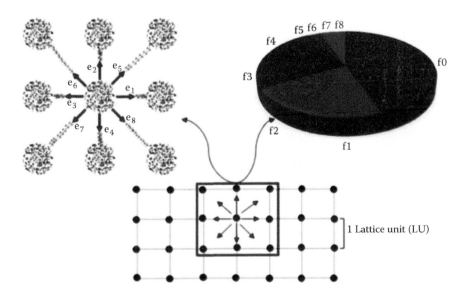

FIGURE 7.1   D2Q9 model showing discrete space and velocities.

The equilibrium distribution function can be estimated by density and velocity of the flow. On the other hand, macroscopic flow variables are obtained using the distribution function, for example,

$$\rho = \sum_i f_i, \tag{7.2}$$

$$\boldsymbol{u} = \frac{1}{\rho} \sum_i f_i \boldsymbol{e}_i. \tag{7.3}$$

LBM algorithm consists of three different steps as (1) *collision*, (2) *propagation*, and (3) *macroscopic quantity (flow variable) calculation*. The standard (first order) discretized Boltzmann equation is

$$f_i\left(\boldsymbol{x} + e_i \delta x, t + \delta t\right) - f_i\left(\boldsymbol{x}, t\right) = \frac{f_i\left(\boldsymbol{x}, t\right) - f_i^{eq}\left(\boldsymbol{x}, t\right)}{\tau}, \tag{7.4}$$

where
   τ is called the relaxation time, which is proportional to the time required by particles to reach equilibrium condition
   δx is the lattice unit and δt is the time step

Thus, distribution function can be estimated at each time step, using the distribution function at equilibrium and the distribution function of previous time step. Then, the macroscopic quantities are obtained, using Equations 7.2 and 7.3. Note the collision term in Equation 7.4 is called *Bhatnagar-Gross-Krook (BGK) collision operator* and implies that collision leads to the equilibrium state.

7.3   LBM FOR INTERFACIAL PROBLEMS

The first multiphase LBM model was proposed by Rothman and Keller [9] in their LGA simulations, which was later developed by Gunstensen et al. [10] for LBM. Such models are generally categorized in four different classes, including color gradient models, interparticle potential models, mean-field theory models, and free energy models. In the following sections, the basics of these models are reviewed.

7.3.1   Color Gradient Models

Color gradient model of Gunstensen et al. [10], followed by Alexander et al. [11], is known as the first LBM interfacial model. In this model, two distribution functions are used: one for the blue fluid (f_{bi}) and one for the red (f_{ri}). A mixture distribution function, which is the combination of red and blue ones, is utilized for simulation as

$$f_i(\boldsymbol{x} + \boldsymbol{e}_i \delta t, t + \delta t) - f_i(\boldsymbol{x}, t) = \Omega_i, \tag{7.5}$$

where Ω_i is defined as

$$\Omega_i = \Omega_i^c + \Omega_i^p, \tag{7.6}$$

where
   Ω_i^c represents the collision part
   Ω_i^p represents the effects of interfacial tension

The macroscopic values of density and velocity for each fluid are then defined as

$$\rho_k = \sum_i f_{ki}, \quad \rho_k u_k = \sum_i e_i f_{ki}, \quad k = r, b. \tag{7.7}$$

Also, the density and velocity of the mixture is calculated as

$$\rho = \rho_r + \rho_b, \quad \rho u = \rho_r u_r + \rho_b u_b. \tag{7.8}$$

Gunstensen et al. [10] modeled particle collisions similar to the modeling introduced earlier by McNamara and Zanetti [7] in LGA and later by Alexander et al. [11].

In order to determine Ω_i^p due to interfacial effects, the following relation has been used:

$$\Omega_i^p = A|F|\cos(2\theta_i), \tag{7.9}$$

where F is the local color gradient defined by the order variable (η) as

$$F(x, t) = \sum_i e_i \eta(x + e_i \delta t, t), \tag{7.10}$$

and θ_i is the angle between F and e_i. The order variable (η) is defined as

$$\eta(x, t) = \rho_r(x, t) - \rho_b(x, t). \tag{7.11}$$

It is obvious that F is zero everywhere, except at the interface.

In this step, $f_i'' = f_i' + \Omega_i^p$ is calculated. Finally, after determination of collision and interfacial parts, the distribution function of f_i is calculated. However, another step (called recoloring) is needed to adjust red and blue color distribution functions. In other words, in order to separate the two phases and maintain the interface, recoloring (segregation) step is required. Thus, the following maximum problem is solved for post recoloring the values of color distribution functions (R_i'', B_i'') as

$$W(R_i', B_i') = \max\left[\sum_i (R_i'' - B_i'')e_i\right] \cdot F, \tag{7.12}$$

where the following constraints must be considered:

$$\rho_r'' = \sum_i R_i'' = \rho_r, \quad B_i'' = f_i - R_i''. \tag{7.13}$$

In other words, the local color gradient momentum ($h = \sum_i (R_i'' - B_i'')e_i$) is aligned along the color gradient (F). In fact, the recoloring step forces each color distribution toward the same color by maximizing $-h \cdot F$.

Following the original model proposed by Gunstensen et al. [10], Grunau et al. [12] modified the color gradient method by performing collisions for each color distribution functions as

$$R_i(\boldsymbol{x}+\boldsymbol{e}_i\delta t,t+\delta t)-R_i(\boldsymbol{x},t)=\Omega_{ri}+\Omega_{ri}^p, \tag{7.14}$$

$$B_i(\boldsymbol{x}+\boldsymbol{e}_i\delta t,t+\delta t)-B_i(\boldsymbol{x},t)=\Omega_{bi}+\Omega_{bi}^p, \tag{7.15}$$

where Ω_{ri}, Ω_{bi}, and Ω_{ki}^p are defined, respectively, as

$$\Omega_{ri}=-1/\tau_r\left[R_i-R_i^{eq}\right], \tag{7.16}$$

$$\Omega_{bi}=-1/\tau_b\left[B_i-B_i^{eq}\right], \tag{7.17}$$

$$\Omega_{ki}^p=\frac{A_k}{2}|\boldsymbol{F}|\left[\frac{\boldsymbol{e}_i\cdot\boldsymbol{F}}{|\boldsymbol{F}|}-\frac{1}{2}\right],\quad k=r,b. \tag{7.18}$$

This model still needs the recoloring step to maintain the sharpness of the interface. Some later studies have been conducted on recoloring step, for example, d'Ortona et al. [13] have presented a new method for recoloring step.

Thereafter, this model has been further investigated and was improved significantly. Lishchuk [14] used the continuum surface force (CSF) model, originally introduced by Brackbill et al. [15], and implemented surface tension as a body force in momentum conservation equation. Lishchuk applied CSF model for previous immiscible LBM models and substantially reduced the artifacts of color gradient model proposed by Gunstensen et al. [14].

This idea was developed further by Halliday et al. [16], who represented the interface by the continuum approximation. They applied the CSF body force (Φ) exactly across the interface according to the following equation:

$$\Phi=\frac{3}{2}k\gamma w_i\nabla\left(V_F\right)\boldsymbol{n}_i\cdot e_i, \tag{7.19}$$

where
k is the local interface curvature
γ is the surface tension coefficient
w_i is the weighting factor associated with i direction
$\boldsymbol{n}_i$ is the normal vector to the interface
V_F is the volume fraction of fluid, defined as

$$V_F\left(x,t\right)=\frac{\rho_r\left(x,t\right)-\rho_b\left(x,t\right)}{\rho_r\left(x,t\right)+\rho_b\left(x,t\right)}. \tag{7.20}$$

Besides, Latva-Kokko and Rothman [17], concerning the lattice pinning in color gradient models, introduced the diffusion scheme. They changed the recoloring step and allowed generation of wider interfaces. This way, color distribution is symmetric around the color gradient and lattice pining problem is diminished considerably. In this scheme, postrecoloring distribution functions are as follows:

$$R_i=\frac{\rho_r}{\rho_r+\rho_b}f_i+\beta\frac{\rho_r\rho_b}{\left(\rho_r+\rho_b\right)^2}f_i^{eq,0}\cos\theta_i, \tag{7.21}$$

$$B_i = \frac{\rho_b}{\rho_r + \rho_b} f_i - \beta \frac{\rho_r \rho_b}{(\rho_r + \rho_b)^2} f_i^{eq,0} \cos\theta_i,$$ (7.22)

where

β is the parameter that determines the tendency of two fluids to be separated

θ_i is the angle between color gradient F and direction e_i

$f_i^{eq,0}$ is the zero velocity equilibrium distribution function [17]

7.3.2 Interparticle Potential Models

LBM was first developed for perfect gases and in order to improve its capability, Shan and Chen [18,19] proposed a model in which interparticle attractions and also other body forces (e.g., gravitational force) were considered. The effects of these forces are added to the velocity after each time step, and the modified velocity is used to calculate the equilibrium distribution function. Thus, the velocity would be

$$u^{eq} = u + \Delta u = u + \frac{\tau F}{\rho},$$ (7.23)

where

F is the summation of interparticle attractions and other external forces

u^{eq} is the modified velocity

Shan and Chen proposed the interparticle forces as presented in the equation that follows, in which the equation of state approaches the van der Waals equation of state as

$$F(x,t) = -G\Psi(x,t) \sum_i w_i \Psi(x + e_i \delta t, t) e_i,$$ (7.24)

where G is the strength of interparticle interactions and potential function Ψ is a function of density. Different potential functions are used, while one of the well-known potential functions is

$$\Psi(\rho) = \Psi_0 e^{-\rho_0/\rho},$$ (7.25)

where ρ_0 and Ψ_0 are arbitrary and depend on fluid characteristics. Equation 7.24 yields the equation of state as

$$P = \rho RT + \frac{GRT}{2}(\Psi(\rho))^2,$$ (7.26)

where the second term indicates the real gas effects. Since the attraction forces between particles are limited to the nearest neighboring particles (and the repulsion between particles are neglected), special attention is needed when using this model. Note Shan–Chen model leads to inconsistent thermodynamics, unless a particular equation of state is used [20].

7.3.3 Mean-Field Models

The Boltzmann equation with interparticle forces is used to describe nonideal fluids [21,22] as

$$\frac{Df}{Dt} \equiv \frac{\partial f}{\partial t} + \xi \cdot \nabla f + F \cdot \nabla_\xi f = -\frac{f - f^{eq}}{\lambda},$$ (7.27)

where $\lambda = \delta t \cdot \tau$ and the Maxwell–Boltzmann distribution function (f^{eq}) depends on the flow characteristics as

$$f^{eq} = \frac{\rho}{(2\pi RT)^{D/2}} \exp\left(-\frac{(\xi - u)^2}{2RT}\right). \tag{7.28}$$

One can estimate $F \cdot \nabla_\xi f$ as

$$F \cdot \nabla_\xi f \simeq F \cdot \nabla_\xi f^{eq} = -\frac{F \cdot (\xi - u)}{RT} f^{eq}. \tag{7.29}$$

Equations 7.27 and 7.29 are combined to obtain

$$\frac{\partial f}{\partial t} + \xi \cdot \nabla f = -\frac{f - f^{eq}}{\lambda} + \frac{(F + G) \cdot (\xi - u)}{\rho RT} f^{eq}, \tag{7.30}$$

where

G is the gravity

F is the effective molecular interaction force expressed as [22]

$$F = -\nabla \psi + F_s, \tag{7.31}$$

where ψ is a function of density as

$$\psi(\rho) = b\rho^2 RT\chi - a\rho^2. \tag{7.32}$$

Here, $b = 2\pi\sigma^3/3m$, where σ is the effective diameter and m is the mass of a single molecule. Parameter a depends on the interparticle pair-wise potential [11] and parameter χ is defined as [21]

$$\chi(\rho) = 1 + \frac{5}{8}b\rho + 0.2869(b\rho)^2 + 0.1103(b\rho)^3 + \cdots. \tag{7.33}$$

Equation 7.30 is integrated during a single time step and is assumed that the BGK operator has a constant value over each time step as

$$\int_t^{t+\delta t}\left(\frac{\partial f}{\partial t} + \xi \cdot \nabla f\right)dt = \int_t^{t+\delta t} -\frac{f - f^{eq}}{\lambda} dt + \int_t^{t+\delta t}\frac{(F + G) \cdot (\xi - u)}{\rho RT} f^{eq} dt. \tag{7.34}$$

The second integration is estimated, using a trapezoidal rule, as

$$f(x + \xi\delta t, t + \delta t) - f(x, t) = -\frac{f(x, t) - f^{eq}(x, t)}{\tau}$$
$$+ \frac{\delta t}{2}\left(\left.\frac{(F + G) \cdot (\xi - u)}{\rho RT} f^{eq}\right|_{t+\delta t} + \left.\frac{(F + G) \cdot (\xi - u)}{\rho RT} f^{eq}\right|_t\right). \tag{7.35}$$

We define a new variable as

$$\bar{f} = f - \frac{(\boldsymbol{F}+\boldsymbol{G})\cdot(\xi-\boldsymbol{u})}{2\rho RT} f^{eq}\delta t, \tag{7.36}$$

and then will have

$$\bar{f}(\boldsymbol{x}+\xi\delta t,t+\delta t)-\bar{f}(\boldsymbol{x},t) = -\frac{\bar{f}(\boldsymbol{x},t)-\bar{f}^{eq}}{\tau}+\frac{(\boldsymbol{F}+\boldsymbol{G})\cdot(\xi-\boldsymbol{u})}{\rho RT} f^{eq}\delta t, \tag{7.37}$$

$$\bar{f}^{eq} = \left(1-\frac{(\boldsymbol{F}+\boldsymbol{G})\cdot(\xi-\boldsymbol{u})}{2\rho RT}\delta t\right)f^{eq}. \tag{7.38}$$

Then, the macroscopic density and velocity are obtained as

$$\sum \bar{f}_i = \rho, \tag{7.39}$$

$$\sum \bar{f}_i e_i = \rho \boldsymbol{u} - \frac{1}{2}(\boldsymbol{F}+\boldsymbol{G})\delta t. \tag{7.40}$$

Since this algorithm is usually unstable, He et al. [21] defined a new variable to solve its problem as

$$g = fRT + \psi(\rho)\Gamma(0), \quad \text{where} \quad \Gamma(\text{u}) = \frac{f^{eq}}{\rho}. \tag{7.41}$$

Then, we have

$$\frac{Dg}{Dt} = RT\frac{Df}{Dt}+\Gamma(0)\left[(\xi-\boldsymbol{u})\cdot\nabla\psi(\rho)\right]. \tag{7.42}$$

Combining Equations 7.30 and 7.42, we obtain

$$\frac{Dg}{Dt} = -\frac{g-g^{eq}}{\tau}+(\xi-\boldsymbol{u})\cdot\{\Gamma(\boldsymbol{u})(\boldsymbol{F}_s+\boldsymbol{G})-[\Gamma(\boldsymbol{u})-\Gamma(0)]\nabla\psi(\rho)\}, \tag{7.43}$$

where

$$g^{eq} = \rho RT\Gamma(\boldsymbol{u})+\psi(\rho)\Gamma(0). \tag{7.44}$$

The idea of using two distinct distribution functions was first introduced by He et al. [22,23]. In order to capture the interface, He et al. [22] have defined a new parameter, which identifies the two fluids. The new parameter is called order parameter (ϕ) and can be evaluated using LBM (just like any other flow parameter). To have a sharp interface, surface tension and gravity forces can be ignored, but interparticle interactions must be included in the order parameter evaluation formula.

Since φ has been treated as other flow parameters, one can use a relation similar to Equation 7.30 (without surface tension and gravity forces) as

$$\frac{Df}{Dt} = -\frac{f - f^{eq}}{\tau} - \frac{(\xi - \boldsymbol{u}) \cdot \nabla \psi(\rho)}{RT} \Gamma(\boldsymbol{u}), \tag{7.45}$$

where

$$f^{eq} = \phi\, \Gamma(\boldsymbol{u}). \tag{7.46}$$

It is easy to translate continuous Boltzmann equation to lattice Boltzmann domain. First, we have to obtain the corresponding variables of distribution functions f and g, similar to Equation 7.36, as

$$\bar{f}_i = f_i + \frac{(\boldsymbol{e}_i - \boldsymbol{u}) \cdot \nabla \psi(\rho)}{2RT} \Gamma_i(\boldsymbol{u}) \delta t, \tag{7.47}$$

$$\bar{g}_i = g_i - \frac{1}{2}(\boldsymbol{e}_i - \boldsymbol{u}) \cdot \{\Gamma_i(\boldsymbol{u})(\boldsymbol{F}_s + \boldsymbol{G}) - [\Gamma_i(\boldsymbol{u}) - \Gamma_i(0)]\nabla \psi(\rho)\}\delta t, \tag{7.48}$$

where

$$\Gamma_i(\boldsymbol{u}) = w_i \left[1 + 3\frac{\boldsymbol{e}_i \cdot \boldsymbol{u}}{c^2} + \frac{9}{2}\frac{(\boldsymbol{e}_i \cdot \boldsymbol{u})^2}{c^4} - \frac{3}{2}\frac{\boldsymbol{u}^2}{c^2} \right], \tag{7.49}$$

and eventually,

$$\bar{f}_i(\boldsymbol{x} + \boldsymbol{e}_i\delta t, t + \delta t) - \bar{f}_i(\boldsymbol{x},t) = -\frac{\bar{f}_i(\boldsymbol{x},t) - f_i^{eq}(\boldsymbol{x},t)}{\tau}$$
$$-\frac{2\tau - 1}{2\tau}\frac{(\boldsymbol{e}_i - \boldsymbol{u}) \cdot \nabla \psi(\phi)}{RT} \Gamma_i(\boldsymbol{u})\delta t, \tag{7.50}$$

$$\bar{g}_i(\boldsymbol{x} + \boldsymbol{e}_i\delta t, t + \delta t) - \bar{g}_i(\boldsymbol{x},t) = -\frac{\bar{g}_i(\boldsymbol{x},t) - g_i^{eq}(\boldsymbol{x},t)}{\tau}$$
$$+\frac{2\tau - 1}{2\tau}(\boldsymbol{e}_i - \boldsymbol{u}) \cdot \{\Gamma_i(\boldsymbol{u})(\boldsymbol{F}_s + \boldsymbol{G}) - [\Gamma_i(\boldsymbol{u}) - \Gamma_i(0)]\nabla \psi(\rho)\}\delta t. \tag{7.51}$$

Macroscopic density, velocity, and order variables are obtained as

$$\phi = \sum \bar{f}_i, \tag{7.52}$$

$$p = \sum \bar{g}_i - \frac{1}{2}\boldsymbol{u}.\nabla \psi(\rho)\delta t, \tag{7.53}$$

$$\rho RT\boldsymbol{u} = \sum \boldsymbol{e}_i\bar{g}_i + \frac{RT}{2}(\boldsymbol{F}_s + \boldsymbol{G})\delta t. \tag{7.54}$$

7.3.4 FREE ENERGY MODELS

Swift et al. [20] presented a thermodynamic consistent model by introducing equilibrium pressure tensor for a nonideal fluid, directly into the collision operator, using a modified equilibrium distribution function. Considering BGK collision operator, we have

$$\Omega_i = -\frac{1}{\tau}(f_i - f_i^{eq}). \tag{7.55}$$

Equilibrium distribution function is proposed as follows:

$$f_i^{eq} = A + Be_{i\alpha}u_\alpha + Cu_\alpha^2 + Du_\alpha u_\beta e_{i\alpha}e_{i\beta} + F_\alpha e_{i\alpha} + G_{\alpha\beta}e_{i\alpha}e_{i\beta}, \tag{7.56}$$

$$f_0^{eq} = A_0 + C_0 u_\alpha^2, \tag{7.57}$$

where parameters A, B, C, D, E, F, and G have been found, using some constraints as

$$\sum_i f_i^{eq} = \rho, \tag{7.58}$$

$$\sum_i f_i^{eq} e_{i\alpha} = \rho u_\alpha, \tag{7.59}$$

$$\sum_i f_i^{eq} e_{i\alpha}e_{i\beta} = P_{\alpha\beta} + \rho u_\alpha u_\beta, \tag{7.60}$$

where $P_{\alpha\beta}$ is the pressure tensor that can be derived using the Helmholtz free energy functional as

$$F(\rho) = \int dx \left\{ f_0(\rho) + \frac{\kappa}{2}|\nabla\rho(x)|^2 \right\}, \tag{7.61}$$

where $f_0(\rho)$ is the Helmholtz free energy density of a uniform fluid of density ρ specifying the bulk energy and the second term is the gradient energy related to surface tension.

The nonideal equations of state and pressure have been derived as

$$p_0 = \rho\frac{df_0}{d\rho} - f_0, \tag{7.62}$$

$$p(x) = \rho\frac{\delta F}{\delta\rho} - F(x) = p_0 - \kappa\rho\nabla^2\rho - \frac{\kappa}{2}|\nabla\rho|^2. \tag{7.63}$$

Then, the pressure tensor is

$$P_{\alpha\beta}(x) = p(x)\delta_{\alpha\beta} + \kappa\frac{\partial\rho}{\partial x_\alpha} : \frac{\partial\rho}{\partial x_\beta}, \tag{7.64}$$

where $\delta_{\alpha\beta}$ is the Dirac delta function [20,24]. In equilibrium, surface tension has been introduced in components of pressure tensor [20]. Using constraints presented in Equations 7.58 through 7.60, one can estimate the equilibrium distribution function parameters as

$$\begin{cases} A_0 = \rho - 2\left(p_0 - \kappa\rho\nabla^2\rho\right) \\ A = \left(p_0 - \kappa\rho\nabla^2\rho/3\right) \end{cases} \begin{cases} C_0 = -\rho \\ C = \dfrac{-\rho}{6} \end{cases}, \tag{7.65}$$

$$B = \frac{\rho}{3}, \quad D = \frac{2\rho}{3}, \quad F_\alpha = 0, G_{xx} = -G_{yy} = \frac{\kappa}{3}\left[\left(\frac{\partial\rho}{\partial x}\right)^2 - \left(\frac{\partial\rho}{\partial y}\right)^2\right], \quad G_{xy} = 2\frac{\kappa}{3}\frac{\partial\rho}{\partial x}\frac{\partial\rho}{\partial y}.$$

Using the Swift model, Inamuro et al. [23] have used two distribution functions f_i and g_i to solve interface and flow field, respectively. They used the standard lattice Boltzmann equation with BGK collision operator as

$$f_i(x + e_i\delta t, t + \delta t) - f_i(x,t) = -\frac{1}{\tau_f}\left[f_i(x,t) - f_i^{eq}(x,t)\right], \tag{7.66}$$

$$g_i(x + e_i\delta t, t + \delta t) - g_i(x,t) = -\frac{1}{\tau_g}\left[g_i(x,t) - g_i^{eq}(x,t)\right]. \tag{7.67}$$

Using the Swift model, they introduced the equilibrium distribution functions as

$$f_i^{eq} = H_i\phi + F_i\left[p_0 - \kappa_f\phi\nabla^2\phi - \frac{\kappa_f}{6}|\nabla\phi|^2\right] + 3E_i\phi e_{i\alpha}u_\alpha + E_i\kappa_f G_{\alpha\beta}(\phi)e_{i\alpha}e_{i\beta}, \tag{7.68}$$

$$g_i^{eq} = E_i\left[3p + 3e_{i\alpha}u_\alpha - \frac{3}{2}u_\alpha u_\alpha + \frac{9}{2}e_{i\alpha}e_{i\beta}u_\alpha u_\beta + A\Delta x\left(\frac{\partial u_\beta}{\partial x_\alpha} + \frac{\partial u_\alpha}{\partial x_\beta}\right)e_{i\alpha}e_{i\beta}\right] + E_i\kappa_g G_{\alpha\beta}(\rho)e_{i\alpha}e_{i\beta}, \tag{7.69}$$

where

$$E_1 = \frac{2}{9}, \quad E_2 = E_3 = \cdots = E_7 = \frac{1}{9}, \quad E_8 = E_9 = \cdots = E_{15} = \frac{1}{72},$$

$$H_1 = 1, \quad H_2 = H_3 = \cdots = H_{15} = 0, \tag{7.70}$$

$$F_1 = \frac{-7}{3}, \quad F_i = 3E_i (i = 2,3,\ldots,15),$$

and

$$G_{\alpha\beta}(\chi) = \frac{9}{2}\frac{\partial\chi}{\partial x_\alpha}\frac{\partial\chi}{\partial x_\beta} - \frac{3}{2}\frac{\partial\chi}{\partial x_\eta}\frac{\partial\chi}{\partial x_\eta}\delta_{\alpha\beta}. \tag{7.71}$$

One can use van der Walls, free energy density f_0 and calculate p_0 using Equation 7.62 as

$$p_0(\phi) = \phi T \frac{1}{1-b\phi} - a\phi^2. \tag{7.72}$$

Macroscopic velocity, pressure, and order variables can finally be evaluated as

$$\boldsymbol{u} = \sum_i e_i g_i, \tag{7.73}$$

$$p = \frac{1}{3} \sum_i g_i, \tag{7.74}$$

$$\phi = \sum_i f_i. \tag{7.75}$$

One of LBM challenges arise when density ratio between two phases is large. Inamuro et al. presented a model based on projection method, in which continuity equation at the interface is satisfied at every time step by solving the pressure Poisson equation [25]. They used two distribution functions f_i and g_i to estimate the order parameter and predicted velocity without a pressure gradient, respectively, as

$$f_i(\boldsymbol{x} + e_i \delta t, t + \delta t) - f_i(\boldsymbol{x},t) = -\frac{1}{\tau_f} \left[f_i(\boldsymbol{x},t) - f_i^{eq}(\boldsymbol{x},t) \right], \tag{7.76}$$

$$g_i(\boldsymbol{x} + e_i \delta t, t + \delta t) - g_i(\boldsymbol{x},t) = -\frac{1}{\tau_g} \left[g_i(\boldsymbol{x},t) - g_i^{eq}(\boldsymbol{x},t) \right]$$

$$+ 3 E_i e_{i\alpha} \left[\frac{\partial}{\partial x_\beta} \left\{ \mu \left(\frac{\partial u_\beta}{\partial x_\alpha} + \frac{\partial u_\alpha}{\partial x_\beta} \right) \right\} \right] \Delta x \tag{7.77}$$

and

$$g_i^{eq} = E_i \kappa_g G_{\alpha\beta}(\rho) e_{i\alpha} e_{i\beta} - \frac{2}{3} F_i \frac{\kappa_g}{\rho} \left(\frac{\partial \rho}{\partial x_\alpha} \right)^2$$

$$+ E_i \left[1 + 3 e_{i\alpha} u_\alpha - \frac{3}{2} u_\alpha u_\alpha + \frac{9}{2} e_{i\alpha} e_{i\beta} u_\alpha u_\beta + \frac{3}{2} \left(\tau_g - \frac{1}{2} \right) \Delta x \left(\frac{\partial u_\beta}{\partial x_\alpha} + \frac{\partial u_\alpha}{\partial x_\beta} \right) e_{i\alpha} e_{i\beta} \right]. \tag{7.78}$$

The other equilibrium distribution function f_i^{eq} is the same as Equation 7.68 and its coefficients are presented in (7.70). Then we find order variable and predicted velocity as

$$\boldsymbol{u}^* = \sum_i e_i g_i, \tag{7.79}$$

$$\phi = \sum_i f_i. \tag{7.80}$$

The current velocity can be estimated using the following equation:

$$\text{Sh}\frac{\boldsymbol{u}-\boldsymbol{u}^*}{\Delta t} = -\frac{\nabla p}{\rho}, \tag{7.81}$$

where $\text{Sh} = \Delta t/\Delta x$ is the Strouhal number. The divergence of the this equation yields

$$\text{Sh}\frac{\nabla\cdot\boldsymbol{u}-\nabla\cdot\boldsymbol{u}^*}{\Delta t} = -\nabla\cdot\left(\frac{\nabla p}{\rho}\right). \tag{7.82}$$

The current velocity satisfies the continuity equation, while the predicted velocity does not necessarily do so. Then,

$$\text{Sh}\frac{\nabla\cdot\boldsymbol{u}^*}{\Delta t} = \nabla\cdot\left(\frac{\nabla p}{\rho}\right). \tag{7.83}$$

This (the pressure Poisson equation) is solved and then the current velocity is estimated using Equation 7.81. Inamuro et al. [25] translated the Poisson equation into LBM domain as

$$h_i^{n+1}(\boldsymbol{x}+\boldsymbol{e}_i\Delta x) = h_i^n(\boldsymbol{x}) - \frac{1}{\tau_h}\left[h_i^n(\boldsymbol{x}) - E_i p^n(\boldsymbol{x})\right] - \frac{1}{3}E_i\frac{\partial u_i^*}{\partial x_i}\Delta x, \tag{7.84}$$

where n is the iteration number and $\tau_h = 1/\rho + 0.5$. Pressure is evaluated at every iteration, using the following equation:

$$p = \sum_{i=1} h_i. \tag{7.85}$$

One may use a breaking condition as

$$\frac{\left|p^{n+1}-p^n\right|}{\rho} < \varepsilon. \tag{7.86}$$

Inamuro's model has some shortcomings, because it does not give cutoff values of order parameter and also surface tension coefficient analytically, and then the minimum and maximum of order parameter have not been attenuated and changed during simulation. On the other hand, the projection method is computationally expensive [25,26].

Zheng et al. [26] have introduced a free energy–based model, which does not have the aforementioned difficulties. In this model, the order parameter is used in a way that does not change the density of phases and surface tension coefficient has been also introduced analytically. Also, in order to capture the interface, Cahn–Hilliard equation has been considered [26]. This equation can be brought to the LBM domain and then the order parameter can be estimated. First, a lattice Boltzmann equation is presented to solve the flow field as

$$f_i(\boldsymbol{x}+\boldsymbol{e}_i\delta t, t+\delta t) - f_i(\boldsymbol{x},t) = \Omega_i, \tag{7.87}$$

where

$$\Omega_i = \frac{f_i^{eq}(\boldsymbol{x},t) - f_i(\boldsymbol{x},t)}{\tau_n} + \left(1 - \frac{1}{2\tau_n}\right)\frac{w_i}{c_s^2}\left[(\boldsymbol{e}_i - \boldsymbol{u}_i) + \frac{\boldsymbol{e}_i \cdot \boldsymbol{u}_i}{c_s^2}\boldsymbol{e}_i\right](\mu_\phi \nabla \phi + \boldsymbol{F}_b)\delta t, \qquad (7.88)$$

and the equilibrium distribution function is expressed as

$$f_i^{eq} = w_i A_i + w_i n\left(3e_{i\alpha}u_\alpha - \frac{3}{2}u_\alpha^2 + \frac{9}{2}u_\alpha u_\beta e_{i\alpha}e_{i\beta}\right), \qquad (7.89)$$

where

$$A_1 = \frac{9}{4}n - \frac{15\left(\phi\mu_\phi + \frac{1}{3}n\right)}{4}, \quad A_i = 3\left(\phi\mu_\phi + \frac{1}{3}n\right), \quad i = 2,3,\ldots,9,$$

$$w_1 = \frac{4}{9}, \quad w_i = \frac{1}{9}, \quad i = 2,\ldots,5, \quad w_i = \frac{1}{36}, \quad i = 6,\ldots,9. \qquad (7.90)$$

If fluid densities are denoted by ρ_A and ρ_B for fluids A and B, respectively, n and ϕ may be defined as

$$n = \frac{\rho_A + \rho_B}{2}, \quad \phi = \frac{\rho_A - \rho_B}{2}. \qquad (7.91)$$

In order to capture the interface, the Cahn–Hilliard equation has been transported into the LBM domain and an equilibrium distribution function and also a lattice Boltzmann equation are allocated to estimate the order parameter. The lattice Boltzmann equation is

$$g_i(\boldsymbol{x} + \boldsymbol{e}_i\delta t, t + \delta t) = g_i(\boldsymbol{x},t) + (1-q)\left[g_i(\boldsymbol{x} + \boldsymbol{e}_i\delta t, t) - g_i(\boldsymbol{x},t)\right] + \frac{\left[g_i^{eq}(\boldsymbol{x},t) - g_i(\boldsymbol{x},t)\right]}{\tau_\phi}, \qquad (7.92)$$

where

$$q = \frac{1}{\tau_\phi + 0.5}, \qquad (7.93)$$

and

$$g_i^{eq} = A_i + B_i\phi + C_i\phi e_i \cdot u_i. \qquad (7.94)$$

The coefficients of Equation 7.94 are obtained as

$$B_1 = 1, \quad B_i = 0, \quad i = 2,\ldots,5, \quad C_i = \frac{1}{2q},$$

$$A_1 = -2\Gamma\mu_\phi, \quad A_i = \frac{1}{2}\Gamma\mu_\phi, \quad i = 2,\ldots,5, \qquad (7.95)$$

and the mobility has been controlled by Γ (see Ref. [26] for details).

The interface thickness can involve some nodes to reduce density gradients. Jacqmin [27] suggested, considering the interface locally flat, the order parameter is then set as Equation 7.81 through the interface

$$\phi = \phi^* \tan h\left(\frac{2\zeta}{W}\right),$$

(7.96)

where
   ϕ^* is the initial order parameter
   ζ is the coordinate located at the middle of the flat interface
   W is the interface thickness defined as

$$W = \frac{\sqrt{2\kappa/A}}{\phi^*}.$$

(7.97)

On the other hand, the surface tension for the flat interface can be calculated as [26–28]

$$\gamma = \int \kappa \left(\frac{\partial \phi}{\partial \zeta}\right)^2 d\zeta.$$

(7.98)

Substituting Equation 7.96 into 7.98 yields

$$\gamma = \frac{4\sqrt{2\kappa A}}{3} \phi^{*3},$$

(7.99)

where A and κ are two coefficients related to interface thickness and surface tension. Parameter A is an amplitude parameter to control the amount of free energy density in the interface. One can find pressure, using Equation 7.100, as

$$p = A\left(3\phi^4 - 2\phi^{*2}\phi^2 - \phi^{*4}\right) - \kappa\phi\nabla^2\phi + \frac{|\nabla\phi|^2}{2} + \frac{n}{3}.$$

(7.100)

Finally, macroscopic quantities can be estimated as

$$\phi = \sum_i g_i,$$

(7.101)

$$n = \sum_i f_i,$$

(7.102)

$$u = \frac{\frac{1}{2}\left(\mu_\phi \nabla \phi + F_b\right) + \sum_i f_i e_i}{n}.$$

(7.103)

As Fakhari and Rahimian [29] pointed out, Zheng's model has some drawbacks, that is, a falling drop will travel more distance in a fluid with high-density contrast (compared to a fluid with low-density contrast). They have simulated falling drop, using Zheng's model for two cases (high- and low-density contrast), but with the same average density (*n*). They have shown that drop travels the same distance at two cases, but this distance is less than the one passed by the falling drop when it is simulated by the kinetic-based LBM. Then, they concluded that Zheng's model is only appropriate for cases in which fluids have similar densities. However, it seems that further studies are required to be able to make such rigid conclusion.

7.4 EXAMPLES

Recently, LBM has been used to study basic multiphase phenomena. In these studies, the interface was captured using a second distribution function in order to recover the interface equation or to represent the presence of another fluid. One can investigate the effect of surface tension or other parameters on the interface. These parameters are categorized by some dimensionless parameters and studies are focused on multiphase phenomena using these dimensionless numbers.

Here, we present some results to verify the efficiency of LBM in multiphase simulations and interface capturing. Figure 7.2 illustrates deformation and breakup of a falling drop in time. Such simulations provide fundamental physical insight into interfacial problems that can be helpful for further studies on hydrodynamic effects on drops and bubbles. These studies are also useful for some industrial applications, such as secondary atomization, spray cooling, and painting. Case (a) presents a falling drop without an initial linear or rotational velocity, but in cases (b) and (c), drop has rotational speeds of 15 and 150 rad/s counterclockwise, respectively. We do not want to discuss a specific physic in detail here, but it is a wonderful physics showing high capabilities of LBM.

In LBM, there are several ways to capture an interface. In present simulations, we used Zheng's model, in which two distribution functions are used: one for solving the flow field and the other for capturing the interface. Unlike level set, VOF, or front tracking methods, LBM does not track the interface, but the collision and propagation of particles form the interface at each time step. This is one of the advantages of LBM with respect to other numerical methods.

In these simulations, all boundary conditions are periodic. There are various methods to impose boundary conditions in LBM. It must be noted that the interface is not a boundary and no boundary condition is applied for interfaces.

Figure 7.3 shows the impact of a drop with a fluid film at nine different time steps. The drop affects the surface of the fluid and causes the fluid film to oscillate. The oscillation eventually damps.

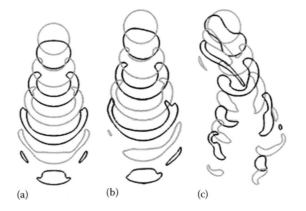

(a)   (b)   (c)

FIGURE 7.2   Deformation and breakup of a falling drop. (a) Zero rad/s, (b) 15 rad/s, and (c) 150 rad/s initial rotational velocity.

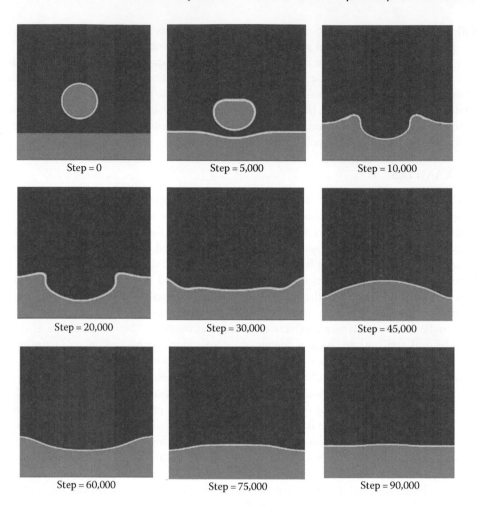

FIGURE 7.3   Impact of a drop with a liquid film at different time steps.

Density differences between two fluids cause propagation of errors, which leads to instabilities. In order to solve this difficulty, the interface must have an initial thickness. Usually, four or five nodes are involved at the interface. Another way is to damp the velocity noises or distribution function noises at each time step.

7.5   CONCLUSION AND FURTHER WORKS

Multiphase problems are present in a wide range of fluid dynamic problems and thus are of great importance to many researchers from different fields of study. This chapter begins with an introduction to multiphase interfacial flows and LBM. Then, basic LBM was presented in Section 7.2 of this chapter and four classes of models proposed for interfacial problems: color gradient models, interparticle potential models, mean-field models, and free energy models. These models were presented in details and their advantages and disadvantages were discussed. In order to verify the efficiency of LBM in interfacial flow simulations, two challenging examples presented. It is shown that LBM is able to maintain the sharpness of the interface without any additional interface tracking algorithm.

Further studies can be conducted on problems with higher density ratios and those in which transportation of mass or energy occurs across the interface.

REFERENCES

1. A.A. Mohamad, *Lattice Boltzmann Method: Fundamentals and Engineering Applications with Computer Codes*. Springer Science & Business Media, New York, 2011.
2. S. Chen and G.D. Doolen, Lattice Boltzmann method for fluid flows, *Annual Review of Fluid Mechanics*, 30, 329–364, 1998.
3. S. Succi, *The Lattice Boltzmann Equation for Fluid Dynamics and Beyond*. Oxford University Press, London, U.K., 2001.
4. R.G. Lerner and G.L. Trigg, *Encyclopaedia of Physics*, 2nd edn., Vol. 129, No. 2, John Wiley & Sons, New York, 1990.
5. A. Prosperetti and G. Tryggvason, *Computational Methods for Multiphase Flow*. Cambridge University Press, New York, 2009.
6. Z. Guo and C. Shu, *Lattice Boltzmann Method and Its Applications in Engineering (Advances in Computational Fluid Dynamics)*. World Scientific, London, U.K., 2013.
7. G.R. McNamara and G. Zanetti, Use of the Boltzmann equation to simulate lattice-gas automata, *Physical Review Letters*, 61(20), 2332, 1988.
8. U. Frisch, D.D. Humieres, B. Hasslacher, P. Lallemand, Y. Pomeau, and J.P. Rivet, Lattice gas hydrodynamics in two- and three-dimensions, *Complex Systems*, 1(4), 649–707, 1987.
9. D.H. Rothman and J.M. Keller, Immiscible cellular-automaton fluids, *Journal of Statistical Physics*, 52(3–4), 1119–1127, 1988.
10. A.K. Gunstensen, D.H. Rothman, S. Zaleski, and G. Zanetti, Lattice Boltzmann model of immiscible fluids, *Physical Review A*, 43(8), 4320–4327, 1991.
11. F.J. Alexander, S. Chen, and D.W. Grunau, Hydrodynamic spinodal decomposition: Growth kinetics and scaling functions, *Physical Review B*, 48(1), 634, 1993.
12. D. Grunau, S. Chen, and K. Eggert, A lattice Boltzmann model for multiphase fluid flows, *Physics of Fluids A: Fluid Dynamics*, 5(10), 2557–2562, 1993.
13. U. d'Ortona, D. Salin, M. Cieplak, R.B. Rybka, and J.R. Banavar, Two-color non-linear Boltzmann cellular automata: Surface tension and wetting, *Physical Review E*, 51(4), 3718, 1995.
14. S.V. Lishchuk, C.M. Care, and I. Halliday, Lattice Boltzmann algorithm for surface tension with greatly reduced microcurrents, *Physical Review E*, 67(3), 36701, 2003.
15. J.U. Brackbill, D.B. Kothe, and C. Zemach, A continuum method for modeling surface tension, *Journal of Computational Physics*, 100(2), 335–354, 1992.
16. I. Halliday, R. Law, C.M. Care, and A. Hollis, Improved simulation of drop dynamics in a shear flow at low Reynolds and capillary number, *Physical Review E*, 73(5), 56708, 2006.
17. M. Latva-Kokko and D.H. Rothman, Diffusion properties of gradient-based lattice Boltzmann models of immiscible fluids, *Physical Review E*, 71(5), 56702, 2005.
18. X. Shan and H. Chen, Simulation of non-ideal gases and liquid-gas phase transitions by lattice Boltzmann equation, *Physical Review E*, 49(4), 2941, 1994.
19. X. Shan and H. Chen, Lattice Boltzmann model for simulating flows with multiple phases and components, *Physical Review E*, 47, 1815–1820, 1993.
20. M.R. Swift, W.R. Osborn, and J.M. Yeomans, Lattice Boltzmann simulation of non-ideal fluids, *Physical Review Letters*, 75(5), 830, 1995.
21. X. He, X. Shan, and G.D. Doolen, Discrete Boltzmann equation model for nonideal gases, *Physical Review E*, 57(1), 13–16, 1998.
22. X. He, S. Chen, and R. Zhang, A lattice Boltzmann scheme for incompressible multiphase flow and its application in simulation of Rayleigh–Taylor instability, *Journal of Computational Physics*, 152(2), 642–663, 1999.
23. T. Inamuro, R. Tomita, and F. Ogino, Lattice Boltzmann simulations of drop deformation and breakup in shear flows, *International Journal of Modern Physics B*, 17(1–2), 21–26, 2003.
24. R. Evans, Advances in physics, the nature of the liquid-vapour interface and other topics in the statistical mechanics of non-uniform, classical fluids, *Advances in Physics*, 28(2), 143–200, 1979.
25. T. Inamuro, T. Ogata, S. Tajima, and N. Konishi, A lattice Boltzmann method for incompressible two phase flows with large density differences, *Journal of Computational Physics*, 198(2), 628–644, 2004.
26. H.W. Zheng, C. Shu, and Y.T. Chew, A lattice Boltzmann model for multiphase flows with large density ratio, *Journal of Computational Physics*, 218(1), 353–371, 2006.
27. D. Jacqmin, Calculation of two phase Navier–Stokes flows, *Using Phase-Field Modeling*, 127, 96–127, 1999.
28. J.S. Rowlinson and B. Widom, *Molecular Theory of Capillarity*. Oxford, London, U.K., 1989.
29. A. Fakhari and M.H. Rahimian, Phase-field modeling by the method of lattice Boltzmann equations, *Physical Review E*, 81(3), 036707, 2010.

8 Mesoscale Computational Techniques in Interfacial Science
Smoothed Particle Hydrodynamics Method

Mohammad Taeibi Rahni, Massoud Rezavand,
Iman Mazaheri, Mohsen Karbaschi, and Reinhard Miller

CONTENTS

8.1   INTRODUCTION

Liquid–fluid interfaces appear in different forms (e.g., simple and complex) and their complete dynamics are not yet fully understood due to their interphase coupling, whereby different phases may strongly affect one another. In addition, the interactions between physicochemical properties and hydrodynamics usually make their dynamics considerably complicated.

Detailed numerical simulations are often computationally too intensive to predict many real-life multiphase flows; however, they can provide necessary insight needed for modeling actual flow problems. Note besides the importance of capturing the motion and deformation of the interface correctly (as well as the transport of mass, energy, etc., through it), one should also be concerned about obtaining sharp enough interfaces (with least numerical dissipation). Many researchers have been conducting interesting simulations regarding liquid–fluid interfaces. Most of these works have not been including much detailed thermodynamical behavior (especially variations of molecular properties along the interface; they have mostly used integrated effects).

As mentioned in Chapter 7, in about the past 15 years, new computational fluid dynamics (CFD) ideas—*meshless* methods— have been developed and widely used to study fluid dynamics, including multiphase flows containing interfaces. Two of the main such methods are the smoothed particle hydrodynamics (SPH) method and the lattice Boltzmann method (LBM). These new techniques have shown to overcome some of the difficulties of conventional CFD. In these methods, the continuum assumption is somehow removed by discretizing the fluid (instead of the media containing it). Then, the mathematical modeling is a bit similar to the one used in molecular dynamics, wherein the particle collisions are also taken into account.

The main difference between SPH and LBM is that the first one uses the Lagrangian form of the nonlinear partial differential full Navier–Stokes equations, while the approximated ordinary differential Boltzmann equation is used in the latter. Even though LBM has become more popular in simulation of many fluid flow problems, there has also been some interesting research utilizing SPH.

8.2 SPH BASICS

SPH is a fully Lagrangian, meshless and particle-based method, which was developed originally by Lucy [1] and independently and simultaneously by Gingold and Monaghan [2] in order to solve astrophysical and compressible problems. The attractive features of this method in simulation of large deformations persuaded researchers to use this method for incompressible fluid flows. Monaghan [3] used SPH for incompressible fluid flows for the first time and in particular simulated free surface flow. Thereafter, in the past one and a half decades, it has been used widely in other fields corresponding to fluid dynamics, such as non-Newtonian fluids [4,5], fluid–solid interaction [6,7], heat transfer problems [8–11], and multiphase flows [12–18].

8.2.1 SPH Formulation

The main concept of SPH method is an interpolation method, which represents any arbitrary function in terms of its values at a number of points (particles) [19]. The integral representation of an arbitrary function $f(x)$ is defined as:

$$f(x) = \int_{\Omega} f(x')W(x - x', h)dx', \qquad (8.1)$$

where
   Ω is the computational domain
   W is a weighting function called Kernel
   h is the effective length of the Kernel function

The Kernel function satisfies the following two properties:

$$\int_{\Omega} W(x - x')dx' = 1, \qquad (8.2)$$

$$\lim_{h \to 0} W(x - x', h) = \delta(x - x'), \qquad (8.3)$$

where δ is the Dirac delta function.

In SPH method, fluid is represented by a finite number of particles, which are the interpolating points, and each has its own mass and volume. In order to approximate the properties of the fluid on these points, the integral interpolation presented in Equation 8.1 is replaced by the following summation:

$$f(x) = \sum_{b=1}^{N} \frac{m(x_b)}{\rho(x_b)} f(x_b) W(x - x_b, h),$$ (8.4)

where

  m is the mass
  ρ is the density
  subscript b labels the particles

This summation is restricted to N particles, which lie within a circle with a radius kh and are centered at x. Here, k is a coefficient that depends on the order of Kernel and for a third-, fourth-, and fifth-order Kernel, their values are 2, 2.5, and 3, respectively. Many examples of various Kernels are presented in the literature. The Kernel function used here is a quintic spline [20] with the form

$$W(x - x_b, h) = \begin{cases} n_d\left[(3-q)^5 - 6(2-q)^5 + 15(1-q)^5\right], & q < 1, \\[2mm] n_d\left[(3-q)^5 - 6(2-q)^5\right], & 1 \le q < 2, \\[2mm] n_d(3-q)^5, & 2 \le q < 3, \\[2mm] 0, & q \ge 3, \end{cases}$$ (8.5)

where

$$q = |x - x_b| / h \text{ and } n_d = \frac{7}{478\pi h^2}.$$

8.2.2 Incompressibility Constraint in SPH

To impose the incompressibility constraint in SPH, there are two general approaches: weakly compressible (WCSPH) and incompressible (ISPH). WCSPH has been more common and was first proposed by Monaghan [3]. In this approach, fluid is considered compressible and pressure is calculated from an appropriate equation of state. One of the most popular equations of state used in WCSPH method has the form [3]

$$P_a = \frac{\rho_{0a} c_a^2}{k}\left[\left(\frac{\rho_a}{\rho_{0a}}\right)^k - 1\right],$$ (8.6)

where

  ρ_0 is the reference density
  c_a is the speed of sound
  k is the polytropic coefficient (gas constant), which is 1.4 for air

In order to ensure the incompressibility constraint (much numbers less than about 0.3), the value of the speed of sound in Equation 8.6 must be at least 10 times greater than the maximum velocity of the fluid and thus density fluctuations should be less than about 1%. On the other hand, since the stability of the numerical method is based on the speed of sound, it will be achieved in smaller time steps (compared to physical time scales of the problem), leading to much more computational costs.

ISPH is a more recent approach and was proposed by Cummins and Rudman [21] for the first time. In this approach, the incompressibility constraint is enforced by the well-known projection method, which has been widely used in Eulerian mesh-based methods [22,23]. In ISPH method, there are different algorithms for enforcing the incompressibility condition, in which the incompressibility criterion, that is, the source term of the pressure Poisson equation, is different. From a theoretical point of view, in an incompressible flow, both variations of density and the divergence of velocity are zero, but in numerical methods, satisfying one of these conditions does not guarantee satisfaction of the other. This is the reason for proposing different methods based on an incompressibility criterion.

In the original model proposed by Cummins and Rudman [21], the divergence of velocity field was considered as the source term for the pressure Poisson equation, and the satisfaction of the continuity equation was achieved by assuming the variations of velocity to be zero. This is why this algorithm is sometimes called a divergence-free method.

In a second algorithm, Shao and Lo [24] calculated the relative variations of density directly as the criterion of incompressibility. Hu and Adams [15] combined these two methods and, as a third method, proposed a new ISPH projection-based method, in which both the divergence-free condition and the density-invariance condition are imposed by a fractional time-step integration algorithm.

These three algorithms are comprehensively compared by Xu et al. [25]. They proposed that the divergence-free algorithm has desirable accuracy features in some cases. However, in the case of highly disordered particles, the method will be unstable. On the other hand, the density-invariance algorithm has good stability characteristics, even when highly disordered particles are presented. However, this algorithm's predictions are not accurate. The algorithm proposed by Hu and Adams [15] benefits both from the accuracy features of the divergence-free algorithm and the stability characteristics of density-invariance algorithm. However, as it has to solve the pressure Poisson equation twice at each time step, it is computationally more expensive.

Xu et al. [25] have pointed out that in divergence-free algorithms, particles move along the streamlines and particle clustering occurs in some specific situations. In order to benefit from the accuracy of divergence-free algorithm and avoid the particles accumulation, they have proposed a new ISPH algorithm that shifts particles slightly from the streamlines and thus corrects the hydrodynamic properties in the new position by a Taylor expansion interpolation. This algorithm, whose detailed steps are presented in Ref. [26], has been used for the simulations presented in the later sections of this work.

8.2.3   PROJECTION ALGORITHM

First, the intermediate position of the particles is predicted, using the current time-step velocities as:

$$\vec{r}_a^{*} = \vec{r}_a^{n} + \Delta t(\vec{u}_a^{n}), \tag{8.7}$$

where
   $\vec{r}$ is the position
   $\vec{u}$ is the velocity
   Δt is the time step
   the superscripts * and n denote intermediate and current time steps, respectively

At this intermediate position, the intermediate velocities are calculated by solving the momentum equation excluding the pressure gradient as:

$$\vec{u}_a^* = \vec{u}_a^n + \left(\vec{g} + \mu \nabla^2 \vec{u} + F_s\right)\Delta t, \tag{8.8}$$

where

$\vec{g}$ is the gravitational acceleration
μ is the dynamic viscosity coefficient
F_s is the surface tension force

In order to discretize the viscous term in Equation 8.8 (in SPH form), the formulation originally proposed by Cleary [27] is used, which conserves both linear and angular momentums as:

$$\left(\frac{\mu}{\rho}\nabla^2 u_a\right) = \sum_b m_b \left(\frac{1}{\rho_b \rho_a}\frac{4\mu_a\mu_a}{\mu_a + \mu_b}\frac{u_{ab}r_{ab}}{|r_{ab}|^2 + \eta^2}\right)\nabla_a W_{ab}, \tag{8.9}$$

where

u_{ab} is the difference between the velocity of particle a and its neighboring particle b ($u_{ab} = u_a - u_b$)
r_{ab} is the relative displacement between particle a and particle b ($r_{ab} = r_a - r_b$)
η is a small number for avoiding denominator to become zero

The pressure at time step $n+1$ can then be calculated by solving the following pressure Poisson equation:

$$\nabla \cdot \left(\frac{1}{\rho_a}\nabla p_a^{n+1}\right) = \frac{1}{\Delta t}\nabla \cdot (u_a^*), \tag{8.10}$$

where p is the pressure. The left-hand side of this equation can be approximated by the formulation proposed by Cummins and Rudman [21] as:

$$\nabla \cdot \left(\frac{1}{\rho_a}\nabla p_a^{n+1}\right) = \sum_b \frac{m_b}{\rho_b}\left(\frac{4}{\rho_a + \rho_b}\right)\frac{(p_a - p_b)\vec{r}_{ab} \cdot \nabla_a W_{ab}}{|\vec{r}_{ab}|^2 + \eta^2}. \tag{8.11}$$

The velocity field at time step $n+1$ can be obtained from correcting the predicted velocities by the pressure field calculated in Equation 8.10 as:

$$\vec{u}_a^{n+1} = \vec{u}_a^* - \Delta t \frac{1}{\rho}\nabla p^{n+1}. \tag{8.12}$$

Finally, all particles are moved to their new position by the following scheme:

$$\vec{r}_a^{n+1} = \vec{r}_a^n + \frac{1}{2}(\vec{u}_a^{n+1} + \vec{u}_a^n)\Delta t. \tag{8.13}$$

8.3   SPH FOR INTERFACIAL PROBLEMS

Various methods for modeling multiphase interfacial flows are presented and also previous multiphase SPH studies are reviewed in this section.

8.3.1 Simulation of Interfacial Multiphase Flows via SPH

Multiphase flows have been widely studied in many experimental and numerical works. Most numerical investigations in this field have been performed by Eulerian mesh-based methods (conventional CFD). Although these methods have been proved to be accurate and have had reliable results in many problems, they need to be more investigated in cases dealing with highly disordered interfaces. Heterogeneity in density and other hydrodynamic complexities near or at the interface make interfacial flows very complicated and thus particular considerations are needed. In the case of interfacial multiphase flows, we need a specific algorithm to track the interface accurately. Volume of fluid [28], level set [29], and front tracking methods are the well-known algorithms for interface tracking in conventional CFD simulation of interfacial multiphase flows. An alternative to Eulerian mesh-based methods for simulation of interfacial flows is the Lagrangian particle-based methods (e.g., LBM or SPH).

SPH is a fully Lagrangian meshless method and does not need any computational grid. It replaces fluid by a finite number of fluid particles and the flow is described by evolution of these particles. Consequently, the Lagrangian nature of SPH usually lets one to track the interface without any need for a specific algorithm. This fact nominates SPH as an important candidate for simulation of such flows.

Monaghan and Kocharyan [12] have simulated some multiphase flow problems with low density ratios, using SPH for the first time. While Colagrossi and Landrini [13] used the method proposed in Ref. [12] for air–water problems (density ratio about 1/1000) and found that the method was unstable and gave very poor results. Note the reason for these instabilities is that when the integral interpolant is performed over two particles with severe density gradients, singularities occur in the solution. To remedy these singularities, Colagrossi and Landrini [13] proposed a new form of particle evolution equation, which enhanced the stability of the method. They also used a periodic reinitialization of the density according to moving-least-square (MLS) density correction [30], a general correction to the standard form of the artificial viscosity used in SPH and a smoothing scheme for velocity field based on XSPH* formalism [31]. MLS is a first-order accurate interpolation method for irregularly distributed particles and requires computation of a special MLS kernel. Thus, it enforces additional computational costs. XSPH is a velocity correction method in which velocities of neighboring particles are used, in order to smooth the velocity field.

Using the concept of particle number density, Hu and Adams [14,15] presented the spatial derivatives of the field variables as a smoothing function, in which the neighboring particles participate just in specific volume (not in density). This way, the influence of severe density gradients is removed. However, their method is not able to handle free surface problems.

Grenier et al. [16] presented a Hamiltonian formulation, as an extension of the ones proposed in Refs. [13,14]. Their method can be adopted for free surface problems. Their proposed method is based on use of a Shepard kernel for density calculation, which enhances the accuracy of the method and requires knowing a spatial distribution of the particle volumes (enforcing computational costs). They have also used a kind of repulsive force at the interface in problems where surface tension effects were negligible.

Shao [32] introduced two different models for interfacial multiphase flows in ISPH approach. First, he developed a coupled model, in which there is no difference between particles of different phases and employs the standard ISPH equations at the interface. In his second model, which is decoupled, particles of different phases are treated separately, and then by employing pressure and viscous force equations at the interface, different phases couple. These two models have been used for different test cases with different density ratios. It is shown that because of more accurate modeling of the interface, the decoupled model has a better performance.

Monaghan and Rafiee [33] presented a simple method for multiphase problems with high density ratios, using a repulsive force between the particles of different phases. In their method, when

---

* XSPH is a special version of SPH where each particle is moved with a smoothed velocity obtained by averaging the velocity of its neighboring particles.

interaction is between two particles of different phases, pressure is increased by a repulsive force acting on the interface. This method has been validated for various test cases.

Szewc et al. [17] simulated the motion of an air bubble in a viscous fluid with a WCSPH formulation similar to the one proposed in Ref. [14]. They have alleviated the effect of large density gradients at the interfaces, using the particle number density concept. At the interface, they have also used a repulsive force similar to one used in Refs. [16,33].

Zainali et al. [18] have presented a new formulation in ISPH approach for Newtonian and non-Newtonian multiphase flows with high density ratios. In their formulation, the concept of particle number density is used in discretization of the functions and their derivatives. In their discretized formulations, density and viscosity are smoothened by a Harmonic mean interpolation. In order to enhance the robustness of the method in problems with high density ratios, density, and viscosity are further smoothened by a smoothed color function.

More recently, Chen et al. [34] have presented a new multiphase WCSPH model with the assumption that pressure and space are continuous across the interface. In their method, when the concerned particle is near the interface, all neighboring particles within the support domain belonging to the other phase are regarded as particles of the same phase, and only the information of position, velocity, volume, and pressure of these particles are involved in solution of acceleration and density of the concerned particle. They have also presented a new corrected density reinitialization algorithm, in order to avoid the pressure instability caused by free movement of SPH fluid particles. This new algorithm leads to a better computational accuracy and to smoother pressure fields. In their simulations, a cutoff value of particles density is also used, in order to avoid negative pressures.

8.3.2 Surface Tension Models

Surface tension is caused by the cohesive forces between molecules of the two fluid phases at the interface. Special attention must be paid in order to accurately take the surface tension force into account. In SPH, there are two general classes of methods used for this purpose. The methods in the first class are microscopic and use the interparticle interactions in order to implement the interparticle cohesive forces. Nugent and Pasch [35] have used the van der Waals (vdW) interparticle attraction potentials to simulate stable drop configurations. The cohesive vdW used in their work leads to an attractive central force acting between particles. These forces vanish outside fringes of the interface and cancel each other at the interface. In their method, the interaction range of attractive forces needs to be larger than the one used for all other forces in SPH formulations.

The cosine model used by Tartakovsky and Meakin [36] is a similar approach, but a combination of attractive and repulsive forces is used (instead of the vdW potentials). Although implementation of microscopic interparticle potentials is straightforward, some difficulties arise when using these methods. One difficulty is that the calculated surface tension requires calibration. In addition, surface tension depends on the resolution of the solution, such that when resolution is increased to more than a certain amount, it does not converge to a fixed value [37]. On the other hand, Breinlinger et al. [38] pointed out that since the interaction between particles are not readily available and must be fitted, using a macroscopic method, in which surface tension and contact angles are given as input parameters, is advantageous.

The second class of SPH methods for surface tension considerations belongs to the models that use a macroscopic point of view. These models are based on the continuum surface force (CSF), originally proposed by Brackbill et al. [39]. In CSF model, a color function is attributed to each fluid and the interface is modeled as a transition region in which the color function varies smoothly from its value in one fluid to its value in the other. Morris [40] has used this approach in SPH for the first time. He presented various formulations, in order to implement surface tension for multiphase flow problems with low density and viscosity ratios (but without free surfaces). Adami et al. [37] have pointed out that all formulations presented in Ref. [40] are based on a smoothed color function, leading to incorrect curvatures for the interface. One reason for the incorrect calculation of divergences for unit normals (the curvature) is that in the transition band, full support of Kernel function is not satisfied.

In order to remedy the aforementioned problems, Hu and Adams [14] proposed the continuum surface stress (CSS) model. In this model, a sharp color function with a discontinuity at the interface is used and the curvature of the interface is calculated by a surface stress tensor, which depends on the gradient of the color function. Adami et al. [37] proposed that in a multiphase problem with high density ratio, surface tension is not distributed uniformly between two different sides of the interface. For example, in an air–water problem, surface tension is dominated in the side of water. For this reason, they have proposed a density weighted formulation for color gradient. They have also proposed a reproducing divergence approximation, in order to achieve accurate calculations of the interface curvature.

In a recent work, Zainali et al. [18] have mentioned that in their ISPH simulations for multiphase problem, in contrary to Hu and Adams [14], they have obtained much more accurate results, using CSF method (compared to CSS) for surface tension. In the present work, the CSF method proposed by Morris [40] is also used.

In CSF method, the surface tension is replaced by a volumetric force as:

$$F_s = f_s \delta_s, \tag{8.14}$$

where

δ_s is the surface delta function that reaches its maximum at the interface
f_s is the surface force defined as:

$$f_s = \gamma \kappa \hat{n} + \nabla_s \gamma, \tag{8.15}$$

where

γ is the surface tension coefficient

$\hat{n}$ is the unit normal vector to the interface
κ is the curvature of interface
∇_s is the surface gradient

The second term in this equation acts tangentially on the interface and forces fluid from the regions with low surface tension to the one with high surface tension. In Ref. [40], surface tension is considered constant and thus the second term of Equation 8.15 is neglected. The unit normal vector to the interface is calculated as:

$$\hat{n} = \frac{\nabla C}{|\nabla C|}, \tag{8.16}$$

where C is the color function as:

$$C = \begin{cases} 1, & \text{in fluid } A, \\ 0, & \text{in fluid } B. \end{cases} \tag{8.17}$$

The curvature of the interface is calculated as:

$$\kappa = -\nabla \cdot \hat{n}. \tag{8.18}$$

Various functions can be employed as δ_s, which has to satisfy the normalization condition. In this work, this function is defined as:

$$\delta_s = |n| = |\nabla C|. \tag{8.19}$$

In order to accurately predict the surface tension, the curvature of the interface must also be calculated accurately. Hence, the unit normal vectors to the interface and their divergence must be calculated. A discretized form for calculation of the unit normal is

$$n_a = \sum_b \frac{m_b}{\rho_b} C_b \nabla_a W_{ab}, \tag{8.20}$$

where C_b is the color index of the neighboring particle, b. In order to accurate achieve more calculations of the unit normal vector, color function can be smoothened by the concept of particle number density as:

$$\tilde{C}_a = \sum_b \frac{m_b}{\rho_b} C_b W_{ab} = \frac{\sum_b C_b W_{ab}}{\Theta_a}, \tag{8.21}$$

where Θ_a is the particle number density and is approximately equal to the inverse of the volume of the corresponding particle a as:

$$\Theta_a = \sum_b W_{ab} = \frac{\rho_a}{m_a}. \tag{8.22}$$

Using Θ_a, a more accurate formulation can be written as:

$$n_a = \sum_b \frac{m_b}{\rho_b} (\tilde{C}_b - \tilde{C}_a) \nabla_a W_{ab}. \tag{8.23}$$

The curvature of the interface can be obtained by calculating the divergence of the unit normal vector as:

$$(\nabla \cdot \hat{n})_a = \sum_b \frac{m_b}{\rho_b} \hat{n}_b \cdot \nabla_a W_{ab}, \tag{8.24}$$

and according to Ref. [19], a more accurate calculation can be obtained, using the difference of unit normal of particles a and b as:

$$(\nabla \cdot \hat{n})_a = \sum_b \frac{m_b}{\rho_b} (\hat{n}_b - \hat{n}_a) \cdot \nabla_a W_{ab}. \tag{8.25}$$

It is noteworthy that unit normal vectors at the fringes of interface might be incorrect leading to incorrect calculations of its curvature. Hence, a constraint is needed for unit normals to be reliable. Morris [40] has proposed a constraint as:

$$|n_a| > \frac{\varepsilon}{h}, \tag{8.26}$$

where ε is a constant, which was set to 0.01 in Ref. [40] and to 0.08 in Ref. [18]. Normals satisfying this condition are considered reliable to participate in calculation of the curvature. Finally, the calculated surface tension can be imposed in the momentum equation as:

$$(F_s)_a = -\frac{\gamma}{\rho_a}(\nabla \cdot \hat{n})_a n_a. \tag{8.27}$$

8.4 EXAMPLES

Two popular examples, wherein ISPH has been used, are presented in this section.

8.4.1 RAYLEIGH–TAYLOR INSTABILITY

One of the well-known examples of multiphase flows with complex interface evolution is Rayleigh–Taylor instability (RTI). This problem has been investigated in several studies, using SPH. For example, Grenier et al. [16] and Monaghan and Rafiee [33] studied it, using WCSPH method, while Cummins and Rudman [21], Hu and Adams [15], and Shadloo et al. [41] studied RTI, using ISPH method. In this problem, computational domain is a rectangle with dimensions of 1×2 (width × height). Particles are initially set on regular positions and particle spacing is $dx = 0.005$. Thus, 200×400 fluid particles are used for simulations. The heavy fluid ($\rho_H = 1.8$) is above the light one ($\rho_L = 1.0$), and they are separated by an interface located at $y = 1 - 0.15\sin(2\pi x)$. Reynolds number is $\mathrm{Re} = \sqrt{H^3 g}/v = 420$, wherein H is the width of the domain, g is the gravitational acceleration, and v is the kinematic viscosity coefficient, which is equal for the two fluids used. The method proposed by Morris et al. [20] is employed to enforce no-slip boundary condition at the solid boundaries (using ghost particles). In Figure 8.1, snapshots of RTI at $t^* = t(g/H)^{1/2} = 5.0$ are compared with those of Colicchio [42], which were presented by Grenier et al. [16]. It is observed that the complex interface evolution is accurately tracked.

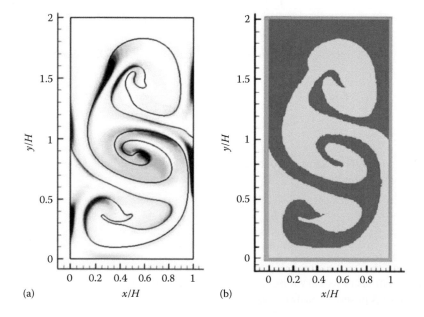

FIGURE 8.1 The position of particles at $t(g/H)^{1/2} = 5.0$: (a) Colicchio (From Colicchio, G., Violent Disturbance and Fragmentation of Free Surface, PhD dissertation, University of Southampton, Southampton, U.K., 2004.), (b) present work.

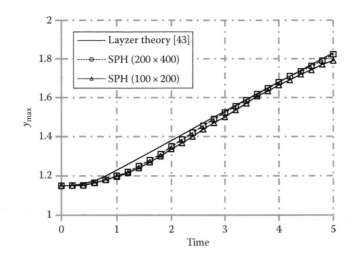

FIGURE 8.2   The time variation of the highest point of the light fluid compared with the Layzer theory. (Layzer, D., *The Astrophys. J.*, 122, 1, 1955.)

Similar to Ref. [33], in order to further analyze the RTI results, the time variation of the highest point of the light fluid is compared to the theoretical results of Layzer [43] in Figure 8.2. Simulations have been performed with two different spatial resolutions ($dx=0.005$ and $dx=0.01$) and in both cases, there are close agreements. Of course, as expected higher resolution leads to closer agreements.

8.4.2   Bubble Rising in a Viscous Flow

Another classical example of interfacial multiphase flows is bubble rising in a viscous quiescent flow. In addition to many Eulerian mesh-based numerical studies, this problem has been investigated in several works, using SPH method. For example, Colagrossi and Landrini [13], Grenier et al. [16], and Szewc et al. [17] studied this problem via WCSPH approach, while Zainali et al. [18] studied it, using ISPH method.

A detailed description of the geometry and the physical properties of this problem is presented in Ref. [44]. A rectangular domain with dimensions 1×2 (width $\times$ height) is considered. The bubble ($\rho_B=0.1$) is initially located a distance equal to its diameter above the bottom of the domain and is surrounded by water ($\rho_L=1$). The dynamic viscosity coefficient is 0.1 for bubble and 1 for liquid particles. The bubble Reynolds number is 35 and Eötvös number is set to 10.

The bubble shapes at $t=3$ time units are shown in Figure 8.3 for two different spatial resolutions ($dx=0.00625$ and $dx=0.0125$) and are compared to the high-resolution results of Ref. [44], which are obtained using finite element method (FEM) analysis.

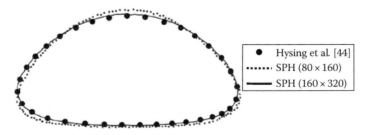

FIGURE 8.3   The shape of the bubble at $t=3$ time units, compared to the FEM results. (Hysing, S. et al., *Int. J. Numerical Methods Fluids*, 60, 1259, 2009.)

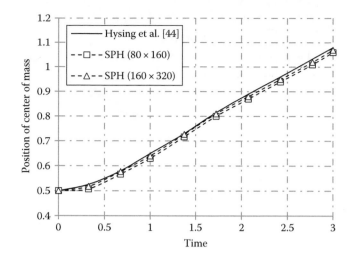

FIGURE 8.4 Time variation of the position of the bubble's center of mass, compared to the high-resolution FEM results. (Hysing, S. et al., *Int. J. Numerical Methods Fluids*, 60, 1259, 2009.)

The time variation of the position of the bubble's center of mass is compared to the aforementioned FEM results in Figure 8.4. Simulations have been performed with two different spatial resolutions, and in both cases, close agreements with high-resolution FEM results are observed. As expected, it can be seen that for higher resolution, the agreements are better.

8.5 SUMMARY, CONCLUSION, AND OUTLOOKS

Liquid–fluid interfaces are present in many industrial and natural phenomena and thus real insight into such problems is very useful. In this chapter, a fully Lagrangian and meshless method (SPH) for modeling interfacial fluid dynamic problems is investigated. The basics of SPH were reviewed first and then SPH formulations were described. Then, two different approaches for enforcing incompressibility constraint were introduced. In the literature, it has been shown that ISPH method employed here predicts smoother pressure fields and is computationally more efficient than WCSPH.

The projection algorithm used here was also described in details. In addition, the advantages of SPH, as a Lagrangian particle-based method in simulation of interfacial problems, were discussed, wherein many multiphase models were reviewed.

Next, two different models for implementation of the surface tension force in SPH computations were reviewed and the CSF model was illustrated. Then, the capabilities of SPH method in simulation of multiphase interfacial problems were verified in two challenging test cases, in which good agreements were achieved in comparison to reliable theoretical results.

Using SPH and its derivatives, many more challenging studies in interfacial flow problems with much higher density and viscosity ratios can be conducted. Also, the transport of mass and energy across the interface can be included in such future investigations.

REFERENCES

1. L.B. Lucy, *The Astronomical Journal*, 82, 1013–1024, 1977.
2. R.A. Gingold and J.J. Monaghan, *Monthly Notices of the Royal Astronomical Society*, 181, 375–389, 1977.
3. J.J. Monaghan, *Journal of Computational Physics*, 110, 399–406, 1994.
4. H. Zhu, N.S. Martys, C. Ferraris, and D.D. Kee, *Journal of Non-Newtonian Fluid Mechanics*, 165, 362–375, 2010.

5. M. Hashemi, R. Fatehi, and M. Manzari, *Journal of Non-Newtonian Fluid Mechanics*, 166, 1239–1252, 2011.
6. C. Antoci, M. Gallati, and S. Sibilla, *Computers & Structures*, 85, 879–890, 2007.
7. A. Rafiee and K.P. Thiagarajan, *Computer Methods in Applied Mechanics and Engineering*, 198, 2785–2795, 2009.
8. P.W. Cleary and J.J. Monaghan, *Journal of Computational Physics*, 148, 227–264, 1999.
9. R. Rook, M. Yildiz, and S. Dost, *Numerical Heat Transfer, Part B: Fundamentals*, 51, 1–23, 2007.
10. K. Szewc, J. Pozorski, and A. Taniere, *International Journal of Heat and Mass Transfer*, 54, 4807–4816, 2011.
11. M. Tong and D.J. Browne, *International Journal of Heat and Mass Transfer*, 73, 284–292, 2014.
12. J. Monaghan and A. Kocharyan, *Computer Physics Communications*, 87, 225–235, 1995.
13. A. Colagrossi and M. Landrini, *Journal of Computational Physics*, 191, 448–475, 2003.
14. X. Hu and N.A. Adams, *Journal of Computational Physics*, 213, 844–861, 2006.
15. X. Hu and N.A. Adams, *Journal of Computational Physics*, 227, 264–278, 2007.
16. N. Grenier, M. Antuono, A. Colagrossi, D. Le Touzé, and B. Alessandrini, *Journal of Computational Physics*, 228, 8380–8393, 2009.
17. K. Szewc, J. Pozorski, and J.P. Minier, *International Journal of Multiphase Flow*, 50, 98–105, 2013.
18. A. Zainali, N. Tofighi, M. Shadloo, and M. Yildiz, *Computer Methods in Applied Mechanics and Engineering*, 254, 99–113, 2013.
19. J.J. Monaghan, *Annual Review of Astronomy and Astrophysics*, 30, 543–574, 1992.
20. J.P. Morris, P.J. Fox, and Y. Zhu, *Journal of Computational Physics*, 136, 214–226, 1997.
21. S.J. Cummins and M. Rudman, *Journal of Computational Physics*, 152, 584–607, 1999.
22. A.J. Chorin, *Mathematics of Computation*, 22, 745–762, 1968.
23. W. Gao, Y. Duan, and R.X. Liu, *Journal of Hydrodynamics, Series B*, 21, 201–211, 2009.
24. S. Shao and E.Y. Lo, *Advances in Water Resources*, 26, 787–800, 2003.
25. R. Xu, P. Stansby, and D. Laurence, *Journal of Computational Physics*, 228, 6703–6725, 2009.
26. A. Ghasemi, B. Firoozabadi, and M. Mahdinia, *European Journal of Mechanics-B/Fluids*, 38, 38–46, 2013.
27. P.W. Cleary, *Applied Mathematical Modelling*, 22, 981–993, 1998.
28. C.W. Hirt and B.D. Nichols, *Journal of Computational Physics*, 39, 201–225, 1981.
29. S. Osher and J.A. Sethian, *Journal of Computational Physics*, 79, 12–49, 1988.
30. G.A. Dilts, *International Journal for Numerical Methods in Engineering*, 44, 1115–1155, 1999.
31. J. Monaghan, *Journal of Computational Physics*, 82, 1–15, 1989.
32. S. Shao, *International Journal for Numerical Methods in Fluids*, 69, 1715–1735, 2012.
33. J. Monaghan and A. Rafiee, *International Journal for Numerical Methods in Fluids*, 71, 537–561, 2013.
34. Z. Chen, Z. Zong, M.B. Liu, L. Zou, H.T. Li, and C. Shu, *Journal of Computational Physics*, 283, 169–188, 2015.
35. S. Nugent and H. Posch, *Physical Review E*, 62, 4968–4975, 2000.
36. A. Tartakovsky and P. Meakin, *Physical Review E*, 72, 1–9, 2005.
37. S. Adami, X. Hu, and N. Adams, *Journal of Computational Physics*, 229, 5011–5021, 2010.
38. T. Breinlinger, P. Polfer, A. Hashibon, and T. Kraft, *Journal of Computational Physics*, 243, 14–27, 2013.
39. J. Brackbill, D. B. Kothe, and C. Zemach, *Journal of Computational Physics*, 100, 335–354, 1992.
40. J. P. Morris, *International Journal for Numerical Methods in Fluids*, 33, 333–353, 2000.
41. M. Shadloo, A. Zainali, and M. Yildiz, *Computational Mechanics*, 51, 699–715, 2013.
42. G. Colicchio, Violent Disturbance and Fragmentation of Free Surface, PhD Dissertation, University of Southampton, Southampton, U.K., 2004.
43. D. Layzer, *The Astrophysical Journal*, 122, 1–12, 1955.
44. S. Hysing, S. Turek, D. Kuzmin, N. Parolini, E. Burman, S. Ganesan, and L. Tobiska, *International Journal for Numerical Methods in Fluids*, 60, 1259–1288, 2009.

9 Macroscale Computational Techniques in Interfacial Science

Mohsen Karbaschi, Mohammad Taeibi Rahni,
Dariush Bastani, and Reinhard Miller

CONTENTS

9.1 INTRODUCTION

Fluids are deformable materials composed of molecules, which show no resistance to shear forces. In principle, even a very small volume of a fluid contains a very large number of molecules associated with the physical behavior of the fluid at different dynamic conditions. As the molecular structures of fluids differ, the behavior of fluids under the action of forces differs considerably. The problem is more complicated when two fluids come close to each other and form an interface. Here, any change in the molecular structure of the interface would affect the flow filed around the interface and vice versa. Therefore, powerful computational techniques are required for comprehensive studies on fluid–liquid interfaces.

Nowadays, computational techniques are powerful tools for the simulation of different complex phenomena. By improving and refining these methods, as well as rapid increase in the speed of computers, it is possible to precisely compute most physical events occurring in fluid flows. In particular, for interfacial science and complex fluid–liquid interfaces, a wide variety of different computational techniques have been developed.

In molecular-scale simulations, the behaviors of each molecule (or a collection of molecules) at a certain point have been computationally studied (see Chapter 6). However, in macroscale simulations, the average of these behaviors over a large number of molecules (in the vicinity of a respective point) has been numerically investigated. While in mesoscale computations, instead of discretizing the fluid media, the fluid itself is hypothetically broken into many small particles whose dynamics are then to be studied. Examples of such techniques are lattice Boltzmann method (LBM), which implements an approximated Boltzmann equation [originally derived for molecular dynamics studies (see Chapter 7) and smoothed particle hydrodynamics (SPH) method, which uses the Lagrangian form of the full Navier–Stokes equations (see Chapter 8)].

This chapter is dedicated to some basics of macroscale computational techniques first for fluid flows in general and then for computations of the dynamics of fluid–liquid interfaces. Section III of this book contains chapters on the advancement and application of relevant computational techniques. Section IV of the book is dedicated to different advanced simulations performed for specific physics containing complex interfaces.

9.2 BASIC DEFINITIONS

There are many natural events occurring at interfaces of soft matters and colloidal systems. These include convection and dissipation of fluids, which affect the rate of mass, heat, and momentum transfer to and from the interface. In multiphase flows, these quantities are quite sensitive to the topology of the dispersed and continuous phases. These phenomena also interact with the boundary layers and, therefore, associate with interfacial thermomechanical properties, such as surface tension, adsorption kinetics, and interfacial rheology [1,2]. From a macroscale point of view, all these phenomena can be governed by transport equations and the equations for interfacial interactions, for example, adsorption isotherms. Many of these fluid-interface interactions are nonlinear, and thus there is no analytical solution to accurately describe their dynamical behavior.

In computational fluid dynamics (CFD), however, many different numerical techniques have been developed to solve the governing partial differential equations (PDE) within acceptable numerical errors, so as to properly describe such complex physical phenomena.

9.2.1 Fluid Dynamics

Multiphase flow refers to any flow consisting of more than one fluid (phase). Unless we deal with problems associated with stability (e.g., microfluidic flows or turbulent flow regimes), the basis of numerical simulations for single-phase and multiphase flows is generally the same. Even though the theoretical background of single-phase flow simulations is well established, there are still many unresolved difficulties with computations of complex multiphase flow problems. To better understand the complexities of such flows on an advanced level, one needs to first have a clear picture of the different fundamental aspects of both fluid dynamics and CFD. Thus, we try to briefly present the basis of these two subjects before getting into the details of interfacial computations.

9.2.2 Continuum Assumption and the Navier–Stokes Equations

For most fluids around us, we can ignore the spaces between the fluid molecules (mean free path) and thus assume there is fluid at all spatial locations. This assumption is called *continuum* and is used when deriving the Navier–Stokes equations. It simply means that the matter is continuously distributed throughout the whole region and the smallest volume of liquid contains a large number of molecules. In addition, with this assumption, we can define different macroscale quantities, such as velocity, density, temperature, pressure, and kinetic energy at each point of the fluid medium [3].

In principle, the dynamical behavior of fluids is determined by conservation laws of quantities such as mass, momentum, and energy. Conservation laws express that the net variation of a quantity

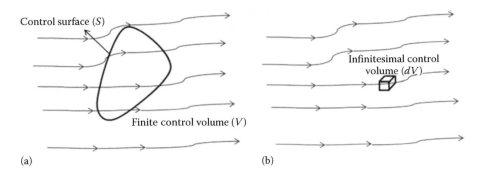

FIGURE 9.1 Schematics of two different control volume approaches: finite control volume approach (a) and infinitesimal control volume approach (b).

inside the *control volume* is equal to the net amount of that quantity being transported across the boundaries of that volume (plus some extra sources, if exist). The mathematical description of the conversation laws results in the *Navier–Stokes, transport,* or *governing* equations of motion in both differential and integral forms. To mathematically describe the conversation laws, one needs to divide the flow field into a number of volumes and describe the behavior of fluid in one finite region. For this purpose, the so-called finite control volume is defined as any arbitrary finite region of the flow, bounded by closed surfaces. Using an infinitesimal control volume approach, the fluid elements are extremely small, and the flow equations are derived in partial differential form. On the other hand, applying the fundamental physical principles to a finite control volume would result in an integral form of the Navier–Stokes equations [4]. Figure 9.1 represents the schematics of these two different control volume approaches.

The integral form of the conservation laws states that the sum of the changes of any extensive property per unit volume, L, inside a control volume is equal to its loss or gain through the boundaries of the control volume plus what is created or consumed by sources and sinks inside the control volume, i.e.:

$$\frac{d}{dt}\int_V L dV = -\int_A L\vec{v} \cdot \hat{n} dA - \int_A D\nabla L \cdot \hat{n} dA - \int_V Q dV, \tag{9.1}$$

where
    V is volume
    A is boundary surface
    D is diffusivity (or diffusion coefficient)
    Q is source term
    $\vec{v}$ is velocity vector
    $\hat{n}$ is unit vector normal to the boundary surface

The differential form of this equation is written as:

$$\frac{\partial L}{\partial t} = -\nabla . L\vec{v} - \nabla \cdot D\nabla L - Q. \tag{9.2}$$

The first terms in the right-hand sides of Equations 9.1 and 9.2 refer to the rate of loss or gain, which are defined by the rate of transport of the quantity through the control volume (entering or leaving) [4]. For different quantities, there are two dominant mechanisms of transport, that is, convection and diffusion. Diffusion refers to the transport due to the presence of flow gradients and represents the transports performed by random thermal motion of molecules within the fluid (it always exists, even in the absence of any flow). On the other hand, convection refers to the transport of quantities due to the motion of the carrier fluid performed by different forces acting on the fluid. In some systems, or for special cases, one of these two mechanisms may dominate, which can result in some simplifications. For example, the so-called Euler equations are the simplified forms of the Navier–Stokes equations, describing the pure convection of flow quantities under inviscid flow condition [5].

For conservation of mass, one can consider density (ρ; mass per unit volume) as the intensive property of interest. If no mass is created or destroyed by any means, for example, a chemical reaction, the source term in the mass conservation equation is zero and thus the integral and differential equations can be written as:

$$\frac{d}{dt}\int_V \rho dV = -\int_A \rho \vec{v} \cdot \hat{n} dA, \tag{9.3}$$

$$\frac{\partial \rho}{\partial t} = -\nabla \cdot \rho \vec{v}. \tag{9.4}$$

Similarly, for the conservation of momentum and energy, considering their corresponding quantities, the related equations can be derived.

For different problems in fluid dynamics, different physical/geometrical constraints or boundary conditions have to be applied. These constraints, together with additional equations applied, for instance, to define a specific chemical reaction or any thermodynamic relation, may be added. This results in a coupled system of equations, describing the complete dynamics of the fluid flow system. Different numerical techniques have been developed in CFD to solve such systems numerically. As follows, the basics of such techniques are briefly described.

9.3 BASICS OF CFD SIMULATIONS

In principle, there are a wide variety of methodologies for computations of fluid flow problems. However, using the right methodology for a specific problem is the key point to ensure convergence to a most accurate solution.

9.3.1 GRIDDING STRATEGY

In order to compute the dynamics of fluid flow with macroscale methods, we divide the physical space into a large number of geometrical elements. These *elements*, also called *grid cells*, can be formed by first creating the grid nodes, which should be properly distributed over the physical space. These grid nodes are then connected by grid lines to make the grid.

When generating the grid, there are specific requirements to be considered, otherwise one may face divergence problems and/or inaccuracy of the solution. For instance, any grid generation should avoid mismatching grid nodes or voids and overlaps of grid cells. In addition, the location of grid nodes should be adjusted carefully so as to increase the grid quality by creating smooth grid cells and reducing grid skewness. The topology of cells is another important issue, as it directly affects the stability, convergence, and accuracy of the solution. Note, for an optimal grid, one needs to reach aspect ratios of the elements (ratio of longest to shortest sides) as close to unity as possible. In addition, local variations in cell sizes should be minimized, that is, the size of the elements must vary gradually.

A grid can be either structured or unstructured, depending on the requirements of the problem. In structured grids, the nodes are distributed in an organized and similar manner along all grid

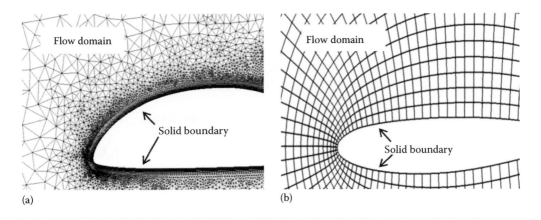

FIGURE 9.2   Samples of unstructured (a) and structured (b) grid types.

lines. Thus, in such grid, the nodes are easily identified by their indices. Such grids consist of a number of grid lines that do not cross each other. Unstructured grids on the other hand are much more complex. Therefore, a list of connectivity definitions is required to specify the set of nodes that make a specific element. Of course, unstructured grids are very useful for complicated geometries. Figure 9.2 represents examples of structured and unstructured grids.

Although in CFD simulations different approaches for grid generation are introduced, there is still no straightforward technique for an optimal gridding and the grid generation strongly depends on the geometry and the physics of the problem of interest.

9.3.2   Discretization Methods

As mentioned in Section 9.2, a coupled system of different forms of equations needs to be solved to numerically model the physical behavior of a fluid flow system. The numerical techniques for solving the governing equations directly depend on the type of equations, boundary geometry, and other properties of the flow system. To better understand the basis of numerical methods, let us first describe the expanded differential form of the Navier–Stokes equations for a 2D incompressible flow of a Newtonian fluid on a typical grid as shown in Figure 9.3. This provides the most common form of PDEs that describe transport phenomena in fluid flows.

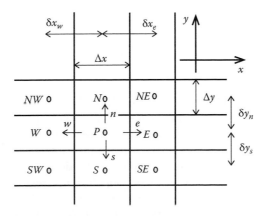

FIGURE 9.3   A schematic of a typical 2D control volume. (From Ferziger, J.H. and Perić, M., *Computational Methods for Fluid Dynamics*, Springer, Berlin, Germany, 2002.)

If we consider u and v as velocity components in x- and y-directions, Equation 9.4 can be expanded as

$$\frac{\partial \rho}{\partial t} = -\frac{\partial(\rho u)}{\partial x} - \frac{\partial(\rho v)}{\partial y}. \tag{9.5}$$

For an incompressible flow, a simplified form of this equation would be

$$\frac{\partial u}{\partial x} + \frac{\partial v}{\partial y} = 0. \tag{9.6}$$

Conservation of momentum in x- and y-directions is presented as

$$\rho\left(\frac{\partial u}{\partial t} + u\frac{\partial u}{\partial x} + v\frac{\partial u}{\partial y}\right) = -\frac{\partial P}{\partial x} + \mu\left(\frac{\partial^2 u}{\partial x^2} + \frac{\partial^2 u}{\partial y^2}\right) + \rho g_x, \tag{9.7}$$

$$\rho\left(\frac{\partial v}{\partial t} + v\frac{\partial u}{\partial x} + v\frac{\partial v}{\partial y}\right) = -\frac{\partial P}{\partial y} + \mu\left(\frac{\partial^2 v}{\partial x^2} + \frac{\partial^2 v}{\partial y^2}\right) + \rho g_y. \tag{9.8}$$

Here, the transports of momentum due to convection (first-order derivative of velocity with respect to position) and viscous diffusion (second-order derivative of velocity with respect to position) are the left-hand sides and the second terms on the right, respectively.

The aforementioned PDEs do not have any analytical solution and hence their numerical solution is the only remaining option. After grid generation, they need to be discretized, that is, one needs to obtain algebraic equations from the original PDEs, the solution of which provides a proper approximation. Here, having a correct strategy in discretization is very crucial to ensure convergence to an accurate solution, with the same requirement as for the grid generation.

In addition, one needs to be careful about stability and consistency of numerical schemes used. Consistency refers to the fact that it must be demonstrated that the obtained algebraic equations are equivalent to the original differential equations, when the grid spacing tends to zero. On the other hand, a numerical scheme is stable if errors do not grow toward very large numbers during computations. Note, for stable schemes, the solution changes only by a small amount, when input data vary negligibly.

In principle, there are a wide variety of methodologies applied for discretization of equations in fluid dynamics. However, they are basically classified into three main categories for discretization with respect to space. These are finite difference schemes, finite volume schemes, and finite element schemes. For discretization with respect to time, we have two main categories of the so-called implicit and explicit schemes. Different discretization methods are briefly explained in the following subsections.

9.3.2.1 Finite Difference Method

Discretization of ordinary differential equations (ODEs) or PDEs, based on finite difference method (FDM), is the oldest and easiest method. In FDM, different terms of the equations are replaced by their corresponding Taylor series expansions, to convert the differential equations to a set of algebraic equations. For example, to approximate a first-order derivative of an arbitrary variable L at point P with respect to x (Figure 9.3), one can develop Taylor series expansions as

$$L\left(x_P + \delta x_e\right) = L\left(x_P\right) + \delta x_e \left.\frac{\partial L}{\partial x}\right|_{x_P} + \frac{1}{2}\delta x_e^2 \left.\frac{\partial^2 L}{\partial x^2}\right|_{x_P} + \frac{1}{6}\delta x_e^3 \left.\frac{\partial^3 L}{\partial x^3}\right|_{x_P} + \cdots, \tag{9.9}$$

$$L\left(x_P - \delta x_w\right) = L\left(x_P\right) - \delta x_w \left.\frac{\partial L}{\partial x}\right|_{x_P} + \frac{1}{2}\delta x_w^2 \left.\frac{\partial^2 L}{\partial x^2}\right|_{x_P} - \frac{1}{6}\delta x_w^3 \left.\frac{\partial^3 L}{\partial x^3}\right|_{x_P} + \cdots. \tag{9.10}$$

By adding and subtracting these, neglecting the differences between δx_w and δx_e (setting them both to be equal to Δx), and neglecting higher-order terms, one can obtain

$$L\left(x_P + \Delta x\right) + L\left(x_P - \Delta x\right) = 2L\left(x_P\right) + \Delta x^2 \left.\frac{\partial^2 L}{\partial x^2}\right|_{x_P} + O(\Delta x^2), \tag{9.11}$$

$$L\left(x_P + \Delta x\right) - L\left(x_P - \Delta x\right) = 2\Delta x \left.\frac{\partial L}{\partial x}\right|_{x_P} L(x_P) + O(\Delta x^2). \tag{9.12}$$

Therefore, a second-order central difference approximation of a first-order derivative and a second-order derivative can be consequently written as

$$\left.\frac{\partial L}{\partial x}\right|_{x_P} = \frac{L\left(x_P + \Delta x\right) - L\left(x_P - \Delta x\right)}{2\Delta x}, \tag{9.13}$$

$$\left.\frac{\partial^2 L}{\partial x^2}\right|_{x_P} = \frac{L\left(x_P + \Delta x\right) + L\left(x_P - \Delta x\right) - 2L(x_P)}{\Delta x^2}. \tag{9.14}$$

Using Equations 9.9 and 9.10, one can also directly obtain first-order approximations of a first-order derivative as

$$\left.\frac{\partial L}{\partial x}\right|_{x_P} = \frac{L(x_P + \Delta x) - L(x_P)}{\Delta x}, \tag{9.15}$$

$$\left.\frac{\partial L}{\partial x}\right|_{x_P} = \frac{L(x_P) - L(x_P - \Delta x)}{\Delta x}. \tag{9.16}$$

These types of approximations of first-order derivatives are called forward and backward difference approximations. Figure 9.4 shows a schematic of the definitions for a first-order derivative and its different approximations.

Using FDMs, such type of approximations can be applied to the derivative terms of the PDEs at each node and thus a coupled system of algebraic equations will be obtained. Of course, to increase the accuracy (decreasing truncation error in computations), one can increase the order of approximations or to refine the grid. Note, FDMs can be applied to any type of grid, but it is mostly used for structured ones.

Stability issues in FDMs depend directly on the chosen method of discretization with respect to time. The same as in first order derivative approximation with respect to space discussed before, one can obtain the following first-order approximation for derivative with respect to time:

$$\left.\frac{\partial L}{\partial t}\right|_t = \frac{L(t + \Delta t) - L(t)}{\Delta t} = \frac{L(t) - L(t - \Delta t)}{\Delta t}. \tag{9.17}$$

Considering a forward difference in time results in the so-called explicit scheme for our numerical solution. This is when dependent variables are explicitly found in terms of known quantities. This simply

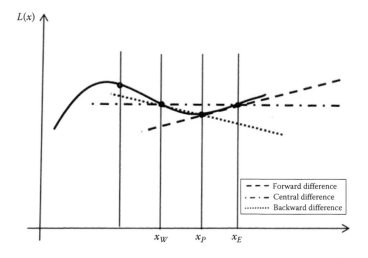

FIGURE 9.4 Different approximations for a first-order derivative. (From Ferziger, J.H. and Perić, M., *Computational Methods for Fluid Dynamics*, Springer, Berlin, Germany, 2002.)

means, for explicit calculations, the state of the system at a later time is calculated from the state of the system at the current time from known quantities. In contrast, considering the backward difference in time leads to the so-called implicit scheme where the dependent variables are defined by a coupled set of equations. In other words, the state of the system at a later time can be calculated from equations that involve both the current state (known quantities) and the later state (unknown quantities) of the system. The Crank–Nicolson scheme, for example, is based on assuming a central difference approximation for derivatives with respect to time [7]. The application of the implicit Crank–Nicolson scheme, is stable and very accurate but results in numerically intensive simulations as some iterative techniques need to be applied for solving the coupled set of equations. On the other hand, in the application of explicit schemes, the time step is constrained by the required accuracy and is sensitive to stability issues. Therefore, for complicated geometries, one needs to apply explicit schemes with a special care about the stability issue for the applied grid and numerical method.

9.3.2.2 Finite Volume Method

Finite volume method (FVM) is another type of methods one can use to discretize the governing equations of fluid flows. Its main difference compared to FDM is that in FVM, the integral form of the equations is used over finite control volumes of the domain. Considering surface and volume integration of different quantities over each control volume or the boundaries to satisfy the conservation laws, it allows the finite volume formulation to be naturally conservative in a discrete sense. For approximation of surface integrals, cell-face values are sometimes approximated by the values at the grid nodes (the same as finite difference approximations). In addition, they are considered as the averaged values over the cell faces. Also, different orders of approximations can be considered, depending on the required accuracy and other physical or numerical specifications of the problem. Therefore, different schemes, such as upwind interpolation, quadratic upwind interpolation, linear interpolation, and skew upwind interpolation have been described [8].

These averaged values can physically represent the rate of transport of quantities through a specific cell face. Therefore, a balance between the rates of transport of each quantity through the control volume can be obtained to ensure conservation of quantities.

9.3.2.3 Finite Element Method

In finite element method (FEM), the residuals of PDEs are converted to weak formulations, using mostly weighted residual methods [9,10]. Thereafter, the volumetric integral of the weak form of equations is satisfied, using the Galerkin method. Then, piecewise-polynomial functions are used to approximate the variations of the solution over each grid line. In this method, more mathematics is involved allowing the application of adaptive procedures for handling singularities or getting higher-order local approximations.

9.4 COMPUTATIONAL APPROACHES TO MULTIPHASE FLOWS

There are many complicated physical flow processes that require much more efforts than just solving the Navier–Stokes equations. For instance, variations of physical properties due to the presence of chemical reactions leading to species production or dissipation or temperature gradients make the flow processes very complex. The situation is the same for multiphase flows wherein interfaces with very complicated configurations exist. Usually, there are multiscale interactions controlling the dynamic/molecular and macroscopic behavior of multiphase flow systems. In gas–liquid systems, for example, a wide range of configurations may exist with consequences for bulk (e.g., hydrodynamics, heat, and mass transfer) and interfacial properties (e.g., physicochemical and rheological properties). This can include bubbly flows at lower velocities, where the dispersed bubbles are within a continuous liquid phase with a complex motion or slug flows at higher velocities, where bubble sizes increase due to coalescence. In such situations, the hydrodynamics does not only affect the flow patterns or the mechanical properties of the interfaces due to the transport of surfactant molecules, but also affect the process of coalescence between bubbles or break-off of large bubbles into smaller ones.

In this book, different chapters are dedicated to advanced numerical simulations applied for special configurations of large interfaces. To better understand the basis of these methodologies, in this subsection, an overview of the main concepts in computations of liquid–fluid interfaces is presented and discussed. For multiphase flow systems including many small particles, drops, or bubbles (e.g., spray systems or bubbly flows), there are too many complex interactions between the elements of the dispersed phase and the continuous phase that must be taken into account. The basis of handling such engineering problems is the same as for computations of large interfaces. However, computational techniques applied to predict the dynamics of such systems include simplifying assumptions in order to allow a decomposition of the whole system into simpler subsystems. Thereafter, the computational results of simpler subsystems would be extended to simulate the original problem. Computation of such systems is briefly explained in Section 9.4.4.

In general, computational methods dealing with large fluid interfaces are categorized into two main techniques: interface-capturing methods and interface-tracking methods (also called front-capturing or front-tracking methods). In interface-capturing methods, there is no explicit element to represent the interface; instead, a high variation of a specific variable is used to specify the regions containing the interface. As a result, the interface location is not precisely computed and may diffuse over several cells, which of course affects the accuracy of computations. However, it makes the computations easy, especially when we deal with drastically contracting interfaces (e.g., fast growing or contractions of the interfaces or interface break-off and coalescence). In interface-tracking methods, a linked set of grid nodes or cell faces represents the interface. Therefore, variations of the interface can be tracked by tracking the motion of the grid cells, since the interface grid follows the fluid flow. Here, the location of the interface is usually computed more precisely, providing a more accurate computation of variables along and across the interface. However, taking a well-curved interface needs intensive refinement of the grid, which results in intensive computations. In addition, large interfacial deformations or changes in the topology of the interface (e.g., the process of breakup or coalescence) result in distortions of the grid cells at the interface and therefore can not be handled with this approach.

9.4.1 VOLUME OF FLUID TECHNIQUE

In interface-capturing methods, a special characteristic function is defined to represent the interface. The volume of fluid (VOF) technique, presented by Nichols and Hirt (1975) [11,12], is one of the methods developed based on the interface-capturing approach. In this method, a color function (ϕ) is used to mark the regions occupied by one of the two fluids. The formulation of this quantity and its transport equation for a two phase flow systems are

$$\phi = \frac{V_{\text{Fluid I}}}{V_{\text{Fluid I}} + V_{\text{Fluid II}}}, \tag{9.18}$$

$$\frac{\partial \phi}{\partial t} = -\nabla \cdot \phi \vec{v}. \tag{9.19}$$

The values of ϕ represent the location of the interface (i.e., $0 < \phi < 1$) and also the individual phases (i.e., $\phi = 0$ or $\phi = 1$). In addition, all molecular properties of the fluids involved can be represented by this quantity. For example, to represent the density or viscosity of the fluids, the following formulations are used:

$$\rho = \rho_{\text{Fluid I}} \phi + \rho_{\text{Fluid II}} (1 - \phi), \tag{9.20}$$

$$\frac{1}{\mu} = \frac{\phi}{\mu_{\text{Fluid I}}} + \frac{1 - \phi}{\mu_{\text{Fluid II}}}. \tag{9.21}$$

Here, the *continuum surface force method* can be considered to model surface tension effects. This can be done by calculating the mean curvature, $K = -\nabla \cdot \left(\dfrac{\nabla \phi}{|\nabla \phi|} \right)$, and then evaluating the volume force F_{sv} (as an alternative for the Laplace surface force). Figure 9.5 represents schematically the interface, using a VOF technique, together with the results of a growing drop simulation. As we can see, a linear piecewise *interface reconstruction approach* is applied.

9.4.2 LEVEL-SET TECHNIQUE

The level-set methods (LSMs) are another important group of interface-capturing approaches, in which a level-set function represents the interface. The basis of this method is similar to the basis of the VOF technique. However, LSMs are better for accurately computing the curvature of an interface. This is very important for the calculation of the Laplace equation and the pressure jump across the interface. The method was first developed by Osher and Sethian [13]. It is very popular for tracking moving interfaces and is well capable in dealing with topology changes. In LSMs, interfaces are explicitly represented as a zero level set of a higher-order function, called level-set

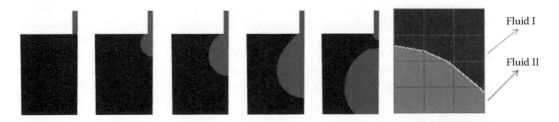

FIGURE 9.5 Interface-capturing technique (volume of fluid) and the results of a growing drop simulation.

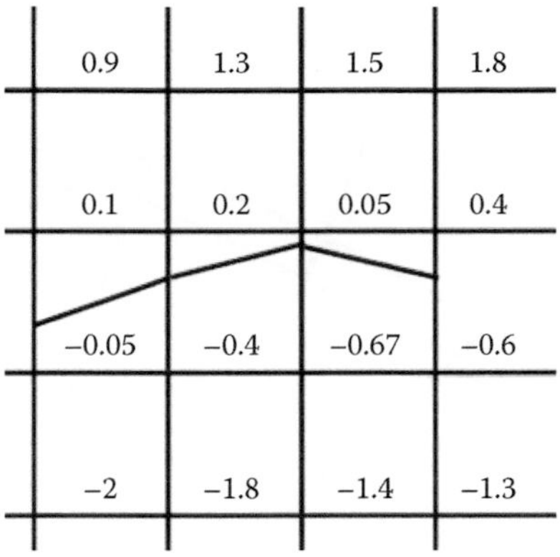

0.9	1.3	1.5	1.8
0.1	0.2	0.05	0.4
−0.05	−0.4	−0.67	−0.6
−2	−1.8	−1.4	−1.3

FIGURE 9.6 Location of the interface with respect to changes of the level-set function at grid points.

function ($\phi(x,t = 0)$ = a given value), as shown in Figure 9.6. This provides information about the initial position of the interface. For the evaluation of the surface with time, a function F gives the normal speed of the surface movements and accompanies the change of the level-set function. The level-set function for the evaluation of the interface is defined as

$$\phi(x,t) + F|\nabla\phi| = 0, \tag{9.22}$$

where $|\nabla\phi|$ is the normalized gradient of the level-set function. The speed function controls the smoothness of the level-set function. In addition, the speed function should be defined in such a way that it is suitable for the given specific application, that is pulling the interface toward the desired position. This function depends on the geometry of the front and also on the physics of the problem.

9.4.3 INTERFACE-TRACKING APPROACHES

In interface-tracking approaches, the Navier–Stokes equations are solved with moving boundaries. The arbitrary Lagrangian–Eulerian approach, also known by its acronym ALE, is one of the main methods developed based on this approach, to avoid grid distortion when the grid follows the fluid flows. Therefore, this method is capable to alleviate many drawbacks of other traditional multi-phase CFD methods. In ALE approach, the grid cells maybe moved with the same velocity as the fluid at the corresponding position (called Lagrangian approach) or might be kept as a fixed grid (called Eulerian approach) or might be moved with an arbitrary velocity to ensure smooth interaction between the phases [14]. Therefore, the equations should be adapted to the grid deformation and the velocity of the moving domain contributes as an additional variable in the equations. For example, the equation of motion would be modified as

$$\frac{\partial(\rho\vec{u})}{\partial t} + \left[(\vec{u} - \vec{u}_m) \cdot \nabla\right](\rho\vec{u}) = -\nabla P + \mu\nabla^2\vec{u} + \rho g, \tag{9.23}$$

where $\vec{u}_m$ is the mesh deformation velocity of the moving domain.

A mathematical brief description of this method together with its application to characterize the process of drop formation based on the profile analysis tensiometry can be found in [15]. For more detailed description on the basics of the method, one can also see [16,17]. Figure 9.7 shows a schematic picture of interface tracking by the movement of grid cells.

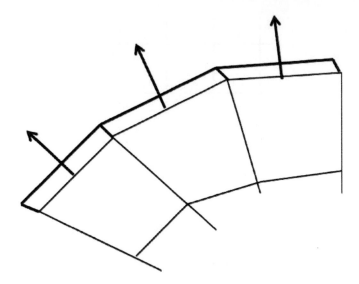

FIGURE 9.7 Schematic picture of the surface-tracking approach by tracking the movement of the cell surfaces.

9.4.4 Flow Computations of Systems Containing Very Small Particles, Bubbles, or Drops

Although many problems dealing with complex fluid interfaces can be computationally studied by using the presented approaches or hybrid schemes in the preceding text, there are many complex configuration of multiphase flows where these techniques can not adequately address the interactions between different phases. Examples are sprays or bubbly liquids, which are widely used to facilitate dispersion of a liquid or gas into a continuous fluid for many thermal or fluid mixing applications. These systems include many complexities of fluid–droplet/bubble/particle, droplet–droplet, bubble–bubble, or particle–particle interactions. The dynamic performance of these two-phase flow systems is directly influenced by many parameters such as geometrical specifications, continuous phase velocity, drop/ bubble/particle size distributions and volume fractions, and temperature. In this context, CFD is a powerful tool to perform a parametric analysis. However, for such problems, we have to deal with complicated and highly nonlinear systems of PDEs that describe interactions at different time scales between different fluid elements. To overcome such complexities, a decomposition of systems into simpler subsystems was introduced by many researchers in order to avoid substantial difficulties inherent in the original systems. There are also schemes proposed to interpret and extend the results of computations of subsystems into the simulation of the whole system. The study of such complex engineering systems is out of the scope of this book. As an example of such systems and introduced simplifications, one can refer to [18–20].

9.5 SUMMARY

This chapter presents the basis of macroscale computational techniques in interfacial science. Here, fundamental aspects of fluid dynamics and its computations together with different common and advanced techniques applied for formulation and discretization of governing equations are introduced. In addition, specific computational techniques for investigation of multiphase fluid flows are presented and discussed.

REFERENCES

1. R. Miller, L. Liggieri, *Interfacial Rheology*, CRC Press, Boca Raton, FL, 2009.
2. R. Miller, L. Liggieri, *Bubble and Drop Interfaces*, Brill, Leiden, 2011.
3. Y.-C. Fung, *A First Course in Continuum Mechanics*, Englewood Cliffs, NJ: Prentice-Hall, Inc., 1977, 351 p., Vol. 1, 1977.
4. R.B. Bird, W.E. Stewart, E.N. Lightfoot, *Transport Phenomena*, John Wiley & Sons, New York, 2007.
5. N. Ron-Ho, A multiple-grid scheme for solving the Euler equations, *AIAA J.*, 20 (1982) 1565–1571.
6. J.H. Ferziger, M. Perić, *Computational Methods for Fluid Dynamics*, Berlin, Germany: Springer, 2002.
7. J. Crank, P. Nicolson, A practical method for numerical evaluation of solutions of partial differential equations of the heat-conduction type, *Mathematical Proceedings of the Cambridge Philosophical Society*, Cambridge, U.K.: Cambridge University Press, pp. 50–67, 1947.
8. H.K. Versteeg, W. Malalasekera, *An introduction to Computational Fluid Dynamics: The Finite Volume Method*, Pearson Education, Harlow, 2007.
9. G. Dhatt, E. Lefrançois, G. Touzot, *Finite Element Method*, John Wiley & Sons, Hoboken, 2012.
10. K.J. Bathe, Finite Element Procedures, Pearson Education, New Jersey, 2006.
11. B.D. Nichols, C.W. Hirt, Methods for calculating multidimensional, transient free surface flows past bodies, *First International Conference on Numerical Ship Hydrodynamics*, Gaithersburg, MD, 1975.
12. C.W. Hirt, B.D. Nichols, Volume of fluid (VOF) method for the dynamics of free boundaries, *J. Comput. Phys.*, 39 (1981) 201–225.
13. S. Osher, J.A. Sethian, Fronts propagating with curvature-dependent speed: algorithms based on Hamilton-Jacobi formulations, *J. Comput. Phys.*, 79 (1988) 12–49.
14. G. Tryggvason, R. Scardovelli, S. Zaleski, *Direct Numerical Simulations of Gas-Liquid Multiphase Flows*, Cambridge, U.K.: Cambridge University Press, 2011.
15. K. Dieter-Kissling, M. Karbaschi, H. Marschall, A. Javadi, R. Miller, D. Bothe, On the applicability of drop profile analysis tensiometry at high flow rates using an interface tracking method, *Colloids Surf. A*, 441 (2014) 837–845.
16. C. Hirt, A.A. Amsden, J. Cook, An arbitrary Lagrangian-Eulerian computing method for all flow speeds, *J. Comput. Phys.*, 14 (1974) 227–253.
17. S. Ushijima, Three dimensional arbitrary Lagrangian–Eulerian numerical prediction method for non-linear free surface oscillation, *Int. J. Numer. Methods Fluids*, 26 (1998) 605–623.
18. A. Kubilay, D. Derome, B. Blocken, J. Carmeliet, CFD simulation and validation of wind-driven rain on a building facade with an Eulerian multiphase model, *Build. Environ.*, 61 (2013) 69–81.
19. Z. Zhang, Q. Chen, Prediction of particle deposition onto indoor surfaces by CFD with a modified Lagrangian method, *Atmos. Environ.*, 43 (2009) 319–328.
20. R.I. Singh, A. Brink, M. Hupa, CFD modeling to study fluidized bed combustion and gasification, *Appl. Therm. Eng.*, 52 (2013) 585–614.

Section III

Applied Multiscale Computational Methodologies

10 Computational Quantum Chemistry Applied to Monolayer Formation at Gas/Liquid Interfaces

Yuri B. Vysotsky, Elena S. Kartashynska, Elena A. Belyaeva, Valentin B. Fainerman, Dieter Vollhardt, and Reinhard Miller

CONTENTS

10.1 INTRODUCTION

Langmuir monolayers have been quite deeply and comprehensively investigated during the previous 80 years. Experimental investigation of the structure of the surfactant monolayers includes such methods as IR spectrometry [1,2], Brewster angle microscopy (BAM) [1], atomic force microcopy (AFM) [3], fluorescence [4] microscopy, and grazing incidence X-ray diffraction (GIXD) [5]. Since the 1990s, quantum chemical and dynamic methods have been juxtaposed with the experimental ones owing to the increase of instrumental possibilities of computers. These methods enable modeling the monolayer unit cells and calculation of their structural and thermodynamic parameters.

The main prerequisite for the development of the quantum chemical model of calculation of the thermodynamic clusterization parameters of surfactants are the following experimental facts:

- GIXD data show that densely packed monolayers at the air/water interface possess crystalline lattice [5].
- Analysis of Π-A isotherms for different surfactants at the air/water interface reveal that molecules of the condensed monolayers are in the maximum extended *all-trans* conformation and tilted to the normal to the interface at angles from 0° to 49° [6,7].
- Molecules are immersed in the water phase on the same number of the methylene links (usually 2–3 links) [8].
- Spontaneous monolayer formation of aliphatic alcohols, acids, and amines starts when the molecules have 11–14 carbon atoms in their hydrocarbon chain (see, e.g., at 15°C [9]).

The described experimental facts should be reproduced in the frameworks of developed quantum chemical approach.

The use of quantum chemical methods in description of the monolayer clusterization at different interfaces tackles a number of challenges, and the main one is the absence of the possibility to calculate structural and thermodynamic parameters of the molecules that are at the interface between two immiscible phases. It is possible to neglect one of the phases presuming that molecule completely belongs to the other phase. However, in this case, one has to estimate the neglected increments, and in some way, to consider the influence of the interface on the calculated parameters. In the frameworks of our quantum chemical model developed to describe the clusterization of the substituted alkanes at the air/water interface, the quantum chemical calculations were carried out for the structures in vacuum. It was decided that the results obtained for molecules and clusters in vacuum are close to the corresponding results obtained in air phase as dielectric permittivities for these two phases are also similar. In the same time, the impact of the water phase was accounted via its stretching and orienting effect. When constructing the cluster structures, we considered that all molecules are maximum extended in *all-trans* conformation and immersed in the water phase on the same number of the methylene links so that the *mattress*-like structure is formed at the water surface (see Figure 10.1).

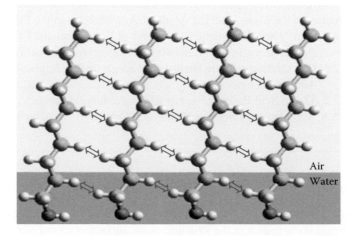

FIGURE 10.1   Orientation of the amphiphilic molecules with respect to the interface.

The methods of molecular dynamics [10] did not give any satisfactory result for the description of such systems, because the interactions between the alkyl chains and those between hydrophilic headgroups were chaotic. The attempts to exploit the COSMO model, which takes into account the presence of the solvent in the system, on the example of aliphatic alcohols [11] showed that dimerization enthalpies calculated for surfactants in water and in vacuum slightly differ. In this connection, the account of the interface is implicit and shows up via its orienting and stretching effect. But the proposed scheme does not fit to the description of the clusterization of ionized surfactants, because it is impossible to disregard the solvation effects.

Undoubtedly, the solvation effects contribute to the clusterization process for nonionic surfactants as well. However, as it was shown in Ref. [12], the impact of the interface change on clusterization of aliphatic alcohols emerges via the shift of the temperature boundaries of existing monolayer phases by only 4°C. With a glance at the calculated *temperature effect* of clusterization, we can say that the alcohol clusterization at the water medium is possible for compounds with 1–2 carbon atoms longer alkyl chain than those at the mercury medium. At the same time, the structure of the monolayer unit cell is the same for both air/water and air/mercury interfaces. In both cases, the monolayer unit cell for $C_{22}H_{45}OH$ is hexagonal with the next parameters: 5.0×7.5 Å at 5°C. In our studies, we consider conventional Langmuir monlayers at the air/water interface. However, one can say that the model proposed by us is applicable for any interface, provided the considered amphiphiles are in nonionized state.

The second hindrance to be surmounted is: the correct consideration of the intermolecular interactions, which contributes to the monolayer formation. Particularly, the description of the intermolecular interactions needs accounting of the electron correlation, because the significant effect of the dispersion attraction for molecular systems with closed molecular shells has merely correlation nature [13–15]. Simple quantum mechanical methods (e.g., Hartree–Fock method) fail to take into account the correlation effects. Their account needs sygnificant computing resources and time expenses. In the same time, the modern quantum chemical approaches include the atom–atom potentials at a certain stage, which implicitly take into consideration the electron correlation. For large amphiphile aggregates with alkyl chain lengths up to 15–20 carbon atoms, the realization of such calculations is difficult in the framework of *ab initio* methods with account of electron correlation even using small basic sets [16] in terms of the existing instrumental facilities. In this connection, the precise structural and thermodynamic parameters of the amphiphile monomer formation are reasonable to calculate using more accurate (*ab initio* or semiempirical) methods, whereas for clusters, such calculations should be done on the basis of the atom–atom potentials. The interactions between larger domains could be accounted in the frameworks of even rougher schemes. For example, Bagatur'yants sticks to such ideology in his studies [17] by modeling thin films of inorganic and complex compounds using the density functional theory.

As our investigations show [18], the quantum chemical semiempirical PM3 method fits best of all for the solution of the described problem. So, optimization and calculation of the thermodynamic parameters for all regarded nonionic surfactant classes was done in the framework of this method. One should use the supramolecular approach to obtain the thermodynamic clusterization parameters of the aggregate determined by intermolecular interactions. So, it is necessary to calculate the required parameters for the aggregate and to subtract the corresponding parameters calculated for the monomer multiplied by the number of the monomers in the cluster [19]. The supramolecular approximation was used in Refs. [20,21] dealing with the calculation of the thermodynamics of formation of semifluorinated alkanes and the mixed system sodium unsaturated carboxylate/dodecyltrimethyl ammonium bromide clusters. However, the authors confined themselves only to calculation of the dimerization enthalpy for these compounds disregarding the entropy factor that makes it impossible to judge about the possibility of such process in terms of Gibbs' energy.

We consider only solid highly ordered monolayers disregarding monolayers in the gaseous and liquid-expanded phases. Therefore, the monolayer translational symmetry needs to be accounted in a calculation that is another complex task. Fortunately, intermolecular interactions are pairwise additive as the force of the intermolecular interaction dramatically decreases with increase of the distance between two interacting molecules (it is inversely proportional to the distance raised to the sixth power). So, it is correct to consider only intermolecular interactions between the nearest molecules in the theoretical description of the monolayer [22,23] neglecting the translational symmetry in the chosen approximation.

Intermolecular interactions between molecules of the substituted alkanes in the monolayers depend significantly on the mutual orientation of the interacting molecules. They appeared to be identified by the mutual disposition of the hydrogen atoms of the hydrocarbon chain and the mutual disposition of the functional groups. The nearest atoms are presumed to make the maximum increment in energy of intermolecular interaction. We showed [11,24–30] that intermolecular interaction between two nearest surfactant molecules can be resolved into increments of the interactions between the nearest methylene groups and the nearest functional groups (see Figure 10.2). We singled out several types of the pairwise CH·HC interactions depending on the mutual orientation of the interacting molecules. The most energetically preferable are the «a» and «f» types of the interactions (see Figure 10.2) [29,31], which are realized in the monolayers of fatty alcohols at the air/water interface. In addition, the study of Ref. [31] revealed that pairwise CH·HC interactions of the «a» and «f» types are isoenergetic. For the description of formation of large clusters and infinite films, the additive scheme should be constructed, which bases on the classification of the interactions between the alkyl chains of the amphiphiles.

If there are the calculated values of the thermodynamic clusterization parameters for small clusters (dimers, trimers, tetramers, and hexamers), one can build the correlation dependencies of the enthalpy, entropy, and Gibbs' energy on the number of the pairwise CH·HC interaction realized in the cluster. It allows defining the thermodynamic clusterization parameters for any cluster using the found correlation dependencies if the cluster structure is specified and the number of the pairwise CH·HC interaction and the interaction between the functional groups in it is counted. In order to get the parameters of the infinite 2D cluster per monomer molecule, one should, first of all, model the fragment of the infinite film and express the number of realized CH·HC interactions using the number of the methylene fragments in the monomer alkyl chain comprising this film. Then, the dependencies of the clusterization thermodynamic parameters on the number of CH·HC interactions should be constructed using the values of the coefficients obtained earlier for small aggregates (dimers, trimers, tetramers, and hexamers). Then, one has to divide the correlation dependencies of clusterization thermodynamic parameters by the number of the monomers in the cluster and calculate the limits of the obtained expressions at the infinite number of molecules in the cluster.

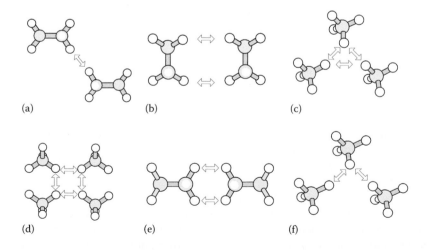

(a)   (b)   (c)

(d)   (e)   (f)

FIGURE 10.2 Types of intermolecular CH·HC interactions. (a)–(f) correspond to the revealed types of interactions.

These allow us to calculate the thermodynamic clusterization parameters for the clusters of any dimensions including 2D monolayers in the framework of the proposed model. The values of the thermodynamic clusterization parameters for surfactants can be estimated on the basis of the increments of the CH·HC interactions and the interactions between the hydrophilic parts of the molecules, because the spontaneous clusterization threshold of amphiphiles belonging to certain homologous series depends on the alkyl chain length (in turn, it defines the number of the CH·HC interactions realized in the cluster). In this case, the monolayer formation is possible only when the increment of the interactions between the methylene units of the alkyl chain exceeds the destabilizing effect contributed by the interactions between the hydrophilic parts of the molecules.

10.2 MODEL AND METHOD DESCRIPTION

10.2.1 MAIN IDEAS OF THE DEVELOPED MODEL FOR CALCULATION OF THERMODYNAMIC CLUSTERIZATION PARAMETERS OF AMPHIPHILES AT THE AIR/WATER INTERFACE

So, from what was said in the preceding section, one can single out the following key points for the quantum chemical model of calculation of the thermodynamic clusterization parameters:

1. The intermolecular CH·HC interactions between the methylene groups of the alkyl chains of the interacting amphiphiphile molecules provide the main contribution to the Gibbs' energy of cluster formation.
2. The supramolecular approximation was used to calculate the thermodynamic parameters of the cluster formation of the considered amphiphile types.
3. The additive scheme was constructed on the basis of the results of direct calculations. This scheme defines the values of the thermodynamic parameters of clusterization as the total contribution of the CH·HC interactions and the interactions between the hydrophilic parts of the amphiphile molecules realized in the cluster.
4. As demonstrated in Figure 10.1, only the CH·HC interactions realized between two methylene groups of the alkyl chains arranged opposite to each other (marked with the arrow) are taken into account and the interactions between the alkyl groups arranged much farther are neglected because of the decrease of the interaction energy inversely proportional to r^6, that is, the CH·HC interactions are additive in pairs.
5. The coefficients found in the framework of the additive scheme were used to obtain the thermodynamic parameter values of the cluster formation of large associates including 2D monolayers.
6. Implicit account of the interface via its orienting and stretching effect.

The calculation of the thermodynamic (enthalpy, entropy, and Gibbs' energy) and structural parameters of monolayer clusterization were carried out according to the quantum chemical model described in compliance with the next procedure:

1. Conformational analysis, which includes the construction of the potential energy dependence of the monomer on the torsion angle values of the functional groups in the hydrophilic part of the amphiphile with respect to the hydrophobic alkyl chain, and calculation of the thermodynamic parameters for the formed conformers
2. Calculation in supramolecular approximation of the clusterization thermodynamic parameters of small aggregates (dimers, trimers, tetramers, etc.) constructed on the basis of the conformers previously found
3. Determination of the tilt angle of the alkyl chains of the amphiphiles in the monolayer with respect to the interface, according to the scheme, which will be described in detail later on and finding the geometric parameters of the unit cell of the 2D lattice

4. Construction of the additive scheme for calculation of the thermodynamic clusterization parameters, which is based on the dependencies of the clusterization enthalpy, entropy, and Gibbs energy for small aggregates of the amphiphile on the number of the intermolecular CH·HC interactions and the interactions between the hydrophilic parts realized in these clusters

5. Calculation of the thermodynamic clusterization parameters of the 2D films per monomer molecule on the basis of the increments introduced by the CH·HC interactions and the interactions between the hydrophilic parts, which are estimated using the additive scheme

6. Analysis of the obtained dependencies of clusterization Gibbs' energy in order to define which of the clusters mainly govern the 2D film formation

10.2.2 JUSTIFICATION OF THE CALCULATION METHOD

Calculation of any multiatomic system is based on the approximate solution of Schrödinger's equation. It can be done using two types of methods: dynamic and quantum mechanical one. Among the dynamic methods, there are the molecular dynamic method (Langevin's dynamic method) and Monte Carlo method. These methods differ from each other by the specification degree for description of the molecular motion. This specification degree decreases from molecular dynamic method to Monte Carlo method. In turn, it leads to the decrease of the computing time expenses and the increase of the number of atoms in the regarded systems.

The existing quantum chemical methods can be classified into nonempirical and semiempirical methods. All the integrals of the Hartree–Fock's set of equations are calculated explicitly in the framework of nonempirical methods without any experimentally found parameters except the fundamental physical constants. In the same time, the accuracy of nonempirical methods depends on the number of functions included in the basic set, which in turn define the computing time and computer resources requests. Therefore, the modern nonempirical quantum chemical methods installed on modern PCs allow the investigation of systems containing no more than 10 atoms. As opposed to nonempirical methods, semiempirical methods allow significant increase in the number of atoms in the considered systems. The faster calculations are possible due to the replacement of some integrals in the Fock's operator by parameters obtained experimentally or expressions obtained using them. It should be mentioned that the accuracy of the semiempirical methods depends on how similar the structure of the investigated compounds is to the structure of compounds used for parametrization of the certain method. The semiempirical methods are arranged chronologically in the line: CNDO, INDO, MINDO/3, MNDO, AM1, and PM3. The CNDO method fits for calculation of molecular properties defined by the electronic distribution (such as dipole moments of molecules), while the lengths of the valence bonds are underestimated. The drawback of CNDO method caused by neglecting the differences in the Coulombic repulsion of electrons with parallel and antiparallel spins is eliminated in the INDO method. It is capable of calculation of the electronic structure of molecules with open shells. However, these two methods do not fit for the construction of the potential energy surfaces because of inappropriate reproduction of the atomization heats and orbital energies. Modification of the INDO method with regard to the correct calculation of the formation heats and molecular geometry leads to the appearance of new methods: MINDO and MNDO. But these methods appeared to be unfit for the description of systems with hydrogen bonds and heteroatoms and overestimated the activation barriers for the reactions. In order to eliminate these drawbacks, Dewar and coworkers developed the new semiempirical AM1 method on the basis of more rigorous approximation Neglecting Diatomic Differential Overlap (NDDO). The AM1 method was the first procedure that included atom–atom potentials. This allowed somehow effectively taking into account the electron correlation, without which it is impossible to describe intermolecular interaction realized in van der Waals' aggregates. The AM1 method reproduces the hydrogen bond and gives calculated data with a higher level of agreement with the experiment than the methods described earlier [32]. The PM3 method is reparametrization of the AM1 method and differs from its predecessor by higher accuracy of reproduction of formation heats for the compounds [33]. This appeared to be useful for calculations of the thermodynamic parameters of formation and clusterization of amphiphiles.

10.3 CALCULATION OF THERMODYNAMIC CLUSTERIZATION PARAMETERS OF SUBSTITUTED ALKANES AT THE AIR/WATER INTERFACE

The assessment of enthalpy, entropy, and Gibbs' energy of the surfactant clusterization is shown in this chapter with the example of 2-hydroxycarboxylic acids. These surfactants possess two functional groups located close to each other with respect to the hydrocarbon chain. Such orientation of functional groups induces the enlargement of all the hydrophilic part of the surfactant molecule, which can affect the geometric parameters of the unit cell of the lattice and the tilt of the hydrophobic chain with respect to the interface. So, 2-hydroxycarboxylic acids are interesting objects for illustration of the calculation of thermodynamic and structural parameters of the surfactant clusterization.

10.3.1 CONFORMATIONAL ANALYSIS OF MONOMERS AND THEIR THERMODYNAMIC PARAMETERS OF FORMATION

The conformational analysis starts with the construction of the potential energy dependence on the values of two torsion angles $\angle\alpha = C_2-C_3-C_4-O_2$ and $\angle\beta = C_2-C_3-O_1-H_1$. These angles correspond to the carboxylic and hydroxylic group location with respect to the alkyl chain of the 2-hydroxycarboxylic acid molecule (Figure 10.3). Both of them were varied in the range of 0°–360° in steps of 15°. The potential energy dependence on these torsion angles is illustrated in the example of the monomer of 2-hydroxytridecanoic acid, as shown in Figure 10.4a. There are five minima in this plot. Additional optimization of the monomer structures in the vicinity of these minima has confirmed that five stable conformations exist. They are characterized by the following values of the α and β angles: (80°, −54°), (−96°, −60°), (67°, 29°), (−61°, −164°); and (−123°, 25°), respectively. It should be noted that the heats of formation for all five conformers are quite similar, but the second monomer ($\alpha = -96°$, $\beta = -60°$) is energetically most advantageous and the forth monomer ($\alpha = -61°$, $\beta = -164°$) is the least preferable.

For the most energetically advantageous conformer of 2-hydroxytridecanoic acid ($\alpha = -96°$, $\beta = -60°$), a dependence of the monomer heat of formation on the value of torsion angle $\angle\gamma = C_1-C_2-C_3-C_4$ was obtained. This angle γ defines the general location of the hydrophilic headgroup of the acid molecule with respect to its alkyl chain (the headgroup of molecule includes both functional groups). As a result, three minima were obtained with the next γ values: 73°, 168°, and 291° (Figure 10.5). These γ values for 2-hydroxycarboxylic acids coincide virtually with those obtained for α-amino acids studied in Ref. [24] (74°, 166°, and 295°). As in the case of α-amino acids, conformers of

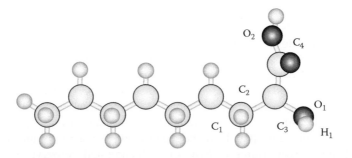

FIGURE 10.3   Torsion angles of the functional groups of 2-hydroxynonanoic acid.

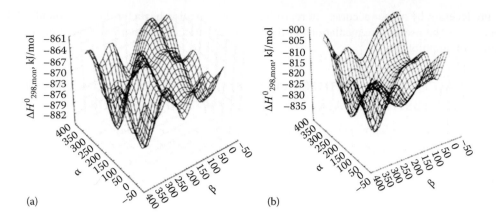

FIGURE 10.4 Potential energy surfaces for the monomer of 2-hydroxytridecanoic acid: (a) γ = 168° and (b) γ = 73°.

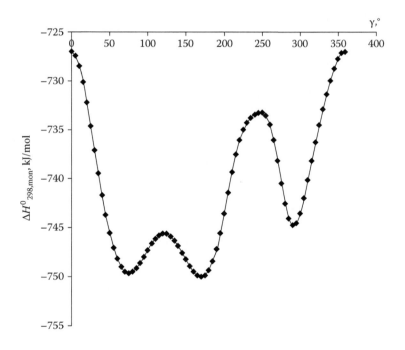

FIGURE 10.5 Dependence of the monomer formation enthalpy of 2-hydroxytridecanoic acid on the value of γ angle.

2-hydroxycarboxylic acids with γ = 168° are slightly more energetically preferable. Despite this fact, it will be shown in the following that clusterization of monomers with γ = 73° is characterized by a great decrease in the Gibbs energy of clusterization. It should be mentioned that the last two conformers of 2-hydroxycarboxylic acids with γ = 73°, described earlier, have different values of the α and β angles: (−176°, 171°) and (98°, −172°) for Monomers 4 and 5, respectively. The potential energy dependence for 2-hydroxytridecanoic acid with γ = 73° is shown in Figure 10.4b.

Optimized geometric structures of the five conformers obtained for 2-hydroxycarboxylic acids with γ = 73° are listed in Figure 10.6. It can be seen that the Monomers 1 and 4 are stabilized with interactions realized between hydrogen atom of the OH unit and hydroxylic oxygen of the COOH unit. In addition, in Monomer 1, there is an interaction between carbonyl oxygen and α-hydrogen

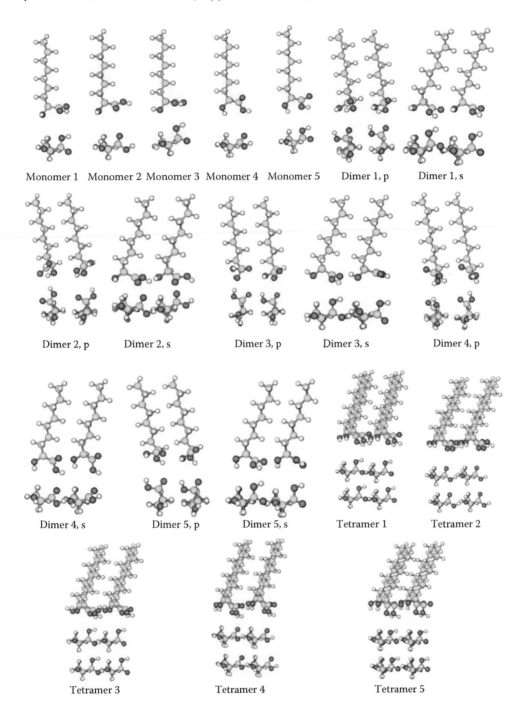

Monomer 1   Monomer 2   Monomer 3   Monomer 4   Monomer 5   Dimer 1, p   Dimer 1, s

Dimer 2, p   Dimer 2, s   Dimer 3, p   Dimer 3, s   Dimer 4, p

Dimer 4, s   Dimer 5, p   Dimer 5, s   Tetramer 1   Tetramer 2

Tetramer 3   Tetramer 4   Tetramer 5

FIGURE 10.6   Geometrical structures of 2-hydroxycarboxylic acid aggregates.

of the alkyl chain. In Monomer 4, this interaction involves carbonyl oxygen and γ-hydrogen of the alkyl chain. Monomer 2 has an intramolecular interaction between hydrogen atom of the OH group and carbonyl oxygen of the COOH group. The structure of Monomer 3 has an intramolecular interaction between carbonyl oxygen and α-hydrogen of the hydrophobic alkyl chain. Monomer 5 is stabilized by the interaction between hydrogen and oxygen atoms of the two OH units and by

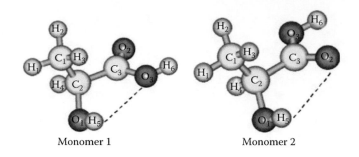

Monomer 1   Monomer 2

FIGURE 10.7   Structure of lactic acid molecule.

the intramolecular interaction between hydrogen atom of carboxylic group and γ-hydrogen of the alkyl chain.

As in the case of α-amino acids described previously [24], the hydrophilic parts of the 2-hydroxycarboxylic acids possess atomic groups, which are capable of formating intramolecular hydrogen bonds. This defines the conformational peculiarities of the mentioned compounds. The assumption about the existence of intramolecular hydrogen bonds in molecules of α-amino and 2-hydroxycarboxylic acids is also proved by the examples of R-lactic acid and S-alanine [34–37]. The values of the torsion angles of the functional groups for four of five conformers of 2-hydroxy-carboxylic acids (apart from Monomer 4) are found to coincide within the range of 50° with corresponding values of the structural parameters for α-amino acids described previously [24]. In addition, the mutual locations of the functional groups in the most energetically advantageous structures of α-amino and 2-hydroxycarboxylic acids are most similar: $\angle\alpha = -81°$, $\angle\beta = -55$ for α-amino acid Monomer 1 and $\angle\alpha = -96°$, $\angle\beta = -60°$ for 2-hydroxycarboxylic acid Monomer 2.

It should be also noted that the structures of the most energetically preferred structures of lactic acid shown in Figure 10.7 agree with the structures found in Ref. [41] using cc-pVDZ basis set in the framework of SCF and MP2 level. The hydrogen interactions $O_3 \cdot H_5$ and $O_2 \cdot H_5$ are present between two functional groups in these conformers and marked with dashed lines in Figure 10.7. The lengths of the bonds obtained in the framework of the semiempirical PM3 method are rather longer than those assessed using *ab initio* methods. Nevertheless, the two regarded structures have lower Gibbs' energy of formation in comparison to the others. At the same time, Monomer 2 is more energetically advantageous among the structures listed in Ref. [34] that agree with our findings.

Table 10.1 contains the calculated thermodynamic parameters of formation (enthalpy, absolute entropy, and Gibbs' energy) at 298 K for five monomer structures of 2-hydroxycarboxylic acids with $\gamma = 73°$. It should be noted that free rotation of the methylene groups in the alkyl chain was not taken into account while calculating the standard entropy values. As shown previously [11,25,28,29], the correction for the free rotation of the methylene groups for different types of amphiphiles is independent of the functional group. It was found to be 7.1, 6.6, 7.0, and 6.1 J/(mol·K) at 298 K for amines, alcohols, thioalcohols, and carboxylic acids, respectively. Corrected values of absolute entropy and free Gibbs' energy of formation for 2-hydroxycarboxylic acids are listed in brackets in Table 10.1.

It should be mentioned that experimental data regarding the standard thermodynamic characteristics of 2-hydroxycarboxylic acid formation are scarce. They concern only the crystalline phase of the first two members of the homologous series. Standard enthalpy, entropy, and Gibbs' energy of formation of 2-hydroxypropanoic acid are −694.1 kJ/mol, 142.3 J/(mol·K), and −522.9 kJ/mol, respectively. For 2-hydroxybutanoic acid, these data are available only for the enthalpy of formation: −676.6 kJ/mol [38,39]. That is why it is impossible to compare the obtained calculated data with experimental values.

The correlation dependences of the standard thermodynamic characteristics of 2-hydroxycarboxylic acids on the alkyl chain length (*n*) were constructed on the basis of the calculated data. These dependences are linear, similar to other classes of amphiphiles studied [11,24–26,28–30,40–43].

TABLE 10.1
Thermodynamic Parameters of 2-Hydroxycarboxylic Acid Formation at 298 K

System	Monomer 1	Monomer 2	Monomer 3	Monomer 4	Monomer 5
			$\Delta H^0_{298,\text{mon}}$, kJ/mol		
$C_3H_6O_3$	−613.44	−618.19	−597.17	−604.02	−597.18
$C_4H_8O_3$	−632.34	−637.45	−620.00	−622.45	−620.05
$C_5H_{10}O_3$	−654.94	−660.40	−642.55	−644.66	−642.54
$C_6H_{12}O_3$	−677.50	−683.37	−665.13	−667.28	−665.11
$C_7H_{14}O_3$	−700.18	−706.42	−687.80	−689.90	−687.78
$C_8H_{16}O_3$	−722.82	−729.46	−710.44	−712.57	−710.44
$C_9H_{18}O_3$	−745.50	−752.53	−733.11	−735.24	−733.12
$C_{10}H_{20}O_3$	−768.17	−775.59	−755.78	−757.91	−755.79
$C_{11}H_{22}O_3$	−790.85	−798.66	−778.46	−780.59	−778.48
$C_{12}H_{24}O_3$	−813.52	−821.73	−801.14	−803.27	−801.15
$C_{13}H_{26}O_3$	−836.21	−844.80	−823.82	−825.96	−823.83
$C_{14}H_{28}O_3$	−858.88	−867.87	−846.50	−848.64	−846.52
$C_{15}H_{30}O_3$	−881.56	−890.95	−869.18	−871.32	−869.20
$C_{16}H_{32}O_3$	−904.25	−914.02	−891.86	−894.00	−891.88
$C_{17}H_{34}O_3$	−926.93	−937.10	−914.54	−916.69	−914.56
			$\Delta S^0_{298,\text{mon}}$, J/(mol·K)		
$C_3H_6O_3$	341.02 (347.72)	334.21 (348.21)	349.39 (356.09)	345.88 (352.58)	350.68 (357.38)
$C_4H_8O_3$	371.86 (385.26)	363.72 (385.78)	379.63 (393.03)	376.07 (389.47)	381.98 (395.38)
$C_5H_{10}O_3$	404.22 (424.32)	394.33 (424.46)	412.22 (432.32)	408.87 (428.97)	413.00 (433.10)
$C_6H_{12}O_3$	436.24 (463.04)	425.52 (463.72)	444.33 (471.13)	440.93 (467.73)	445.43 (472.23)
$C_7H_{14}O_3$	468.43 (501.93)	456.27 (502.53)	476.34 (509.84)	473.23 (506.73)	477.04 (510.54)
$C_8H_{16}O_3$	500.54 (540.74)	487.16 (541.49)	508.33 (548.53)	505.15 (545.35)	509.04 (549.24)
$C_9H_{18}O_3$	532.59 (579.49)	517.35 (579.74)	539.80 (586.70)	537.21 (584.11)	541.28 (588.18)
$C_{10}H_{20}O_3$	564.35 (617.95)	547.85 (618.31)	571.83 (625.43)	568.97 (622.57)	573.27 (626.87)
$C_{11}H_{22}O_3$	595.22 (655.52)	578.35 (656.88)	603.14 (663.44)	600.66 (660.96)	604.00 (664.30)
$C_{12}H_{24}O_3$	627.39 (694.39)	608.43 (695.02)	635.09 (702.09)	631.96 (698.96)	635.45 (702.45)
$C_{13}H_{26}O_3$	659.33(733.03)	638.10 (732.75)	666.66 (740.36)	663.15 (736.85)	667.20 (740.90)
$C_{14}H_{28}O_3$	689.85 (770.25)	668.56 (771.28)	697.94 (778.34)	696.76 (777.16)	699.80 (780.20)
$C_{15}H_{30}O_3$	720.88 (807.98)	698.47 (809.25)	728.28 (815.38)	726.80 (813.90)	729.77 (816.87)
$C_{16}H_{32}O_3$	752.12 (845.92)	727.65 (846.50)	760.95 (854.75)	758.92 (852.72)	762.47 (856.27)
$C_{17}H_{34}O_3$	783.49 (883.99)	757.83 (884.74)	791.78 (892.28)	788.52 (889.02)	791.96 (892.46)
			$\Delta G^0_{298,\text{mon}}$, kJ/mol		
$C_3H_6O_3$	−501.60 (−503.59)	−515.60 (−506.39)	−487.82 (−489.82)	−493.62 (−495.62)	−488.21 (−490.21)
$C_4H_8O_3$	−491.10 (−493.08)	−506.07 (−495.84)	−481.07 (−483.05)	−482.46 (−484.43)	−481.82 (−483.80)
$C_5H_{10}O_3$	−482.74 (−486.71)	−500.55 (−489.32)	−472.73 (−476.70)	−473.84 (−477.82)	−472.95 (−476.92)
$C_6H_{12}O_3$	−474.23 (−480.20)	−495.21 (−482.99)	−464.27 (−470.24)	−465.40 (−471.37)	−464.58 (−470.55)
$C_7H_{14}O_3$	−465.89 (−473.86)	−489.82 (−476.60)	−455.87 (−463.84)	−457.05 (−465.02)	−456.07 (−464.03)
$C_8H_{16}O_3$	−457.50 (−467.46)	−484.47 (−470.26)	−447.44 (−457.41)	−448.62 (−458.59)	−447.65 (−457.62)
$C_9H_{18}O_3$	−449.13 (−461.08)	−478.94 (−463.72)	−438.89 (−450.85)	−440.24 (−452.20)	−439.33 (−451.29)
$C_{10}H_{20}O_3$	−440.65 (−454.61)	−473.50 (−457.28)	−430.50 (−444.45)	−431.78 (−445.73)	−430.94 (−444.89)
$C_{11}H_{22}O_3$	−431.93 (−447.88)	−468.07 (−450.85)	−421.90 (−437.85)	−423.29 (−439.24)	−422.17 (−438.12)
$C_{12}H_{24}O_3$	−423.59 (−441.53)	−462.52 (−444.29)	−413.49 (−431.44)	−414.70 (−432.65)	−413.61 (−431.56)
$C_{13}H_{26}O_3$	−415.18 (−435.12)	−456.86 (−437.60)	−404.97 (−424.92)	−406.07 (−426.01)	−405.15 (−425.10)
$C_{14}H_{28}O_3$	−406.35 (−428.29)	−451.42 (−431.16)	−396.37 (−418.31)	−398.16 (−420.10)	−396.94 (−418.89)
$C_{15}H_{30}O_3$	−397.67 (−421.61)	−445.82 (−424.55)	−387.49 (−411.43)	−389.19 (−413.12)	−387.96 (−411.89)
$C_{16}H_{32}O_3$	−389.06 (−414.99)	−440.03 (−417.73)	−379.30 (−405.24)	−380.84 (−406.77)	−379.78 (−405.71)
$C_{17}H_{34}O_3$	−380.48 (−408.42)	−434.51 (−411.20)	−370.56 (−398.49)	−371.74 (−399.67)	−370.64 (−398.57)

The values of the slopes for enthalpy, absolute entropy, and Gibbs' energy of formation are in the range from −22.50 to −22.67 kJ/mol, from 31.64 to 31.77 J/(mol·K), and from 8.51 to 8.55 kJ/mol, respectively. The deviation of the values of absolute terms from the mean value is also no more than 1.2%. So, it is possible to express these partial correlations in a general form:

$$\Delta H^0_{298,mon} = -(22.68 \pm 0.21) \cdot n - (581.43 \pm 1.91) \ [S = 7.86 \ kJ/mol, N = 75], \qquad (10.1)$$

$$S^0_{298,mon} = (31.41 \pm 0.27) \cdot n + (312.99 \pm 2.41) \ [S = 9.92 \ J/(mol \cdot K), N = 75], \qquad (10.2)$$

where
   S is the standard deviation
   N is the sample size

The values of the slopes in Equations 10.1 and 10.2, which characterize the contributions from the methylene groups, agree well with the values calculated earlier for other classes of amphiphilic compounds at 298 K [11,24–26,28–30,40–43]. Thus, the increments of the methylene units in the values of enthalpy and absolute entropy of formation were in the range from −22.56 to −22.68 kJ/mol and from 30.53 to 39.28 J/(mol·K), respectively. The standard errors for the calculation of enthalpy, entropy, and Gibbs energy of 2-hydroxycarboxylic acid formation do not exceed corresponding values of the amphiphile types previously studied. The correlation coefficients exceed 0.997.

10.3.2 CLUSTERIZATION THERMODYNAMIC PARAMETERS OF SMALL AGGREGATES

The dimers were built from the obtained monomer conformations with $\gamma = 73°$. The structures of these entities based on Monomer 2 are illustrated in Figure 10.8. Here, the vector drawn through the centers of the oxygen atoms of two OH groups was chosen as a direction of the orientation of the monomer head groups in the dimer. According to this definition, as adopted previously [24,43], the structures of the dimers were subdivided into two types, characterized by *parallel* (p) and *serial* (s) relative orientation of the head groups. For example, the definition *Dimer 2, p* indicates that this dimer structure is built on the basis of Monomer 2, and the hydrophilic head groups of the monomers are oriented *parallel* in it (Figure 10.8a).

Note that it is possible to construct dimers, which are the basic units for larger clusters up to 2D films, with the different tilt angles δ and ϕ of their alkyl chains with respect to the normal to the p- and q-directions, respectively, of the monolayer unit cell. Further, on the basis of these δ and ϕ values, it is possible to construct larger clusters, particularly tetramers, which can be regarded as the unit cells of the corresponding monolayers, and thus, define the general tilt angle t

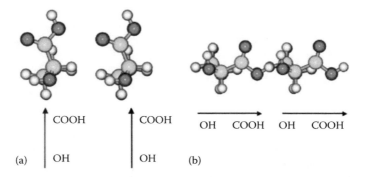

FIGURE 10.8 Relative orientation of 2-hydroxycarboxylic acid monomers in the dimer: (a) *parallel* and (b) *serial*.

of 2-hydroxycarboxylic acid molecules with respect to the normal to the air/water interface. As described in Ref. [43] in detail, the value of the general tilt angle t of the alkyl chain with respect to the normal to the interface can be calculated using the following equations:

$$t = \arcsin\left(\frac{\sin \delta}{\cos \theta_1}\right), \quad \theta_1 = \operatorname{arctg}\left(\frac{\sin \phi}{\sin \delta \cdot \sin \theta} - \operatorname{ctg}\theta\right),$$ (10.3)

where

δ is the tilt angle of amphiphilic molecules with respect to the p-axis of the cluster unit cell
ϕ is the tilt angle of amphiphilic molecules with respect to the q-axis of the cluster unit cell
θ is the angle between the p and q directions (see Figure 10.9)

To determine the tilt angles of 2-hydroxycarboxylic acid molecules with respect to the p and q directions of the cluster unit cell, one should use the procedure described in detail elsewhere [27]. The *parallel* and *serial* types of dimers with the "a" CH·HC interaction type (marked in Figure 10.10 with hollow arrows) have been constructed with two monomers. Applying the parallel shift of one molecule with respect to another one in both of the p and q directions, the dependencies of the dimerization Gibbs energy on the δ and ϕ angles, respectively, were tabulated. The minima of the dimerization Gibbs energy for these associates correspond to optimum δ and ϕ values.

The dependences of the Gibbs energy of dimerization for the structures of Dimer 2, p and Dimer 2, s on the values of the tilt angles of the molecules with respect to the normal to the p and q directions of the monolayer are listed in Table 10.2. These data show that minimum values of the Gibbs energy of dimerization correspond to the structures with the following values of the ϕ_2 angle: 9.8°, 22.0°, and 40.9°. Additional optimization of the dimers with $\phi_2 = 9.8°$ and 22.0° reveals only the existence of the single stable structure with $\phi_2 = 9.8°$. The presence of the second minimum of the Gibbs energy of dimerization with $\phi_2 = 40.9°$ corresponds to dimers, which have one fewer CH·HC contact than the dimer of 2-hydroxytridecanoic acid shown in Figure 10.10. The loss of this CH⋯HC interaction causes an increase in the dimerization Gibbs energy and a lower preference of such structures in comparison with those having the maximum number of CH·HC interactions.

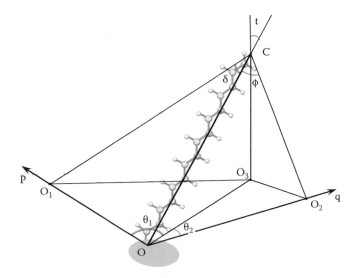

FIGURE 10.9 Amphiphilic molecule orientation with respect to the air/water interface (the oval schematically denotes the hydrophilic part of the amphiphile).

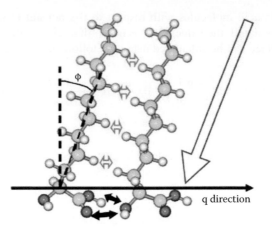

FIGURE 10.10   Determination of the molecular tilt angle with respect to the normal to the q direction.

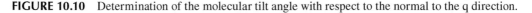

TABLE 10.2
Dependence of the Dimerization Gibbs' Energy of 2-Hydroxycarboxylic Acid Aggregates Built on the Basis of Monomer 2 ($n = 11$) on the δ and ϕ Values

Tilt Angle ϕ,°	ΔH^{dim}_{298}, kJ/mol	ΔS^{dim}_{298}, J/(mol·K)	ΔG^{dim}_{298}, kJ/mol	Tilt Angle δ,°	ΔH^{dim}_{298}, kJ/mol	ΔS^{dim}_{298}, J/(mol·K)	ΔG^{dim}_{298}, kJ/mol
53.7	−41.59	−221.81	24.51	54.3	−43.48	−201.21	16.49
51.1	−43.09	−231.26	25.83	51.1	−50.93	−253.50	24.61
40.9	−50.99	−226.12	16.40	36.6	−53.87	−224.40	13.00
35.7	−54.58	−265.73	24.61	34.0	−61.76	−284.91	23.14
22.0	−57.61	−245.97	15.69	21.0	−60.84	−235.28	9.27
11.5	−51.41	−242.15	20.75	10.2	−60.66	−233.64	8.96
9.8	−58.62	−260.34	18.96	9.8	−55.97	−238.86	15.21

It should be noted that there are two intermolecular interactions in the structure of Dimer 2, s. The first interaction is realized between hydrogen atom of the carboxylic group of one acid molecule and oxygen atom of the OH group of another one. The second interaction occurs between oxygen atom of the carboxylic unit of one molecule and hydrogen atom of the OH unit of another one. The mentioned interactions are marked with double-edged solid black arrows in Figure 10.10. The presence of the described interactions causes the dimer orientation with the corresponding angle $\phi_2 = 9.8°$ to the normal to the q direction.

For Dimers 2, p with *parallel* orientation of the head groups, the minimum values of the dimerization Gibbs energy correspond to the values of the δ_2 angle equal to 10.2°, 21.0°, and 36.6° (Table 10.2). After additional optimization of these structures, two minima were observed. The first minimum corresponds to the structure with $\delta_2 = 21.0°$ and the second one with $\delta_2 = 36.6°$. As in the case of the second structure of Dimer 2, s, the second structure of Dimer 2, p, has one fewer CH⋯HC interaction than the dimer of 2-hydroxytridecanoic acid with $\delta_2 = 21.0°$.

Application of the described procedure to dimers constructed on the basis of other monomer conformations of 2-hydroxycarboxylic acids allows us to determine the molecular tilt angles with respect to the normal to the directions of the spread monolayer. The obtained values of the δ and ϕ

tilt angles were found to be (21.1°, 21.2°), (9.Monomers 1–5, respectively). Using the found values of the δ and φ angles, series of dimer structures with *parallel* Dimer 5, p, there is interaction between hydrogen atom of the carboxylic group of one acid molecule and carbonyl oxygen atom of another one. Whereas the Dimers 1, s, 2, s, and 3, s have interactions between hydrogen atom of the OH unit of one molecule and carbonyl oxygen atom of another one, the *serial* dimers formed using Monomer 5 possess another interaction between hydrogen atom of the OH group of one molecule and hydroxylic hydrogen atom belonging to the COOH unit of the second monomer.

Thermodynamic parameters of formation and clusterization were calculated for all dimer structures described earlier and tetramers built on their basis. Enthalpy, entropy, and Gibbs energy of clusterization were calculated according to the equations: $\Delta H_{T,m}^{Cl} = \Delta H_T^0 - m \cdot H_{T,\text{mon}}^0$; $\Delta S_{T,m}^{Cl} = S_T^0 - m \cdot S_{T,\text{mon}}^0$; $\Delta G_{T,m}^{Cl} = \Delta H_{T,m}^{Cl} - T \cdot \Delta S_{T,m}^{Cl}$, where superscript Cl denotes the parameter belonging to the clusterization process; ΔH_T^0 and S_T^0 are enthalpy and entropy of the aggregates at a certain temperature T, $H_{T,\text{mon}}^0$, and $S_{T,\text{mon}}^0$ are enthalpy and entropy of the corresponding monomers at the same temperature T, and m is the monomer number in the cluster. According to described formulas, the corresponding values of enthalpy, entropy, and Gibbs energy of clusterization are calculated for all the considered aggregates except the tetramer structures comprising Monomer 4 with short alkyl chains of 7 and 9 methylene units. These structures possess edge effects, and hence, they are not included in the further development of the additive scheme for calculation of the thermodynamic clusterization parameters of 2-hydroxycarboxylic acids.

It should be mentioned that here we consider only 2D monolayers comprising monomers with regular structures. However, small aggregates (as usual dimers) can possess distorted structure due to edge effects. However, it is unreasonable to consider such aggregates for constructing the additive scheme, which enables one to calculate the necessary parameters for monolayers with regular structure. The edge effects denote the appearance of such interactions between monomer molecules (both between their functional groups and alkyl chains) that are absent in 2D monolayers. They can be avoided in structures of larger rectangular clusters (tetramers, hexamers) as the increase of the monomer number in the cluster leads to a more ordered optimized structure because of the better regularity of interactions between the atom groups inside of the cluster. Therefore, the structures of Dimer 3, p with edge effects are not included in the further construction of the general correlation for all small associates. The corresponding energetic increments of interactions realized between the functional groups of the 2-hydroxycarboxylic acid molecules were obtained from tetramers 3 having these interactions without the edge effect distortion.

For all homologous series of the calculated thermodynamic parameters of clusterization (enthalpy, entropy, and Gibbs' energy), the correlation dependences on the number of intermolecular CH·HC interactions K_a realized in 2-hydroxycarboxylic acid clusters were built. The obtained data show that the values of the regression slope characterizing the increment of the CH·HC interaction into the clusterization thermodynamic parameters are in the range from −9.80 to −10.83 kJ/mol for enthalpy, from −19.35 to −26.69 J/(mol·K) for entropy, and from −2.56 to −4.56 kJ/mol for Gibbs' energy. These data are rather near to analogous values obtained for amphiphiles (alcohols, saturated and unsaturated carboxylic acids, α-amino acids, amines, thioalcohols, and alkylamides) studied previously. The values of increments for these amphiphiles were found to be in the range from −9.20 to −10.50 kJ/mol for enthalpy, from −17.75 to −24.49 J/(mol·K) for entropy, and from −2.81 to −4.93 kJ/mol for Gibbs' energy of clusterization [11,24–26,28–30,40–43].

The contribution of the interactions of 2-hydroxycarboxylic acid headgroups with *serial* orientation in the enthalpy of clusterization was negative for all dimers apart from the Dimer 4, s structure. Possibly, it can be stipulated by the structural peculiarities of this dimer. Oxygen atom belonging to the OH group of one monomer and carbonyl oxygen atom of another one mutually repulses. In turn, this causes the lower thermodynamic preference of formation of such dimers constructed from Monomer 4 in comparison to other *serial* oriented headgroups. The contribution of the headgroup interactions with *parallel* orientation of the functional groups to the enthalpy of clusterization was statistically insignificant. For the entropy of clusterization, the increment of the hydrophilic group interactions in *serial* dimers was larger (by absolute value) than that for *parallel* dimers. In addition, as noted earlier, the smaller number of intermolecular CH·HC contacts is realized in dimers with *serial* orientation of the headgroups than that in *parallel* dimers (for structures with odd number of carbon atoms in the alkyl chain). As a result, formation of the aggregates with *parallel* orientation of the functional groups in them is more advantageous with respect to the Gibbs energy of clusterization. At the same time, the formation of clusters comprising Monomer 3 is more advantageous among the aggregates built on the basis of all regarded monomer structures.

All partial correlations for dimers and tetramers of 2-hydroxycarboxylic acids are unified into general one, because the values of the slopes in them are quite close:

$$\Delta H_{298,m}^{Cl} = -(10.11\pm0.02)\cdot(K_a+n_{3,s})-(7.97\pm0.34)\cdot n_{1,s}-(5.94\pm0.28)\cdot(n_{2,s}+n_{5,p})$$

$$+(21.11\pm0.48)\cdot n_{4,s}-(2.77\pm0.49)\cdot n_{4,p}, \; [N=152; R=0,9999; S=2.08 \; kJ/mol],$$

$$(10.4)$$

$$\Delta S_{298,m}^{Cl} = -(20.02\pm0.47)\cdot K_a-(89.90\pm4.36)\cdot(n_{1,p}+n_{2,p})-(126.97\pm4.22)\cdot(n_{1,s}+n_{2,s})$$

$$-(65.78\pm8.02)\cdot n_{3,p}-(147.09\pm7.18)\cdot n_{3,s}-(107.15\pm2.67)\cdot\left(n_{4,p}+n_{4,s}+n_{5,p}+n_{5,s}\right),$$

$$[N=152; R=0.9991; S=22.67 \; J/(mol\cdot K)],$$

$$(10.5)$$

$$\Delta G_{298,m}^{Cl} = -(4.24\pm0.12)\cdot K_a+(28.90\pm0.69)\cdot\left(n_{1,p}+n_{1,s}+n_{2,p}+n_{4,p}+n_{5,p}\right)$$

$$+(32.21\pm0.81)\cdot\left(n_{2,s}+n_{3,s}+n_{5,s}\right)+(22.82\pm1.20)\cdot n_{3,p}+(54.73\pm1.08)\cdot n_{4,s},$$

$$[N=152; R=0.9851; S=5.71 kJ/mol],$$

$$(10.6)$$

where K_a is the number of CH·HC interactions realized in the regarded cluster. It can be obtained for *serial* and *parallel* dimers as follows:

$$K_a = \left\{\frac{n-1}{2}\right\} \quad and \quad K_a = \left\{\frac{n}{2}\right\}, \text{respectively,} \quad (10.7)$$

where
   n is the number of methylene groups in the alkyl chain of 2-hydroxycarboxylic acids
   braces denote the integer part of the number
   $n_{i,p}$ and $n_{i,s}$ are the descriptors of *parallel* (p) and *serially* (s) oriented functional headgroups in the structure, with i denoting the number of the corresponding monomer used for the construction of the regarded aggregate

In the case that interactions between the functional groups of the headgroups exist in the aggregate structure, then the value of the corresponding descriptor ($n_{i,s}$ and/or $n_{i,p}$) is equal to the number of such interactions. If the considered interactions are absent, this descriptor is zero.

The values of the standard deviations for enthalpy, entropy, and Gibbs' energy of clusterization for 2-hydroxycarboxylic acids do not exceed the corresponding values for the homologous series studied in Refs. [11,24–26,28–30,40–43].

10.3.3   THERMODYNAMIC CLUSTERIZATION PARAMETERS OF LARGE AND INFINITE CLUSTERS

The correlation dependencies discussed earlier for the calculation of the thermodynamic clusterization parameters (Equations 10.4 through 10.6) enable one to use the found values of the regression coefficients for the construction of the additive scheme. It is possible to calculate the values of the thermodynamic clusterization parameters for 2-hydroxycarboxylic acid homologues of any dimension up to 2D monolayers as a sum of the corresponding increments of the intermolecular CH·HC interactions and the interactions between the hydrophilic parts of the molecules.

Prior to the construction of the additive scheme, we consider the geometric parameters of the unit cells for 2-hydroxycarboxylic acid monolayers built on the basis of the regarded five monomer structures. It should be noted that the available experimental data concerning the structural peculiarities of 2-hydroxycarboxylic acid monolayers with different hydroxylic group position with respect to the alkyl chain show the next dependencies. The closer the OH unit is situated to the COOH group of the hydroxycarboxylic acid molecule, the less ordered is the hydrophobic chain packing [44,45] and the larger is the tilt angle of the alkyl chain with respect to the normal to the interface [45,46]. According to Ref. [44], such observations can be explained as follows. In the case of OH group location in the second, third, and even fourth positions of the alkyl chain, both functional groups act as a single hydrophilic headgroup of the amphiphile and involve 1–2 methylene units located between these functional groups into the water phase. It is natural that the volume of such a hydrophilic part is much larger than the volume of the individual COOH group of unsubstituted carboxylic acid or hydroxycarboxylic acids with OH group position 9, 11, 12, and 16. Correspondingly, although 2-, 3-, or 4-hydroxycarboxylic acids form condensed films, their alkyl chains are less ordered and orientate with a larger tilt angle with respect to the interface than their isomers with more separated functional groups along the alkyl chain.

Applied in this study, the model allows one to assess the values of the geometric parameters of the unit cell for 2-hydroxycarboxylic acid monolayers (see Figure 10.11 on the example of a monolayer comprising Monomer 2). Thus, as a result of the optimization using the PM3 method, the tetramers were obtained (see Figure 10.6), which could be considered as the unit cells of the corresponding monolayers (see Table 10.3). It is possible to compare the calculated values of the geometric parameters for 2-hydroxycarboxylic acids with the experimental data concerning only 9-, 11-, and 12-hydroxycarboxylic acids. One could suggest that the parameters of the unit cell of 2-hydroxycarboxylic acids are similar to those obtained for α-amino acids [7,24], because the structure of these amphiphiles is similar except the NH_2 group that is smaller than the OH group. From Table 10.3, it is seen that the experimentally obtained data for these parameters are quite similar for the mentioned amphiphiles. However, the tilt angle of the alkyl chains of 2-hydroxycarboxylic acids is smaller than for α-amino acids, and it was found in the range of 12.6°–21.5°, whereas for α-amino acids, it is 36°. This is due to the fact that the amino group is rather larger than the OH group and thus, the 2-hydroxycarboxylic acid molecules can be packed with smaller tilt angle values t with respect to the normal to the interface.

The fragment of the 2-hydroxycarboxylic acid monolayers is illustrated in Figure 10.12 in the example of the structure built on the basis of Monomer 2. Here, as in other infinite 2D clusters, two types of interactions between the hydrophilic parts of molecules can be singled out. These interactions were designated previously as *parallel* and *serial* in p and q directions, respectively. In addition, *parallel* interactions of the headgroups possess intermolecular hydrogen bonds, which were described before in the "dimer and tetramer" section. The presence of hydrogen bonds in *parallel* interactions is not typical apart from aggregates comprising Monomer 5. The number of *serial* and *parallel* interactions realized in the cluster can be calculated as follows:

$$n_s = q \cdot (p-1) \text{ and } n_p = p \cdot (q-1), \qquad (10.8)$$

where n is the number of methylene groups in the alkyl chain of 2-hydroxycarboxylic acids and braces denote the integer part of the number.

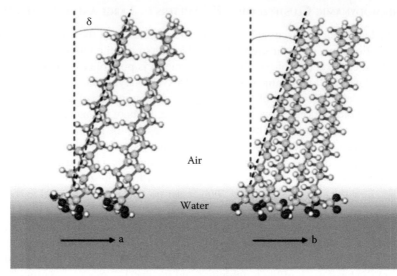

(a)   (b)

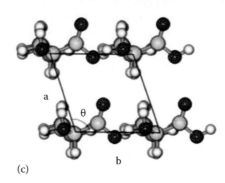

(c)

FIGURE 10.11   Structure of the unit cell of the infinite 2D cluster 2 of 2-hydroxycarboxylic acid: (a) view along the a axis, (b) view along the b axis, and (c) view along the alkyl chain axis.

TABLE 10.3
Geometric Parameters of the Unit Cells of 2-Hydroxycarboxylic Acids in the PM3 Approximation

Structure	a, Å	b, Å	θ,°	t,°
Calculation				
2D cluster 1	4.73	4.87	110	32.1
2D cluster 2	4.86	5.44	94	23.3
2D cluster 3	4.85	5.41	95	21.7
2D cluster 4	4.69	5.20	97	24.4
2D cluster 5	4.70	5.26	97	24.7
Experiment				
9-Hydroxyoctadecanoic acid [47]	4.75	4.89	120.7	12.6
11-Hydroxyoctadecanoic acid [47]	4.95	4.99	120.5	21.5
12-Hydroxyoctadecanoic acid [47]	4.60	4.99	112.3	19.5
α-Aminooctadecanoic acid [7]	4.91	5.25	112.0	36.0

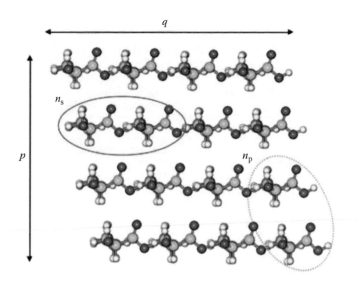

FIGURE 10.12 Fragment of the infinite 2D cluster 2 of 2-hydroxycarboxylic acids: n_s and n_p are schematic denotations of the *serial* and *parallel* interactions between functional groups of monomers, respectively.

To calculate the number of interactions described earlier per monomer molecule of the infinite 2D cluster, one has to divide Equation 10.8 by the number of monomers in the cluster ($m = p \cdot q$) and calculate the limits of the resulting expressions at infinite number of molecules in the cluster. Therefore, for infinite 2D clusters of 2-hydroxycarboxylic acids ($p = \infty$, $q = \infty$), Equation 10.8 becomes

$$n_{s,\infty}/m = n_{p,\infty}/m = 1. \tag{10.9}$$

The number of intermolecular CH·HC interactions per monomer molecule of the 2D monolayers can be calculated using the following expression:

$$K_{a,\infty}/m = n - 1. \tag{10.10}$$

Introducing Equations 10.9 and 10.10 into the correlation equations for enthalpy, entropy, and Gibbs' energy of clusterization (10.4) through (10.6), one obtains the expressions for the clusterization thermodynamic characteristics of 2-hydroxycarboxylic acids per monomer molecule:

$$A_i^{Cl}{}_{(T),\infty}/m = U_i \cdot K_{a,\infty}/m + V_i, \tag{10.11}$$

where the values of the coefficients U_i, and V_i depend on the particular thermodynamic characteristic $A_i^{Cl}{}_{(T),\infty}/m$ (enthalpy, entropy, or Gibbs' energy), on the structure of the monomer (descriptor i corresponds to its serial number) comprising 2D cluster, and on the temperature. The values of these coefficients are listed in Table 10.4.

The dependencies of the thermodynamic parameters of clusterization per 2-hydroxycarboxylic acid molecule on the length of its alkyl chain at 298 K are shown in Figures 10.13 through 10.15 in the example of the energetically most advantageous structure of 2D film 3. Here, the lines correspond to the dependencies calculated according to the additive scheme using Equation 10.11, whereas the points define the results of the direct calculations using the PM3 method. The listed graphs show that the results obtained in the framework of additive scheme agree well with the results of direct calculation.

TABLE 10.4

Values of the Coefficients for Calculation of the Thermodynamic Clusterization Parameters per One 2-Hydroxycarboxylic Acid Molecule

Type of Infinite 2D Cluster	$\Delta H_{i\,(298),\infty}^{Cl}/m,$ kJ/mol		$\Delta S_{i\,(298),\infty}^{Cl}/m,$ J/(mol·K)		$\Delta G_{i\,(298),\infty}^{Cl}/m,$ kJ/mol	
	$V_{\Delta H}$	$U_{\Delta H}$	$V_{\Delta S}$	$U_{\Delta S}$	$V_{\Delta G}$	$U_{\Delta G}$
2D cluster 1	−7.97	−10.14	−216.86	−20.02	56.65	−4.17
2D cluster 2	−5.94	−10.14	−216.86	−20.02	58.68	−4.17
2D cluster 3	−10.14	−10.14	−212.87	−20.02	53.30	−4.17
2D cluster 4	18.34	−10.14	−214.30	−20.02	82.20	−4.17
2D cluster 5	−5.94	−10.14	−214.30	−20.02	57.92	−4.17

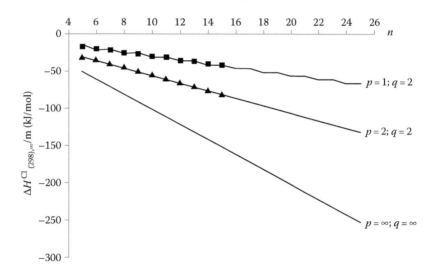

FIGURE 10.13 Dependence of clusterization enthalpy per monomer of 2D film 3 on the alkyl chain length of 2-hydroxycarboxylic acids (p and q are the number of monomer molecules, which define the size of 2D cluster).

It is seen from the dependencies of clusterization Gibbs' energy per monomer molecule (see Figure 10.15) at 298 K that spontaneous monolayer formation of 2-hydroxycarboxylic acids is possible if the alkyl chain length has not less than 14 methylene units for the monolayers built on the basis of Monomer 3. For other monomer structures, this threshold shifts in the region of larger alkyl chains: 15 methylene units for the Monomers 1, 2, 5, and 21 methylene units for Monomer 4. The value of the spontaneous clusterization for the most energetically preferred 2D monolayer 3 is in good agreement with experimental studies and reveals the possibility of monolayer formation of 2-hydroxyhexadecanoic acid [48,49].

The Gibbs' energy is the function of the state. Hence, its variation indicates the preference of the system to be in one or another state, that is, about the direction of the thermodynamic process. If the Gibbs' energy variation is negative, the system transits spontaneously from the initial state to the final state. Therewith in our works dedicated to the surfactant clusterization at the air/water interface [11,24–30,40–43], we regard not only the monolayer formation but also the formation of small aggregates (dimers, trimers, and tetramers). This provides us the possibility to judge about subtle process details

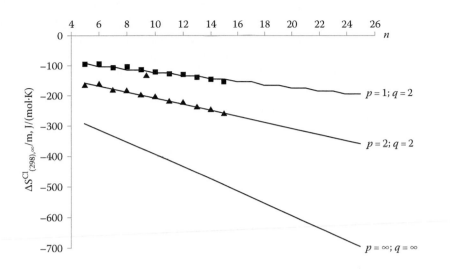

FIGURE 10.14 Dependence of the clusterization entropy per monomer of 2D film 3 on the alkyl chain length of 2-hydroxycarboxylic acids (p and q have the same meaning as in Figure 10.13).

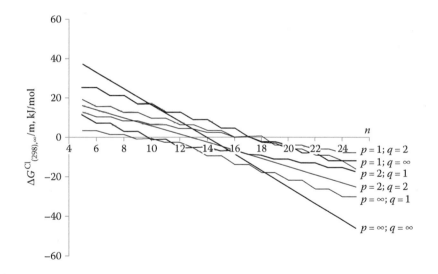

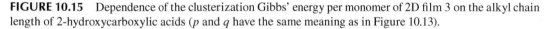

FIGURE 10.15 Dependence of the clusterization Gibbs' energy per monomer of 2D film 3 on the alkyl chain length of 2-hydroxycarboxylic acids (p and q have the same meaning as in Figure 10.13).

of the monolayer formation. Thus, it is possible to make assumptions about the way of the 2-hydroxy-carboxylic acid monolayer formation using the analysis of the graphical data shown in Figure 10.15. One can see that clusterization of 2-hydroxycarboxylic acids with the alkyl chain length of $n = 10$–16 carbon atoms will take place by formation of *parallel* dimers, their further enlargement, and formation of linear clusters, which can aggregate into two-dimensional associates possessing less energetically preferred interactions of the hydrophilic parts as in *serial* dimers. That is, 2D monolayer formation will take place via formation of linear associates stipulating a more friable monolayer structure.

This supposition is supported by the results of the π–A isotherms and BAM studies indicated by a reduced tilt orientational order by the fluid/condensed phase transition region and diminished contrast between condensed and fluid phases [44,50]. Investigations by GIXD provided a strong evidence for the reduced order in condensed 2-hydroxycarboxylic acids monolayers. As noted earlier, the situation is caused by the fact that both functional groups of 2-hydroxycarboxylic acid act as a single hydrophilic head group. GIXD results have shown that if the size of the headgroup exceeds a particular value perpendicular to the tilt direction, a disordered packing results as superposition of lattices with varying tilt azimuth of alkyl chains. Mismatch between the cross-sections of the headgroups and chains results in distortion of the alkyl chain packing in the monolayer [44,50].

10.3.4 COMPARISON WITH EXPERIMENTS

The surface pressure–area (Π–A) isotherms were measured at different temperatures using a computer-interfaced film balance. The surface pressures measured with the Wilhelmy method using a roughened glass plate were reproducible to ±0.1 mN/m and the areas per molecule to ±0.005 nm². The temperature of the experimental system was controlled to within ±0.1°C. The film balance was sheltered in a cabinet to avoid excessive disturbances by convection and contamination by impurities.

2-Hydroxyhexadecanoic (2OH–C16) and 2-hydroxyoctadecanoic (2OH–C18) acids purchased from Nu-Check Prep, Inc., Elysian, MN, were dissolved in a 9:1 (v:v) mixture of hexane (for spectroscopy, Merck) and ethanol (p.a., Merck). After the evaporation of the spreading solvent, the molecules remaining at the air/water interface were continuously compressed at rates of 0.1 nm²/(molecule·min). Ultrapure deionized water with the conductivity of 0.055 μS/cm produced by Purelab Plus was used as subphase for the monolayers of 2OH–C16 acid. 1 M aqueous NaCl subphase adjusted with HCl to pH 3 was used for the monolayers of 2OH–C18 acid.

The experimental surface pressure–area (Π–A) isotherms for the 2-hydroxyhexadecanoic (2-OH C16 acid) and 2-hydroxyoctadecanoic acid (2-OH C18 acid) monolayers at $T = 25$–44°C are shown in Figure 10.16a and b (points). At compression, a kink point characteristic for the beginning of a first-order phase transition is followed by an inclined plateau region, which indicates the two-phase coexistence for the transition from the fluid phase to the condensed phase. The experimental isotherms at higher temperatures indicate that the regarded systems undergo a first-order phase transition from the fluid to condensed monolayer state except the monolayers of 2-OH C16 acid at 25°C and 2-OH C18 acid at 25°C and 30°C.

Using thermodynamic models described in detail elsewhere [51–53], it is possible to describe the equation of state for insoluble monolayers and obtain theoretical Π–A curves. It should be noted that the parameters of the equation of state for the considered monolayers are evaluated using a model, which accounts for nonideality of the mixing entropy [53]. The results calculated according to this thermodynamic model have shown good agreement with the experimental Π–A isotherms obtained for various types of amphiphilic monolayers (e.g., rac-1-(2-methylhexadecanoyl)-2-O-hexadecyl-glycero-phosphoethanolamine and 2,4-di(n-undecylamino)-6-amino-1,3,5-triazine). The values of the molecular areas of amphiphiles estimated from the fitting of the experimental data to the exploited model were found to be quite similar to the real values.

Consider now briefly the main expressions used for the description of the state of insoluble monolayers. The equation of state for monolayers in the fluid (G, LE) state is represented by the Volmer-type equation [51] (see also Equation 1.89 of Chapter 1):

$$\Pi = \frac{kT}{m\cdot(A-\omega)} - \Pi_{coh}, \qquad (10.12)$$

where
  k is the Boltzmann constant
  T is temperature
  ω is the partial molecular area for monomers (or the limiting area of molecule in the gaseous state)
  A is the area per molecule
  Π_{coh} is the cohesion pressure, which takes into account the intermolecular interaction, the m value takes into account the association ($m > 1$) or dissociation ($m < 1$) degree of amphiphilic molecules in the monolayer or the size of these molecules

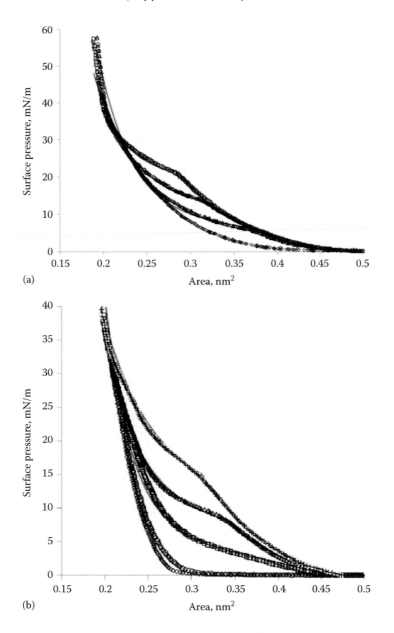

FIGURE 10.16 (a) Experimental (points) and theoretical (lines) Π–A isotherms for 2-OH C16 acid mono-layers at various temperatures: (1) 25°C (◇), (2) 30°C (△), (3) 35°C (○), and (4) 40°C (□). (b) Experimental (points) and theoretical (lines) Π–A isotherms for 2-OH C18 acid monolayers at various temperatures: (1) 25°C (◇), (2) 30°C (○), (3) 36°C (□), (4) 41°C (△), and (5) 44°C (+).

Taking into account the term mentioned earlier for nonideality of mixing entropy for monomers and clusters [53], the equation of state for the surface layer in the fluid (G, LE)/condensed (LC) transition region is as follows:

$$\Pi = \frac{1}{m} \cdot \frac{kT\alpha\beta}{A - \omega\left[1 + \varepsilon(\alpha\beta - 1)\right]} - \frac{kT}{A}\left(1 - \alpha\beta\right) - \Pi_{\text{coh}}. \tag{10.13}$$

In this equation,

    parameter α expresses the dependence of the aggregation constant on the surface pressure

    β is the fraction of the monolayer free from aggregates

    $\varepsilon = 1 - \omega_{cl}/\omega$, ω_{cl} is the area per monomer in a cluster

The ε value should consist of two terms, $\varepsilon = \varepsilon_0 + \eta\Pi$: ε_0 is the relative jump of the area per molecule during phase transition, and η is a relative two-dimensional compressibility of the condensed monolayer. The mentioned parameters have been calculated according to the expressions described in detail in Ref. [53].

The value of the standard Gibbs' energy of clusterization ΔG^{Cl}, calculated per mole of the monomers or oligomers, can be written in the following form [52,54]:

$$\Delta G^{Cl} = RT \ln\left(\frac{\omega_{(n)}}{A_c}\right). \tag{10.14}$$

The results obtained for ΔG^{Cl} at various temperatures can be used to estimate the standard enthalpy (ΔH^{Cl}) and standard entropy of clusterization (ΔS^{Cl}):

$$\Delta G^{Cl} = \Delta H^{Cl} - T\Delta S^{Cl}, \tag{10.15}$$

$$-\frac{\Delta H^{Cl}}{T^2} = \frac{\partial}{\partial T}\left(\frac{\Delta G^{Cl}}{T}\right). \tag{10.16}$$

According to Equations 10.12 and 10.13, the theoretical Π–A isotherms obtained for both considered hydroxycarboxylic acids are depicted in Figure 10.16a and b with solid lines. The model parameters are listed in Table 10.5, where A_c is the molecular area, which corresponds to the onset of the phase transition (i.e., at $\Pi = \Pi_c$), ε_0 is the relative jump of the area per molecule during phase transition, and η is a relative two-dimensional compressibility of the condensed monolayer. Here, for the 2-OH C16 acid at 25°C and 2-OH C18 acid at 25°C and 30°C, the surface pressure of the phase transition is $\Pi_c = 0$ mN/m. Hence, the theoretical dependences are absent for these cases. The model that assumes a dimerization (and some quantity of monomers as $m = 1.5$ for 2-OH C16 acid and $m = 1.8$ for 2-OH C18 acid) for $A \geq A_c$ and equilibrium between monomers and dimers for $A \leq A_c$ results in a perfect agreement with the experiment. For 2-hydroxycarboxylic acids with the regarded alkyl chains, large aggregates are formed mainly from dimers in good agreement with assumptions expressed in the previous section about the possible way of the clusterization process via *parallel* dimers.

Table 10.5 summarizes the thermodynamic characteristics of the aggregation process of 2-hydroxycarboxylic acid dimers. To estimate the parameters of dimerization (or the formation of large aggregates) from monomers, the oligomer–large aggregate transition characteristics should be multiplied by the oligomer aggregation number (as described in our previous paper) [54], which in the present case is $m = 1.5$ for 2-OH C16 acid and $m = 1.8$ for 2-OH C18 acid. These data enable a comparison with the quantum chemical calculation results, which yield the quantities per mole of monomers.

For all systems studied, the values of the experimental enthalpy and entropy are smaller (by absolute value) than those calculated for dimers from the quantum chemical semiempirical PM3 method. This can possibly be ascribed to the effect of intermolecular interactions within the aggregates, which was not accounted for in the theory. The agreement between experimental and quantum chemical data can be improved by introducing the additive corrections ΔH_{add} and ΔS_{add}, which account for this effect. The values of these additive corrections as estimated from a least squares method are (50.79 ± 2.85) kJ/mol and (142.72 ± 4.58) J/(mol·K), respectively. It should be mentioned that the described additive corrections were introduced for agreement between calculated

TABLE 10.5

Parameters of the Equation of State for α-Hydroxycarboxylic Acid in Spread Monolayers and Thermodynamic Characteristics for the Cluster Formation Out of Oligomers

Parameter	Temperature, °C							
	30	35	40	30–40	36	41	44	36–44
	α-Hydroxyhexadecanoic acid				α-Hydroxyoctadecanoic acid			
ω, nm^2	0.20	0.19	0.19		0.21	0.20	0.20	
A_c, nm^2	0.365	0.313	0.287		0.38	0.34	0.312	
Π_{coh}, mN/m	10.1	9.74	10.96		9.32	9.28	10.71	
ε_0	0.0	0.0	0.01		0.0	0.0	0.0	
η, m/mN	0.0025	0.002	0.0025		0.0025	0.003	0.004	
m	1.625	1.48	1.38		2.0	1.76	1.52	
ΔG^0, kJ/mol	−1.52[a]	−1.28[a]	−1.07[a]	−1.92[b]	−1.51[a]	−1.38[a]	−1.16[a]	−2.48[b]
ΔH^0, kJ/mol		−15.2[a]		−22.8[b]		−15.1[a]		−27.2[b]
ΔS^0, J/(mol·K)		−45.2[a]		−67.8[b]		−43.7[a]		−78.7[b]

[a] Formation of large aggregates out of oligomers.
[b] Formation of large aggregates out of monomers, approximately (** = * m)

TABLE 10.6

Thermodynamic Parameters for the Formation of Large Aggregates Out of Monomers

System (Temperature)	ΔH^0, kJ/mol		ΔS^0, J/(mol·K)		ΔG^0, kJ/mol	
	Calc.	Exp.	Calc.	Exp.	Calc.	Exp.
$C_{16}H_{32}O_3$ (35°C)	−19.95	−22.8	−63.22	−67.8	−0.48	−1.92
$C_{18}H_{38}O_3$ (41°C)	−30.05	−27.2	−83.28	−78.7	−3.90	−2.48

and experimental data for aliphatic alcohols, as well [54]. However, in case of alcohols, these additive corrections were somewhat smaller than for 2-hydroxycarboxylic acids. This could be affected by the nature of different amphiphiles. The interactions between functional groups of different surfactants contribute different increments into the values of the thermodynamic clusterization parameters. Table 10.6 summarizes a comparison of the experimental data with the calculated thermodynamic parameters for the formation of dimers from monomers with the correction for the dimerization parameters introduced as explained earlier.

Assuming all aspects mentioned earlier, one can see that all the theoretical models exploited for the description of the 2D clusterization process of 2-hydroxycarboxylic acids agree well. Although the quantum chemical model predicts somewhat higher values of the clusterization thermodynamic parameters, nevertheless, it agrees qualitatively with the experimental data concerning the spontaneous clusterization threshold of the regarded compounds and a possible path of the first-order phase transition.

10.4 COMPARISON OF THE SPONTANEOUS CLUSTERIZATION THRESHOLDS FOR NONIONOGENIC AMPHIPHILES

According to the quantum chemical model described in the second section and illustrated on the example of α-hydroxycarboxylic acids, the thermodynamic clusterization parameters were calculated for seven classes of monosubstituted alkanes and four classes of disubstituted ones. So, the surfactant list contains fatty alcohols, saturated and unsaturated carboxylic acids, amines, nitriles, thioalcohols, monoethoxylated alcohols, α-amino and α-hydroxycarboxylic acids, amides of carboxylic acids, and disubstituted melamines.

10.4.1 MONOSUBSTITUTED ALKANES

One homologous series of the classical amphiphiles are aliphatic alcohols. This type of compounds along with carboxylic acids has been frequently used as a standard model for experimental studies of monolayer properties. Therefore, aliphatic alcohols have been selected for quantum chemical simulations because of the considerable experimental data array for the comparison of the theoretical model results with experimental data. For aliphatic alcohols, $C_nH_{2n+1}OH$ ($n = 6$–16), two types of monolayers were considered: with oblique and hexagonal unit cell. The formation of both monolayer types is almost isoenergetic, but the structural parameters of the modeled hexagonal film are in better agreement with the existing experimental data. They are calculated to be: $a = 4.3$ Å and $b = 7.4$ Å and agree well with the corresponding experimental values: $a = 5.0$ Å and $b = 7.5$ Å. The tilt angle of alcohol molecules with respect to the normal to the interface is $t = 4°$, whereas the corresponding experimental values were found in the range of $0°$–$9°$ [12,55–58]. It was found that the spontaneous clusterization process is possible for alcohols having more than 10 carbon atoms in the alkyl chain, agreeing well with the results of numerous experimental studies [55,59–66]. The analysis of the calculated dependencies of the thermodynamic clusterization parameters per monomer for small aggregates and 2D films suggests the possible way of the formation of the hexagonal monolayer phase via formation of alcohol trimers and their further aggregation up to the infinite monolayer structure. This assumption corresponds well with the small cluster aggregation numbers $m = 2.3$–3.0 for the LE phase obtained by thermodynamic analysis of experimental Π–A isotherms of homologues C_{12}–C_{14} alcohols [18,11,54].

The results of the quantum chemical simulations obtained for aliphatic alcohols suggest the study of the effect of chemical substitution. So far, we have considered ethoxylated alcohols C_nE_m possessing only one ethoxy-unit C_nE_1 ($n = 6$–16). It was determined that spontaneous clusterization of the regarded compounds occurs for more than 14 carbon atoms in the alkyl chain. This agrees well with the analysis of the Π–A isotherms obtained experimentally for a number of C_nE_1 with $n = 14$, 16, and 18 [67]. The comparison of the threshold chain lengths for aliphatic and monoethoxylated alcohols at 298 K reveals that the introduction of the $-O-CH_2-CH_2-$ unit into the hydrophilic part of the amphiphile causes the shift of the spontaneous clusterization threshold to homologues with longer alkyl chains by three methylene units in agreement with results described in Refs. [68–74]. Monoethoxylated alcohol monolayers along with aliphatic alcohols have hexagonal unit cell with quite similar values of the calculated parameters: $a = 4.02$ Å and $b = 7.94$ Å, and $t = 4°$.

The second classical objects in the studies of amphiphiles behavior are carboxylic acids. We considered carboxylic acids with 7–16 carbon atoms of the alkyl chain. On the basis of quantum chemical calculations, it was shown that molecules in the highly ordered monolayer can be oriented at the angle ~16° (tilted monolayer), or at the angle ~0° to the normal to the air/water interface (untilted monolayer). The structural parameters of both tilted and untilted monolayers correspond well to the experimental data [75,76]. The parameters of the unit cell of the modeled tilted monolayer are: $a = 8.0$–8.2 Å and $b = 4.2$–4.5 Å (with the corresponding experimental data 8.4–8.7 Å and 4.9–5.0 Å [75,76]). For the modeled untilted monolayer, these parameters are: $a = 7.7$–8.0 Å and $b = 4.6$ Å (with the corresponding experimental data 8.4 Å and 4.8–4.9 Å [75,76]). Enthalpy, entropy, and Gibbs' energy of clusterization were calculated for both structures. The correlation dependencies of the calculated parameters on the number of pair-wise intermolecular CH·HC interactions in the clusters and the pair-wise interactions between functional groups were obtained. It was shown that spontaneous clusterization of carboxylic acids at the air/water interface under standard conditions is energetically preferable for molecules with 13 or more carbon atoms in the alkyl chain, and also this result agrees with the corresponding experimental parameters [77].

Good agreement of calculated results with the experimental data for saturated carboxylic acids encouraged us to investigate acids with double bonds in the alkyl chains. For this purpose, we considered both *cis*- and *trans*-unsaturated carboxylic acids with different position of the double bond. The model systems used are unsaturated fatty acids of the composition $\Delta = 12-15$ and $\omega = 6-11$, where Δ and ω refer to the number of carbon atoms between the functional group and double bond, and that between the double bond and methyl group, respectively. The calculations showed that for the monomers with equal alkyl chain lengths, the thermodynamic parameters (enthalpy, entropy, and Gibbs' free energy) do not depend on the position of the double bond. According to the exploited quantum chemical model, the thermodynamic parameters of clusterization were calculated for dimers, trimers, and tetramers of *cis*- and *trans*-unsaturated carboxylic acids of different structures. These clusterization parameters were used to construct the additive scheme capable of the prediction of the spontaneous clusterization threshold.

The calculations showed that the spontaneous clusterization threshold for *cis*-unsaturated carboxylic acids with $\Delta = 12$, 13, 14, and 15 corresponds to the total alkyl chain length of 18–19, 17–18, again 18–19, and 17–18 carbon atoms, respectively, that is, 18–19 carbon atoms for molecules with even Δ, and 17–18 carbon atoms for molecules with odd Δ. These results could be expected to be applicable for any *cis*-unsaturated carboxylic acids. In particular, they agree with the experimental data for *cis*-16-heptadecenoic acid (Δ16, ω1) [49,78], *cis*-9-hexadecenoic acid (Δ7, ω9) [49,79], *cis*-11-eicosenoic acid (Δ11, ω9) [49,80], and *cis*-9-octadecenoic acid (Δ9, ω 9) [49,80,81]. From the comparison of the calculated results with corresponding parameters of clusterization for saturated carboxylic acids [82], it is shown that for *cis*-unsaturated carboxylic acids with even and odd Δ values, the spontaneous clusterization threshold corresponds to the number of carbon atoms in the alkyl chain by six and seven units higher, respectively, than for saturated acids. This fact is attributed to the differences in the structures of *cis*-unsaturated fatty acids and saturated fatty acids, namely, to the presence of sp^3- and sp^2-hybridized carbon atoms in the *cis*-unsaturated fatty acids that contribute to additional intermolecular CH·HC-interactions [26]. Also, the number of carbon atoms in the saturated carboxylic acids alkyl chain is more than in the alkyl chain of unsaturated acids. This difference is due to the presence of an additional carbon atom involved in the acid group (because, according to the current classification of organic compounds, the carbon atom, which belongs to the carboxylic group of unsaturated acid, is accounted for in the Δ value, whereas for saturated acids, this atom is not considered to be the constituent of the fatty radical). Moreover, for even Δ values, one hydrogen atom in the alkyl chain participates in the formation of an additional intermolecular hydrogen bond what does not exist for saturated acids and *cis*-unsaturated acids with odd Δ. Also, the number of "a" type intermolecular CH·HC interactions [11,18,25,28,29,83] corresponding to the spontaneous clusterization threshold is 5–6 for both saturated and unsaturated acids, showing that this type of interaction governs the clusterization.

In comparison with the *cis*-unsaturated carboxylic acids, the *trans*-unsaturated acids start to form spontaneous clusters, and thus monolayers, with shorter hydrocarbon chains of 16–18 carbon atoms, whereas the corresponding threshold for *cis*-unsaturated carboxylic acids was found for 18–19 carbon atoms in the alkyl chain. These results are in agreement with experimental data [84]. Also, the calculated structural parameters of *trans*-unsaturated carboxylic acids monolayer for the unit cell with $a = 6.98$ Å, $b = 8.30$ Å, and for the molecular tilt angle with ~25° agree with the experimental parameters (for *trans*-13-docosenoic acid: $a = 5.30$ Å, $b = 8.48$ Å, and $t = 25.25°$ [84]). So, the proposed approach can be applied to the description of the clusterization process in monounsaturated acids with disconnected double bonds, and then, generalized to polyunsaturated acids with disconnected double bonds.

Other classes of amphiphiles, which were investigated in the frameworks of the described scheme, are thioalcohols [28], amines [29], and nitriles [82]. The spontaneous clusterization thresholds for these compounds are 14–15, 17–18, and 18–19 carbon atoms in the alkyl chain, respectively. Obviously, this can be ascribed to the fact that intermolecular hydrogen bonds N–H·N and S–H·S occurring between the monomer functional groups of amines and thioalcohols are less strong than intermolecular hydrogen bonds O–H·O appearing between amphiphilic molecules and water molecules of the interface. That is why the decrease of the amphiphile solubility is probably the reason enabling their cluster formation.

The proposed quantum chemical model was also tested on the amphiphiles having two alkyl chains in their structure. The melamine-type series of 2,4-di(n-alkylamino)-6-amino-1,3,5-triazine ($2C_nH_{2n+1}$-melamine) with $n = 9 - 16$ were considered [42]. The four most stable conformations were determined, which then were used to construct the clusters. The peculiar feature of these structures is the existence of a bend at one of the alkyl chains. Thus, the formation of infinite films becomes possible because of their spatial arrangement. From the calculation of the relative amount of various conformers in the mixture, it follows that, if the alkyl chain length is lower than 11–12 carbon atoms, the mixture is composed mainly of the monomers that do not contain any intramolecular interactions, whereas for higher alkyl chain lengths, the monomers that involve such interactions prevail in the mixture. Thus, for all clusters considered, the thermodynamic parameters (enthalpies, entropies, and Gibbs' energies) of clusterization were calculated. It was shown that the dependencies of these parameters on the alkyl chain length either exhibit stepwise shape or are represented by the combination of a linear and stepwise function. This depends on the different number of CH·HC interactions in the structures considered. Five types of clusters that are capable of the formation of infinite 2D films were considered in detail. For each of these types, the dependencies of the clusterization enthalpy, entropy, and Gibbs' energy on the alkyl chain length in the constituting monomers were obtained. Using these dependencies, it became possible to calculate these thermodynamic characteristics for clusters of any size, and also for infinite 2D films. It was shown that the spontaneous clusterization of $2C_nH_{2n+1}$-melamine became possible if the alkyl chain length exceeds nine carbon atoms. The assessed value of the spontaneous clusterization threshold agrees well with corresponding experimental value obtained by Π–A measurements [85–90].

10.4.2 Disubstituted Alkanes

The interfacial behavior of disubstituted alkanes is of interest, because a large number of compounds involved in the biological assemblies have more than one functional group. For example, monolayers of amphiphiles having a second polar group that are substituted at the lipophilic alkyl chain have been frequently studied as model systems, because they are constituents of the structural architecture of biological membranes and, thus, abundant in biological systems [91–98]. The biomembrane characteristics are modified in a unique way by the positions of the secondary polar OH group in the alkyl chain of the lipid [92]. Therefore, hydroxycarboxylic fatty acids are good candidates for the characterization of the effect of a second polar group substituted in different positions of the alkyl chain [48,75,99,100]. Because of the small size of the OH group, the steric influence of the substituent on the cross-section area of the amphiphile, and thus on the alkyl chain lattice, is minimized. So far, studies of the behavior of long chain hydroxycarboxylic acids have focused on the fundamental subject concerning the bipolar nature of the amphiphiles. Substitution of the hydroxy group at the C2 position of the alkyl chain gives rise to completely different behavior from that of midchain and terminal substitutions [46,101–103]. In the case of midchain position of the OH substitution, the molecules show bipolar behavior and the two polar groups act independently from each other as polar entities [102,103].

In contrast, OH substitution in 2-position leads to the formation of an enlarged monopolar unit [46,101]. It was interestingly found that this gives rise to a loss of ordering attributed to a size mismatch between the enlarged headgroup and the alkyl chain [50,104]. The alkyl chains afford greater lateral freedom, reducing the possibility for tilt ordering. Correspondingly, large differences exist also between the main monolayer characteristics of hydroxycarboxylic acids OH substituted in C2 position and midchain in mesoscopic and microscopic scale using surface pressure–area per molecule (Π–A) isotherms, BAM and GIXD [102,103].

As illustrated in the third section, the quantum chemical approach is also capable of predicting the spontaneous clusterization threshold for disubstituted alkanes. Let us remind that the spontaneous clusterization threshold for the 2-hydroxycarboxylic acid at the air/water interface at standard conditions is 14 carbon atoms in the alkyl chain that is in good agreement with the existing experimental data [48,49]. Comparing the Gibbs' energy values of clusterization for small aggregates, a possible scheme of 2D clusterization is proposed based on the formation of linear aggregates constructed by dimers with *parallel* orientation of the functional groups for 2-hydroxycarboxylic acid monomers with $n \leq 16$ carbon atoms in the alkyl chain. The following parameters of the unit cell for the most energetically preferred structure of 2D film were found: $a = 4.85$ Å, $b = 5.41$ Å, and the angle between them $\theta = 95°$. The molecular tilt angle of 2-hydroxycarboxylic acids with respect to the air/water interface was calculated to be ~20°.

In the frameworks of exploited quantum chemical model, thermodynamics and structure of monolayer formation for aliphatic α-amino acids were considered as well. These surfactants have more big headgroup than α-hydroxycarboxylic acids. The spontaneous clusterization thresholds for both homo- and heterochiral α-amino acids at the air/water interface are estimated in the range 13–14 carbon atoms in the alkyl chain at standard conditions. It was shown that the geometric parameters of the unit cell for the homochiral infinite 2D film, which corresponds to the most advantageous conformation of the monomer, were calculated to be: $a = 4.57$–4.71 Å and $b = 5.67$–5.75 Å, with the angle between the axes $\theta = 100°$–$103°$ and a general tilt angle of the alkyl chains with respect to the normal to the interface $t = 30°$. These values agree well with the available experimental data: $a = 4.91$ Å and $b = 5.25$ Å, $\theta = 112°$, and $t = 36°$ [7]. For heterochiral α-amino acid monolayers, the rectangular unit cell is typical with the calculated geometric parameters: $a = 4.62$ Å and $b = 10.70$ Å and the tilt angle of the alkyl chain with respect to the interface $t = 35°$, which is in good agreement with the GIXD data: $a = 4.80$ Å, $b = 9.67$ Å, and $t = 37°$ [7]. In addition, possible relative orientations of the monomers in the heterochiral clusters were considered. It was shown that for the racemic mixtures of α-amino acids, the formation of heterochiral 2D films is most energetically preferable with the alternating (rather than *checkered*) packing of the enantiomers with opposite specific rotation.

It is interesting to compare the spontaneous clusterization thresholds for disubstituted alkanes, which have the second functional group in 2-position. The OH group exchange by the NH_2 group slightly affects the values of the spontaneous clusterization thresholds for α-substituted carboxylic acids. This indicates the comparability of the increments of interactions of the hydrophilic parts by the values of the clusterization in Gibbs' energies.

The amides of carboxylic acid are another homologous series of amphiphiles investigated in the framework of the proposed quantum chemical model. The functional amide group $CONH_2$ can be represented as a combination of the ketonic C=O group and the amine NH_2 group. Calculations revealed that among all classes of amphiphiles described earlier, aliphatic amides have almost the same spontaneous clusterization threshold as saturated and α-substituted carboxylic acids and thioalcohols with 14–15 carbon atoms in the alkyl chain. Molecules of modeled amide monolayers are tilted by 22°–23° with respect to normal to the interface in reasonable agreement with existing experimental data (18°) [6].

The comparison of the spontaneous clusterization thresholds for amphiphiles that differ only by one functional group such as saturated carboxylic acids—amides or alcohols—amines is also of interest. The objective of this comparison is to define the relative influence of functional groups OH and NH_2 on the clusterization thermodynamics of these amphiphilic compounds. As already mentioned, the spontaneous clusterization threshold for amines is 18–19 carbon atoms in the alkyl chain, whereas for alcohols, it is 11–12 carbon atoms. Calculated values agree well with experimental data, which shows the presence of stable monolayers for alcohols starting from dodecanol [55,59–66] and for amines starting from hexadecylamine [105]. This fact indicates that intermolecular interactions of the functional groups realized between amine molecules provide a larger destabilizing contribution to the Gibbs' energy of clusterization than the interactions between alcohol molecules with the same alkyl chain length. It is also proved by the fact that tetradecylamine is not capable of stable monolayer formation [29,105]. On the other hand, the exchange of the hydroxyl group in carboxylic acids by the amino group does not lead to any significant shift in the spontaneous clusterization threshold to longer alkyl chains. It can be perhaps explained by the comparability in the increments of intermolecular hydrogen bonds O·H–O, which are realized between carboxylic acid molecules, and intermolecular hydrogen bonds O·H–N [106], which are realized between amide molecules.

In addition, taking into consideration the increase of the hydrophilic headgroup volume, molecules of the disubstituted alkanes orientate with larger tilt angle with respect to the normal to the interface: 0°–9° for alcohols, carboxylic acids, and amines; 18° for amides; 13°–22° for 2-hydroxycarboxylic acids; and 36° for α-amino acids. This gives rise to the decrease of the number of intermolecular CH·HC interactions between hydrophobic alkyl chains of the disubstituted alkanes, whereas the monolayers of monosubstituted alkanes have a maximal number of CH·HC interactions.

10.5 TEMPERATURE EFFECT OF SURFACTANT CLUSTERIZATION

Amphiphilic assemblies range from simple structures, such as monolayers, bilayers, micelles, and vesicles, to highly complex biological architectures. A variety of recent experimental studies revealed numerous interesting observations that required new theoretical insights about the detailed orientation and distance dependence of the intermolecular interaction between molecules within the assemblies. Bottom-up studies of amphiphilic monolayers, which have been often used as simplest model systems for more complex architectures, have been directed toward understanding the driving forces for the aggregation (condensation) processes. The results have been correlated to the formation of the condensed phase at the main-phase transition of the amphiphiles at the air/water interface and took into account conclusions on the properties of cluster formation, ordering at clusterization, and reorganization of clusters. Using semiempirical quantum chemical methods, thermodynamic quantities (enthalpy, entropy, and Gibbs' energy) of the dimerization and clusterization in finite and infinite clusters of amphiphilic compounds have been calculated. It was shown that the clusterization process is very sensitive to temperature, pH, and different additives in the subphase [6,64,77,107–120].

Despite numerous experimental studies of different factors, particularly, of the temperature on the process of monolayer formation, there are less theoretical models capable of the prediction of the aggregation behavior of amphiphiles (see, e.g., Ref. [121]). This requires further thorough investigations not only on structural and thermodynamic parameters of clusterization of substituted alkanes at the air/water interface, but also temperature conditions that allow one to obtain condensed 2D films of the corresponding amphiphiles with given alkyl chain lengths. The use of quantum chemical semiempirical methods allows one to obtain these parameters with appropriate accuracy without carrying out additional experimental studies.

According to our model, the values of enthalpy and entropy of cluster formation of infinite 2D monolayers can be defined using Equation 10.11. Let us remind that here the values of the coefficients U_i, and V_i depend on the particular thermodynamic characteristic $A_i^{Cl}{}_{(T),\infty}/m$ (enthalpy or entropy, or Gibbs' energy), on the structure of the monomer (descriptor i corresponds to its serial number) comprising 2D cluster, and on the temperature. $K_{a,\infty}/m$ is the number of the "a" CH·HC interactions type realized in the cluster (e.g., for α-hydroxycarboxylic acids, it can be calculated using Equation 10.10). It depends on the number of methylene fragments (n) in the alkyl chain and the structural features of the considered clusters. The coefficients $V_{\Delta H}$ and $V_{\Delta S}$ define the contributions of interactions between the hydrophilic parts of interacting amphiphile molecules in the enthalpy and entropy of cluster formation, respectively, $U_{\Delta H}$ and $U_{\Delta S}$ are the coefficients that define the contributions of the intermolecular CH·HC interactions between the alkyl chains of amphiphilic compounds in the enthalpy and entropy of cluster formation, respectively. The values of the regression coefficients that enter into the expressions for the calculation of enthalpy and entropy of clusterization per monomer for all considered types of amphiphilic compounds at the temperature 298 K are listed in Table 10.7.

TABLE 10.7

Values of the Coefficients for Calculation of the Thermodynamic Parameters of Clusterization per One Monomer Molecule of the Infinite 2D Cluster of Substituted Alkanes at 298 K

Long-Chain Amphiphile	$\Delta H^{Cl}_{(298),\infty}$, kJ/mol		$\Delta S^{Cl}_{(298),\infty}$, J/(mol·K)	
	$V_{\Delta H} \pm \Delta V_{\Delta H}$	$U_{\Delta H} \pm \Delta U_{\Delta H}$	$V_{\Delta S} \pm \Delta V_{\Delta S}$	$U_{\Delta S} \pm \Delta U_{\Delta S}$
Alcohols	−3.65 ± 1.38	−9.15 ± 0.07	−156.28 ± 6.8	−19.79 ± 0.2
Monoethoxylated alcohols	−35.09 ± 12.0	−8.51 ± 0.18	−295.82 ± 18.0	−19.30 ± 0.4
Amines	−4.96 ± 0.94	−10.11 ± 0.08	−166.0 ± 11.9	−24.5 ± 1.0
Thioalcohols	−7.40 ± 1.45	−9.78 ± 0.10	−211.9 ± 8.9	−17.8 ± 0.6
Carboxylic acids	0.07 ± 1.74	−10.39 ± 0.15	−160.0 ± 7.6	−23.1 ± 0.6
cis-Unsaturated carboxylic acids	−47.42 ± 0.96	−9.20 ± 0.08	−341.7 ± 16.3	−18.4 ± 0.6
trans-Unsaturated carboxylic acids	−21.54 ± 0.72	−9.20 ± 0.08	−316.0 ± 5.4	−18.4 ± 0.6
Amides of carboxylic acids	−4.23 ± 0.99	−10.24 ± 0.08	−175.8 ± 15.1	−22.3 ± 1.2
α-Amino acids	−17.98 ± 0.52	−10.28 ± 0.04	−217.1 ± 9.42	−21.7 ± 0.5
α-Hydroxycarboxylic acids	−10.14 ± 0.02	−10.14 ± 0.02	−212.87 ± 7.6	−20.02 ± 0.7

For the calculation of the temperature of the spontaneous cluster formation threshold of amphiphiles (T), the equation for definition of the Gibbs' energy of cluster formation is used:

$$\Delta G^{Cl}_{(T),\infty}/m = \Delta H^{Cl}_{(T),\infty}/m - T \cdot \Delta S^{Cl}_{(T),\infty}/m = V_{\Delta H} + U_{\Delta H} \cdot K_{a,\infty}/m - T(V_{\Delta S} + U_{\Delta S} \cdot K_{a,\infty}/m). \quad (10.17)$$

From this expression, if $\Delta G^{Cl}_{(T),\infty}/m$ is taken as zero, it is easy to get the equation for the temperature of the spontaneous cluster formation as the ratio of enthalpy to entropy of cluster formation including the corresponding thermodynamic properties of cluster formation per monomer of the infinite 2D films:

$$T = \frac{V_{\Delta H} + U_{\Delta H} \cdot K_{a,\infty}/m}{V_{\Delta S} + U_{\Delta S} \cdot K_{a,\infty}/m}, \quad (10.18)$$

where the coefficients $U_{\Delta H}$, $U_{\Delta S}$, $V_{\Delta H}$, and $V_{\Delta S}$ have the same meaning as in Equation 10.17, their values calculated at 298 K are listed in Table 10.7, and $K_{a,\infty}/m$ is the number of the intermolecular CH·HC interactions per monomer of the infinite 2D clusters of the considered types of substituted alkanes.

It should be noted that the values of the coefficients $U_{\Delta H}$, $U_{\Delta S}$, $V_{\Delta H}$, and $V_{\Delta S}$ depend on the temperature and there are a few approaches to take this dependence into account, which differ from each other by the degree of their theoretical validity. The main idea of scheme 1 is the use of the temperature dependencies of the enthalpy and entropy of clusterization of the small associates (dimers, tetramers, and hexamers) determined by using the expansion coefficients of heat capacity (a, b, and c'). Note that the values of ΔH^{Cl}_{298}, ΔS^{Cl}_{298}, a, b, and c' should be factorized contributions of the hydrophilic head groups and the CH·HC interactions between alkyl chains of the interacting molecules. Then, the terms T, T^2, T^3, $T\ln T$, $K_a T$, $K_a T^2$, and $K_a T \ln T$ can be inserted into the formula for calculating ΔG^{Cl}_{298}. Due to the lack of reference data for the values of the expansion coefficients of the heat capacity (a, b, and c'), it is impossible to exploit this scheme. Therefore, the simplification of Scheme 1 can be implemented in different ways: the corresponding variables T, T^2, T^3, $T\ln T$, $K_a T$, $K_a T^2$, and $K_a T \ln T$ as constants are found by linear regression using all the temperature data array available for ΔG^{Cl}_{298} (Scheme 2); by linear regression for the enthalpy and entropy of clusterization at each of the considered temperatures (Scheme 3); and neglecting the dependence of the coefficients of enthalpy and entropy of clusterization on the temperature and using the values of these coefficients found for one of the temperature values, for example, 298 K (Scheme 4).

Here we regard the simplest and consequently the roughest approach in detail. The use of the other more complicated schemes is discussed in Refs. [122] and [123]. The investigations shows that parameters used in these more complex schemes are independent of the amphiphile type used in the developed schemes. In the simplest scheme the values of the coefficients $U_{\Delta H}$, $U_{\Delta S}$, $V_{\Delta H}$, and $V_{\Delta S}$ are taken from the correlation expressions for the enthalpy and the entropy of clusterization calculated at one definite temperature value (e.g., 298 K, see Table 10.7). That is, we consider the dependency of the values of the coefficients $U_{\Delta H}$, $U_{\Delta S}$, $V_{\Delta H}$, and $V_{\Delta S}$ on the temperature in Equation 10.18 as negligible. It can be done if the process of the cluster formation is carried out in a narrow temperature range. Then one can estimate the temperature of the spontaneous cluster formation threshold of different types of substituted alkanes versus their alkyl chain length.

The calculated temperature values of the spontaneous cluster formation for eight types of amphiphilic compounds with the alkyl chain length of 10–25 carbon atoms are listed in Table 10.8. The relative error of the T obtained according to the error theory depends on the alkyl chain length of the considered types of surfactants. It decreases with lengthening of the alkyl chain. One can see that this value for the regarded alkyl chain lengths ($n = 10$–25) and for all considered types of amphiphilic compounds varies within the ranges of 3%–8%.

Consider now the comparability of the clusterization threshold temperature using the simplest Scheme 4 with experimental data in the example of aliphatic alcohols that form hexagonal monolayers (see Figure 10.17). It is seen that even the simplest scheme reproduces well the experimental data. The existing deviations are rather small and are in the range of 2–5 K for alcohols with alkyl chain of 9–14 carbon atoms. That is no more than 1.9% of relative error.

It should be noted that the fractionally linear dependency of the temperature of the spontaneous clusterization threshold on the alkyl chain length can be fitted within the linear function $T = e + f \cdot n$ in case of compounds with a narrow range of alkyl chain lengths (e.g., $n = 12$, 14, and 16 carbon atoms). Therefore, the calculated values of the coefficients e and f are: 230, 202, 188, 199, 199, 201, 176, and 209 K and 6.78, 6.07, 9.21, 7.19, 7.45, 7.45, 8.80, and 5.76 K, respectively, per carbon atom for alcohols, amines, thioalcohols, carboxylic acids, α-amino acids, amides of carboxylic acids, α-hydroxycarboxylic acids, and monoethoxylated alcohols. The calculated values of e and f coefficients for the considered types of substituted alkanes agree reasonably with those obtained for lipid series with alkyl chain lengths of 12, 14, and 16 carbon atoms: 225 and 7.10 K per carbon atom [125]. This confirms yet again the same nature of the intermolecular CH·HC interactions between alkyl chains of different types of amphiphilic compounds and their crucial contribution to the process of the formation of amphiphilic monolayers.

It should be mentioned that the influence of the alkyl chain length of amphiphilic molecules on the temperature of the spontaneous clusterization threshold can be clearly shown in the example: dependence of the temperature variation of the spontaneous clusterization threshold ΔT on the alkyl chain lengthening by two methylene groups. This choice is induced by the structural features of the infinite amphiphilic monolayers that are the result of different dependencies of the number of the intermolecular CH·HC interactions per monomer of substituted alkanes on the alkyl chain length. As proved in different papers [11,24–26,28,29], the intermolecular CH·HC interactions realized between alkyl chains of different amphiphile classes play a crucial role in the process of their cluster formation, while the interaction between their hydrophilic head groups makes a destabilizing contribution to the Gibbs' energy of clusterization, that is, in the clusterization process. This affects the values of the thermodynamic parameters of clusterization, in turn T and ΔT values are linearly or stepwise dependent on the alkyl chain length of one or another amphiphile. Thus, for example, for long-chain amines, carboxylic acids, thioalcohols, and cis-unsaturated acids, the dependencies of the number of the intermolecular CH·HC interactions per monomer on the alkyl chain length are stepwise; for the homochiral α-substituted carboxylic acids and trans-unsaturated carboxylic acids, these dependencies are linear; and for aliphatic and monoethoxylated alcohols comprising hexagonal monolayers, these dependencies are the combination of linear and stepwise. However, the

TABLE 10.8
Temperature of the Cluster Formation Threshold of Substituted Alkanes

System / Alkyl Chain Length, n	T, K																Relative Error, %
	10	11	12	13	14	15	16	17	18	19	20	21	22	23	24	25	
Alcohols[a]	292	300	311	317	326	331	338	342	348	352	357	360	364	367	371	373	5
Carboxylic acids	—	—	285	285	301	301	314	314	325	325	334	334	342	342	349	349	5–6
Unsaturated carboxylic acids: cis–trans	—	—	—	—	—	—	277	281	281	294	294	306	306	317	317	326	5
	—	—	—	—	—	—		285	292	299	305	311	316	322	327	331	4
Amides of carboxylic acids	—	—	287	287	303	303	316	316	327	327	336	336	345	345	352	352	7–8
α-Amino acids	—	278	288	296	304	311	317	323	329	334	339	343	348	352	355	359	3–4
α-Hydroxycarboxilic acids	—	—	281	291	300	308	316	323	330	336	342	347	352	357	361	366	4
Amines	—	—	275	275	288	288	299	299	308	308	316	316	323	323	328	328	6–7
Thioalcohols	—	—	294	294	313	313	330	330	345	345	358	358	369	369	380	380	5–6
Monoethoxylates alcohols[a]	—	—	278	283	290	295	301	305	310	314	318	321	326	328	332	334	7–8

[a] The monolayer with hexagonal unit cell is considered for the marked surfactants.

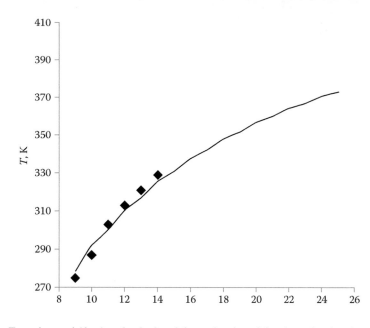

FIGURE 10.17 Experimental (dots) and calculated dependencies of the clusterization threshold temperature (T) on the alkyl chain length of aliphatic alcohol. (From Nagle, J.F., *Annu. Rev. Phys. Chem.*, 31, 157, 1980.)

structure of all considered infinite 2D clusters [11,24–26,28,29] of substituted alkanes with oblique unit cell is likely that lengthening of the alkyl chain by two CH_2 units causes the increase of the number of the CH·HC interactions per monomer by two interactions. In case of films with hexagonal unit cell elongation of surfactant chain by two CH_2 units leads to the increase of the number of the CH·HC interactions per monomer by 2.5 interactions. Then, having defined ΔT as the difference of the temperatures of the spontaneous clusterization threshold of amphiphiles with ($n + 2$) and n carbon atoms and using Equation 5.1, one can obtain

$$\Delta T = \frac{2 \cdot (U^0_{\Delta H} \cdot (V_{\Delta H(T)} - V^0_{\Delta S}) - V^0_{\Delta H} \cdot (U_{\Delta H(T)} - U^0_{\Delta S}))}{((V_{\Delta H(T)} - V^0_{\Delta S}) + (U_{\Delta H(T)} - U^0_{\Delta S}) \cdot (K_{a,\infty} / m + 2)) \cdot ((V_{\Delta H(T)} - V^0_{\Delta S}) + (U_{\Delta H(T)} - U^0_{\Delta S}) \cdot (K_{a,\infty} / m))}$$

(10.19)

or

$$\Delta T = \frac{2.5 \cdot (U^0_{\Delta H} \cdot (V_{\Delta H(T)} - V^0_{\Delta S}) - V^0_{\Delta H} \cdot (U_{\Delta H(T)} - U^0_{\Delta S}))}{((V_{\Delta H(T)} - V^0_{\Delta S}) + (U_{\Delta H(T)} - U^0_{\Delta S}) \cdot (K_{a,\infty} / m + 2.5)) \cdot ((V_{\Delta H(T)} - V^0_{\Delta S}) + (U_{\Delta H(T)} - U^0_{\Delta S}) \cdot (K_{a,\infty} / m))}$$

(10.20)

for amphiphiles with oblique and hexagonal unit cells, respectively.

Figure 10.18 illustrates the dependencies of ΔT on the alkyl chain length for infinite 2D clusters of substituted alkanes. These dependencies are plotted for amphiphiles with alkyl chain lengths of 10–25 carbon atoms and the water subphase in liquid state. From the data presented in Figure 10.18, it is seen that larger variations of the temperature for the spontaneous clusterization threshold are realized for molecules with shorter alkyl chain lengths than for molecules with longer alkyl chains. That means, the shorter the alkyl chain length of the considered amphiphilic compound, the more necessary is it to decrease the subphase temperature to enable the process of spontaneous cluster formation. For instance, in case of fatty alcohols (for the monolayer with hexagonal unit cell), the lengthening of the alkyl chain of decanol ($n = 10$) by two CH_2 units is equal to decrease of the temperature of the cluster formation threshold by 15 K, whereas for hexadecanol ($n = 16$), this effect is

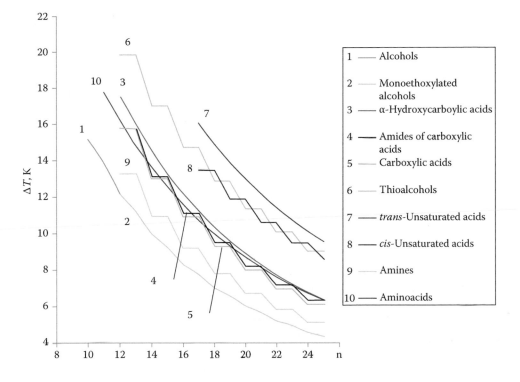

FIGURE 10.18   Dependence of the ΔT variation on the alkyl chain length (n).

essentially smaller by 8 K, and in the case of docosanol ($n = 22$), this effect is essentially smaller by 5 K. It should be noted that the dependencies of the temperature variation for the spontaneous cluster formation threshold ΔT on the alkyl chain length are stepwise for all considered types of amphiphilic compounds except for α-substituted carboxylic acids and *trans*-unsaturated carboxylic acids. As indicated earlier, this fact is caused by the structural features of the considered monolayers, which are the result of different dependencies of the number of the intermolecular CH·HC interactions per monomer of the substituted alkanes on the alkyl chain length.

Nevertheless, as shown in Figure 10.18, the types of the dependencies of the *temperature effect* are comparable for different classes of amphiphilic compounds. Thus, the addition of two CH_2 units to the alkyl chain has a similar effect for reducing the temperature of the spontaneous cluster formation by 10–20 K (or 5–10 K per CH_2 unit) well agreeing with the available experimental data [67,77]. The longer the hydrophobic alkyl chain of the considered amphiphile, the smaller is the difference between the values of the temperature variation ΔT. In addition, the values of the coefficients defining the contribution of the CH·HC interactions in the values of enthalpy and entropy for cluster formation of regarded types of substituted alkanes are quite similar (see Table 10.7). This allows the conclusion that the difference in the ΔT values is determined by the nature of the hydrophilic part of the surfactant and its contribution to the ΔT variation becomes less significant with lengthening of the alkyl chain [126,127].

10.6   ASSESSMENT OF STRUCTURAL AND THERMODYNAMIC PARAMETERS OF AMPHIPHILIC MONOLAYERS

10.6.1   ESTIMATION OF THE MOLECULAR TILT ANGLE OF AMPHIPHILES IN THE FILMS

The possibility of wide application of amphiphilic mono- and multilayers in electronics, optics, and the creation of artificial biomembranes and biosensors provokes the necessity of detailed investigation of the structure of the obtained films. Particularly, one of the important questions is the detection of the factors influencing the condensed monolayer thickness, which in turn depends on the tilt angle of the surfactant molecules with respect to the interface. As the experimental studies of amphiphilic compounds show [128], their molecules in the condensed monolayer phase are in linear conformation and almost in upright orientation with respect to the interface, opposite to the weakly ordered molecules of the monolayers in liquid-expanded state. Concerning the simulation of the amphiphiles behavior in the monolayers of different density, the data obtained in Refs. [129–132] show that the alkyl chains tend to align and orient vertically at the interface with the increase of the surface pressure. Nevertheless, it should be noted that different orientations of the molecules are possible in these monolayers. Thus, in Ref. [103], the GIXD observations found that the molecules of 9-, 11-, and 12-hydroxyoctadecanoic acids are tilted with respect to the normal to the air/water interface at angles from 13° to 22°, whereas the α-amino acids are even more tilted with 36° [7]. The authors in Ref. [133] studied α-amino acid derivatives and revealed that surfactant molecules can be tilted with angles from 18° to 40° with respect to the normal to the interface and up to 49° for N-alkyl substituted α-amino acids [134].

The results of X-ray diffraction analysis of different nonionic surfactants [6,7,44,75,76,103,133–137] prove that alcohols and carboxylic acids are almost in upright position with respect to the interface, whereas amphiphiles having a bigger hydrophilic part are significantly tilted with respect to the normal to the interface. In addition, alcohols, carboxylic acids, and their derivatives can form crystalline monolayers with a hexagonal unit cell, whereas the monolayers of homochiral α-amino acids and their derivatives have oblique unit cells. In the case of carboxylic acids, there is the possibility to form two general types of monolayers with upright and inclined orientation of molecules with respect to the interface. In the studies [44,136,137], the authors explain this fact in such a way that inclined orientation of surfactant molecules is realized in monolayers with lower surface pressure values than with vertically oriented molecules. The authors of Ref. [138] investigating the monolayer structure of chiral amphiphiles suggest that the value of the molecular tilt angle in the monolayer depends on the size of the functional group located around the chiral center of the molecule and can vary in range from 15° to 45°. Several authors [139,140] consider that inclined position is inherent for amphiphilic compounds with the cross-section area of the hydrophilic part larger than that of the alkyl chain. Researches of the studies of surfactant films at solid surfaces assume that the molecular tilt angle depends to a large extent on the commensurability of the volume of the surfactant headgroups and the volume of the atoms belong to the solid surface [141]. If the surfactant head group is commensurable to the volume of the atom of the solid surface, then the amphiphilic molecule orients vertically to this surface. In case of significant differences between sizes of the surfactant head groups and atoms of the surface, the inclined orientation of molecules is realized in the monolayer. In the judgment of Ref. [142], the position of amphiphilic molecules at the interface is caused by two mutually opposed forces: the force of sterical repulsion between the surfactant molecules and the force of Coulombic attraction between the dipoles and the liquid surface. The first force favors the upright orientation of amphiphilic molecules, whereas the second one favors the inclined orientation. It should be noted that the value of the tilt angle for the same surfactant class in the monolayer depends on the type of the crystal lattice formed: the molecules arranged in hexagonal packing possess smaller tilt angle with respect to the normal to the interface than those packed in the oblique lattice. According to our results, the position of the amphiphiles with respect to the interface is defined by structure and orientation of the hydrophilic parts of the interacting molecules and the CH·HC interactions realized between the hydrophobic chains [143].

On the basis of the quantum chemical calculations, we have shown that the molecular tilt angle of amphiphiles in the monolayer with respect to the normal to the interface changes discontinuously. It has discrete values dependent on the size of the hydrophilic part and the structural peculiarities of the intermolecular CH·HC interactions realized between the molecules in the monolayer. It is explained by the fact that the geometrical parameters of the alkyl chains of the monomers and their aggregates obtained using quantum chemical calculations are the same for different classes of amphiphiles and are defined by the mutual orientation of molecules in the clusters. The size of the amphiphile hydrophilic part determines two quantum numbers for the tilt angles ϕ and δ with respect to the normal to the propagation directions of the 2D film (which correspond to the dimerization Gibbs' energy minimum). In turn, these two angles define the option for a certain value of the general tilt angle of molecule in the monolayer from the discrete set of possible values (see Figure 10.9 and Equation 10.3).

In order to illustrate the proposed approach, we investigated the dimer structures of six amphiphile types with increasing size of the hydrophilic parts in the sequence: saturated alcohols, carboxylic acids, amides of carboxylic acids, α-hydroxycarboxylic acids, and α-amino acids, N-acyl-substituted alanine. It was analytically found that the tangent of the angles ϕ and δ depends linearly on the number of CH·HC interaction lost because of sterical hindrances appearing during the disposition of the bigger hydrophilic parts of the molecules:

$$tg\phi = \pm 0.095 + 0.265 \cdot i + 0.531 \cdot n_{lost}^{q}, \quad tg\delta = \pm 0.095 + 0.265 \cdot i + 0.531 \cdot n_{lost}^{p}. \qquad (10.21)$$

Here, n_{lost}^{p} and n_{lost}^{q} are the numbers of CH·HC interaction lost as a result of mutual repulsion of bigger hydrophilic parts of the amphiphile in dimers in p and q-directions of the monolayer propagation (see Table 10.9); parameter i is an identifier of the structural peculiarities of the dimers: i can have the value of 1 or –1 in the case when the corresponding dimer in p or q-direction is formed by lower or upper shift of the given monomer molecule with respect to another one. The «–» sign before the free term indicates that in the regarded dimer, the upper shift of the CH·HC interactions is realized; otherwise, this free term is positive. The values of the increment and free term in the described linear dependence (10.21) by $tg\phi$ (or $tg\delta$) are in good agreement with those obtained using the correlation analysis of the considered angle values for dimers of the chosen amphiphile classes optimized in the framework of the PM3 method.

The values of the quantum numbers n_{lost}^{p} and n_{lost}^{q}, which stand for the number of lost CH·HC interactions in amphiphile dimers of both p- and q-directions, can be determined, as described in Table 10.9. For dimers in q-direction, length and height of the hydrophilic part are limiting parameters, whereas for dimers in p-direction, these are width and height. In this connection, only the values of quantum number n_{lost}^{q} are listed in Table 10.9, depending on the linear sizes of the hydrophilic part of the amphiphile. The same values result for n_{lost}^{p} with the only difference that the criterion parameters are width and height. During the amphiphile aggregation with small hydrophilic parts, no sterical hindrances appear during their mutual location in the dimer. So, the maximal number of the intermolecular CH·HC interactions is realized in the dimer formed by the monomer of a certain alkyl chain length. In case of amphiphilic compounds with more bigger hydrophilic parts, it is necessary

---

TABLE 10.9

Values of Quantum Numbers n_{lost}^{q} and n_{lost}^{p} for Lost CH·HC Interactions in Dimers Formed in p and q-Direction of Monolayer Propagation Depending on the Linear Sizes of the Hydrophilic Parts

Linear Size	$n_{lost}^{q} \cdot (n_{lost}^{p})$
Height: 0.8–1.7Å	
Length (width for n_{lost}^{p}) up to 2.8 Å	0
Height: 1.1–2.0 Å	1
Length (width for n_{lost}^{p}) more than 2.8 Å	
Height: 3.6–4.5 Å	2
Length (width for n_{lost}^{p}) more than 2.8 Å	

TABLE 10.10

Tilt Angles Values of Amphiphilic Molecules with Respect to the Interface in Condensed Monolayers

Surfactant Class	Calculated Values According Equation 10.21			Direct Calculation in PM3 Method		Exp.	Refs.
	ϕ	δ	t	ϕ	δ	t	t
Saturated alcohols:						0–9	
With hexagonal unit cell	—	—	—	3	4	4	[135]
With oblique unit cell	9	9	13	8	10	13	
Carboxylic acids:							
With hexagonal unit cell	—	—	—	7	0	0	0	[44,75,76, 136,137]
With oblique unit cell	20	9	22	13	13	18	20	
Amides of carboxylic acids	20	9	22	23	10	23	18	[7]
α-Hydroxycarboxylic acids	10	9	14	19	7	21	13–22	[47]
α-Amino acids	20	10	25	19	16	30	36	[7]
N-acylsubstituted alanine	35	9	41	37	12	43	46	[134]

to shift the monomer molecules (upward or downward) with respect to each other in p- or q-direction in order to avoid the sterical hindrances. This leads to the loss of one or more CH·HC interactions in the formed dimer but allows the orientation of the given hydrophilic parts of molecules.

In order to assess the value of the amphiphile tilt angle with respect to the normal to the interface, one has to define the value of the θ angle between the p and q-directions of the spread monolayer. In the simplest case $\theta = 90°$ when the amphiphile dimers form a squared tetramer. The values of the θ angle obtained for the optimized tetramer structures are close to 90° and make up 87° for saturated alcohols, 86° for carboxylic acids, 98° for amides of carboxylic acids, 95° for α-hydroxycarboxylic acids, 103° for α-amino acids, and 105° for N-acyl-substituted alanine. The calculated values of the molecular tilt angles for the regarded amphiphile monolayers agree with the corresponding experimental values obtained using the GIXD method (see Table 10.10). It should be noted that for alcohols and carboxylic acids, the listed values of the tilt angle correspond with the monolayers with oblique unit cell. During the formation of the monolayers with hexagonal unit cell, the amphiphilic molecules orientate themselves in such a way that the mutual shift of the molecules in both directions of the monolayer propagation is minimal and the corresponding values of the ϕ or δ angles are less than 9°. It is also necessary to mention that the hydrophilic parts of the amphiphilic molecule can orientate themselves not along the p and q directions of the monolayer, but along the diagonal of the square. In such case, the values of ϕ or δ angles can be lower than 9° (as, e.g., for carboxylic acids studied in Ref. [25]). In addition, the increase in atom number with different electronegativity in the surfactant headgroup leads to the probable origin of intermolecular hydrogen bonds between the monomer headgroups in the cluster. It can in turn affect the realization of one or another shift of hydrophobic chains of monomers during the formation of CH·HC interactions.

It can be clearly seen from Table 10.10 that the experimental data are quite well reproduced by the quantum chemical calculations of the PM3 method and can be predicted on the basis of the geometrical parameters of the amphiphile structure and the intermolecular CH·HC interactions formed in the aggregation process. The presented data make it possible to state that the tilt angles of the amphiphiles with respect to the normal to the interface have discontinuous values. One of the possible values depends on the number of CH·HC interactions realized between the hydrocarbon chains of the amphiphile which in turns is caused by the size of the hydrophilic parts of the amphiphiles participating in the aggregation. The proposed approach provides prior assessment of the molecular tilt angle with respect to the interface even without preliminary quantum chemical calculations and time-consuming experiments.

10.6.2 ASSESSMENT OF THERMODYNAMIC PARAMETERS OF FORMATION AND CLUSTERIZATION OF DISUBSTITUTED ALKANES IN THE FRAMEWORK OF SUPERPOSITION-ADDITIVE APPROACH

10.6.2.1 Introduction into the Historical Background of Application of Additive Schemes

The postulate about the existence of atoms in molecules [144] is the rigorous basis not only for the development of the conception about functional groups but also a tool for the assessment of the transferability of their properties [145]. The degree of transferability for atomic groups could be directly calculated using corresponding values of the atomic properties or the bond properties between them. In addition, the transferability to a certain extent could be estimated using similarity indices [146] on the basis of the comparability of the atomic charge densities in the molecules or the critical points for corresponding bonds [147–149].

Such an approach was implemented to the calculation of the atomic transferability in molecules for different classes of organic compounds, particularly hydrocarbons [150,151], alcohols [152], aldehydes [153], ketones [153,154], ethers [155–157], and nitriles [158]. These works show that the energetic properties of oxygen, nitrogen, and carbon atoms depend significantly on the nearest environment of the fragment in which regarded atoms are included. In this connection, the calculations according to the different additive schemes performed previously [159–161] should be carried out, accounting for the properties of the nearest molecular environment for the considered molecular fragment, that is, justifying the properties of molecular graphs. One more display of atomic transferability based on the postulate about the existence of atoms in molecules is the superposition-additive approach (SAA). Now, it is widespread and generally recognized [162–168].

For the first time, this approach was developed for the calculations of the electronic structure and physicochemical properties of conjugated systems [162]. Later [163], it was applied to the calculation of the thermodynamic parameters of formation and atomization of conjugated systems, their dipole electric polarizabilities, molecular diamagnetic receptivity, π-electronic ring current, etc. Earlier the superposition-additive method was only applied to a few types of compounds (conjugated systems and radicals). However, it would be interesting not only to generalize this approach to saturated alkanes and their derivatives, but also to study its applicability to van der Waals molecules. In this respect, substituted alkanes are especially promising, because they possess a nonbranched alkyl chain, and are capable of the formation of 2D films (clusters or van der Waals molecules) at the interface. For these systems, relevant experimental data (enthalpy, entropy, Gibbs' energy of formation, critical temperatures, and standard heat capacities) are available [38,39,169–174], and the thermodynamic parameters were calculated according to the quantum chemical semiempirical PM3 method [11,25,26,28,29,40].

The implementation of the superposition-additive approach offers a new and unique possibility to obtain parameters, which are unknown for some members of homologous series of surfactants and amphiphiles, in the case that corresponding experimental or theoretically calculated parameters (e.g., thermodynamic parameters at the formation of monolayers) are available for other members of homologous series. Also it suggests the option to calculate the thermodynamic parameters in the case that corresponding values for other classes of surfactants are available. The present section deals with the extension of the superposition-additive approach onto the description of the thermodynamic parameters (enthalpy and Gibbs' energy) of formation, absolute entropy, and thermodynamic parameters of clusterization of saturated alkanes and their substituents at the water/air interface. The studied entities are molecules and clusters of mono- and disubstituted alkanes considered previously.

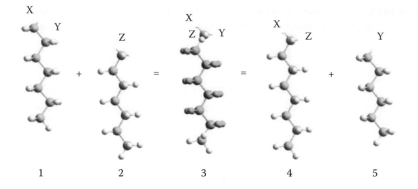

FIGURE 10.19 Generalized superposition-additive scheme for the calculation of thermodynamic parameters of surfactant monomers with alkyl chain length n = 8 taken as an example: X, Y, and Z are schematic denotations of the functional groups.

10.6.2.2 Theoretical Basis of Superposition-Additive Approach

The postulate about the way in which atoms exist in molecules is the theoretical basis for different superposition-additive schemes [144]. Its general idea is that each atom in a molecule retains its individuality in various chemical combinations (i.e., in various molecules). This refers to the transferability of atomic properties. In addition, the atomic values, being summed over all atoms in the molecule, yield the molecular average so that the corresponding molecular characteristics are additive.

The main idea of the superposition-additive approach is based on transferability of atomic properties and additivity of molecular properties. The essence of the procedure is the assumption that when two molecular graphs are superimposed, the properties of the constituent atoms remain unchanged. Then, if the same superposition can be constructed in two different ways, each one involving two entities, it becomes possible to calculate structure and properties of one of these entities, provided the structure and properties of the remaining three entities are known.

This principle is graphically illustrated in Figure 10.19. The molecules (1), (2), (4), and (5) are structures, which involve the alkyl chain and the functional groups X, Y, and Z (these groups can be the same or different). Structure (3) is the result of superposition of the structures (1) and (2), and also of the structures (4) and (5). As these two superpositions lead to the same result, the properties of any of the four molecules could be expressed as the algebraic sum of the corresponding properties of the other three molecules. Thus, for example, to calculate any thermodynamic parameter of molecule (4), one has to add the corresponding values of the molecules (1) and (2) and to subtract the value of the molecule (5).

10.6.2.3 Application of SAA to Mono- and Disubstituted Alkanes

Previously [163–167], we have shown that it is possible to use reasonably a quite wide range of superposition-additive schemes. However, the best results are given by schemes possessing maximal superimposition of molecular graphs. Here, we illustrate the use of SAA on the example of schemes that correspond to maximal mutual overlapping of alkyl chains (C_nH_{2n-4}). The first scheme for monosubstituted alkanes can be presented as follows:

$$\text{SAS\_1: } A(C_nH_{2n}XY) = A(C_{n-1}H_{2n-2}XZ) + A(C_{n-1}H_{2n-1}Y) - A(C_{n-2}H_{2n-3}Z), \qquad (10.22)$$

where

 A is the thermodynamic parameter (absolute entropy, enthalpy, or Gibbs' energy of the formation of the compound from elementary substances) at normal conditions (=298.15 K)

 n is the number of atoms in the hydrocarbon chain

 X, Y, and Z are the schematic definitions of the structural fragments of the functional groups (e.g., H for alkanes, COOH for carboxylic acids, OH for alcohols, NH_2 for amines, etc.)

It is possible to use a number of schemes with the same mutual overlapping of the alkyl chains but using molecules of the same amphiphile class (X=Y=Z) or different ones (X≠Y≠Z). To calculate the thermodynamic parameters of a molecule using SAS_1, containing n carbon atoms in the alkyl chain, one should add the corresponding parameters of two molecules with $(n-1)$ carbon atoms in the alkyl chain, and subtract the corresponding parameter of the molecule with $(n-2)$ carbon atoms in the alkyl chain. For example, to calculate the thermodynamic parameter for octane (X=Y=Z=H), one should add the values of this parameter for two heptane molecule and subtract from the calculated sum the corresponding thermodynamic parameter of hexane.

The thermodynamic formation parameters for monosubstituted alkanes were calculated using data obtained in the framework of the PM3 method and compared with directly calculated values. In particular, for enthalpy and Gibbs' energy of the formation of substituted alkane monomers from the elementary substances, and their absolute entropy, the standard deviations of the values calculated according to the SAS_1 from the results of PM3 calculations for alkanes, alcohols, thioalcohols, amines, fatty acids, nitriles, and cis-unsaturated carboxylic acids are: 0.05, 0.004, 2.87, 0.02, 0.01, 0.77, and 0.01 kJ/mol for enthalpy; 2.32, 5.26, 4.49, 0.53, 1.22, 1.02, and 5.30 J/(mol · K) for absolute entropy; 0.69, 1.56, 3.82, 0.15, 0.37, 0.69, and 1.58 kJ/mol for Gibbs' energy, whereas the deviations from the experimental data are 0.52, 5.75, 1.40, 1.00, and 4.86 kJ/mol; 0.52, 0.63, 1.40, 6.11, and 2.21 J/(mol · K); 2.52, 5.76, 1.58, 1.78, and 4.86 kJ/mol, respectively (for nitriles and cis-unsaturated carboxylic acids, experimental data are not available). The proposed approach provides also quite accurate estimates of enthalpy, entropy, and Gibbs' energy of boiling and melting, critical temperatures, and standard heat capacities for several classes of substituted alkanes.

Good predictive abilities of SAA revealed in the example of monosubstituted alkanes gave us an idea to adjust it to the disubstituted ones, such as α-substituted carboxylic acids and carboxylic acid amides. For this purpose, we regarded only the most interesting SAS, which use the data for three classes of amphiphiles, that is, when X≠Y≠Z. It should be mentioned that in the exploited SAS, the energetically most favored monomer and cluster structures of bifunctional amphiphiles are used (except 2-hydroxycarboxylic acids) [24,27,175]. The second favored structures of 2-hydroxy-carboxylic acid monomers and clusters have the most similar structure of the hydrophilic part of the molecule to that of α-amino acids. This enables maximal overlapping of the molecular graphs for the regarded compounds. This is the reason for using these particular 2-hydroxycarboxylic acid structures. The geometry of the small aggregates (dimers and tetramers) of α-substituted carboxylic acids is quite similar, as well. The tilt angle of amphiphilic molecules with respect to the normal to the interface was calculated earlier [24,175] to be 20°–30° for structures built on the basis of the chosen acid monomers. In addition, there are equal number of intermolecular CH·HC interactions realized in the regarded aggregates of α-substituted carboxylic acids up to 2D films. It should be noted that mismatch of the number of the mentioned interactions in the superposition of the molecular graphs for the regarded structures contradicts the main principles of the superposition-additive approach. In light of the aforementioned fact, such mismatch does not give any satisfactory result.

Note, that in SAS exploited here for calculation of thermodynamic parameters of formation and clusterization of bifunctional amphiphiles we use the monomers and clusters of mono- and unsubstituted alkanes of different structures for α-substituted acids and amides. Also, the exploited structures of monomers, aggregates, and 2D films for mono- and unsubstituted alkanes could not be most preferable according to the Gibbs' energy, and they cannot be realized in experiment. However, the use of such structures is necessary for ensuring a maximal molecular graph overlapping for the structures participating in the superposition. This is the key point for implementation of different superposition-additive schemes.

The use of the SAS_1 for bifunctional amphiphiles reveals systematic errors for the description of the thermodynamic parameters of formation and can be attributed to the incomplete

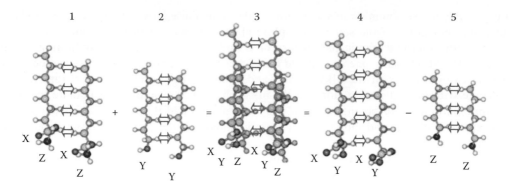

FIGURE 10.20 SAS 2 for the calculation of the formation and clusterization parameters of disubstituted alkanes illustrated in the example of 2-hydroxycarboxylic acid dimers (X=COOH, Y=OH, and Z=NH$_2$).

reproduction of the values for the increments of intra- and intermolecular interactions between functional groups in the molecules of bifunctional alkanes by the corresponding values of the increments for monosubstituted alkanes. These errors can be taken into account using an additive correction. Consideration of this correction allows the assessment of the values of the thermodynamic parameters of formation for 2-hydroxycarboxylic acids, α-amino acids, and alkyl amides with the following values of standard deviations: 0.01, 0.03, and 0.02 kJ/mol for enthalpy; 0.88, 0.50, and 1.16 J/(mol · K) for absolute entropy; and 0.31, 0.29, and 0.37 kJ/mol for Gibbs' energy, respectively.

As noted recently [164] regarding the application of the proposed approach to the calculation of the thermodynamic parameters of formation and clusterization of amphiphile aggregates, it should be kept in mind that, in contrast to the individual monomers, these systems involve also intermolecular CH·HC interactions (see Figure 10.20). Thus, the CH·HC-interactions realized in clusters of amphiphiles with odd number of carbon atoms in the monomer chains do not overlap with those realized in the clusters of amphiphiles with even number of carbon atoms in the monomer chains. This provokes inapplicability of the superposition-additive scheme 1 to the calculation of the homologue with n carbon atoms out of two homologues with $(n - 1)$ and one with $(n - 2)$ carbon atoms in their chains. Therefore, in this case, the schemes provided with the principle of mutual molecular graphs overlapping would be correct. Molecular graph overlapping is possible if one uses thermodynamic properties of clusters with odd (even) number of methylene units for the calculation of the corresponding parameters only for clusters having odd (even) number of CH$_2$ groups.

To calculate the parameters of the aggregate formation and its clusterization, one should apply SAS_2 with maximal overlapping of molecular graphs of the structures involved in the mentioned scheme (see Figure 10.20):

$$\text{SAS\_2: } A(C_nH_{2n}XY)_m/m = A(C_{n-2}H_{2n-4}XZ)_m/m + A(C_{n-2}H_{2n-3}Y)_m/m - A(C_{n-4}H_{2n-7}Z)_m/m,$$
$$(10.23)$$

where

A is the thermodynamic parameter of formation or clusterization of associates per monomer molecule

X, Y, and Z are the schematic definitions of the structural fragments of the functional groups as in expression (10.22)

m is the number of monomers in a cluster ($m = 2$ for dimers, $m = 4$ for tetramers)

n is the number of carbon atoms in the hydrocarbon chain of monomer

Application of the SAS_2 gives the opportunity to calculate the thermodynamic formation parameters of small aggregates of monosubstituted alkanes (total for alcohols, amines, and thioalcohols) with the next mean values of the standard deviations (from the directly calculated data in the framework of the PM3 method): 0.17, 1.09, and 0.36 kJ/mol for the formation enthalpy of dimers, trimers, and tetramers respectively, and 4.49, 5.76, and 6.14 J/(mol · K) for entropy of cluster formation respectively [164]. The mentioned Scheme 2 is also capable of the assessment of the clusterization thermodynamic functions for monosubstituted alkanes. For carboxylic acids taken as an example, the standard deviations for enthalpy, entropy, and Gibbs' energy of clusterization were 0.28–0.74 kJ/mol, 1.85–7.24 J/(mol · K), and 0.77–1.47 kJ/mol for the dimers, and 0.23–1.26 kJ/mol, 2.17–23.18 J/(mol · K), and 0.29–6.95 kJ/mol for tetramers. In the frameworks of this approach, it is also possible to assess the thermodynamic clusterization parameters of the 2D monolayers.

The use of SAS_2 for the calculation of the thermodynamic parameters of aggregate formation and clusterization for bifunctional amphiphiles reveals the next peculiarities. There are systematic errors of the description of the thermodynamic formation and clusterization parameters comparatively to the directly calculated values. As noted by the authors of Ref. [104], for amphiphiles with two closely located functional groups (not further than at a distance of four methylene units), these functional groups act as a single hydrophilic part of the molecule. In this connection, the presence of the mentioned systematic errors for the thermodynamic parameters of the aggregate formation, as in the case of monomers, is caused by the following fact. The sum of the increments for individual functional groups of the amphiphiles involved in superposition is only partially reproduced by the increments of the interactions between the single bifunctional hydrophilic parts of molecule. The accounting of these systematic errors leads to significant decrease of the values of the standard deviations. They are 0.42–1.35 and 0.51–3.21 kJ/mol for formation enthalpy of dimers and tetramers of α-substituted carboxylic acids and amides, and 3.44–5.33 and 3.26–6.95 J/(mol · K) for entropy, respectively. The systematic errors are the highest for alkyl amides among all the classes of bifunctional alkanes considered. As shown in Ref. [27], it is caused by incomplete overlapping of the molecular graphs of the structures involved in the scheme. This can be attributed to the impossibility to represent intermolecular interactions of the hydrophilic head groups in the amide associates only as a combination of the corresponding interactions realized in the clusters of carboxylic acids, amines, and alcohols. The p-π-conjugated system realized in the functional group of amide molecule differs from that present in carboxylic acids, and it is absent in the molecules of amines and alcohols.

The reason of the systematic errors for assessment of the clusterization thermodynamic parameters for bifunctional alkanes is analogous to the above described one. But it is possible to account it and decrease the values of the standard deviations. Thus, the values of the thermodynamic parameters of clusterization per monomer molecule of 2D films for the considered bifunctional amphiphiles are estimated with the deviation in the range of 0.10–0.61 kJ/mol, 1.96–2.99 J/(mol · K), and 0.03–0.87 kJ/mol for clusterization enthalpy, entropy, and Gibbs' energy, respectively.

Eventually, the superposition-additive approach at first developed and tested on monofunctional nonionic amphiphiles can be updated and used for calculation of the thermodynamic clusterization parameters for amphiphiles with two functional groups. It provides a high accuracy and enables one to obtain correct results in a simpler and much more expedient way than experimental and quantum chemical studies. The proposed approach can find its application in the estimation of monolayer parameters, especially when the amphiphiles in question are uncommon, expensive, or dangerous.

10.7   CONCLUSIONS

In this chapter, the model of the quantum chemical description of the thermodynamic parameters (enthalpy, entropy, and Gibbs' energy) of the amphiphile clusterization at the air/water interface is developed. This model is tested in the example of ten classes of amphiphiles: saturated and monoethoxylated alcohols, thioalcohols, saturaterd and monoienic carboxylic acids, amines, amides of carboxylic acids, α-substituted carboxylic acids, and disubstituted melamines. The obtained

TABLE 10.11

Threshold Alkyl Chain Lengths of Nonionic Surfactants Enabling the Spontaneous Clusterization at 25°C

Surfactant Class	Calculated Value	Experimental Value
Fatty alcohols	11	10–12
Monoethoxylated alcohols	14–15	14
Saturated carboxylic acids	13	13
cis-Monoienic carboxylic acids	17–19	18–19
trans-Monoienic carboxylic acids	16–18	18
Fatty amides of carboxylic acids	14–15	14
α-Amino acids	13–14	14
α-Hydroxycarboxylic acids	14	14
Amines	18	16–18
Thioalcohols	14–15	16
N-alkylsubstituted alanine (at 15°C)	15	14
Dialkylsubstituted melamine	10	10–12

thermodynamic and structural parameters of clusterization of the regarded amphiphiles are in good agreement with the existing experimental data.

The developed model includes implicit account for the interface impact via its orienting and stretching effect on the amphiphilic molecules. The water phase retracts the hydrophilic part of the amphiphile and two-three nearest to it methylene units, whereas the hydrophobic chain of the amphiphile is repulsed from the interface and is located in the gaseous phase. All hydrogens of the methylene units in the hydrocarbon chain are in *trans*-conformation, ensuring the most extended linear form. The quantum chemical calculation was done in the framework of supermolecular approximation only for amphiphilic aggregates with monomers of certain spatial orientation. Taking into consideration the pair-wise additivity of interactions realized in the cluster, it is possible to limit oneself only to the calculation of the small aggregates (dimers, trimers, tetramers, and hexamers). The values of their thermodynamic parameters of formation and clusterization depend significantly on the mutual orientation of monomers in the clusters. In the long term, the values of clusterization thermodynamic parameters for large and infinite clusters (2D films) can be described using the additive scheme, which define the required parameters as a sum of the increments of the corresponding CH·HC interactions realized between the alkyl chains and the interactions between the functional groups of the amphiphile.

The main results concerning the thermodynamics of clusterization for the regarded amphiphile classes can be illustrated in the example of the threshold chain length that enables spontaneous clusterization (see Table 10.11).

The applied model fits the determination of the dependence of the geometric parameters of the monolayer unit cells on the size of the hydrophilic part of the amphiphile. The compounds with smaller functional group (alcohols, carboxylic acids) form crystalline monolayers with a hexagonal unit cell, whereas amphiphiles with a bigger hydrophilic part (α-substituted carboxylic acids and their derivatives) have oblique unit cells. It is found that the geometric parameters of the monolayer unit cells for the regarded amphiphiles are defined by the intermolecular CH···HC interactions of the "a" type, whereas the tilt angle of the alkyl chains with respect to the interface is determined by the volume and structural peculiarities of hydrophilic parts, as well (see Table 10.12).

In addition, the superposition-additive approach using the calculated data of the PM3 method provides an opportunity to assess the values of the thermodynamic parameters of formation and clusterization for different mono- and disubstituted alkanes without performing time-consuming experiments. The approach to the assessment of the *temperature effect* (developed on the basis of

TABLE 10.12

Geometric Parameters of the Unit Cells of Surfactant Films

	Parameters of the Monolayer Unit Cell							
	a, Å		b, Å		θ,°		t,°	
Surfactant	Calc.	Exp.	Calc.	Exp.	Calc.	Exp.	Calc.	Exp.
Alcohols [135]	4.30	5.00	7.4	7.50	87	90	4	0–9
Carboxylic acids [44,75,76,136,137]	4.60	4.80	8.00	8.40	86	90	0	0
α-Hydroxycarboxylic acids: homochiral [47]	4.85	4.99	5.41	4.99	95	121	20	13–22
α-Amino acids [7]:								
Homochiral	4.71	4.91	5.67	5.25	103	112	30	36
Heterochiral	4.62	4.80	10.70	9.67	90	90	35	36
N-alkylsubstituted alanine [134]	4.65	4.93	6.20	5.64	105	104	43	46

the array of calculated data) has a big prognostic importance. It allows the calculation of the temperature, enabling spontaneous clusterization of the considered nonionic amphiphiles with given alkyl chain length. A further application of the model developed here can include the direct account of the water phase impact on the clusterization process and the influence of subphase pH on its thermodynamics.

REFERENCES

1. Baszkin, A., Norde, W. *Physical Chemistry of Biological Interfaces*. New-York: Marcel Dekker, Inc., 2000, p. 836.
2. Skryshevsky, V. A., Tolstoy, V. P., Chernyshova, I. V. *Handbook of Infrared Spectroscopy of Ultrathin Films*. Hoboken, NJ: John Wiley & Sons, Inc., 2003, p. 710.
3. Cohen, S. H., Bray, M. T., Lightbody, M. L. *Atomic Force Microscopy. Scanning Tunneling Microscopy*. New York: Plenum Press, 1994, p. 468.
4. Valeur, B. *Molecular Fluorescence. Principles and Applications*. Gamburg, Germany: Wiley-VCH, 2002, p. 402.
5. Dutta, P. *Curr. Opp. Solid Mater. Sci.*, 1997, *2*, 557–562.
6. Weinbach, S. P., Jacquemain, D., Leveiller, F., Kjaer, K., Als-Nielsen, J., Leiserowitz, L. *J. Am. Chem. Soc.*, 1993, *115*, 11110–11118.
7. Weissbuch, I., Berfeld, M., Bouwman, W., Kjaer, K., Als-Nielsen, J., Lahav, M., Leiserowitz, L. *J. Am. Chem. Soc.*, 1997, *119*, 933–942.
8. Kadam, M. M., Sawant, M. R. *J. Dispers. Sci. Technol.*, 2006, *27*, 861–868.
9. Vollhardt, D., Fainerman, V. B., Emrich, G. *J. Phys. Chem. B*, 2000, *104*, 8536–8543.
10. Tarek, M., Bandyopadhyay, S., Klein, M. L. *J. Mol. Liq.*, 1998, *78*, 1–6.
11. Vysotsky, Y. B., Bryantsev, V. S., Fainerman, V. B., Vollhardt, D. *J. Phys. Chem. B*, 2002, *106*, 11285–11294.
12. Kraack, H., Ocko, B. M., Pershan, P. S., Sloutskin, E., Tamam, L., Deutsch, M. *Langmuir.*, 2004, *20*, 5386–5395.
13. Kaplan, I. G., Rodimova, O. V. *Sov. Phys. Usp.*, 1978, *21*, 918–943.
14. Tsuzuki, S., Tanabe, K. *J. Phys. Chem.*, 1991, *95*, 2272–2278.
15. Williams, D. E., Craycroft, D. J. *J. Phys. Chem.*, 1987, *91*, 6365–6373.
16. Tsuzuki, S., Honda, K., Uchimaru, T., Mikami, M. *J. Chem., Phys.*, 2006, *124*, 114304.
17. Vladimirova, K. G., Freidzon, A. Y., Kotova, O. V., Vaschenko, A. A., Lepnev, L. S., Bagatur'yants, A. A., Vitukhnovskiy, A. G., Stepanov, N. F., Alfimov, M. V. *Inorg. Chem.*, 2009, *48*, 11123–11130.
18. Vysotsky, Y. B., Bryantsev, V. S., Fainerman, V. B. *J. Phys. Chem. B*, 2002, *106*, 121–131.
19. Hobza, P. *Annu. Rep. Prog. Chem., Sect. C Phys. Chem.*, 1997, *93*, 257–287.

20. Ferreira, M. L., Sierra, M. B., Morini, M. A., Schulz, P. C. *J. Phys. Chem. B,* 2006, *110,* 17600–17606.
21. Dynarowicz-Latka, P., Perez-Moralez, M., Monoz, E., Broniatowski, M., Martin-Romero, M. T., Camacho, L. *J. Phys. Chem. B,* 2006, *110,* 6095–6100.
22. Dai, L. *Intelligent Macromolecules for Smart Devices: From Materials Synthesis to Device Applications.* London, U.K.: Springer-Verlag, 2004, p. 496.
23. Lee, Y. S. *Self-Assembly and Nanotechnology a Force Balance Approach.* Columbus, OH: John Wiley & Sons, Inc., Publication, 2008, p. 344.
24. Vysotsky, Y. B., Fomina, E. S., Belyaeva, E. A., Aksenenko E. V., Vollhardt, D., Miller, R. *J. Phys. Chem. B,* 2009, *113,* 16557–16567.
25. Vysotsky, Y .B., Muratov, D. V., Boldyreva, F. L., Fainerman, V. B., Vollhardt, D., Miller, R. *J. Phys. Chem. B,* 2006, *110,* 4717–4730.
26. Vysotsky, Y .B., Belyaeva, E. A., Vollhardt, D., Aksenenko, E. V., Miller, R. *J. Phys. Chem. B,* 2009, *113,* 4347–4359.
27. Vysotsky, Y. B., Fomina, E. S., Belyaeva, E. A., Fainerman, V. B., Vollhardt, D., Miller, R. *J. Phys. Chem. C,* 2012, *116,* 26358–26376.
28. Vysotsky, Y. B., Belyaeva, E. A., Fainerman, V. B., Vollhardt, D., Miller, R. *J. Phys. Chem. C,* 2007, 111, 5374–5381.
29. Vysotsky, Y .B., Belyaeva, E. A., Fainerman, V. B., Aksenenko, E. V., Vollhardt, D., Miller, R. *J. Phys. Chem. C,* 2007, *11,* 15342–15349.
30. Vysotsky, Y. B., Belyaeva, E. A., Fomina, E. S., Vollhardt, D., Fainerman, V. B., Miller R. *J. Phys. Chem. B,* 2012, *116,* 2173–2282.
31. Fomina, E. S., Vysotsky, Y. B., Belyaeva E. A., Vollhardt, D., Fainerman, V. B., Miller, R. *J. Phys. Chem. C,* 2014, *118,* 4122–4130.
32. Minkin, V. I., Simkin, B. Y., Minyaev, R. M. *The Theory of Molecular Structure (in Russian).* Rostov-on-Don, Russia: Phoenix, 1997, p. 560.
33. Soloviov, M. E., Soloviov, M. M. *Computational Chemistry (in Russian).* Moscow, Russia: Solon-Press, 2005, p. 536.
34. Pecul, M., Rizzo, A., Leszczynski. *J. Phys Chem. A,* 2002, *106,* 11008–11016.
35. Csaszar, A. G. *J. Phys. Chem.* 1996, *100,* 3541–3551.
36. Blanco, S., Lesarri, A., Lopez, J. C., Alonso, J. *J. Am. Chem. Soc.,* 2004, *126,* 11675–11683.
37. Stepanian, S. G., Reva, I. D., Radchenko, A. D., Adamowicz, L. *J. Phys. Chem. A,* 1998, *102,* 4623–4629.
38. Stull, D. R., Westrum, E. F. Jr., Sinke, G.C. *The Chemical Thermodynamics of Organic Compounds.* New York: John Wiley & Sons, 1969, p. 536.
39. Speight, J. G. *Lange's Handbook of Chemistry.* New York: McGraw-Hill, Inc., 1999, p. 1623.
40. Vysotsky, Y .B., Bryantsev, V. S., Fainerman, V. B., Vollhardt, D., Miller, R. *J. Phys. Chem. B,* 2002, *106,* 121–131.
41. Vysotsky, Y. B., Bryantsev, V. S., Boldyreva, F. L., Fainerman, V. B., Vollhardt D. *J. Phys. Chem. B,* 2005, *109,* 454–462.
42. Vysotsky, Y. B., Shved, A. A., Belyaeva, E. A., Aksenenko, E. V., Fainerman, V. B., Vollhardt, D., Miller, R. *J. Phys. Chem. B,* 2009, *113,* 13235–13248.
43. Vysotsky, Y. B., Fomina, E. S., Belyaeva, E. A., Aksenenko, E. V., Vollhardt, D., Miller, R. *J. Phys. Chem. B,* 2011, *115,* 2264–2281.
44. Weidemann, G., Brezesinski, G., Vollhardt, D., Bringezu, F., De Meijere, K., Möhwald, H. *J. Phys. Chem. B,* 1998, *102,* 148–153.
45. Siegel, S., Vollhardt, D. *Progr. Colloid Polym. Sci.,* 2002, 121, 67–71.
46. Vollhardt, D., Fainerman, V. B. *J. Phys. Chem. B,* 2004, *108,* 297–302.
47. Vollhardt, D., Siegel, S., Cadenhead, D. A. *J. Phys. Chem. B,* 2004, *108,* 17448–17456.
48. Kellner, B. M., Cadenhead, D.A. *J. Colloid Interface Sci.,* 1978, *63,* 452–460.
49. Mingotaud, A. F., Mingotaud, C., Patterson, L. K. *Handbook of Monolayers.* San Diego, CA: Academic Press, Inc., 1993, p. 1385, vol. 1.
50. Weidemann, G., Brezesinski, G., Vollhardt, D., DeWolf, C., Möhwald, H. *Langmuir,* 1999, *15,* 2901–2910.
51. Fainerman, V. B., Vollhardt, D. *J. Phys. Chem. B,* 1999, *103,* 145–150.
52. Vollhardt, D., Fainerman, V. B., Siegel, S. *J. Phys. Chem. B,* 2000, *104,* 4115–4121.
53. Fainerman, V. B., Vollhardt, D. *J. Phys. Chem. B,* 2008, *112,* 1477–1481.
54. Vysotsky, Y. B., Bryantsev, V. S., Fainerman, V. B., Vollhardt, D., Miller, R. *Colloid Surf. A Physicochem. Eng. Asp.,* 2002, *209,* 1–14.
55. Lenne, P.-F., Bonosi, F., Renault, A., Bellet-Amalric, E., Legrand, J.-F., Petit, J.-M., Rieutord, F., Berge, B. *Langmuir,* 2000, *16,* 2306–2310.

56. Weinbach, S. P., Weissbuch, I., Kjaer, K., Bouman, W. G., Als-Nielsen, J., Lahav, M., Leiserowitz, L. *Adv. Mater.,* 1995, *7*, 857–862.
57. Majewski, J., Popovitz-Biro, R., Bouwman, W. G., Kjaer, K., Als-Nielsen, J., Lahav, M., Leiserowitz, L. *Chem. Eur. J.,* 1995, *1*, 304–311.
58. Vollhardt, D., Fainerman, V. B. *Adv. Interface Sci.,* 2010, *154*, 1–19.
59. Aveyard, R., Carr, N., Slezok, H. *Can. J. Chem.,* 1985, *63*, 2742–2746.
60. Tsay, R.-Y., Wu, T.-F., Lin, S.-Y. *J. Phys. Chem. B,* 2004, *108*, 18623–18629.
61. Braun, R., Casson, B. D., Bain, C. D. *Chem. Phys. Lett.,* 1995, *245*, 326–334.
62. Melzer, V., Vollhardt, D., Weidemann, G., Brezesinski, G., Wagner, R., Möhwald, H. *Phys. Rev. E,* 1998, *57*, 901–907.
63. Aratono, M., Takiue, T., Ikeda, N., Nakamura, A., Motomura, K., *J. Phys. Chem.* 1992, *96*, 9422–9424.
64. Vollhardt, D. *Adv. Colloid Interface Sci.* 1999, *79*, 19–57.
65. Bonosi, F., Renault, A., Berge, B. *Langmuir,* 1996, *12*, 784–787.
66. Berge, B., Renault, A. *Europhys. Lett.,* 1993, *21*, 773–777.
67. Islam, N., Kato, T. *J. Colloid Interface Sci.,* 2006, *294*, 288–294.
68. Prime, E. L., Tran, D. N. H., Plazzer, M., Sunario, D., Leug, A. H. M., Yiapanis, G., Baoukina, S., Yarowsky, I., Qiao, G. G., Solomon, D. H. *Colloid Surf. A Physicochem. Eng. Asp.,* 2012, *415*, 47–58.
69. Lu, J. R., Thomas, R. K., Penfold, J. *Adv. Colloid Interface Sci.,* 2000, *84*, 143–304.
70. Islam, N., Kato, T. *Langmuir,* 2003, *19*, 7201–7205.
71. Pollard, M. L., Pan, R., Steiner, C., Maldarelli, C. *Langmuir,* 1998, *14*, 7222–7234.
72. Shukla, R. N., Gharpurey, M. K., Biswas, A. B. *J. Colloid Interface Sci.,* 1967, *23*, 1–8.
73. Ramirez, P., Perez, L. M., Trujillo, L. L., Ruiz, M., Munoz, J., Miller, R. *Colloid Surf. A Physicochem. Eng. Asp.,* 2011, *375*, 130–135.
74. Chanda, J., Bandyopadhyay, S. *J. Phys. Chem. B,* 2006, *110*, 23482–23488.
75. Kellner, B. M. J., Cadenhead, D. A. *Chem. Phys. Lipids,* 1979, *23*, 41–48.
76. Durbin, M. K., Malik, A., Ghaskadvi, R., Shih, M. C., Zschack, P., Dutta, P. *J. Phys. Chem.,* 1994, *98*, 1753–1755.
77. Dynarowicz-Łatka, P., Dhanabalanb, A., Oliveira, O. N. Jr. *Adv. Colloid Interface Sci.,* 2001, *91*, 221–293.
78. O'Brien, K. C., Rogers, C. E., Lando, J. B. *Thin Solid Films,* 1983, *102*, 131–140.
79. Bishop, D. G., Kenrick, J. R., Bayston, J. H., Macpherson, A. S., Johns, S. R. *Biochim. Biophys. Acta,* 1980, *602*, 248–259.
80. Peltonen, J. P., Rosenholm, J. B. *Thin Solid Films,* 1989, *179*, 543–547.
81. Tomoasia-Cotisel, M., Zsako, J., Mocanu, A., Lupea, M., Chifu, E. *J. Colloid Interface Sci.,* 1987, *117*, 464–476.
82. Glazer, J., Goddard, E. D. *J. Chem. Soc.,* 1950, 3406–3408.
83. Vysotsky, Y. B., Belyaeva, E. A., Vollhardt, D., Aksenenko, E. V., Miller, R. *J. Colloid Interface Sci.,* 2008, *326*, 339–346.
84. Vollhardt, D. *J. Phys. Chem. C,* 2007, *111*, 6805–6812.
85. Vollhardt, D. *Adv. Colloid Interface Sci.,* 2005, *116*, 63–80.
86. Koyano, H., Bissel, P., Yoshihara, K., Ariga, K., Kunitake, T. *Chem. A Eur. J.,* 1997, *3*, 1077–1082.
87. Vollhardt, D., Liu, F., Rudert, R. *Chem. Phys. Phys. Chem.,* 2005, *6*, 1246–1250.
88. Vollhardt, D., Liu, F., Rudert, R., He, W. *J. Phys. Chem. B,* 2005, *109*, 10849–10857.
89. Fainerman, V. B., Vollhardt, D., Aksenenko, E. V., Liu, F. *J. Phys. Chem. B,* 2005, *109*, 14137–14143.
90. Vollhardt, D., Fainerman, V. B., Liu, F. *J. Phys. Chem. B.* 2005, *109*, 11706–11711.
91. Weissbuch, I., Leiserowitz, L., Lahav, M. *Curr. Opin. Colloid Interface Sci.,* 2008, *13*, 12–22.
92. Pascher, I. *Biochim. Biophys. Acta,* 1976, *455*, 433–451.
93. Baret, J. F., Hasmonay, H., Firpo, J. L., Dupin, J. J., Dupeyrat, M. *Chem. Phys. Lipids,* 1982, *30*, 177–187.
94. Menger, F. M., Richardson, S. D., Wood, M. G., Sherrod, M. J. *Langmuir,* 1989, *5*, 833–838.
95. Huda, M. S., Fujio, K., Uzu, Y. *Bull. Chem. Soc. Jpn.,* 1996, *69*, 3387–3394.
96. Jacobi, S., Chi, L. F., Plate, M., Overs, M., Schäfer, H. J., Fuchs, H. *Thin Solid Films,* 1998, *327*, 180–184.
97. Overs, M., Fix, M., Jacobi, S., Chi, L. F., Sieber, M., Schäfer, H. J., Fuchs, H., Galla, H. J. *Langmuir,* 2000, *16*, 1141–1148.
98. Bergström, S., Aulin-Erdtman, G., Rolander, B., Stenhagen, E., Östling, S. *Acta Chem. Scand.,* 1952, *6*, 1157–1174.
99. Matuo, H., Rice, D.K., Balthasar, D.M., Cadenhead, D.A. *Chem. Phys. Lipids,* 1982, *30*, 367–380.
100. Asgharian, B., Cadenhead, D.A. *Langmuir,* 2000, *16*, 677–681.
101. Siegel, S., Vollhardt, D., Cadenhead, D. A. *Colloids Surf. A,* 2005, *256*, 9–15.

102. Vollhardt, D., Siegel, S., Cadenhead, D. A. *Langmuir,* 2004, *20*, 7670–7677.
103. Yim, K. S., Rahaii, B., Fuller, G.G., *Langmuir,* 2002, *18*, 6597–6601.
104. Wiedemann, G., Brezesinski, G., Vollhardt, D., Möhwald, H. *Langmuir,* 1998, *14*, 6485–6492.
105. Jarvis, N. L. *J. Phys. Chem.,* 1965, *69*, 1789–1797.
106. Taylor, R., Kennard, O., Versichel, W. *Acta Cryst. B,* 1984, *B40*, 280–288.
107. Henry, D. J., Dewan, V. I., Prime, E. L., Qiao, G. G., Solomon, D. H., Yarovsky, I. *J. Phys. Chem. B,* 2010, *114*, 3869–3878.
108. Gutierrez-Campos, A., Diaz-Leines, G., Castillo, R. *J. Phys. Chem. B,* 2010, *114*, 5034–5046.
109. Flores, A., Corvera-Poire, E., Garza, C., Castillo, R. *J. Phys. Chem. B,* 2006, *110*, 4824–4835.
110. Ben-Naim, A. *Hydrophobic Interactions.* New York: Springer, 1980, p. 320.
101. Du, X., Liang, Y. *J. Phys. Chem. B,* 2000, *104*, 10047–10052.
112. Hoffmann, F., Hühnerfuss, H., Stine, K. J. *Langmuir,* 1998, *14*, 4525–4534.
113. Harvey, N. G., Rose, P. L., Mirajovsky, D., Arnett, E. M. *J. Am. Chem. Soc.,* 1990, *112*, 3547–3554.
114. Hühnerfuss, H., Neumann, V., Stine, K. J. *Langmuir,* 1996, *12*, 2561–2569.
115. Schwartz, D. *Annu. Rev. Phys. Chem.,* 2001, *52*, 107–137.
116. Conboy, J. C., Messmer, M. C., Walker, R. A., Richmond, G. L. *Progr. Colloid Polym. Sci.,* 1997, *103*, 10–20.
117. Oishi, Y., Takashima, Y., Suehiro, K., Kajiyama, T. *Langmuir,* 1997, *13*, 2527–2532.
118. Gang, Z., Kun, F., Xia, S., Pingsheng, H. *Langmuir,* 2002, *18*, 6602–6605.
119. Huilin, Z., Weixing, L., Shufang, Y., Pingsheng, H. *Langmuir,* 2000, *16*, 2797–2801.
120. Pallas, N. R., Pethica, B. A. *Langmuir,* 1985, *1*, 509–513.
121. Baoukina, S., Monticelli, L., Marrink, S. J., Tieleman, D. P. *Langmuir,* 2007, *23*, 12617–12623.
122. Vysotsky, Y. B., Fomina, E. S., Belyaeva, E. A., Vollhardt, D., Fainerman, V. B., Miller, R. *J. Phys. Chem. B,* 2012, *116*, 8996–9006.
123. Vysotsky, Y. B., Fomina, E. S., Fainerman, V. B., Vollhardt, D., Miller, R. *Phys. Chem. Chem. Phys.,* 2013, *15*, 11623–11628.
124. Nagle, J. F. *Annu. Rev. Phys. Chem.,* 1980, *31*, 157–196.
125. Birdi, K. S. *Self-Assembly Monolayer Structures of Lipids and Macromolecules at Interfaces.* New York: Springer, 1999.
126. Schreiber, F. *Prog. Surf. Sci.,* 2002, *65*, 151–257.
127. Peltonen, L., Hirvonen, J., Yliruusi, J. *J. Colloid Interface Sci.,* 2001, *239*, 134–138.
128. Guyot-Sionnest, P., Hunt, J. H., Shen, Y. R. *Phys. Rev. Lett.,* 1987, *59*, 1597–1600.
129. McMullen, R. L., Kelty, S. P. *J. Phys. Chem. B,* 2007, *111*, 10849–10852.
130. Knock, M. M., Bell, G. R., Hill, E. K., Turner, H. J., Bain, C. D. *J. Phys. Chem. B,* 2003, *107*, 10801–10814.
131. Howes, A. J., Radke, C. J. *Langmuir,* 2003, *23*, 1835–1844.
132. Shi, L., Tummala, N. R., Striolo, A. *Langmuir,* 2010, *26*, 5462–5474.
133. Eliash, R., Weissbuch, I., Weygund, M. J., Kjaer, K., Als-Nielsen, J., Lahav, M., Leiserowitz, L. *J. Phys. Chem. B,* 2004, 108, 7228–7240.
134. Nandi, N., Vollhardt, D. *Chem. Rev.,* 2003, 103, 4033–4075.
135. Wang, J.-L., Leveiller, F., Jacquemain, D., Kjaer, K., Als-Nielsen, J., Lahav, M., Leiserowitz, L. *J. Am. Chem. Soc.,* 1994, 116, 1192–1204.
136. Riviere, S., Henon, S., Meunier, J., Schwartz, D. K., Tsao, M.-W., Knobler, C. M. *J. Chem. Phys.,* 1994, 101, 10045–10051.
137. Hönig, D., Overbeck, G. A., Möbius, D. *Adv. Mater.,* 1992, 4, 419–424.
138. Nandi, N., Bacchi, B. *J. Am. Chem. Soc.,* 1996, 118, 11208–11216.
139. Morpeth, F. F. (Ed.) *Preservation of Surfactant Formulation.* New York: Blackie Academic & Professional, 1995, p. 367.
140. Förster, G., Meister, A., Blume, A. *Curr. Opin. Colloid Interface Sci.,* 2001, *6*, 294–302.
141. Ulman, A. *Chem. Rev.,* 1996, *96*, 1533–1554.
142. Iwamoto, M., Wu, C.-X., Zhong-can, O.-Y. *Chem. Phys. Lett.,* 1999, *312*, 7–13.
143. Kuzmenko, I., Kjaer, K., Als-Nielsen, J., Lahav, M., Leiserowitz, L. *J. Am. Chem. Soc.,* 1999, *121*, 2657–2661.
144. Bader, R. F. W. *Atoms in Molecules: A Quantum Theory.* Oxford, U.K.: Caledon Press, 2001, p. 438.
145. Mandado, M., Vila, A., Grana, A. M., Mosquera, R. A., Cioslowski, J. *Chem. Phys. Lett.,* 2003, 371, 739–741.
146. Luzanov, A.V. *Int. J. Quant. Chem.,* 2011, *111*, 2196–2220.
147. Popelier, P. L. A. *J. Phys. Chem. A,* 1999, *103*, 2883–2890.
148. Cioslowski, J., Nanayakkara, A. *J. Am. Chem. Soc.,* 1993, *115*, 11213–11215.

149. Marti, J. A. *Chem. Phys.,* 2001, *265*, 263–271.
150. Lorenzo, L., Mosquera, R. A. *Chem. Phys. Lett.,* 2002, *356*, 305–312.
151. Wilberg, K. B., Bader, R. F. W., Lau, C. D. H. *J. Am. Chem. Soc.,* 1987, *109*, 1001–1012.
152. Mandado, M., Grana, A. M., Mosquera, R. A. *J. Mol. Struct. (Theochem.),* 2002, *584*, 221–234.
153. Grana, A. M., Mosquera, R. A. *J. Chem. Phys.,* 1999, *110*, 6606–6617.
154. Grana, A. M., Mosquera, R. A. *J. Chem. Phys.,* 2000, *113*, 1492–1501.
155. Vila, A., Mosquera, R. A. *Chem. Phys. Lett.,* 2000, *332*, 474–480.
156. Vila, A., Caraballo, E., Mosquera, R. A. *Can. J. Chem.,* 2000, *78*, 1535–1543.
157. Vila, A., Mosquera, R. A. *J. Chem. Phys.,* 2001, *115*, 1264–1274.
158. Lopez, J. L., Mandado, M., Grana, R. A., Mosquera, R. A. *Int. J. Quant. Chem.,* 2002, *86*, 190–198.
159. Davis, M. I., Douheret, G. *J. Chem. Soc. Faraday Trans.,* 1998, *94*, 2389–2394.
160. Hoiland, H., Vikingstad, E. *Acta Chem. Scand. A,* 1976, *30 A*, 182–186.
161. Zhuo, S., Wei, J., Si, W., Ju, G. J. *J. Chem. Phys.,* 2004, 120, 2575–2581.
162. Vysotskii, Y. B., Zaikovskaya, Y. V., Solonskii, I. N. *Russ. J. Org. Chem.,* 2001, *37*, 101–118.
163. Vysotsky, Y. B., Bryantsev, V. S. *Int. J. Quant. Chem.,* 2004, *96*, 123–135.
164. Vysotsky, Y. B., Belyaeva E. A., Fomina, E. S., Vasylyev, A. O., Vollhardt, D., Fainerman, V. B., Aksenenko, E. V., Miller, R. *J. Colloid Interface Sci.,* 2012, *387*, 162–174.
165. Vysotsky, Y. B., Belyaeva, E. A., Fainerman, V. B., Aksenenko, E. V., Vollhardt, D., Miller, R. *Phys. Chem. Chem. Phys.,* 2011, *13*, 20927–20932.
166. Vysotsky, Y. B., Vasylyev, A. O., Belyaeva, E. A., Eilasyan, E. G. *Effects of Substituents in Chemical Thermodynamics. Quantum Chemical Approach (in Russian).* Donetsk, Ukraine: DonNUET, 2009, p. 200.
167. Vysotsky, Y. B., Belyaeva, E. A., Vasylyev, A. O., Fainerman, V. B., Aksenenko, E. V., Vollhardt, D., Miller, R. *Colloids Surf. A: Physicochem. Eng. Asp.,* 2012, *413*, 303–306.
168. Rybachenko, V. I. (Ed.) *From Molecules to Functional Architecture. Supramolecular Interactions.* Donetsk, Ukraine: East Publisher House, 2012, p. 103–127.
169. Daubert, T. E., Danner, R. P., Sibul, H. M., Stebbins, C. C. *Physical and Thermodynamic Properties of Pure Chemicals: Data Compilation.* Bristol, PA: Taylor & Francis, 1998, p. 9860.
170. Alberty, R. A., Gehrig, C. A. *J. Phys. Chem. Ref. Data,* 1985, 14, 803–820.
171. Davidovits, P. *Chem. Rev.,* 2006, *106*, 1323–1354.
172. Mundy, C. J., Kuo, I.- F. W. *Chem. Rev.,* 2006, *106*, 1282–1304.
173. Helgaker, T. *Chem. Rev.* 1999, *99*, 293–352.
174. Knunyants, I. L. (Ed.) *Khimicheskaja Enciklopedija (Encyclopaedia of Chemistry) (in Russian).* Moscow, Russia: Bolshaja Rossijskaja Enciklopedija, 1988–1998, vol. 1–5.
175. Fomina, E. S. *Chem. Phys. Technol. Surf.,* 2012, *3*, 405–418.

11 Molecular Dynamics Simulation of Surfactant Monolayers

Bin Liu, Jirasak Wong-Ekkabut, and Mikko Karttunen

CONTENTS

11.1 INTRODUCTION

In this chapter, we provide a brief introduction to molecular simulations of lipid/surfactant monolayers. We do not aim to provide a comprehensive review. Instead, we first discuss the very timely problem of nanoparticle interactions with the lung surfactant and how that can be studied by simulations. After that, we provide a detailed introduction on the various aspects of building a monolayer simulation and show a case study using simulations of cationic surfactants and zwitterionic lipids. The aim is to provide the reader with a detailed view of how to build simulations, what aspects are important, and what kind of properties can be analyzed. In our other contributions of this volume, we discuss electrostatic interactions in detail (see Chapter 6). That discussion is also valid here and we refer the reader to Chapter 6 regarding the details of how the important electrostatic interactions must be accounted for in interfacial systems.

11.2 EFFECT OF CARBON NANOPARTICLES ON LUNG SURFACTANT

Continuous combustion of fossil fuel produces airborne pollutants into the atmosphere. To a large degree, pollutants consist of carbonaceous particles with a broad size distribution [1–4]. Usually, larger particles can be trapped and removed from the respiratory system, while the smaller ones, especially those in the nanometer range, can reach the alveoli and get transferred into the blood circulatory system [5–7]. Therefore, these combustion-generated particles are responsible for various respiratory and cardiovascular diseases [5,8,9]. There have been growing public health concerns regarding nanomaterials, and the potential risk issues associated with carbon nanoparticles (CNPs) have been intensely studied. It has been known for some time that CNPs deposit in the lung and can induce pneumoconiosis [10–15].

Computer simulations provide an approach to investigate the molecular-level interactions of nanoparticles with the lung surfactant. Importantly, simulations results can predict how nanoparticles influence hydrogen bonding, hydrophobic interactions, and ordering of the lung surfactant.

This allows for detailed analysis of the structural and dynamic properties of lung surfactant in the presence of varying concentrations of CNPs. In reality, the lung surfactant consists of various biological molecules, for example, phospholipids, cholesterol, and surfactant protein [16]. The inclusion of all of the components is impossible, since simulations are limited by the number of molecules they can handle as well as accessible time scales. To reduce the complexity of the system, pure lipid monolayer of dipalmitoylphosphatidylcholine (DPPC) and fullerene C60 have been used in a number of studies as a simple model of lung surfactant and CNPs, respectively. In our own study using such a system, coarse-grained molecular dynamics (CGMD) simulations with the Martini force field [17] were performed with a series of constant particle number, volume and temperature (NVT) simulations with various box sizes at C60:DPPC ratios up to 0.3 [18]. CGMD allow us to reach a reasonably large length and long time scales (the systems consisted of 1600 molecules per monolayer and run for 5–10 µs) so that monolayer collapse and pore formation could be observed [19,18]. In the presence of fullerenes, the surface tensions of monolayers significantly decreased at high compression (at small area per molecule), resulting in collapse of the monolayer, as shown in Figure 11.1. This is in agreement with the results from simulations of ternary lipid mixture monolayers [20]. On the other hand, at low compression (at a large area per molecule), fullerenes increase the surface tension of a monolayer, leading to pore formation, as shown in Figure 11.2. Interestingly, the monomeric fullerene had been suggested to be a stable form in both lipid monolayer and bilayer [14,21]. This result is, however, reliable only at low compression. Study of aggregation of fullerenes in monolayer showed that monomeric fullerene becomes significantly less especially at high compression and high concentration of fullerenes, Figure 11.3; see Ref. [18]. Aggregation of fullerenes causes a decrease of the effective area per molecule and lower surface tension. When fullerene clusters become larger in size than the monolayer thickness; the monolayer bends and folds into a bilayer/a hemispherical budding structure to prevent exposure to the vapor and water phase as shown in Figure 11.1. The free energy calculations [18] of a single fullerene transferring across the monolayer suggest that fullerene can easily penetrate into lipid monolayers and spontaneous translocation of fullerene out of the monolayer is rather difficult.

In conclusion, the simulations were able to suggest that the potentially harmful effects of the deposited CNPs on the respiratory system might be related to the difficulty of CNP clearance from lung surfactant. In addition, simulations suggest that CNPs may alter the physical and mechanical properties of lung surfactant [20,18], which are responsible for respiratory distress syndrome [20,22–25].

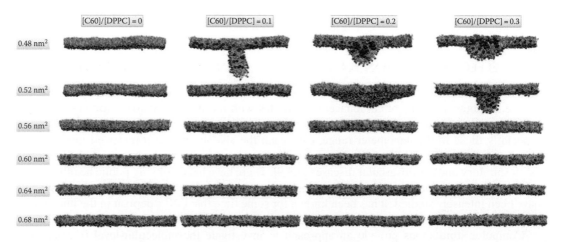

FIGURE 11.1 Snapshots illustrating the equilibrium systems of 1600 molecules/monolayer in the *xz*-plane. Cyan: lipid tails, purple: phosphate group in lipid heads, and red: fullerene. Water was omitted for clarity.

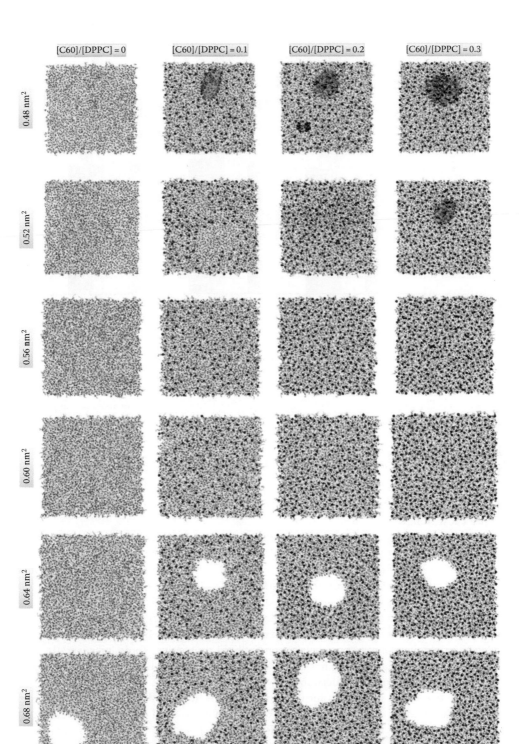

FIGURE 11.2   Snapshots illustrating the equilibrium systems of 1600 molecules/monolayer in the *xy*-plane. Colors and simulation time are the same as in Figure 11.1.

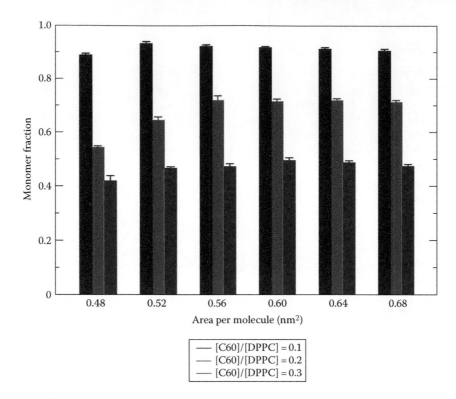

FIGURE 11.3 The monomer fraction of fullerenes as a function of the area per molecule at [C60]/[DPPC] ratios of 0.1 (red), 0.2 (green), and 0.3 (blue).

11.3 PARAMETRIZING LIPID MOLECULES

Section 11.2 demonstrated the utility of molecular simulations. In this section, we discuss the details of how to obtain parameters for lipid or surfactant molecules when they are not readily available from prior research.

Before being able to perform any MD simulations, one must obtain force field parameters, or in case they do not exist, parametrize the molecules of interest so that the MD program can understand their compositions, structures, and interactions with each other and other components in simulation. This is an essential step as it determines the simulation's correctness, quality, and value. Although lots of lipid and surfactant molecules have been parameterized, for many lipid molecules, significant manual work, including *ab initio* calculations for partial charges and constructing topology files which describe the modeling or parametrization understandable to a MD program, is still mandatory to obtain a quality parametrization for lipid molecules. Tools for generating topology files automatically from a structure file such as a PDB file exist [26]. But the quality of the generated topology files is usually far from desirable.

There are two broad categories for the force fields in lipid modeling: the atomistic approach and the coarse-grained (CG) approach. The CG force fields, as used in the lung surfactant study in the previous section, such as the famous MARTINI model [17,27], are known for their speed and larger system sizes. Atomistic force fields, on the other hand, are often able to provide quantitative predictions that can be verified by experiments, and are versatile. The atomistic force fields can be classified into two flavors: the all-atom ones and the united-atom ones. In an all-atom force field, such as OPLS [28,29], AMBER [30–33], and CHARMM [34–36], all atoms are explicitly present in the simulation. In a united-atom force field, such as the GROMOS force field [37–39], the nonpolar hydrogens bonded to the carbons in an acyl chain are absorbed into the carbons to which they are bonded, to form a *united atom* to reduce the number of interacting sites. Here, we focus on some practical issues and skills useful in obtaining good quality parametrization for lipid molecules by using one of the most widely used atomistic force fields for lipid simulations, the GROMOS force field and its derivative, the Berger lipid model [40]. It is also worth noting that the OPLS force field has also found many applications in lipid simulations (yet work remains to be done for OPLS peptide parameterization [41]). In addition, there is currently a very interesting open collaboration platform called *Matching lipid force fields with NMR data* (available at http:// nmrlipids.blogspot.ca). This approach is pioneered by Markus Miettinen and Samuli Ollila that is groundbreaking and may lead to completely new developments and integration of experimental and computational lipid data.

From the practical point of view, a good way to obtain a quality parametrization for a lipid molecule is by studying and reusing mature, well-tested parametrization for other lipid molecules, which share notable amounts of parts as the lipid to be parametrized. It may, of course, be the case that no such parametrization exists. This shortcut approach has been applied to many lipid molecules with success. One of the most famous *baseline* lipid parametrizations from which many other parametrizations were derived is the DPPC parametrization [42] based on the GROMOS force field and the Berger lipid model. Many other saturated dichained lipids, including DMPC [43] and DLPC, can be easily parametrized by adding or removing repeating hydrocarbons. Borrowing the parametrization for the double-bonded hydrocarbons, this DPPC parametrization can be adapted to construct parametrization for unsaturated dichained lipids, such as DOPC, POPC, and SOPC in principle. A word of warning should be given, however: Double bonds can be tricky to parametrize and it has been shown that old parametrizations are wrong and can influence the observed physical properties of lipids and their interactions with others [44,45]. Similarly, the PC headgroup can also be substituted by other parametrized headgroups, such as the PG headgroup, to obtain parametrizations for the corresponding PG lipids [46].

If the headgroup of the lipid of interest has not been parametrized, one can use the parameter set of a force field to parametrize it. The parameter set includes equilibrium position and force constant for bonded interactions such as bond stretching, bond angle bending, proper and improper dihedral interactions, and van der Waals, radii and constants. What is usually missing in a force field for a specific headgroup is the partial charges. Quantum chemistry calculations are needed to obtain the partial charges to parametrize a headgroup. Ideally, one should apply quantum chemistry approaches to calculate the partial charges for the entire lipid. However, as the computational cost of a quantum chemistry calculation scales as the third order or even more of the number of electrons in the system, it becomes quickly intractable as the size of the lipid increases. Fortunately, the *locality* of partial charges and the *insulating* property of hydrocarbons can be employed to reduce the computational cost. The *locality* of partial charges means the partial charge of a specific site (an atom or a united atom) is only influenced heavily by its first and second bonded neighbors. One important exception is aromatic rings. In any case, an aromatic ring must be treated as a whole. The *insulating* property of hydrocarbons means one hydrocarbon can be essentially regarded as a neutrally charged dividing point to separate two independent partial charge regions. Therefore, one can perform quantum chemistry calculations for a pseudomolecule composed of a headgroup and a methyl or ethyl group. If the headgroup contains one or more hydrocarbons, one can divide the headgroup again into smaller parts and cap them with methyl or ethyl groups to form pseudomolecules. Usually the accuracy of partial charges obtained from such a pseudomolecule is within the tolerance of an MD simulation for lipids.

Quantum chemistry calculations for partial charges can be performed by using the well-known Gaussian package [47] and some open source packages such as the GAMESS family, which includes GAMESS-US [48,49] and Firefly [48,50] as its two major variants. One of the most popular basis set, for example, 6-31G* and 6-31G(d,p) [51–53], which offer both decent accuracy and acceptable computational cost, is often used for calculating the partial charges for a lipid. 6-31G* has also been employed to obtain the partial charges in the AMBER force field [54]. These basis sets usually work well for neutral and cationic lipids. But for anionic lipids, 6-31+G* or 6-31+G(d,p) [55,56], which include diffuse functions to account for the presence of significant charge density that are distant from the atomic nuclei, are needed to get accurate results at the cost of slower or even difficult convergence. To take the effect of electron correlation on partial charges into account, post-Hartree–Fock (HF) methods or density functional theory (DFT) methods are usually employed as they are generally superior to the plain HF level calculation in which electron correlation is totally neglected. The Moeller–Plesset level 2 (MP2) method, which is a post-HF method, is usually preferable as it can take most of electron correlation into account at affordable computational cost. DFT methods can also work well, provided one chooses an appropriate exchange-correlation functional (E_{xc}). The quantum chemistry packages mentioned earlier can offer four sets of partial charges, that is, those by Mulliken population analysis [57,58], Löwdin population analysis [59,60], electrostatic potential analysis (ESP) [61,62], and natural population analysis (NPA) [63–65]. NPA can only be done by the natural bond orbital (NBO) module [65], which exists as a plug-in for all the major quantum chemistry packages. Once one obtains the four sets of partial charges, one should first use one's chemical instinct to judge which set is most reasonable. Our experience shows that usually, the NPA scheme is the choice as it is not sensitive to the choice of basis set, theory level, or initial structure of the molecule being investigated. But the choice of partial charge scheme could differ from case to case. In principle, one should also employ polarized continuum model (PCM) [66–69] to reflect the influence of the aqueous environment on partial charges. However, our experience shows the use of PCM makes negligible difference for partial charges of the molecules for biological or physiological simulations.

Quantum chemistry calculations for partial charges usually take two steps. First, one uses the plain HF level calculation to perform geometry optimization for the molecule being investigated and obtain the *equilibrium* structure. We put a double quote to encompass the word equilibrium, because the geometry optimization usually ends up in a local minimum or even a saddle point on the potential energy surface as the global minimum is extremely difficult to reach if possible at all. This is caused by the high dimensional and very complex potential energy surface landscape of any molecule of decent size. The point of performing this step is getting a structure reasonably close to the real equilibrium structure for the second quantum chemistry calculation step, and in real MD trajectories, molecules are always close to their equilibrium structures but seldom sit there. The second step involves using the optimized structure obtained in the first step to perform single point calculation at either MP2 level or with a DFT method to account for electron correlation. Our experience shows the partial charges obtained by using a MP2 level or DFT calculation are distinguishable from those from the plain HF level calculation, but reasonably close.

One may encounter the difficult situation in which one bonded interaction in the molecule being parametrized has not been parametrized in a specific force field. One obvious approach is to look for experimental results to find the equilibrium position and force constant for it. One can also resort to quantum chemistry calculations to perform a scan of the potential energy surface on the dimension of interest. In the following example (Figure 11.4), the angle bending interaction between CH_2–(C=O)–C(benzene) is not parametrized in the GROMOS force field [39]. The first step to parametrize it using

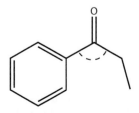

FIGURE 11.4 Diagram for a pseudomolecule for parametrizing the angle bending interaction between CH_2–(C=O)–C(benzene).

the harmonic oscillator approximation is capping CH_2 with CH_3 (methyl group) to form a pseudo-molecule. To facilitate the parametrization, one would better convert the structure representation of this pseudomolecule from Cartesian to internal coordinates (Z-matrix representation), which can be accomplished by using chemical visualization programs such as MacMolPlt [70]. Then one can generate a set of input files for quantum chemistry calculations with single point structures represented by Z-matrix and with varying CH_2–$(C{=}O)$–C angles, which should cover the guessed equilibrium angle. This set of quantum chemistry calculations usually can give a set of system energies, which can be almost perfectly fitted to a parabola against the varying CH_2–$(C{=}O)$–C angles. From the fitting, one can retrieve the equilibrium angle and force constant for this angle bending interaction.

11.4 SIMULATION BOX SETUP

Once parametrization for all molecules has been obtained; the next step for the simulation is to construct a simulation box, which consists of all the components needed and has the appropriate geometric configuration. In theory, any box type that can fill up the entire space with periodic boundary condition can be used for monolayer simulations, including some perhaps bizarre sounding box types like rhombic dodecahedron or truncated octahedron. For the easy of analysis and practical reasons, the simplest rectangular box type is almost always used unless there are some special requirements.

For monolayer at air/water interface simulations, there are two popular geometrical configuration setups (see Figure 11.5). In Figure 11.5a, the simulated monolayer is placed at the interface between the water and air phases, and a wall potential is applied to the bottom of the water phase to prevent molecules from escaping. The water slab should be thick enough to allow recovery of bulk water property for the region that has a direct effect on the monolayer [71]. Another very popular configuration is displayed in Figure 11.5b where two symmetrical monolayers are separated by a water slab thick enough to restore bulk water property in the middle and hence prevent interactions between these two monolayers [72]. Caution should be taken when one uses the setup in Figure 11.5a with the constant particle number, pressure and temperature (NPT) ensemble, or in a situation in which severe buckling may develop, as the varying box size or monolayer geometry may interfere with the wall potential and cause artifacts.

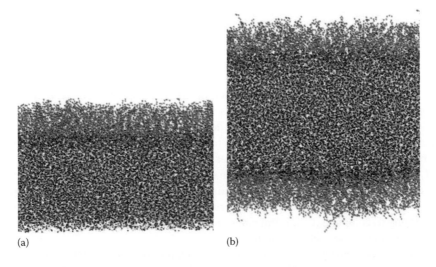

(a) (b)

FIGURE 11.5 Two types of simulation box setup for monolayers at the air/water interface. (a) A wall potential is applied to prevent water from escaping. (b) Symmetrical monolayers separated by a thick water slab. Monolayers displayed here consist of DPPC lipids modeled by the GROMOS force field and the Berger lipid model.

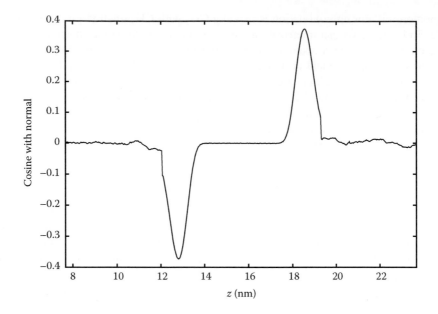

FIGURE 11.6 Water dipole orientation for the symmetrical configuration in Figure 11.5b. The two symmetrical peaks correspond to the phosphate region in the DPPC headgroup, which significantly reorients water dipole. Bulk water property rapidly resumes away from the headgroup region. The ripples in the lipid chain region and in the air phase are caused by small number of water molecules that have escaped from the water slab.

One way to check if the water slab in Figure 11.5b is thick enough is to calculate water dipole orientation along the z-axis and calculate the Debye screening length. In the vicinity of the polar headgroups of the monolayers, water dipole orientation is distinctively different from that in bulk water region, which should be isotropic. With either configuration, the air phase (essentially vacuum in most simulations) needs to be thick enough to prevent the interactions between the simulated system and its periodic images in z-direction (Figure 11.6).

11.5 RUNNING MONOLAYER SIMULATIONS

Once the simulation box has been constructed, the production simulations are usually the least complicated step compared to parametrization or analysis (which will be discussed in detail in the following later). Modern MD packages, such as GROMACS [73], NAMD [74], and AMBER [75], usually provide reliable default parameter settings and excellent documentation. One must, however, always pay attention to the particular demands of the system and verify that the behavior is physically correct [76]. Typically, one needs to conduct trial simulations to verify the choice of parameters against existing experiments or other simulation results.

Usually monolayer simulations start with the energy minimization step. Steep descent and conjugate gradient (CG) methods are the most popular choices. This step relaxes the energy introduced by the artificial system setup, which could otherwise make the following dynamic simulation steps unstable. Failing to complete the minimization step usually indicates serious issues in either parametrization or simulation box setup or both.

Depending on the goals one wants to achieve with a monolayer simulation, the next step could be either a constant temperature (*NVT*) simulation that comprises of both the equilibration stage and the production stage, or a series of *NVT* simulations or constant temperature constant pressure (*NPT*) simulations for equilibration followed by a production of *NPT* simulation. The choice of thermostat and/or barostat determines the quality of *NVT* or *NPT* simulations to a large extent.

Popular thermostats include Nosé-Hoover [77,78] or Nosé-Hoover chains [79], Berendsen thermostat or its variants [80], Andersen thermostat [81], and the increasingly popular V-Rescale thermostat [82], which has proven to be suitable for both equilibration and production simulations [83]. Popular barostats include Berendsen coupling [80], which is very useful for situations where the system is far from equilibration as it provides first-order decaying toward equilibrium, and the Parrinello–Rahman coupling [84] that serves the production stage very well and is generally the recommended method.

In the past, treating long-ranged Coulombic interactions were computationally intensive and tricky to handle. The particle mesh Ewald (PME) algorithm [85] is becoming the *de facto* standard treatment for Coulombic interactions as it offers both satisfactory accuracy and very decent efficiency [86–90], provided one chooses appropriate cutoff ranges. The choice of real-space range is usually less important when PME is used than with other algorithms that treat Coulombic interactions such as reaction field [91], since in PME, the real-space cutoff is no more than a division of computational burden into a real-space part and a reciprocal space part. The lower sensitivity to the choice of cutoff range in PME is another advantage. Recent reviews are provided in Refs. [89,90]. Our other contribution in this book also contains a detailed discussion of electrostatic interactions when interfaces are present (see Chapter 6).

11.6   ANALYSIS AND A CASE STUDY FOR DPPC/CTAB MONOLAYERS

In this section, we discuss both the conventional analysis that can be relatively easily done and some advanced analysis techniques, which have been recently developed in the context of monolayer simulations. As shown in Figures 11.5b and 11.7, both pure DPPC monolayers and DPPC/CTAB mixtures were simulated by employing the symmetrical configuration setup [72]. Each monolayer in the simulation box consists of 128 lipids. Cetyltrimethylammonium bromide (CTAB) is a cationic surfactant [92]. It has a trimethyl ammonium headgroup and a lipid chain of 16 hydrocarbons. A series of *NVT* simulations with various simulation box sizes at various CTAB molar fractions were conducted.

Snapshots along the trajectory are often an intuitive and important way to gauge how the simulation evolves in time. Figure 11.7 displays snapshots at the end of 1 μs trajectories for three monolayers with various CTAB molar fractions. The visualizations were obtained by using the VMD [93,94] software, possibly the most popular MD visualization tool. Severe buckling occurs in the

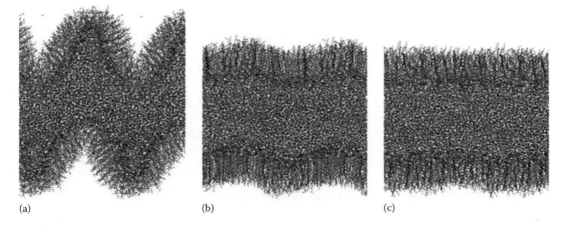

(a)   (b)   (c)

FIGURE 11.7   Pure DPPC monolayer and DPPC/CTAB mixtures at area per lipid 0.4 nm². (a) High surface pressure at very low area per lipid induces buckling in pure DPPC monolayer. (b) 20% cationic CTAB (deep blue) with 80% DPPC mixture has only very wild surface undulation. (c) 30% Cationic CTAB (deep blue) with 70% DPPC mixture resumes flat geometry even at very low area per lipid.

pure DPPC monolayer (Figure 11.7a) at a low area per lipid 0.4 nm^2, which indicates high surface pressure. The monolayer with 20% CTAB displays much milder buckling and with 30% CTAB, buckling almost disappears. This indicates CTAB stabilizes the flat geometry of DPPC monolayers, especially with high surface pressure.

The goals of conducting an MD simulation can be categorized into studying statistical properties and investigating dynamical processes. Before taking statistics, one must ensure equilibrium has been reached and that the trajectories from the equilibration have been discarded from the analysis. The most common approach to judge if the system has entered equilibrium is to investigate the trend of various energies, including total, kinetic, potential, and other energies belonging to various degrees of freedom. If at least one of them is still displaying a systematic increase or decrease, the system is still not in equilibrium. This is, however, not a sufficient criterion and other quantities, for example, the number of hydrogen bonds, must be monitored. In addition, analysis of fluctuations is often a useful way to analyze equilibrium. Analysis of lateral diffusion of lipids (effective mixing) is another important quantity. The lateral diffusion coefficient can be evaluated by

$$D_\alpha = \lim_{t \to \infty} \frac{1}{4t} \langle r^2(t) \rangle = \lim_{t \to \infty} \frac{1}{4tN_\alpha} \sum_{i=1}^{N_\alpha} \langle r_i^2(t) \rangle, \tag{11.1}$$

where
   the subscript α denotes a specific type of lipid. In this case study, it is either DPPC or CTAB.
   $\langle r_i^2(t) \rangle$ is the average squared lateral displacement of the ith lipid belonging to type α at time t
   N_α is the total number of lipids of type α in the system

The motion of the center of mass of the corresponding leaflet needs to be removed from $r_i^2(t)$.

One of the most important characterizations for the behavior of monolayers is the surface tension/pressure to area per lipid isotherms. From the pressure tensor in the simulation box, the surface tension of a monolayer can be evaluated [95] as

$$\gamma = \langle (P_N - P_L) \cdot L_z \rangle / 2 = \langle (P_N - P_L) \rangle \cdot L_z / 2, \tag{11.2}$$

where
   L_z is the box size in z-direction
   $P_N = P_{zz}$ is the normal pressure and the third diagonal component of the pressure tensor
   $P_L = (P_{xx} + P_{yy})/2$ is the lateral pressure
   P_{xx} and P_{yy} are the first and second components of the pressure tensor

The brackets denote averaging over time. The second equality applies only to NVT simulations where the box size in z-direction is a constant, which applies to the case study here.

To get a more direct comparison between simulations and experimental data, the surface pressure of a monolayer can be evaluated, which can be accomplished by deducting the surface tension of the monolayer from the bare air/water surface tension under the same condition:

$$\Pi(A_L, T) = \gamma_0(T) - \gamma(A_L, T), \tag{11.3}$$

where $\gamma_0 \equiv \gamma_0(T)$ is the bare water/air surface tension, which is a function of temperature, and both Π and γ are functions of the area per lipid A_L and temperature.

However, no water model can reproduce the real bare air/water surface tension for a broad range of temperature, which might be used in biological or physiological simulations.

Therefore, instead of using experimental data for bare air/water surface tension, the simulated values by the specific water model used in a simulation should be employed to ensure consistency. In addition, density profiling is a valuable tool to investigate the relative positioning of all relevant components in the simulation box and the change of it caused by other factors (Figures 11.8 through 11.10).

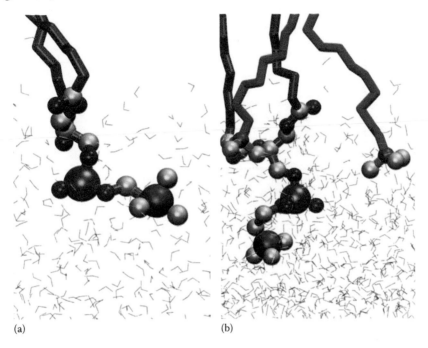

(a)   (b)

FIGURE 11.8 Phosphorus (the large tan atom)–nitrogen (the large blue atom) vector in PC headgroups reoriented by neighboring cationic CTAB. (a) The P-N vector of DPPC is oriented almost parallel to the monolayer plane. (b) The cationic CTAB (green lipid tail) essentially reorients the P-N vector of DPPC.

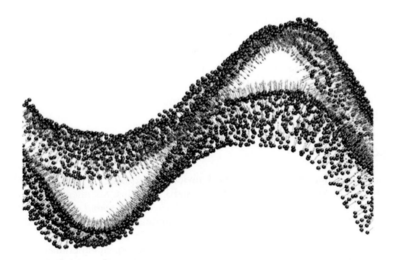

FIGURE 11.9 The normal vectors (arrows) for highly buckled DPPC monolayers separated by a water slab. Each monolayer has 2048 DPPCs. Water is disabled in visualization for clarity. Phosphates were chosen to approximate the interface between water and DPPC monolayers. The normal vectors always point toward water.

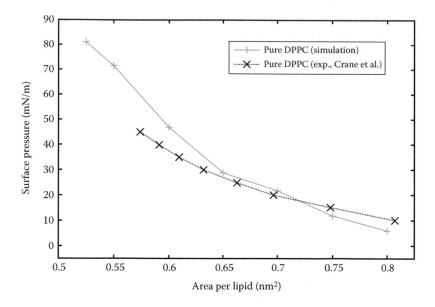

FIGURE 11.10   Surface pressure-area per lipid isotherms for pure DPPC monolayers simulated at 323 K. (The experimental data were obtained by Crane, J.M. et al., *Biophys. J.*, 77, 3134, 1999.)

11.7   DISCUSSION AND CONCLUSIONS

In this chapter, we have provided a detailed discussion of how to set up monolayer simulations, the caveats and various technical details as well as software commonly used for such simulations. Our aim was not to provide a comprehensive review of the vast literature on the topic but rather to provide a fairly hands-on approach to help the interested readers to set up, run, and analyze their own simulations.

ACKNOWLEDGMENTS

We thank Dr. Toby Zeng for fruitful discussion on quantum chemistry calculations and Prof. Pierre-Nicolas Roy for providing some of the computational resources. Computational resources were provided by SharcNet (https://www.sharcnet.ca/), the WestGrid HPC Consortium, the SciNet Consortium, and Compute Canada. Financial support was provided by the Waterloo Institute for Nanotechnology Nanofellowship program (BL), the University of Waterloo (MK), the Ontario Government (BL), and Natural Sciences and Engineering Research Council of Canada (NSERC; MK).

REFERENCES

1. D. B. Kittelson. Engines and nanoparticles: A review. *J Aerosol Sci*, 29:575–588, 1998.
2. Z. D. Ristovski, L. Morawska, N. D. Bofinger, and J. Hitchins. Submicrometer and supermicrometer particulate emission from spark ignition vehicles. *Environ Sci Technol*, 32:3845–3852, 1998.
3. M. M. Maricq, D. H. Podsiadlik, and R. E. Chase. Examination of the size-resolved and transient nature of motor vehicle particle emissions. *Environ Sci Technol*, 33:1618–1626, 1999.
4. L. Morawska, N. D. Bofinger, L. Kocis, and A. Nwankwoala. Submicrometer and supermicremeter particles from diesel vehicle emissions. *Environ Sci Technol*, 32:2033–2042, 1998.
5. C. Buzea, I. I. Pacheco, and K. Robbie. Nanomaterials and nanoparticles: Sources and toxicity. *Biointerphases*, 2:Mr17–Mr71, 2007.
6. G. Oberdorster, E. Oberdorster, and J. Oberdorster. Nanotoxicology: An emerging discipline evolving from studies of ultrafine particles. *Environ Health Perspect*, 113:823–839, 2005.

7. G. Oberdorster, Z. Sharp, V. Atudorei, A. Elder, R. Gelein, W. Kreyling, and C. Cox. Translocation of inhaled ultrafine particles to the brain. *Inhal Toxicol*, 16:437–445, 2004.

8. C. W. Lam, J. T. James, R. McCluskey, S. Arepalli, and R. L. Hunter. A review of carbon nanotube toxicity and assessment of potential occupational and environmental health risks. *Crit Rev Toxicol*, 36:189–217, 2006.

9. A. Seaton, W. MacNee, K. Donaldson, and D. Godden. Particulate air pollution and acute health effects. *Lancet*, 345:176–8, 1995.

10. G. Oberdorster, J. Ferin, and B. E. Lehnert. Correlation between particle size, in vivo particle persistence, and lung injury. *Environ Health Perspect*, 102(Suppl. 5):173–179, 1994.

11. D.F. Merlo, S. Garattini, U. Gelatti, C. Simonati, L. Covolo, M. Ceppi, and F. Donato. A mortality cohort study among workers in a graphite electrode production plant in Italy. *Occup Environ Med*, 61:e9, 2004.

12. C.G. Uragoda. A cohort study of graphite workers in Sri Lanka. *Occup Med (Lond)*, 47:269–272, 1997.

13. A. Huczko, H. Lange, E. Calko, H. Grubek-Jaworska, and P. Droszcz. Physiological testing of carbon nanotubes: Are they asbestos-like? *Fullerene Sci Technol*, 9:251–254, 2001.

14. Z. Wang and S. Yang. Effects of fullerenes on phospholipid membranes: A langmuir monolayer study. *Chemphyschem*, 10:2284–2289, 2009.

15. C. C. Chiu, W. Shinoda, R. H. DeVane, and S. O. Nielsen. Effects of spherical fullerene nanoparticles on a dipalmitoyl phosphatidylcholine lipid monolayer: A coarse grain molecular dynamics approach. *Soft Matter*, 8:9610–9616, 2012.

16. J. Goerke. Pulmonary surfactant: Functions and molecular composition. *Biochim Biophys Acta*, 1408:79–89, 1998.

17. S. J. Marrink, A. H. de Vries, and A. E. Mark. Coarse grained model for semiquantitative lipid simulations. *J Phys Chem B*, 108(2):750–760, January 2004.

18. N. Nisoh, M. Karttunen, L. Monticelli, and J. Wong-Ekkabut. Lipid monolayer disruption caused by aggregated carbon nanoparticles. *Rsc Adv*, 5:11676–11685, 2015.

19. S. Baoukina, L. Monticelli, S. J. Marrink, and D. P. Tieleman. Pressure-area isotherm of a lipid monolayer from molecular dynamics simulations. *Langmuir*, 23:12617–12623, 2007.

20. J. Barnoud, L. Urbini, and L. Monticelli. C60 fullerene promotes lung monolayer collapse. *J R Soc Interface*, 12:20140931, 2015.

21. J. Wong-Ekkabut, S. Baoukina, W. Triampo, I. M. Tang, D. P. Tieleman, and L. Monticelli. Computer simulation study of fullerene translocation through lipid membranes. *Nat Nanotechnol*, 3:363–368, 2008.

22. F. R. Poulain and J. A. Clements. Pulmonary surfactant therapy. *West J Med*, 162:43–50, 1995.

23. J. A. Clements. Pulmonary edema and permeability of alveolar membranes. *Arch Environ Health*, 2:280–283, 1961.

24. J. A. Clements. Surface phenomena in relation to pulmonary function. *Physiologist*, 5:11–28, 1962.

25. B. Robertson and H. L. Halliday. Principles of surfactant replacement. *Biochim Biophys Acta*, 1408:346–361, 1998.

26. A. W. Schüttelkopf and D. M. F. van Aalten. PRODRG: A tool for high-throughput crystallography of protein-ligand complexes. *Acta Crystallogr D Biol Crystallogr*, 60(Pt 8):1355–1363, August 2004.

27. S. J. Marrink and D. Peter Tieleman. Perspective on the martini model. *Chem Soc Rev*, 42:6801, 2013.

28. W. L. Jorgensen, D. S. Maxwell, and J. Tirado-Rives. Development and testing of the OPLS all-atom force field on conformational energetics and properties of organic liquids. *J Am Chem Soc*, 118(45):11225–11236, January 1996.

29. G. A. Kaminski, R. A. Friesner, J. Tirado-Rives, and W. L. Jorgensen. Evaluation and reparametrization of the OPLS-AA force field for proteins via comparison with accurate quantum chemical calculations on peptides †. *J Phys Chem B*, 105(28):6474–6487, July 2001.

30. W. D. Cornell, P. Cieplak, C. I. Bayly, I. R. Gould, K. M. Merz, D. M. Ferguson, D. C. Spellmeyer, T. Fox, J. W. Caldwell, and P. A. Kollman. A second generation force field for the simulation of proteins, nucleic acids, and organic molecules. *J Am Chem Soc*, 117(19):5179–5197, May 1995.

31. J. Wang, P. Cieplak, and P. A. Kollman. How well does a restrained electrostatic potential (RESP) model perform in calculating conformational energies of organic and biological molecules? *J Comput Chem*, 21(12):1049–1074, September 2000.

32. E. J. Sorin and V. S. Pande. Exploring the helix-coil transition via all-atom equilibrium ensemble simulations. *Biophys J*, 88(4):2472–2493, April 2005.

33. Y. Duan, C. Wu, S. Chowdhury, M. C. Lee, G. Xiong, W. Zhang, R. Yang et al. A point-charge force field for molecular mechanics simulations of proteins based on condensed-phase quantum mechanical calculations. *J Comput Chem*, 24(16):1999–2012, December 2003.

34. E. Neria, S. Fischer, and M. Karplus. Simulation of activation free energies in molecular systems. *J Chem Phys*, 105(5):1902, 1996.

35. A. D. MacKerell, D. Bashford, M. Bellott, R. L. Dunbrack, J. D. Evanseck, M. J. Field, S. Fischer et al. All-atom empirical potential for molecular modeling and dynamics studies of proteins. *J Phys Chem B*, 102(18):3586–3616, April 1998.

36. A. D. Mackerell, M. Feig, and C. L. Brooks. Extending the treatment of backbone energetics in protein force fields: Limitations of gas-phase quantum mechanics in reproducing protein conformational distributions in molecular dynamics simulations. *J Comput Chem*, 25(11):1400–1415, August 2004.

37. W. R. P. Scott, P. H. Hünenberger, I. G. Tironi, A. E. Mark, S. R. Billeter, J. Fennen, A. E. Torda, T. Huber, P. Krüger, and W. F. van Gunsteren. The GROMOS biomolecular simulation program package. *J Phys Chem A*, 103(19):3596–3607, May 1999.

38. X. Daura, A. E. Mark, and W. F. van Gunsteren. Parametrization of aliphatic CHn united atoms of GROMOS96 force field. *J Comput Chem*, 19(5):535–547, 1998.

39. C. Oostenbrink, A. Villa, A. E. Mark, and W. F. van Gunsteren. A biomolecular force field based on the free enthalpy of hydration and solvation: The GROMOS force-field parameter sets 53A5 and 53A6. *J Comput Chem*, 25:1656–1676, 2004.

40. O. Berger, O. Edholm, and F. Jahnig. Molecular dynamics simulations of a fluid bilayer of dipalmitoylphosphatidylcholine at full hydration, constant pressure, and constant temperature. *Biophys J*, 72:2002–2013, 1997.

41. E. A. Cino, W.-Y. Choy, and M. Karttunen. Comparison of secondary structure formation using 10 different force fields in microsecond molecular dynamics simulations. *J Chem Theory Comput*, 8:2725–2740, 2012.

42. P. Tieleman, Biocomputing Group, University of Calgary, http://wcm.ucalgary.ca/tieleman/downloads. Accessed on February 22, 2015.

43. A. A. Gurtovenko, M. Patra, M. Karttunen, and I. Vattulainen. Cationic DMPC/DMTAP lipid bilayers: Molecular dynamics study. *Biophys J*, 86(6):3461–3472, June 2004.

44. M. Bachar, P. Brunelle, D. Peter Tieleman, and A. Rauk. Molecular dynamics simulation of a polyunsaturated lipid bilayer susceptible to lipid peroxidation. *J Phys Chem B*, 108:7170–7179, 2004.

45. H. Martinez-Seara, T. Rog, M. Karttunen, R. Reigada, and I. Vattulainen. Influence of cis double-bond parametrization on lipid membrane properties: How seemingly insignificant details in force-field change even qualitative trends. *J. Chem. Phys.*, 129:105103, 2008.

46. W. Zhao, T. Róg, A. A. Gurtovenko, I. Vattulainen, and M. Karttunen. Atomic-scale structure and electrostatics of anionic palmitoyloleoylphosphatidylglycerol lipid bilayers with Na+ counterions. *Biophys J*, 92(4):1114–1124, February 2007.

47. M. J. Frisch, G. W. Trucks, H. B. Schlegel, G. E. Scuseria, M. A. Robb, J. R. Cheeseman, J. A. Montgomery, Jr. et al. Gaussian 03, Revision C.02. Gaussian, Inc., Wallingford, CT, 2004.

48. M. W. Schmidt, K. K. Baldridge, J. A. Boatz, S. T. Elbert, M. S. Gordon, J. H. Jensen, S. Koseki et al. Montgomery. General atomic and molecular electronic structure system. *J Comput Chem*, 14(11):1347–1363, November 1993.

49. M. S. Gordon and M. W. Schmidt. Advances in electronic structure theory: GAMESS a decade later. In C. E. Dykstra, G. Frenking, K. S. Kim, and G. E. Scuseria, eds., *Theory and Application of Computational Chemistry: The First Forty Years*, pp. 1167–1189. Elsevier, Amsterdam, the Netherlands, 2005.

50. A. A. Granovsky. Firefly quantum chemistry package version 8. http://classic.chem.msu.su/gran/firefly/index.html. Accessed on November 10, 2014.

51. R. Ditchfield, W. J. Hehre, and J. A. Pople. Self-consistent molecular-orbital methods. IX. An extended gaussian-type basis for molecular-orbital studies of organic molecules. *J. Chem. Phys.*, 54(2):724, 1971.

52. W. J. Hehre, R. Ditchfield, and J. A. Pople. Self-consistent molecular orbital methods. XII. Further extensions of gaussian-type basis sets for use in molecular orbital studies of organic molecules. *J. Chem. Phys.*, 56(5):2257, 1972.

53. P. C. Hariharan and J. A. Pople. The influence of polarization functions on molecular orbital hydrogenation energies. *Theor. Chem. Acc.*, 28:213–222, 1973.

54. V. Hornak, R. Abel, A. Okur, B. Strockbine, A. Roitberg, and C. Simmerling. Comparison of multiple Amber force fields and development of improved protein backbone parameters. *Proteins*, 65(3):712–725, November 2006.

55. G. A. Petersson, A. Bennett, T. G. Tensfeldt, M. A. Al-Laham, W. A. Shirley, and J. Mantzaris. A complete basis set model chemistry. I. The total energies of closed-shell atoms and hydrides of the first-row elements. *J. Chem. Phys.*, 89(4):2193, 1988.

56. G. A. Petersson and M. A. Al-Laham. A complete basis set model chemistry. II. Open-shell systems and the total energies of the first-row atoms. *J. Chem. Phys.*, 94(9):6081, 1991.
57. R. S. Mulliken. Electronic population analysis on LCAO [Single Bond] MO molecular wave functions. I. *J. Chem. Phys.*, 23(10):1833, 1955.
58. R. S. Mulliken. Criteria for the construction of good self-consistent-field molecular orbital wave functions, and the significance of LCAO-MO population analysis. *J. Chem. Phys.*, 36(12):3428, 1962.
59. P.-O. Löwdin. On the non-orthogonality problem. *Adv. Quantum Chem.*, 5:185–199, 1970.
60. L.C. Cusachs and P. Politzer. On the problem of defining the charge on an atom in a molecule. *Chem. Phys. Lett.*, 1(11):529–531, January 1968.
61. D. E. Williams. *Net Atomic Charge and Multipole Models for the Ab Initio Molecular Electric Potential*, pp. 219–271. John Wiley Sons, Inc., Weinheim, Germany, 2007.
62. B. Wang and G. P. Ford. Atomic charges derived from a fast and accurate method for electrostatic potentials based on modified AM1 calculations. *J. Comput. Chem.*, 15(2):200–207, February 1994.
63. K. C. Gross, P. G. Seybold, and C. M. Hadad. Comparison of different atomic charge schemes for predicting pKa variations in substituted anilines and phenols. *Int. J. Quantum Chem.*, 90(1):445–458, August 2002.
64. R. S. Vareková, S. Geidl, C.-M. Ionescu, O. Skrehota, M. Kudera, D. Sehnal, T. Bouchal, R. Abagyan, H. J. Huber, and J. Koca. Predicting pK(a) values of substituted phenols from atomic charges: Comparison of different quantum mechanical methods and charge distribution schemes. *J. Chem. Inf. Model.*, 51(8):1795–1806, August 2011.
65. E. D. Glendening, J. K. Badenhoop, A. E. Reed, J. E. Carpenter, J. A. Bohmann, C. M. Morales, and F. Weinhold. Nbo 5.9. Theoretical Chemistry Institute, University of Wisconsin, Madison, WI, 2012. http://www.chem.wisc.edu/~nbo5, 2012.
66. S. Miertuš, E. Scrocco, and J. Tomasi. Electrostatic interaction of a solute with a continuum. A direct utilizaion of ab initio molecular potentials for the prevision of solvent effects. *Chem. Phys.*, 55(1):117–129, February 1981.
67. J. Tomasi and M. Persico. Molecular interactions in solution: An overview of methods based on continuous distributions of the solvent. *Chem. Rev.*, 94(7):2027–2094, November 1994.
68. R. Cammi and J. Tomasi. Remarks on the use of the apparent surface charges (ASC) methods in solvation problems: Iterative versus matrix-inversion procedures and the renormalization of the apparent charges. *J. Comput. Chem.*, 16(12):1449–1458, 1995.
69. J. Tomasi, B. Mennucci, and R. Cammi. Quantum mechanical continuum solvation models. *Chem. Rev.*, 105(8):2999–3093, August 2005.
70. B. M. Bode and M. S. Gordon. Macmolplt: A graphical user interface for GAMESS. *J. Mol. Graph. Model.*, 16(3):133–138, June 1998.
71. Y. N. Kaznessis, S. Kim, and R. G. Larson. Simulations of zwitterionic and anionic phospholipid monolayers. *Biophys. J.*, 82(4):1731–1742, April 2002.
72. B. Liu, M. I. Hoopes, and M. Karttunen. Molecular dynamics simulations of DPPC/CTAB monolayers at the air/water interface. *J. Phys. Chem. B*, 118(40):11723–11737, October 2014.
73. B. Hess, C. Kutzner, D. van der Spoel, and E. Lindahl. GROMACS 4: Algorithms for highly efficient, load-balanced, and scalable molecular simulation. *J. Chem. Theory Comput.*, 4:435–447, 2008.
74. J. C. Phillips, R. Braun, W. Wang, J. Gumbart, E. Tajkhorshid, E. Villa, C. Chipot, R. D. Skeel, L. Kalé, and K. Schulten. Scalable molecular dynamics with NAMD. *J. Comput. Chem.*, 26(16):1781–1802, December 2005.
75. D. A. Case, T. E. Cheatham, T. Darden, H. Gohlke, R. Luo, K. M. Merz, A. Onufriev, C. Simmerling, B. Wang, and R. J. Woods. The Amber biomolecular simulation programs. *J. Comput. Chem.*, 26(16):1668–1688, December 2005.
76. J. Wong-Ekkabut, M. S. Miettinen, C. Dias, and M. Karttunen. Static charges cannot drive a continuous flow of water molecules through a carbon nanotube. *Nat. Nanotechnol.*, 5:555–557, 2010.
77. S. Nose. A unified formulation of the constant temperature molecular dynamics methods. *J. Chem. Phys.*, 81(1):511, 1984.
78. W. G. Hoover. Canonical dynamics: Equilibrium phase-space distributions. *Phys. Rev. A*, 31:1695–1697, March 1985.
79. G. J. Martyna, M. L. Klein, and M. Tuckerman. Nose–Hoover chains: The canonical ensemble via continuous dynamics. *J. Chem. Phys.*, 97(4):2635, 1992.
80. H. J. C. Berendsen, J. P. M. Postma, W. F. van Gunsteren, A. DiNola, and J. R. Haak. Molecular dynamics with coupling to an external bath. *J. Chem. Phys.*, 81(8):3684, 1984.
81. H. C. Andersen. Molecular dynamics simulations at constant pressure and/or temperature. *J. Chem. Phys.*, 72(4):2384, 1980.

82. G. Bussi, D. Donadio, and M. Parrinello. Canonical sampling through velocity rescaling. *J. Chem. Phys.*, 126:014101, 2007.

83. J. Wong-Ekkabut and M. Karttunen. Assessment of common simulation protocols for simulations of nanopores, membrane proteins, and channels. *J. Chem. Theory Comput.*, 8(8):2905–2911, August 2012.

84. M. Parrinello. Polymorphic transitions in single crystals: A new molecular dynamics method. *J. Appl. Phys.*, 52(12):7182, 1981.

85. U. Essmann, L. Perera, M. L. Berkowitz, T. Darden, H. Lee, and L. G. Pedersen. A smooth particle mesh Ewald method. *J. Chem. Phys.*, 103(19):8577, 1995.

86. D. M. York, T. A. Darden, and L. G. Pedersen. The effect of long-range electrostatic interactions in simulations of macromolecular crystals: A comparison of the Ewald and truncated list methods. *J. Chem. Phys.*, 99(10):8345, 1993.

87. M. Patra, M. Karttunen, M. T. Hyvönen, E. Falck, P. Lindqvist, and I. Vattulainen. Molecular dynamics simulations of lipid bilayers: Major artifacts due to truncating electrostatic interactions. *Biophys. J.*, 84(6):3636–3645, June 2003.

88. M. Patra, M. T. Hyvönen, E. Falck, M. Sabouri-Ghomi, I. Vattulainen, and M. Karttunen. Long-range interactions and parallel scalability in molecular simulations. *Comput. Phys. Commun.*, 176(1):14–22, January 2007.

89. M. Karttunen, J. Rottler, I. Vattulainen, and C. Sagui. Electrostatics in Biomolecular Simulations: Where Are We Now and Where Are We Heading? *Current Topics in Membranes,* 60:49–89, 2008.

90. G. A. Cisneros, M. Karttunen, P. Ren, and C. Sagui. Classical electrostatics for biomolecular simulations. *Chem. Rev.*, 114(1):779–814, January 2014.

91. J.A. Barker and R.O. Watts. Monte Carlo studies of the dielectric properties of water-like models. *Mol. Phys.*, 26(3):789–792, August 1973.

92. Ch. J. Johnson, E. Dujardin, S. A. Davis, C. J. Murphy, and S. Mann. Growth and form of gold nanorods prepared by seed-mediated, surfactant-directed synthesis. *J. Mater. Chem.*, 12(6):1765–1770, May 2002.

93. W. Humphrey, A. Dalke, and K. Schulten. VMD: Visual molecular dynamics. *J. Mol. Graph.*, 14(1):33–38, 27–28, February 1996.

94. J. Stone. An efficient library for parallel ray tracing and animation. Technical report, In *Intel Supercomputer Users Group Proceedings*, Sandia National Laboratory, Albequerque NM, 1995.

95. S. Baoukina, S. J. Marrink, and D. P. Tieleman. Structure and dynamics of lipid monolayers: Theory and applications. In R. Faller, T. Jue, M. L. Longo, and S. H. Risbud, eds., *Biomembrane Frontiers: Nanostructures, Models and the Design of Life*, pp. 75–99. Humana Press, New York, 2009.

96. J. M. Crane, G. Putz, and S. B. Hall. Persistence of phase coexistence in disaturated phosphatidylcholine monolayers at high surface pressures. *Biophys. J.*, 77(6):3134–3143, December 1999.

12 Molecular Dynamics Simulation of Droplet Spreading and Wettability Issues

Joël De Coninck

CONTENTS

12.1 INTRODUCTION

Wetting has a very long and fascinating history. The solid–liquid interactions are related to physics, but chemistry is deeply involved at the interface. The flow of the liquid in contact with a solid refers to fluid mechanics, and the solid itself is described in terms of material science. More than 200 years ago, Young and Laplace derived from macroscopic arguments the basic equations describing the equilibrium configuration of partially wetting liquids. Starting about 50 years ago, the dynamics of wetting became of central interest. Nowadays, still fundamental question remain unresolved. Most of the open questions are related to the microscopic characteristics of the considered materials for wetting. Young's equation, for example, has been validated at the microscopic scale only recently [1].

Apart from the fundamental problem of whether a given solid is wetted by a given liquid, many of the practical applications require the precise knowledge of the rates of wetting processes. Particularly, one is interested to know how fast a liquid can wet a given area of a solid surface. To understand the mechanisms controlling the dynamics of wetting, let us here consider the simplest case: the spreading of a liquid drop on top of a flat solid substrate. When such a liquid drop is placed in contact with the solid, capillary forces drive the interface spontaneously toward equilibrium. As the drop spreads, the contact angle θ relaxes from its initial maximum value of 180° at the moment of contact to its equilibrium angle θ_0 in the case of partial wetting or $0°$ if the liquid wets the solid completely. At this stage, a unifying approach describing the dynamics of wetting is still missing, but interesting progress has been achieved rather recently in this direction.

Since the shape of the drop will change versus time, it is clear that dissipation occurs both on a macroscopic scale, due to the reorganization of molecules as described by viscosity, and near the solid, as described by some kind of friction. Considering different channels of dissipation, different macroscopic theories have been developed, and upon certain assumption about the various microscopic characteristics, the behavior of the liquid can be described [2–7]. On the other hand, the macroscopic behavior can very well be measured and the proposed theories can be verified experimentally. However, these indirect methods do not lead to a full understanding of the processes that take place at the microscopic level. It is, for instance, still impossible to predict the exact wetting behavior of an arbitrary solid–liquid system before it is actually measured. Moreover, the main existing models to describe wetting seem to be incompatible with each other, although both can be fitted equally well to experimental data in many cases [8].

It is the goal of this work to review some recent molecular dynamics (MD) simulations to study the details of drop spreading and to clarify the physical mechanisms controlling the dynamic of wetting.

12.2   ONE WORD ABOUT THE TECHNIQUE

The MD algorithm and interactions used in this study are standard [9,10] except that the fluid is made of 8- or 16-atom chains of Lennard-Jones atoms to control the viscosity of the liquid. In this simplistic but consistent way, we attempt to come closer to the experiments. The basic interaction between pairs of atoms is of Lennard-Jones type:

$$V_{ij}(r) = C_{ij}\left(\left(\frac{\sigma}{r}\right)^{12} - \left(\frac{\sigma}{r}\right)^{6}\right)$$

where i,j represent the solid or fluid and the fluid atoms are grouped into chains with the additional pairwise confining potential

$$U(r) = A\left(\frac{\sigma}{r}\right)^{6}$$

between adjoining atoms. The constant A is usually fixed to 1. Aside from the issue of molecular size, the chain structure has the effect of strongly reducing the volatility of the fluid. To model the substrate, we consider a lattice made of a few layers of fcc unit cells, at each site of which is an atom of mass 50 so as to have comparable time scales for the atomic motion in fluid and solid. For computational convenience, the tail of the potentials are cut off at $r_c = 2.5$, in units of the fluid core size σ.

In our simulations, the interaction parameters are chosen as $C_{ff} = 1$ and $C_{ss} = 1$. The affinity between the solid and the liquid is explored herewith by varying the amplitude of C_{sf}. Given the potential, the motion follows from integrating Newton's equations, using a fifth-order predictor–corrector algorithm. In the remainder of this chapter, we nondimensionalize by using σ,ε and the fluid monomer mass as the units of distance, energy, and mass, respectively. The resulting natural time unit is $\tau = \sigma\sqrt{m/\varepsilon}$. Our typical time step is 5×10^{-3} in reduced units.

With these interactions, the liquid and solid considered here have the following properties for $T = 33.33°K$ and $r_c = 2.5\sigma$ (Table 12.1).

TABLE 12.1

Physical Properties of the Considered Liquid and Solid

Liquid Density	ρ_L	18.26 ± 0.07	10^{-3} atom/Å³
Solid density	ρ_S	17.12	10^{-3} atom/Å³
Surface tension	γ	2.493 ± 0.646	10^{-3} N/m
Shear viscosity	η	0.248 ± 0.004	10^{-3} Pa·s

12.3 SOME THEORETICAL CONSIDERATIONS

As demonstrated by de Gennes [7], the out-of-balance interfacial tension forces $F = \gamma(\cos\theta_0 - \cos\theta_d)$ (where γ is the liquid/air interfacial tension and θ_d and θ_0 are the dynamic and equilibrium contact angles, respectively) can be compensated by three channels of dissipation: viscous dissipation, dissipation in the precursor film associated with the complete wetting case ($\theta_0 = 0$), and dissipation in the close vicinity of the solid near the contact line. According to this view, when partial wetting is considered (Figure 12.1), the viscous dissipation described by the hydrodynamic (HD) approach [4–7] and the dissipation near the contact line described by the molecular-kinetic theory (MKT) [2–3] are the dominant channels.

Both theories predict the dynamic contact angle θ_d versus the contact-line velocity, V.

For the HD theory, we have

$$V = \frac{\gamma}{9\eta}\left(\theta_d^3 - \theta_0^3\right)\left[\ln\left(\frac{L}{L_s}\right)\right]^{-1} \tag{12.1}$$

where

L is a characteristic length scale of the droplet (m)
L_s is the slip length (m) near the wetting line

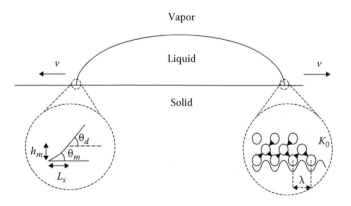

FIGURE 12.1 A liquid drop partially wetting a solid surface. The hydrodynamics parameters are illustrated on the left of the droplet. θ_m designs a microscopic contact angle, measured at a microscopic height h_m over a slip length L_s; quite often, $\theta_m = \theta_0$. The molecular-kinetic parameters are described on the right of the droplet. They refer to K_0, the equilibrium frequency of molecular displacements occurring at the contact line, and λ, the average distance of each displacement.

The MKT, however, gives

$$V = 2K_0\lambda \sin h\left(\frac{\gamma\lambda^2\left(\cos\theta_0 - \cos\theta_d\right)}{2k_BT}\right)$$ (12.2)

where

k_B is the Boltzmann's constant
T is the temperature (K)

The density of absorption site is set to λ^{-2}. When dealing with contact angles close to θ_0, the MKT can be simplified since the argument of the sin h function becomes sufficiently small:

$$V = \frac{\gamma}{\xi}\left(\cos\theta_0 - \cos\theta_d\right)$$ (12.3)

with ξ the friction at the contact line given by $\xi = k_BT/K_0\lambda^3$ (Pa·s). These approaches lead to different predictions for small dynamic contact angles and for a vanishing θ_0 on a flat solid surface: HD leads to $R_d \sim t^{1/10}$ and $\theta_d \sim t^{-3/10}$, whereas MKT gives $R_d \sim t^{1/7}$ and $\theta_d \sim t^{-3/7}$ where R_d is the dynamic drop radius and t the time.

Brochard-Wyart and de Gennes [11] have pointed out that for small angles, the dynamics is more likely to be controlled by viscous dissipation, whereas for large angles, contact-line friction would be the governing channel of dissipation. To overcome the inconsistency between these two models, Petrov and Petrov [12] and de Ruijter et al. [13] argued that both channels of dissipation should coexist and therefore combined these two approaches but in different ways.

Petrov considered that the microscopic contact angle appearing in the HD approach is the dynamic contact angle described by the MKT model and that both dissipations occur at the same time but in different regions. De Ruijter et al. used a purely mechanical scheme to combine both channels of dissipation. Within this model, dissipations can occur over different time scales; the dissipation in the vicinity of the triple line (MKT) usually first predominates, then the dissipation due to the friction in the inner layers of the liquid (viscosity) controls the spreading of the droplet as the contact angle falls recovering in that way the results of [11]. This model is specific to droplet geometry where the spherical cap approximation is valid. Nevertheless, it can be adapted to other geometries such as meniscus rise around a fiber or imbibition by porous substrates.

The idea that the dynamic contact angle depends on the contact-line velocity rather than the overall flow field of the liquid has been criticized by Shikhmurzaev in his interface formation theory; see [14]. Very recent results [15] seem to support this.

Most of these theoretical developments have been widely tested against experimental data [16]. To evaluate the effectiveness of a given model, the first criterion is to minimize the error between the experimental data and the fit by the relevant equation. It is important to keep in mind that the so-called best fit does not guarantee that the model is physically adequate; it just indicates that the mathematical function modeling the physics is able to adjust to the experimental data. A way to validate the fitted parameters would be to determine them by some independent measurement. If the agreement with the fitted ones were good, the model would be validated for the liquid and solid under study. Unfortunately, every model includes some molecular or at least microscopic parameters like the jump length, the jump frequency, or the slip length, which cannot be measured directly. As we will see, some numerical simulations such as the MD are able to measure some microscopic parameters, but no experimental method has so far been devised to measure such quantities. Therefore, it is not easy to validate the results of one fitting analysis. Moreover, although the HD and MKT approaches are physically different, macroscopic and molecular, their mathematical expression leads to behaviors that are very similar, as shown by the scaling laws. It is thus possible to analyze the results with either theory. This is one of the reasons why these models still coexist in spite of their differences. Another problem is a tendency to measure the dynamics of the contact angle (in spontaneous spreading) or the variation of the dynamic contact angle as a function of the contact-line velocity (in forced wetting) over a too narrow range of time or velocity. For instance, if the number of experimental points does not cover at least several decades of time, it may be possible to fit equally well the experimental data with any of the theoretical approaches [8]. This makes the analysis even more complex.

12.4   SOME EXPERIMENTAL RESULTS

Of course, wetting science is not limited to the study of droplet spreading on top of a flat surface. Many other systems, most of the time more complex, are of both fundamental and industrial importance. The solid can be a pore, a fiber, or a network of these entities, and the liquid can be of finite or effectively infinite size. To illustrate the differences between a flat surface and a fiber, it is first straightforward to consider the radii of curvature. If gravity is negligible, then when a droplet wets a flat surface, its radii of curvatures are equal. These become greater and reach an infinite value in the extreme case of the formation of a flat film. In the fiber geometry, the radii of curvature are of opposite sign, and one of them is determined by the fiber radius. This simple observation outlines the fundamental differences between the two geometries and therefore suggests that the study of the liquid–fiber system should provide distinctive results. This statement is valid both when a droplet of liquid spreads on a horizontal fiber and when a fiber is wetted by a bath of liquid. It is rather easy to reconsider theoretically the spontaneous dynamics of wetting around a fiber to test the possible existence of the different dissipation mechanisms that may appear in the process. It can easily be shown in this case that HD predicts $\theta_d \sim t^{-1/2}$, whereas the MKT leads to $\theta_d \sim t^{-1}$. This has been achieved experimentally in [17] where the spontaneous wetting of a fiber is studied using a PET fiber (poly(ethylene terephthalate) of 0.8 mm of diameter put in contact with silicon oils [poly(dimethylsiloxane) (PDMS)] of different viscosities. Some corresponding results are given in Figure 12.2.

When these data are put in a logarithmic scale (Figure 12.3), we see directly that the contact angle dynamics present two different trends during the latest stages of the meniscus rise as the meniscus approaches equilibrium (i.e., small contact angles). On one hand, the liquid with a low viscosity has a slope around −1.0, and on the other hand, the liquids with higher viscosity have a slope around −0.5 as shown more precisely in Table 12.2.

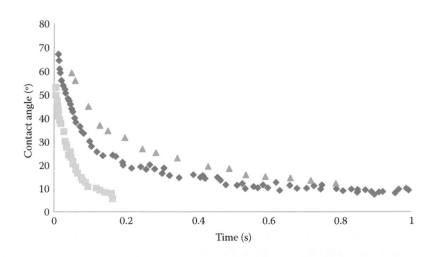

FIGURE 12.2   Contact angle dynamics versus time for PDMS5 (□), PDMS20 (◊), and PDMS50 (Δ).

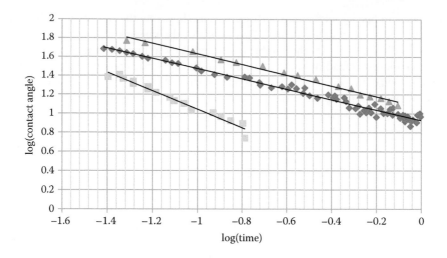

FIGURE 12.3 Late stage contact angle dynamics versus time for PDMS5 (□), PDMS20 (◊), and PDMS50 (Δ) on a logarithmic scale. The solid line corresponds to the best linear fits.

TABLE 12.2
Slopes of the Different Contact Angle Dynamics

Silicon Oils	Slope	R^2
PDMS5	-0.928 ± 0.012	0.986
PDMS20	-0.528 ± 0.014	0.964
PDMS50	-0.513 ± 0.006	0.982

These two trends can be related to the two channels of dissipation considered before, viscous dissipation and friction. These simple experiments clearly prove their experimental existence during spreading.

12.5 PARTIAL WETTING CASE

Let us now study how we can use MD simulations to check the microscopic details of these channels of dissipation. Having defined our system in terms of potentials and parameters previously, we can proceed to study the spreading process by examining the behavior of a liquid drop when it is brought into contact with the solid. In what follows, we present the results of large-scale simulations of spreading drops carried out for a wide range of solid–liquid interactions and corresponding values of θ_0.

To build our system in a consistent way, we construct a spherical drop with a large number of molecules above the center of a planar circular solid substrate comprising also a large number of atoms. The droplet radius is obtained by locating the L–V interface in the radial density profile. The radius of the solid is a certain number of times larger than that of the drop (4 here) to allow sufficient space for the liquid to spread without reaching the edges of the solid as illustrated in Figure 12.4.

The drops are formed by first equilibrating a large cube of 216,320 atoms for $3 \cdot 10^5$ iterations and then extracting the required drop from the center of the cube. This allows us to have a large equilibrated drop at our disposal.

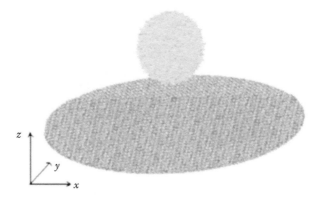

FIGURE 12.4 The initial configuration with an equilibrated drop and a solid surface. The images were generated using the VMD software. (From Humphrey, W. et al., *J. Mol. Graph.*, 14, 33, 1996.)

Next, we allow the drop and the solid to equilibrate independently for 10^5 iterations at constant temperature, with the drop placed in the vacuum above the substrate at a distance of 100 Å. The drop is then moved into contact with the solid and spreading starts. From this moment, only the temperature of the solid is controlled. The position of each atom of the liquid is measured every 1000 iterations, allowing us to extract the contact angle and the base radius from each snapshot. Drops are followed for $2 \cdot 10^6$ to $8 \cdot 10^6$ iterations (10–40 ns), until spreading stops or the contact angle ceases to change with respect to the noise of the simulation.

12.6   COMPLETE WETTING CASE

When the coupling becomes large enough $C_{sf} > 1$, the drop wets completely the substrate and finally forms a monolayer on top of it. The required time is of course quite huge in terms of simulation time steps. Within this kind of simulation, it is clearly observable that there appear several layers during spreading, as observed in real experiments [19,20]. A typical snapshot is given in Figure 12.5.

12.7   HOW TO ANALYZE SIMULATION DATA

12.7.1   IN THE PARTIAL WETTING CASE

To determine the contact angle and radius at any instant, we must first locate the $L–t$ interface. As described in details in [21,22], we divide the drop into horizontal layers of a certain thickness (small

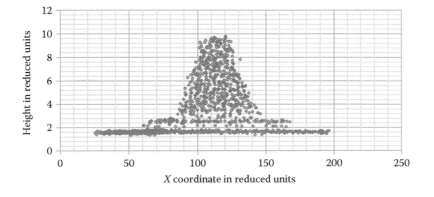

FIGURE 12.5 Side view of a drop of 1600 16-atom molecules during complete spreading.

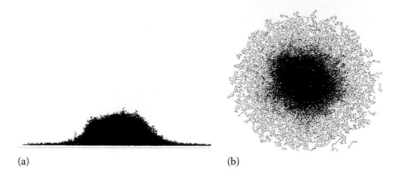

(a)   (b)

FIGURE 12.6   Side view of a drop of 2000 16-atom molecules during spreading (a). The corresponding top view (b).

enough to give sufficient layers while retaining a uniform density within each layer, 3 Å here). By symmetry, we determine the center of these layers and calculate the density of atoms as a function of the radial distance from the center. The radius of each layer is then given as the distance from the center at which the density falls below a selected cutoff value, typically 85% of the central density. Here, we have of course to check that the results do not depend on the choice of this cutoff value. This has been successfully verified. Once we have the profile of the drop, we can study its evolution with time in a way that mimics real experiments.

To measure the contact angle, we find the best circular fit to the measured profile since there is no gravity taken into account here. The first two layers in contact with the solid are omitted from the fit, since this part of the drop deviates from its otherwise spherical form for entropic reasons [23]. The circular fit is then extrapolated to the position of the first liquid atoms. The tangent to the circle at this position gives the contact angle and the base radius. We have also verified that this method ensures a constant drop volume.

In our simulations, several levels of solid–liquid interaction are investigated, by varying the coupling coefficients in the Lennard-Jones potential C_{sf} from 0.2 to 1.1.

12.7.2   In the Complete Wetting Case

A typical drop undergoing terraced spreading is shown in Figure 12.6 from side and top views. In this case, the drop is made of 2000 16-atom chains, with solid–liquid attraction coefficients $C_{sf} = 1$ [24–26]. One may note the absence of vapor and the distinct layering near the wall. The layering may be studied versus time, and we observe that the corresponding radii of the first few layers vary as $\sqrt{t}$.

This pseudodiffusive spreading with $\sqrt{t}$ dependence has been widely observed for experimental precursor films of completely wetting liquids such as PDMS on silicon wafers [19,20].

12.8   RESULTS

Versus time, we get the contact angles for different couplings given in Figure 12.7.

From the contact angle and the base radius of the drop versus time, the Gdynia facility described in [8] computes the speed of the contact line versus time $V(t)$ and fits these data using different theories such as the MKT or the HD. This systematic statistical tool (available free of charge at http://www.lpsi.be) evaluates dynamic contact angle data from spontaneous and forced wetting experiments and compares them with the current theories. The method has several advantages over conventional fitting procedures. It not only minimizes the weighted sum of the squares of differences between fitted values and experimental data using a combination of the simplex and Levenberg–Marquardt algorithms, it also explores the robustness of these fits by using the bootstrap technique

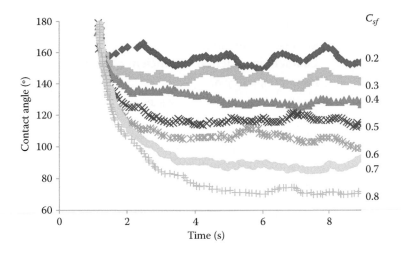

FIGURE 12.7 Contact angle versus time for various couplings C_{sf}.

and systematically varying the initial values of the parameters of the fit. This gives access to a distribution of the fitted parameters and to their possible correlation. Having such a method is important, as we can improve our understanding of the physical mechanism of wetting.

12.9 PHYSICAL MECHANISMS

We now fit the data $\theta d(t)$, $V(t)$ to Equation 12.2 for each solid–liquid coupling using the known values of γ and θ_0. Typical results are given in Figure 12.8.

At the lowest couplings (0.2–0.3), spreading is too fast and the statistical noise too great to yield stable fits. For the remaining couplings (0.3–0.8), the results in terms of the contact-line friction coefficient ξ and the jump frequency K_0 are listed in Table 12.3, together with their standard

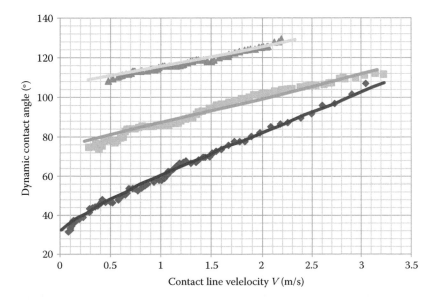

FIGURE 12.8 Typical behavior of $\theta_d(V)$ during the spreading of a 40,000-atom drop at couplings C_{sf} of 1.0, 0.8, and 0.6 from bottom to top. The fits to Equation 12.2 are shown by the smooth curves.

TABLE 12.3

Results of Spreading in Terms of the
Contact-Line Friction Coefficient ξ and the
Jump Frequency K_0

Coupling C_{sf}	θ_0 (°)	ξ (mPa·s)	K_0 (10^9 Hz)
0.2	165.9 ± 1.6		
0.3	147.2 ± 0.4		
0.4	134.8 ± 1.1	0.157 ± 0.043	37.0 ± 20.4
0.5	121.3 ± 0.8	0.212 ± 0.056	27.4 ± 14.9
0.6	105.8 ± 2.4	0.364 ± 0.096	16.0 ± 8.7
0.7	91.2 ± 1.9	0.429 ± 0.114	13.5 ± 7.3
0.8	74.6 ± 1.7	0.528 ± 0.140	11.0 ± 6.0
0.9	55.4 ± 1.6	0.747 ± 0.197	7.77 ± 4.22
1.0	31.8 ± 0.8	0.868 ± 0.227	6.67 ± 3.61
1.1	0	1.100 ± 0.291	5.28 ± 2.87

deviations. The values of K_0 are calculated using the constant value for the jump distance $\lambda = 4.3 \pm 0.4$ Å. The jumps start at the minimum value of 3.93 Å, which is the interdistance between two solid atoms. The corresponding jump distribution in the simulation peaks rapidly at about 4.0 Å and then decays more slowly toward zero with a long tail. Taking the mean value given by the distribution as the characteristic value for the collective motion of the contact line in our simulations gives $\lambda = 4.3 \pm 0.4$ Å. This value is interestingly independent of the coupling C_{sf}.

As we have seen in Figure 12.8, the MKT accounts well for the dynamics of spreading in the simulations. However, as pointed out before, the fact that a theory can fit some data is not sufficient to prove its validity.

One of the advantages of MD simulations is that they allow us to look very closely at the atomic motions. In what follows, we therefore study the random thermal displacements of the atoms at equilibrium to determine K_0 directly and compare them with the fitted values. Strategies for the direct measurement of K_0 and λ have already been developed by de Ruijter and coworkers [10,21,22,27]. The approach taken here is similar.

The first step is to determine the equilibrium density profile in the z-direction normal to the solid surface. This is done by computing the atomic density within thick horizontal slices starting at the solid and working outward into the bulk. As previously observed in MD simulations [28–30], the liquid is strongly layered near the solid at high couplings. As the coupling is reduced, the layering becomes progressively less pronounced, until at $C_{sf} = 0.2$ the profile resembles that of the $L–V$ interface. Far away from the solid (>25 Å), any layering decays and the liquid has a uniform density equal to that of the bulk.

At equilibrium, liquid atoms adjacent to the solid surface undergo random thermal displacements, jumping from one potential well to the next and from one layer to another, but with no net flux. The equilibrium frequency K_0 will be some average of all possible types of atom displacement. Nevertheless, due to the layering, these movements are likely to be anisotropic, with greater mobility within a given layer than between layers. We therefore divide the displacements into two categories:

1. Parallel jumps within the first layer
2. Perpendicular jumps between the first and second layers

Ideally, we are interested by the three-phase zone (TPZ) where the density profile $\rho(z)$ is inevitably modified by the proximity of the $L–V$ interface. Since the TPZ is only a few atoms thick and subject to large fluctuations, this is not easy. However, motion of the contact line during spreading is possible only because of a net flux into the first layer. We therefore apply our analysis to the entire first layer, assuming a uniform behavior throughout [22,27]. From the density profile, we locate the first two layers at each coupling. A parallel jump is then defined as a jump within the first layer (in the xy plane) of not less than $d_{\parallel}$, the distance between adjacent atoms in the solid surface, that is, 3.93 Å. Smaller displacements are considered as thermal agitation or noise.

Similarly, a perpendicular jump is defined as a jump from the first layer to the second one, with a displacement along z of not less than $d\perp$, the distance of separation between the two first peaks in the density profile. Jumps between layer 1 and layer 3 or beyond are also neglected. The aforementioned definitions are essentially those used by de Ruijter and coworkers and could equally be applied here. Once we have these definitions, we can analyze the XYZ file of all the atoms initially in the first layer (up to 5000) and follow them during the simulation to find the time necessary to make a jump. This analysis gives us distributions of jump distances and frequencies. To improve our precision, we average over a few hundred different starting times $t > t_0$ (the time at which equilibrium is achieved). For the highest frequencies, the simulations are relaunched from equilibrium for a further 2×10^5 iterations with more frequent recording of the positions (every 100 time steps). More details are available in [31].

Using these criteria, we can easily determine the cumulative percentage of atoms initially in the first layer that have made a parallel or perpendicular jump as a function of time. By differentiating the cumulative data, we obtain the distribution of the time required to make a jump. All have pronounced maxima but are skewed to long times. Ideally, we should consider the complete distribution to calculate the characteristic mean value of $K_{0\parallel}$ or $K_{0\perp}$ by inversion. To simplify the procedure, we assume that the average time for the collective movement of all the atoms is the time required for half the atoms of the first layer to have made a parallel or perpendicular jump [10,22]. The characteristic frequency is then the inverse of this time. By repeating the counting procedure for each of the 500 starting times, we obtain average values of $K_{0\parallel}$ and $K_{0\perp}$ for each coupling together with their standard deviations. The results, together with the values of K_0 obtained by fitting the spreading dynamics with the MKT, are plotted in Figure 12.9 as a function of the coupling.

As can be seen, there is broad agreement between $K_{0\perp}$ and K_0 over a wide range of couplings, which confirms the result obtained by de Ruijter and coworkers [10]. Spreading is possible only if there is a net flux into the first layer. Because adsorption and desorption of atoms from the solid surface is less frequent than a displacement along the surface, the former becomes the rate-determining event for spreading in our system. The upward drift of K_0 at the lowest coupling is also logical.

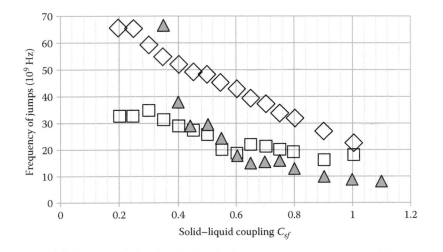

FIGURE 12.9 Frequencies of jumps $K_{0\parallel}$ (◊), $K_{0\perp}$ (□), and K_0 (▲) as a function of the coupling.

As the density of atoms in the first layer is reduced, adsorption–desorption events will play a decreasing role in dynamic wetting.

Inevitably, our definitions of parallel and perpendicular jumps are somewhat arbitrary. We have therefore checked the robustness of this approach by varying the criteria. As shown in [31], whichever definition we use, the agreement between the measured rate of jumping and that found by applying the MKT to droplet spreading remains very good at least for this very low-viscosity liquid.

The results of Figure 12.9 predict also that the coefficient of contact-line friction ζ will be dependent on the strength of solid–fluid interactions. In [32], it has been shown that ζ is in fact proportional to the viscosity and to the exponential of the equilibrium work of adhesion by analyzing a very large set of dynamic wetting data drawn from more than 20 publications and representative of a very wide range of systems, from MD-simulated Lennard-Jones liquids and substrates, through conventional liquids and solids, to molten glasses and liquid metals on refractory solids. The combined set spans 9 decades of viscosity and 11 decades of contact-line friction. These results suggest that the broad agreement between the MKT and such a wide range of data is strong evidence that the dynamic contact angle is directly dependent on the velocity of the contact line and confirm that equilibrium wetting properties dictate the corresponding dynamics.

In the complete wetting regime, as pointed out earlier, a pseudodiffusive spreading with $\sqrt{t}$ dependence has been widely observed for precursor films of completely wetting liquids. This remarkable property has also been observed for lipid films on water [33] and metals on metal surfaces [34]. Interesting theoretical explanations have been advanced combining micro- and macroarguments [35]. These results also support a molecular-kinetic interpretation in which the driving force for spreading is the lateral pressure π in the monolayer; see [36] for more details, leading to the prediction

$$V = 2K_0 \lambda \sin h\left(\frac{\lambda^2 \pi}{2k_B T} \right) \tag{12.4}$$

Near equilibrium, it can easily be shown that spreading is indeed pseudodiffusive and follows the square root of time. In this regime, it can be shown that the controlling factor is here again the equilibrium frequency of molecular displacements within the monolayer.

12.10 CONCLUSIONS

The results presented here illustrate the value of large-scale MD in modeling wetting and spreading at both global and molecular scales. Simulation is clearly an important tool to help us understand the underlying physics. However, any conclusions must be tempered by an acknowledgment of the limitations imposed by current computational power and the relative simplicity of the systems we are able to investigate. Nevertheless, we can expect that all the physics observed here will be present in a real system.

This review confirms that the dynamics of spreading and wetting in our system is described rather well by the MKT and that contact-line friction is the dominant channel of dissipation for this low-viscosity liquid. By applying the MKT to the spreading data and also analyzing the detailed behavior on the atomic scale, it has been possible not only to model the global behavior of the dynamic contact angle as a function of contact-line speed but to demonstrate that the key parameters in the model, K_0 and λ, are an accurate reflection of the molecular motions at the atomic scale. Thus, the MKT appears valid at some fundamental level.

Many more questions still remain open, however, within this subject currently: What will happen for more viscous liquids? For heterogeneous substrates? For rough substrates? For two or complex liquid systems? We hope that the combination of tools now available such as experimental physics, theories, and numerical simulations will help us to understand these mechanisms, but very much remains to be done, which is the best conclusion after all for a review.

ACKNOWLEDGMENTS

The author wishes to thank all the members of the LPSI team for fruitful discussions specially Michel de Ruijter, Emilie Bertrand, David Seveno, Anne-Marie Cazabat, and Terry Blake. This research has been partially funded by the Interuniversity Attraction Poles Programme (IAP 7/38 MicroMAST) from the Belgian Science Policy Office, the Région Wallonne, and the FNRS.

REFERENCES

1. D. Seveno, T. D. Blake, and J. De Coninck, *Phys. Rev. Lett.*, 2013, 111, 096101-1.
2. T.D. Blake and J. Haynes, *J. Colloid Interface Sci.*, 1969, 30, 421.
3. T.D. Blake, *J. Colloid Interface Sci.*, 2006, 299, 1–13.
4. O.V. Voinov, *Fluid Dyn.*, 1976, 11, 714–721.
5. R.G. Cox, *J. Fluid Mech.*, 1986, 16, 169–194.
6. D. Bonn, J. Eggers, J. Indekeu, and J. Meunier, *Rev. Mod. Phys.*, 2009, 81, 739–805.
7. P.G. de Gennes, *Rev. Mod. Phys.*, 1985, 57, 827–863.
8. D. Seveno, A. Vaillant, R. Rioboo, H. Adão, J. Conti, and J. De Coninck, *Langmuir*, 2009, 25, 13034–13044.
9. M.P. Allen and D.J. Tildesley, *Computer Simulation of Liquids*, Oxford University Press, New York, 1991.
10. J. De Coninck and T.D. Blake, *Annu. Rev. Mater. Res.*, 2008, 38, 1.
11. F. Brochard-Wyart and P.G. de Gennes, *Adv. Colloid Interface Sci.*, 1992, 39, 1–11.
12. P.G. Petrov and J. Petrov, *Langmuir*, 1992, 8, 1762–1767.
13. M. de Ruijter, J. De Coninck, and G. Oshanin, *Langmuir*, 1999, 15, 2209–2216.
14. Y.D. Shikhmurzaev, *Capillary Flows with Forming Interfaces*, Chapman & Hall CRC, London, U.K., 2008.
15. T.D. Blake, J.-C. Fernandez-Toledano, G. Doyen, and J. De Coninck, *Forced Wetting and Hydrodynamic Assist*, 2015, to appear in Physics of Fluids, 2015.
16. M. Schneemilch, R.A. Hayes, J.G. Petrov, and J. Ralston, *Langmuir*, 1998, 14, 7047.
17. M.S. Vega, D. Seveno, G. Lemaur, M.H. Adão, and J. De Coninck, *Langmuir*, 2005, 21, 9584–9590.
18. W. Humphrey, A. Dalke, and K. Schulten, *J. Mol. Graph.*, 1996, 14, 33.
19. F. Heslot, N. Fraysse, and A.M. Cazabat, *Nature*, 1989, 338, 640.
20. F. Heslot, A.M. Cazabat, P. Levinson, and N. Fraysse, *Phys. Rev. Lett.*, 1990, 65, 599.
21. T.D. Blake, A. Clarke, J. De Coninck, and M.J. de Ruijter, *Langmuir*, 1997, 13, 2164.
22. M.J. de Ruijter, T.D. Blake, and J. De Coninck, *Langmuir*, 1999, 15, 7836.
23. J. De Coninck, F. Dunlop, and F. Menu, *Phys. Rev. E*, 1993, 47, 1820–1823.
24. J. De Coninck, U. D'Ortona, J. Koplik, and J. Banavar, *Phys. Rev. Lett.*, 1995, 74, 928–932.
25. U. D'Ortona, J. De Coninck, J. Koplik, and J. Banavar, *Phys. Rev. E*, 1996, 53, 562–569.
26. J. De Coninck, N Fraysse, M.P. Valignat, and A.M. Cazabat, *Langmuir*, 1993, 9, 1906–1909.
27. D. Seveno, G. Ogonowski, and J. De Coninck, *Langmuir*, 2004, 20, 8385.
28. G. Saville, *J. Chem. Soc. Faraday Trans.*, 1977, 273, 1122.
29. J.R. Grigera, S.G. Kalko, and J. Fischbarg, *Langmuir*, 1996, 12, 154.
30. N.V. Priezjev, A.A. Darhuber, and S.M. Troian, *Phys. Rev. E*, 2005, 71, 041608.
31. E. Bertrand, T. Blake, and J. De Coninck, *J. Phys. Condens. Matter*, 2009, 21, 464124.
32. D. Duvivier, T.D. Blake, and J. De Coninck, *Langmuir*, 2013, 29, 10132–10140.
33. S. He and J.B. Ketterson, *Philos. Mag. B*, 1998, 77, 831.
34. M.N. Popescu and S. Dietrich, *Phys. Rev. E*, 2004, 69, 061602.
35. P.G. de Gennes and A.M. Cazabat, *C. R. Acad. Sci. II*, 1990, 310, 1601.
36. E. Bertrand, T.D. Blake, and J. De Coninck, *Langmuir*, 2005, 21, 6628–6635.

13 Implementation of the Immersed Boundary Method to Study Interactions of Fluids with Particles, Bubbles, and Drops

Efstathios E. Michaelides and Zhi-Gang Feng

CONTENTS

13.1 INTRODUCTION

There are three fundamental numerical approaches for the study of the movement of particles, bubbles, and drops in fluids. The first one is the continuum model (or two-fluid model), in which both the fluid phase and the dispersed phase are treated as continuous media. The continuum model requires the knowledge of interactions at the boundaries of the phases, such as drag and heat transfer coefficients, and makes use of the apparent viscosity of the dispersed phase. The second approach is the discrete particle model. It treats the discrete phase as separate particles that interact with the flow and traces the position and velocity of all the elements of the dispersed phase by solving Lagrangian equations of motion. The fluid phase is considered a continuum phase, and the effects of the elements of the dispersed phase are included by adding the interaction terms (for mass, momentum, and energy) into the governing equations of the fluid by empirical equations. The third approach is the direct numerical simulation (DNS). This is also a Eulerian–Lagrangian method that resolves the fluid flow around particles, bubbles, and drops. The method accounts for the interactions of the fluid with the dispersed phase by solving the Navier–Stokes equations for the fluid phase and the initial value problem for the motion of the dispersed phase simultaneously. With the significant increase of computational power, the DNS method is becoming a more enabling and popular approach to study complex particulate flow problems.

The conventional DNS methods, such as the finite volume (FVM) and finite element methods (FEM), are not very efficient in the simulations of particulate flows with a large number of particles.* The main obstacle with these methods is the need to generate new, geometrically adapted grids, a very time-consuming task in 3D flows. Methods such as Stokesian dynamics [1] and boundary element method essentially neglect the fluid inertia effects and can only be applied to particulate flows under creeping flow conditions. Kalthoff et al. [2] proposed a method that incorporates analytical solutions for the near-particle region, with some parameters determined by matching the outer flow conditions. Hence, they are able to determine the forces on particles based on the analytical flow solutions. Zhang and Prosperetti adopted the same concept and used the Stokes' analytical solution in the vicinity of the particles in the *Physalis* method [3]. The method requires the knowledge of accurate analytical solutions for the motion of particles, which are available for particles with simple shapes and under creeping flow conditions (e.g., Stokes' flow). Ladd successfully applied the lattice Boltzmann method (LBM) to particle–fluid suspensions [4,5]. The LBM overcame the limitations of the conventional FVM and FEM by using a fixed, nonadaptive (Eulerian) grid system to represent the flow field. The LBM has proven to be a robust and efficient method to accurately simulate particulate flows with small or large numbers of particles [6,7]. The method has also undergone several modifications and improvements and has been applied to particles of irregular shapes and aggregates of particles [8,9].

When the LBM is used to simulate particle–fluid interaction problems, the no-slip condition on the particle–fluid interface is realized by the so-called *bounce-back* rule (see [5]), and the particle surface is represented by the so-called boundary nodes, which are essentially the set of the midpoints of the links between two fixed grids, in which one of the grid points is within the fluid domain and the other is within the solid domain. This arrangement causes the computational boundary of a particle to be defined by a stepwise scheme. In order to represent a smooth boundary and to accurately represent the shape of any particle, it is necessary to use a large number of lattice points. In addition, when a particle moves, its computational boundary changes and varies in each time step. This introduces fluctuations in the forces that act on the particle and limits the ability of the LBM to solve particle–fluid interaction problems at high Reynolds numbers.

One of the difficulties in several numerical methods that were developed for the simulation of the flow of particles is the treatment of the boundaries and interfaces: mass, momentum, and energy are exchanged in the boundaries, which needs to be accounted for. Typically, this is accomplished by prescribing as boundary conditions (BCs) closure equations for the mole (or mass) fractions, the velocity, and the temperature. This practice introduces two problems: (1) the correct stipulation of the closure equations and (2) the correct description of the interface within the numerical grid of the fluid.

A moment's reflection proves that the prescription of the interfacial BCs is rather peculiar within the framework of the governing equations the numerical codes solve: For example, in our Newtonian framework, the governing equations that are solved for the transport of momentum are given in terms of forces (rates of change of momentum), while velocity interfacial conditions need to be prescribed at the interface. A more consistent approach within the Newtonian framework would be to describe the interfaces and other boundaries in terms of forces. Peskin in 1977 [10] used such an approach and developed the immersed boundary method (IBM) in order to model the flow of blood in the heart. The method uses a fixed Cartesian mesh for the fluid, which is composed of Eulerian nodes. For the solid boundaries, which are immersed in the fluid, the IBM uses a set of Lagrangian points that are advected according to the rules of fluid–solid interactions. The boundaries are modeled as continuous force fields acting on the fluid. Fogelson and Peskin in 1988 [11] showed that this method may also be employed to simulate flows with a group of suspended particles, both rigid and deformable. Mohd-Yusof applied this method with a system of discrete forces, which are calculated

---

* While the terms *particles* and *particulate flow* will be used for brevity in the rest of this chapter, the immersed boundary method may be easily converted to be used for bubble and droplet flows.

from the discretized Eulerian equations at a group of points in the computational domain [12]. Several researchers refer to this method as the direct forcing (DF) scheme.

Feng and Michaelides in 2004 [13] combined the IBM and the LBM for the first time to model the flow and interactions of groups of particles in flow systems. They computed the boundary force density at the interfaces using a penalty method. The advantage of the IBM is that instead of remeshing the fluid domain to account for the flow of particles, the method uses a fixed mesh to represent the fluid field. The moving boundaries of the particles are represented by a set of Lagrangian boundary points, which are advected by the fluid. Feng and Michaelides [14] later in 2005 developed the *Proteus* code, using a numerical method that combines DF and the IBM. *Proteus* makes use of Eulerian lattice nodes for the fluid flow field and Lagrangian boundary nodes to represent particles or moving-boundary surfaces. The method may be applied to both deformable and rigid particles. Several months later, Uhlman developed a similar computational method that combines the IBM with a finite-difference-based fluid solver [15]. The following section describes in more detail the implementation of the IBM in particulate flows.

13.2 FUNDAMENTALS OF THE IBM

The key attribute of the IBM is that the existence of a solid body is represented by its effect on the fluid. This is enforced by introducing a fluid *body force density* term into the momentum equations. Let us consider a particle with a boundary surface, Γ, immersed in a 3D incompressible viscous fluid with a domain, Ω. The particle boundary surface, Γ, is represented by the Lagrangian parametric coordinates, s, and the flow domain, Ω, is represented by the Eulerian coordinates x. Hence, the positions of the surface may be written as $\mathbf{x} = \mathbf{X}(s,t)$. The no-slip BC at the particle surface is satisfied by enforcing the velocity at all interfaces to be equal to the velocity of the fluid at the same location:

$$\frac{\partial \mathbf{X}(s,t)}{\partial t} = \mathbf{u}(\mathbf{X}(s,t),t) \tag{13.1}$$

where $\mathbf{u}$ is the fluid velocity.

Let $\mathbf{F}(s,t)$ and $\mathbf{f}(\mathbf{x},t)$ represent the particle surface force density and the fluid body force density, respectively. The governing equations for the fluid–particle mixture are as follows:

$$\rho\left(\frac{\partial \mathbf{u}}{\partial t} + \mathbf{u}\cdot\nabla\mathbf{u}\right) = \mu\nabla^2\mathbf{u} - \nabla p + \mathbf{f} \tag{13.2}$$

$$\nabla\cdot\mathbf{u} = 0 \tag{13.3}$$

with

$$\mathbf{f}(\mathbf{x},t) = \int_\Gamma \mathbf{F}(s,t)\delta(\mathbf{x}-\mathbf{X}(s,t))ds \tag{13.4}$$

and

$$\frac{\partial \mathbf{X}}{\partial t} = \int_\Omega \mathbf{u}(\mathbf{x},t)\delta(\mathbf{x}-\mathbf{X}(s,t))d\mathbf{x} \tag{13.5}$$

where
    $p(\mathbf{x},t)$ is the fluid pressure
    ρ is the fluid density
    μ is the fluid viscosity

Equations 13.2 and 13.3 are the governing equations for viscous, isothermal, incompressible flow, which include the effect of the particle interface as a force density. Equation 13.4 shows how the force density of the fluid, $f(x,t)$, may be calculated from the immersed boundary force density, $F(s,t)$. Equation 13.5 is essentially the no-slip condition at the interface, since the particle moves the same velocity as the neighboring fluid.

In the numerical implementation of the IBM, the entire fluid domain, including the parts that are occupied by immersed bodies, is divided into a set of fixed regular nodes. The fluid nodes are not moving with the flow and are Eulerian nodes. The immersed boundary of the particles is discretized by a group of interface boundary points that move under the action of the fluid. We will call these boundary nodes Lagrangian nodes. It must be noted that the Lagrangian nodes do not necessarily coincide with the Eulerian nodes. For this reason, the forces on the nodes of the Lagrangian grid are projected onto the nodes of the Eulerian grid using Equation 13.5.

The force density at the Lagrangian nodes that represent the fluid–particle interfaces may be written as follows:

$$\mathbf{f} = \rho\left(\frac{\partial \mathbf{u}}{\partial t} + \mathbf{u}\cdot\nabla\mathbf{u}\right) - \mu\nabla^2\mathbf{u} + \nabla p \tag{13.6}$$

With the known velocity and pressure fields at the time step $t = t_n$, we have an explicit scheme to determine the force term at these Lagrangian boundary points at time $t = t_{n+1}$, which is

$$f_i^{(n+1)} = \rho\left(\frac{u_i^{(n+1)} - u_i^{(n)}}{\Delta t} + u_j^{(n)}u_{i,j}^{(n)}\right) - \mu u_{i,jj}^{(n)} + p_{,i}^{(n)} \tag{13.7}$$

In the last equation, the Einstein notations for subscripts and comma notation for the derivatives are used.

In order to impose the BC that at $t = t_{n+1}$, the velocity on the immersed Lagrangian boundary points is equal to the calculated velocity of the particle at the same point, $U_i^{P(n+1)}$, the density force at these points is given as

$$f_i^{(n+1)} = \rho\left(\frac{U_i^{P(n+1)} - u_i^{(n)}}{\Delta t} + u_j^{(n)}u_{i,j}^{(n)}\right) - \mu u_{i,jj}^{(n)} + p_{,i}^{(n)} \tag{13.8}$$

The last equation is often called *direct forcing*, since it may be used to evaluate the force density at the Lagrangian boundary points without introducing any predefined parameters. Mohd-Yusof [12] and Fadlun et al. [16] computed this force at the Eulerian nodes. In the case of particulate flows where the particle interfaces are constantly moving, the task to directly compute these forces at the Eulerian nodes becomes formidable, and for this reason, Feng and Michaelides introduced the spreading of the calculated force density from the Lagrangian points to the neighboring Eulerian points [14]. The projection of the force from the Lagrangian to the Eulerian points, using Equation 13.4, speeds the computations significantly.

The spreading techniques are accurate to the first order. This is the same order of accuracy as in most of the numerical techniques that treat interfaces including the *bounce-back rule* used in the LBM. Our experience has shown that the method provides accurate and reliable results for particulate flows.

Figure 13.1 is a schematic diagram of the main steps of the IBM: Part (a) shows the boundary of an ellipsoidal particle, which is carried by the fluid, within the Eulerian grid where the fluid computations take place. Part (b) shows the discretization of the fluid–particle interface using a fixed number of points. If the interface is deformable, the interfacial points are connected by springs with finite spring constant (see [13]). If the interface is rigid, the spring constant is set to be very large. Part (c) depicts the substitution of the ellipsoidal particle surface and its effect on the fluid domain by a system of forces acting on the nodes of the Lagrangian grid. Part (d) is the final step in the interface modeling and shows the spreading of the forces from the Lagrangian points that move with the particle to the nodes of the Eulerian grid, which are stationary.

For the solution of the governing equations in the Eulerian grid nodes, the velocities at the Lagrangian points at the current time step $t = t_n$ may be computed using Equation 13.5. Alternatively, one may employ a bilinear interpolation using the velocity values of four neighboring grid points for 2D flows or eight neighboring grid points for 3D flows. This scheme was adopted in the *Proteus* code. The calculated force density using Equation 13.8 is at a Lagrangian boundary point. For this reason, it needs to be spread into the neighboring Eulerian nodes, where the computations take place, using a spreading function. Goldstein et al. explained this spreading as smoothing the boundary surface and used a Gaussian function to smoothen the boundary within one grid spacing [17]. In the *Proteus* code, a delta function was used to spread the force density to the nearby Eulerian nodes. The spreading procedure that was employed may be summarized as follows in two dimensions.

Two-dimensional particle boundary, Γ, is considered, as shown in Figure 13.2. A small area, ε, around a Lagrangian boundary point s_i, is also considered. The latter is the coordinate corresponding

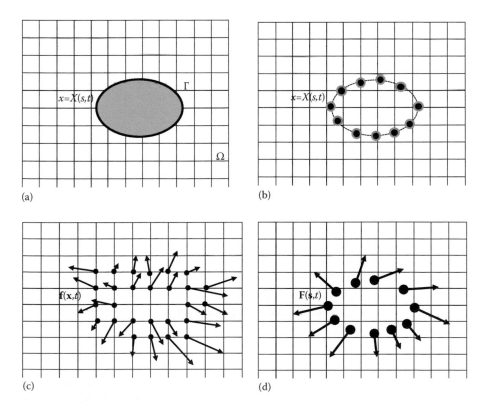

(a)
(b)
(c)
(d)

FIGURE 13.1 The four main steps in the implementation of the immersed boundary method. Clockwise from top left: (a) the Eulerian grid and the (ellipsoidal) particle in the Lagrangian domain, (b) the discretization of the Lagrangian particle surface, (c) the substitution of the Lagrangian interface with a system of forces, and (d) the spreading of the forces in the Eulerian grid.

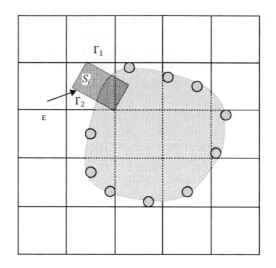

FIGURE 13.2 Schematic diagram for the force spreading.

to a particular Lagrangian boundary node *i*. By integrating the force density within the small area, we obtain the following expression:

$$\int_{\varepsilon} \mathbf{f}\big(\mathbf{X}(\mathbf{s},t)\big)dA = \int_{\Gamma} \mathbf{F}(\mathbf{s},t)\left(\int_{\varepsilon} \delta\big(\mathbf{x}-\mathbf{X}(\mathbf{s},t)\big)dA\right)d\mathbf{s} = \int_{\Gamma_{\varepsilon}} \mathbf{F}(\mathbf{s},t)d\mathbf{s} \qquad (13.9)$$

The relation between the flow force density, $\mathbf{f}(\mathbf{X}(\mathbf{s},t))$, and the surface force density, $\mathbf{F}(\mathbf{s},t)$, of Equation 13.4 was used to derive Equation 13.9. The latter implies that the flow force density integral for a small area ε is equal to the boundary force density integral over the boundary element, Γ_{ε}, which is the intersection of this small area and the whole particle boundary Γ as shown in Figure 13.2 [14].

We stipulate that the force of the flow on the particle is distributed in a small band along the particle boundary, Γ, with a force density given by the function $\mathbf{f}(\mathbf{X}(\mathbf{s},t))$. In the numerical implementation of the IBM, we consider a small area of this band represented by a Lagrangian boundary node $\mathbf{s}_i$. Then, the flow force density in this area is approximated by its value at this boundary node, $\mathbf{f}(\mathbf{X}(\mathbf{s}_i,t))$. For a uniform grid with grid spacing $\delta x = \delta y = \delta$, it is reasonable to assume that this *area of influence* represented by the node $\mathbf{s}_i$ is $\delta_s\delta$ (where δ_s is the length of the small boundary element Γ_e). Then, the force acting on this area is $\mathbf{f}(\mathbf{X}(\mathbf{s}_i,t))\delta_s\delta$ and the integral of the surface force density over the small area may be approximated as

$$\int_{\Gamma_{\varepsilon}} \mathbf{F}(\mathbf{s},t)d\mathbf{s} \approx \mathbf{f}\big(\mathbf{X}(\mathbf{s}_i,t)\big)\delta_s\delta \qquad (13.10)$$

For the 3D case, the area around a Lagrangian point is substituted by a small volume around this point. Hence, the line integral is replaced by a surface integral over a small boundary surface area intersected by the small volume. Also, the line distance δ_s is replaced by δA, the area of this boundary surface element.

Using Equation 13.10 as an approximation for the surface force density integral, the flow force density at each Eulerian node is determined from Equation 13.5. This is the same procedure for the force spreading as if a delta function were used for the spreading of the force. Equation 13.10 also provides an approach for the computation of the force acting on the particles, since the total force acting on a particle is equal to the sum of the forces acting on each surface element.

An alternative method to compute the force acting on the particles is by integrating the fluid stress over the particle surface:

$$F_i^{P(n)} = \oint_{\Gamma} \sigma_{ij}^{(n)} n_j dS = \oint_{\Gamma} \left(-p^{(n)} \delta_{ij} + \mu \left(u_{i,j}^{(n)} + u_{j,i}^{(n)} \right) \right) n_j dS \qquad (13.11)$$

When the DF method is also implemented, the surface integral is taken only over the external surface. Feng and Michaelides determined that there is a negligible difference between the computed values by using the two different approaches [14].

One of the drawbacks of using a delta function to spread the force field emanating from an interface is that the distributed forces cover an area that is larger than the surface area of the particle, as it may be seen in Figure 13.1. Consequently, the *size* of the particle that influences the flow field is larger and this may introduce errors in the interfacial transport of mass, momentum, and energy. This is due to the fact that the density force computed from Equation 13.8 at each boundary node is a body force. To achieve better accuracy, the boundary node has to be placed beneath the particle surface (e.g., at 0.5 grid step when a very fine grid is used) rather than on the particle surface. Yu and Shao [18] and Breugem [19] suggested corrections for this numerical diffusion of surfaces by applying the set of boundary forces slightly inward from the actual surfaces. They both suggested that the boundary forces be applied at distances that are approximately one-third of the Eulerian grid size and inward from the actual surface. Deen et al. [20] suggested calculations using the extrapolation of the variables of the Eulerian grid inside the boundary of the particle in a way that is similar to the *ghost cell* method that have been used elsewhere.

13.3 NUMERICAL IMPLEMENTATION OF THE IBM IN 3D ISOTHERMAL FLOWS

The first issue arising in the implementation of the IBM is how to set up the Lagrangian boundary points in order to accurately represent the particle surface. For a spherical particle in three dimensions, it is impossible to find evenly distributed boundary nodes that represent the surface of the sphere. Feng and Michaelides [14] used a number of strips, with the width of a strip being comparable to the grid spacing. Each strip is composed of a number of evenly distributed points. The number of the points in each strip is chosen in a way that the spacing between two neighboring points is approximately equal to the width of the strips.

The forces are first computed at the Lagrangian boundary nodes and then are spread to the Eulerian nodes, where the computations for the fluid take place. The force corresponding to each Lagrangian boundary node is spread into all its neighboring Eulerian nodes. The choice of the spreading function must meet certain criteria, which were outlined by Peskin [21]. For the 3D simulations, the following equations are used:

$$\delta(r) = \begin{cases} \dfrac{1}{4} \left(1 + \cos\left(\dfrac{\pi |r|}{2} \right) \right), & |r| \le 2 \\ 0, & |r| > 2 \end{cases} \qquad (13.12)$$

and

$$D(\mathbf{x} - \mathbf{x}_{mn}) = \delta(x - x_{mn}) \delta(y - y_{mn}) \delta(z - z_{mn}) \qquad (13.13)$$

With the use of these expressions, the spreading occurs within a distance of two lattice units from each Lagrangian node. When the spreading process is performed for each Lagrangian boundary node, the force on each Eulerian node is the summation of all the contributions from the surrounding Lagrangian boundary nodes. The spreading process can be stated succinctly as follows: For a system of N number of particles, with M Lagrangian boundary points used to represent each particle surface, the flow force density at an Eulerian node $\mathbf{x}_i$ is computed as follows:

$$\mathbf{f}\left(\mathbf{x}_i, t\right) = \sum_{n=1}^{N} \sum_{m=1}^{M} \mathbf{f}\left(\mathbf{x}_{mn}\left(t\right), t\right) \delta A_{mn} D\left(\mathbf{x}_i - \mathbf{x}_{mn}\left(t\right)\right) \tag{13.14}$$

where

$\quad$ $\mathbf{x}_{mn}$ is the position of the mth Lagrangian boundary point for the nth particle at the time t
$\quad$ δA_{mn} is the surface represented by boundary point, $\mathbf{x}_{mn}$; and the function $D(\mathbf{r})$ is given by Equation 13.13

During the simulations of the particle motion, it is necessary to compute the total force and torque acting on each particle. The force on a particle includes forces such as the gravity/buoyancy force as well as the particle collision forces, $\mathbf{F}_i^{col}$, as well as the hydrodynamic force exerted by the fluid. Hence, the total force exerted on the ith particle is

$$\mathbf{F}_i = \left(1 - \frac{\rho_f}{\rho_s}\right) M_i \mathbf{g} + \oint_{S_i} \sigma \cdot \mathbf{n} dS + \mathbf{F}_i^{col} \tag{13.15}$$

where M_i is the mass of the ith particle. The torque on a particle is computed by using the following expression:

$$\mathbf{T}_i = \oint_{S_i} \left(\mathbf{x} - \mathbf{x}_i\right) \times (\sigma \cdot \mathbf{n}) dS \tag{13.16}$$

In order to minimize the skewing effect (drift) due to the accumulation of the numerical error when one uses the rotation matrix of the particle, a unit quaternion is used to represent the rotation of the particles. In the 3D implementation, *Proteus* used a Runge–Kutta scheme to solve the following set of differential equations:

$$\frac{d}{dt} \begin{bmatrix} \mathbf{x}(t) \\ \mathbf{q}(t) \\ \mathbf{P}(t) \\ \mathbf{L}(t) \end{bmatrix}_i = \begin{bmatrix} \mathbf{v}(t) \\ \frac{1}{2}\varpi(t)\mathbf{q}(t) \\ \mathbf{F}(t) \\ \mathbf{T}(t) \end{bmatrix}_i \tag{13.17}$$

with initial conditions

$$\begin{bmatrix} \mathbf{x}(0) \\ \mathbf{q}(0) \\ \mathbf{P}(0) \\ \mathbf{L}(0) \end{bmatrix}_i = \begin{bmatrix} \mathbf{x}_{i,0} \\ \mathbf{q}_{i,0} \\ m_i \mathbf{v}_{i,0} \\ I_i \varpi_{i,0} \end{bmatrix} \tag{13.18}$$

where

> $\mathbf{q}_i(t)$ is the unit quaternion form for the rotation matrix of ith particle. The initial value is deter-
>   mined from the initial rotation matrix
> $\mathbf{P}_i(t)$ is the linear momentum of the ith particle
> $\mathbf{L}_i(t)$ is the angular momentum of the ith particle

At any given time, the translational and rotational velocities of particles are computed by using the following expressions:

$$\mathbf{v}_i(t) = \frac{\mathbf{P}_i(t)}{M_i}; \quad \boldsymbol{\omega}_i(t) = \mathbf{I}_i(t)^{-1}\mathbf{L}_i(t) \tag{13.19}$$

where $\mathbf{I}_i(t)$ is the inertia matrix of the ith particle in the current coordinate system. This is related to the moment of inertia matrix $\mathbf{I}_{i0}$ of the initial coordinate system of the ith particle and the rotation matrix of the ith particle, $\mathbf{R}_i(t)$, by the expression

$$\mathbf{I}_i(t) = \mathbf{R}_i(t)\mathbf{I}_{i0}\mathbf{R}_i(t)^T \tag{13.20}$$

This numerical scheme is capable to simulate the motion and rotation not only of spherical particles but of particles with irregular shapes. In the special case of spherical particles, $\mathbf{I}_i(t)$ is a constant diagonal matrix

$$\mathbf{I}_i = \begin{bmatrix} \frac{2}{5}M_i r_i^2 & & \\ & \frac{2}{5}M_i r_i^2 & \\ & & \frac{2}{5}M_i r_i^2 \end{bmatrix} \tag{13.21}$$

where r_i is the radius of ith particle.

It must be noted that in several DNS techniques, for example, the LBM, the rigid body motion inside the particle is not enforced *a priori*. The problem that is actually solved with these techniques is the interaction between a fluid and a solid shell filled with the same fluid. The shell has the same boundary as the particle and carries the total mass of the particle. In general, the contribution from the internal flow to the resulting motion of the particle is not significant for low-frequency flows [22]. When the IBM is used with the DF scheme, the particles are treated as shells filled with the same fluid. However, since the spreading of the force at the boundary nodes is not limited to the flow close to the boundaries of the particles, the internal flow is allowed to develop. To remedy this deficiency, one may enforce the rigid body motion for the internal fluid by setting up a few internal Lagrangian points. The velocity at these points is determined by the rigid body motion of the particle. Therefore, one of the advantages of *Proteus* is that it enforces the rigid body motion in the interior of the particles in a straightforward way. In [23], Feng and Michaelides suggested improvements to this numerical method by using a more robust method for the implementation of velocity updating. Wu and Shu [24] also suggested an alternative way to enforce the no-slip condition at interfaces by transferring the force density function to the unknown velocity correction. The latter is determined by enforcing the nonslip BC at the interface.

13.4   IMPLEMENTATION OF THE IBM WITH HEAT TRANSFER

The typical process to conduct numerical studies in nonisothermal flows in the past was to solve independently the energy equation, using the calculated velocity data from the Navier–Stokes equations. Yu et al. [25] changed this practice for DNS by extending the distributed-Lagrange-multiplier/fictitious-domain method, which was developed by Glowinski et al. [26], to include heat transfer calculations in 2D particulate flow systems. Thus, they solved simultaneously the fluid mass, momentum, and energy-governing equations.

A glance at the implementation of the IBM proves that the method is ideally suited for the simultaneous solution of multiphase flow with mass and energy transfer. While for the momentum exchange, the boundaries of particles are substituted with a system of forces, for the energy interactions, the boundaries of particles may be substituted by a system of heat sources or heat sinks. Feng and Michaelides recognized this analogy and extended the IBM from isothermal systems to systems that include heat transfer in a straightforward way [23,27]. The presence of particles at different temperatures is equivalent to a system of heat sources (for hotter particles) or heat sinks (for colder particles). Hence, in analogy to the momentum interactions, the temperature field in the Eulerian domain of the fluid is governed by the following modified energy equation:

$$\rho_f c_f \frac{\partial T}{\partial t} + \rho_f c_f \vec{u} \cdot \nabla T = k_f \nabla^2 T + q_p + \lambda, \quad \vec{x} \in \Omega \tag{13.22}$$

where

c_f and k_f are the specific heat and the thermal conductivity of the fluid
q_p represents any energy sources inside the particle
The additional term, λ, is an energy density term that is added to enforce the temperature field in the region occupied by the particles

As it happens with the force balance equations, this term is zero in the field occupied by the fluid alone. In the solid region, this term is computed by the equation

$$\lambda = \rho_f c_f \frac{\partial T}{\partial t} + \rho_f c_f \vec{u} \cdot \nabla T - k_f \nabla^2 T - q_p, \quad \vec{x} \in \sum \Gamma_i \tag{13.23}$$

where $T(t)$ is the temperature of each particle. In the original publications, the temperature of each particle was assumed to be uniform, that is, $T = T_p(t)$, which essentially implies that the Biot number vanishes ($\text{Bi} \ll 1$). This assumption may be relaxed by solving the energy equation inside the particles to obtain the temperature field, $T_p(t,x_i)$.

The transient temperature of the particles, $T_p(t)$, is determined by solving the following differential equation, which is obtained from the energy balance for each particle:

$$\rho_p V_p c_p \frac{dT_p}{dt} = \oint_{\partial s} k_f \vec{\nabla} T_f \cdot \vec{n} ds + \int_s q_p dv \tag{13.24}$$

where $\vec{n}$ is the outward normal vector.

Whenever particles have different temperatures than the fluid, the heat sources of sinks create a temperature gradient within the fluid, which would modify the fluid properties. This effect may be computed in a conceptually easy manner by using empirical equations for the properties of the fluid at every point of the Eulerian grid. For relatively small temperature differences between the particles and fluid, the Boussinesq approximation has been often used successfully for the coupling of the energy and momentum equations. Hence, the momentum Equation 13.6 is rewritten as follows:

$$\rho_{f0} \frac{\partial \vec{u}}{\partial t} + \rho_{f0} \vec{u} \cdot \nabla \vec{u} = -\vec{\nabla} p + \mu_f \nabla^2 \vec{u} + \beta_f \left(T - T_{f0} \right) \vec{g} + \vec{f}, \quad \vec{x} \in \Omega \tag{13.25}$$

where

the subscript 0 denotes values far from the particles

β_f is the thermal expansion coefficient of the fluid

Deen and Kuipers [28] extended the IBM method to study the coupled mass and heat transfer in dense particulate systems and connected the results of the study to the operation of a simple 1D heterogeneous reactor model. The computational results based on this method agree well with the results from a 1D model, which is based on empirical correlations for the fluid–particle mass and heat transfer coefficients.

13.5 PARTICLE–PARTICLE AND PARTICLE–WALL INTERACTIONS

Because it treats the fluid–particle interfaces in a computationally efficient manner, the IBM is ideally suited for studies of large groups of particles, particle clusters, and particle aggregates. In such studies, interparticle collisions and collisions between particles and flow boundaries are unavoidable and need to be modeled appropriately. In general, the Eulerian grid used in a DNS study is not fine enough to handle the lubrication force that develops between the particles or between particles and a solid boundary, when the boundaries almost touch. Therefore, a mechanism is necessary to be introduced in the IBM numerical scheme in order to account for the repulsive forces that develop before and during the collision processes. In the absence of such a mechanism, the motion of the particles will result in the partial penetration of the surfaces. When several such particle boundary penetrations occur simultaneously, the computational results may become meaningless.

Glowinski et al. suggested a collision technique by introducing a repulsive force when the gap between two particles is lower than a given threshold, the so-called safe zone [26]. This artificial short-range repulsive force is added as an external force in the IBM. Thus, when two particles with radii R_i and R_j approach very closely, the collision force exerted on each particle is

$$\mathbf{F}_{ij}^P = \begin{cases} 0, & \left\| \mathbf{x}_i - \mathbf{x}_j \right\| > R_i + R_j + \zeta \\ \dfrac{c_{ij}}{\varepsilon_P} \left(\dfrac{\left\| \mathbf{x}_i - \mathbf{x}_j \right\| - R_i - R_j - \zeta}{\zeta} \right)^2 \left(\dfrac{\mathbf{x}_i - \mathbf{x}_j}{\left\| \mathbf{x}_i - \mathbf{x}_j \right\|} \right), & \left\| \mathbf{x}_i - \mathbf{x}_j \right\| \le R_i + R_j + \zeta \end{cases} \tag{13.26}$$

where

$\mathbf{x}_i$ is the position of the particle i

the parameter, c_{ij}, is the force scale in the problem that is studied, for example, a good choice for this scale in sedimentation problems is the gravity/buoyancy force on a single particle

ε_P is the stiffness parameter for collisions

ζ is the threshold or *safe zone*, which is specified in advance

Glowinski et al. [26] provide an extensive discussion on the choice of the stiffness parameters. Similarly, the repulsive force between a particle and a wall is given by the reflection method assuming that an identical, fictitious particle approaches the wall from the other side with a normal velocity that is opposite to that of the particle under consideration. This collision technique allows particles to interpenetrate when the stiffness parameter ε_P is large. Because of this, it is said that the particles undergo *soft collisions*.

Several other soft collision schemes that are more physically meaningful have been proposed in the recent past, most of them emanating from granular flow applications (see [29]). The contact forces, which are composed of the normal and tangential forces, are evaluated from the small overlapping displacement between the particles and the relative velocities of the colliding particles. Using the linear spring-dashpot model, the normal contact force may be written as

$$F_{ij}{}^n = -k_n \delta_{ij}^n - \eta_n v_{ij}^n \tag{13.27}$$

where

δ_{ij}^n is the normal overlapping displacement

k_n and η_n are the normal spring stiffness and damping coefficient

v_{ij}^n is the relative velocity component in the normal direction of particle j with respect to particle i

Similarly, the tangential collision force is given by the expression

$$F_{ij}^t = -k_t \delta_{ij}^t - \eta_t v_{ij}^t \qquad (13.28)$$

where

δ_{ij}^t is the tangential overlapping displacement

k_t and η_t are the tangential spring stiffness and damping coefficient, respectively

v_{ij}^t is the relative tangential velocity component at the contact point

This velocity is computed as follows:

$$\vec{v}_{ij}^t = \vec{v}_{ij} - \left(\vec{v}_{ij} \cdot \hat{n}_{ij} \right) \hat{n}_{ij} + \left[\vec{\omega}_i \times r_i \vec{n}_{ij} - \vec{\omega}_j \times r_j \left(-\vec{n}_{ij} \right) \right] \qquad (13.29)$$

Feng et al. [30] applied this soft sphere collision method in the case of particle–wall collisions, which are special cases of the interparticle collisions when the radius of one of the particle approaches infinity. They assumed that the particles are homogeneous and isotropic, with $k_n = k_t = k$ and $\eta_n = \eta_t = \eta$.

In the soft sphere model, the effect of the friction contact between the colliding particles on the tangential force may be taken into account by the following expression:

$$f_{ij}^t = \begin{cases} -k\delta_{ij}^t - \eta v_{ij}^t, & \text{if} \quad \left| f_{ij}^t \right| \le \mu_s \left| f_{ij}^n \right| \\ \mu_k \left| f_{ij}^n \right| \dfrac{\delta_{ij}^t}{\left| \delta_{ij}^t \right|}, & \text{if} \quad \left| f_{ij}^t \right| > \mu_s \left| f_{ij}^n \right| \end{cases} \qquad (13.30)$$

where μ_s and μ_k are the coefficients of static and dynamic friction. Most studies do not differentiate between μ_s and μ_k and use a single coefficient of friction, μ. The sliding distance or tangential displacement can be computed by integrating the relative tangential velocity during the contact time of the particles.

The normal spring stiffness for dry collisions has been well studied both theoretically, for example, [29], and experimentally, for example, [31]. In general, it is determined by the bulk material properties of the particles. The range of the spring stiffness used in gas–solid flows is from 200,000 dyn/cm (200 N/m) to 50,000,000 dyn/cm (50,000 N/m). Using smaller values for the spring stiffness allows a larger time step to be used in the simulation. However, such weaker springs may cause significant particle–particle overlapping, which is physically unrealistic. The range of the damping coefficient, η, is between 0 and 100 dyn*s/m (0.1 N*s/m), and typical values of the coefficient of friction are close to $\mu = 0.3$ for particulate flows in gases [30].

13.6 EXAMPLES OF THE APPLICATION OF IBM

The IBM is an accurate method to perform numerical calculations with single solid and deformable particles or with ensembles of particles. The method yields reliable and accurate local and averaged information for all types of dispersed flows. We provide here two examples on two types of problems the IBM may be used to provide needed data for engineering and scientific applications. The first pertains to an isothermal application and the second to a fundamental problem in heat transfer.

13.6.1 Determination of the Boundary Conditions for Two-Fluid Models

Practicing engineers working with typical industrial applications, such as the operation of fluidized bed reactors, boilers, oil refinement, and chemical reactor columns, do not have the need for the detailed information the IBM (and all the other DNS methods) provide. For this reason, they prefer to use numerical codes and models that yield globally averaged properties, such as models based on Eulerian–Eulerian techniques (two-fluid models). The models treat the dispersed phases as separate continua (within the fluid continuum) that interact with the fluid. One of the requirements for these models is the *a priori* knowledge of the BCs of the dispersed phase. While the no-slip velocity BC for the fluid phase has been proven to be both convenient and accurate from a very large volume of experimental and analytical results, the condition is neither justified nor accurate to be used for the dispersed phase. Both physical experiments and numerical simulations show that there is finite tangential slip of the particulate phase at a boundary. The normal slip is always zero, because neither the particles nor the fluid are allowed to penetrate the boundary.

Physical experiments for the behavior of a dispersed phase close to a wall are expensive and difficult to perform. In addition, they are laden with high uncertainty, especially when the dispersed phase is not dilute. Since numerical methods, such as the IBM, may provide accurate data for the behavior of particles close to boundaries, for both dilute and dense flow regimes, the IBM would be ideally suited to provide this type of information for the two-fluid models. Davis et al. [32] performed such a numerical study and determined the trajectories and interactions of a large ensemble of particles in the vicinity of a vertical wall. They determined the tangential velocity slip (relative tangential velocity) of the spherical particles when they are in contact with the vertical wall. The results of this study are given in Figure 13.3 as the dimensionless tangential slip velocity at the wall versus a Reynolds number, which is defined in terms of the terminal velocity of the spheres, when they fall in a stagnant fluid. It is evident that this type of information that emanates from the implementation of the IBM will significantly improve the accuracy of two-fluid models.

13.6.2 Natural Convection from Particles and Aggregates

An accurate numerical method, such as the IBM, may also be used to solve fundamental problems. One such problem is the natural convection from single particles of irregular shapes and aggregates of particles. Natural convection from single spheres has been studied analytically, experimentally and numerically by several researchers (e.g., [33–35]). Novel numerical methods enable us to determine the heat transfer, not only from a single sphere but from particles of any shape as well as

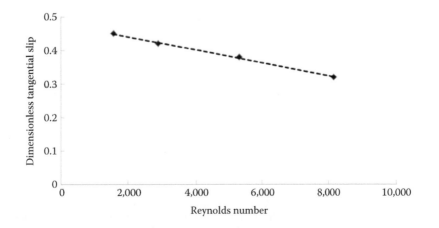

FIGURE 13.3 Dimensionless tangential slip velocity of a sphere at a plane boundary.

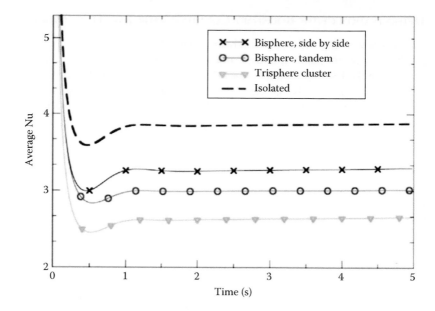

FIGURE 13.4   Average Nu for spherical clusters in contact. Gr = 100 and Pr = 0.72.

particle aggregates. As it is apparent from Section IV, the IBM is an especially suitable method to study problems associated with heat transfer, because the method allows the simultaneous solution of the momentum and energy equations. Feng et al. [36] used the IBM to determine the natural convection heat transfer coefficients from single isolated spheres; bispheres, in two orientations—side by side—and in tandem; and clusters of three touching spheres. The latter are oriented in such a way that their centers form an equilateral triangle in a vertical plane with a horizontal base. Some of the results of this study are summarized in Figure 13.4, which depicts the evolution of the average Nusselt number versus time for Prandtl number equal to 0.72 and Grashof number equal to 100. The average Nusselt number represents the heat transfer per sphere in the aggregate. It is observed in this figure that (a) the steady natural convection is established relatively fast and (b), as the spheres aggregate, the heat transfer per sphere decreases.

ACKNOWLEDGMENTS

This work was partly supported by the Tex Moncrief Chair of Engineering at Texas Christian University, by a grant from the U.S. Department of Energy (award number DE-FE0011453), with Mr. Steven Seachman as the project administrator, and by the UTSA/SwRI CONNECT grant.

REFERENCES

1. Brady, J. F. and Bossis, G., 1988, Stokesian dynamics, *Ann. Rev. Fluid. Mech.*, 20, 111–157.
2. Kalthoff, W., Schwarzer, S., and Herrmann, H. J., 1996, An algorithm for the simulation of particulate suspensions with inertia effects, *Phys. Rev. E*, 56, 2234–2242.
3. Zhang, Z. and Prosperetti, A., 2005, A second-order method for three-dimensional particle simulation, *J. Comput. Phys.*, 210, 292–324.
4. Ladd, A. J. C., 1994, Numerical simulations of particulate suspensions via a discretized Boltzmann equation Part I. Theoretical foundation, *J. Fluid Mech.*, 271, 285–310.
5. Ladd, A. J. C., 1994, Numerical simulations of particulate suspensions via a discretized Boltzmann equation Part II. Numerical results, *J. Fluid Mech.*, 271, 311–339.
6. Aidun, C.K., Lu, Y.-N., and Ding, E.-J., 1998, Direct analysis of particulate suspensions with inertia using the discrete Boltzmann equation, *J. Fluid Mech.*, 373, 287.

7. Feng, Z.-G. and Michaelides, E. E., 2002, Inter-particle forces and lift on a particle attached to a solid boundary in suspension flow, *Phys. Fluids*, 14, 49–60.
8. Michaelides, E. E., 2003 Hydrodynamic force and heat/mass transfer from particles, bubbles and drops—The Freeman scholar lecture, *J. Fluids Eng.*, 125, 209–238.
9. Ernst, M, Dietzel, M., and Sommerfeld, M., 2013, A lattice Boltzmann method for simulating transport and agglomeration of resolved particles, *Acta Mech.*, 224, 2425–2449.
10. Peskin, C.S., 1977, Numerical analysis of blood flow in the heart, *J. Comput. Phys.*, 25, 220–252.
11. Fogelson, A.L. and Peskin, C.S., 1988, A fast numerical method for solving the three-dimensional stokes equation in the presence of suspended particles, *J. Comput. Phys.*, 79, 50–69.
12. Mohd-Yusof, J., 1997, *Combined Immersed Boundaries/B-Spline Methods for Simulations of Flows in Complex Geometries, Annual Research Briefs*, Center for Turbulence Research, Stanford University, Stanford, CA.
13. Feng, Z.-G. and Michaelides, E. E., 2004, An immersed boundary method combined with Lattice Boltzmann method for solving fluid and particles interaction problems, *J. Comput. Phys.*, 195, 602–628.
14. Feng, Z.-G. and Michaelides, E.E., 2005, Proteus—A direct forcing method in the simulation of particulate flows, *J. Comput. Phys.*, 202, 20–51.
15. Uhlmann, M., 2005, An immersed boundary method with direct forcing for the simulation of particulate flows, *J. Comput. Phys.*, 209, 448–476.
16. Fadlun E.A., Verisco R., Orlandi, P., and Mohd-Yusof J., 2000, Combined immersed boundary finite-difference methods for three-dimensional complex flow simulations, *J. Comput. Phys.*, 161, 35–60.
17. Goldstein, D., Handler, R., and Sirovich, L., 1993, Modeling a no-slip flow boundary with an external force field, *J. Comput. Phys.*, 105, 354–366.
18. Yu, Z. and Shao, X., 2007, A direct-forcing fictitious domain method for particulate flows, *J. Comput. Phys.*, 227, 292–314.
19. Breugem, W.P., 2012, A second-order accurate immersed boundary method for fully resolved simulations of particle-laden flows, *J. Comput. Phys.*, 231, 4469–4498.
20. Deen, N.G., Kriebitzsch, S.H.L., van der Hoef, M.A., and Kuipers, J.A.M., 2012, Direct numerical simulation of flow and heat transfer in dense fluid–particle systems, *Chem. Eng. Sci.*, 81, 329–344.
21. Peskin, C.S., 2002, The immersed boundary method, *Acta Numerica*, 11, 479–517.
22. Ladd, A.J.C. and Verberg, R., 2001, Lattice-Boltzmann simulations of particle-fluid suspensions, *J. Statist. Phys.*, 104, 1191–1251.
23. Feng, Z.-G. and Michaelides, E. E., 2009, Heat transfer in particulate flows with direct numerical simulation (DNS), *Int. J. Heat Mass Transf.*, 52(3–4), 777–786.
24. Wu, J. and Shu, C., 2010, Particulate flow simulation via a boundary condition-enforced immersed boundary-Lattice Boltzmann scheme, *Commun. Comput. Phys.*, 7, 793–812.
25. Yu, Z., Shao, X., and Wachs, A., 2006, A fictitious domain method for particulate flows with heat transfer, *J. Comput. Phys.*, 217, 424–452.
26. Glowinski, R., Pan, T.-W., Hesla, T.I., Joseph, D.D., and Periaux, J., 2001, A fictitious domain approach to the direct numerical simulation of incompressible viscous flow past moving rigid bodies: Application to particulate flow, *J. Comput. Phys.*, 169, 363–426.
27. Feng, Z.-G. and Michaelides, E. E., 2008, Inclusion of heat transfer computations for particle laden flows. *Phys. Fluids*, 20,675–684.
28. Deen N.G. and Kuipers J.A.M., 2014, Direct numerical simulation of fluid flow accompanied by coupled mass and heat transfer in dense fluid–particle systems, *Chem. Eng. Sci.*, 116, 6, 645–656.
29. Johnson, K.L., 1985, *Contact Mechanics*, Cambridge University Press, Cambridge, U.K.
30. Feng, Z.G., Michaelides, E.E., and Mao, S.L., 2010, A three-dimensional resolved discrete particle method for studying particle-wall collision in a viscous fluid, *J. Fluids Eng.*, 132, 967–973.
31. Mullier, M.U., Tüzün, U., and Walton, O.R., 1991, A single-particle friction cell for measuring contact frictional properties of granular materials, *Powder Technol.*, 65, 61–74.
32. Davis, A.P., Michaelides, E.E., and Feng, Z.-G., 2012, Particle velocity near vertical boundaries—A source of uncertainty in two-fluid models, *Powder Technol.*, 220, 15–23.
33. Mahony, J. J., 1956, Heat transfer at small Grashof Number, *Proc. R. Soc. Lond.*, A238, 412–423.
34. Hossain, M.A. and Gebhart, B., 1970, Natural convection about a sphere at low Grashof number, in *Proceedings of the Fourth International Heat Transfer Conference*, Vol. 5, Versailles, France.
35. Riley, N., 1986, The heat transfer from a sphere in free convective flow, *Comput. Fluids*, 14(3), 225–237.
36. Feng, Z.G., Musong, S.G., and Michaelides, E.E., 2016, A three-dimensional immersed boundary method for free convection from single spheres and aggregates, *J. Fluids Eng.*, in print.

14 Mesoscale Lattice Boltzmann Model of Dispersions Stabilized by Surfactants

Ruud G.M. Van der Sman and Marcel B.J. Meinders

CONTENTS

14.1 INTRODUCTION

In the past decade, the lattice Boltzmann (LB) method has evolved as the simulation technique of choice for complex fluids and soft matter [2,3,13,16,18,19,22,24,27]. Consequently, we have been developing LB models for complex food materials, which are often complex fluids or soft matter [14,15,17,31,32,35–37,41,42]. LB is a very versatile method, able to model both macroscopic phenomena, like hydrodynamics [28] and convection-diffusion phenomena [33,34,38], as mesoscopic phenomena occurring in complex fluids and soft matter. Essentially, the LB method solves the classical Boltzmann equation, but where space, time, *and* particle velocity are discretized. If one adheres to the conservation laws of mass, momentum, and energy, and if the lattice has sufficient symmetry, the LB method would produce correct physical behavior, as classically described by the Navier–Stokes equation and Fick's law. This is not surprising, taking into account that the Boltzmann equation is fundamental to these classical laws. Next to these macroscopic phenomena, the Boltzmann equation connects the LB method also to the microscopic scale and thermodynamics. Henceforth, it has become the natural vehicle to simulate complex fluids and soft matter, whose behavior has both macroscopic as microscopic traits. These materials consist largely of a fluid continuum, with microscopic elements dispersed into it, such as immiscible droplets, bubbles, or solid particles.

From a numerical point of view, the LB method is very simple [28]. For almost all calculations, the information is locally available. Communication with neighboring lattice sites is only during the propagation step, where particles move to their next sites according to their discrete velocity. These properties make the LB method very efficient in parallel computing, which is often required due to the often large difference in length scales present in the addressed physical problems concerning complex fluids or soft matter. Because of the numerical simplicity of the method, modelers can focus on the complexity of the physics of their problems.

In this chapter, we will particularly focus on the physical problem of dispersions of immiscible fluids, such as emulsions or foams that are stabilized by surfactants. In these models, we make use of the connection of the LB method to both the macroscopic and microscopic scales. At the macroscale, these dispersions are subject to hydrodynamics, as they occur in the processing of food materials. With thermodynamics, we describe the microscopic phenomena happening at the scale of the interface between the immiscible fluids, where the surfactants are preferentially absorbed. The thermodynamics of these complex fluids is described using a free energy functional, which includes surface free energy terms [39]. In the free energy–based LB model, the interface is thought to be have a finite width, in the spirit of van der Waals, Ginzburg-Landau, and Cahn-Hilliard [29]. Similar *diffuse* interface models have been developed in material science for binary alloys, which are known as the phase field method [43].

In the next section, we will explain the essentials of the LB method—and how it can be related to the thermodynamic description of a free energy functional. We will give the generalized Navier–Stokes and convection-diffusion equations. The section will be concluded with a brief description of boundary conditions.

Subsequently, we focus on the free energy functional that should describe the desired physics of the complex fluid. We will increase the complexity step by step, starting with a binary immiscible fluid, without surfactants. After introducing nonionic surfactants, we discuss the extension of the free energy functional toward ionic surfactants.

After the presentation of the theory, we present some results obtained for emulsion droplets, and thin films between foam bubbles. The model for ionic surfactants is still under development, but we show some promising early results for the situation of a water film having two parallel interfaces. We conclude with a short section with an outlook toward further extensions of the free energy functional.

14.2   LATTICE BOLTZMANN METHOD

The LB method has the particle distribution functions $f_i(\mathbf{x}, t)$ as a state variable. $f_i(\mathbf{x}, t)$ is the number density of particles at lattice site $\mathbf{x}$ moving with velocity $\mathbf{c}_i$. Particles will collide with each other, and they will subsequently propagate to neighboring lattice site. $f_i(\mathbf{x}, t)$ will move to $\mathbf{x} + \mathbf{c}_i \Delta t$. The collision follows a relaxation toward the equilibrium particle distribution $f_i^{eq}(\mathbf{x}, t)$, see the classical Boltzmann equation. The combined action of the collision and propagation step can be represented with the discrete Boltzmann equation:

$$f_i(\mathbf{x} + \mathbf{c}_i \Delta t, t + \Delta t) = f_i^{eq}(\mathbf{x}, t) + \sum_j \Lambda_{ij}[f_j(\mathbf{x}, t) - f_j^{eq}(\mathbf{x}, t)] \tag{14.1}$$

Λ is the so-called collision matrix, with the eigenvalues determining the transport coefficients such as the shear viscosity. The velocity moments of the equilibrium distributions are related to the macroscopic variables, that is, conserved quantities and the driving forces for transport [32]. For fluid flow, the following moments hold:

$$\sum_i f_i^{eq} = \rho$$

$$\sum_i f_i^{eq} c_{i,\alpha} = \rho u_\alpha \tag{14.2}$$

$$\sum_i f_i^{eq} c_{i,\alpha} c_{i,\beta} = P_{\alpha\beta} + \rho u_\alpha u_\beta$$

where

$\rho = \sum_i f_i$ is the mass density of the fluid

$\rho u_\alpha = \sum_i f_i c_{i,\alpha}$ is the momentum in the Cartesian direction α

$P_{\alpha\beta}$ is the Korteweg–deVries stress tensor, which is derived from a free energy functional in case of a complex fluid

For a simple fluid, the stress tensor is a diagonal tensor, with $P_{\alpha\beta} = p\delta_{\alpha\beta}$. The pressure p follows the ideal gas law: $p = \rho c_s^2$, with c_s the speed of sound. α and β indicate Cartesian directions, and $\delta_{\alpha\beta}$ is the Kronecker delta.

The phase field method uses an order parameter to distinguish the immiscible fluids, $\phi = \pm 1$. At the interfacial region, the order parameter changes continuously between the extremal values. The order parameter evolves according to a convection-diffusion equation, with diffusion driven by the chemical potential μ_ϕ, which is also derived from the free energy functional. For the order parameter, another particle distribution function is used: g_i. The governing physics is imposed by the velocity moments of the equilibrium distribution:

$$\sum_i g_i^{eq} = \phi$$

$$\sum_i g_i^{eq} c_{i,\alpha} = \phi u_\alpha \qquad (14.3)$$

$$\sum_i g_i^{eq} c_{i,\alpha} c_{i,\beta} = \Gamma_\phi \mu_\phi \delta_{\alpha\beta} + \phi u_\alpha u_\beta$$

The velocity field is obtained from the moment of the fluid particle distribution function f_i.

For the formulation of boundary conditions in the LB method, one makes use of the microscopic picture of moving particles [19]. The no-slip boundary condition for fluid flow, and the no-flux boundary condition for convection-diffusion, just follows bounce-back conditions. Particles are just reflected back in their opposite propagation direction. The boundary between fluid and wall is positioned just midway two lattice sites. For fluid flow, there is no simple equivalent of the bounce-back in traditional computation fluid dynamics (CFD) solvers. This makes the LB method very efficient in modeling fluid flow in complex geometries like porous media, and also in solid particle suspensions.

A similar order parameter exists for the surfactant ψ, which is modeled with the particle distribution function h_i. ψ is the volume fraction of the surfactant. Via application of the Chapman–Enskog expansion method, one can derive the governing equations [34,40]. They are the following:

$$\partial_t \phi + \partial_\alpha \phi u_\alpha = \partial_\alpha M_\phi \partial_\alpha \mu_\phi$$

$$\partial_t \psi + \partial_\alpha \psi u_\alpha = \partial_\alpha M_\psi \partial_\alpha \mu_\psi \qquad (14.4)$$

$$\partial_t \rho u_\alpha + \partial_\beta \rho u_\alpha u_\beta = -\partial_\beta P_{\alpha\beta} + \partial_\alpha \rho \nu (\partial_\alpha u_\beta + \partial_\beta u_\alpha)$$

where

ν is the viscosity

M_ϕ and M_ψ are mobilities of the order parameters

μ_ψ and μ_ϕ are chemical potentials

All the intricacies of the dispersed flow problem with surfactants will emerge from the specifics of the free energy density functional $F = F(\rho,\psi)$, which is discussed in the next section. The chemical potentials, and the trace of the pressure tensor p, follow from

$$\mu_\phi = \frac{\partial F}{\partial \phi}$$

$$\mu_\psi = \frac{\partial F}{\partial \psi} \qquad (14.5)$$

$$p = \phi\mu_\phi + \psi\mu_\psi - F$$

14.3   FREE ENERGY FUNCTIONALS

14.3.1   IMMISCIBLE FLUIDS

For a binary system of immiscible fluids, the free energy functional is given by Cahn–Hilliard [1,10,21]. It is a function of the order parameter only, and consists of a bulk free energy $F_{1,\phi}$ and a surface free energy $F_{0,\phi}$. The bulk free energy is the well-known double well potential:

$$F_{1,\phi} = A(\phi^2 - 1)^2 \qquad (14.6)$$

which has two minima at $\phi = \pm 1$, which are the values of the order parameter in the bulk phases of the two immiscible fluids.

The surface free energy has a squared gradient term in the spirit of van der Waals:

$$F_{0,\phi} = \kappa(\nabla\phi)^2 \qquad (14.7)$$

The surface free energy term makes it possible for the two immiscible phases to coexist, with an interfacial region where the order parameter continuously changes between the two extremal values. The interface is thought to be located at $\phi = 0$. For planar interfaces, the order parameter will follow a $\phi(x) = \tan h(x/\zeta)$. The width of the interface ζ follows from $\zeta^2 = 2\kappa/A$. Because of the smooth transition of the order parameter, within the finite interfacial region, the Cahn–Hilliard models have also been called *diffuse interface* models.

The tanh profile of the order parameter implies that the squared gradient term, $(\nabla\phi)^2$, is an approximation of a delta-function. In thermodynamics, delta-functions are used to combine surface free energy with bulk free energy into a single expression.

Because of the squared gradient term in the free energy functional, there will also be differential terms in the chemical potential μ_ψ and the Korteweg–deVries tensor $P_{\alpha\beta}$. These differential terms represent nonlocal interactions between the two immiscible fluids in the interfacial region.

14.3.2   IMMISCIBLE FLUIDS AND NONIONIC SURFACTANTS

For the inclusion of the contribution of surfactants to the free energy, we have built upon the functional for immiscible fluids. The surfactant contributions follow from the theory from Diamant and Andelman [6]. In their original theory, the surface free energy has been described by a sharp Dirac-delta function—making their theory a sharp interface model. We have transformed their theory into a diffuse interface model, via replacing the Dirac-delta function with a smooth approximation, like the squared gradient term: $\hat{\delta} = (\nabla\phi)^2$ [32].

The surfactant has both a bulk free energy contribution $F_{1,\psi}$ and a surface free energy contribution $F_{0,\psi}$. The bulk free energy contribution follows regular solution theory, and includes an extra term to model differential solubility of the surfactant for the two immiscible phases. The bulk free energy contribution is

$$F_{1,\psi} = \theta[\psi \ln(\phi) + (1-\psi)\ln(1-\psi)] + \beta\psi^2 + \gamma\psi\phi \tag{14.8}$$

The first two terms represent ideal mixing, with θ the reduced temperature. The third term is due to self-interaction of the surfactant [20]. The last term is the solvation energy of the surfactant, leading to differential solubility [39].

The surface free energy contribution is linear with the surfactant concentration:

$$F_{0,\psi} = \alpha\psi\hat{\delta} \tag{14.9}$$

with $\hat{\delta}$ as the approximation of the Dirac-delta function. Recent studies have shown that one can also take an analytical expression for $\hat{\delta}$, instead of the squared gradient term [7,39]. This greatly improves the numerical stability of the scheme. Examples of these analytical expressions are $\hat{\delta} = 1 - \phi^2$ or $\hat{\delta} = (1-\phi^2)^2$. The latter has identical shape as the squared gradient term, in case of a planar interface.

14.3.3 IMMISCIBLE FLUIDS AND IONIC SURFACTANTS

Thin films between bubbles or droplets are often stabilized via ionic surfactants. Subsequently, we have extended our model with the contribution of ionic surfactants and their counterions. Again, we can build upon the sharp interface model of Diamant and Andelman [5], which we have extended with the solvation energy for the ions, see Onuki [23]. In practice, the surfactants are anionic, and have cationic counterions. Hence, they are represented with the following order parameter ψ_-, and ψ_+. In the bulk phase, both the surfactant is expected to follow regular solution theory, but the counterions follow only ideal mixing. Their bulk free energy contributions is

$$\begin{aligned} F_{1,\psi_{\pm}} = \theta[\nu_+^{-1}\psi_+ \ln\psi_+ &+ \nu_-^{-1}\psi_- \ln\psi_- \\ &+ (1 - \psi_- - \psi_+)\ln(1 - \psi_- - \psi_+)] \\ &+ \beta\psi_-^2 + \gamma_+\psi_+ + \gamma_-\psi_- \end{aligned} \tag{14.10}$$

We have accounted for the difference in size of the anionic surfactant and counterions, via their ratio of molar volume with that of the solvent $\nu_{\pm}$.

The surface free energy is only due to the anionic surfactant:

$$F_{0,\psi,-} = \alpha\psi_-\hat{\delta} \tag{14.11}$$

with $\hat{\delta}$ one of the earlier mentioned approximations of the Dirac-delta function.

At the interfacial region, there will be an inhomogeneous distribution of charge, leading to the development of an electric field U, which also contributes to the free energy. There are two contributions: the electrostatic contribution of the ions F_q, and the self-energy of the electric field, F_U:

$$\begin{aligned} F_q &= \nu_+^{-1}q_+\psi_+ - \nu_-^{-1}q_-\psi_-]U \\ F_U &= \frac{1}{2}(\epsilon_0 + \phi\epsilon_1)(\nabla U)^2 \end{aligned} \tag{14.12}$$

where
  $q_{\pm}$ are the charges of the ions
  $\epsilon = \epsilon_0 + \phi\epsilon_1$ is the dielectric permittivity of the fluid, which is a function of the order parameter
    ϕ [23,26]

The free energy functional also needs to minimize with respect to the electric field. By doing so, one obtains the Poisson equation:

$$\left(\frac{q_+ \psi_+}{v_+} - \frac{q_- \psi_-}{v_-} \right) = \epsilon(\phi) \nabla^2 U \tag{14.13}$$

The Poisson equation has to be solved for every time step.

14.4 NUMERICAL ISSUES

To obtain a LB implementation of the phase field method, one needs to obtain the chemical potentials and Korteweg–deVries pressure tensor from the free energy functionals. These expressions can be inserted in the equilibrium distributions. Consequently, one obtains a thermodynamic model of a dispersion of immiscible fluids, with (anionic) surfactants—which can be driven by fluid flow, diffusion, and electric fields. For the latter, the Poisson equation needs to be solved. The LB method does not have a particular advantage for solving the Poisson equation, and hence, any efficient Poisson solver can be used.

Subsequent research has shown that the thermodynamic consistent LB method is not the most efficient model from a numerical point of view. The efficiency and also the numerical stability can be improved if one takes a nonvariational method, see [8]. Here, one relaxes the constraint of thermodynamic consistency, and still obtain the desired physics. This can be done, because the diffuse interface is a numerical construct, having a width much larger than the physical width of the interface. Several terms in the chemical potentials and pressure, related to the finite size of the interface, can be excluded. These terms can be identified via comparison of the expression of the chemical potential and pressure with those of the sharp interface model, see Diamant and Andelman [6]. More details on the nonvariational model will be presented in an upcoming paper.

Furthermore, numerical analysis has shown it is more efficient to solve the convection-diffusion equation for the surfactants and counterions with the finite volume method. The LB method has problems with the divergence of the chemical potential for $\psi = 0$ and $\psi = 1$, due to the ideal-mixing terms in the free energy functionals. The results presented in the following are computed with the new hybrid LB/finite volume scheme, using a nonvariational model for the chemical potentials and pressure.

In the studies concerning ionic surfactants, the Poisson equation is solved with red-black successive overrelaxation method [12]. The solution of the previous timestep is used as an initial guess. After an initial period, the Poisson solver will converge within six steps.

14.5 RESULTS

14.5.1 IMMISCIBLE FLUIDS WITHOUT SURFACTANTS

We have applied the Cahn–Hilliard/phase field model earlier to investigate emulsion droplet formation in microfluidic devices [31]. This is a nice illustration of how fluid dynamics is linked to mesoscale phenomena, like droplet break up, and moving contact lines. Droplet breakup involves a change of the topology of the interface between the two immiscible interfaces, which presents quite a challenge for traditional CFD methods. In the Cahn–Hilliard/phase field, the change of topology is just a consequence of the minimization of the free energy functional, and need not to be computed explicitly. Also wetting boundary conditions are easily implemented, just via assigning a certain value of the order parameter to lattice sites representing solid walls [31,37]. Moving contact lines are also an immediate consequence of the free energy functional. Contact angle dynamics just follows from the coupling of fluid dynamics to the thermodynamics, as described by the free energy functional. Contact line dynamics correctly follows the Cox–Voinov theory [37].

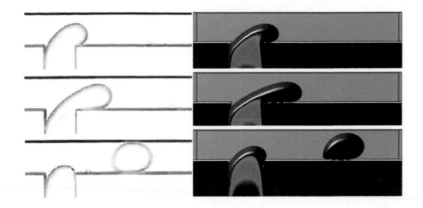

FIGURE 14.1 Snapshots of LB simulation of emulsion droplet breakup in T-junction of microchannels, compared with experimental observations at comparable times.

Snapshots of droplet breakup in a T-junction of microchannels are shown in Figure 14.1, and are compared to experimental observations at similar times [31]. Quantitative analysis has shown that the LB method can accurately predict the droplet breakup without any parameter fitting. All physical parameters were known beforehand from literature, or independent experiments.

14.5.2 EQUILIBRIUM PROPERTIES OF NONIONIC SURFACTANTS

The free energy approach has a big advantage that equilibrium profiles and properties can be derived analytically, and provides a good check for the validity of the numerical scheme [32]. Examples of simulated profiles of the order parameters ψ and ϕ are shown in Figure 14.2, and compared to the

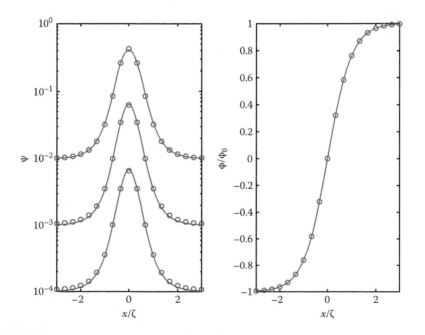

FIGURE 14.2 Profiles for surfactant concentration ψ and order parameter ϕ for a planar interface. Simulations are performed for $\beta = 0$ and $\gamma = 0$, rendering Langmuir sorption behavior.

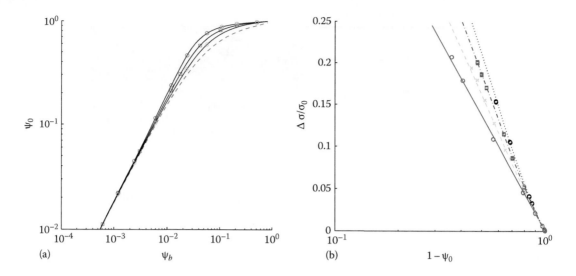

FIGURE 14.3   (a) shows that simulation results (symbols) follow the Frumkin sorption isotherm (solid lines) for $-\beta/\theta = \{0.5, 1.0, 1.5\}$, $\alpha/\theta = 9.0$. The dashed line gives the Langmuir isotherm for $\beta = 0$. (b) shows the relative surface pressure as function of the surfactant loading ψ_0 for $-\beta/\theta$ from the range $\{0, 1, 2, 3\}$, compared to analytical predictions (lines).

analytical solutions. Observe that the surfactant profile has its maximum right at the interface, where $\phi = 0$. This is due to the free energy functional, stating that surface energy is lowered if surfactant is absorbed at the interface.

The adsorption isotherm is derived from the equality of the chemical potential of the surfactant at the interface, where $\phi = 0$, and in the bulk phase. Analysis shows that Langmuir or Frumkin isotherms are obtained, with the latter in case of $\beta \neq 0$. The surface pressure follows from the grand potential, see [23]. In Figure 14.3, we show the sorption isotherm, and surface pressure for several model parameters. They compare favorably with the analytical solutions for the Frumkin sorption isotherms.

14.5.3   Thin Film Stability

We are currently using the nonionic surfactant model to investigate the stability of thin films of foam. First we have put two elongated bubbles, covered with surfactant, in a square box, and we let them relax toward their equilibrium shape. During the relaxation, the two interfaces will meet, and a thin water film is formed. This flow problem is quite comparable with the problem of liquid draining from a foam film, or two coalescing emulsion droplets. Snapshots of this simulation are shown in Figure 14.4. More detail of the coalescence process is shown in Figure 14.5. Here, the contour lines of the two interfaces at different times are drawn in the same figure. Here, one can clearly observe the dimple formation: the two films approach each other at two locations off center. Dimple formation is a well-known phenomenon occurring during coalescence [4], and the simulations are clearly able to reproduce that. Coalescence occurs because of high flow resistance in the thin film. It is energetically more favorable for the films to approach each other at the edges of the thin film. After collapse of the film, a satellite liquid droplet is formed. In the right pane of Figure 14.5, we have shown the velocity field at various times just prior to coalescence. Here, one can clearly see there is hardly any flow in the middle of the thin film. The interface is not fully loaded with surfactants, and displays partial mobility. In prior simulations of coalescence with insoluble surfactants using the lubrication approximation, one either assumes fully mobile or fully immobile (in case of fully loaded interface). We have analyzed the time evolution distance between the thin films as

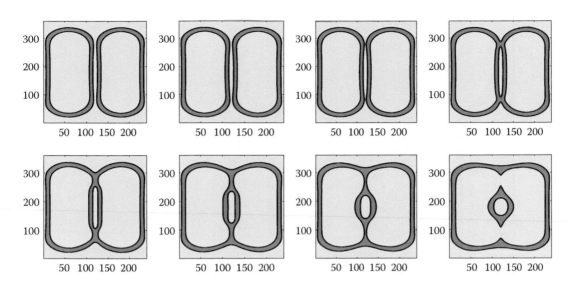

FIGURE 14.4   Collapse of thin film in between two fluid blobs confined in a square box with perfect wetting walls. The interface is loaded with surfactant, soluble in the continuous phase. Initially, the film has a configuration of two parallel planes. Bubbles deform due to minimization of interfacial energy, requiring the thin film to drain. We show contour plots of the surfactant concentration at equidistant times.

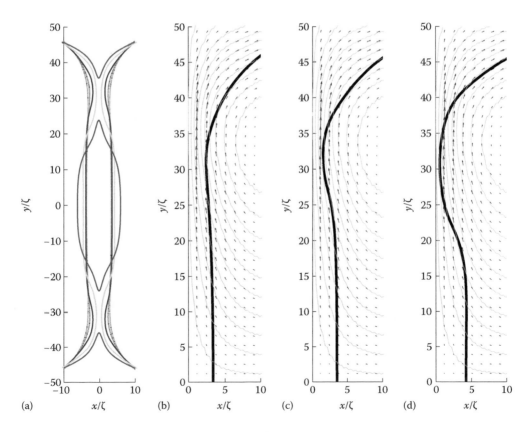

FIGURE 14.5   (a) Evolution of the interfaces in the thin film region at equidistant times, around the coalescence event. (b–d) Show the velocity field, streamlines, and interface (black solid line) in the upper quadrant of the interfacial region, at equidistant time just before coalescence.

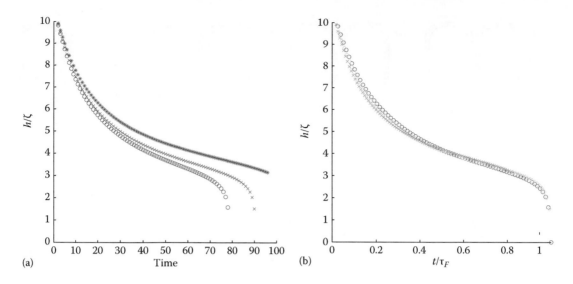

FIGURE 14.6   (a) Closest distance between two interfaces of a thinning film, as function of time for differ-ent (initial) bulk surfactant concentrations: $\psi_b = 1,2,5 \times 10^{-3}$. (b) Self-similar pattern of dimensionless gap between coalescing bubbles h/ζ as function of time, normalized to the lifetime of the film τ_F.

function of surfactant loading. If the time is rescaled by the lifetime of the film (τ_F), all curves col-lapse to a similar master curve—which can be seen in Figure 14.6.

14.5.4   PLANAR FILM WITH IONIC SURFACTANTS

We have tested the model for ionic surfactants with the equilibrium properties, which again follow from the equality of the chemical potentials for the interface and bulk regions. It immediately fol-lows that the ion concentrations in the bulk phases follow a Poisson–Boltzmann distribution:

$$\psi_{\pm}(x) = \psi_{b,\pm} \exp\left(-\frac{q_{\pm}U(x)}{\theta}\right) \tag{14.14}$$

The electrostatic potential is assumed to follow from the linearized Poisson–Boltzmann equation:

$$-\epsilon_1 \nabla^2 U = -\left(\frac{q^+}{v^+}\psi^+ - \frac{q^-}{v^-}\psi^-\right) \tag{14.15}$$

Substitution of the Poisson-Boltzmann distributions, $q_+ = q_- = q$, $v_+ = v_- = v$, the electroneutrality condition, and the assumption $qU/\theta \ll 1$ give

$$-\epsilon_1 \nabla^2 U = -2\frac{q^2\psi_b}{\theta v}U \tag{14.16}$$

Introducing the Debye–Hueckel screening length L_κ:

$$\frac{1}{L_\kappa^2} = \frac{2q^2\psi_b}{\theta v \epsilon_1} \tag{14.17}$$

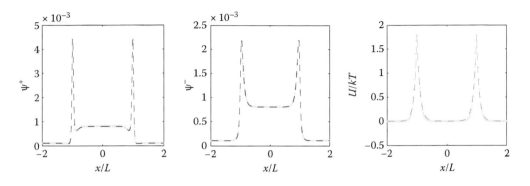

FIGURE 14.7   Steady-state solution of thin liquid film, with ions preferentially soluble in liquid phase, with the cationic surfactant absorbing at the interface. Solvation energy is the same for both ions. Differential sorption leads to the development of a double layer at the interfaces. The numerical solution follows the analytical solutions, with the dashed lines following the previous analytical expressions.

The solution for the potential is

$$U(\tilde{x}) = U_{ref} + U_1 \sin h\left(\frac{x}{L_\kappa}\right) \tag{14.18}$$

with $x = 0$ in the middle of the thin film.

Equating the chemical potential at the interface equal to the chemical potential of the bulk, and defining $\psi_c = \exp(-\alpha/\theta)$, and $\psi_U = \exp(-qU_0/\theta)$, we obtain the Davies adsorption isotherm:

$$\Psi = \frac{\psi_b}{\psi_b + \psi_c/\psi_U} \tag{14.19}$$

From the same equality of chemical potentials, we can deduce an analytical expression for the profiles of the ion concentrations $\psi_\pm(x)$.

We have compared simulations results with the previous analytical relations, which are shown in Figure 14.7. We observe that the ions' concentrations in the bulk phases are consistent with the Poisson–Boltzmann distribution, the electric field is consistent with the solution of the Poisson–Boltzmann equation, and the amount of ionic surfactant absorbed at the interface is in compliance with the Davies sorption isotherm. If we make the solvation energy different for surfactant and counterion, our model can reproduce effects of ions at interfaces as described by Onuki [23].

14.6   CONCLUSIONS AND OUTLOOK

We have shown that the free energy approach is a powerful framework for describing complex fluids stabilized by surfactants. The complex physics is an emerging field from the coupling of the thermodynamics, as described by the free energy functional, with macroscale transport as hydrodynamics and diffusion. Without any special numerical algorithm, our numerical scheme can cope with challenges on the mesoscale as droplet breakup, moving contact line dynamics, coalescence with dimple formation, formation of electric field, and disjoining pressure in thin films with ionic surfactants.

Because of the large amount of parameter, and multiple coupled physics, it requires careful analysis to obtain an efficient and stable numerical scheme. Recent research has shown that these properties can be improved greatly by using the finite volume scheme for diffusion of solutes like the surfactants and counterions. Furthermore, it is beneficial to use a nonvariational approach, such that redundant terms can be removed from the free energy functional, without altering the physics. Furthermore, we have seen broadening of the interface at high surfactant loadings. This will increase the computational demand of the scheme. The interface thickness must be smaller than any relevant physical length scale in the investigated problem. For problems such as the thin films in foams, as investigated earlier, the interface thickness must be significantly smaller than the film thickness and the Debye length. Hence, it is advantageous to go beyond simple uniform grids, as in standard LB scheme. For efficient numerical solution, adaptive, hierarchical grids will be required [30]. These techniques are becoming mainstay in the field of high-performance computing, and must not be seen as any hurdle.

Despite the aforementioned numerical challenges, we think the free energy approach still has huge potential. Furthermore, we see much potential in its application to other complex physical problems, such as (1) the effects of surfactants on spreading of droplets [11], (2) the effects on salts on surface tension [25], (3) the stabilization of water films by ions [25], (4) the effect of multiple (ionic) surfactant on the stabilization of dispersions [6], and (5) stabilization of dispersions by fat crystals [9]. The latter can be described by a phase field model [43], where the fat crystal can grow out of a liquid oil phase upon cooling.

REFERENCES

1. DM Anderson, GB McFadden, and A A Wheeler. Diffuse-interface methods in fluid mechanics. *Annual Review of Fluid Mechanics*, 30:139–165, 1998.
2. ME Cates, O Henrich, D Marenduzzo, and K Stratford. Lattice Boltzmann simulations of liquid crystalline fluids: Active gels and blue phases. *Soft Matter*, 5(20):3791–3800, 2009.
3. ME Cates and PS Clegg. Bijels: A new class of soft materials. *Soft Matter*, 4(11):2132–2138, 2008.
4. DYC Chan, E Klaseboer, and R Manica. Film drainage and coalescence between deformable drops and bubbles. *Soft Matter*, 7(6):2235–2264, 2011.
5. H Diamant and D Andelman. Kinetics of surfactant adsorption at fluid-fluid interfaces. *The Journal of Physical Chemistry*, 100(32):13732–13742, 1996.
6. H Diamant, G Ariel, and D Andelman. Kinetics of surfactant adsorption: The free energy approach. *Colloids and Surfaces A: Physicochemical and Engineering Aspects*, 183:259–276, 2001.
7. S Engblom, M Do-Quang, G Amberg, and AK Tornberg. On diffuse interface modeling and simulation of surfactants in two-phase fluid flow. *Communications in Computational Physics*, 14:879–915, 2013.
8. R Folch and M Plapp. Towards a quantitative phase-field model of two-phase solidification. *Physical Review E*, 68(1):010602, 2003.
9. S Ghosh and D Rousseau. Fat crystals and water-in-oil emulsion stability. *Current Opinion in Colloid & Interface Science*, 16(5):421–431, 2011.
10. ME Gurtin, D Polignone, and J Viñals. Two-phase binary fluids and immiscible fluids described by an order parameter. *Mathematical Models and Methods in Applied Sciences*, 6(6):815–831, 1996.
11. M Hanyak, DKN Sinz, and AA Darhuber. Soluble surfactant spreading on spatially confined thin liquid films. *Soft Matter*, 8(29):7660–7671, 2012.
12. MJ Holst and F Saied. Numerical solution of the nonlinear Poisson–Boltzmann equation: Developing more robust and efficient methods. *Journal of Computational Chemistry*, 16(3):337–364, 1995.
13. F Jansen and J Harting. From bijels to pickering emulsions: A lattice Boltzmann study. *Physical Review E*, 83(4):046707, 2011.
14. J Kromkamp, A Bastiaanse, J Swarts, GBPW Brans, RGM Van Der Sman, and RM Boom. A suspension flow model for hydrodynamics and concentration polarisation in crossflow microfiltration. *Journal of Membrane Science*, 253(1):67–79, 2005.
15. J Kromkamp, D van den Ende, D Kandhai, RGM van Der Sman, and RM Boom. Shear-induced self-diffusion and microstructure in non-Brownian suspensions at non-zero Reynolds numbers. *Journal of Fluid Mechanics*, 529:253–278, 2005.
16. T Krüger, S Frijters, F Günther, B Kaoui, and J Harting. Numerical simulations of complex fluid-fluid interface dynamics. *The European Physical Journal Special Topics*, 222(1):177–198, 2013.
17. T Kulrattanarak, RGM van der Sman, CGPH Schroën, and RM Boom. Analysis of mixed motion in deterministic ratchets via experiment and particle simulation. *Microfluidics and Nanofluidics*, 10(4):843–853, 2011.

18. H Kusumaatmaja and JM Yeomans. Modeling contact angle hysteresis on chemically patterned and superhydrophobic surfaces. *Langmuir*, 23(11):6019–6032, 2007.
19. AJC Ladd and R Verberg. Lattice-Boltzmann simulations of particle-fluid suspensions. *Journal of Statistical Physics*, 104(5):1191–1251, 2001.
20. H. Liu and Y. Zhang. Phase-field modeling droplet dynamics with soluble surfactants. *Journal of Computational Physics*, 229(24):9166–9187, 2010.
21. J Lowengrub and L Truskinovsky. Quasi-incompressible Cahn–Hilliard fluids and topological transitions. *Proceedings of the Royal Society of London. Series A: Mathematical, Physical and Engineering Sciences*, 454(1978):2617–2654, 1998.
22. D Marenduzzo, E Orlandini, and JM Yeomans. Hydrodynamics and rheology of active liquid crystals: A numerical investigation. *Physical Review Letters*, 98(11):118102, 2007.
23. A Onuki. Ginzburg-Landau theory of solvation in polar fluids: Ion distribution around an interface. *Physical Review E*, 73(2):021506, 2006.
24. I Pagonabarraga. Lattice Boltzmann modeling of complex fluids: Colloidal suspensions and fluid mixtures. In *Novel Methods in Soft Matter Simulations*, M Karttunen, A Lukkarinen, and I Vattulainen (Eds), pp. 279–309. Springer, Berlin, Germany, 2004.
25. PB Petersen and RJ Saykally. On the nature of ions at the liquid water surface. *Annual Review of Physical Chemistry*, 57:333–364, 2006.
26. B Rotenberg, I Pagonabarraga, and D Frenkel. Coarse-grained simulations of charge, current and flow in heterogeneous media. *Faraday Discussions*, 144:223–243, 2010.
27. SA Setu, I Zacharoudiou, GJ Davies, D Bartolo, S Moulinet, AA Louis, JM Yeomans, and DGAL Aarts. Viscous fingering at ultralow interfacial tension. *Soft Matter*, 9(44):10599–10605, 2013.
28. S Succi. *The Lattice-Boltzmann Equation*. Oxford University Press, Oxford, U.K., 2001.
29. MR Swift, E Orlandini, WR Osborn, and JM Yeomans. Lattice Boltzmann simulations of liquid-gas and binary fluid systems. *Physical Review E*, 54(5):5041, 1996.
30. J Tölke, S Freudiger, and M Krafczyk. An adaptive scheme using hierarchical grids for lattice Boltzmann multi-phase flow simulations. *Computers & Fluids*, 35(8):820–830, 2006.
31. S Van der Graaf, T Nisisako, CGPH Schroen, RGM Van Der Sman, and RM Boom. Lattice Boltzmann simulations of droplet formation in a T-shaped microchannel. *Langmuir*, 22(9):4144–4152, 2006.
32. RGM van der Sman and S van der Graaf. Diffuse interface model of surfactant adsorption onto flat and droplet interfaces. *Rheologica Acta*, 46(1):3–11, 2006.
33. RGM van der Sman. Lattice-Boltzmann scheme for natural convection in porous media. *International Journal of Modern Physics C*, 8(04):879–888, 1997.
34. RGM van der Sman. Galilean invariant lattice Boltzmann scheme for natural convection on square and rectangular lattices. *Physical Review E*, 74(2):026705, 2006.
35. RGM van der Sman. Simulations of confined suspension flow at multiple length scales. *Soft Matter*, 5(22):4376–4387, 2009.
36. RGM van der Sman. Soft matter approaches to food structuring. *Advances in Colloid and Interface Science*, 176:18–30, 2012.
37. RGM van der Sman. Investigation of lattice Boltzmann wetting boundary conditions for capillaries with irregular polygonal cross-section. *Computer Physics Communications*, 2013.
38. RGM van der Sman and MH Ernst. Convection-diffusion lattice Boltzmann scheme for irregular lattices. *Journal of Computational Physics*, 160(2):766–782, 2000.
39. RGM van der Sman and MBJ Meinders. Mesoscale models of dispersions stabilized by surfactants and colloids. *Advances in Colloid and Interface Science*, 211:63–76, 2014.
40. RGM van der Sman and S van der Graaf. Emulsion droplet deformation and breakup with lattice Boltzmann model. *Computer Physics Communications*, 178(7):492–504, 2008.
41. HM Vollebregt, RGM van der Sman, and RM Boom. Suspension flow modelling in particle migration and microfiltration. *Soft Matter*, 6(24):6052–6064, 2010.
42. HM Vollebregt, RGM van der Sman, and RM Boom. Model for particle migration in bidisperse suspensions by use of effective temperature. *Faraday Discussions*, 158(1):89–103, 2012.
43. AA Wheeler, WJ Boettinger, and GB McFadden. Phase-field model for isothermal phase transitions in binary alloys. *Physical Review A*, 45(10):7424, 1992.

15 Hybrid Numerical Method for Interfacial Flow with Soluble Surfactant and Its Application to an Experiment in Microfluidic Tip Streaming

Michael R. Booty, Michael Siegel, and Shelley L. Anna

CONTENTS

15.1 INTRODUCTION

This chapter describes a hybrid numerical method for solving problems of two-phase flow with soluble surfactant in the limit of large bulk Peclet number.

Most numerical studies of the effect of surfactant on the deformation and breakup of a single bubble or drop in an imposed flow are for insoluble surfactant, that is, for surfactant that is confined to the interface alone. A soluble surfactant advects and diffuses in the bulk fluid as a passive scalar, but when local interfacial surface area is changed during deformation, the surface and bulk surfactant concentrations are brought out of equilibrium. This causes an exchange or transfer of surfactant between its dissolved form in the bulk and its adsorbed form on the interface that acts so as to restore equilibrium between the two surfactant phases.

Since typical surfactant molecules are large relative to solvent molecules, their bulk diffusivity is small and is of the order of 10^{-10} m^2 s^{-1} [1]. For an oil–water interface and a length scale of 100 μm, this leads to a bulk Peclet number $Pe \simeq 10^6$. As a result, exchange of surfactant between the surface and bulk phases occurs in a narrow layer of dimensionless width of order $Pe^{-1/2} \simeq 10^{-3}$, which we refer to as a transition layer, that lies adjacent to the fluid interface.

The hybrid method combines a leading order matched asymptotic or singular perturbation reduction of the transition layer dynamics with a numerical method or underlying flow solver for two-phase flow with insoluble surfactant. Part of the success of this formulation is that, for zero Reynolds number Stokes flow, a highly accurate boundary integral method can be adapted to solve the moving boundary value problem, including the influence of surfactant solubility. Without the large bulk Peclet number treatment, this and other surface-based methods would not easily apply.

The method has been developed to a point that it has led to what we believe are the first fully resolved simulations of surfactant-mediated tip streaming of an isolated drop with a surfactant that is soluble. We also describe ongoing experiments with a microfluidic flow device that has the potential, via tip streaming, to form a controlled stream of monodisperse droplets of micron size for use in a variety of applications. Surfactant solubility effects are fundamental to the experiments and the successful operation of the device.

There is a long history and an extensive literature on drop deformation and breakup. Some of the references cited here touch on this. For a recent review, see, for example, [2].

This chapter begins with a derivation of the hybrid asymptotic–numerical method. This is followed by review of its validation against simulations at large but finite bulk Peclet number that utilize a complex variable formulation for study of the deformation of a 2D inviscid bubble, which first appeared in [3]. The general solution of the initial boundary value problem for the transition layer dynamics can be expressed in terms of a convolution integral in the limit of infinite Biot number. This leads to an alternative, mesh-free formulation for the influence of surfactant solubility effects using the hybrid method that was presented in [4] and is reviewed here. The mesh-free formulation has the potential to greatly improve the computational efficiency of the hybrid method and provides further validation of the method and its implementation.

Sample results for the study of an isolated drop that is strained in an axisymmetric uniaxial extension are given next and include representative simulations of tip streaming that were described in [5]. This is followed by description of the microfluidic flow experiments for surfactant-mediated tip streaming in a flow-focusing geometry that were presented in [6,7]. A comparison between preliminary results from the numerical simulation of conditions for tip streaming of an isolated drop of [5] and an operating diagram for tip streaming in the experiments of [7] follows, before the concluding remarks.

15.2 DERIVATION OF THE HYBRID ASYMPTOTIC–NUMERICAL METHOD IN THE LARGE BULK PECLET NUMBER LIMIT

In dimensionless form, the governing equations and boundary conditions begin with the condition of mass conservation for incompressible flow and the Navier–Stokes equation for momentum conservation of a Newtonian fluid

$$\nabla \cdot \boldsymbol{u} = 0 \quad \text{and} \quad Re(\partial_t + \boldsymbol{u} \cdot \nabla)\boldsymbol{u} = -\nabla p + \lambda \nabla^2 u, \quad \boldsymbol{x} \in \Omega_i. \tag{15.1}$$

The geometrical setup is of a single drop in a surrounding fluid, where the interior drop domain is Ω_1 and the surrounding fluid domain is Ω_2. For ease of notation, we omit subscripts on quantities in the separate domains unless ambiguity can result, as occurs, for example, at a fluid interface. The quantity $\lambda = \mu_i/\mu_2$ is the ratio of the local fluid viscosity to the exterior fluid viscosity, so that $\lambda = 1$ in Ω_2 and $\lambda = \mu_1/\mu_2$ is the viscosity contrast in Ω_1. Lengths are made nondimensional by the radius a of the undeformed spherical drop, and velocity is made nondimensional by the capillary velocity of a surfactant-free interface, $U = \sigma_0/\mu_2$, where σ_0 is the constant surface tension in the clean or surfactant-free case of two pure fluids. Time is made nondimensional by a/U and pressure by $\mu_2 U/a$. The presentation below is given in the Stokes flow limit $Re = 0$.

At the interface S, continuity of velocity and the kinematic condition for evolution of the interface shape hold, so that

$$[\boldsymbol{u}] = 0 \quad \text{and} \quad \frac{d\boldsymbol{x}}{dt} = (\boldsymbol{u} \cdot \boldsymbol{n})\boldsymbol{n}, \tag{15.2}$$

where

[·] denotes the jump in a quantity across S
$\boldsymbol{x}$ is a point on S for all t
$\boldsymbol{n}$ is the outward unit normal on Ω_1

The condition for stress balance across S is

$$-(p_2 - p_1)\boldsymbol{n} + 2(\boldsymbol{e}_2 - \lambda\boldsymbol{e}_1) \cdot \boldsymbol{n} = \sigma\kappa\boldsymbol{n} - \nabla_s\sigma, \tag{15.3}$$

where

$\boldsymbol{e}_i$ is the rate of strain tensor
σ is the interfacial surface tension
κ is the sum of the principal normal curvatures of S
∇_s is the surface gradient

The last term, $\nabla_s\sigma$, is the Marangoni stress. This is zero in the surfactant-free case $\sigma = \sigma_0$ but is the cause of most of the modification to the dynamics of a surfactant-laden interface. The drop is deformed by an imposed linear flow, such as a 2D pure strain, 2D simple shear, or axisymmetric 3D strain (i.e., a uniaxial extension), given, respectively, by

$$\boldsymbol{u}^\infty = Ca(x_1, -x_2), \quad \boldsymbol{u}^\infty = Ca(x_2, 0), \quad \boldsymbol{u}^\infty = Ca\left(-\frac{r}{2}\boldsymbol{e}_r + x\boldsymbol{e}_x\right), \tag{15.4}$$

as $|\boldsymbol{x}| \to \infty$. Here, $Ca = Ga/U$ is the capillary number and G is the dimensional imposed strain rate.

The addition of interfacial surfactant with surface concentration Γ requires the inclusion of a surface equation of state and a conservation law for the evolution of Γ, namely,

$$\sigma = 1 + E\ln(1 - \Gamma), \tag{15.5}$$

$$\left.\frac{\partial\Gamma}{\partial t}\right|_n + \nabla_s \cdot (\Gamma\boldsymbol{u}_s) + \kappa u_n\Gamma = \frac{1}{Pe_s}\nabla_s^2\Gamma + J\boldsymbol{n} \cdot \nabla C|_s, \quad \boldsymbol{x} \in S. \tag{15.6}$$

The first of these is the Frumkin equation of state [1]; its linearization $\sigma = 1 - E\Gamma$ is used in some of the results described in the succeeding text. Here, $E = RT\Gamma_\infty/\sigma_0$ is the elasticity number, and Γ_∞ is the maximum theoretical monolayer packing density, which is the scale for nondimensionalization of Γ. In (15.6), the time derivative is taken in the direction normal to S when the interface is in motion, and the interfacial fluid velocity is expressed in terms of its tangential and normal projections as $\boldsymbol{u} = \boldsymbol{u}_s + u_n\boldsymbol{n}$. On the right-hand side, the first term represents surface diffusion of surfactant with surface Peclet number $Pe_s = Ua/D_s$ and surface diffusivity D_s. The second term is absent when the surfactant is considered to be insoluble but acts as a source term for the exchange of surfactant between its bulk and surface forms when the influence of solubility is included. Here, C is the bulk surfactant concentration made nondimensional by a constant ambient concentration C_∞, and the mechanism for exchange is a Fickian diffusive flux in the sublayer immediately adjacent to S.

The exchange parameter $J = DC_\infty/\Gamma_\infty U$ is a measure of the bulk–interface exchange relative to advection of surfactant on S.

Additional relations are introduced to describe the evolution of dissolved or bulk surfactant. Away from the interface, dissolved surfactant is advected with the flow and diffuses through it as a passive scalar, and, to describe the bulk–interface exchange dynamics, an adsorption–desorption model needs to be introduced. Here,

$$(\partial_t + \boldsymbol{u} \cdot \nabla)C = \frac{1}{Pe}\nabla^2 C, \quad \boldsymbol{x} \in \Omega_2, \tag{15.7}$$

$$\boldsymbol{Jn} \cdot \nabla C \mid_s = Bi(K(1-\Gamma)C\mid_s - \Gamma), \quad \boldsymbol{x} \in S, \tag{15.8}$$

$$C(\boldsymbol{x},0) = 1 \quad \text{and} \quad C \to 1 \quad \text{as} \quad |\boldsymbol{x}| \to \infty. \tag{15.9}$$

In Equation 15.7 for the evolution of the bulk surfactant concentration, $Pe = Ua/D$ is the Peclet number for the bulk concentration, and D is its bulk diffusivity. Bulk surfactant is taken to be present in the external flow region Ω_2 alone. On the right-hand side of the bulk–interface surfactant exchange boundary condition (15.8), the Biot number $Bi = \kappa_d a/U$ is the ratio of the flow time scale a/U to the timescale for desorption kinetics κ_d^{-1} and $K = \kappa_a C_\infty/\kappa_d$ is a dimensionless equilibrium partition coefficient, which is the ratio of the bulk-to-interface adsorption rate $\kappa_a C_\infty$ to the interface-to-bulk desorption rate κ_d. This is referred to as the mixed-kinetic form of the interface boundary condition.

Under equilibrium conditions, the bulk surfactant concentration $C \equiv 1$ throughout Ω_2, and from boundary condition (15.8), the associated equilibrium surface surfactant concentration is $\Gamma_e = K/(1 + K)$. Any local change in interfacial area following a fluid particle on the interface, such as can occur during deformation or in a steady deformed state, tends to move the surface and ambient bulk surfactant concentrations away from equilibrium. However, in the limit $Pe \to \infty$, Equation 15.7 for the evolution of C is singularly perturbed, so that any departure from equilibrium between the surface and bulk concentrations is accompanied by a large normal gradient of the bulk concentration C in a narrow transition layer immediately adjacent to the interface. This rapid spatial variation is resolved by introducing a leading order *inner* or local expansion that is formed in the limit $\epsilon = Pe^{-1/2} \to 0$.

An intrinsic or surface-fitted coordinate system (ξ_1,ξ_2,n) is introduced, where the tangential coordinates ξ_1 and ξ_2 are aligned with the principal directions of curvature of S and n is distance along the outward normal measured from S. To transform (15.7) from the Eulerian to the intrinsic frame, the Eulerian fluid velocity is written as $\boldsymbol{u} = \boldsymbol{u}_t + u_p\boldsymbol{n}$, where $\boldsymbol{u}_t$ and $u_p\boldsymbol{n}$ are its tangential and normal projections. With a well-defined choice of origin O' on S for the intrinsic frame, $\boldsymbol{U}_s$ is the velocity of O' relative to the origin O of the Eulerian coordinate system projected onto the tangent plane locally, so that $\boldsymbol{U}_s$ varies on S. In the intrinsic frame, Equation 15.7 becomes

$$(\partial_t + \boldsymbol{v}_t \cdot \nabla_t + v_p\partial_n)C = \epsilon^2\nabla_s^2 C, \tag{15.10}$$

where
   $\boldsymbol{v}_t = \boldsymbol{u}_t - \boldsymbol{U}_s$ is the fluid velocity relative to O' projected onto the local tangent plane
   $v_p = u_p - u_n$ is the normal component of the fluid velocity relative to S
   $u_n = \boldsymbol{u} \cdot \boldsymbol{n}$ is the Eulerian normal velocity of S
   ∇_t is the tangential projection of the gradient
   ∇_s^2 is the Laplacian in the intrinsic frame

When the Reynolds number Re is not large, there is no mechanism to support a boundary layer or local region of large shear of the fluid velocity $\boldsymbol{u}$ adjacent to S. To a first approximation, in terms of a local normal coordinate $N = n/\varepsilon = O(1)$, (15.10) therefore becomes

$$(\partial_t + \boldsymbol{v}_s \cdot \nabla_s + \partial_n v_p \mid_s N \partial_N)C = \partial_N^2 C. \tag{15.11}$$

Here, the tangential fluid velocity profile in the transition layer is of plug-flow or zero-shear type at leading order, so that $\boldsymbol{u}_t = \boldsymbol{u}_s$ is the tangential velocity of a fluid particle on S, as it appears in Equation 15.6, and $\boldsymbol{v}_s = \boldsymbol{u}_s - \boldsymbol{U}_s$. On the interface, the relative normal velocity $v_p = 0$, so that nearby its leading order approximation is the first nonzero term of its Taylor series, $\partial_n v_p \mid_s n$. When the incompressibility condition of (15.1) is written in intrinsic coordinates and then evaluated in the limit $n \to 0$, the normal derivative $\partial_n v_p \mid_s$ on the interface can be expressed in terms of surface data alone by

$$\partial_n v_p \mid_s = -\kappa u_n - \nabla_s \cdot \boldsymbol{u}_s. \tag{15.12}$$

The large bulk Peclet number limit requires that the bulk–interface exchange coefficient is rescaled so that $J_0 = J/\varepsilon = O(1)$ in (15.6) and (15.8), and the transfer flux becomes $J_0 \partial_N C \mid_s$.

Outside the transition layer, the small ε-large Pe limit of (15.7) implies that C is constant on particle paths. With the equilibrium initial data of (15.9), this leads to a far-field boundary condition that $C \to 1$ as $N \to \infty$ in regions where the flow direction enters or stays within the transition layer, which is where $\partial_n v_p \mid_s \leq 0$. In regions where the flow leaves the transition layer, $\partial_n v_p \mid_s > 0$ and $\lim_{N \to \infty} C$ is left to be free or determined by the solution of (15.11) for large N. This occurs in a neighborhood of any stagnation points on the interface where the flow on S is toward the point and the flow in the bulk leaves S. Here, an excess bulk concentration develops with time, $C > 1$ with $\lim_{N \to \infty} \partial_N C = 0$, and at large times, the outward flow advects a plume of high bulk surfactant concentration away from the interface. This form for the far-field boundary condition was addressed in [5].

To summarize, in the transition layer, the bulk surfactant concentration is governed by

$$(\partial_t + \boldsymbol{v}_s \cdot \nabla_s + \partial_n v_p \mid_s N \partial_N)C = \partial_N^2 C \quad \text{in} \quad \Omega^r \tag{15.13}$$

$$\text{where} \quad \partial_n v_p \mid_s = -\kappa u_n - \nabla_s \cdot \boldsymbol{u}_s, \tag{15.14}$$

$$J_0 \partial_N C \mid_s = Bi(K(1-\Gamma)C \mid_s - \Gamma) \quad \text{on} \quad S, \tag{15.15}$$

$$\text{as} \quad N \to \infty \begin{cases} C \to 1 & \text{when} \quad \partial_n v_p \mid_s \leq 0, \\ \partial_N C \to 0 & \text{when} \quad \partial_n v_p \mid_s > 0, \end{cases} \tag{15.16}$$

$$C(\xi_1, \xi_2, N, 0) = 1, \tag{15.17}$$

where Ω^r is the transition layer domain. The surface surfactant concentration is governed by

$$\left. \frac{\partial \Gamma}{\partial t} \right|_n + \nabla_s \cdot (\Gamma \boldsymbol{u}_s) + \kappa u_n \Gamma = \frac{1}{Pe_s} \nabla_s^2 \Gamma + J_0 \left. \frac{\partial C}{\partial N} \right|_s \quad \text{on} \quad S. \tag{15.18}$$

Since the asymptotic reduction that leads to (15.13) to (15.18) has been made to leading order in the limit of infinite bulk Peclet number, the parameter Pe does not appear—it has been scaled out. However, when the interface profile and bulk surfactant concentration are shown together, a finite value of Pe needs to be chosen or reintroduced so as to give the transition layer a nonzero width and distinguish the bulk surfactant concentration there from the ambient far-field value.

The bulk and surface surfactant diffusivities are of the same order [1], so that the bulk and surface Peclet numbers Pe and Pe_s are both large. However, the surface diffusion term of (15.18) is not given the same asymptotic treatment. In most situations, there is no tendency for the surface concentration Γ to form near discontinuities, although near discontinuities in its gradient are observed in some of the examples in the succeeding text. A possible exception to this occurs in capillary-induced pinch off, see, for example, [8–10]. When it does occur, the dimensionality of a near discontinuity of Γ on the interface is one order lower than that of a near discontinuity of C in the bulk flow, so that it is less computationally demanding to resolve it by reducing the mesh size. In the examples in the succeeding text, the influence of surface diffusion is found to be sufficiently small that it is neglected unless stated otherwise.

The reduced model for surfactant solubility just given is quite general. Any numerical means for solving problems of two-phase flow with insoluble surfactant that is governed by Equations 15.1 through 15.6 can be used, together with a suitable solution method for the reduced, infinite Peclet number initial boundary value problem for solubility effects that is given by (15.13) through (15.17). It is important to note that this does not require the input of any off-surface data from the underlying flow solver for its solution. In the results described in the succeeding text, the zero Reynolds number limit of (15.1) has been used, which affords a surface-based method for the underlying flow solver.

15.3   VALIDATION OF THE METHOD

In the context of a 2D bubble with an inviscid interior in the zero Reynolds number or Stokes flow limit, complex variable methods offer a highly accurate means for validating the large bulk Peclet number asymptotic model. They also happen to give a solution that has a simple elliptical shape, which can be used for further validation.

The solution technique relies on classical complex variable methods for solving the biharmonic equation for the stream function. This is based on a Goursat representation for a complex stress–stream function that is written in terms of two functions that are analytic in the exterior flow domain Ω_2, [11]. The interface and far-field conditions are sufficient to determine the two analytic functions together with a conformal map $z(\zeta,t)$ that maps $z = x_1 + ix_2 \in \Omega_2$ back to the interior of the unit circle $|\zeta| < 1$ in the ζ-plane. The analysis has been developed for bubble dynamics in a number of studies since the early work of, for example, Richardson [12]. It was generalized by Tanveer and Vasconcelos [13], and later by Siegel to include the influence of insoluble surfactant [14]. These references cite additional studies in the area.

Omitting all detail of this part of the analysis, which can be found for soluble surfactant in [3], the result of the underlying flow solver is that, for an initially circular or elliptical bubble that evolves in either a pure strain or a simple shear flow, the interface is described exactly by a Joukowski map

$$z = \frac{a(t)}{\zeta} + b(t)\zeta. \tag{15.19}$$

This holds when the flow is surfactant-free [13], when there is insoluble surfactant on the interface [14], or with soluble surfactant at arbitrary Pe and other parameters. As a consequence, the interface remains elliptical for all time, with a changing aspect ratio that is influenced by surfactant effects. For a pure strain $\boldsymbol{u}^\infty = Ca(x_1,-x_2)$, the evolution of the map parameters is given by

$$\frac{d}{dt}(ab) = -2I_0ab + 2Caa^2,$$

$$\text{where} \quad I_0 = \frac{1}{4\pi}\int_0^{2\pi}\frac{\sigma(v,t)}{|z_\zeta(e^{iv},t)|}dv, \quad \text{and} \quad a^2 - b^2 = 1. \tag{15.20}$$

The surface tension σ appears in the integral and expresses the influence of surfactant on the dynamics. For an initially circular bubble, $a(0) = -1$ and $b(0) = 0$.

The evolving surface tension in (15.20) can be found in two distinct ways:

1. *A "traditional" method with Pe finite.* The surfactant dynamics with $Pe < \infty$ is governed by (15.5) through (15.9), which is solved by finite differences on an orthogonal curvilinear coordinate system in the z-plane that is coupled to the map (15.19) and flow solver (15.20).
2. *The hybrid method with Pe infinite.* The surfactant dynamics is governed by (15.5) and the infinite Pe transition layer reduction of (15.13) through (15.18). This is solved by finite differences on a rectangular grid in the intrinsic coordinate frame and coupled to (15.19) and (15.20).

Away from the interface, the grid of the traditional method is generated by the time-dependent map (15.19) acting on grid points (v_i, r_j) that are fixed in the ζ-plane, where $\zeta = re^{iv}$ and

$$v_i = \frac{2\pi i}{M}, \quad i = 0,1,\ldots,M-1, \quad r_j = 1 - \frac{j}{P-1}r_m, \quad j = 0,1,\ldots,P-1. \tag{15.21}$$

The preimage in the ζ-plane is the annular region $1 - r_m \le r \le 1$ for some r_m, which can be chosen to confine the grid closer to the interface and refine the mesh by decreasing r_m.

The need to evaluate the fluid velocity u in (15.7) away from the interface greatly increases the computational overhead of a surface-based solver for insoluble surfactant when it is adapted to include the effects of solubility at finite Pe. In 2D, to find u at each of MP grid points, an integral of boundary data around the M grid points on the interface needs to be performed, requiring $O(M^2P)$ operations. Although this can be reduced to $O(MP)$ operations by a fast-multipole method, the complex variable formulation allows the same improvement to be obtained with relative ease by analytic continuation of the Goursat functions from their known values on the unit disk $|\zeta| = 1$ into the disk interior in the ζ-plane, combined with knowledge of the map $z(\zeta,t)$. The Goursat function f, for example, is first represented by the M Fourier coefficients $\hat{f}_k$ of its discrete Fourier transform (DFT) at M points $\zeta_{i0} = e^{iv_i}$ on $|\zeta| = 1$. The continuation of f away from the interface is given by multiplying $\hat{f}_k$ by r_j^k to give the inverse fast Fourier transform (FFT) at the grid point $\zeta_{ij} = r_je^{iv_i}$,

$$f(\zeta_{ij}) = \sum_{k=-1}^{M-2}\hat{f}_k r_j^k e^{ikv_i}. \tag{15.22}$$

Since $r_j < 1$, this operation is well posed. At each of the P values of j, the FFT is performed for the M values of i in $O(M \log M)$ operations, so that neglecting the logarithmic correction f is evaluated at all grid points on and away from the interface in $O(MP)$ operations. The sum in (15.22) begins at $k = -1$ since f is analytic except for a simple pole at $\zeta = 0$, which corresponds to an imposed far-field flow that is linear in z.

These aspects of the conformal mapping method have been emphasized, since without a simple means to refine the mesh close to the interface and a fast evaluation of u off the interface, it is not feasible to solve the finite Peclet number advection–diffusion Equation 15.7 accurately at values of Pe that are of the order of 10^3 or larger.

The finite difference scheme for the traditional method used an alternating direction implicit method, with upwinding for the advection terms in the equations for C and Γ. The implementation was first-order accurate in space and time and found to be stable for a time step $\Delta t \lesssim 10^{-3}$ over a wide range of Pe for mesh sizes up to 1024×1024.

In the hybrid method, (15.13) is solved on a fixed rectangular grid (v_i, N_j), where $N_j = jN_m/(P-1)$ for $j = 0,1,\ldots,P-1$, and the far-field boundary condition (15.16) is applied at $N = N_m$. There is no need to remesh the grid as the interface evolves and no need to resolve large gradients of the scaled normal gradient $\partial_N C$ in the hybrid method. Surface quantities are computed via DFTs, centered finite differences are used to evaluate C and its derivatives for $N > 0$, and a semi-implicit method is used for the time step, to produce a method for evaluation of Γ and C that is second-order accurate in space and time. The time update of $\boldsymbol{u}$ and Γ is performed in $O(M)$ operations, and the evaluation and time update of C is performed in $O(MP)$ operations.

For both methods, computation of the underlying flow solver is second-order accurate in time. Since its integrand is periodic, evaluation of the integral I_0 in (15.20) by the trapezoidal rule readily gives spectral accuracy in space.

Figure 15.1 shows the bubble shape and bulk surfactant concentration C in the first quadrant for an initially circular bubble that is stretched in a pure strain with capillary number $Ca = 0.25$ at $t = 1.0$, as found by the hybrid method (a) and traditional method (b). Computational parameters are $M = P = 256$, $\Delta t = 5 \times 10^{-4}$, $r_m = 0.9$, and $N_m = 10$; physical parameter values are $E = 0.1$, $K = 1.5$, $J_0 = 1$ (so that $J = Pe^{-1/2}$), and $Bi = \infty$ so that $Cl_s = \Gamma/K(1 - \Gamma)$, with a nonequilibrium initial surface concentration $\Gamma(v,0) = 0.5$ that initiates strong bulk–interface surfactant exchange at $t = 0^+$.

The straining flow advects surface surfactant toward the bubble ends and contracts the interface there, causing a relatively high surface concentration of surfactant, so that surfactant tends to desorb or leave the interface for the bulk flow. Conversely, near the middle section of the bubble, advection and local generation of surface area cause a relatively low surface surfactant concentration that is replenished by adsorption from the far-field bulk flow.

At this value of $Pe = 10^3$, the profiles and other quantities computed by the two methods are close, and sufficiently so that there is no discernible difference between the data shown in the two parts of Figure 15.1. With the hybrid method, the mesh size can be reduced to $M = P = 64$ with no change in the data to within plotting resolution, while at larger Pe, the traditional method begins to lose accuracy due to inadequate resolution of the narrow transition layer.

Figure 15.2 shows the surface surfactant concentration $\Gamma(v,t)$ versus v at $t = 1.0$ computed by the hybrid method (dotted curves) and by the traditional method (solid curves) for $Pe = 50$, 10^2, 10^3, 10^4, and 5×10^4. The number of grid points is fixed with $M = P = 256$ and other parameters are as in Figure 15.1. Since the transition layer is constructed to be exact in the limit $Pe \to \infty$, solutions computed by the traditional method are expected to approach solutions computed by the hybrid method as Pe increases. This is seen in the upper panel when the Peclet number is less than about 10^3. However, when $Pe > 10^3$, at this fixed resolution, the traditional method does not resolve the large normal gradient of bulk surfactant near S, especially at the bubble poles or ends where $v \simeq 0$, π, 2π, and the profile shown for the surface concentration Γ is seen to diverge from the expected limiting solution there when Pe is increased beyond 10^3.

The accuracy of the traditional method at $Pe = 10^4$ with the same number of grid points is improved when the parameter r_m of (15.21) is reduced to $r_m = 0.1$, so that the grid points are clustered in a more narrow annular region near the interface in the ζ-plane. The lower panel of Figure 15.2 shows that at this greatly increased resolution, the traditional and hybrid methods give results that are again nearly indistinguishable. At yet larger, more realistic Pe, the traditional method becomes prohibitively expensive due to the separation of spatial scales in the transition layer and the need to accurately resolve its dynamics.

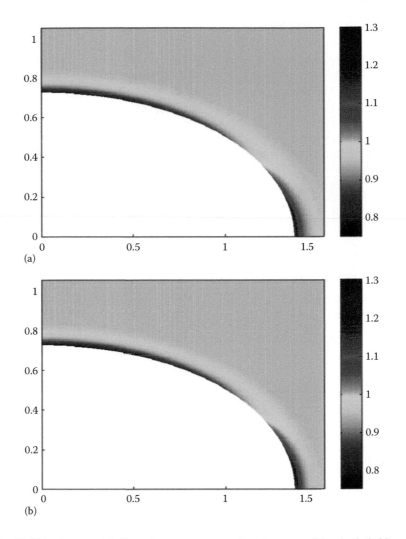

FIGURE 15.1 Bubble shape and bulk surfactant concentration C computed by the hybrid method (a) and the traditional method (b) at time $t = 1.0$. The main physical parameter values are $Pe = 10^3$ and $Ca = 0.25$; for other parameters, see text. Good agreement between the two methods is found at this low Pe. (Reprinted from *J. Comput. Phys.*, 229, Booty, M.R. and Siegel, M., A hybrid numerical method for interfacial fluid flow with soluble surfactant, 3864–3883, Copyright 2010, with permission from Elsevier.)

A complex variable formulation in terms of a pair of Goursat functions on each fluid domain was used in our subsequent study [4] for a viscous drop in 2D with viscosity contrast $\lambda = \mu_1/\mu_2 \geq 0$. However, the underlying flow solver was changed to a Sherman–Lauricella formulation of the boundary integral method. This is more computationally efficient as a means of solving the biharmonic equation than conformal mapping when the interior fluid is viscous, $\lambda > 0$, although it is not apparent or built in to the Sherman–Lauricella formulation that the exact elliptical profile will be retained in the limit when $\lambda = 0$, for which the integral equation is rank deficient. This provides an additional means of validation that the flow solver, together with the method for including surface and bulk surfactant effects, has been implemented correctly.

The Sherman–Lauricella formulation is familiar in classical 2D problems of elasticity [15] but has also been applied to two-phase Stokes flow [16,17]. It is an *indirect* method, in that all primitive variables can be expressed via the Goursat functions in terms of a single complex density that is defined on the drop interface and satisfies a Fredholm second kind integral equation.

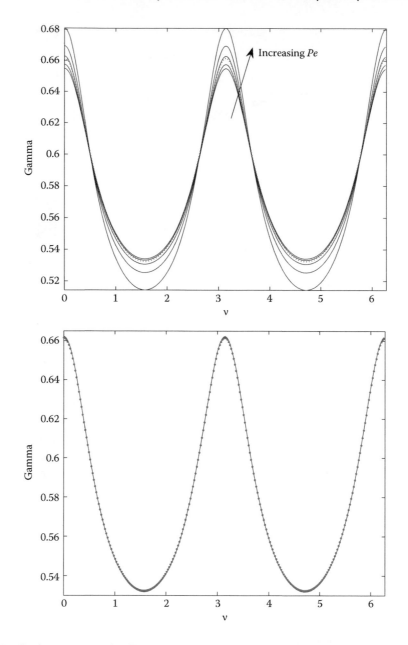

FIGURE 15.2 Surface concentration $\Gamma(v,t)$ at $t = 1.0$ computed by the hybrid method (dotted) and by the traditional method (solid curves) with $Pe = 50$, 10^2, 10^3, 10^4, and 5×10^4. Other parameters are as in Figure 15.1. In the lower panel, $Pe = 10^4$ and the accuracy of the traditional method has been increased by setting $r_m = 0.1$. (Reprinted from *J. Comput. Phys.*, 229, Booty, M.R. and Siegel, M., A hybrid numerical method for interfacial fluid flow with soluble surfactant, 3864–3883, Copyright 2010, with permission from Elsevier.)

In [4], a mesh-free method was presented that gives the solution of the initial boundary value problem (15.13) to (15.17) for the bulk surfactant concentration in the transition layer as a convolution integral, which is reviewed here.

With the initial parameterization for the interface in terms of arc-length, $\xi_1 = s$, a coordinate transformation to a Lagrangian surface coordinate s_0 is made by setting $s = s(s_0, t)$ where $\partial_t s = v_s(s,t)$ and $s = s_0$ at $t = 0$. The dependence on s_0 then only appears parametrically, and on introducing

$$u = C - 1, \quad \partial_n v_p \big|_s (s_0, t) = \psi_0(t),$$ (15.23)

$$\Gamma(s_0, t) = \Gamma_0(t), \quad h_0(t) = \frac{\Gamma_0(t)}{K(1 - \Gamma_0(t))} - 1,$$ (15.24)

the boundary value problem (15.13) to (15.17) becomes

$$\partial_t u + \psi_0(t) N \partial_N u = \partial_N^2 u,$$ (15.25)

$$u(N = 0, t) = h_0(t); \quad u \to 0 \quad \text{as} \quad N \to \infty; \quad u(N, t = 0) = 0.$$ (15.26)

Here, the diffusion-controlled version of the boundary condition (15.15), which is with $Bi = \infty$, has been used, and the far-field condition (15.16) has been simplified to $C \to 1$.

The solution can be found in terms of a convolution integral by Duhamel's principle. This gives an expression for C throughout the transition layer, if needed, and also gives an expression for its normal derivative at the interface, which can be substituted directly in the bulk–interface exchange term of (15.18), namely,

$$\frac{\partial C}{\partial N}\bigg|_s = \frac{-2h_0(0)}{\sqrt{\pi}\gamma(t, 0)} - \frac{2}{\sqrt{\pi}} \int_0^t \frac{\partial_\tau h_0(\tau)}{\gamma(t - \tau, \tau)} d\tau,$$ (15.27)

$$\text{where} \quad \gamma(t - \tau, \tau) = 2e^{\frac{1}{2}a(t - \tau, \tau)} \left(\int_0^{t - \tau} e^{-a(\bar{t}, \tau) d\bar{t}} \right)^{1/2},$$ (15.28)

$$\text{and} \quad a(t, \tau) = 2 \int_0^t \psi_0(t' + \tau) dt'.$$ (15.29)

In (15.27), $h_0(0) = 0$ if the bulk and surface surfactant concentrations are initially in equilibrium.

The results of this mesh-free version of the hybrid method were compared with those of a mesh-based version. For the mesh-free method, the Lagrangian surface coordinate s_0 is the natural coordinate for parameterization of the interface. Interface fluid markers were therefore used in the boundary integral equation of the underlying flow solver, and evaluation of the integrals with fixed step size by the trapezoidal rule gave spectral accuracy in space. An accelerated evaluation of the convolution integral in (15.27) was not introduced, and the implementation was first-order accurate in time.

In the mesh-based version of the hybrid method, the parameterization for the interface was arc-length, normalized in time to be $\alpha \in [0, 2\pi)$. With a fixed step size in α, this maintains spectral spatial accuracy of the underlying flow solver, and to retain this, Chebyshev–Lobatto collocation points were introduced in the normal N-direction to compute C in (15.13) through (15.17). The second derivative in (15.13) was evaluated implicitly via a Chebyshev differentiation matrix, and first derivative terms in (15.13) and (15.18) were found explicitly by FFTs. The time step was a semi-implicit two-step variation of the Crank–Nicolson method, giving second-order accuracy in time.

Figure 15.3 shows data from the mesh-based method (solid lines) and the mesh-free method (dotted) for an inviscid bubble in a simple shear flow, with $Ca = 0.5$, $E = 0.1$ in the linearized equation of state, $Pe_s = \infty$, $K = 1.5$, $J_0 = 1.0$, $Bi = \infty$, and an initial surface surfactant concentration $\Gamma(s, 0) = 0.6$

that is in equilibrium with the ambient bulk concentration. Panel (a) of Figure 15.3 shows the inter-face profile from time $t = 0.0$ up to $t = 4.0$ in increments of $\Delta t = 0.5$. A least squares fit of the data shows that the exact elliptical shape of the interface is maintained by both methods with an error that is close to round off. The profile appears to be close to a steady state at the final time $t = 4.0$. Panel (b) shows the surface surfactant concentration Γ versus the scaled arc-length α at the same times. This has local maxima near the bubble poles that become highly peaked as t increases. A close-up of this is shown in panel (c). Agreement between the results of the two methods is close, but at its worst near the maxima of Γ at later times. Better agreement is found when the constant number of mesh-free Lagrangian markers is increased. Panel (d) shows the bulk surfactant concentration at $t = 4.0$ as

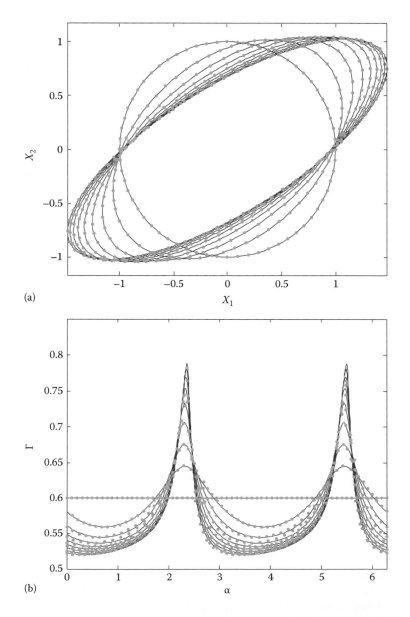

FIGURE 15.3 An inviscid bubble is deformed by a simple shear flow. Data of the mesh-based method (solid curves) and of the mesh-free method (markers) are superimposed. (a) Interface profile at times $t = 0$, (0.5), 4.0. (b) Surface surfactant concentration Γ versus scaled arc-length α at the same time. (*Continued*)

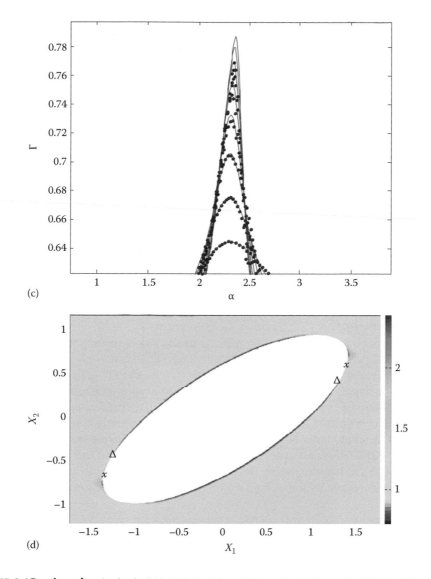

FIGURE 15.3 (*Continued*)   An inviscid bubble is deformed by a simple shear flow. Data of the mesh-based method (solid curves) and of the mesh-free method (markers) are superimposed. (c) Close-up of data from (b). (d) Bulk surfactant concentration C at time $t = 4.0$ with $Pe = 8000$. Stagnation points on the interface are indicated by × (interface contraction) or Δ (interface expansion). (Reprinted from Kuan, X.U., Booty, M.R., and Siegel, M., Analytical and computational methods for two-phase flow with soluble surfactant, *SIAM J. Appl. Math.*, 73, 523–548. Copyright 2013 Society for Industrial and Applied Mathematics. All rights reserved. With permission.)

computed by the mesh-based method with Pe set to 8000 to display the data. The four markers in the panel indicate the location of stagnation points on the interface, which steadily approach each other and coalesce at a time $t \simeq 7.0$, after which the direction of flow is everywhere clockwise around the interface.

For these runs, the mesh-based computations begin with $M = 128$ equally spaced grid points on the interface, which is increased by doubling to $M = 512$ before the final time $t = 4.0$. In the normal direction, there are $P = 64$ Chebyshev–Lobatto points, and the far-field boundary condition is imposed at $N_m = 20$. The time step is fixed at 10^{-3}, and the CPU time is 276 min on a 2.4 GHz AMD

Opteron model 180 processor. The computations with the mesh-free method have a constant number of $M = 256$ fluid markers on the interface throughout, the time step is 2.5×10^{-3}, and the CPU time is 332 min. In both methods, the bubble area is conserved with a relative error of less than 10^{-8}.

15.4 SAMPLE RESULTS

In a recent study [5], the mesh-based version of the hybrid method was extended to a 3D axisymmetric geometry to study the influence of surfactant solubility effects when the imposed flow is the uniaxial extension given by cylindrical polar coordinates in (15.4). The flow solver is based on the boundary integral equation

$$
u_\alpha(\boldsymbol{x}_0) + \frac{\lambda - 1}{4\pi(\lambda + 1)} \int_I^{PV} Q_{\alpha\beta\gamma}(\boldsymbol{x}, \boldsymbol{x}_0) u_\beta(\boldsymbol{x}) n_\gamma(\boldsymbol{x}) ds(\boldsymbol{x})
$$

$$
= \frac{2\boldsymbol{u}_\alpha^\infty(\boldsymbol{x}_0)}{1 + \lambda} + \frac{1}{4\pi(\lambda + 1)} \int_I M_{\alpha\beta}(\boldsymbol{x}, \boldsymbol{x}_0)(\nabla_s \sigma - \sigma\kappa\boldsymbol{n})_\beta(\boldsymbol{x}) ds(\boldsymbol{x}), \tag{15.30}
$$

see, for example, [18,19].

In (15.30), the subscripts are the coordinate symbols, r or x, and I traces the contour of the interface S in the first quadrant of the meridional plane. The kernels $Q_{\alpha,\beta,\gamma}$ and $M_{\alpha,\beta}$ are the axisymmetric stresslet and Stokeslet distributions, respectively, and PV denotes the principal value. Integrals with nonsingular kernels are evaluated by 8- or 16-point Gauss–Legendre quadrature, and weakly singular integrals that appear in the Stokeslet distributions are calculated by Gauss-log quadrature. Cubic splines are used to interpolate the interface position at quadrature points that lie between the computational grid points of I and to compute the normal direction and curvature of S. Cubic splines are also used to implement an adaptive grid technique that redistributes grid points to have arc-length spacing that is inversely proportional to the interface curvature, see [20]. This improves resolution of high curvature regions such as drop tips or necking regions close to pinch off.

The implementation has overall second-order accuracy in space and first order in time. In a typical computation, the interface profile is represented with $M = 160$–240 points in the first quadrant.

The method was validated by comparison with studies for drop deformation in axisymmetric flow with insoluble surfactant [21,22]. In a typical computation of this type, total surfactant mass was conserved to within about 1%, and the drop volume was conserved to within 0.001%.

Figure 15.4a shows comparison between data of the hybrid method (dashed lines) and an asymptotic solution (symbols) that is valid for small Ca in the steady state when the elasticity number $E = 0$ in (15.5) and the interface boundary condition (15.15) is replaced by $Cl_s = 0$. In this case, the surface surfactant concentration becomes completely uncoupled, the bulk concentration C does not influence the underlying flow, but the flow is coupled to the evolution of C, and the interface is a perfect sink for C. The asymptotic solution is given in [5]. The comparison is shown at equally spaced values of the declination ϕ from the axis of extension or drop tip, $\phi = 0$, at which $C \equiv 0$ for all N, to the drop equator, $\phi = \pi/2$. In the computations with the hybrid method, $Ca = 0.015$ and $\lambda = 1$ in the same limit for E and Cl_s.

Figure 15.4b shows the computed bulk concentration profiles at times that are sufficiently large that the interface shape is steady. However, bulk surfactant is seen exiting the transition layer in a plume that is beginning to form near the drop tip, where $\partial_n v_p|_s > 0$. The far-field boundary condition (15.16) must be applied to show this. In all other computations shown earlier, the simpler far-field

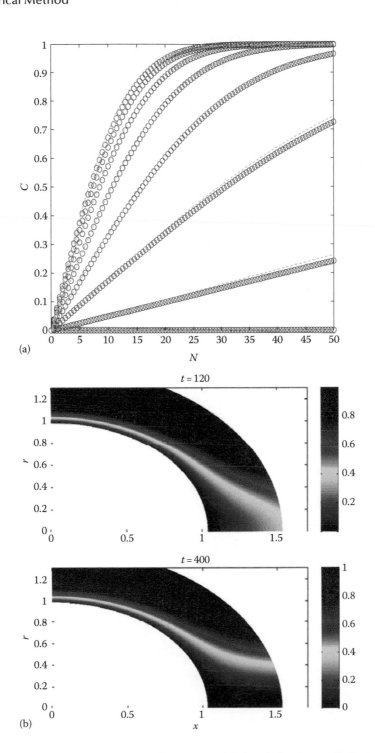

FIGURE 15.4 (a) Comparison between a steady-state asymptotic solution for small Ca (symbols) and the solution computed by the hybrid method (dashed lines) when $Ca = 0.015$, $\lambda = 1$, and $t \gg 1$, showing the bulk concentration $C(N,\phi_i)$ versus N at different locations ϕ_i. (b) The bulk concentration at times $t = 120$ and 400, plotted with $Pe = 10^4$. (Reprinted with permission from Wang, Q., Siegel, M., and Booty, M.R., Numerical simulation of drop and bubble dynamics with soluble surfactant, *Phys. Fluids*, 26, 052102. Copyright 2014, AIP Publishing LLC.)

condition $C \to 1$ was used, but the times are not sufficiently large for significant leaching of bulk surfactant out of the transition layer to occur.

Figure 15.5a shows an example of a tip-streaming filament at a sequence of times. The physical parameters are $\lambda = 0.025$, $Ca = 0.127$, $E = 0.2$, $Pe_s = 10^3$, $J_0 = 0.05$, $Bi = 0.02$, and $K = 1.5$, and the initial surfactant concentration $\Gamma(x,0) = 0.6$ is in equilibrium with the ambient bulk concentration. The viscosity ratio must be sufficiently small for tip streaming to occur, and the capillary number must be large enough for there to be no steady drop shape. The imposed flow simultaneously stretches the drop in the x-direction of the uniaxial extension and advects the surface surfactant concentration Γ toward the drop tip. Geometric focusing caused by decreasing radius near the tip causes a high surface concentration to develop there. In the absence of both surface diffusion ($Pe_s = \infty$) and solubility effects, surface surfactant is frozen or anchored into the flow, and the high concentration in the vicinity of the tip allows a thin thread or filament to be drawn out by the imposed flow. This persists with small surface diffusion and solubility. The combination of low surface tension on the thread and small radius at a longitudinal section $x = $ constant allow the thread to be in near equilibrium via the Young–Laplace relation $\Delta p = \sigma/R$ along its length for some time. At some point, however, the tip-streaming thread pinches off or breaks via capillary instability.

Panel (b) of Figure 15.5 shows the bulk surfactant concentration just before pinch off at the latest time $t = 26.0$ of this computational run. On the main, parent, or central part of the drop surface, the surface concentration has been decreased by stretching and advection toward the tip, so that solubility tends to replenish surface surfactant by adsorption from the bulk. Conversely, near the thread tip where the surface concentration is at its largest, surfactant desorbs from the interface to enter the bulk, as Figure 15.5b shows.

An indication that the computation is resolved is given by the profile in Figure 15.5a at time $t = 26.0$, which is shown with the data overlaid for computation with 200 nodes (dashed curve), 250 nodes (dotted curve), and 300 nodes (solid curve) distributed on the boundary in the first quadrant. Convergence is seen to be slowest near the drop tip.

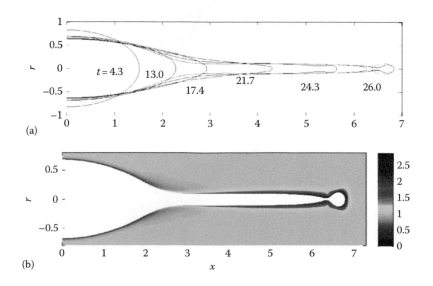

FIGURE 15.5 (a) Time evolution of a tip-streaming filament, with viscosity ratio $\lambda = 0.025$ and other parameters held fixed, shown at the indicated times. For the profile at $t = 26.0$, the overlaid dashed, dotted, and solid curves show convergence of the data near the drop tip as the mesh size is decreased. (b) Instantaneous bulk surfactant concentration at $t = 26$. (Reprinted with permission from Wang, Q., Siegel, M., and Booty, M.R., Numerical simulation of drop and bubble dynamics with soluble surfactant, *Phys. Fluids*, 26, 052102. Copyright 2014, AIP Publishing LLC.)

15.5   EXPERIMENT ON MICROFLUIDIC TIP STREAMING

Experiments using a microfluidic flow-focusing device to form a thin tip-streaming filament that subsequently breaks up into near-monodisperse droplets have been reported in [6,7] and are briefly reviewed here. A schematic diagram of the device is shown in Figure 15.6a. The exterior or continuous phase liquid, which is a light mineral oil, enters from the two outside channels, while the interior or dispersed phase liquid is aqueous and contains well-characterized dissolved surfactants with known bulk concentration. The viscosity ratio is fixed at $\lambda = 1/40 = 0.025$. The flow is focused toward an orifice in a baffle, downstream of the interior phase channel exit. This region is outlined by a dashed line in Figure 15.6a, and a close-up image of the orifice is shown to the right. Notation and the values of device dimensions and volume flow rates are introduced in the caption of Figure 15.6, with Q_C being the volume flow rate of the exterior phase in both side channels.

The volume flow rate of the two phases is controlled, and their ratio $\phi = Q_C/Q_D$ can be varied in the range $40 \leq \phi \leq 320$. The Reynolds number based on either phase flow rate is in the range from 10^{-3} to 3. The surfactant used in [6] was $C_{12}E_8$, which is a nonionic polyethoxylated surfactant that is soluble in water. Solutions of $C_{12}E_8$ in deionized water with bulk concentration C_∞ in the range from 0 to 14.5 times the critical micelle concentration were used. The three quantities that can be varied for experiments with a specific surfactant are the capillary number, the flow rate ratio ϕ, and the bulk surfactant concentration C_∞.

A well-defined capillary number was introduced in [6] that is based on the surface tension when the surface and bulk surfactant concentrations are in equilibrium, that is, with $\Gamma_e = K/(1 + K)$, and

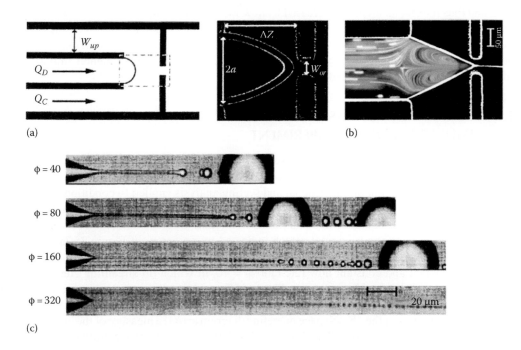

FIGURE 15.6   (a) Schematic diagram (left) denoting flow of both phases from left to right and an image (right) of the orifice region with interface not tip streaming. The dimensions shown are $W_{up} = 280$ μm, $a = 90$ μm, $\Delta Z = 180$ μm, and $W_{or} = 34$ μm. (b) Streakline image of the interior, aqueous fluid during formation of a tip-streaming thread (vertical length bar 50 μm). (c) Representative images of thread formation and subsequent breakup at different values of the flow rate ratio $\phi = Q_C/Q_D$. (Reprinted with permission from Anna, S.L. and Mayer, H.C., Microscale tipstreaming in a microfluidic flow focusing device, *Phys. Fluids*, 18, 121512. Copyright 2006, AIP Publishing LLC.)

on an elongation rate $G = \Delta V/\Delta Z$, where $\Delta V = V_{constr} - V_{up}$ is the difference between the average flow velocity in the orifice and in the upstream side channels.

Under suitable conditions, the combination of flow focusing and surfactant cause the interface to emit a tip-streaming filament or thread, which is drawn downstream through the orifice as shown in Figure 15.6b. The device geometry is planar, with depth 98 μm, but the interface becomes conical toward the tip.

Figure 15.6c shows a mode of operation in which a primary drop with diameter comparable to the orifice size is produced, followed by a time interval during which a very thin thread is drawn out from the interior fluid cone tip. The thin thread grows to a maximum length, after which it pinches off at the end of the interior fluid cone; the cone tip then becomes more rounded before subsequently emitting a second large primary drop, and the process repeats in a more or less periodic manner.

With other conditions fixed, as the flow rate ratio φ increases, the maximum length of the intact thread increases and its radius appears to decrease. Droplet sizes resulting from capillary breakup of the thin thread decrease accordingly. This is seen with φ = 40, 80, and 160 in Figure 15.6c, which are shown when the thread has reached its maximum length. When the flow rate ratio is increased above φ = 160, the time interval between primary droplet formation increases further, until a state of nearly continuous tip streaming is reached when φ = 320, as shown. Droplets of 1–2 μm diameter are produced in a nearly continuous stream for a relatively long time interval of 10–100 ms compared with 1–10 ms at lower φ.

The studies [6,7] and references given there contain much more discussion on the experimental conditions and on the dynamics that result. Here, we point out that there is a continual generation of fresh interface at the exit from the interior channel, see Figure 15.6b, with initial adsorption of surfactant to the interface from the interior bulk flow. This is followed by a stage in which the cone radius and thus interfacial area are decreased by flow focusing as it approaches the cone tip near the orifice. During this stage, the interface and bulk surfactant concentrations are more or less close to local equilibrium, with exchange occurring between the interface and the recirculating cone eddy. The dynamics of surfactant adsorption and desorption is clearly of fundamental importance to understanding the operation of the device.

15.6 PRELIMINARY RESULTS FROM NUMERICAL SIMULATION OF THE EXPERIMENT

In the more recent study [7], bounds are developed that define a region in a dimensionless phase space in which the microfluidic device is expected to operate in a tip-streaming mode. The bounds are derived by a combination of dimensional and scaling analysis applied to conditions at the interface of the tip-streaming cone and thread, so that, for example, an interfacial surfactant balance is maintained, and the stress-balance boundary condition (15.3) is satisfied. Combined with physical constraints, this leads to the triangular region indicated by an arrow in Figure 15.7a. The figure also shows data from experimental runs using a device of dimensions that are close to those of [6], with $C_{12}E_8$ surfactant and constant flow rate ratio φ = 40, in which tip-streaming conditions were observed. These are indicated by squares in Figure 15.7a, and good agreement is seen between the predicted bounds and experimental data.

The state parameters of the phase-space operating diagram are (1) a measure of the bulk surfactant concentration and (2) the flow rate. The dimensionless bulk surfactant concentration is given by

$$\bar{C} = \frac{\mu_2 a \kappa_a C_\infty}{RT\Gamma_\infty}, \tag{15.31}$$

which is the ratio of a surface relaxation time $\tau_s = \mu_2 a/RT\Gamma_\infty$ to the timescale for surfactant adsorption $\tau_a = 1/\kappa_a C_\infty$. The dimensionless flow rate $\bar{Q}$, up to a numerical prefactor that depends on the

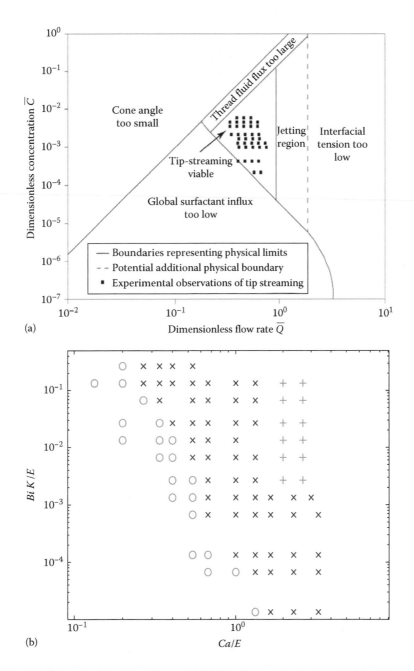

FIGURE 15.7 Conditions for tip streaming and other breakup modes in terms of the dimensionless bulk concentration $\bar{C}$ and flow rate $\bar{Q}$ (note log–log scales). (a) Experimental data and region within which tip streaming is expected to be viable. (Reprinted with permission from Moyle, T.M., Walker, L.M., and Anna, S.L., Predicting conditions for microscale surfactant mediated tip streaming, *Phys. Fluids*, 24, 082110. Copyright 2012, AIP Publishing LLC.) (b) Simulation data with an unbounded geometry. The "o" markers denote steady-state drop shapes, "×" markers denote tip-streaming drops, and "+" markers denote a nontip-streaming breakup mode. For parameter values and other details, see text. (Reprinted with permission from Wang, Q., Siegel, M., and Booty, M.R., Numerical simulation of drop and bubble dynamics with soluble surfactant, *Phys. Fluids*, 26, 052102. Copyright 2014, AIP Publishing LLC.)

device geometry, is given by the surface relaxation time τ_s times a convection rate that is proportional to the flow rate of the interior fluid and also depends on the device geometry and dimensions. See [7] for the precise definition of $\bar{Q}$.

In the numerical study [5], we reported on simulations of tip streaming and states near tip streaming in an unconfined geometry, so as to make a comparison with these experimental results—despite there being apparently important distinctions between the experiment and simulation setup. The simulations use the hybrid numerical method of the infinite bulk Peclet number limit with the finite Biot number, mixed-kinetic form of the surfactant exchange boundary condition (15.15), an example of which is shown in Figure 15.5. The drop is initially at equilibrium with its surroundings and is then stretched by an imposed uniaxial extension. Parameter values of the simulations are generally close to those in the experiments of [7], and estimates based on the experimental data were given in [5]; fixed parameter values of the simulations are $\lambda = 1/40$, $E = 0.15$, $Pe_s = 10^3$, and $Bi = 2 \times 10^{-4}$. The experimentally relevant bulk surfactant concentration C_∞ of (15.31) is changed with other parameters held fixed by varying J_0 and K with J_0/K fixed in the simulations. In the present notation,

$$\bar{C} = \frac{BiK}{E} \quad \text{and} \quad \bar{Q} \simeq \frac{\mu_2 aG}{RT_\infty} = \frac{Ca}{E}, \tag{15.32}$$

where $\bar{C}$ is defined at (15.31), per [7]. However, with the unconfined geometry of the simulations, it is not possible to exactly reproduce the definition of $\bar{Q}$ in the flow-focusing experiments, so that the experimental convection rate is approximated by the imposed strain rate G. This gives the axes of the simulation data for conditions of tip streaming that are shown for comparison in Figure 15.7b. Note the difference in the axis scales between parts (a) and (b) of Figure 15.7.

Conditions for tip streaming are denoted by "×" in the simulation data of Figure 15.7b, and conditions that produce a steady drop shape are denoted by "o." At relatively large ambient surfactant concentration and flow rate, in the simulations, tip streaming transitions to a different breakup mode denoted by "+," in which the drop is drawn out rapidly into a near-cylindrical shape. This shape has a small decrease in radius and small increase in Γ with increasing axial distance x, relative to tip streaming, in which there is a clearer distinction between a large parent drop and thin tip-streaming thread. In the experiments, the corresponding region is described as jetting in Figure 15.7a. Jetting is characterized by a broader finger or jet of the interior fluid that breaks up relatively quickly downstream of the orifice into drops that are less monodisperse and with diameter comparable to the orifice width. This can occur with flow focusing in the absence of surfactant, see Figure 4d of [6].

The differences in setup between the experiments and simulations are (1) the flow focusing versus unconfined geometry and (2) the continual supply of surfactant from the interior bulk flow to an initially clean interface at the exit from the interior fluid channel versus an initial equilibrium surface concentration of surfactant and bulk–interface exchange with the exterior flow.

Given these differences, the level of agreement in location of the tip-streaming region seen in Figure 15.7 is surprisingly good or may partly be a consequence of chance. The region of the simulations shows some dependence on the choice of parameters, and if the Biot number is increased from $Bi = 2 \times 10^{-4}$ to $Bi = 2 \times 10^{-2}$ in the simulations, the phase diagram is similar to that of Figure 15.7b but with each region translated upward and to the right. The distinction between tip streaming and other dynamic breakup modes is also to some extent subjective, so that location of transition boundaries is liable to some interpretation. In ongoing work, the simulations have been adapted to include axisymmetric, annular flow-focusing baffles in an attempt to explore the influence of flow focusing on the tip streaming of an otherwise isolated drop.

15.7 CONCLUSION

The main theme of this chapter is the formulation and development of a hybrid numerical method that has been developed for studies of two-phase flow to include the effects of surfactant solubility. In most circumstances, the bulk Peclet number in the equation for evolution of bulk surfactant concentration is large, and of the order of 10^6 or more. This causes large gradients in bulk concentration to develop adjacent to a fluid interface when the surface and bulk concentrations are out of equilibrium. A leading order asymptotic reduction for the evolution of bulk surfactant in a surface-fitted coordinate system in the limit $Pe \to \infty$ is introduced to resolve the fine spatial scale of this transition layer. Solution of the initial boundary value problem for the transition layer dynamics can be solved in parallel with a numerical method for solving two-phase flow with insoluble surfactant. Here, we have used surface-based boundary integral methods.

In studies with the hybrid method that are described here, the bulk surfactant concentration away from the transition layer is taken to be spatially uniform and effectively diffusion-free. The first of these assumptions can be relaxed by incorporating, for example, a relatively coarse mesh semi-Lagrangian flow solver in the bulk flow, which requires the flow velocity away from the interface.

In presenting the development of the hybrid method, attention has centered on its validation in the large Peclet number limit and on validation of the numerical implementation. The references cited include results from more detailed parameter studies in various cases and limits that seek to provide insight into the mechanism of surfactant solubility, and how this depends on, for example, dimensionless parameters such as the Biot number and exchange parameter that appear in the mixed-kinetic form of the boundary condition (15.8), or (15.15) in the large Pe limit.

Results from recent, ongoing experimental studies on tip streaming in a microfluidic flow-focusing device have also been reviewed. The experiments aim to provide a monodisperse stream of micron-size droplets for use in a variety of emerging applications. The experiments directly motivate a fundamental understanding of surfactant solubility effects, and their numerical simulation provides a significant challenge. Comparison of the modeling and experimental results with the results of numerical simulations in an unconfined geometry that take first steps in this direction have been made and reviewed here.

ACKNOWLEDGMENT

This work was supported by the National Science Foundation (NSF) under grant nos. DMS-1009105, DMS-1412789, and CBET-1033814.

REFERENCES

1. C.-H. Chang and E.I. Franses. Adsorption dynamics of surfactants at the air/water interface: A critical review of mathematical models, data, and mechanisms. *Colloid Surf. A*, 100, 1–45, 1995.
2. H.A. Stone. Dynamics of drop deformation and breakup in viscous fluids. In *Annual Reviews of Fluid Mechanics*, J. L. Lumley, M. Van Dyke, and H. L. Reed (Eds). Vol. 26, pp. 65–102, Annual Reviews, Palo Alto, CA, 1994.
3. M.R. Booty and M. Siegel. A hybrid numerical method for interfacial fluid flow with soluble surfactant. *J. Comput. Phys.*, 229, 3864–3883, 2010.
4. K. Xu, M.R. Booty, and M. Siegel. Analytical and computational methods for two-phase flow with soluble surfactant. *SIAM J. Appl. Math.*, 73, 523–548, 2013.
5. Q. Wang, M. Siegel, and M.R. Booty. Numerical simulation of drop and bubble dynamics with soluble surfactant. *Phys. Fluids*, 26, 052102, 2014.
6. S.L. Anna and H.C. Mayer. Microscale tipstreaming in a microfluidic flow focusing device. *Phys. Fluids*, 18, 121512, 2006.

7. T.M. Moyle, L.M. Walker, and S.L. Anna. Predicting conditions for microscale surfactant mediated tipstreaming. *Phys. Fluids*, 24, 082110, 2012.

8. B. Ambravaneswaran and O.A. Basaran. Effects of insoluble surfactants on the nonlinear deformation and breakup of stretching liquid bridges. *Phys. Fluids*, 11, 997–1015, 1999.

9. R.V. Craster, O.K. Matar, and D.T. Papageorgiou. Pinchoff and satellite formation in surfactant covered viscous threads. *Phys. Fluids*, 14, 1364–1376, 2002.

10. Y.-N. Young, M.R. Booty, M. Siegel, and J. Li. Influence of surfactant solubility on the deformation and breakup of a bubble or capillary jet in a viscous fluid. *Phys. Fluids*, 21, 072105, 2009.

11. W.E. Langlois. *Slow Viscous Flow*. MacMillan, New York, 1964.

12. S. Richardson. Two-dimensional slow viscous flows with time-dependent free boundaries driven by surface tension. *Euro. J. Appl. Math.*, 3, 193–207, 1992.

13. S. Tanveer and G.L. Vasconcelos. Time-evolving bubbles in two-dimensional Stokes flow. *J. Fluid Mech.*, 301, 325–344, 1995.

14. M. Siegel. Influence of surfactant on rounded and pointed bubbles in two-dimensional Stokes flow. *SIAM J. Appl. Math.*, 59, 1998–2027, 1999.

15. N.I. Muskhelishvili. *Some Basic Problems of the Mathematical Theory of Elasticity*. P. Noordhoff, Groningen, the Netherlands, 1963.

16. L. Greengard, M.C. Kropinski, and A. Mayo. Integral equation methods for Stokes flow and isotropic elasticity in the plane. *J. Comput. Phys.*, 125, 403–414, 1996.

17. M.C.A. Kropinski and E. Lushi. Efficient numerical methods for multiple surfactant-coated bubbles in a two-dimensional Stokes flow. *J. Comput. Phys.*, 230, 4466–4487, 2011.

18. C. Pozrikidis. *Boundary Integral and Singularity Methods for Linearized Viscous Flow*. Cambridge University Press, Cambridge, U.K., 1992.

19. H.A. Stone and L.G. Leal. Relaxation and breakup of an initially extended drop in an otherwise quiescent fluid. *J. Fluid Mech.*, 198, 399–427, 1989.

20. J.R. Lister and H.A. Stone. Capillary breakup of a viscous thread surrounded by another viscous fluid. *Phys. Fluids*, 10, 2758–2764, 1998.

21. H.A. Stone and L.G. Leal. The effects of surfactants on drop deformation and breakup. *J. Fluid Mech.*, 220, 161–186, 1990.

22. C.D. Eggleton, T.M. Tsai, and K. Stebe. Tip streaming from a drop in the presence of surfactant. *Phys. Rev. Lett.*, 87, 048302, 2001.

16 Finite Element Computations for Dynamic Liquid–Fluid Interfaces

Sashikumaar Ganesan, Andreas Hahn,
Kristin Simon, and Lutz Tobiska

CONTENTS

16.1 INTRODUCTION

The influence of surface active agents (surfactants) on the deformation of droplets and the structure of the surrounding flow field is an active research area with numerous applications. The convective transport of surfactants induced by the flow field generates a local accumulation resulting in a non-uniform concentration of surfactants at the liquid–fluid interface. The appearing Marangoni forces may lead to a destabilization of the interface with essential consequences for the flow structure. This is a complex process, whose tailored use in applications requires a fundamental understanding of the mutual interplay.

We present advanced finite element techniques that allow for a robust and accurate numerical solution of the underlying system of partial differential equations (PDEs). Two test examples demonstrate the potential of the proposed method.

16.2   TWO-PHASE AND FREE-SURFACE FLOWS

We derive a system of PDEs describing two-phase flows and free surface flows and complete it by initial and boundary conditions. Finally, we derive a variational formulation of the problem, which is needed for finite element computations.

16.2.1   GOVERNING EQUATION SYSTEMS

We consider an incompressible two-phase flow with surfactant in a fixed bounded domain $\Omega \subset \mathbb{R}^d, d = 2,3$. We assume that the liquid filling $\Omega_1(t)$ is completely surrounded by another liquid filling $\Omega_2(t) = \Omega \backslash \Omega_1(t)$ and the liquids are immiscible. Here, $t \in [0,T]$ denotes the time. The interface between the liquids is denoted by $\Gamma_F(t) = \partial\Omega_1(t)$.

The mathematical model consists of the time-dependent incompressible Navier–Stokes equations in each phase

$$\rho_k(\partial_t \mathbf{u} + (\mathbf{u} \cdot \nabla)\mathbf{u}) - \nabla \cdot \mathbb{S}_k(\mathbf{u}, p) = \rho_k g\mathbf{e}, \quad \nabla \cdot \mathbf{u} = 0 \text{ in } \Omega_k(t) \times (0,T], \tag{16.1}$$

$k = 1, 2$, completed by the initial conditions,

$$\Omega_k \big|_{t=0} = \Omega_{k,0}, \quad \mathbf{u} \big|_{t=0} = \mathbf{u}_0,$$

the kinematic and force balancing conditions,

$$\mathbf{w} \cdot \mathbf{n} = \mathbf{u} \cdot \mathbf{n}, \quad [\mathbf{u}] = \mathbf{0}, \quad [\mathbb{S}(\mathbf{u}, p)] \cdot \mathbf{n} = \sigma(c_\Gamma)\mathcal{K}\mathbf{n} + \nabla_\Gamma \sigma(c_\Gamma) \quad \text{on } \Gamma_F(t),$$

and homogeneous Dirichlet-type boundary conditions on the fixed (in time) boundary $\partial\Omega$. Here, $\mathbf{u} = (u_1,\ldots,u_d)$ is the fluid velocity, p is the pressure, ρ_k is the density of the associated fluid phase, g is the gravitational constant, and $\mathbf{e}$ is a unit vector in the direction of the gravitational force. Further, $\mathbf{w}$ on $\Gamma_F(t)$ denotes the interface velocity, $\mathbf{n}$ is the unit normal vector (on $\Gamma_F(t)$ directed outward of $\Omega_1(t)$), $\mathcal{K}$ is the sum of principal curvatures, $[\cdot]$ is the jump across the interface $\Gamma_F(t)$, c_Γ is the surfactant concentration on the interface, and $\sigma(c_\Gamma)$ is the surface tension coefficient depending on c_Γ. Note that $\sigma(c_\Gamma)\mathcal{K}\mathbf{n} + \nabla_\Gamma \sigma(c_\Gamma)$ can be written as surface divergence of a surface stress tensor; see Section 16.4.4.

The stress tensor $\mathbb{S}_k(\mathbf{u}, p)$ for a Newtonian incompressible fluid and the velocity deformation tensor $\mathbb{D}(\mathbf{u})$ are given by

$$\mathbb{S}_k(\mathbf{u}, p) = 2\mu_k \mathbb{D}(\mathbf{u}) - p\mathbb{I}, \quad \mathbb{D}(\mathbf{u}) = \frac{1}{2}\left(\nabla\mathbf{u} + (\nabla\mathbf{u})^T\right), \quad k = 1,2,$$

where

μ_k is denoting the dynamic viscosity of fluid k

$\mathbb{I}$ is the identity tensor

In the case of two-phase flows, the pressure is unique up to an additive constant only. Indeed, we have for $\tilde{p} = p + \text{const}$

$$\nabla \cdot \mathbb{S}(\mathbf{u}, \tilde{p}) = \nabla \cdot \mathbb{S}(\mathbf{u}, p),$$

and the force balancing condition only includes the pressure jump over the interface. Therefore, the pressure has to be fixed by an additional condition, for example, setting the mean value to zero.

This is different for free surface flows, where the normal component of the normal stress on Γ_F satisfies

$$\mathbf{n} \cdot \mathbb{S}(\mathbf{u}, \tilde{p}) \cdot \mathbf{n} - \mathbf{n} \cdot \mathbb{S}(\mathbf{u}, p) \cdot \mathbf{n} = p - \tilde{p} \neq 0.$$

It remains to add equations describing the surfactant transport. Let us assume that the surfactant is insoluble in $\Omega_1(t)$ and soluble in $\Omega_2(t)$. We denote the surfactant concentration in the outer phase by C. For the transport in the bulk phase, we have

$$\partial_t C + \mathbf{u} \cdot \nabla C = \nabla \cdot [D_C \nabla C] \quad \text{in } \Omega_2(t) \times (0, T], \tag{16.2}$$

completed by the initial condition

$$C\big|_{t=0} = C(0) \quad \text{in } \Omega_2(0),$$

and the boundary conditions

$$\mathbf{n} \cdot D_C \nabla C = -S(C, c_\Gamma) \quad \text{on } \Gamma_F(t), \quad \mathbf{n} \cdot D_C \nabla C = 0 \quad \text{on } \partial \Omega.$$

Here
   D_C is the diffusive coefficient of the outer phase surfactant concentration C
   S is the source term depending on C and c_Γ

The surfactant concentration c_Γ on Γ_F is described by the initial condition $c_\Gamma\big|_{t=0} = c_\Gamma(0)$ and the PDE on the evolving surface $\Gamma_F(t)$

$$\dot{c}_\Gamma + (\nabla_\Gamma \cdot \mathbf{u}) c_\Gamma = \nabla_\Gamma \cdot (D_\Gamma \nabla_\Gamma c_\Gamma) + S(C, c_\Gamma) \quad \text{on } \Gamma_F \times (0, T], \tag{16.3}$$

with
   $\dot{c}_\Gamma$ denoting the material derivative
   D_Γ is the diffusion coefficient

The surface gradient ∇_Γ of a function φ living in a neighbouhood of Γ_F is given by $\nabla_\Gamma \varphi := \nabla \varphi - (\nabla \varphi \cdot \mathbf{n})\mathbf{n}$. It can be proven that the restriction of $\nabla_\Gamma \varphi$ on Γ_F depends only on the values of φ on Γ_F. Therefore, $\nabla_\Gamma c_\Gamma$ can be calculated by extending c_Γ to a neighborhood of the surface. The surface divergence of the velocity $\nabla_\Gamma \cdot \mathbf{u}$ is given as the trace of the projected velocity deformation tensor onto the tangential plane, that is,

$$\nabla_\Gamma \cdot \mathbf{u} = tr((\mathbb{I} - \mathbf{n} \otimes \mathbf{n})\mathbb{D}(\mathbf{u})).$$

Note that we assume the free surface $\Gamma_F(t)$ to be closed. Thus no boundary conditions are needed for the surface PDE.

Finally, we need to model the dependencies of the surface tension σ on c_Γ and of the source term S on C and c_Γ. In applications, these are often deduced from the linear Henry or the nonlinear Langmuir equation of state [1]. In the linear case, we get

$$S(C, c_\Gamma) = k_a c_\Gamma^\infty C - k_d c_\Gamma \quad \text{and} \quad \sigma(c_\Gamma) = \sigma_0 \left(1 + E \left(1 - \frac{c_\Gamma}{c_\Gamma^\infty} \right) \right),$$

Here

c_Γ^∞ denotes a reference surfactant concentration

σ_0 is the reference surface tension for a surfactant concentration of c_Γ^∞

$E = RT_\infty c_\Gamma^\infty / \sigma_0$ is the corresponding surface elasticity, R is the ideal gas constant, T_∞ is the temperature of the system

k_a and k_d are the adsorption and desorption constants

The nonlinear equation of state leads to

$$S(C, c_\Gamma) = k_a C \left(c_\Gamma^\infty - c_\Gamma \right) - k_d c_\Gamma \quad \text{and} \quad \sigma(c_\Gamma) = \sigma_0 \left(1 + E \ln \left(1 - \frac{c_\Gamma}{c_\Gamma^\infty} \right) \right),$$

with c_Γ^∞ set to the maximum surface surfactant packing concentration.

Two-phase flows with different boundary conditions besides $\Gamma_F(t)$ and free-surface flows can be modeled in a similar way. Only a few changes have to be made. For example, if we want to model a freely oscillating droplet, there is only one (time-dependent) fluid phase $\Omega(t)$ to be considered. It has a closed free boundary $\Gamma_F(t) = \partial \Omega(t)$ and the surfactant is soluble in the fluid phase.

In the following, we will refer to the stated two-phase problem. Nevertheless, all of the presented techniques can be transferred to other two-phase or free surface flows easily.

The model described previously is a coupled nonlinear system of PDEs with respect to the unknown velocity $\mathbf{u}$, pressure p, concentration of surfactant in the bulk phase C and on the surface c_Γ, and the position of the free surface $\Gamma_F(t)$.

Some of the challenges to solve the PDE system numerically are the accurate tracking or capturing of the unknown interface Γ_F, the handling of jumps in the fluid properties, the fluid and surfactant mass conservation over time, the inclusion of capillary effects, and the handling of Marangoni forces.

16.2.2 VARIATIONAL FORM OF THE INCOMPRESSIBLE NAVIER–STOKES EQUATIONS IN FIXED DOMAINS

Next, we derive the variational form of the Navier–Stokes equations in a fixed domain Ω. The immiscible two-phase flow Navier–Stokes equations can be written into a one-field formulation by defining

$$\rho(\mathbf{x}) = \begin{cases} \rho_1 & \text{for } \mathbf{x} \text{ in } \Omega_1 \\ \rho_2 & \text{for } \mathbf{x} \text{ in } \Omega_2 \end{cases}, \quad \mu(\mathbf{x}) = \begin{cases} \mu_1 & \text{for } \mathbf{x} \text{ in } \Omega_1 \\ \mu_2 & \text{for } \mathbf{x} \text{ in } \Omega_2 \end{cases}.$$

Let $V(\Omega) := (H_0^1(\Omega))^3$ and $Q(\Omega) := L_0^2(\Omega)$ be the usual Sobolev spaces and its vector-valued versions, and let $(\cdot, \cdot)_\Omega$ be the inner product in $L^2(\Omega)$. We suppress the argument Ω and write V and Q, respectively, if the context is clear.

To obtain a variational form of Equation 16.1, we multiply the momentum equation by a test function $\mathbf{v} \in V$ and the mass balance equation by a test function $q \in Q$. Then, integration by parts in each sub-domain Ω_k, $k = 1, 2$, is applied. Incorporating the force balancing conditions on the interface, the variational form of Equation 16.1 reads as follows.

For given Ω and $\mathbf{u}_0 = \mathbf{u}(0)$, find $(\mathbf{u}, p) \in V \times Q$ such that

$$(\rho \partial_t \mathbf{u}, \mathbf{v})_\Omega + a(\mathbf{u}, \mathbf{u}, \mathbf{v})_\Omega - b(p, \mathbf{v})_\Omega = f(c_\Gamma, \mathbf{v})_\Omega, \tag{16.4}$$

$$b(q, \mathbf{u})_\Omega = 0, \tag{16.5}$$

for all $(\mathbf{v},q) \in V \times Q$ and $t \in (0,T]$, where

$$a(\mathbf{w},\mathbf{u},\mathbf{v})_\Omega = 2(\mu\mathbb{D}(\mathbf{u}),\mathbb{D}(\mathbf{v}))_\Omega + (\rho(\mathbf{w}\cdot\nabla)\mathbf{u},\mathbf{v})_\Omega,$$

$$b(p,\mathbf{v})_\Omega = (p,\nabla\cdot\mathbf{v})_\Omega,$$

$$f(c_\Gamma,\mathbf{v})_\Omega = g(\rho\mathbf{e},\mathbf{v})_\Omega + \int_{\Gamma_F} \mathbf{v}\cdot(\sigma(c_\Gamma)\mathcal{K}\mathbf{n} + \nabla_\Gamma\sigma(c_\Gamma))\,\mathrm{d}s.$$

16.2.3 Variational Form of the Surfactant Transport Equations in Fixed Domains

The variational form of the transport equations in the bulk and on the interface can be derived in a similar way. Let $G(\Omega_2):= H^1(\Omega_2)$ and $M(\Gamma_F):= H^1(\Gamma_F)$ be the solution spaces for the bulk and the interface equations. Again, we skip the arguments and write G and M if the context is clear.

The variational form of the surfactant transport equations can be obtained by multiplying the bulk Equation 16.2 with $\phi \in G$ and the interface Equation 16.3 with $\psi \in M$, integrating over Ω_2 and Γ_F, respectively, and incorporating the boundary conditions for C. Let $\langle\cdot,\cdot\rangle_{\Gamma_F}$ be the inner product in $L^2(\Gamma_F)$; then the variational formulation reads

For given Ω_2, C_0, $c_{\Gamma,0}$, $\mathbf{u}$, find $(C,c_\Gamma) \in G \times M$ such that

$$(\partial_t C,\phi)_{\Omega_2} + a_c(\mathbf{u},C,\phi)_{\Omega_2} = f_c(C,c_\Gamma,\phi)_{\Gamma_F}, \tag{16.6}$$

$$\frac{\mathrm{d}}{\mathrm{d}t}\langle c_\Gamma,\psi\rangle_{\Gamma_F} + a_\Gamma(c_\Gamma,\psi)_{\Gamma_F} = f_\Gamma(C,c_\Gamma,\psi)_{\Gamma_F}, \tag{16.7}$$

for all $(\phi,\psi) \in G \times M$ and $t \in (0,T]$, where

$$a_c(\mathbf{u},C,\phi)_{\Omega_2} = D_c(\nabla C,\nabla\phi)_{\Omega_2} + ((\mathbf{u}\cdot\nabla)C,\phi)_{\Omega_2},$$

$$f_c(C,c_\Gamma,\phi)_{\Gamma_F} = -\langle S(C,c_\Gamma),\phi\rangle_{\Gamma_F},$$

$$a_\Gamma(c_\Gamma,\psi)_{\Gamma_F} = D_s\langle\nabla_\Gamma c_\Gamma,\nabla_\Gamma\psi\rangle_{\Gamma_F},$$

$$f_\Gamma(C,c_\Gamma,\phi)_{\Gamma_F} = \langle S(C,c_\Gamma),\psi\rangle_{\Gamma_F}.$$

16.3 INTERFACE CAPTURING AND TRACKING METHODS

There are well-established numerical methods for solving systems of PDEs in fixed domains. However, in case an unknown interface (or a free boundary) is part of the problem a new level of complexity appears. We need an appropriate representation of the interface and techniques to model its movement. We mainly distinguish between interface capturing and interface tracking methods.

In the first group, the interface is described implicitly by an indicator function defined on a domain containing the bulk phases. In volume of fluid methods, the volume fraction of each phase in a cell is used (see, e.g., [2,3]), and the position of the interface has to be reconstructed. In finite element approaches, the interface often is captured by the level set method described in Section 16.3.1.

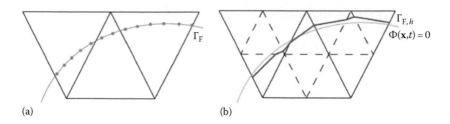

FIGURE 16.1 In front tracking methods, Γ_F is represented by markers • (a) and reconstruction of approximate interface $\Gamma_{F,h}$ in level set methods (b).

Interface tracking methods are based on an explicit representation of the interface. In its simplest form a set of markers is put on the interface $\Gamma_F(t)$ and transported by the flow field **u** to obtain the markers on the interface $\Gamma_F(t + \Delta t)$ (see Figure 16.1 (a) and, e.g., [4]). We consider two representatives in which the markers are the vertices of a triangulation of $\Gamma_F(t)$, the direct parametric surface tracking in Section 16.3.2 and the arbitrary Lagrangian–Eulerian approach in Section 16.3.3, respectively.

16.3.1 Level Set Techniques

Level set techniques can capture the motion of an interface $\Gamma_F(t) \in \mathbb{R}^d$ embedded in a fixed domain Ω. They have been developed in 1988 by Sethian and Osher [5]. According to the name, the interface is described by the zero level of a scalar field $\Phi(\mathbf{x},t) : \Omega \times [0,T] \to \mathbb{R}$.

Initially, $\Phi(\mathbf{x},0)$ is given by the signed distance function to the initial interface $\Gamma_F(0)$. Therefore, $\Gamma_F(0)$ is represented by the zero level of $\Phi(\mathbf{x},0)$ and $|\nabla\Phi(\mathbf{x},0)| = 1$ is ensured in the whole domain Ω. In a velocity field $\mathbf{u}(\mathbf{x}, t)$, the motion of the interface is given by the transport of the scalar field Φ, modeled by

$$\partial_t\Phi + \mathbf{u} \cdot \nabla\Phi = 0 \quad \text{in } \Omega \times (0,T].$$

This ensures that every point belonging to the interface remains at the zero level of $\Phi(\mathbf{x},t)$. The big advantage of level set techniques is their ability to inherently handle topological changes.

The transport equation is solved numerically on a fixed mesh. Its hyperbolic type requires stabilization and higher-order methods to gain appropriate numerical solutions. Hence, the level set method provides challenges regarding mass and shape conservation. To improve the shape conservation, the property $|\nabla\Phi(\mathbf{x},t)| = 1$ has to be ensured over time. Reinitializations of $\Phi(\mathbf{x},t)$ as an approximate signed distance function are needed [6].

Another disadvantage is the implicit description of the interface. Its position at time t has to be reconstructed as zero level of $\Phi(\mathbf{x},t)$, which can be rather complex. Although, the transport equation is solved by piecewise polynomial functions using higher-order methods, to gain the approximate interface $\Gamma_{F,h}(t)$ piecewise linear interpolations of $\Phi(\mathbf{x},t)$ (on a finer mesh) are often used [compare Figure 16.1 (b)].

A surface mesh is given directly by the restriction of the given bulk mesh to $\Gamma_{F,h}(t)$. These meshes are no longer shape regular. Nonetheless, shape regularity of the underlying bulk mesh can compensate most of the upcoming problems. This kind of meshes is the object of our current research, see, for example, [7–9].

16.3.2 Interface Tracking by Direct Parametric Surfaces

Having in mind that we have to solve PDEs in the bulk and on the interface, it is natural to use separate meshes to discretize Ω and $\Gamma_F(t)$. The position of the interface is described by the vertices of the surface mesh (markers). A Eulerian formulation of the Navier–Stokes equation on the bulk mesh gives the fluid velocity $\mathbf{u}$ in each time step.

Note that the motion of the interface is only driven by the normal component of the interface velocity $\mathbf{w}$. The tangential component does not change the shape of the interface but the distribution of the markers on the interface and, hence, influences the mesh quality. There are many ways to move the surface according to the fluid velocity, that is, satisfying

$$\mathbf{w} \cdot \mathbf{n} = \mathbf{u} \cdot \mathbf{n} \quad \text{on } \Gamma_{\mathrm{F}}(t).$$

Two standard approaches are to move the markers of the interface with the fluid velocity $\mathbf{w} = \mathbf{u}$ or with its normal projection $\mathbf{w} = (\mathbf{u} \cdot \mathbf{n})\mathbf{n}$ leading to

$$\frac{d\mathbf{X}}{dt} = \mathbf{u} \quad \text{or} \quad \frac{d\mathbf{X}}{dt} = (\mathbf{u} \cdot \mathbf{n})\mathbf{n}, \tag{16.8}$$

for the position vector $\mathbf{X}$.

Recently, a new approach has been developed [10–12], which exploits the tangential part of the surface velocity to improve the mesh quality. The idea combines a weak formulation of the mesh movement Equation 16.8 with the computation of curvature quantities based on a weak formulation of the identity $\Delta_\Gamma id = \mathcal{K}\mathbf{n}$. The weak form of the identity is derived by partial integration on the closed surface Γ_{F}. The first equation in Equation 16.8 is tested against vector-valued functions $\chi\mathbf{n}$ with $\chi \in L^2(\Gamma_{\mathrm{F}})$. The resulting system,

$$\left\langle \frac{d\mathbf{X}}{dt} - \mathbf{u}, \chi\mathbf{n} \right\rangle_{\Gamma_{\mathrm{F}}} = 0 \quad \text{for all } \chi \in L^2(\Gamma_{\mathrm{F}}),$$

$$\langle \mathcal{K}\mathbf{n}, \eta \rangle_{\Gamma_{\mathrm{F}}} + \langle \nabla_\Gamma id, \nabla_\Gamma \eta \rangle_{\Gamma_{\mathrm{F}}} = 0 \quad \text{for all } \eta \in H^1(\Gamma_{\mathrm{F}})^d,$$

is used—in a discrete setting—to compute the position vector $\mathbf{X}$ and the scalar mean curvature $\mathcal{K}$ based on a given velocity field $\mathbf{u}$. Continuous, piecewise linear functions have been used in [11] to represent the position of $\Gamma_{\mathrm{F}}(t)$ and the scalar mean curvature. Furthermore, appropriate quadrature rules have been applied. For the semi-discrete continuous-in-time problem, an equidistribution property of the vertices has been proven in the 2D case in [13]. An extension of the method to three dimensions in [14] shows that the induced tangential motion gives rise to conformal polyhedral surfaces, which demonstrates the good mesh quality of the method. The extension to higher-order approximations of Γ_{F} and $\mathcal{K}$ is an open problem.

16.3.3 ARBITRARY LAGRANGIAN–EULERIAN FEM

The ALE method is a direct parametric surface tracking method. But instead of having two separate meshes for Ω and Γ_{F}, the purpose is to treat both with one mesh.

The time-dependent domain $\Omega(t)$ and its motion are described by a function that gives the position of every point of the domain dependent on time.

Let $\mathcal{A}_t$ be a family of mappings that maps a point $\hat{\mathbf{x}}$ of the parameter space (or reference domain) $\hat{\Omega}$ to a point $\mathbf{x}$ in the domain $\Omega(t)$ at each time point $t \in (0,T)$

$$\mathcal{A}_t : \hat{\Omega} \to \Omega(t), \quad \mathbf{x}(\hat{\mathbf{x}},t) = \mathcal{A}_t(\hat{\mathbf{x}}).$$

In the following, we name $\hat{\mathbf{x}} \in \hat{\Omega}$ the ALE (or parametric) coordinate and $\mathbf{x} \in \Omega(t)$ the Eulerian (or spatial) coordinate. Further, we refer to $\hat{\Omega}$ as the ALE frame and to $\Omega(t)$ as the Eulerian frame.

A function v given on $\Omega(t)$, that is, in the Eulerian frame, is carried over to the parameter space, or ALE frame, by

$$\hat{v}(\hat{\mathbf{x}}, t) = v(\mathbf{\mathcal{A}}_t(\hat{\mathbf{x}}), t),$$

where $\hat{v}$ denotes the function given on the parameter space. With this definition, one does not need to distinguish between hat and nonhat functions, since the context becomes clear when the function is evaluated. For example, if we write $v(\hat{\mathbf{x}}, t)$, actually $v(\mathbf{\mathcal{A}}_t(\hat{\mathbf{x}}), t)$ is meant and the other way around.

However, concerning partial derivatives in time there are two possibilities, now. One can take the derivative in the Eulerian frame, denoted by $\partial_t v$, or in the ALE frame, denoted by $\partial_t \hat{v}$. Both are different. In order to express the partial time derivative given in the ALE frame in the Eulerian frame as well, we introduce the following notation:

$$(\hat{\partial}_t v)(\mathbf{x}, t) := (\partial_t \hat{v})(\mathbf{x}, t).$$

The time derivative in the Eulerian frame is connected to the time derivative in the ALE frame via the chain rule

$$\hat{\partial}_t v = \partial_t v + \partial_t \mathbf{\mathcal{A}}_t \cdot \nabla v = \partial_t v + \mathbf{w} \cdot \nabla v. \tag{16.9}$$

Note that $\hat{\partial}_t v$ is the material derivative of v with respect to the domain velocity $\mathbf{w}$ given by $\mathbf{w} = \partial_t \mathbf{\mathcal{A}}_t$ or $\mathbf{w} = \hat{\partial}_t \mathbf{x}$. We say a PDE is given in ALE form, if the evolution of a quantity in time is given in the ALE frame. In order to transform a PDE given in the Eulerian frame into the ALE form the connection of the time derivatives (Equation 16.9) is used. That is, a PDE given in the form

$$\partial_t u + \mathcal{L}(u) = 0, \tag{16.10}$$

where $\mathcal{L}$ denotes a differential operator in the spatial coordinate, becomes

$$\hat{\partial}_t u - \mathbf{w} \cdot \nabla u + \mathcal{L}(u) = 0. \tag{16.11}$$

The difference of Equations 16.11 and 16.10 is that the time derivative is now in the ALE frame. As a consequence, an additional convective term appears [15].

The evolution of the bulk domain, or bulk grid in the discrete case, is determined by giving $\Omega(0)$ and $\mathbf{w}$. Having an interface resolving grid for $\Omega(0)$, that is, a grid where the phase change happens across a facet of an element and not in the interior, one has an interface resolving grid for $\Omega(t)$ at time t, too. Thus, the discrete interface $\Gamma_F(t)$ at time t and its triangulation are given by the evolution of the initial interface and its triangulation. Therefore, there is no need to reconstruct the interface, it is given explicitly. At each time step, the interface triangulation consists of facets from the bulk mesh. Figure 16.2 shows a cut through a bulk mesh to illustrate this fact.

Our model does not determine the motion of the internal bulk mesh. Instead, just the motion of the interface or surface is given by the condition $\mathbf{w} \cdot \mathbf{n} = \mathbf{u} \cdot \mathbf{n}$. Setting $\mathbf{w} = \mathbf{u}$ in Ω a fully Lagrangian description would be achieved. However, this leads to large distortions of the mesh. Therefore, the additional degree of freedom in extending $\mathbf{w}$ is used to maintain a good mesh quality. Often, a simple harmonic extension of the surface velocity is used, that is, by solving a

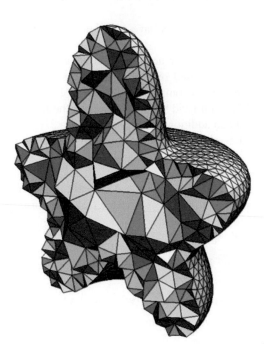

FIGURE 16.2 Cut through a bulk mesh.

Laplace problem for **w**. A more advanced, extension of the surface velocity is the elastic mesh update. For that, the equations for a linear elastic body are solved. In both cases, a finite element method is used in [16].

16.4 FINITE ELEMENT METHOD FOR INTERFACE FLOWS

In this section, we present the discretization of the Navier–Stokes equations and the surfactant transport equations in the bulk and on the interface. The interface motion is tracked using the ALE method introduced in Section 16.3.3. In Section 16.4.1, different temporal discretization methods are discussed, and Section 16.4.2 introduces an efficient decoupling technique for the upcoming equation system. Proper finite elements for the space discretization are specified in Section 16.4.3. Section 16.4.4 is devoted to a technique to incorporate the surface tension terms into the Navier–Stokes equation, and mass conservation properties are discussed in Section 16.4.5.

16.4.1 TEMPORAL DISCRETIZATION

Some of the popular temporal discretization methods are discussed in this section. Mostly, implicit schemes are preferred over explicit ones to avoid a time step restriction, which depends on the unknown velocity of the fluid. One simple and strongly A-stable method is the backward Euler method. Even though the Euler method is only first-order accurate, it is often used due to its stability. Nevertheless, a second-order accurate solution in time can be obtained with almost the same computational cost as the backward Euler scheme, when the Crank–Nicolson method is used. Unfortunately, the Crank–Nicolson method is not strongly A-stable and occasionally it suffers from instabilities. Other higher-order methods, such as implicit Runge–Kutta and backward multi-step methods, are well known for ordinary differential equations. Discontinuous Galerkin methods are common for space discretization. Recently, they are also used for time discretization

[17–19]. However, all mentioned methods have a higher complexity and storage requirement in common.

Another scheme that is second-order accurate and strongly A-stable is the fractional-step-ϑ scheme (FS). It has first been proposed in [20] as a three-step operator-splitting scheme, which separates the nonlinearity and the incompressibility in the Navier–Stokes equations. Due to the advances in computing and fast multigrid solvers, nowadays the full Navier–Stokes system is solved in all three sub-steps, to avoid an additional splitting error [16,21].

Let $0 = t_0 < t_1 < \cdots < t_N = T$ be a decomposition of the given time interval $[0, T]$ and $\Delta t_n = t_{n+1} - t_n$, $0 \le n \le N-1$, be a sequence of time step sizes and a superscript n on a time-dependent function names the function at t_n, for example, $u^n = u(t_n)$. We use the ALE technique to track the motion of the interface. Thereby, for every time step $t_n \to t_{n+1}$ the ALE frame is given by $\hat{\Omega} = \Omega^n$ and the Eulerian frame is set to $\Omega = \Omega^{n+1}$. A generalized semi-discrete (in time) formulation of the two-phase flow problem (Equations 16.4 through 16.7) reads:

For given $\Omega_k^n (k=1,2)$, $\mathbf{u}^n$, $\mathbf{w}^n$, p^n, c_Γ^n, and C^n find $(\mathbf{u}^{n+1}, p^{n+1}, C^{n+1}, c_\Gamma^{n+1}) \in V(\Omega^{n+1}) \times Q(\Omega^{n+1}) \times G(\Omega_2^{n+1}) \times M(\Gamma_F^{n+1})$ such that for all $(\mathbf{v}, q, \phi, \psi) \in V(\Omega^{n+1}) \times Q(\Omega^{n+1}) \times G(\Omega_2^{n+1}) \times M(\Gamma_F^{n+1})$,

$$(\rho \mathbf{u}^{n+1}, \mathbf{v})_{\Omega^{n+1}} + \alpha \Delta t_n \left(a((\mathbf{u}^{n+1} - \mathbf{w}^{n+1}), \mathbf{u}^{n+1}, \mathbf{v})_{\Omega^{n+1}} - f\left(c_\Gamma^{n+1}, \mathbf{v}\right)_{\Omega^{n+1}} \right)$$

$$+ \Delta t_n \left(b(q, \mathbf{u}^{n+1})_{\Omega^{n+1}} - b(p^{n+1}, \mathbf{v})_{\Omega^{n+1}} \right)$$

$$= (\rho \mathbf{u}^n, \mathbf{v})_{\Omega^n} - \beta \Delta t_n \left(a((\mathbf{u}^n - \mathbf{w}^n), \mathbf{u}^n, \mathbf{v})_{\Omega^n} - f(c_\Gamma{}^n, \mathbf{v})_{\Omega^n} \right),$$

$$(C^{n+1}, \phi)_{\Omega_2^{n+1}} + \alpha \Delta t_n \left(a_c((\mathbf{u}^{n+1} - \mathbf{w}^{n+1}), C^{n+1}, \phi)_{\Omega_2^{n+1}} - f_c\left(C^{n+1}, c_\Gamma^{n+1}, \phi\right)_{\Omega_2^{n+1}} \right)$$

$$= (C^n, \phi)_{\Omega_2^{n+1}} - \beta \Delta t_n \left(a_c((\mathbf{u}^n - \mathbf{w}^n), C^n, \phi)_{\Omega_2^{n+1}} - f_c\left(C^n, c_\Gamma^n, \phi\right)_{\Omega_2^{n+1}} \right),$$

$$\left\langle c_\Gamma^{n+1}, \psi \right\rangle_{\Gamma_F^{n+1}} + \alpha \Delta t_n \left(a_\Gamma\left(c_\Gamma^{n+1}, \psi\right)_{\Gamma_F^{n+1}} - f_\Gamma\left(C^{n+1}, c_\Gamma^{n+1}, \psi\right)_{\Gamma_F^{n+1}} \right)$$

$$= \left\langle c_\Gamma^n, \psi \right\rangle_{\Gamma_F^n} - \beta \Delta t_n \left(a_\Gamma\left(c_\Gamma^n, \psi\right)_{\Gamma_F^n} - f_\Gamma\left(C^n, c_\Gamma^n, \psi\right)_{\Gamma_F^n} \right),$$

where $\mathbf{w}^{n+1} = \mathbf{w}(\mathbf{u}^{n+1})$ is an extension of the interface velocity into the bulk, as described in Section 16.3.3.

Note that the pressure and the mass conservation equations in the Navier–Stokes equation are handled fully implicit. The surfactant equation in the bulk is discretized on the new domain Ω_2^{n+1}, whereas the discretized surface equation includes integrals over Γ_F^n and Γ_F^{n+1}.

Based on the choice of the coefficient α and β, different time stepping methods can be obtained, see Table 16.1. The FS scheme uses different values of α and β in the three sub-steps of $(t_{3n}, t_{3(n+1)})$, see Table 16.2.

16.4.2 Decoupling Strategies

For a start, the surfactant concentration in the force balancing condition is usually treated explicitly, that is, c_Γ^n is used in (t^n, t^{n+1}). Thus, the Navier–Stokes equation decouples from the transport

TABLE 16.1

Choice of α and β for Different Temporal Schemes

Coefficient	Forward Euler	Backward Euler	Crank-Nicolson
α	0	1	0.5
β	1	0	0.5

TABLE 16.2

Choice of α and β in the Three Sub-Steps of Fractional-Step-ϑ Scheme

Coefficient	FS Step 1	FS Step 2	FS Step 3
α	0.585786	0.414214	0.585786
β	0.414214	0.585786	0.414214

equations. The nonlinear Navier–Stokes equations are commonly approximated by linearized equations using a fixed point iteration [22]. The linearized form of the Navier–Stokes equations reads as follows.

For given Ω^n, c_Γ^n, $\mathbf{u}_0^{n+1} := \mathbf{u}^n$, and $\mathbf{w}_0^{n+1} := \mathbf{w}^n$, find $(\mathbf{u}_i^{n+1}, p_i^{n+1}) \in V(\Omega^n) \times Q(\Omega^n)$ such that for all $(\mathbf{v}, q) \in V(\Omega^n) \times Q(\Omega^n)$,

$$\left(\rho \mathbf{u}_i^{n+1}, \mathbf{v} \right)_{\Omega^n} + \alpha \Delta t_n \left(a \left(\left(\mathbf{u}_{i-1}^{n+1} - \mathbf{w}_{i-1}^{n+1} \right), \mathbf{u}_i^{n+1}, \mathbf{v} \right)_{\Omega^n} - f \left(c_\Gamma^n, \mathbf{v} \right)_{\Omega^n} \right)$$

$$+ \Delta t_n \left(b \left(q, \mathbf{u}_i^{n+1} \right)_{\Omega^n} - b \left(p_i^{n+1}, \mathbf{v} \right)_{\Omega^n} \right)$$

$$= \left(\rho \mathbf{u}_{i-1}^{n+1}, \mathbf{v} \right)_{\Omega^n} - \beta \Delta t_n \left(a \left(\left(\mathbf{u}_{i-1}^{n+1} - \mathbf{w}_{i-1}^{n+1} \right), \mathbf{u}_{i-1}^{n+1}, \mathbf{v} \right)_{\Omega^n} - f \left(c_\Gamma^n, \mathbf{v} \right)_{\Omega^n} \right),$$

for $i = 1,2,\ldots.$

A linear elasticity or a Laplace problem has to be solved in order to determine the domain velocity $\mathbf{w}_i^{n+1}$ in every iteration step [22,23]. The iteration has to be continued until the residual of the Navier–Stokes system becomes less than a prescribed tolerance. After stopping the fixed point iteration set $\mathbf{u}^{n+1} := \mathbf{u}_i^{n+1}$, $\mathbf{w}^{n+1} := \mathbf{w}_i^{n+1}$, and $p^{n+1} := p_i^{n+1}$.

Since the updated domain velocity $\mathbf{w}^{n+1}$ is known, the new position of the computational domains Ω_k^{n+1}, $k = 1, 2$, can be obtained before solving the surfactant concentration equations. Even though the Navier–Stokes equations and the concentration equations are decoupled, the updated fluid velocity $\mathbf{u}^{n+1}$ and the domain velocity $\mathbf{w}^{n+1}$ are needed in the convective terms of the transport equations.

Further, the concentration equations in the bulk phase and on the interface are still coupled. An explicit treatment of the source term $S(C, c_\Gamma)$ in each transport equation would decouple them into two linear equations. However, the adsorption and desorption of surfactants on the interface are strongly coupled with the solutions of the transport equations. An implicit treatment of the source term $S(C, c_\Gamma)$ seems to be necessary to capture the dynamics of the fluid flow induced by the Marangoni forces. A Gauss–Seidel-type fixed point iteration can be used to efficiently decouple the transport equations. The decoupled semi-discrete (in time) form of Equations 16.6 and 16.7 at time t_n reads as follows.

For given C^n, c_Γ^n, $\mathbf{u}^{n+1}$, $\mathbf{w}^{n+1}$, $c_{\Gamma,0}^{n+1} := c_\Gamma^n$, and $C_0^{n+1} := C^n$, find $(C_i^{n+1}, c_{\Gamma,i}^{n+1}) \in G(\Omega_2^{n+1}) \times M(\Gamma_F^{n+1})$ such that for all $(\phi, \psi) \in G(\Omega_2^{n+1}) \times M(\Gamma_F^{n+1})$,

$$\left(C_i^{n+1}, \phi \right)_{\Omega_2^{n+1}} + \alpha \Delta t_n \left(a_c \left((\mathbf{u}^{n+1} - \mathbf{w}^{n+1}), C_i^{n+1}, \phi \right)_{\Omega_2^{n+1}} - f_c \left(C_i^{n+1}, c_{\Gamma,i-1}^{n+1}, \phi \right)_{\Omega_2^{n+1}} \right)$$

$$= (C_{i-1}^{n+1}, \phi)_{\Omega_2^{n+1}} - \beta \Delta t_n \left(a_c ((\mathbf{u}^{n+1} - \mathbf{w}^{n+1}), C_{i-1}^{n+1}, \phi)_{\Omega_2^{n+1}} - f_c (C_{i-1}^{n+1}, c_{\Gamma,i-1}^{n+1}, \phi)_{\Omega_2^{n+1}} \right),$$

$$\left\langle c_{\Gamma,i}^{n+1}, \psi \right\rangle_{\Gamma_F^{n+1}} + \alpha \Delta t_n \left(a_\Gamma \left(c_{\Gamma,i}^{n+1}, \psi \right)_{\Gamma_F^{n+1}} - f_\Gamma \left(C_i^{n+1}, c_{\Gamma,i}^{n+1}, \psi \right)_{\Gamma_F^{n+1}} \right)$$

$$= \left\langle c_{\Gamma,i-1}^{n+1}, \psi \right\rangle_{\Gamma_F^{n+1}} - \beta \delta t_n \left(a_\Gamma \left(c_{\Gamma,i-1}^{n+1}, \psi \right)_{\Gamma_F^{n+1}} - f_\Gamma \left(C_i^{n+1}, c_{\Gamma,i-1}^{n+1}, \psi \right)_{\Gamma_F^{n+1}} \right),$$

for $i = 1,2,\ldots$.

In computations, the iteration is stopped when the residual of these equations becomes less than a given tolerance.

16.4.3 FINITE ELEMENT DISCRETIZATION

The choice of finite elements for the Navier–Stokes equations is very crucial in computations of two-phase flows. One of the inherent properties of two-phase flows is the discontinuity of the pressure across the interface. Thus, the natural choice for approximating the pressure is the use of discontinuous finite elements. As shown in [11,24], deviating from this choice induces spurious velocities, also known as parasitic currents. Errors in interface approximation and curvature calculation will also lead to spurious velocities in the computation of two-phase flows (see [24] for more details).

Additionally, the choice of finite elements for the pressure approximation influences the possibilities for the velocity approximation because for stability, an inf-sup condition has to be satisfied. For example, the choice of continuous piecewise polynomials of degree less than or equal to k for the velocity (P_k) and discontinuous piecewise polynomials of degree less than or equal to $k-1$ for the pressure (P_{k-1}^{disc}) leads to a finite element method of order k. Unfortunately, stability of this method is established for $k \geq d$ in the d-dimensional case only on special refinements of macro-element meshes [25–27]. Since the mesh in Lagrangian methods moves, guaranteeing the refinement rule is very challenging. Therefore, we prefer continuous piecewise polynomials of degree less than or equal to k enriched with cell bubble (polynomial) functions (P_k^{bubble}) for the velocity together with the discontinuous pressure, that is, the inf-sup stable pair ($P_k^{bubble}, P_{k-1}^{disc}$), for $k = 2, 3$ [28–30]. On quadrilateral and hexahedral meshes, alternatively the stable pair (Q_k, P_{k-1}^{disc}) can be used in [31].

Note that despite the choice of discontinuous elements for the pressure, spurious velocities might be generated in Eulerian methods due to an interface unresolved mesh [24]. For such meshes, the interface across which the pressure is discontinuous is located within a finite element cell. A discontinuity within a cell can be represented by extending the finite element space with discontinuous basis functions. This approach is called XFEM [32]. The inf-sup stability of such modified finite element pairs is an open problem.

Standard finite elements can be used for the spatial discretization of the surfactant transport equations in the bulk and on the interface. Taking into account that for ALE methods, the surface triangulation consists of a subset of facets of the bulk mesh, a natural choice for the surface finite element space is the restriction of the bulk finite element space. Furthermore, this combination needs no interpolation of the bulk concentration on the surface or vice versa and, hence, reduces computational costs. Since the derivatives of the concentration are needed in the force balancing condition of the Navier–Stokes equations, polynomials of degree two or more are preferred.

16.4.4 Discretizing Surface Tension and Marangoni Forces

At the interface Γ_F, we have the boundary condition

$$[\mathbb{S}(\mathbf{u}, p)]\mathbf{n} = \sigma(c_\Gamma)\mathcal{K}\mathbf{n} + \nabla_\Gamma \sigma(c_\Gamma).$$

This boundary condition is a special case of the more general Boussinesq–Scriven model taking viscosity properties of the interface into consideration [33,34]. This model uses the boundary condition

$$[\mathbb{S}(\mathbf{u}, p)]\mathbf{n} = \nabla_\Gamma \cdot \mathbb{S}_\Gamma, \quad \mathbb{S}_\Gamma = (\sigma + (\lambda_\Gamma - \mu_\Gamma)\nabla_\Gamma \cdot \mathbf{u})\mathbb{P} + \mu_\Gamma \mathbb{D}_\Gamma \mathbf{u},$$

where
  $\mathbb{S}_\Gamma$ stands for the surface stress tensor
  $\mathbb{D}_\Gamma$ denotes the surface deformation tensor
  $\mathbb{P} = \mathbb{I} - \mathbf{n}\mathbf{n}^T$ is the projection into the tangent plane of the interface

Setting $\lambda_\Gamma = \mu_\Gamma = 0$, viscous effects of the interface are neglected, and we obtain

$$\nabla_\Gamma \cdot \mathbb{S}_\Gamma = \nabla_\Gamma \cdot (\sigma\mathbb{P}) = \sigma\mathcal{K}\mathbf{n} + \nabla_\Gamma \sigma.$$

In the weak formulation, we get the surface tension functional on the right-hand side of the problem in the form

$$\int_{\Gamma_F} \mathbf{v} \cdot [\mathbb{S}(\mathbf{u}, p)] \cdot \mathbf{n}\, ds = \int_{\Gamma_F} \mathbf{v} \cdot \nabla_\Gamma \cdot (\sigma\mathbb{P})ds.$$

Note that for the surface gradient of the identity mapping we have

$$(\nabla_\Gamma(id_\Gamma)_i)_j = \delta_{ij} - n_i n_j = \mathbb{P}_{ij}, \quad i, j = 1, \ldots, d.$$

Setting $\Phi = \sigma\mathbb{P}v = \sigma\nabla_\Gamma id_\Gamma \mathbf{v}$, we use the Gauß theorem on closed surfaces (compare [35, Theorem 2.10])

$$\int_\Gamma \nabla_\Gamma \cdot \Phi\, ds = \int_\Gamma \mathcal{K}\Phi \cdot \mathbf{n}\, ds, \tag{16.12}$$

and the product rule

$$\nabla_\Gamma \cdot \Phi = \nabla_\Gamma \cdot (\sigma\nabla_\Gamma id_\Gamma) \cdot \mathbf{v} + \sigma\nabla_\Gamma id_\Gamma : \nabla_\Gamma \mathbf{v}.$$

Since $\Phi = \sigma\mathbb{P}\mathbf{v}$ is in the tangential plane, we have $\Phi \cdot \mathbf{n} = 0$, the integral on the right-hand side of Equation 16.12 vanishes, and the surface tension functional becomes

$$\int_{\Gamma_F} \mathbf{v} \cdot [\mathbb{S}(\mathbf{u}, p)] \cdot \mathbf{n} ds = \int_{\Gamma_F} \nabla_\Gamma \cdot (\sigma \nabla_\Gamma id_\Gamma) \cdot \mathbf{v} ds = -\int_{\Gamma_F} \sigma \nabla_\Gamma id_\Gamma : \nabla_\Gamma \mathbf{v} ds. \qquad (16.13)$$

In the discretized setting, Equation 16.13 cannot be handeled fully implicit because Γ^{n+1} is not known a priori. To increase stability compared to a fully explicit inclusion, the surface tension functional is treated semi-implicitly [36,37]. The integration domain stays Γ_F^n, but the identity term is evaluated at Γ_F^{n+1} by using $id_{\Gamma^{n+1}} = id_{\Gamma^n} + \Delta t_n \mathbf{u}^{n+1}$.

16.4.5 MASS CONSERVATION OF FLUID AND SURFACTANTS

Volume conservation of the inner phase $\Omega_1(t)$ in the continuous model is a consequence of

$$\frac{d}{dt} \int_{\Omega_1(t)} dx = \int_{\Gamma_F} \mathbf{w} \cdot \mathbf{n} d\gamma = \int_{\Gamma_F} \mathbf{u} \cdot \mathbf{n} d\gamma = \int_{\Omega_1(t)} \nabla \cdot \mathbf{u} dx = 0,$$

where we used the Reynolds transport theorem, the domain velocity property $\mathbf{w} \cdot \mathbf{n} = \mathbf{u} \cdot \mathbf{n}$ on the interface Γ_F, and the incompressibility constraint in the bulk phases (Equation 16.5). Due to $\Omega = \Omega_1(t) \cup \Omega_2(t)$ being a fixed domain, there is volume conservation of the outer phase $\Omega_2(t)$, too. With $\rho = \rho_k$ in $\Omega_k(t)$, mass conservation in both phases can be concluded. It should be mentioned that for the mass conservation of the fluid phases, we do not need $\nabla \cdot \mathbf{u} = 0$ pointwise. It is sufficient to satisfy the incompressibility constraint for the integral mean of each phase.

The conservation of surfactants follows from the weak form of the bulk and surface equation by setting the test functions Φ and ψ in Equations 16.6 and 16.7 equal to one. Then, we have for the change of surfactant mass

$$\frac{d}{dt} \left[(C, 1)_{\Omega_2} + \langle c_\Gamma, 1 \rangle_{\Gamma_F} \right] = \left(\frac{\partial C}{\partial t} + \mathbf{u} \cdot \nabla C, 1 \right)_{\Omega_2} + \langle S(C, c_\Gamma), 1 \rangle_{\Gamma_F} = 0.$$

Here, we used the Reynolds transport theorem and the divergence constraint $\nabla \cdot \mathbf{u} = 0$ pointwise in Ω_2.

It is desirable to have the mass conservation of fluid phases and surfactants in the discretized model, too. However, discretization errors in space and time and the use of stabilized schemes to handle the convection dominated transport equations do not allow the same conclusions for the discretized schemes.

The standard mixed finite elements used for the Navier–Stokes equations satisfy the incompressibility constraint only approximatively. Instead of $\nabla \cdot \mathbf{u}_h = 0$ pointwise, the discrete velocity field $\mathbf{u}_h$ is only discretely divergence-free, that is, $(q_h, \nabla \cdot \mathbf{u}_h)_\Omega = 0$ for all test functions q_h from the pressure space. Therefore, mass conservation of the fluid phases can only be concluded if the pressure space contains a discontinuous test function q_h with $q_h|_{\Omega_1(t)} = 1$ and $q_h|_{\Omega_2(t)} = 0$. For discontinuous pressure approximations, this is the case, but continuous pressure approximations as the well-known Taylor–Hood element do not satisfy such a property.

For the semi-discrete continuous-in-time discretization of coupled flow transport problems in fixed domains, the mass conservation of transported species has been discussed in [38]. Mass conservation can be achieved using a higher-order discretization for the flow problem compared to the transport problem. This concept has been applied in [39]. Indeed, a second-order Taylor–Hood approach to approximate velocity and pressure has been combined with a first-order method for discretizing the surfactant transport equations in the bulk and on the surface. Other possibilities are the use of divergence-free elements [38] or reconstruction operators [40] transferring discretely divergence-free functions into divergence-free ones.

Particular time discretization schemes are required to retain the conservation properties over time. Based on discrete versions of the Reynolds transport theorem, unconditionally stable higher order discontinuous Galerkin discretizations in time are discussed in [17–19].

16.5   NUMERICAL EXPERIMENTS

In the computations, a dimensionless model is used. By introducing dimensionless variables, the model equations become parameterized by 11 dimensionless numbers: the Reynolds numbers Re_2, the Weber number We, the Froude number Fr, the Peclet number for the bulk surfactant transport Pe_C, the Peclet number for the surface surfactant transport Pe_Γ, the Biot number Bi, the Damköhler number Da, the surface elasticity E, the surfactant scaling β, and the dimensionless scaling factors ρ_1/ρ_2 and μ_1/μ_2.

The dimensionless variables are chosen in the following way:

$$\tilde{\mathbf{x}} = \frac{\mathbf{x}}{L}, \quad \tilde{\mathbf{u}} = \frac{\mathbf{u}}{U_\infty}, \quad \tilde{\mathbf{w}} = \frac{\mathbf{w}}{U_\infty}, \quad \tilde{t} = \frac{U_\infty}{L}t, \quad \tilde{p} = \frac{p}{\rho_2 U_\infty^2}, \quad \tilde{c}_\Gamma = \frac{c_\Gamma}{c_\Gamma^\infty}, \quad \tilde{C} = \frac{C}{C_\infty},$$

where

L is a characteristic length
U_∞ is a characteristic velocity
C_∞ is a characteristic bulk concentration

With these variables, the dimensionless numbers are given as

$$Re = \begin{cases} Re_2\mu_2/\mu_1 & \mathbf{x} \in \Omega_1(t) \\ Re_2 & \mathbf{x} \in \Omega_2(t) \end{cases}, \text{where } Re_2 = \frac{\rho_2 U_\infty L}{\mu_2},$$

$$We = \frac{\rho_2 U_\infty^2 L}{\sigma_0}, \quad Fr = \frac{U_\infty^2}{Lg}, \quad Pe_C = \frac{U_\infty L}{D_C}, \quad Pe_\Gamma = \frac{U_\infty L}{D_\Gamma},$$

$$Bi = \frac{k_d L}{U_\infty}, \quad Da = \frac{k_a c_\Gamma^\infty}{U_\infty}, \quad E = \frac{RT_\infty c_\Gamma^\infty}{\sigma_0}, \quad \beta = \frac{c_\Gamma^\infty}{LC_\infty},$$

and

$$\varrho = \begin{cases} \rho_1/\rho_2 & \mathbf{x} \in \Omega_1(t) \\ 1 & \mathbf{x} \in \Omega_2(t) \end{cases}.$$

After dropping the tilde, the model equations in the dimensionless form read as follows:

$$\varrho(\partial_t \mathbf{u} + (\mathbf{u} \cdot \nabla)\mathbf{u}) - \nabla \cdot \mathbb{S}(\mathbf{u}, p) = \frac{\varrho}{Fr}e, \quad \nabla \cdot \mathbf{u} = 0 \text{ in } \Omega \times (0,T],$$

$$\mathbf{w} \cdot \mathbf{n} = \mathbf{u} \cdot \mathbf{n}, [\mathbf{u}] = 0, [\mathbb{S}(\mathbf{u}, p)] \cdot \mathbf{n} = \frac{1}{We}\nabla_\Gamma \cdot (\sigma(c_\Gamma)\mathbb{P}) \quad \text{on } \Gamma_F(t) \times (0,T],$$

$$\partial_t C + \mathbf{u} \cdot \nabla C = \frac{1}{\mathrm{Pe}_C} \Delta C \quad \text{in } \Omega_2(t) \times (0,T],$$

$$\frac{1}{\mathrm{Pe}_C} \nabla C \cdot \mathbf{n} = -\beta S(C, c_\Gamma) \quad \text{on } \Gamma_F(t) \times (0,T],$$

$$\dot{c}_\Gamma = \frac{1}{\mathrm{Pe}_\Gamma} \Delta_\Gamma c_\Gamma + S(C, c_\Gamma) \quad \text{on } \Gamma_F(t) \times (0,T],$$

where the dimensionless stress tensor $\mathbb{S}$ is given by

$$\mathbb{S}(\mathbf{u}, p) = \frac{2}{\mathrm{Re}} \mathbb{D}(\mathbf{u}) - p\mathbb{I},$$

and the dimensionless source term is given as

$$S(C, c_\Gamma) = \frac{\mathrm{Da}}{\beta} C - \mathrm{Bi} c_\Gamma \quad \text{or} \quad S(C, c_\Gamma) = \frac{\mathrm{Da}}{\beta} C(1 - c_\Gamma) - \mathrm{Bi} c_\Gamma,$$

for the case of the Henry sorption isotherm and the Langmuir sorption isotherm, respectively.

16.5.1 Rising Bubble

The influence of soluble surfactants on the dynamics of rising bubbles driven by buoyancy force in quiescent water is studied in this section. We extend the benchmark of rising bubble considered in [41] as test case 1 by taking into account soluble surfactants in the outer phase. A 3D axisymmetric setting (see [22]), with a bubble rising in a liquid column is considered. Initially, the bubble is in rest and its surface is clean, that is, $c_\Gamma(0) = 0$. The initial surfactant concentration in the outer phase is $C(0) = 1$. Further, the initial shape of the bubble is assumed to be spherical with the diameter $d = 0.5$.

The dimensionless numbers used in the test cases are: $\mathrm{Re}_2 = 99$, $\mathrm{Re}_1 = \mathrm{Re}_2 \mu_2/\mu_1 = 9.9$, $\mathrm{We} = 40$, $\mathrm{Fr} = 1$, $\mathrm{Bi} = 0.01$, $\mathrm{Da} = 100$, $\beta = 200$, $E = 0.5$, $\rho_1/\rho_2 = 0.1$, and $\mathrm{Pe}_\Gamma = 10$. We consider a clean bubble and two settings with $\mathrm{Pe}_C = 100$ and $\mathrm{Pe}_C = 1000$, respectively. We write shortly Pe for Pe_C.

Even though the bubble is in rest initially, it starts to rise due to the buoyancy force. To study the effect of surfactants, the rise velocity is computed using

$$\text{Rise velocity} = \frac{1}{|\Omega_1(t)|} \int_{\Omega_1(t)} \mathbf{u}_z \, r dr dz.$$

Further, the sphericity is computed as the ratio between the surface area of volume-equivalent sphere and the surface area of the bubble. The rise velocity and the sphericity of the clean bubble and the bubbles with surfactants are presented in Figure 16.3. The rise velocity starts to increase in all test cases and reaches a terminal velocity. In the case of rising bubble with surfactants, a retarding effect is observed. Since $c_\Gamma(0) = 0$, the retarding effect increases gradually when c_Γ increases due to absorption. The dimensionless terminal velocities of the bubbles with and without surfactants are 0.51 and approximately 0.48, respectively. This clearly indicates that the presence of surfactants reduces the terminal rising velocity. Although the presence of surfactants decreases the surface tension, interestingly, the surfactant bubble remains spherical in comparison to the clean bubble due to the lower terminal velocity.

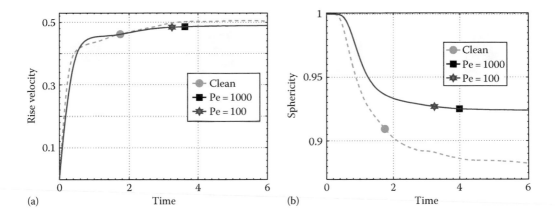

FIGURE 16.3 Rise velocity (a) and the sphericity (b) of the rising bubble with and without surfactants.

The concentration of surfactant on the interface at different times and the gradual increase in the total mass of surfactant on the interface for both cases, Pe = 100 and Pe = 1000, are shown in Figure 16.4. The surfactant concentration on the interface increases gradually due to absorption when the bubble passes through the soluble surfactant liquid column. In addition, the advection of the surfactant along the interface toward the rear stagnation point (arclength = 0) leads to an accumulation of the surfactant there. This can clearly be seen in Figure 16.4 (a). The increase of the total surface surfactant mass is higher for Pe = 100 than for Pe = 1000, due to the higher diffusion coefficient for the surfactant transport in the bulk phase toward the interface [see Figure 16.4 (b)].

Because of the dominance of the absorption over the desorption, the bulk phase surfactant in the vicinity of the interface is adsorbed onto the interface. This behavior can clearly be seen in Figure 16.5. The bulk surfactant concentration C adjacent to the interface, especially at the bottom region of the rising bubble, is lower due to absorption onto the interface. In a desorption dominated case, the bulk surfactant concentration would be higher near the interface, especially at the bottom region due to c_Γ being increased at the rear stagnation point.

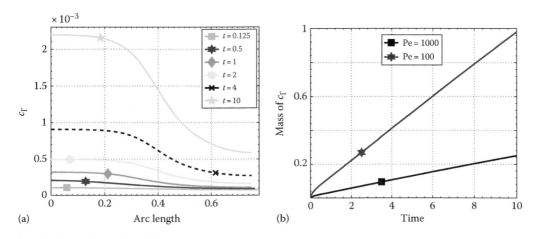

FIGURE 16.4 c_Γ on the interface at different times for Pe = 1000 (a) and the total mass of surfactant on the interface over time for Pe = 100 and Pe = 1000 (b).

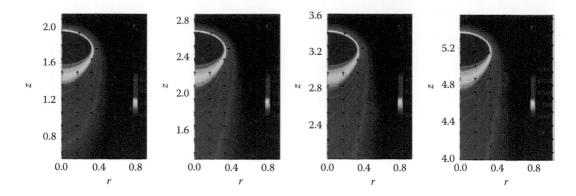

FIGURE 16.5 Contour plots of the bulk surfactant concentration at different times, $t = 3, 4.5, 6, 10$, for the rising bubble with Pe = 1000.

16.5.2 FREELY OSCILLATING DROPLET

In this section, we consider the influence of surfactants on the oscillation of a droplet in a gravitation free environment. The surrounding of the droplet is neglected; thus, we consider a free-surface flow problem. In this example, we perform a full 3D computation.

We consider four different cases, with different amounts of surfactant present. The dimensionless numbers are fixed for all cases to Re = 1000, We = 13.5, $\text{Pe}_C = 100$, $\text{Pe}_\Gamma = 100$, Bi = 1, Da = 1, $\beta = 1$, and $E = 1$. The Froude number is considered to be infinity, since no gravitational forces are present.

A droplet of nonspherical shape will start to oscillate due to the surface tension force. Here, we distort a sphere with the second-order spherical harmonic $Y_{2,0}$, that is, we can write the smooth initial surface $\Gamma(0) = \partial\Omega(0)$ as the zero set

$$\Gamma(0) = \{(r, \theta, \varphi) \in \mathbb{R}_0^+ \times [0, \pi] \times [0, 2\pi] : r - \epsilon Y_{2,0}(\theta, \varphi) - 1 = 0\},$$

where $\epsilon = 0.2$ and spherical coordinates were used.

The first case is a clean bubble (clean), where no surfactant is present. Next, we add a little amount of surfactant into the system, case (i). The following cases, case (ii) and case (iii), are obtained by doubling the amount of surfactant each. The initial condition for the surfactant concentration in the bulk and on the surface is set such that the surfactant sorption is in equilibrium, that is, $S(C(0), c_\Gamma(0)) = 0$. The resulting ratio

$$\frac{\text{Da}}{\beta\text{Bi}} = \frac{c_\Gamma}{C(1 - c_\Gamma)},$$

together with the given total surfactant mass gives the initial concentrations $C(0)$ and $c_\Gamma(0)$. Table 16.3 gives an overview of the resulting values. The initial velocity is zero for all cases, such that the droplet starts at rest.

In Figure 16.6, the tip position over time is shown. The tip position is the maximum elongation of the droplet in z-direction. We see the oscillations for all four cases. The oscillation frequency is largest for the clean case and decreases with the increasing amount of surfactant in the different cases. The damping rates first increase with higher amount of surfactant, however for the last case (iii), the damping rate decreases slightly. The different frequencies and damping rates can be

TABLE 16.3

Initial Surfactant Concentrations

Case	$C(0)$	$c_\Gamma(0)$	Total Mass
Clean	—	—	—
i	3.0391×10^{-2}	2.9495×10^{-2}	0.5
ii	6.2170×10^{-2}	5.8531×10^{-2}	1
iii	1.3013×10^{-1}	1.1515×10^{-1}	2

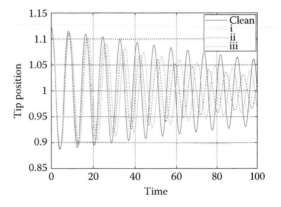

FIGURE 16.6 Tip position of the oscillating bubble.

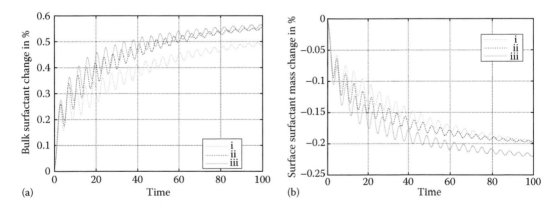

FIGURE 16.7 Mass of the bulk surfactant (a) and mass of the surface surfactant (b).

explained by the effect of the surfactant on the surface tension force, that is, on one hand, a reduction of the surface tension force with increasing surface surfactant concentration, which results in lower frequencies, and, on the other hand, the Marangoni forces, which increase the damping rates. Figure 16.7 shows the mass change in percent over time for the bulk surfactant (a) and the surfactant on the surface (b). The overall trend for the surface surfactant mass is decreasing, because the surface area gets smaller with the droplet approaching rest. Since the total amount of surfactant is a constant, the trend for the bulk surfactant mass has to be increasing. This is clearly seen in Figure 16.7 for all three cases.

ACKNOWLEDGMENTS

The authors thank the Council of Scientific and Industrial Research in India for financial support within the project 25(0228)/14/EMR-II and the German Research Foundation (DFG) for financial support within the Priority Program SPP 1506 "Transport Processes at Fluidic Interfaces" with the project To143/11-2 and within the graduate program Micro-Macro-Interactions in Structured Media and Particle Systems (GK 1554).

REFERENCES

1. J. Eastoe and J. Dalton, Dynamic surface tension and adsorption mechanisms of surfactants at the air–water interface, *Advances in Colloid and Interface Science*, 85(2), 103–144, 2000.
2. A. Alke and D. Bothe, 3D Numerical modeling of soluble surfactant at fluidic interfaces based on the volume-of-fluid method, *Fluid Dynamics & Materials Processing*, 5(4), 345–372, 2009.
3. A. J. James and J. Lowengrub, A surfactant-conserving volume-of-fluid method for interfacial flows with insoluble surfactant, *Journal of Computational Physics*, 201(2), 685–722, 2004.
4. M. Muradoglu and G. Tryggvason, A front-tracking method for computation of interfacial flows with soluble surfactants, *Journal of Computational Physics*, 227, 2238–2262, 2008.
5. S. Osher and J. A. Sethian, Fronts propagating with curvature-dependent speed: Algorithms based on Hamilton-Jacobi formulations, *Journal of Computational Physics*, 79(1), 12–49, 1988.
6. S. Gross and A. Reusken, *Numerical Methods for Two-Phase Incompressible Flows*, Vol. 40 of Springer Series in Computational Mathematics. Springer-Verlag, Berlin, Germany, 2011.
7. S. Gross, V. Reichelt, and A. Reusken, A finite element based level set method for two-phase incompressible flows, *Computing and Visualization in Science*, 9(4), 239–257, 2006.
8. M. A. Olshanskii, A. Reusken, and X. Xu, A stabilized finite element method for advection-diffusion equations on surfaces, *IMA Journal of Numerical Analysis*, 34(2), 732–758, 2014.
9. M. A. Olshanskii, A. Reusken, and J. Grande, A finite element method for elliptic equations on surfaces, *SIAM Journal of Numerical Analysis*, 47(5), 3339–3358, 2009.
10. J. W. Barrett, H. Garcke, and R. Nürnberg, A stable parametric finite element discretization of two-phase Navier–Stokes flow, *ArXiv e-prints*, Aug. 2013. arXiv:1308.3335 [math.NA].
11. J. W. Barrett, H. Garcke, and R. Nürnberg, Eliminating spurious velocities with a stable approximation of viscous incompressible two-phase Stokes flow, *Computer Methods in Applied Mechanics and Engineering*, 267, 511–530, 2013.
12. J. W. Barrett, H. Garcke, and R. Nürnberg, On the stable numerical approximation of two-phase flow with insoluble surfactant, *ArXiv e-prints*, Nov. 2013. arXiv:1311.4432 [math.NA].
13. J. W. Barrett, H. Garcke, and R. Nürnberg, A parametric finite element method for fourth order geometric evolution equations, *Journal of Computational Physics*, 222(1), 441–467, 2007.
14. J. W. Barrett, H. Garcke, and R. Nürnberg, On the parametric finite element approximation of evolving hypersurfaces in $\mathbb{R}^3$, *Journal of Computational Physics*, 227(9), 4281–4307, 2008.
15. F. Nobile, Numerical approximation of fluid-structure interaction problems with application to haemodynamics. PhD thesis, SB, Lausanne, Switzerland, 2001.
16. G. Matthies, Finite element methods for free boundary value problems with capillary surfaces. PhD thesis, Otto-von-Guericke-Universität, Fakultät für Mathematik, Magdeburg, Germany, 2002.
17. A. Bonito, I. Kyza, and R. H. Nochetto, Time-discrete higher-order ALE formulations: Stability, *SIAM Journal of Numerical Analysis*, 51(1), 577–604, 2013.
18. A. Bonito, I. Kyza, and R. H. Nochetto, Time-discrete higher order ALE formulations: A priori error analysis, *Numerische Mathematik*, 125(2), 225–257, 2013.
19. A. Bonito, I. Kyza, and R. H. Nochetto, A DG approach to higher order ALE formulations in time, in *Recent Developments in Discontinuous Galerkin Finite Element Methods for Partial Differential Equations*, X. Feng et al., (Eds), Vol. 157 of IMA Volumes in Mathematics and its Applications, pp. 223–258. Springer International Publishing, Switzerland, 2014.
20. M. O. Bristeau, R. Glowinski, and J. Periaux, Numerical methods for the Navier-Stokes equations. Application to the simulation of compressible and incompressible flows, *Computer Physics Reports*, 6, 73–188, 1987.
21. S. Turek, *Efficient Solvers for Incompressible Flow Problems. An Algorithmic and Computational Approach*. Springer-Verlag, Berlin, Germany, 1999.

22. S. Ganesan and L. Tobiska, An accurate finite element scheme with moving meshes for computing 3D-axisymmetric interface flows, *International Journal of Numerical Methods Fluids*, 57(2), 119–138, 2008.

23. S. Ganesan and L. Tobiska, Arbitrary Lagrangian-Eulerian finite-element method for computation of two-phase flows with soluble surfactants, *Journal of Computational Physics*, 231(9), 3685–3702, 2012.

24. S. Ganesan, G. Matthies, and L. Tobiska, On spurious velocities in incompressible flow problems with interfaces, *Computer Methods in Applied Mechanics and Engineering*, 196(7), 1193–1202, 2007.

25. L. R. Scott and M. Vogelius, Conforming finite element methods for incompressible and nearly incompressible continua, in *Large-Scale Computations in Fluid Mechanics, Proceedings of 15th AMS-SIAM Summer Seminar* (B. E. Enquist, S. Osher, and R. C. J. Somerville, eds.) Vol. 22, pp. 221–244, 1985. Lectures in Applied Mathematics. American Mathematical Society, Providence, Rhode Island.

26. L. R. Scott and M. Vogelius, Norm estimates for a maximal right inverse of the divergence operator in spaces of piecewise polynomials, *RAIRO, Modélisation Mathématique et Analyse Numérique*, 19, 111–143, 1985.

27. S. Zhang, A new family of stable mixed finite elements for the 3D Stokes equations, *Mathematics of Computation*, 74, 543–554, 2005.

28. D. Boffi, F. Brezzi, and M. Fortin, *Mixed Finite Element Methods and Applications*, Vol. 44 of Springer Series in Computational Mathematics. Springer, Heidelberg, Germany, 2013.

29. M. Crouzeix and P.-A. Raviart, Conforming and nonconforming finite element methods for solving the stationary Stokes equations I, *RAIRO Analyse Numérique*, 7, 33–76, 1973.

30. V. Girault and P.-A. Raviart, *Finite Element Methods for Navier-Stokes Equations: Theory and Algorithms*, Vol. 5 of Springer Series in Computational Mathematics. Springer-Verlag, Berlin, Germany, 1986.

31. G. Matthies and L. Tobiska, The inf-sup condition for the mapped Q_k/P_{k-1}^{disc} element in arbitrary space dimensions, *Computing*, 69(2), 119–139, 2002.

32. C. Lehrenfeld and A. Reusken, Analysis of a Nitsche-XFEM-DG discretization for a class of two-phase mass transport problems, *SIAM Journal of Numerical Analysis*, 51(2), 958–983, 2013.

33. A. Reusken and Y. Zhang, Numerical simulation of incompressible two-phase flows with a Boussinesq-Scriven interface stress tensor, *International Journal for Numerical Methods in Fluids*, 73(12), 1042–1058, 2013.

34. J. C. Slattery, L. Sagis, and E.-S. Oh, *Interfacial Transport Phenomena*, 2nd edn. Springer, New York, 2007.

35. G. Dziuk and C. M. Elliott, Finite element methods for surface PDEs, *Acta Numerica*, 22, 289–396, 2013.

36. E. Bänsch, *Numerical Methods for the Instationary Navier-Stokes Equations with a Free Capillary Surface*. Habilitationsschrift, Albert-Ludwigs-Universität, Freiburg i. Br., Germany, 1998.

37. G. Dziuk, An algorithm for evolutionary surfaces, *Numerische Mathematik*, 58, 603–611, 1991.

38. G. Matthies and L. Tobiska, Mass conservation of finite element methods for coupled flow-transport problems, *International Journal of Computing Science and Mathematics*, 1(2–4), 293–307, 2007.

39. J. W. Barrett, H. Garcke, and R. Nürnberg, Stable finite element approximations of two-phase flow with soluble surfactant, Preprint 14, Universität Regensburg, Regensburg, Germany, 2014.

40. A. Linke, G. Matthies, and L. Tobiska, Robust arbitrary order mixed finite element method for the incompressible stokes equations, ESAIM M2AN, doi:10.1051/m2an/2015044.

41. S. Hysing, S. Turek, D. Kuzmin, N. Parolini, E. Burman, S. Ganesan, and L. Tobiska, Quantitative benchmark computations of two-dimensional bubble dynamics, *International Journal of Numerical Methods in Fluids*, 60, 1259–1288, 2009.

17 Finite Element Techniques for the Numerical Simulation of Two-Phase Flows with Mass Transport

Christoph Lehrenfeld and Arnold Reusken

CONTENTS

17.1 INTRODUCTION

Two-phase incompressible flows with surface tension forces are usually modeled by either a diffusive interface or a sharp interface model. In this chapter, we restrict to the numerical simulation of the latter class of models. For numerical simulations based on a diffusive interface model, we refer to the literature, for example, [1–4]. In systems with incompressible fluids, a sharp interface model for the fluid dynamics typically consists of the Navier–Stokes equations for the bulk fluids with an interfacial surface tension force term on the right-hand side in the momentum equation. This model is combined with a convection-diffusion equation for mass transport (of a solute). We consider flow regimes with a low Reynold's number (laminar flow), a small capillary number (significant surface tension forces), and a moderate Schmidt number, so that concentration boundary layers can be resolved. We explain why these models typically have a very high numerical complexity. Such flow models cannot be solved reliably and accurately by the commercial codes that are available today. There is a need for more efficient and reliable numerical techniques for this class of models. The development, analysis, and application of such tailor-made numerical simulation methods are an important field in numerical analysis and computational engineering science. In this chapter, we restrict to a certain class of finite element discretization methods for two-phase incompressible flows, which has been developed in recent years. We treat three important and rather general finite element techniques that can be used for an accurate discretization of the mass transport equation: (1) the extended finite element method (XFEM) for the approximation of discontinuities; (2) Nitsche-method for a convenient handling of interface conditions (e.g., Henry condition); and (3) the space-time finite element technique. We restrict ourselves to an explanation of the main ideas of these methods and refer to the literature for more detailed information. Some results of numerical experiments with these methods that were obtained using the DROPS solver [5] are presented. This finite element code is specifically developed for testing, improving, and validating (new) tailor-made numerical methods for the simulation of sharp interface models for two-phase incompressible flows.

The two main contributions in this chapter are a discussion of major numerical challenges for this problem class and an explanation of the three finite element techniques mentioned before.

17.2   MATHEMATICAL MODEL

17.2.1   FLUID DYNAMICS AND MASS TRANSPORT

We introduce a standard mathematical model for the fluid dynamics in a two-phase flow with mass transport between the bulk phases. This model is often used in the literature, for example, [6–9]. We restrict ourselves to *isothermal conditions*, *incompressible* fluids and assume that there is *no change of phase*.

The given domain $\Omega \subset \mathbb{R}^3$, contains two different immiscible incompressible phases (liquid–liquid or liquid–gas), which may move in time and have different material properties ρ_i (density) and μ_i (viscosity), $i = 1, 2$. The density and viscosity, ρ_i and μ_i, $i = 1, 2$ are assumed to be constant in each phase. For each point in time, $t \in [0,T]$, Ω is partitioned into two open bounded subdomains $\Omega_1(t)$ and $\Omega_2(t)$, $\overline{\Omega} = \overline{\Omega_1}(t) \cup \overline{\Omega_2}(t)$, $\Omega_1(t) \cap \Omega_2(t) = \emptyset$, each of them containing one of the phases. These phases are separated from each other by the interface $\Gamma(t) = \overline{\Omega_1}(t) \cap \overline{\Omega_2}(t)$. For convenience we assume that $\Omega_1(t)$ is strictly contained in Ω, that is, does not touch $\partial\Omega$. The normal velocity $V_\Gamma = V_\Gamma(x,t) \in \mathbb{R}$ denotes the magnitude of the velocity of the interface Γ at $x \in \Gamma(t)$ in normal direction. $\mathbf{n}_\Gamma$ denotes the unit normal on Γ pointing from Ω_1 to Ω_2. To model interfacial forces we use the following standard (Cauchy) ansatz. The interface is considered to be a 2D continuum and on each (small) connected surface segment $\gamma \subset \Gamma$ there is a contact line force on $\partial\gamma$ of the form $\sigma_\Gamma n$. This σ_Γ is called the *interface stress tensor* and constitutive laws for σ_Γ have to be provided by surface rheology. Examples will be given in Remark 1. Based on the basic conservation laws of mass and momentum, the following standard model (in strong formulation) for the *fluid dynamics* of a two-phase incompressible flow can be derived:

$$\begin{cases} \rho_i \left(\dfrac{\partial \mathbf{u}}{\partial t} + (\mathbf{u} \cdot \nabla)\mathbf{u} \right) = \mathrm{div}\,\sigma_i + \rho_i \mathbf{g} \\ \mathrm{div}\,\mathbf{u} = 0 \end{cases} \quad \text{in } \Omega_i(t), \quad i = 1,2, \tag{17.1}$$

$$[\sigma\mathbf{n}_\Gamma]_\Gamma = \mathrm{div}_\Gamma \sigma_\Gamma \quad \text{on } \Gamma(t), \tag{17.2}$$

$$[\mathbf{u}]_\Gamma = 0 \quad \text{on } \Gamma(t), \tag{17.3}$$

$$V_\Gamma = \mathbf{u} \cdot \mathbf{n}_\Gamma \quad \text{on } \Gamma(t), \tag{17.4}$$

with the bulk phase stress tensor $\sigma_i = -p\mathbf{I} + \mu_i(\nabla\mathbf{u} + (\nabla\mathbf{u})^T)$, that is, we consider Newtonian bulk fluids. The vector $\mathbf{g}$ denotes an external (gravity) force. The operator div_Γ denotes the tangential divergence, see [9]. The notation $[\cdot]_\Gamma$ is used to denote the jump of a quantity across Γ. The assumption that there is no change of phase leads to the dynamic interface condition (Equation 17.4). Viscosity of the fluids leads to the continuity condition in Equation 17.3. Momentum conservation in a (small) material volume that intersects the interface leads to the stress balance condition in

Equation 17.2. To make the problem well-posed one needs suitable initial conditions for $\Gamma(0)$ and $\mathbf{u}(x, 0)$ and boundary conditions for $\mathbf{u}$ or $\sigma\mathbf{n}$ on $\partial\Omega$.

These Navier–Stokes equations model the fluid dynamics. Note that the evolution of the interface $\Gamma(t)$ is *implicitly* defined by this model.

Remark 1: We mention three important choices for the interface stress tensor σ_Γ. For an extensive treatment of constitutive models for the surface stress tensor, we refer to [6]. For $x \in \Gamma$ we define the projection $\mathbf{P}(x) = I - \mathbf{n}_\Gamma(x)\mathbf{n}_\Gamma(x)^T$. The operator $\nabla_\Gamma = \mathbf{P}\nabla$ is the tangential gradient. A so-called *clean interface* model for surface tension is given by $\sigma_\Gamma = \tau\mathbf{P}$, with a *constant* surface tension coefficient $\tau > 0$. This results in the surface force

$$\text{div}_\Gamma\sigma_\Gamma = \text{div}_\Gamma(\tau\mathbf{P}) = -\tau\kappa\mathbf{n}_\Gamma \tag{17.5}$$

used in Equation 17.2. Here κ is the mean curvature of Γ, that is, $\kappa(x) = \text{div } \mathbf{n}_\Gamma(x)$ for $x \in \Gamma$. *In the remainder of this chapter we restrict to this case.* In certain systems, for example, with surfactants, the surface tension coefficient τ cannot be assumed to be constant. This then gives rise to so-called Marangoni forces and the surface tension force in Equation 17.2 takes the more general form $\text{div}_\Gamma\sigma_\Gamma = \text{div}_\Gamma(\tau\mathbf{P}) = -\tau\kappa\mathbf{n}_\Gamma + \nabla_\Gamma\tau$. A third example for the interface stress tensor is one that is used to model viscous forces in the interface. With $\mathbf{D}_\Gamma(\mathbf{u}) := \mathbf{P}(\nabla_\Gamma\mathbf{u} + (\nabla_\Gamma\mathbf{u})^T)\mathbf{P}$ the *Boussinesq-Scriven* constitutive law is given by

$$\sigma_\Gamma = \tau\mathbf{P} + (\lambda_\Gamma - \mu_\Gamma)\text{div}_\Gamma\mathbf{u}\mathbf{P} + \mu_\Gamma\mathbf{D}_\Gamma(\mathbf{u}).$$

The constants $\mu_\Gamma > 0$ and $\lambda_\Gamma > 0$ are called the *interface shear viscosity* and *interface dilatational viscosity*, respectively. Other examples of interface stress tensors σ_Γ, for example, for the important case of visco-elastic interfacial behavior, are treated in [10].

We assume that one or both phases contain a dissolved species that is transported due to convection and molecular diffusion and does not adhere to the interface. The concentration of this species is denoted by $c(x, t)$. A standard mathematical model for the mass transport is as follows:

$$\frac{\partial c}{\partial t} + \mathbf{u} \cdot \nabla c = \text{div}(D_i\nabla c) \quad \text{in } \Omega_i(t), i = 1,2, \tag{17.6}$$

$$[D_i\nabla c \cdot \mathbf{n}_\Gamma]_\Gamma = 0 \quad \text{on } \Gamma(t), \tag{17.7}$$

$$c_1 = C_H c_2 \quad \text{on } \Gamma(t). \tag{17.8}$$

The diffusion coefficient D_i is piecewise constant and in general $D_1 \neq D_2$. The interface condition in Equation 17.7 comes from mass conservation, which implies flux continuity across the interface. The condition in Equation 17.8 is Henry's law, see Remark 2. The model has to be completed with suitable initial and boundary conditions for the concentration c.

Remark 2: Henry's law is a constitutive law describing the balance of chemical potentials at the interface. Under the assumption that kinetic processes at the interface are sufficiently fast, an instantaneous thermodynamical equilibrium is obtained. Henry's law states a linear dependence of the concentration at the interface on the partial pressure in the fluids $p = \beta_i c_i$. The constants $\beta_i, i = 1,2$ depend on the solute, the solvent, and the temperature. The partial pressures from both sides have to coincide. This yields

$$\beta_1 u_1 = \beta_2 u_2, \tag{17.9}$$

which results in Equation 17.8 with $C_H = \beta_2/\beta_1$. For further details on the modeling, we refer to [7,11].

17.2.2 Example of a Two-Phase Flow with Mass Transport

In this section, we describe a concrete example of a two-phase system. The fluid dynamics and mass transport in this system can be accurately described by the models introduced in the previous section. This example will be used in the numerical experiments in Section 17.8. The example has also been considered in [12–15].

In a large container filled with a water-glycerol mixture, the rise of a 4 mm air bubble is considered. Initially, the shape of the bubble is spherical and the bubble is placed close to the bottom of the container. Both fluids are at rest. Due to buoyancy forces, the bubble rises up and deforms during this process until a quasi-stationary state with a fixed ellipsoidal shape and a constant rise velocity is reached. We consider the dissolution process of oxygen from the gaseous phase to the liquid during the rise of the bubble. At the initial state, the oxygen concentration inside the liquid mixture is set to zero while the concentration inside the bubble is constant u_0.

In Table 17.1, the material parameters for the considered substance systems are listed. The mixture of water and glycerol consists of 18%(volume) water and 82%(volume) glycerol. The corresponding material data are taken from the literature, see [12,15]. In [15], experimental results for the considered setup are given, however, only with respect to the fluid dynamics.

In order to characterize the predominant effects in flows and specifically two-phase flows the dimensionless numbers Re, Ca, and Sc are introduced. The Reynold's number Re describes the ratio between inertia forces and viscous forces within a fluid. For the case of a rising bubble it is defined as

$$Re = \frac{\rho_L U_B d_B}{\mu_L}, \tag{17.10}$$

where U_B is a characteristic velocity of the fluid, here the terminal rise velocity of the bubble and d_B is the (initial) bubble diameter.

Note that in the definition of the Reynold's number the fluid properties of the gas phase do not enter. In the considered case, we have $Re \approx 7.6$ which says that inertia forces are important for the fluid behavior, but viscosity is still sufficiently high such that the flow can be considered as *laminar*.

Related to the surface tension force the dimensionless capillary number Ca is often used to characterize the ratio between viscous and surface tension forces:

$$Ca = \frac{\mu_L U_B}{\tau}, \tag{17.11}$$

with τ the constant surface tension coefficient of a clean interface. For the system considered here, we have a capillary number $Ca \approx 0.14$ which states that the surface tension forces are predominant. The dominance of the surface tension forces implies that the deformation of the bubble shape is moderate and that the resulting bubble shape is stable.

TABLE 17.1

Material Parameters for the Example Considered in this Chapter

	Liquid Phase ($\Omega_2 = \Omega_l$)	Disperse Phase ($\Omega_1 = \Omega_B$)
Density (kg/m³)	1205	1.122
Dynamic viscosity (Pa·s)	0.075	$1.824 \cdot 10^{-4}$
Henry weight β (1)	1	33
Diffusion coeff. (m²/s)	$6.224 \cdot 10^{-6}$	$1.916 \cdot 10^{-5}$
Surface tension (N/m)	0.063	
Bubble diameter (m)	0.004	

The Reynold's and capillary numbers describe the main characteristics of the fluid flow behavior. For the description of the most important effects of the mass transport in such a two-phase flow system, one typically uses the Schmidt number

$$Sc = \frac{\mu}{\rho D},$$

(17.12)

which describes the ratio between kinematic viscosity $v = \mu/\rho$ and diffusion D in a fluid. In liquids, the diffusion is typically in orders of magnitude smaller than the kinematic viscosity leading to high Schmidt numbers in liquid phases. Typical values for Sc in liquids are about 1000. In gaseous phases the Schmidt number is typically of order one. For the dissolution process in the considered example, the Schmidt number in the gaseous phase is significantly smaller than in the liquid phase. This is reflected in considerably smaller layers attached to the interface in the liquid phase than in the gaseous phase. The presence of thin boundary layers renders the accurate simulation of mass transport extremely difficult. In [12] instead of the physically correct diffusivity in the liquid, an artificial diffusivity is used to be able to prescribe the Schmidt number. Several Schmidt numbers have been considered to investigate the dependency of the dissolution process on the Schmidt number Sc. Here, we restrict to the case $Sc = 10$ in the liquid phase, for which a thin boundary layer exists the resolution of which is, however, still possible. This corresponds to the material parameters of Table 17.1 (with an artificial value $D_1 = 6.224 \cdot 10^{-6}\,\mathrm{m^2/s}$ for the diffusion coefficient in the liquid phase).

In Figure 17.1a the rise of the bubble is illustrated. One can capture the main characteristics of the dissolution of oxygen in the liquid phase and the velocity field in terms of streamlines. The results shown are simulation results obtained with discretization techniques discussed later.

Remark 3 The characteristics of a two-phase flow problem as introduced before depend on the values of the dimensionless numbers Re, Ca, and Sc. A further relevant dimensionless number is the Peclet number Pe, which is the ratio of rate of advection of a physical quantity by the flow to the rate of diffusion of the same quantity. The relation $Pe = Re \cdot Sc$ holds. In certain flow regimes there are phenomena that cause a (strong) increase in the numerical complexity of the simulation. We briefly address a few important examples of such phenomena.

Flows with a low Reynold's number (diffusion dominates) are laminar and typically show a much smoother and more stable behavior than flows with a (very) large Reynold's number (convection

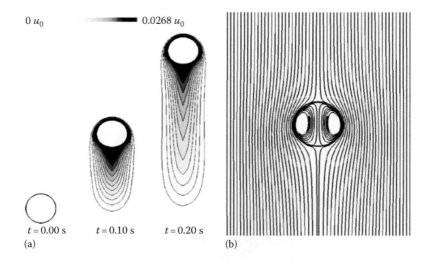

FIGURE 17.1 Concentration contours in the fluid phase at several times for the dissolution process of oxygen from a rising air bubble in a water-glycerol mixture for Schmidt number $Sc = 10$ (a) and streamlines around the bubble at $t = 0.2$ s (b).

dominates), which may even become turbulent. Already for one-phase incompressible flows, it is well-known that the numerical simulation of laminar flows is significantly easier than that of flows with a large Reynold's number. This obviously carries over to two-phase incompressible flows. Note that in two-phase flows, depending on the fluid parameters, the Reynold's numbers within the separate phases can be very different.

For a large capillary number Ca, surface tension is small compared to viscous or inertia forces acting on the interface. This typically leads to large deformations of the interface. Such a strong dynamics of the interface is numerically difficult to handle. A small capillary number on the other hand causes other numerical difficulties. These are due to the fact that a small value for Ca corresponds to a large surface tension force, which strongly influences the fluid dynamics but is singular in the sense that it acts only on the (evolving) interface.

If in the mass transport problem the Peclet number Pe is very large, this corresponds to a convection-dominated transport problem. Then typically, very sharp interior or boundary layers occur, which are difficult to deal with numerically. For laminar flows with a small Reynold's number, say $Re = \mathcal{O}(1)$, the magnitude of the Peclet number is of the same order as that of the Schmidt number, due to the relation $Pe = Re \cdot Sc$. Some specific numerical challenges for two-phase flow simulations are discussed in Section 17.3.

17.2.3 ONE-FLUID MODEL

As a basis for numerical simulations of two-phase flows one typically does *not* use a model with two Navier–Stokes equations in the two subdomains, as in Equation 17.1, with coupling conditions as in Equations 17.2 and 17.3 and a dynamic condition as in Equation 17.4. Instead one very often uses a *one-fluid* model, which we explain in this section. For the numerical *interface representation* different techniques are used in the literature, for example, the *volume of fluid* (VOF) method, the *level set* (LS) method, and the *phase field* method. For a treatment of these methods, we refer to the literature, for example, [8,9,16,17]. In this chapter, we restrict to the level set method, which we briefly explain. At $t = 0$ a *smooth* function $\phi_0(x), x \in \Omega$, is chosen, which characterizes the initial interface $\Gamma(0)$ in the following way:

$$\phi_0(x) < 0 \Leftrightarrow x \in \Omega_1(0), \quad \phi_0(x) > 0 \Leftrightarrow x \in \Omega_2(0), \quad \phi_0(x) = 0 \Leftrightarrow x \in \Gamma(0).$$

A popular choice is to take ϕ_0 (approximately) equal to a signed distance function to the initial interface, see Figure 17.2.

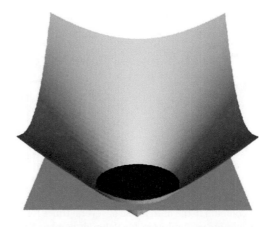

FIGURE 17.2 Initial level set function ϕ_0 equals a signed distance function, 2D example.

For $t > 0$ the level set function values $\phi(x,t)$ are defined by keeping the values constant along characteristics, induced by the velocity field $\mathbf{u}$. This results in the transport equation

$$\frac{\partial \phi}{\partial t} + \mathbf{u} \cdot \nabla \phi = 0 \quad \text{in } \Omega, t \geq 0, \tag{17.13}$$

which is called the *level set equation*. The interface $\Gamma(t)$ can be characterized by values of the level set function at time t:

$$\Gamma(t) = \{x \in \Omega : \phi(x,t) = 0\}. \tag{17.14}$$

The level set equation describes the evolution of the interface, hence the condition in Equation 17.4 is not needed anymore. The jumps in the coefficients ρ and μ can be described using the level set function ϕ in combination with the Heaviside function $H : \mathbb{R} \to \mathbb{R}$; $H(\zeta) = 0$ for $\zeta < 0$, and $H(\zeta) = 1$ for $\zeta > 0$. For ease one can set $H(0) = 1/2$. We define

$$\begin{aligned}
\rho(\phi) &:= \rho_1 + (\rho_2 - \rho_1)H(\phi), \\
\mu(\phi) &:= \mu_1 + (\mu_2 - \mu_1)H(\phi).
\end{aligned} \tag{17.15}$$

The continuity condition in Equation 17.3 is easy to satisfy by restricting to numerical approximations for the velocity that are continuous. The important stress balance condition in Equation 17.2 can be reformulated as a *localized force term in the momentum equation*. Based on these observations, the model (Equations 17.1 through 17.4) can be reformulated as follows:

$$\begin{cases} \rho(\phi)\left(\dfrac{\partial \mathbf{u}}{\partial t} + (\mathbf{u} \cdot \nabla)\mathbf{u}\right) = \operatorname{div}\sigma(\phi) + \rho(\phi)\mathbf{g} + \delta_\Gamma \operatorname{div}_\Gamma \sigma_\Gamma \text{ in } \Omega, \\ \operatorname{div}\mathbf{u} \qquad\qquad\qquad = 0 \end{cases} \tag{17.16}$$

$$\frac{\partial \phi}{\partial t} + \mathbf{u} \cdot \nabla \phi = 0 \quad \text{in } \Omega, \tag{17.17}$$

with $\sigma(\phi) := -p\mathbf{I} + \mu(\phi)(\nabla\mathbf{u} + (\nabla\mathbf{u})^\mathrm{T})$, that is, $\sigma = \sigma_i$ in $\Omega_i(t)$, and δ_Γ a suitable Dirac delta function that localizes the force $\operatorname{div}_\Gamma \sigma_\Gamma$ on the interface. Note that in Equation 17.16, we now have *one* Navier–Stokes equation in the whole domain Ω. Hence, this model is called the one-fluid model. Compared to the two Navier–Stokes equations in Equation 17.1 the Navier–Stokes equation in Equation 17.16 is more complicated, due to the discontinuities in viscosity μ and density ρ and the localized interface force term $\delta_\Gamma \operatorname{div}_\Gamma \sigma_\Gamma$. To obtain a well-posed model, one has to add suitable boundary and initial conditions for ϕ and $\mathbf{u}$. Note that the initial condition for ϕ determines the initial interface $\Gamma(0)$, due to Equation 17.14. This one-fluid model forms the basis for many numerical simulations of the fluid dynamics in two-phase incompressible flow systems in the literature.

In this chapter, we restrict to *finite element discretization methods*. For these one needs a suitable *variational formulation* of the partial differential equations. Since in the part on numerical methods later we focus on the mass transport equation, we only treat a variational formulation of the mass transport model (Equations 17.6 through 17.8). This is done in Section 17.5. For suitable variational formulations of the one-fluid model, consisting of the Navier–Stokes equations (Equation 17.16) and the level set Equation 17.17, we refer to the literature, for example, [9].

17.3 NUMERICAL CHALLENGES

The two-phase flow models introduced previously pose enormous challenges to numerical simulation tools. Such flow models cannot be solved reliably and accurately by the commercial codes that are available today. We address a few causes of the very high numerical complexity of this problem class later.

- *Evolving unknown interface.* The interface evolution is determined by the dynamic condition $V_\Gamma = \mathbf{u} \cdot \mathbf{n}_\Gamma$. The interface is a geometric object and it turns out that an accurate numerical approximation of this object and its evolution is a big challenge. In case of geometric singularities (e.g., droplet break-up or collision) it becomes even more complicated. An accurate interface approximation is of major importance for the two-phase flow simulation, in particular in systems where the surface tension is a driving force.

- *Discretization of surface tension force.* In many systems, the surface tension force has a strong effect on the fluid dynamics. Hence, an accurate discretization of $\mathrm{div}_\Gamma \sigma_\Gamma$ is essential for accurate simulation results. This surface tension force acts only on the (unknown) interface and depends on the curvature of the interface, see Equation 17.5. An accurate discretization of the surface tension force turns out to be a very difficult task.

- *Moving discontinuities.* Many quantities are discontinuous across the (evolving) interface. For example, the density and viscosity values have jumps across the interface. Due to surface tension forces, the pressure is discontinuous across the interface. The force balance (Equation 17.2) and a jump in the viscosity across the interface typically induce a discontinuity across the interface of the normal derivative of the velocity. Due to the Henry condition (Equation 17.8), the concentration c is discontinuous across the interface. For an accurate numerical treatment of such moving discontinuities, special numerical techniques are required.

- *Transport processes on moving manifolds.* If surfactants are considered, this leads to a transport equation on the (evolving) interface. The numerical solution of partial differential equations on moving manifolds is a difficult topic, which has been addressed in the literature only recently.

- *Resolution of boundary layers.* In many applications, boundary layers form at the interface of the two-phase system. The resolution of these boundary layers is very demanding. In [12] the dependency of the boundary layer thickness on the Schmidt number has been investigated for the setting described earlier and a correlation of the form $Sc^{-0.65} \cdot 0.5$ mm has been observed. The findings in that paper imply a ratio between the bubble diameter and the thickness of the boundary layer of 8 for $Sc = 1$ up to 400 for $Sc = 1000$. Hence, for realistic Schmidt numbers ($Sc \approx 1000$), the resolution of boundary layers with three-dimensional simulation tools is extremely difficult. For these realistic Schmidt numbers, the Peclet number $Pe = Re \cdot Sc$ is very large and thus the transport problem is strongly convection dominated, which causes problems with respect to the stability of the finite element discretization. Typically, this requires some form of convection stabilization.

- *Numerical solution of discrete problems.* After implicit discretization in time and discretization in space one obtains a very large nonlinear system of equations (for the discrete quantities) in each time step. In simulations, by far most of the total computing time is needed for solving these large nonlinear systems. For efficiency reasons, one has to use iterative solvers that take advantage of certain structural properties of the problem. Related to the development of such iterative solvers, there are many open problems.

- *Linearization.* The flow model contains several strongly nonlinear couplings. In the model given in Equations 17.16 and 17.17, the transport of the level set function depends on the flow field $\mathbf{u}$. The latter is determined from the Navier–Stokes Equation (Equation 17.16), but in this equation there is a strong dependence on (the zero level of) the level set function ϕ. This coupling between fluid dynamics (Equation 17.16) and interface evolution

(Equation 17.17) is strongly nonlinear. If mass transport is considered and if there is a dependence of τ on the dilute concentration, that is, $\tau = \tau(c)$, this coupling in two directions between fluid dynamics and mass transport is also in general strongly nonlinear. The same holds for a coupling in two directions between fluid dynamics and surfactant transport, that is, $\tau = \tau(S)$, where S is the surfactant concentration on the evolving interface. The topic of development of efficient and robust linearization techniques is an important one.

For dealing with these challenges, special numerical approaches have been developed. In the remainder of this chapter, we treat three numerical techniques in more detail:

- XFEM for nonaligned discontinuities (Section 17.4)
- Nitsche method for interface conditions (Section 17.5)
- Space-time FEM (Section 17.6)

All three techniques are based on rather general numerical concepts. Our aim is to explain the basic ideas of these concepts and avoid technical details. In Section 17.7, these techniques are combined and result in a very efficient numerical method for the simulation of the mass transport equation. This type of advanced finite element methods for the mass transport equations have been introduced in the literature only very recently; see, for example [18,19].

17.4   XFEM FOR NONALIGNED DISCONTINUITIES

In interface capturing methods, like VOF and the level set method, the interface is given implicitly and typically *not aligned* with the grid or triangulation that is used in the discretization of the flow variables. This *nonalignment* causes severe difficulties for the discretization methods, because certain important quantities (pressure, solute concentration) are discontinuous across the interface. In the past decade, in the finite element literature the extended finite element method (XFEM), also called cut finite element method, has been developed to overcome these difficulties. There is an extensive literature on the XFEM, see [20–24] and the references therein. We outline the basic idea of this technique for the approximation of the discontinuous solutions such as the pressure p or the solute concentration c.

It is known that the approximation of a smooth function with a discontinuity across the non-aligned interface with piecewise polynomials gives only poor approximation results. The approximation error (in the L^2 norm) scales with $\sqrt{h}$ independent of the polynomial degree. To overcome this, we introduce a special finite element space. The main idea is depicted in Figure 17.3. We first

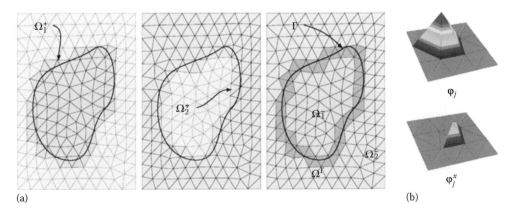

(a)   (b)

FIGURE 17.3   Extended finite element space as a sum of fictitious domain spaces (a) and sketch of a standard and an extended basis function (b), 2D example.

consider the simpler problem of approximating a function w_1 in the domain Ω_1. The function in Ω_1 has a smooth extension $\tilde{w}_1$ into Ω_1^+, the domain of all elements (triangles in Figure 17.3) which have some part in Ω_1. For the approximation of this smooth extension $\tilde{w}_1$, we use standard piecewise linears in Ω_1^+ and obtain the usual good approximation of $\tilde{w}_1$ and thereby also of w_1. We apply the same procedure on Ω_2^+ and can thus construct a good approximation of a smooth function w which is discontinuous across Γ. For this procedure, we need one degree of freedom per vertex except for vertices in Ω^{Γ}, the domain of elements which are intersected by the interface. Here, we need two degrees of freedom per vertex, one for each domain (Ω_1^+ and Ω_2^+). The corresponding finite element space is

$$Q_h^{\Gamma} = \mathcal{R}_1 Q_h \oplus \mathcal{R}_2 Q_h,$$

where Q_h is the finite element space of piecewise linears and $\mathcal{R}_i$ is the restriction operator to Ω_i.

Note that $dim(\mathcal{R}_i Q_h)$ coincides with the number of vertices in Ω_i^+. This finite element space and especially its dimension depends on the location of the interface Γ.

The finite element space Q_h^{Γ} can also be characterized as an extension of the standard finite element space Q_h, that is, $Q_h^{\Gamma} = Q_h \oplus Q_h^x$, with a suitable space Q_h^x, which is spanned by so-called enrichment functions. We explain how the basis functions of Q_h^{Γ} are constructed. To this end, consider a basis function $\varphi_j \in Q_h$ corresponding to a vertex $\mathbf{x}_j$ in Ω^{Γ}. Without loss of generality assume the vertex $\mathbf{x}_j$ lies in Ω_1. Then we define the enrichment function $\varphi_j^x = \mathcal{R}_2 \varphi_j$. For a vertex in Ω_2 we accordingly define $\varphi_j^x = \mathcal{R}_1 \varphi_j$. This choice gives the locality property of the new basis functions $\varphi_j^x(\mathbf{x}_k) = 0$ for all vertices $\mathbf{x}_k$. The basis functions φ_j and φ_j^x corresponding to a vertex $\mathbf{x}_j$ in Ω^{Γ} are depicted in Figure 17.3 (b) for a 2D example. In practice, the interface Γ is replaced by a numerical approximation Γ_h.

This XFEM has been successfully applied in the simulation of two-phase flow fluid dynamics. We illustrate this for a toluene-water rising droplet system. The results are taken from [9] and illustrate the effect of using the XFEM instead of a standard FEM for discretization of the pressure variable. We refer to section 7.11.2 in [9] for a precise description of the flow system and the numerical solver components used. The pressure is either discretized using the XFEM space $Q_h^{\Gamma_h}$ or the standard finite element space Q_h consisting of piecewise linears. Figure 17.4 shows the initial shape of the droplet and the droplet shapes after 10 time steps for both cases.

While the interface is smooth using the extended pressure finite element space, it shows many *spikes* in the case of the standard pressure space. These spikes are of course nonphysical and only caused by numerical oscillations at the interface, so-called *spurious velocities*, which are shown in Figure 17.5. The velocity field for the XFEM case is smooth showing the characteristic vortices, see Figure 17.6. Note that the scaling of the color coding in both figures is very different, with a maximum velocity of $5 \cdot 10^{-3}$ m for the extended pressure space compared to $5 \cdot 10^{-1}$ m for the standard pressure space. These results clearly show, that for this realistic two-phase flow example the

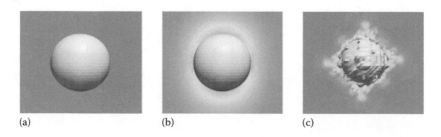

(a)   (b)   (c)

FIGURE 17.4 Initial droplet shape (a) and after 10 time steps for the XFEM case (b) and the standard FEM case (c).

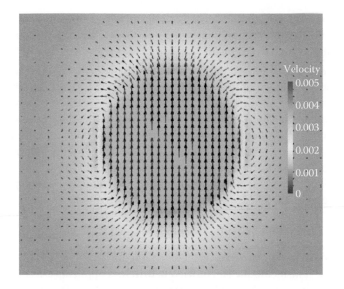

FIGURE 17.5 Velocity field at interface for the standard FEM case.

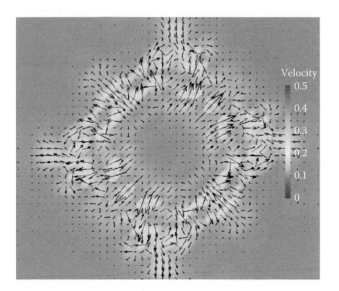

FIGURE 17.6 Velocity field at interface for the XFEM case.

standard pressure space Q_h is not suitable, whereas the extended pressure space $Q_h^{\Gamma_h}$ yields satisfactory results.

In the numerical simulations of two-phase flows presented in this chapter, we use the XFEM for the discretization of the (discontinuous) pressure variable and the (discontinuous) solute concentration c. In the literature one can also find extended finite element techniques for the discretization of the velocity, which is nonsmooth across the interface because it has a kink behavior. Such a so-called kink enrichment method for the velocity is treated in, for example, [21]. In the simulation of the two-phase fluid dynamics, however, such an enrichment of the velocity finite element space is not often used in practice, see [21] for an explanation of this.

Remark 4 The finite element space $Q_h^{\Gamma_h}$ depends on the location of the interface Γ_h. Thus, if the interface evolves the finite element space Q_h^{Γ} and its dimension changes. This causes difficulties for the numerical treatment of unsteady problems. A solution to this is discussed in Section 17.6.

Remark 5 The use of the XFEM enrichment functions, which results in an accurate discretization, causes problems for the solution of the linear system arising from the discretization. The key problem is that the conditioning of the resulting matrix depends on the position of the interface with respect to the triangulation. To overcome this problems stabilization techniques have been introduced in [24–26]. For the discretization of a (Navier–)Stokes problem with an enrichment of the pressure space, a stabilization is used to guarantee LBB-stability for the pair of finite element spaces in [24]. In [25,26], stabilized discretizations for interface problems as they appear for mass transport problems are discussed. In [27], it is shown that for the solution of mass transport problems *optimal* iterative strategies can be designed without additional stabilizations.

17.5 NITSCHE METHOD FOR INTERFACE CONDITIONS

The Nitsche method is a general finite element technique for enforcing boundary or interface conditions in a weak sense. The technique can be used in different applications, see, for example, [23,28]. In this section, we explain the basic idea of the Nitsche method for the mass transport problem (Equations 17.6 through 17.8) for the case that the interface is *stationary*. The method allows an accurate and convenient treatment of the Henry interface condition. The generalization to the case of an evolving interface is briefly addressed in Section 17.7.

The Nitsche method can only be applied in a variational setting. Therefore, we first need a well-posed variational formulation of the mass transport equation. The Henry condition can be reformulated in the form $[\beta c]_{|\Gamma} = 0$, with a suitable piecewise constant $\beta_i > 0$ in Ω_i, $i = 1, 2$. We assume that the velocity field **u** is given. Furthermore, for simplicity we assume that the dilute concentration has to satisfy the homogeneous boundary condition $c = 0$ on $\partial\Omega$. For the variational formulation, we need suitable scalar products and (Sobolev) spaces. Here, we outline the key components; details are given in section 10.2 in [9]. We define a function space

$$V := \{v \in L^2(\Omega) : v_{|\Omega_i} \in H^1(\Omega_i), i = 1, 2, v_{|\partial\Omega} = 0, [\beta v]_{|\Gamma} = 0\}.$$

It is important to note that the Henry condition is put as an essential condition in the definition of this space. On V, we use the weighted scalar products

$$(u, v)_0 := (\beta u, v)_{L^2} = \int_{\Omega} \beta u v \, dx, (u, v)_{1, \Omega_1 \cup \Omega_2} := \sum_{i=1}^{2} \beta_i \int_{\Omega_i} \nabla u_i \cdot \nabla v_i \, dx.$$

Related to the transport problem, we define the bilinear form

$$a(u, v) := (Du, v)_{1, \Omega_1 \cup \Omega_2} + (\mathbf{u} \cdot \nabla u, v)_0, \quad u, v \in V. \tag{17.18}$$

Here D denotes the piecewise constant diffusion coefficient. A suitable variational formulation of the mass transport problem, with a source term f, is as follows: determine $c = c(\cdot, t) \in V$ such that for $t \in [0, T]$

$$\left(\frac{\partial c}{\partial t}, v\right)_0 + a(c, v) = (f, v)_0 \quad \text{for all } v \in V, \tag{17.19}$$

holds. The derivative $(\partial c / \partial t)$ is defined in a suitable weak sense (not specified here). It can be shown that this is a well-posed weak formulation of the masstransport equation.

We introduce a Nitsche based finite element discretization of this problem. We assume that the interface is nonaligned, and therefore one has to be careful with the choice of the finite element space. In Section 17.4, it is explained that the XFEM space Q_h^Γ is a suitable finite element space for the discontinuous dilute concentration c. It is very difficult (often even impossible), in particular, for the case of an evolving interface, to incorporate the Henry interface condition $[\beta c]_{|\Gamma} = 0$ in this finite element space. The Nitsche technique allows us to use the space Q_h^Γ *without* the Henry interface condition. The Henry condition is enforced, in a weak sense, by changing the bilinear form in the variational problem. Before we introduce this modified bilinear form, we need a suitable averaging of functions across the interface. We consider a triangulation $\mathcal{T}_h$ consisting of tetrahedra T, with $h_T := \text{diam}(T)$ and $h := \max\{h_T : T \in \mathcal{T}_h\}$. For any tetrahedron $T \in \mathcal{T}_h$ let $T_i := T \cap \Omega_i$ be the part of T in Ω_i. We define the weighted average

$$\{v\} := \kappa_1(v_1)_{|\Gamma} + \kappa_2(v_2)_{|\Gamma}, \quad (\kappa_i)_{|T} := \frac{|T_i|}{|T|}, \tag{17.20}$$

where v_i denotes the restriction of a function v to Ω_i. We now introduce the modification of the bilinear form in Equation 17.18:

$$a_h(u,v) := (Du,v)_{1,\Omega_1 \cup \Omega_2} + (\mathbf{u} \cdot \nabla u, v)_0 + N_h^\Gamma(u,v) \quad \text{with}$$
$$N_h^\Gamma(u,v) := -([\beta u],\{D\nabla v \cdot \mathbf{n}_\Gamma\})_\Gamma - (\{D\nabla u \cdot \mathbf{n}_\Gamma\},[\beta v])_\Gamma + \lambda h^{-1}([\beta u],[\beta v])_\Gamma, \tag{17.21}$$

with $\lambda > 0$ a parameter. Here $(\cdot,\cdot)_\Gamma$ denotes the L^2 scalar product on the interface Γ (in practice replaced by an approximation Γ_h). In this modified bilinear form, the term $(\{D\nabla u \cdot \mathbf{n}_\Gamma\},[\beta v])_\Gamma$ is added for consistency reasons, the term $([\beta u],\{D\nabla v \cdot \mathbf{n}_\Gamma\})_\Gamma$ is included to maintain symmetry and the term $\lambda h^{-1}([\beta u],[\beta v])_\Gamma$ is used because we want to satisfy the Henry condition $[\beta c]_{|\Gamma} = 0$ (approximately). The latter term penalizes the jump $[\beta u]_{|\Gamma}$. An important property of this Nitsche method is that the penalty term is consistent: it vanishes if we insert the solution c of the continuous problem. The Nitsche-XFEM discretization of the mass transport problem (Equation 17.19) is as follows: determine $c_h = c_h(\cdot,t) \in Q_h^\Gamma$ such that for $t \in [0,T]$

$$\left(\frac{\partial c_h}{\partial t}, v_h\right)_0 + a_h(c_h,v_h) = (f,v_h)_0 \quad \text{for all } v_h \in Q_h^\Gamma. \tag{17.22}$$

In practice, the integrals (and average) over Γ that occur in this equation are replaced by the corresponding ones over Γ_h. Note that Equation 17.22 is a discretization only w.r.t. the space variable. The time discretization can be realized by some finite differencing (method of lines). In Section 17.6, we explain an alternative for this *first space then time discretization* approach which is also suitable for the case where the interface is *not stationary*. Again, we emphasize that the Henry condition is not built into the finite element space Q_h^Γ, but weakly enforced by a suitable modification of the bilinear form. The performance of the method turns out to be rather robust w.r.t. the choice for the stabilization parameter λ. Error analysis for this method has been developed and yields that for the discrete solution c_h an optimal error bound of the form $\| c(\cdot,t) - c_h(\cdot,t) \|_{L^2(\Omega)} \leq ch^2$ holds [29]. Also bounds for the error in the Henry condition are available and of the form $\| [\beta c_h(\cdot,t)]_\Gamma \|_\Gamma \leq ch^{3/2}$.

17.6   SPACE-TIME FORMULATION OF PARABOLIC PROBLEMS

The solution of problems with moving interfaces in an Eulerian framework is challenging. Between time steps some degrees of freedom switch phases. The derivation of appropriate discrete equations for these unknowns is not straightforward. The solution can be discontinuous across the interface. Hence, an approximation of the time derivative by some kind of finite difference stencil, as it is typically done in the method of lines, is not feasible. One way to deal with this problem is a formulation in space-time. The domain and the time interval define a so-called time slab on which the new formulation of the problem is derived. In the space-time setting, the (space-time) interface is stationary and the time derivate can naturally be defined within the separate space-time subdomains.

We explain the basic concept of the space-time finite element method for a simpler problem with a *smooth* solution, namely the parabolic model problem

$$\frac{\partial u}{\partial t} - \Delta u = f \quad \text{in } \Omega, t \in [0,T],$$

$$u(\cdot,0) = u_0 \quad \text{in } \Omega, \tag{17.23}$$

$$u(\cdot,t) = 0 \quad \text{on } \partial\Omega.$$

For simplicity we assume f to be independent of t. We use a partitioning of the time domain $0 = t_0 < t_1 < \cdots < t_N = T$, with a fixed time step size $\Delta t = T/N$, that is, $t_j = j\Delta t$. This assumption of a fixed time step is made to simplify the presentation, but is *not* essential for the method. Corresponding to each time interval $I_n := (t_{n-1}, t_n)$, we have a consistent triangulation $\mathcal{T}_n$ of the domain Ω. This triangulation may vary with n. Let V_n be a finite element space of continuous piecewise polynomial functions corresponding to the triangulation $\mathcal{T}_n$, with boundary values equal to zero. For $1 \le n \le N$ and a nonnegative integer k we define, on each time slab $Q^n := \Omega \times I_n$, a space-time finite element space as follows:

$$V_{kn} := \left\{ v : Q^n \to \mathbb{R} : v(x,t) = \sum_{j=0}^{k} t^j \phi_j(x), \phi_j \in V_n \right\}, \tag{17.24}$$

for $1 \le n \le N$. The corresponding space-time discretization of Equation 17.23 reads: determine u_h such that for all $n = 1,2,\ldots,N$, $(u_h)_{|Q^n} \in V_{kn}$ and

$$\int_{t_{n-1}}^{t_n} \left(\frac{\partial u_h}{\partial t}, v_h \right)_{L^2} + \left(\nabla u_h, \nabla v_h \right)_{L^2} dt + \left([u_h]^{n-1}, v_h^{n-1,+} \right)_{L^2} = \int_{t_{n-1}}^{t_n} (f,v_h)_{L^2} dt \quad \text{for all } v_h \in V_{kn}, \tag{17.25}$$

where

$$(\cdot,\cdot)_{L^2} = (\cdot,\cdot)_{L^2(\Omega)}$$

$$[w_h]^n = w_h^{n,+} - w_h^{n,-}, \quad w_h^{n,+(-)} = \lim_{s \to 0^{+(-)}} w_h(\cdot,t_n + s),$$

$u_h^{0,-} \in V_1$ an approximation of the initial data u_0

Such space-time finite element methods for parabolic problems are well-known in the literature. For an analysis and further explanation of this discretization method we refer to the literature, for example, [30].

As examples, we consider two important special cases: $k = 0$, $k = 1$. If $k = 0$ then $v_h \in V_{kn}$ does not depend on t. Define $u_h^n(x) := u_h(x,t)$, $t \in I_n$. The method shown in Equation 17.25 for determining $u_h^n \in V_n$ reduces to the implicit Euler scheme:

$$\frac{1}{\Delta t}\left(u_h^n - u_h^{n-1}, v_h \right)_{L^2} + \left(\nabla u_h^n, \nabla v_h \right)_{L^2} = (f,v_h)_{L^2} \quad \text{for all } v_h \in V_n.$$

We now consider $k = 1$. Then on Q^n, the function u_h^n can be represented as $u_h^n(x,t) = \hat{u}_h^n(x) + (1/\Delta t)(t - t_{n-1})\tilde{u}_h^n(x)$, with $\hat{u}_h^n, \tilde{u}_h^n \in V_n$. These unknown functions are uniquely determined by the coupled system

$$\left(\hat{u}_h^n + \tilde{u}_h^n, v_h\right)_{L^2} + \Delta t \left(\nabla \hat{u}_h^n + \frac{1}{2}\nabla \tilde{u}_h^n, \nabla v_h\right)_{L^2} = \left(u_h^{n-1,-}, v_h\right)_{L^2} + \Delta t(f, v_h)_{L^2},$$

$$\frac{1}{2}\left(\tilde{u}_h^n, v_h\right)_{L^2} + \Delta t \left(\frac{1}{2}\nabla \hat{u}_h^n + \frac{1}{3}\nabla \tilde{u}_h^n, \nabla v_h\right)_{L^2} = \frac{1}{2}\Delta t(f, v_h)_{L^2}, \qquad (17.26)$$

for all $v_h \in V_n$, see [30]. Note that although the discretization is in $d + 1$ dimension, if d is the dimension of the spatial domain, the method has a time-stepping structure such that the computational complexity essentially depends (only) on the dimension of the underlying spatial finite element space V_n. The considered method is often called a discontinuous Galerkin (DG) method (in time) as the space-time finite element space does not ensure continuity in time in a strong sense, but only in a weak sense. As a time integration method, the space-time DG method is stable and has good smoothing properties. The big advantage, however, is the great flexibility that comes with the space-time formulation. This flexibility is used to incorporate the more difficult case where for instance diffusion coefficients are discontinuous across a moving interface and the finite element space is no longer independent of time. We explain this in the next section.

17.7 SPACE-TIME NITSCHE-XFE METHOD

The discretization of a parabolic problem with a moving interface is significantly more difficult than for the model problem discussed before. We apply the concepts of the XFEM space and the Nitsche technique in a *space-time setting*, which results in a suitable discretization method for the mass transport problem with a moving interface. We only sketch the components of such a discretization and refer to [19] for details.

In order to capture the moving discontinuities, a generalization of the enrichment procedure of the XFEM space as described in Section 17.4 is needed. We take $k = 1$ (linear finite elements) in Equation 17.24 and apply the enrichment procedure on a space-time slab, see Figure 17.6. In the enrichment special discontinuous space-time basis functions, which depend on the interface position in space-time, are added to the original finite element space V_{1n}. This is depicted in Figure 17.7: the marked space-time elements are cut by the (space-time) interface and corresponding to *every* degree of freedom in these elements an enrichment function is added. This can lead to a different

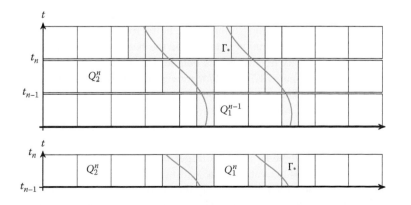

FIGURE 17.7   Sketch of space time slabs and relevant areas for the space-time XFEM.

number of unknowns on different time slabs. This, however, is not a problem due to the weak coupling introduced by the DG formulation in time.

The (weak) enforcement of the interface condition for the stationary interface as described in Section 17.5 has a straightforward extension to the space-time domain. The corresponding contribution of the Nitsche integrals takes the form $\int_{t_{n-1}}^{t_n} N_h^{\Gamma(t)}(u,v)\mathrm{d}t$, with $N_h^{\Gamma(t)}$ as in Equation 17.21.

The resulting discrete problem does not allow for a block structure as in Equation 17.26 where only *space* integrals have to evaluated. This is due to the fact that the problem parameters D and β as well as the ansatz and test function depend on space *and time*. Instead, integration on *space-time geometries* has to be used.

Remark 6 Due to the discontinuous coefficients and shape functions the space-time formulation is posed on space-time subdomains $Q_i^n = \bigcup_{t\in[t_{n-1},t_n]} \Omega_i(t)$. To implement a method with integrals of the form $\int_{Q_i^n} \cdot\,\mathrm{d}x\mathrm{d}t$ one needs special strategies for the numerical integration. We do not discuss this here but refer to [18] for more information on this.

To summarize, a very efficient and accurate finite element discretization of the mass transport problem (Equations 17.6 through 17.8) is obtained, by using a variational formulation in *space-time* in which the Henry condition is treated by the *Nitsche technique* and the finite element space is built by the *space-time XFEM*. A rigorous error analysis of this method, resulting in second-order error bounds both with respect to the space and time mesh size, is given in [19].

17.8   NUMERICAL SIMULATION OF MASS TRANSPORT PROBLEM

We discuss the results of numerical simulations of the two-phase flow system described in Section 17.2.2. The simulation of the fluid dynamics of this flow problem has also been considered in [12–14]. The numerical simulation including mass transport, with $Sc = 10$, has further been investigated in [12,13]. For the fluid dynamics of the system experiments have been carried out in [15]. The methods described before are implemented in the software package DROPS, see [5]. For the rise velocity and the pathlength of the barycenter of the bubble, we get values of 0.112 m/s and 21.44 mm, respectively. The final shape of the bubble has an aspect ratio of maximum height to maximum width of 0.88. All three values are in good agreement with the numerical and experimental results in [12–15]. Based on this and a systematic validation study of fluid dynamics simulations with DROPS in [31,32] we conclude that the flow field obtained with DROPS is sufficiently accurate to serve as a reliable input for the simulation of the mass transport problem, which we discuss next.

We consider the numerical solution of the described system with a discretization combining the space-time formulation, the extended finite element technique (in space-time) and the Nitsche method. Piecewise linear functions in space and time are used and a Nitsche stabilization parameter $\lambda = 20$. In Figure 17.1, the evolution of concentration isolines in the liquid for the simulation at different time stages is shown. The corresponding concentration fields inside the bubble at different times and the streamlines corresponding to the velocity (relative to the bubble rise velocity) at $T = 0.2$ are shown in Figure 17.8. Note that the scalings in Figures 17.1 and 17.8 are very different due to the large Henry number.

In Figure 17.9, we consider the concentration along straight lines which are crossing the center of the bubble at $T = 0.2$. We consider the line through the tip of the bubble ($0°$), a line through the equator ($90°$) and one line at a $135°$ angle from the tip. On those lines, we plotted the concentration. Due to the Henry interface condition, the concentration has jumps across the interface such that the concentration inside the bubble is 33 times larger than outside. We adapted the scaling for the concentration inside and outside the bubble. The scaling is chosen such that a continuous line in the plot corresponds to a concentration field fulfilling the Henry interface condition. We observe in the plot that this condition is fulfilled very accurately. We considered the data given in [13, Figure 9.35] for a comparison. The results are in good agreement.

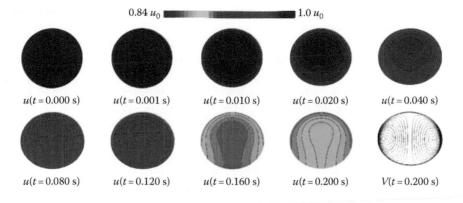

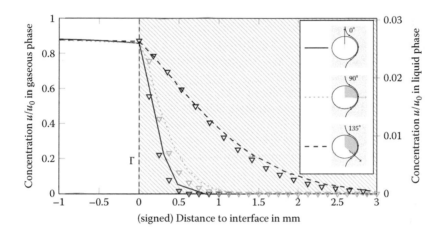

FIGURE 17.8   Concentration distribution inside the gaseous bubble.

FIGURE 17.9   Concentration along lines for angles 0°, 90°, and 135° computed with DROPS (lines) and comparison data from [13] (triangles).

The two-phase flow problem considered in this section has a dynamically evolving and deforming interface with a complex flow structure and a convection-dominated transport of the solute. The finite element techniques used in the simulation of the fluid dynamics and the solute transport provide an efficient solution strategy which has been proven to be robust with respect to the interface position an the moving discontinuities. This can be concluded from the results presented earlier and from other simulation studies with the DROPS package, see [5]. The techniques are flexible, in the sense that these can be combined with adaptivity concepts (local grid refinement close to the interface) and with higher order discretizations. This flexibility and the fact that for the space-time Nitsche-XFEM for the mass transport problem there is a rigorous second-order error bound available, are advantages of this approach compared to the finite volume-based method used in [13].

17.9   CONCLUSION AND OUTLOOK

In this chapter, we outlined important numerical challenges in the numerical simulation of sharp interface models for two-phase incompressible flows. Three important and relatively new finite element techniques (XFEM, Nitsche, space-time FEM) are explained and combined for the discretization of the mass transport equation. As far as we know, the resulting space-time Nitsche-XFE method is the only Eulerian discretization method for this type of solute transport equation which has a proven second order accuracy.

In our opinion, the development of satisfactory numerical simulation tools for two-phase incompressible flows is still in its infancy. Hence, an outlook discussion can be very extensive. Here, we restrict ourselves to a few brief comments on topics for further research.

We applied the space-time XFEM approach to the solute transport equation. The application of the space-time technique for the simulation of the fluid dynamics problem (Navier–Stokes model) appears to be a promising approach. This, however, has not been studied in the literature so far and is a topic of current research.

Almost nothing is known concerning accurate simulation of two-phase flows with more complex interfacial rheology, for example, Newtonian bulk fluids with interfaces that have (due to surfactants or nano-particles) viscous or visco-elastic behavior. Such systems are very important in applications, see [6,10].

We are not aware of any literature in which numerical methods for a realistic model of solute transport combined with reaction processes are studied.

A general key difficulty in this field is the validation of numerical methods. Only very few papers are available, see [32], in which benchmark results for two-phase flow simulations are presented.

For laminar *one*-phase incompressible flows there is an extensive literature on error analyses of finite element discretization methods. For two-phase incompressible flows (almost) no rigorous discretization error analysis is known.

Time and space discretization results in a very large and strongly nonlinear system of equations for the discrete unknowns. If one applies standard (black box) iterative solvers, the computing times, even on modern architectures, are unacceptable. Hence, special solution methods, that make use of problem specific structures are needed. Such solvers are not available, yet.

Discrete linear systems arising from discretizations of two-phase flows are already challenging due to high contrasts in material parameters (density, viscosity, and diffusivity). If additionally non-standard components such as extended finite element or space-time finite element spaces come into play, the iterative solution of linear systems gets extremely difficult. Only for special cases, satisfying solution strategies are known in the literature, for example, [24,25,27].

ACKNOWLEDGMENT

The authors gratefully acknowledge funding by the German Science Foundation (DFG) within the Priority Program (SPP) 1506 "Transport Processes at Fluidic Interfaces."

REFERENCES

1. D. Anderson, G. McFadden, and A. Wheeler, Diffusive-interface methods in fluid mechanics, *Annu. Rev. Fluid Mech.*, 30, 139–165, 1998.
2. D. Jacqmin, Calculation of two-phase Navier–Stokes flows using phase-field modeling, *J. Comput. Phys.*, 155, 96–127, 1999.
3. K. Teigen, X. Li, J. Lowengrub, F. Wang, and A. Voigt, A diffusion-interface approach for modelling transport, diffusion and adsorption/desorption of material quantities on a deformable interface, *Commun. Math. Sci.*, 7, 1009–1037, 2009.
4. S. Aland, J. Lowengrub, and A. Voigt, A continuum model of colloid stabilized interfaces, *Phys. Fluids*, 23, 062103, 2011.
5. S. Gross, J. Berger, J. Grande, P. Esser, O. Fortmeier, C. Lehrenfeld, E. Loch, T.H. Nguyen, V. Reichelt, L. Zhang, and Y. Zhang, DROPS package for simulation of two-phase flows. http://www.igpm.rwth-aachen.de/DROPS/, 2015.
6. L. Sagis, Dynamic properties of interfaces in soft matter: Experiments and theory, *Rev. Mod. Phys.*, 83(4), 1367, 2011.
7. J. Slattery, L. Sagis, and E.-S. Oh, *Interfacial Transport Phenomena*, 2nd edn. New York: Springer, 2007.

8. G. Tryggvason, R. Scardovelli, and S. Zaleski, *Direct Numerical Simulations of Gas-Liquid Multiphase Flows*. Cambridge, U.K.: Cambridge University Press, 2011.

9. S. Gross and A. Reusken, *Numerical Methods for Two-Phase Incompressible Flows*. Heidelberg, Germany: Springer, 2011.

10. L. Sagis, Modeling interfacial dynamics using nonequilibrium thermodynamics frameworks, *Eur. Phys. J.-Spec. Top.*, 222, 105–137, 2013.

11. M. Ishii, *Thermo-Fluid Dynamic Theory of Two-Phase Flow*. Paris, France: Eyrolles, 1975.

12. D. Bothe, M. Koebe, and H.-J. Warnecke, VOF-simulations of the rise behavior of single air bubbles with oxygen transfer to the ambient liquid, in *Transport Phenomena with Moving Boundaries* (F.-P. Schindler, ed.), pp. 134–146. Düsseldorf, Germany: VDI-Verlag, 2004.

13. M. Koebe, *Numerische Simulation aufsteigender Blasen mit und ohne Stoffaustausch mittels der Volume of Fluid (VOF) Methode*. PhD thesis. Paderborn, Germany: Universität Paderborn, 2004.

14. A. Onea, *Numerical Simulation of Mass Transfer with and without First Order Chemical Reaction in Two-Fluid Flows*. PhD thesis. Karlsruhe, Germany: Forschungszentrum Karlsruhe, September 2007.

15. F. Raymond and J.-M. Rosant, A numerical and experimental study of the terminal velocity and shape of bubbles in viscous liquids, *Chem. Eng. Sci.*, 55, 943–955, 2000.

16. J. Sethian, *Level Set Methods and Fast Marching Methods*. Cambridge, U.K.: Cambridge University Press, 1999.

17. S. Osher and R. Fedkiw, *Level Set Methods and Dynamic Implicit Surfaces*, Applied Mathematical Sciences, Vol. 153, Berlin, Germany: Springer, 2003.

18. C. Lehrenfeld, The Nitsche XFEM-DG space-time method and its implementation in three space dimensions, *SIAM J. Scientific Comput.*, 37(1), A245–A270, 2015.

19. C. Lehrenfeld and A. Reusken, Analysis of a Nitsche XFEM-DG discretization for a class of two-phase mass transport problems, *SIAM J. Numer. Anal.*, 51, 958–983, 2013.

20. J. Chessa and T. Belytschko, An extended finite element method for two-phase fluids, *ASME J. Appl. Mech.*, 70, 10–17, 2003.

21. T. Fries and T. Belytschko, The generalized/extended finite element method: An overview of the method and its applications, *Int. J. Numer. Method Eng.*, 84, 253–304, 2010.

22. H. Sauerland and T. Fries, The extended finite element method for two-phase and free-surface flows: A systematic study, *J. Comput. Phys.*, 230, 3369–3390, 2011.

23. A. Hansbo and P. Hansbo, An unfitted finite element method, based on Nitsche's method, for elliptic interface problems, *Comput. Methods Appl. Mech. Eng.*, 191(47–48), 5537–5552, 2002.

24. P. Hansbo, M. Larson, and S. Zahedi, A cut finite element method for a Stokes interface problem, *Appl. Numer. Math.*, 85, 90–114, 2014.

25. E. Burman and P. Hansbo, Fictitious domain finite element methods using cut elements: II. A stabilized Nitsche method, *Appl. Numer. Math.*, 62(4), 328–341, 2012; *Third Chilean Workshop on Numerical Analysis of Partial Differential Equations (WONAPDE 2010)*, Concepción.

26. S. Zahedi, E. Wadbro, G. Kreiss, and M. Berggren, A uniformly well-conditioned, unfitted Nitsche method for interface problems: Part I, *BIT Numer. Math.*, 53, 791–820, September 2013.

27. C. Lehrenfeld and A. Reusken, Optimal preconditioners for Nitsche-XFEM discretizations of interface problems, Technical Report 406. Aachen, Germany: IGPM, RWTH Aachen, August 2014. Submitted to *Numer. Math.*

28. P. Hansbo, Nitsche's method for interface problems in computational mechanics, *GAMM-Mitt.*, 28(2), 183–206, 2005.

29. A. Reusken and T. Nguyen, Nitsche's method for a transport problem in two-phase incompressible flows, *J. Fourier Anal. Appl.*, 15, 663–683, 2009.

30. V. Thomee, *Galerkin Finite Element Methods for Parabolic Problems*. Berlin, Germany: Springer, 1997.

31. E. Bertakis, S. Gross, J. Grande, O. Fortmeier, A. Reusken, and A. Pfennig, Validated simulation of droplet sedimentation with finite-element and level-set methods, *Chem. Eng. Sci.*, 65, 2037–2051, 2010.

32. H. Marschall, S. Boden, C. Lehrenfeld, C. Falconi, U. Hampel, A. Reusken, M. Worner, and D. Bothe, Validation of interface capturing and tracking techniques with different surface tension treatments against a Taylor bubble benchmark problem, *Comput. Fluids*, 102, 336–352, 2014.

18 Finite Volume/Finite Area Interface-Tracking Method for Two-Phase Flows with Fluid Interfaces Influenced by Surfactant

Chiara Pesci, Kathrin Dieter-Kissling,
Holger Marschall, and Dieter Bothe

CONTENTS

18.1 INTRODUCTION

This chapter is concerned with continuum physical and numerical modeling of free-surface and two-phase flows under the influence of surfactant. There are two general numerical model categories to describe systems comprising fluidic interfaces: *Eulerian interface-capturing methods* and *Lagrangian interface-tracking methods*. The first describe two-phase systems on a Eulerian grid by introducing an additional variable, describing the position of the interface [level-set (LS) functions] or of one of the continuous phases (color or phase-field functions). The evolution of the interface is tracked by solving a corresponding advection equation. Commonly known representatives of interface-capturing methods are volume-of-fluid (VOF) methods [1], LS methods [2], hybrid coupled LS/VOF methods [3], or diffuse-interface methods [4]. Concerning surfactant transport, all methods have been enhanced to capture insoluble or soluble surfactants [5–8].

In Lagrangian methods, the fluidic interface is represented by either a deformable computational grid (boundary integral methods [9]) or arbitrary Lagrangian–Eulerian (ALE) interface-tracking methods [10] or a set of Lagrangian marker particles aligned to the interface [11].

The present work bases on the ALE model originally presented by Muzaferija and Perić [10] to model free-surface flows and enhanced by Tuković and Jasak [12]. Our method combines an unstructured collocated finite volume method (FVM) supporting a moving computational mesh with automatic mesh deformation [13] and remeshing [14–16] algorithm. The surfactant transport along the free surface is taken into account adopting the unstructured collocated finite area method (FAM) [17].

18.2 CONTINUUM PHYSICAL MODELING

18.2.1 MATHEMATICAL MODEL

Consider a fluid domain Ω containing two immiscible fluids, separated by a deformable interface. The mathematical model for two-phase flows employs a sharp interface representation, meaning that the interface is presented as a mathematical surface of zero thickness. The fluid domain has impermeable boundary, and it is divided by a free and deformable interface $\Sigma(t)$ into two subdomains, $\Omega^+(t)$ and $\Omega^-(t)$, corresponding to the two bulk phases. The presence of surfactant species both in the bulk phases and on the interface is taken into account. Within the model, the governing equations are based on the conservation of mass, momentum, and surfactant molar mass. Furthermore, we restrict this description to incompressible Newtonian bulk fluids under isothermal conditions and in the absence of phase change and chemical reactions. Surfactant concentrations are hence assumed as dilute in the bulk, but can be nondilute on the interface. For simplicity, each surfactant species is considered to be soluble in a single bulk phase, only, which can be different for different surfactant. This avoids the occurrence of two simultaneous one-sided sorption processes.

In order to formulate the governing equations, let $V(t)$ be a control volume, inside the total fluid domain Ω, moving with velocity **w**. The control volume is bounded by $\partial V(t)$, with the outward pointing normal field on $\partial V(t)$ denoted as **n** (see Figure 18.1). Moving control volumes are required by the numerical model. In fact, as we will see later, the equations will be expressed in ALE formulation, which employs comoving domain. The intersection $\Sigma(t) \cap V(t)$ is denoted as $S(t)$, and its bounding curve is denoted as $\partial S(t) = C(t)$. The normal field on $\Sigma(t)$ (and on $S(t)$ as well) is $\mathbf{n}_\Sigma$, pointing into the domain Ω^-, say, and **m** is the normal vector to $C(t)$ being also tangential to $\Sigma(t)$.

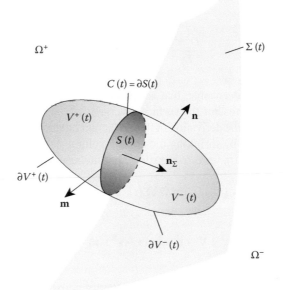

FIGURE 18.1 Domain representation for a two-phase system, moving control volume and area.

The velocity field is denoted as $\mathbf{v}$, the interface velocity is $\mathbf{v}^{\Sigma}$ with $\mathbf{v}^{\Sigma} = \mathbf{v}_{|\Sigma}$, where the notation $\cdot_{|\Sigma}$ denotes the trace of a quantity defined in $\Omega^{\pm}$ on the interface.

The core part of the mathematical model is the integral balance of surfactant molar mass over a moving control volume $V(t)$. Its general form reads as

$$\frac{d}{dt}\left[\int_{V(t)} c_i\,dV + \int_{S(t)} c_i^{\Sigma}\,dS\right] = -\int_{\partial V(t)} c_i(\mathbf{v}-\mathbf{w})\cdot\mathbf{n}\,dS - \int_{\partial V(t)} \mathbf{j}_i\cdot\mathbf{n}\,dS + \int_{V(t)} r_i\,dV$$

$$-\int_{C(t)} c_i^{\Sigma}(\mathbf{v}^{\Sigma}-\mathbf{w})\cdot\mathbf{m}\,dl - \int_{C(t)} \mathbf{j}_i^{\Sigma}\cdot\mathbf{m}\,dl + \int_{S(t)} r_i^{\Sigma}\,dS, \tag{18.1}$$

where

c_i is the molar concentration of surfactant in the bulk (with $i = 1,\ldots,N$)
c_i^{Σ} is the surface molar concentration of surfactant on the interface
$\mathbf{j}_i$ and $\mathbf{j}_i^{\Sigma}$ are the diffusive fluxes in the bulk phase and on the interface, respectively

The source terms due to chemical reactions in the bulk phase and on the interfaces are denoted as r_i and r_i^{Σ}.

Applying the transport and the divergence theorems for moving volumes and for moving surface areas to the left-hand side of the balance equation, the following equation results:

$$\int_{V(t)\backslash S(t)} \left(\partial_t c_i + \nabla\cdot(c_i\mathbf{v}+\mathbf{j}) - r_i\right)dV$$

$$+\int_{S(t)} \left(\frac{D^{\Sigma}c_i^{\Sigma}}{Dt} + c_i^{\Sigma}\nabla_{\Sigma}\cdot\mathbf{v}^{\Sigma} + \nabla_{\Sigma}\cdot\mathbf{j}_i^{\Sigma} + [c_i(\mathbf{v}-\mathbf{v}^{\Sigma})\cdot\mathbf{n}_{\Sigma}+\mathbf{j}_i\cdot\mathbf{n}_{\Sigma}] - r_i^{\Sigma}\right)dS = 0, \tag{18.2}$$

where $\dfrac{D^{\Sigma}}{Dt}$ denotes the associated Lagrangian derivative. For a generic surface quantity ϕ^{Σ}, it can be expressed as

$$\frac{D^{\Sigma}\phi^{\Sigma}}{Dt} = \partial_t^{\Sigma}\phi^{\Sigma} + \mathbf{v}_{\parallel}^{\Sigma} \cdot \nabla_{\Sigma}\phi^{\Sigma}, \tag{18.3}$$

with $\partial_t^{\Sigma}(\cdot)$ being the Thomas derivative, that is, the derivative following the normal motion of the interface. The symbol $\nabla_{\Sigma}\cdot$ represents the surface divergence.*

For more details regarding the derivation of the surfactant balance equations, we refer to [7,18–21]. Localization of Equation 18.2 by shrinking the control volume $V(t)$ to points in $\Omega\backslash\Sigma(t)$, and then the control area $S(t)$ to points on $\Sigma(t)$, leads to the surfactant molar mass balance in local form:

$$\partial_t c_i + \nabla \cdot (c_i \mathbf{v} + \mathbf{j}_i) = r_i \quad \text{in } \Omega\backslash\Sigma(t), \tag{18.4}$$

$$\frac{D^{\Sigma}c_i^{\Sigma}}{Dt} + c_i^{\Sigma}\nabla_{\Sigma} \cdot \mathbf{v}^{\Sigma} + \nabla_{\Sigma} \cdot \mathbf{j}_i^{\Sigma} + \left[\!\left[c_i(\mathbf{v} - \mathbf{v}^{\Sigma}) \cdot \mathbf{n}_{\Sigma} + \mathbf{j}_i \cdot \mathbf{n}_{\Sigma} \right]\!\right] = r_i^{\Sigma} \quad \text{on } \Sigma(t). \tag{18.5}$$

The notation $[\!\![\cdot]\!\!]$ stands for the jump of a physical quantity across the interface, where the jump of ϕ is defined as

$$[\![\phi]\!](t,\mathbf{x}) = \lim_{h \to 0+} (\phi(t,\mathbf{x} + h\mathbf{n}_{\Sigma}) - \phi(t,\mathbf{x} - h\mathbf{n}_{\Sigma})), \quad \mathbf{x} \in \Sigma(t). \tag{18.6}$$

To understand the meaning of the jump term in Equation 18.5, we consider a second form of the partial mass balance that follows from balancing solely the interface. This form involves an additional term, the sorption source term s_i^{Σ}, that takes into account the molar mass exchange between the bulk and the interface. The corresponding integral balance reads as

$$\frac{d}{dt}\int_{S(t)} c_i^{\Sigma}\, dS = -\int_{C(t)} c_i^{\Sigma}\left(\mathbf{v}^{\Sigma} - \mathbf{w}\right) \cdot \mathbf{m}\, dl - \int_{C(t)} \mathbf{j}_i^{\Sigma} \cdot \mathbf{m}\, dl + \int_{S(t)} \left(r_i^{\Sigma} + s_i^{\Sigma}\right) dS. \tag{18.7}$$

Applying the surface transport and the surface divergence theorems to Equation 18.7, we obtain the local form as

$$\frac{D^{\Sigma}c_i^{\Sigma}}{Dt} + c_i^{\Sigma}\nabla_{\Sigma} \cdot \mathbf{v}^{\Sigma} + \nabla_{\Sigma} \cdot \mathbf{j}_i^{\Sigma} = r_i^{\Sigma} + s_i^{\Sigma} \quad \text{on } S(t). \tag{18.8}$$

From the comparison between the two formulations of the local balance (18.5) and (18.8), the relation

$$s_i^{\Sigma} + \left[\!\left[c_i\left(\mathbf{v} - \mathbf{v}^{\Sigma}\right) \cdot \mathbf{n}_{\Sigma} + \mathbf{j}_i \cdot \mathbf{n}_{\Sigma} \right]\!\right] = 0 \tag{18.9}$$

between the diffusive fluxes and the sorption term follows.

---

* In brief, the surface gradient of a quantity $\phi(\mathbf{x})$ is defined as $\nabla_{\Sigma}\phi(\mathbf{x}) = \nabla\phi(\mathbf{x}) - \mathbf{n}_{\Sigma}(\mathbf{x})(\nabla\phi(\mathbf{x}) \cdot \mathbf{n}_{\Sigma}(\mathbf{x}))$ at $\mathbf{x} \in \Sigma$, where ϕ is extended to a neighborhood of ϕ as a differentiable function. Then, the surface divergence of a vector $\mathbf{f}$ is defined as $(\nabla_{\Sigma} \cdot \mathbf{f})(\mathbf{x}) = \mathrm{tr}(\nabla_{\Sigma}\mathbf{f})(\mathbf{x})$.

Collecting the local balance equations for surfactant transport in the bulk phase and on the interface, using the Thomas derivative according to Equation 18.3, we have

$$\partial_t c_i + \nabla \cdot (c_i \mathbf{v} + \mathbf{j}_i) = r_i \quad \text{in } \Omega \setminus \Sigma(t), \tag{18.10}$$

$$\partial_t^{\Sigma} c_i^{\Sigma} + \nabla_{\Sigma} \cdot \left(c_i^{\Sigma} \mathbf{v}^{\Sigma} + \mathbf{j}_i^{\Sigma} \right) = r_i^{\Sigma} + s_i^{\Sigma} \quad \text{on } \Sigma(t). \tag{18.11}$$

The system of equations (18.10) and (18.11) is not closed, that is, additional relations are needed to determine the diffusive fluxes and the source terms as functions of the primitive variables. In particular, the sorption source term will be modeled via a constitutive equation.

Before continuing in the derivation of the mathematical model, we need some assumptions. First, we impose that no chemical reaction inside the bulk and on the interface occur; therefore, the terms r_i and r_i^{Σ} are equal to zero. We also assume that the velocity field can be obtained solving the standard two-phase Navier–Stokes equations for incompressible Newtonian fluids in the bulk with constant density. This is closely related to the assumption that the surfactant species are dilute within the bulk.

In local formulation, the continuity equation and the momentum balance in the bulk phases $\Omega^{\pm}$ read

$$\nabla \cdot \mathbf{v} = 0, \tag{18.12}$$

$$\frac{\partial (\rho \mathbf{v})}{\partial t} + \nabla \cdot (\rho \mathbf{v} \otimes \mathbf{v}) = -\nabla p + \eta \Delta \mathbf{v} + \rho \mathbf{g}, \tag{18.13}$$

where
  $\mathbf{v}$ is the barycentric velocity
  p is the pressure
  ρ is the density
  η is the dynamic viscosity
  $\mathbf{g}$ is the acceleration due to gravity

The two bulk phases are separated by the moving interface $\Sigma(t)$, where at least one of the intensive quantities has a jump discontinuity. Therefore, transmission (or jump) conditions are needed at Σ.

On the interface, the model accounts for the molar mass of surfactant, while the transfer of mass in total is neglected, that is,

$$[\![\dot{m}]\!] = 0 \Leftrightarrow \rho^+ \left(\mathbf{v}^+ - \mathbf{v}^{\Sigma} \right) \cdot \mathbf{n}_{\Sigma} = \rho^- \left(\mathbf{v}^- - \mathbf{v}^{\Sigma} \right) \cdot \mathbf{n}_{\Sigma}. \tag{18.14}$$

Even more, the transfer of phase that is induced by the sorption processes is not considered. Thus,

$$\dot{m}^+ = \dot{m}^- = 0 \Rightarrow \mathbf{v}^+ \cdot \mathbf{n}_{\Sigma} = \mathbf{v}^- \cdot \mathbf{n}_{\Sigma} = V_{\Sigma}, \tag{18.15}$$

where V_{Σ} denotes the speed of normal displacement of $\Sigma(\cdot)$ given by

$$V_{\Sigma} = \mathbf{v}^{\Sigma} \cdot \mathbf{n}_{\Sigma}. \tag{18.16}$$

We assume no-slip at the interface, that is, the tangential velocity components satisfy

$$\mathbf{v}_{\|}^+ = \mathbf{v}_{\|}^- = \mathbf{v}_{\|}^{\Sigma}. \tag{18.17}$$

Considering Equations 18.15 and 18.17, the following conditions at the interface are obtained:

$$[\![\mathbf{v}]\!] = 0 \quad \text{and} \quad V_\Sigma = \mathbf{v} \cdot \mathbf{n}_\Sigma. \tag{18.18}$$

In the same way, the inertia of the adsorbed surfactant is neglected; hence, the interfacial momentum jump condition reads as

$$[\![-\mathbf{S}]\!] \cdot \mathbf{n}_\Sigma = \nabla_\Sigma \sigma + \sigma \kappa_\Sigma \mathbf{n}_\Sigma, \tag{18.19}$$

where
  σ is the surface tension
  $\mathbf{S}$ is the stress tensor

Its constitutive equation in case of incompressible Newtonian fluids is

$$\mathbf{S} = -p\mathbf{I} + \tau, \tag{18.20}$$

where
  $-p\mathbf{I}$ is the isotropic component
  τ is the deviatoric part, that is, the viscous stress tensor

Under the assumptions made before,

$$\tau = \eta(\nabla \mathbf{v} + (\nabla \mathbf{v})^T), \tag{18.21}$$

with dynamic viscosity $\eta \geq 0$. The quantity κ earlier is twice the mean interface curvature, that is, the sum of the local principal curvature of the interface:

$$\kappa = -\nabla_\Sigma \cdot \mathbf{n}_\Sigma. \tag{18.22}$$

Hence, in the stagnant case and with constant surface tension, the momentum jump condition reduces to the Laplace pressure jump.

The surface gradient of the surface tension represents the so-called Marangoni stress. Generally, in the presence of surfactants, the surface tension is a function of temperature and concentrations of the adsorbed surfactants $\sigma = \sigma\left(T, c_1^\Sigma, c_2^\Sigma, \ldots, c_N^\Sigma\right)$, and it can be significantly lowered by the adsorption of these chemical compounds. For simplicity, we consider the system under isothermal conditions; thus, the equation of state for the surface tension, which needs to be modeled, is of the form $\sigma = \sigma\left(c_1^\Sigma, c_2^\Sigma, \ldots, c_N^\Sigma\right)$. Moreover, under this assumption, there is no need for an energy equation. Then, the Marangoni stress can be calculated from the distribution of adsorbed surfactant according to

$$\nabla_\Sigma \sigma = \sum_{i=1}^{N} \frac{\partial \sigma}{\partial c_i^\Sigma} \nabla_\Sigma c_i^\Sigma. \tag{18.23}$$

To close the problem, constitutive equations for the sorption source term and diffusive fluxes in the bulk and on the interface are needed. Moreover, a surface equation of state, that is, a concrete relation $\sigma = \sigma\left(T, c_1^\Sigma, c_2^\Sigma, \ldots, c_N^\Sigma\right)$, is required. These relations depend on the material, that is, the fluids and the dissolved species. Some possible models will be listed in Sections 18.2.2 and 18.2.3.

Two more equations related to mesh motion need to be taken into account. In fact, in the chosen numerical approach, the mesh is deforming under the constraint

$$\mathbf{v}^\Sigma \cdot \mathbf{n}_\Sigma = \mathbf{w} \cdot \mathbf{n}_\Sigma, \tag{18.24}$$

at the interface, where $\mathbf{w}$ is the mesh velocity, ensuring consistency at the interface such that the same set of cell faces represents the interface. Then, referring to the integral form of the balance equations for a comoving control volume with the transported quantity equal to one, the so-called space conservation law is introduced, which describes the change in volume of a control volume due to its motion:

$$\frac{d}{dt} \int_{V(t)} dV - \int_{\partial V(t)} \mathbf{w} \cdot \mathbf{n}_\Sigma \, dS = 0. \tag{18.25}$$

The space conservation law is a central aspect of the interface-tracking method presented in Section 18.3.1.4, since it describes the relationship between the volume rate of change of a control volume V and the corresponding surface velocity $\mathbf{v}^\Sigma$.

18.2.2 Constitutive Equations: Diffusive Fluxes

The modeling of the diffusive fluxes depends on the fact whether cross effect or other effects due to high solute concentrations are relevant. We therefore distinguish between two cases, one referring to dilute systems in which the solute molecules only interact with solvent molecules but not with other dissolved constituents and one accounting for high concentration phenomena that can become relevant for the transport on the interface.

In the present work, surfactant concentrations in the bulk phase are always assumed to be dilute. Therefore, the diffusive fluxes can be modeled applying the Fick's law, that is, the closure relation

$$\mathbf{j}_i = -D_i \, \nabla c_i, \tag{18.26}$$

with diffusion coefficients $D_i > 0$ that are independent of the composition of the mixture. In fact, as we are interested in an isothermal description, we assume the diffusivities to be constant. Let us note in passing that this simplest closure does not account for the constraint

$$\sum_{i=1}^{N} M_i \mathbf{j}_i = 0, \tag{18.27}$$

with M_i denoting the molar mass of species i, which is the link between the partial mass balances, that is, Equation 18.4, and the continuity equation. This inconsistency is not problematic for dilute solutions but needs to be removed for the general case. Indeed, for nondilute surfactant concentrations at the interface, Fick's law is no longer appropriate to describe interfacial diffusive fluxes. Cross effects between different surfactant species and nonideal interactions of the surfactant mixture cannot be neglected. Examples of these effects are countergradient diffusion, osmotic diffusion, or diffusion barriers. To take into account these effects, the interfacial diffusive fluxes are then to be modeled through the Maxwell–Stefan equations [22,23], that is,

$$-\sum_{\substack{j \neq i}}^{N} \frac{c_j^\Sigma \mathbf{j}_i^\Sigma - c_i^\Sigma \mathbf{j}_j^\Sigma}{D_{ij}^\Sigma} = \mathbf{d}_i^\Sigma, \tag{18.28}$$

where
   D_{ij}^Σ are the binary Maxwell–Stefan diffusion coefficients
   $\mathbf{d}_i^\Sigma$ is the surface thermodynamical driving force

Assuming isothermal conditions and neglecting individual body forces, $\mathbf{d}_i^\Sigma$ is given as

$$\mathbf{d}_i^\Sigma = \frac{c_i^\Sigma}{RT}\nabla_\Sigma \mu_i^\Sigma + \frac{y_i^\Sigma}{RT}\nabla_\Sigma \sigma, \qquad (18.29)$$

with the mass fractions $y_i^\Sigma = \rho_i^\Sigma/\rho^\Sigma$ and the molar-based surface chemical potentials μ_i^Σ. The latter are defined as $\mu_i^\Sigma = M_i \dfrac{\partial\left(\rho^\Sigma \psi^\Sigma\right)}{\partial \rho_i^\Sigma}$, where $\rho^\Sigma \psi^\Sigma$ is the interfacial free energy that has to be modeled as a material-dependent quantity.

To solve for the diffusive fluxes $\mathbf{j}_i^\Sigma$, the linear (in the fluxes) system (18.28) is to be inverted, taking into account the constraint

$$\sum_i \mathbf{j}_i^\Sigma = 0. \qquad (18.30)$$

As in the bulk, this constraint reflects the conservation of total mass. The invertibility of the system (18.28 and 18.30) is guaranteed for strictly positive concentrations and diffusivities as shown (for the bulk equations) in [24]. The numerical solution employs an iterative procedure due to Giovangigli [25] and will be described in Section 18.3.

In concrete applications, instead of modeling the free energy of the interface, one usually assumes constitutive relations for the chemical potentials. But note that the definition of a consistent free energy is somewhat involved. For so-called elastic mixtures, a general procedure is described in [26].

For adsorbed species on the solid surface of a porous medium, Krishna [27] gives a constitutive equation for nonideal mixtures as

$$\mu_i^\Sigma = \mu_i^{\Sigma,0}(T,p) + RT \ln(\gamma_i x_i^\Sigma), \qquad (18.31)$$

where
  $\mu_i^{\Sigma,0}$ is the chemical potential at an appropriate reference state
  T is the absolute temperature
  R is the universal gas constant
  γ_i is the activity coefficient
  x_i^Σ is the molar mass fraction of component i

Here, the surface pressure or, equivalently, the surface tension is not included as a variable for the reference chemical potential. This might be reasonable for such solid substrates, but in the fluid interface case, the dependence of Σ needs to be considered. Concrete models for the surface chemical potential of species adsorbed on a fluid interface are given, for example, in Kralchevsky [28], where also free energies and corresponding surface equations of state are listed.

18.2.3 Constitutive Equations: Sorption Processes

A constitutive equation for the sorption source term s_i^Σ in Equation 18.8 is required. To model this term, two different situations can be considered: diffusion-controlled sorption (fast) and kinetically controlled sorption (slow) [29,30]. In the first case, the sorption process is much faster than the diffusive transport, while in the latter case, the diffusive transport is faster than the sorption process, typically due to the presence of a kinetic barrier. Thus, the transfer rate will be determined in two different ways:

1. *Diffusion-controlled (fast) sorption*:

$$s_{i,\text{fast}}^{\Sigma} = -\llbracket \mathbf{j}_i \cdot \mathbf{n}_{\Sigma} \rrbracket = \mathbf{j}_i^{+} \cdot \mathbf{n}^{+} + \mathbf{j}_i^{-} \cdot \mathbf{n}^{-}, \tag{18.32}$$

where $\mathbf{n}^{\pm}$ denotes the outer normals to the two bulk phases at the interface; note that only one of the two terms will be nonzero if the surfactant is only soluble in one of the bulk phases. This relation (18.32) follows from local balance of mass exchange (18.9), but with simplifications according to the aforementioned assumptions. In the case of fast (as compared to diffusive transport) sorption, the adsorption and desorption rates are locally equilibrated, that is,

$$s_i^{\text{ads}}(c_{i|\Sigma}, c_i^{\Sigma}) = s_i^{\text{des}}(c_i^{\Sigma}). \tag{18.33}$$

This equality leads to an additional local relationship between c_i^{Σ} and $c_{i|\Sigma}$, which needs to be accounted for in the numerical solution. In this case of fast sorption, the interfacial surfactant mass balance can be viewed as a dynamical coupling condition between the Dirichlet and the Neumann data of the bulk concentration field. The numerical method needs to map this into a discretized form.

2. *Kinetically controlled (slow) sorption*:

$$s_{i,\text{slow}}^{\Sigma} = s_i^{\text{ads}}\left(c_{i|\Sigma}, c_i^{\Sigma}\right) - s_i^{\text{des}}\left(c_i^{\Sigma}\right), \tag{18.34}$$

where
$s_i^{\text{ads}}\left(c_{i|\Sigma}, c_i^{\Sigma}\right)$ describes the rate of adsorption
$s_i^{\text{des}}\left(c_i^{\Sigma}\right)$ is the desorption rate

Note that the rate of adsorption is a function of the bulk concentration near the interface and the concentration of the adsorbed species, while the desorption rate is usually assumed to be a function of the adsorbed species only.

The starting point of the concrete sorption modeling is the Gibbs adsorption equation. In the considered isothermal case, this reads as

$$d\sigma = -\sum_{i=1}^{N} c_i^{\Sigma} d\mu_i^{\Sigma}, \tag{18.35}$$

where μ_i^{Σ} is the (model based) surface chemical potential of species i. In the case of fast sorption, that is, for local adsorption/desorption equilibrium, we have $\mu_i = \mu_i^{\Sigma}$, and, hence, the Gibbs adsorption equation becomes

$$d\sigma = -\sum_{i=1}^{N} c_i^{\Sigma} d\mu_i. \tag{18.36}$$

Since the bulk concentration of the surfactants can be considered dilute, the chemical potential is given as

$$\mu_i(T, p, x_k) = g_i(T, p) + RT \ln x_i, \tag{18.37}$$

where $g_i(T, P)$ is the Gibbs free energy of the pure constituent i at the temperature and the pressure of the mixture.

For simplicity, a single surfactant system is considered. Then, the Gibbs equation (18.36) simplifies to

$$d\sigma = -c^\Sigma d\mu. \tag{18.38}$$

Introducing Equation 18.37 into Equation 18.38, the relationship between the bulk molar fraction $x = c/c_{tot}$, the concentration of the adsorbed form c^Σ, and the change of surface tension $d\sigma$ is obtained as

$$d\sigma = -RTc^\Sigma d(\ln x). \tag{18.39}$$

At this point, in order to describe the relationship between the surfactant bulk concentration c_i and the concentration of the adsorbed species c_i^Σ, the distinction between diffusion-controlled (fast) sorption and kinetically controlled (slow) sorption is made.

18.2.3.1 Diffusion-Controlled Sorption (Fast)

Fast sorption processes are characterized by a quasi-instantaneous sorption. In the following paragraphs, a collection of sorption model is recalled, which can be found in [28], starting from the classical ones such as Henry, Langmuir, and Frumkin models to close with more recent ones such as reorientation and aggregation models. A detailed description of the latter, more elaborated models can be found in [31,32].

18.2.3.1.1 Henry Model

The simplest way to describe the sorption process is to apply the Henry model, where a linear relationship between the surfactant bulk concentration c and the concentration of the adsorbed one c^Σ is assumed. Thus, the correlation between these two quantities reads

$$c^\Sigma = Kc, \tag{18.40}$$

where K is the Henry constant.* The corresponding equation of state for the surface tension becomes

$$\sigma = \sigma_0 - RTc^\Sigma, \tag{18.41}$$

where
  σ is the surface tension in N/m
  R is the universal gas constant with $R = 8.3144$ J/(mol K)
  T is the absolute temperature of the system in K

In the Henry model, interactions between molecules are not considered, and the model is only applicable in case of low surface coverages by the surfactant, that is, small values of c^Σ.

18.2.3.1.2 Langmuir Model

The Langmuir adsorption isotherm takes into account the interactions between molecules at the interface by means of the Langmuir equilibrium constant a, expressed in mol/m³, and the maximum number of adsorbed molecules per area c_∞^Σ. The adsorption isotherm reads as

$$c^\Sigma = c_\infty^\Sigma \frac{c/a}{1 + c/a}. \tag{18.42}$$

---

* If we consider the molar concentration fields $[c^\Sigma]$ = mol/m², $[c]$ = mol/m³, then $[K]$ = m.

The Langmuir equilibrium constant corresponds to a situation in which half of the surface is covered by surfactant, that is, $c^\Sigma = c_\infty^\Sigma/2$. Thus, this constant can be related to the Henry constant K through the relation $K = c_\infty^\Sigma/a$.

The Langmuir surface equation of state follows as

$$\sigma = \sigma_0 + RT_\infty^\Sigma \ln\left(1 - \frac{c^\Sigma}{c_\infty^\Sigma}\right),$$

(18.43)

or, equivalently,

$$\sigma = \sigma_0 - RT\, c_\infty^\Sigma \ln\left(1 + \frac{c}{a}\right).$$

(18.44)

With increasing chain length of the hydrophobic part of the surfactant (e.g., for fatty acids), the Langmuir model fails to predict the surface tension and the concentration of the adsorbed form of the surfactant. Deviations are due to interaction between the adsorbed molecules.

18.2.3.1.3   Frumkin Model

The starting point of this model is the Langmuir one, but with an additional term describing interactions between the adsorbed molecules that affect c^Σ. As a result, the so-called Frumkin isotherm reads as

$$bc = \frac{c^\Sigma\omega}{1 - c^\Sigma\omega} e^{-2ac^\Sigma\omega},$$

(18.45)

where
   b is the adsorption constant, $[b]$ = m³/mol
   $\omega = 1/c_\infty^\Sigma$ is the partial molar surface area, $[\omega]$ = m²/mol
   a is a dimensionless constant that describes the intermolecular interactions between the adsorbed molecule

For this adsorption isotherm, the surface tension is given as

$$\sigma = \sigma_0 + \frac{RT}{\omega}\left[\ln(1 - c^\Sigma\omega) + a(c^\Sigma\omega)^2\right],$$

(18.46)

or, using the maximum surface coverage c_∞^Σ, as

$$\sigma = \sigma_0 + RTc_\infty^\Sigma\left[\ln\left(1 - \frac{c^\Sigma}{c_\infty^\Sigma}\right) + a\left(\frac{c^\Sigma}{c_\infty^\Sigma}\right)^2\right].$$

(18.47)

Note that Equations 18.46 and 18.47 are identical to the Langmuir equation of state in case $a = 0$.

18.2.3.1.4   Reorientation Model

This model admits the existence of molecules of surfactant in two states, state 1 and 2, in the adsorption layer. Thus, we need to consider two molar areas of an adsorbed molecule, ω_1 and ω_2, with the assumption $\omega_1 > \omega_2$. ω_1 is the molar area required at an empty interface, and ω_2 is the minimum area needed at a covered interface. The relations between the molar areas and surface concentration fields of the two states are described by

$$c^\Sigma = c_1^\Sigma + c_2^\Sigma,$$ (18.48)

$$\omega c^\Sigma = \omega_1 c_1^\Sigma + \omega_2 c_2^\Sigma.$$ (18.49)

The model involves the ratio between surface concentrations in the two possible adsorption states of the molecules calculated through the generalized Joos' adsorption equation:

$$\frac{c_1^\Sigma}{c_2^\Sigma} = e^{(\omega_1 - \omega_2)/\omega} \left(\frac{\omega_1}{\omega_2} \right)^\alpha e^{-(\sigma_0 - \sigma)(\omega_1 - \omega_2)/(RT)},$$ (18.50)

where the coefficient α is usually taken as equal to zero. The associated surface equations of state and the relative adsorption isotherm are

$$\sigma = \sigma_0 + \frac{RT}{\omega} \ln(1 - c^\Sigma \omega),$$ (18.51)

$$bc = \frac{c_2^\Sigma \omega}{(1 - c^\Sigma \omega)^{\omega_2/\omega}},$$ (18.52)

where b is still the adsorption constant as in the Frumkin model. For $\omega_1 = \omega$, this model falls back into the Langmuir model with $b = 1/a$, where a is the Langmuir equilibrium constant.

It has been shown that the reorientation model is superior to the Langmuir and Frumkin ones especially for surfactants with small mutual interaction at the interface, for instance, molecules with a rather large molar area demand.

18.2.3.1.5 Aggregation Model

This model has been developed to contain only one molar area ω_1 of an adsorbed surfactant at a covered interface, as the Langmuir and Frumkin models, but including a critical surface adsorption concentration c_{crit}^Σ and a nondimensional surface adsorption number n.

The mean molar area can be obtained from a weighted average that involves the molar area of the adsorbed molecule and the current and critical adsorption concentrations:

$$\omega = \omega_1 \frac{1 + n \left(\dfrac{c^\Sigma}{c_{crit}^\Sigma} \right)^{n-1}}{1 + \left(\dfrac{c^\Sigma}{c_{crit}^\Sigma} \right)^{n-1}}.$$ (18.53)

Then, the adsorption isotherm and the surface equation of state for the aggregation model read as

$$bc = \frac{c^\Sigma \omega}{\left[1 - c^\Sigma \omega \left(1 + \left(\dfrac{c^\Sigma}{c_{crit}^\Sigma} \right)^{n-1} \right) \right]^{\omega_1/\omega}},$$ (18.54)

$$\sigma = \sigma_0 + \frac{RT}{\omega} \ln \left[1 - c^\Sigma \omega \left(1 + \left(\frac{c^\Sigma}{c_{crit}^\Sigma} \right)^{n-1} \right) \right].$$ (18.55)

18.2.3.1.5.1 Generalized Frumkin Model (Surfactant Mixtures). As an example of sorption model for surfactant mixtures, we recall the generalized Frumkin model. The other sorption models, presented before for a single species, are straightforward to apply to mixtures, too. For a binary mixture of species 1 and 2, the sorption isotherms read as

$$b_1 c_1 = \frac{\theta_1}{(1 - \theta_1 - \theta_2)^{n_1}} \, e^{-2a_1\theta_1 - 2a_{12}\theta_2} \, e^{(1 - n_1)(a_1\theta_1^2 + a_2\theta_2^2 + 2a_{12}\theta_1\theta_2)}, \tag{18.56}$$

$$b_2 c_2 = \frac{\theta_2}{(1 - \theta_1 - \theta_2)^{n_2}} \, e^{-2a_2\theta_2 - 2a_{12}\theta_1} \, e^{(1 - n_2)(a_1\theta_1^2 + a_2\theta_2^2 + 2a_{12}\theta_1\theta_2)}, \tag{18.57}$$

where θ is the surface coverage. The corresponding surface equation of state is

$$\sigma = \sigma_0 + \frac{RT}{\omega_0} \left[\ln(1 - \theta_1 - \theta_2) + \theta_1 \left(1 - \frac{1}{n_1}\right) + \theta_2 \left(1 - \frac{1}{n_2}\right) + a_1\theta_1^2 + a_2\theta_2^2 + 2a_{12}\theta_1\theta_2 \right], \tag{18.58}$$

where n_i are the relative molar areas of the respective species, the coefficients a_i are the interaction parameters of the ith component among each other, while a_{12} is the mixed interaction coefficient. Simplifications of the equation of state lead to the following relation, easily applied to experimental data:

$$e^{(\sigma_0 - \sigma) \, \omega/(RT)} = e^{(\sigma_0 - \sigma)|_1 \, \omega/(RT)} + e^{(\sigma_0 - \sigma)|_2 \, \omega/(RT)} - 1, \tag{18.59}$$

where the average molar area is given by

$$\omega = \frac{\omega_1 c_1^{\Sigma} + \omega_2 c_2^{\Sigma}}{c_1^{\Sigma} + c_2^{\Sigma}}. \tag{18.60}$$

18.2.3.2 Kinetically Controlled Sorption (Slow)

In case of slow sorption processes, that is, in the presence of a kinetic barrier, the surfactant transfer from the sublayer to the interface is much slower than the diffusive transport. The barrier can be due to sterical hindering, spatial reorientation, or conformational changes accompanying the adsorption process. In these cases, a kinetic modeling of the transfer process is needed.

18.2.3.2.1 Henry Model

For the Henry isotherm, the net rate of the reversible adsorption reads as

$$s_{slow}^{\Sigma} = \kappa^{ads} c_{|\Sigma} - \kappa^{des} \frac{c^{\Sigma}}{c_\infty^{\Sigma}}. \tag{18.61}$$

The first term on the right-hand side represents the rate of adsorption, and the second one is the desorption rate. Here, $c_{|\Sigma}$ is the concentration profile in the direct vicinity of the interface.* The parameters k^{ads} and k^{des} describe the rate constants for the adsorption and desorption processes.† The corresponding surface equation of state is still (18.41), that is,

$$\sigma = \sigma_0 - RT c^{\Sigma}.$$

---

* This quantity is a bulk concentration; thus, its physical unit is $[c_{|\Sigma}] = \text{mol/m}^3$.
† The physical units of the two constant are $[k^{ads}] = \text{m/s}$ and $[k^{des}] = \text{mol/(m}^2\text{/s)}$.

18.2.3.2.2 Langmuir Model

The net rate of surfactant adsorption in the Langmuir sorption model reads as

$$s_{\text{slow}}^{\Sigma} = \kappa^{\text{ads}} c_{|\Sigma} \left(1 - \frac{c^{\Sigma}}{c_{\infty}^{\Sigma}} \right) - \kappa^{\text{des}} \frac{c^{\Sigma}}{c_{\infty}^{\Sigma}}, \tag{18.62}$$

where the two constants for adsorption and desorption can be related to the Langmuir equilibrium constant for fast sorption:

$$\frac{1}{a} = \frac{\kappa^{\text{ads}}}{\kappa^{\text{des}}}. \tag{18.63}$$

The corresponding surface tension equation of state is still Equation 18.44 or 18.43, that is,

$$\sigma = \sigma_0 + RT\, c_{\infty}^{\Sigma} \ln\left(1 - \frac{c^{\Sigma}}{c_{\infty}^{\Sigma}} \right).$$

18.2.4 Summary

The full model of a two-phase flow with soluble surfactants under the simplifying assumptions explained earlier reads as follows:

- *Bulk equations*

$$\nabla \cdot \mathbf{v} = 0,$$

$$\partial_t(\rho \mathbf{v}) + \nabla \cdot (\rho \mathbf{v} \otimes \mathbf{v}) = -\nabla p + \eta \Delta \mathbf{v} + \rho \mathbf{g}, \quad \text{in } \Omega^{\pm}(t),$$

$$\partial_t c_i + \nabla \cdot (c_i \mathbf{v} + \mathbf{j}_i) = 0.$$

- *Interface equations*

$$[\![\mathbf{v}]\!] = 0,$$

$$[\![p\mathbf{I} - \tau]\!] \cdot \mathbf{n}_{\Sigma} = \sigma \kappa \mathbf{n}_{\Sigma} + \nabla_{\Sigma} \sigma,$$

$$\partial_t^{\Sigma} c_i^{\Sigma} + \nabla_{\Sigma} \cdot \left(c_i^{\Sigma} \mathbf{v}^{\Sigma} + \mathbf{j}_i^{\Sigma} \right) = s_i^{\Sigma}, \quad \text{on } \Sigma(t),$$

$$\mathbf{v}_{\|}^{+} = \mathbf{v}_{\|}^{-} = \mathbf{v}_{\|}^{\Sigma},$$

$$s_i^{\Sigma} + [\![\mathbf{j}_i \cdot \mathbf{n}_{\Sigma}]\!] = 0.$$

- *Initial conditions*

$$\mathbf{v}(0, \mathbf{x}) = \mathbf{v}_0(\mathbf{x}), \quad \forall\, \mathbf{x} \in \Omega_{\pm}(0),$$

$$c_i(0, \mathbf{x}) = c_{i,0}(\mathbf{x}), \quad \forall\, \mathbf{x} \in \Omega_{\pm}(0),$$

$$c_i^{\Sigma}(0, \mathbf{x}) = c_{i,0}^{\Sigma}(\mathbf{x}), \quad \forall \mathbf{x} \in \Sigma(0).$$

- *Bulk diffusive fluxes*

$$\mathbf{j}_i = -D_i \, \nabla c_i, \quad \text{in } \Omega \text{ (dilute solution)}.$$

- *Interface diffusive fluxes*

$$-\sum_{j \neq i}^{N} \frac{c_j^\Sigma \mathbf{j}_i^\Sigma - c_i^\Sigma \mathbf{j}_j^\Sigma}{D_{ij}^\Sigma} = \mathbf{d}_i^\Sigma, \quad \text{on } \Sigma(t) \text{ (nondilute solution)},$$

$$\mathbf{d}_i^\Sigma = \frac{c_i^\Sigma}{RT} \nabla_\Sigma \mu_i^\Sigma + \frac{y_i^\Sigma}{RT} \nabla_\Sigma \sigma.$$

- *Sorption modeling*

$$s_{i,\text{fast}}^\Sigma = [\![-\mathbf{j}_i \cdot \mathbf{n}_\Sigma]\!], \quad s_i^{\text{ads}}\left(c_{i|\Sigma}, c_i^\Sigma\right) = s_i^{\text{des}}\left(c_i^\Sigma\right) \quad \text{diffusion-controlled sorption (fast)},$$

$$s_{i,\text{slow}}^\Sigma = s_i^{\text{ads}}\left(c_{i|\Sigma}, c_i^\Sigma\right) - s_i^{\text{des}}\left(c_i^\Sigma\right) \quad \text{kinetically controlled sorption (slow)}.$$

Refer to Tables 18.1 and 18.2 for the details on sorption modeling.

18.3 NUMERICAL MODEL

Direct numerical simulation (DNS) can give valuable information about interfacial transport processes in two-phase systems with surfactant that would not be easily accessible through experimental investigations. The numerical methods available for such simulations can be divided into two general categories with respect to the treatment of the fluid/fluid boundary: *interface-capturing* methods and *interface-tracking* methods. All of them rely on the assumption of a sharp interface representation, that is, on a sharp interface model.

Eulerian interface-capturing methods are based on the numerical solution of two-phase Navier–Stokes equations in single field formulation, mostly on a fixed mesh; the interface is captured from a volumetric marker field. The evolution of this volumetric marker field is tracked by solving an advection equation, which is the reason for also calling these approximations volume-tracking methods.

On the other hand, in interface-tracking methods, the fluid interface is explicitly represented by a surface mesh or by a set of connected marker particles. These methods allow for very accurate description of interfacial transport processes and calculation of surface tension force. The major drawback here is the limitation to moderately deformed interfaces and no changes in mesh topology, if it is not accompanied by an automatic remeshing technique, for example [33]. In this category falls our solution strategy: the ALE interface-tracking method. The interface is represented by a boundary-fitted surface mesh holding surfactant molar mass that is subject to interfacial transport. The boundary mesh is advected in a Lagrangian manner under the enforcement of appropriate jump conditions at the interface (18.18 and 18.19), whereas the mesh away from the interface is updated through automatic mesh motion with Laplacian smoothing or remeshing in order to guarantee mesh validity.

The reason for choosing this numerical strategy is justified by its high accuracy due to the fact that the interface is represented by a computational mesh boundary. This means that the local surface tension calculation is very precise, and parasitic currents around the interface are significantly reduced. Moreover, the *force-conservative* approach implies that the condition for zero net surface tension force on closed surfaces is satisfied exactly.

The model equations are discretized by means of collocated FVM [34–36] for transport processes in the bulk and FAM [17] for transport processes on the interface. Such a practice fulfills the

TABLE 18.1

Fast Sorption Models

Model	Adsorption Isotherm	Equation of State (σ)
Henry	$c^{\Sigma} = Kc$	$\sigma = \sigma_0 - RTc^{\Sigma}$
Langmuir	$c^{\Sigma} = c_{\infty}^{\Sigma} \dfrac{c/a}{1 + c/a}$	$\sigma = \sigma_0 - RT\, c_{\infty}^{\Sigma} \ln\left(1 + \dfrac{c}{a}\right)$, or Equation 18.43
Frumkin	$bc = \dfrac{c^{\Sigma}\omega}{1 - c^{\Sigma}\omega}\, e^{-2ac^{\Sigma}\omega}$	$\sigma = \sigma_0 + \dfrac{RT}{\omega}\left[\ln(1 - c^{\Sigma}\omega) + a(c^{\Sigma}\omega)^2\right]$
Additional relation	$\omega = 1/c_{\infty}^{\Sigma}$	
Reorientation	$bc = \dfrac{c_2^{\Sigma}\omega}{(1 - c^{\Sigma}\omega)^{\omega_2/\omega}}$	$\sigma = \sigma_0 + \dfrac{RT}{\omega}\ln(1 - c^{\Sigma}\omega)$
Additional relations	$c^{\Sigma} = c_1^{\Sigma} + c_2^{\Sigma}, \quad \omega c^{\Sigma} = \omega_1 c_1^{\Sigma} + \omega_2 c_2^{\Sigma}$	
	$\dfrac{c_1^{\Sigma}}{c_2^{\Sigma}} = e^{\frac{\omega_1 - \omega_2}{\omega}} \cdot \left(\dfrac{\omega_1}{\omega_2}\right)^{\alpha} \cdot e^{-\frac{(\sigma_0 - \sigma)(\omega_1 - \omega_2)}{RT}}$	
Aggregation	$bc = \dfrac{c^{\Sigma}\omega}{\left[1 - c^{\Sigma}\omega\left(1 + \left(\dfrac{c^{\Sigma}}{c_{\text{crit}}^{\Sigma}}\right)^{n-1}\right)\right]^{\omega_1/\omega}}$	$\sigma = \sigma_0 + \dfrac{RT}{\omega}\ln\left[1 - c^{\Sigma}\omega\left(1 + \left(\dfrac{c^{\Sigma}}{c_{\text{crit}}^{\Sigma}}\right)^{n-1}\right)\right]$
Additional relation	$\omega = \omega_1 \dfrac{1 + n\left(\dfrac{c^{\Sigma}}{c_{\text{crit}}^{\Sigma}}\right)^{n-1}}{1 + \left(\dfrac{c^{\Sigma}}{c_{\text{crit}}^{\Sigma}}\right)^{n-1}}$	

Model	Adsorption Isotherm	Equation of State (σ)
Generalized Frumkin	$b_1 c_1 = \dfrac{\theta_1}{(1 - \theta_1 - \theta_2)^{n_1}}\, e^{\alpha_1}$	$\sigma = \sigma_0 + \dfrac{RT}{\omega_0} F(\theta_i, n_i, a_i)$
	$b_2 c_2 = \dfrac{\theta_2}{(1 - \theta_1 - \theta_2)^{n_2}}\, e^{\alpha_2}$	
Additional relation	See Equations 18.56 and 18.57 for $\alpha_{1,2}$	See Equation 18.58 for $F(\theta_i, n_i, a_i)$
	$\omega = \dfrac{\omega_1 c_1^{\Sigma} + \omega_2 c_2^{\Sigma}}{c_1^{\Sigma} + c_2^{\Sigma}}$	

TABLE 18.2
Slow Sorption Models

Model	Source Term	Equation of State (σ)
Henry	$s_{slow}^{\Sigma} = \kappa^{ads} c_{\Sigma} - \kappa^{des} \dfrac{c^{\Sigma}}{c_{\infty}^{\Sigma}}$	$\sigma = \sigma_0 - RT c^{\Sigma}$
Langmuir	$s_{slow}^{\Sigma} = \kappa^{ads} c_{\Sigma}\left(1 - \dfrac{c^{\Sigma}}{c_{\infty}^{\Sigma}}\right) - \kappa^{des} \dfrac{c^{\Sigma}}{c_{\infty}^{\Sigma}}$	$\sigma = \sigma_0 - RT\, c_{\infty}^{\Sigma} \ln\left(1 + \dfrac{c}{a}\right)$, or Equation 18.43
Additional relation	$\dfrac{1}{a} = \dfrac{\kappa^{ads}}{\kappa^{des}}$	

requirements of conservativeness, boundedness, stability, and consistency of a numerical method [37]. Particular attention will be given to multicomponent mixture of surfactants diffusing and adsorbing at the interface. The description of the transport of surfactant in the bulk and on the interface is included in the model. To account for sorption processes, several sorption models are included.

To have a complete overview of the numerical method, the overall solution algorithm is presented (Algorithm 18.1). Details regarding the most important steps will be described in the following sections. The method is implemented in OpenFOAM, an open source object-oriented C++ library for numerical simulation of computational continuum mechanics and computational fluid dynamics.

Algorithm 18.1 Overall Solution Algorithm

for $t \le T$ **do** {Start time loop}

Copy the values of bulk species concentration c_i, interfacial species concentrations c_i^{Σ}, velocity $\mathbf{v}$ and pressure p from the previous time level to the new time level.

for $n \le nOuterCorrector$ **do** {Start outer loop}
    Calculate the displacement of the interface to compensate the net mass flux (see Section 18.3.1.4)
    Update the interfacial mesh (see Section 18.3.1.4)
    Update the volumetric meshes by automatic mesh motion, eventual remeshing (see Section 18.3.1.4)
    Calculate the new face fluxes
    Update the pressure and velocity boundary conditions at the interface (see Section 18.3.1.5)
    Solve the discretized Navier–Stokes equations applying the PISO loop (see Section 18.3.1.3)
    Calculate the new net fluxes
    Check convergence and mass conservation
end for {End outer loop}

Assemble and solve the discretized bulk species transport equations

Account for sorption processes according to the sorption library (see Section 18.2.3)

Interfacial species transport (see Section 18.3.2):
if Single Surfactant or dilute concentrations **then**
    Assemble and solve the discretized species transport equations

else if Surfactant Mixtures **then**

    Calculate effective diffusivities according to iterative inversion algorithm of Giovangigli

        Read run time selectable properties from dictionaries

        Calculate mixture values of molar masses

        Assemble diffusivity matrix **B** according to Equation 18.94

        Invert **B** according to Equation 18.96

    Insert effective diffusivities into species transport equations (see Section 18.3.2.2)

    Assemble block structured coefficient matrix and source vector

    Solve coupled algebraic equation system

end if

Update the local surface tension values according to the chosen surface tension equation of state (see Sections 18.3.1.6 and 18.3.2.4)

end for{End time loop}

18.3.1 TWO-PHASE HYDRODYNAMICS

18.3.1.1 Governing Equations

The governing equations are presented for moving control volumes in their integral form. To account for mesh motion, an ALE reference frame is chosen. In the field far away from the interface, the mesh is arbitrarily moving with velocities between the interfacial velocity and the velocity of the outer domain boundaries.

- *Mass*

$$\frac{d}{dt}\int_{V(t)} \rho\, dV + \int_{\partial V(t)} ((\mathbf{v}-\mathbf{w})\rho)\cdot \mathbf{n}\, dS = 0. \tag{18.64}$$

- *Momentum*

$$\frac{d}{dt}\int_{V(t)} (\rho\mathbf{v})\, dV + \int_{\partial V(t)} ((\mathbf{v}-\mathbf{w})\otimes(\rho\mathbf{v}))\cdot \mathbf{n}\, dS = \int_{\partial V(t)} (-p\mathbf{I}+\mathbf{S})\cdot \mathbf{n}\, dS + \int_{V(t)} \rho\mathbf{g}\, dV. \tag{18.65}$$

- *Space conservation law*

$$\frac{d}{dt}\int_{V(t)} dV - \int_{\partial V(t)} \mathbf{w}\cdot \mathbf{n}\, dS = 0. \tag{18.66}$$

Here, **v** denotes the fluid velocity and **w** the interfacial mesh velocity. Equation 18.66 is the so-called space conservation law (see Section 18.2.1).

18.3.1.2 Domain and Equation Discretization

18.3.1.2.1 Domain Discretization

The discretization process consists of two general steps: the discretization in space and time of the computational domain and the discretization of the governing partial differential equations, which will be described for a generic transport equation for the transported property ϕ, for example, the surfactant molar concentration c_i^{Σ}.

The time interval is split into a finite number of time steps Δt, with the possibility of adaptive time stepping during the solution process.

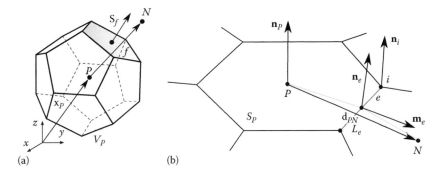

FIGURE 18.2 Computational domain discretization: (a) polyhedral control volume and (b) polygonal control area.

The volumetric domain is subdivided into a finite number of convex polyhedral control volumes V_P, also denoted as computational cells (see Figure 18.2a). The control volumes are nonoverlapping and completely filling the computational domain. The centroid of the control volume is denoted by P and the centroid of a neighboring control volume by N. The cell faces are of polygonal shape and denoted by f. Each face is characterized by its area S_f and its area normal vector $\mathbf{S}_f$.

For the spatial discretization of the interfacial domain, polygonal control areas are introduced, which are consistent with the boundary faces of the volumetric domain at the interfacial boundary. In analogy to the notation in the volumetric domain, the center of a control area is denoted by P and the center of a prototype neighboring control area by N (see Figure 18.2b). The edge e, separating two control areas, is characterized by the edge vector $\mathbf{e}$, its length L_e, and the binormal vector $\mathbf{m}_e$, perpendicular to $\mathbf{e}$ as well as to the edge normal vector $\mathbf{n}_e = \frac{1}{2}(\mathbf{n}_1 + \mathbf{n}_2)$.

18.3.1.2.2 Finite Volume Method

A FVM is applied to discretize the governing equations for the bulk transport processes. For the transport property ϕ, the generic transport equation in integral form reads

$$\frac{d}{dt}\int_{V(t)} \phi \, dV + \int_{\partial V(t)} (\phi(\mathbf{v}-\mathbf{w}) + \mathbf{j}) \cdot \mathbf{n} \, dS = \int_{V(t)} s \, dV. \tag{18.67}$$

The basic strategy for the discretization of a volumetric transport equation is to numerically evaluate the volume integrals transformed into sums of face integrals employing the midpoint rule. The face-centered values are computed through central differences with exception of the convective terms. For these terms, the Gamma discretization scheme [12,38] is employed to ensure boundedness and stability. The diffusive term is discretized implicitly with explicit nonorthogonal correction. Time integration is achieved by means of a second-order implicit three time-level scheme, also known as backward scheme or Gear's method [39].

The fully discretized volumetric transport equation for the arbitrary transported quantity then reads

$$\frac{3\phi_P^n V_P^n - 4\phi_P^o V_P^o + \phi_P^{oo} V_P^{oo}}{\Delta t} + \sum_f F_f \phi_f = \sum_f \Gamma_f \mathbf{S}_f \cdot (\nabla \phi)_f + s_u V_P + s_P V_P \phi_P, \tag{18.68}$$

where
$F_f = S_f \cdot u_f$ is the volumetric face flux
s_u and s_P are the result of linearization of $s(\phi)$ [39,40]

We denote the discrete velocity as $\mathbf{u}$ to distinguish between the discrete and the continuous quantity. The superscripts n, o, and oo represent values evaluated at the new time instance t^n and two previous time instances $t^o = t^n - \Delta t$ and $t^{oo} = t^o - \Delta t$.

18.3.1.2.3 Finite Area Method

The same strategy as for the FVM is applied to discretize the surface transport equation. Polygonal cell faces of the control volumes constitute the finite area mesh. Transport processes at the interface are described by the interfacial transport equation, which, for an arbitrary surface quantity ϕ^Σ, reads

$$\frac{d}{dt} \int_{S(t)} \phi^\Sigma \, dS + \int_{\partial S(t)} (\phi^\Sigma (\mathbf{v} - \mathbf{w})_\parallel + \mathbf{j}^\Sigma) \cdot \mathbf{m} \, dl = \int_{S(t)} s^\Sigma \, dS. \tag{18.69}$$

Area and edge integrals are transformed into sums of edge integrals employing the midpoint rule, and then central differences are applied to calculate edge-centered values. The resulting discretized surface transport equation for the arbitrary discretized quantity ϕ_S reads

$$\frac{3(\phi_P^S)^n (S_P)^n - 4(\phi_P^S)^o (S_P)^o + (\phi_P^S)^{oo} (S_P)^{oo}}{\Delta t} + \sum_e \left(F_e \phi_e^S \right)$$

$$= \sum_e \left(\Gamma_\phi^S \right)_e (\mathbf{m}_e L_e) \cdot \left(\nabla_S \phi^S \right)_e + s_u^S S_P + s_P^S S_P \phi_P^S, \tag{18.70}$$

with the relative edge flux $F_e = (\mathbf{m}_e L_e) \cdot (\mathbf{u} - \mathbf{w})_{\parallel,e}$.

18.3.1.3 Fluid Flow Solution

The pressure–velocity coupling is solved applying the iterative pressure implicit with splitting of operators (PISO) algorithm [41]. Details on the segregated solution of the Navier–Stokes equations can be found in [37].

18.3.1.4 Mesh Motion and Adaption

18.3.1.4.1 Space Conservation Law

After the flow field has been updated by utilizing the PISO algorithm based on the old mesh configuration, the volumetric fluxes over the mesh boundary faces representing the interface are different from zero: $\dot{V}_f \neq 0$. The starting point for the calculation of these fluxes is the space conservation law [10,42], introduced in its integral form in Equation 18.25:

$$\frac{d}{dt} \int_{V(t)} dV - \int_{\partial V(t)} \mathbf{w} \cdot \mathbf{n} \, dS = 0,$$

with $\mathbf{w}$ as the interfacial mesh velocity. Applying a second-order backward time discretization scheme, Gear's method, the discretized space conservation law reads

$$\frac{3V_P^n - 4V_P^o + V_P^{oo}}{2\Delta t} = \sum_f (\mathbf{w} \cdot \mathbf{n})_f S_f \equiv \sum_f \dot{V}_f^n. \tag{18.71}$$

So-called swept volumes δV_f are introduced that describe the volume swept by each face during the update of the mesh. For the different time steps, the swept volumes read as

$$V_P^n - V_P^o = \sum_f \delta V_f^n, \quad V_P^o - V_P^{oo} = \sum_f \delta V_f^o. \tag{18.72}$$

Introducing the definition of swept volumes (18.72) into the discretized space conservation law (18.71) yields

$$\frac{3 \sum_f \delta V_f^n - \sum_f \delta V_f^o}{2 \Delta t} = \sum_f \dot{V}_f^n, \tag{18.73}$$

which provides the advantage of evaluating $\dot{V}_f^n$ without explicitly knowing the mesh velocity. Thus, each flux can now be described as

$$\dot{V}_f^n = \frac{3}{2} \frac{\delta V_f^n}{\Delta t} - \frac{1}{2} \frac{\delta V_f^o}{\Delta t}. \tag{18.74}$$

18.3.1.4.2 Interface Mesh Update

After the PISO iteration on a spatially fixed mesh, the net volume flux $\dot{V}_f'$ over the faces representing the interface $f \in S_\Sigma$ is not zero:

$$\frac{\dot{m}_f}{\rho} - \dot{V}_f = \dot{V}_f' \neq 0, \tag{18.75}$$

where

$\dot{m}_f$ represents the mass flux over the faces f that have to satisfy the continuity equation
$\dot{V}_f$ are the net volumetric fluxes that have to satisfy the discretized space conservation law

If there is no phase change occurring, the mesh has to be updated such that $\dot{V}_f' \equiv 0$, that is, the vertices describing the position of the interface need to be updated accordingly. Note that this procedure is valid only under the constraints of no phase change and no mass transfer across the interface. The mesh update involves the introduction of the so-called control points $\mathbf{x}_C$ at the centers of the boundary faces (Figure 18.3). The control points are necessary, since a simple movement of

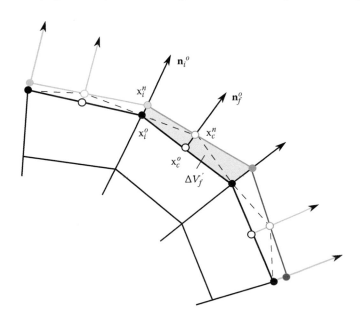

FIGURE 18.3   Control points and displacement directions at the interface.

the face centers would result in a disjoint set of faces. The mesh motion algorithm then involves the following steps:

1. Definition of directions of displacement for the control points $\mathbf{f}_c$
2. Calculation of the swept volume $\delta V_f' = \frac{2}{3} \dot{V}_f' \Delta t$ from Equations 18.73 and 18.75, accounting only for the time step between n and o
3. Calculation of the displacement of the control points according to

$$\mathbf{x}_C^n = \mathbf{x}_C^o + \frac{\delta V_f'}{\mathbf{S}_f \cdot \mathbf{f}_c} \mathbf{f}_c \qquad (18.76)$$

4. Update of the mesh vertices through linear interpolation or least square point interpolation

18.3.1.4.3 Mesh Smoothing versus Remeshing

After the displacement of the interfacial mesh, the volumetric mesh needs to be updated. For this purpose, automatic mesh motion is employed. In cases of low mesh distortion, a Laplacian mesh smoothing algorithm [13] is applied according to a vertex-based solution technique using a minielement finite element method with polyhedral cell support. Here, the mesh topology is not changed; the mesh points are moved ensuring sufficiently high mesh quality based on a mesh quality metrics. In cases when the mesh is subject to large distortions or severe changes, smoothing algorithm cannot guarantee mesh validity. Thus, local remeshing in combination with conservative field remapping algorithms needs to be applied [33]. In order to apply the FAM on top of topologically changing surface meshes, the method has been enhanced to support conservative remapping of area fields based on a supermesh approach [30,37].

18.3.1.5 Interface Boundary Conditions

The computational domain consists of two volumetric meshes, A and B, representing the bulk phases and the interfacial mesh. These two domains need to be coupled by iteratively enforcing appropriate boundary conditions, in particular at the boundary representing the interface.

The denser phase, that is, the liquid phase, is denoted as A; the other phase, that is, the gas phase, is denoted as B. To achieve good convergence during the semi-implicit iterative evaluation, as described in [12], it is necessary to treat the denser phase as phase A. Otherwise, further stabilization would be necessary. The two phases are separated by the interface Σ. The interfacial normal $\mathbf{n}$ is pointing from fluid A to fluid B (Figure 18.4).

On side A, pressure value and normal velocity gradient are specified and on side B, vice versa. The boundary conditions at the interfacial mesh boundaries are reported in Equations 18.77 through 18.80. Note that the absolute pressure p is decomposed as $p = \rho \mathbf{g} \cdot \mathbf{r} + p_m$, where $\mathbf{r}$ is the position

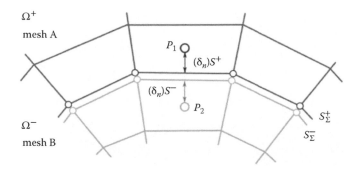

FIGURE 18.4 Two-fluid interface.

vector of the face center. The boundary conditions are written with respect to p_m, which will be denoted as p in the following:

$$[p] = -2[\mu]\nabla_S \cdot \mathbf{v} + \sigma\kappa - 2[\rho\mathbf{g} \cdot \mathbf{r}], \tag{18.77}$$

$$[\mu\, \mathbf{n} \cdot \nabla\mathbf{v}] = -\nabla_S\sigma - [\mu(\nabla_S\mathbf{v}) \cdot \mathbf{n}] - [\mu]\mathbf{n}(\nabla_S \cdot \mathbf{v}), \tag{18.78}$$

$$[\mathbf{v}] = 0, \tag{18.79}$$

$$\mathbf{n} \cdot \left[\frac{1}{\rho}\nabla p\right] = \mathbf{n} \cdot \left[\frac{D\mathbf{v}}{Dt}\right] + \mathbf{n} \cdot \left[\frac{\mu}{\rho}\Delta\mathbf{v}\right]. \tag{18.80}$$

Note that we arrived at four conditions at the interface and not only two, as one would expect. This is due to the fact that the whole domain is divided into two disconnected subdomains, and each of them needs a complete set of boundary conditions, specifying the values of velocity, pressure, and their normal gradient. Specifying these conditions, we obtain, on side A,

$$p_A = p_B - 2(\mu_A - \mu_B)\nabla_S \cdot \mathbf{v} - \sigma\kappa - 2(\rho_A - \rho_B)\mathbf{g} \cdot \mathbf{r}_A,$$

$$(\mathbf{n} \cdot \nabla\mathbf{v})_A = \frac{1}{\mu_A}\nabla_S\sigma + \frac{\mu_B - \mu_A}{\mu_A}\mathbf{n} \cdot (\nabla_S\mathbf{v}) + \frac{\mu_B}{\mu_A}[(\mathbf{I} - \mathbf{nn}) \cdot (\mathbf{n} \cdot \nabla\mathbf{v})_B] - \mathbf{n}(\nabla_S \cdot \mathbf{v}),$$

and, on side B,

$$\mathbf{v}_B = \mathbf{v}_A,$$

$$\mathbf{n} \cdot (\nabla p)_B = -\frac{\rho_B}{\rho_A}\mathbf{n} \cdot (\nabla p)_A + \rho_B\mathbf{n} \cdot \left(\frac{D\mathbf{v}}{Dt}\right)_B + \rho_B\mathbf{n} \cdot \left(\frac{D\mathbf{v}}{Dt}\right)_A + \mu_B\mathbf{n} \cdot (\Delta\mathbf{v})_B + \rho_B\frac{\mu_A}{\rho_A}\mathbf{n} \cdot (\Delta\mathbf{v})_A.$$

18.3.1.6 Surface Tension Treatment

The surface tension calculation is a delicate step, since in our problems, the presence of surfactant directly affects this quantity and its accurate evaluation is fundamental to obtain high-quality results. The surface tension force is treated as in [12], where it is derived from fundamental continuum physics. The total surface tension force for a surface control area of the interface $S \subset \Sigma$ can be formulated as

$$\mathbf{F}^\sigma = \int_{\partial S} \mathbf{m}\sigma\, dl. \tag{18.81}$$

This force can be decomposed into its normal and tangential component

$$\mathbf{F}^\sigma = \int_S \kappa\sigma\mathbf{n}\, dS + \int_S \nabla_S\sigma\, dS. \tag{18.82}$$

Discretization of Equation 18.81 on an unstructured polygonal surface mesh leads to

$$\mathbf{F}^\sigma = \sum_e \int_{L_e} \mathbf{m}\sigma\, dl = \sum_e (\mathbf{m}\sigma)_e L_e, \tag{18.83}$$

where the product $(\mathbf{m}\sigma)_e$ is evaluated applying the trapezoidal rule with $\hat{\mathbf{e}}$ the unit edge vector and $\mathbf{n}_i, \mathbf{n}_j$ the normal vectors located at the edge's vertices:

$$(\mathbf{m}\sigma)_e = \frac{\sigma_i(\hat{\mathbf{e}} \times \mathbf{n}_i) + \sigma_j(\hat{\mathbf{e}} \times \mathbf{n}_j)}{2}. \tag{18.84}$$

After decomposition, the discretized normal and tangential surface tension forces read, respectively,

$$\mathbf{F}_{\text{normal}}^{\sigma} = \mathbf{n}_P \otimes \mathbf{n}_P \sum_e (\mathbf{m}\sigma)_e L_e, \tag{18.85}$$

$$\mathbf{F}_{\text{tangential}}^{\sigma} = (I - \mathbf{n}_P \otimes \mathbf{n}_P) \sum_e (\mathbf{m}\sigma)_e L_e, \tag{18.86}$$

with $\mathbf{n}_P$ is the face normal vector.

If surfactants are present in the system, the surface tension strongly depends on the surfactant surface concentration $\sigma = \sigma(c_1^{\Sigma}, \ldots, c_N^{\Sigma})$. The effects of surfactants will be described in Section 18.3.2.4.

18.3.2 SURFACTANT TRANSPORT

The central part of the model is the surfactant transport equations in the bulk and on the interface. In the following paragraphs, the modeling equations will be rephrased into the ALE reference frame in order to account for the mesh motion introduced by the interface-tracking approach.

18.3.2.1 Surfactant Transport in the Bulk

Transport processes of surfactant in the bulk phase are described through an integral governing equation in ALE reference frame. Thus, the equation for surfactant concentrations in the bulk reads

$$\frac{d}{dt} \int_{V(t)} c_i \, dV + \int_{\partial V(t)} ((\mathbf{v} - \mathbf{w})c_i) \cdot \mathbf{n} \, dS + \int_{\partial V(t)} \mathbf{j}_i \cdot \mathbf{n} \, dS = \int_{V(t)} s_i \, dV. \tag{18.87}$$

18.3.2.2 Surfactant Transport on the Fluid Interface

More attention is given to the derivation of the surfactant transport equations on the interface in ALE formulation. Consider the surface transport equation in local form (18.11), see Section 18.2.1, that is,

$$\partial_t^{\Sigma} c_i^{\Sigma} + \nabla_{\Sigma} \cdot \left(c_i^{\Sigma} \mathbf{v}^{\Sigma} + \mathbf{j}_i^{\Sigma} \right) = s_i^{\Sigma}.$$

Decomposing the fluid velocity at the interface into its normal and tangential components and recalling the interface curvature $\kappa_{\Sigma} = \nabla_{\Sigma} \cdot (-\mathbf{n}_{\Sigma})$, the surface transport equation becomes

$$\partial_t^{\Sigma} c_i^{\Sigma} + \nabla_{\Sigma} \cdot \left(c_i^{\Sigma} \mathbf{v}_{\|} + \mathbf{j}_i^{\Sigma} \right) - c_i^{\Sigma} \kappa_{\Sigma} V_{\Sigma} = s_i^{\Sigma}, \tag{18.88}$$

where $V_{\Sigma} = \mathbf{v}^{\Sigma} \cdot \mathbf{n}_{\Sigma}$ is the speed of normal displacement of the interface Σ. The last term on the left-hand side of Equation 18.88 accounts for the change of surface concentration due to the change of interfacial area. From a numerical point of view, it implies the computation of a second-order derivative because of the presence of curvature κ_{Σ}. The ALE formulation alleviates this problem.

Since we are dealing with moving meshes, to describe transport processes on the interface, a suitable integral balance equation needs to be derived for the moving control areas $S(t)$, bounded by the curve $C(t) = \partial S(t)$. The mesh velocity is denoted as $\mathbf{w}$. The consistency between the physics of the problem and surface mesh motion is reached through the constraint (18.24)

$$\mathbf{w} \cdot \mathbf{n}_{\Sigma} = \mathbf{v}^{\Sigma} \cdot \mathbf{n}_{\Sigma}, \quad \text{on } \Sigma.$$

The motion of the points due to the mesh motion is described by initial value problems $\dot{\mathbf{y}}(t) = \mathbf{w}(t, \mathbf{y}(t))$ with initial condition $\mathbf{y}(t_0) = \mathbf{y}_0$. The moving control area is $S(t) = \{\mathbf{y}(t; t_0, y_0) : \mathbf{y} \in S_0\}$.

The change of surface concentration due to mesh motion is described by the equation

$$\frac{d}{dt} \int_{S(t)} c_i^\Sigma \, dS = \int_{S(t)} \left(\frac{D^w(c_i^\Sigma)}{Dt} + c_i^\Sigma \nabla_\Sigma \cdot \mathbf{w} \right) dS, \tag{18.89}$$

where $\frac{D^w}{Dt}$ is the comoving derivative associated to the motion induced by $\mathbf{w}$. For a surface quantity ϕ^Σ, it is defined as

$$\frac{D^w}{Dt} \phi^\Sigma(t_0, \mathbf{y}_0) = \frac{d}{dt} \phi^\Sigma(t, \mathbf{y}(t; t_0, \mathbf{y}_0))|_{t=t_0}.$$

Employing the Thomas derivative $\partial_t^\Sigma c_i^\Sigma$, we obtain

$$\frac{d}{dt} \int_{S(t)} c_i^\Sigma \, dS = \int_{S(t)} \left(\partial_t^\Sigma c_i^\Sigma + \nabla_\Sigma \cdot \left(c_i^\Sigma \mathbf{w} \right) \right) dS. \tag{18.90}$$

From the surfactant transport equation (18.5), we obtain an expression for $\partial_t^\Sigma c_i^\Sigma$ to substitute into Equation 18.90:

$$\frac{d}{dt} \int_{S(t)} c_i^\Sigma \, dS = \int_{S(t)} \left(s_i^\Sigma - \nabla_\Sigma \cdot \left(c_i^\Sigma (\mathbf{v} - \mathbf{w}) \right) - \nabla_\Sigma \cdot \mathbf{j}_i^\Sigma \right) dS. \tag{18.91}$$

Under the constraint (18.24), the normal component of Equation 18.91 cancels and it remains:

$$\frac{d}{dt} \int_{S(t)} c_i^\Sigma \, dS + \int_{S(t)} \nabla_\Sigma \cdot \left(c_i^\Sigma (\mathbf{v} - \mathbf{w})_\parallel + \mathbf{j}_i^\Sigma \right) dS = \int_{S(t)} s_i^\Sigma \, dS, \tag{18.92}$$

or, in flux formulation with $\mathbf{m}$ being the outer normal to $\partial S(t) \equiv C(t)$,

$$\frac{d}{dt} \int_{S(t)} c_i^\Sigma \, dS + \int_{\partial S(t)} \left(c_i^\Sigma (\mathbf{v} - \mathbf{w})_\parallel + \mathbf{j}_i^\Sigma \right) \cdot \mathbf{m} \, dl = \int_{S(t)} s_i^\Sigma \, dS. \tag{18.93}$$

In Equation 18.93, the effect of area change due to mesh motion on the surfactant concentration is already included, without the necessity of computing the surface curvature κ_Σ. Thus, we obtained a more accurate expression that involves only a first-order derivative (i.e., the surface tangent). Equation 18.93 is the starting point for the finite area discretization. Note that we have dropped the superscript Σ for the fluid velocity $\mathbf{v}$ (see Equation 18.18).

18.3.2.3 Multicomponent Diffusive Fluxes on the Interface

To calculate the diffusive fluxes for multicomponent surfactant systems, the Maxwell–Stefan equations (see Section 18.2.2) are reformulated in matrix form as follows:

$$\mathbf{B}^\Sigma \mathbf{j}^\Sigma = \mathbf{d}^\Sigma, \tag{18.94}$$

where the coefficient matrix $\mathbf{B}^\Sigma$ is a function of the local concentration values and the Maxwell–Stefan diffusivities. The diffusive fluxes of the single species $\mathbf{j}_i^\Sigma$ are condensed into the flux matrix $\mathbf{j}^\Sigma$. In analogy, the species driving forces $\mathbf{d}_i^\Sigma$ are merged into the overall driving force matrix $\mathbf{d}^\Sigma$. The system is not directly invertible, but it has a unique solution [24] under the additional constraint (18.30), that is, $\sum_i \mathbf{j}_i^\Sigma = 0$. For this reason, the iterative inversion algorithm of Giovangigli [25] is adopted, utilizing the group inverse of the coefficient matrix that is identical to the inverse on the solution subspace. Applying a regular splitting method, the diffusive fluxes can be expressed as

$$\mathbf{j}^{\Sigma} = \mathbf{D}\mathbf{d}^{\Sigma}, \tag{18.95}$$

where $\mathbf{D}$ is the matrix of heterogeneous diffusivities, also denoted in this procedure as the group inverse $\mathbf{D} = \mathbf{B}^{\#}$ of the coefficient matrix $\mathbf{B}$. The matrix $\mathbf{D}$ is given as

$$\mathbf{D} = \sum_{\kappa=0}^{\infty} (\mathbf{Q}\mathbf{S})^{\kappa} \mathbf{Q}\mathbf{L}^{-1}\mathbf{Q}, \tag{18.96}$$

where
$\mathbf{S} = \mathbf{I} - \mathbf{L}^{-1}\mathbf{B}^{\Sigma}$
$\mathbf{L}$ is a diagonal matrix with $\mathbf{L} = \mathrm{diag}(L_1,...,L_N)$ such that $L_{\kappa} > B_{\kappa\kappa}$ for $y_{\kappa} > 0$ and $L_{\kappa} \geq B_{\kappa\kappa}$ for $y_{\kappa} = 0$

Usually, good convergence is obtained for $\kappa = 4$. The projection matrix along $\mathbf{U}^{\perp}$ onto the solution subspace is denoted as $\mathbf{Q} = \mathbf{Q}_{\mathbf{U}^{\perp},\mathbb{R}\mathbf{y}^{\Sigma}}$.

Inserting Equation 18.95 into the integral surfactant balance for varying control areas (18.93) leads to the coupled system

$$\frac{d}{dt} \int_{S(t)} c_i^{\Sigma} \, dS + \int_{\partial S(t)} \left(c_i^{\Sigma}(\mathbf{v} - \mathbf{w})_{\|} + \sum_j D_{ij} \nabla c_j^{\Sigma} \right) \cdot \mathbf{m} \, dl = \int_{S(t)} s_i^{\Sigma} \, dS. \tag{18.97}$$

More information on the iterative inversion algorithm and the strategy to solve surface transport equations can be found in [25,37]; the main ideas of the block-coupled solution are reported in the following section. Note that the diffusive term in Equation 18.97 needs a specialized discretization scheme, since the diffusivities are strongly heterogeneous; details can be found in [37].

18.3.2.4 Sorption Modeling
The source term s_i^{Σ} present in Equation 18.97 is model dependent and implemented in a sorption library. The constitutive modeling of sorption processes has been introduced in Section 18.2.3, and it differs considerably for fast and slow sorption. Common to both types is the evaluation of the surface tension as a function of the concentration of the adsorbed surfactant.

Consider a single surfactant for simplicity. For diffusion-controlled (fast) sorption processes, the adsorption and desorption rates are locally equilibrated, as stated in Equation 18.33. This equality leads to an additional local relationship between c^{Σ} and $c_{|\Sigma}$, say $c^{\Sigma} = f(c_{|\Sigma})$. Various sorption models are available to describe this relation, $f(c_{|\Sigma})$ (see Table 18.1). From the sorption model, the value of $c_{|\Sigma} = f^{-1}(c^{\Sigma})$ is taken as a Dirichlet boundary condition for the surfactant bulk equation (18.87).

After the solution of the bulk equation with this Dirichlet data, the source term for the surface concentration equation is computed as stated in Equation 18.32, that is,

$$s_{\mathrm{fast}}^{\Sigma} = \mathbf{j} \cdot \mathbf{n}_{\Sigma} = -D(\mathbf{n} \cdot \nabla c)_{\Sigma}. \tag{18.98}$$

This way, the Neumann data from the new bulk field are transferred to the interfacial molar mass balance. Then, the surface transport equation is solved to obtain the new surface concentration field of the surfactant species.

For kinetically controlled (slow) sorption processes, the source term s^{Σ} is given directly by the sorption model adopted, as a balance between adsorption and desorption, see Equation 18.34, that is,

$$s_{\mathrm{slow}}^{\Sigma} = s^{\mathrm{ads}}(c_{|\Sigma}, c^{\Sigma}) - s^{\mathrm{des}}(c^{\Sigma}). \tag{18.99}$$

In this case, coupling between the interface and the bulk flow is achieved by enforcing the discretized boundary condition (18.9) as a Neumann boundary condition for the bulk equation, that is,

$$\mathbf{n} \cdot \nabla c = -\frac{s^{\Sigma}}{D} \quad \text{at } \Sigma. \tag{18.100}$$

After interfacial and bulk surfactant transport equations are solved, the surface tension $\sigma = \sigma(c^{\Sigma})$ is updated according to the sorption model chosen. The resulting surface tension force is calculated as described in Section 18.3.1.6.

18.3.2.5 Block-Coupled Solution of the Species Transport Equations

The discretized species transport equations (18.97) can be written as a system of linear equations in the form

$$\mathbf{a}_P \phi_P^n + \sum_N \mathbf{a}_N \phi_N^n = \mathbf{b}_P, \tag{18.101}$$

where $\mathbf{a}_P$ and $\mathbf{a}_N$ are $n \times n$ matrices, where n is the number of species in the system. The source term $\mathbf{b}_P$ holds all the explicit terms. The solution quantity ϕ is a vectorial quantity composed by the species concentrations c_i^{Σ}, treated as independent variables. The resulting partial differential system is strongly coupled because of the cross effects between surfactant species (see diffusion term in Equation 18.97). To take into account this coupling and to be consistent with the physics of the problem, a coupled solution method is applied. With the iterative inversion algorithm (see Section 18.3.2.3) and the block-coupled solution procedure, the constraint (18.30), $\sum_i \mathbf{j}_i^{\Sigma} = 0$, is inherently fulfilled. A block-coupled GMRES solver with an incomplete Cholesky preconditioner is applied.

18.4 SIMULATION RESULTS

In this section, the setup for free-surface and two-phase flow DNSs is outlined. More details on the method can be found in [30] for free-surface flows and in [12,17] for two-phase flows. In Section 18.4.1, only the results for kinetically controlled sorption are presented.

18.4.1 Free-Surface Flow under Surfactant Influence

As a test case for single-phase flows, the droplet formation process under the influence of single surfactants and surfactant mixtures is shown. The case setup is depicted in Figure 18.5, and the starting mesh is shown in Figure 18.6. The orifice is represented by a cylindrical tube. At the inlet, an inflow boundary condition is applied to mimic the parabolic velocity profile in the orifice. This condition corresponds to an inlet flow rate of $\dot{V} = 25$ mm^3/s. The inner diameter of the capillary is $d_i = 0.45$ mm and the other diameter is $d_o = 0.7$ mm.

To be consistent with the experimental observation for droplet formation process during a profile analysis tensiometry protocol, the free surface is pinned at the outer edge of the capillary. A water/air system is used.

For a pure system, the surface tension is given by $\Sigma_0 = 0.072$ N/m, and the fluid density is $\rho = 1000$ kg/m^3. Isothermal conditions are assumed with $T = 298$ K. The Reynold's number for this system is Re = 28.3.

The solver validation for the pure hydrodynamics can be found in [43]. The solution method related to surfactants is presented in [30,37] (multicomponent surfactant system).

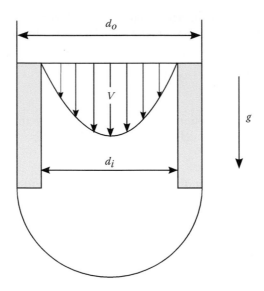

FIGURE 18.5 Simulation setup of the drop profile analysis tensiometry (PAT) capillary.

FIGURE 18.6 Starting mesh for the PAT capillary.

18.4.1.1 Single Surfactant

The effects of a single surfactant on the system are described through two different slow sorption models: the Henry sorption model and the Langmuir one. The data characterizing the models are summarized in the following text. The maximum interfacial surfactant concentration for both cases is given as $c_\infty^\Sigma = 2.65 \cdot 10^{-2}$ **mol/m²**.

18.4.1.1.1 Henry Sorption

This case adopts the kinetically controlled Henry sorption model described by Equations 18.61 and 18.41. The bulk concentration is initialized as $c = 0.001$ mol/L. The surface and bulk diffusivities are taken as $D^\Sigma = 1 \cdot 10^{-7}$ m²/s and $D = 2 \cdot 10^{-5}$ m²/s, respectively. The adsorption coefficient is $\kappa^{ads} = 30$ 1/s, while the desorption coefficient is $\kappa^{des} = 0.2$ mol/(m³s).

FIGURE 18.7 Concentration profile on the free surface, $[c^\Sigma] = \text{mol/m}^2$, during droplet formation process assuming Henry sorption at $t_1 = 0.0$ s, $t_2 = 0.03$ s, and $t_3 = 0.11$ s.

FIGURE 18.8 Surface tension profile, $[\sigma] = \text{N/m}$, during droplet formation process assuming Henry sorption at $t_1 = 0.0$ s, $t_2 = 0.03$ s, and $t_3 = 0.11$ s.

In Figure 18.7, the temporal evolution of the concentration of the adsorbed surfactant is shown, and the corresponding surface tension profiles are reported in Figure 18.8. It can be seen that the concentration of the adsorbed surfactant is higher at the apex of the drop, while in the vicinity of the orifice, where the surface is stretched, the concentration is diluted.

18.4.1.1.2 Langmuir Sorption
The governing equations for this model are (18.62) and (18.43). In order to obtain results comparable with the Henry model, the settings are chosen in analogy with the case earlier.

In Figure 18.9, the temporal evolution of the concentration of the adsorbed surfactant is depicted, and the corresponding surface tension profiles are given in Figure 18.10. As for the Henry sorption model, the concentration of the adsorbed surfactant rises at the apex of the drop due to convective transport mechanisms and change in the surface area, while in the vicinity of the orifice, the concentration is diluted.

With respect to the Henry sorption model, the Langmuir sorption model takes into account the interactions between molecules at the interface by means of the Langmuir equilibrium constant a and the maximum number of adsorbed molecules per area c_∞^Σ. This results into slightly different values of the surfactant concentration and surface tension on the free surface. The higher the surfactant concentration on the free surface, the more pronounced the differences between the two models.

Another important aspect to underline is the difference in the shape of the droplets between the contaminated system and the clean one; see Figure 18.11. Contaminated droplets are longer and

FIGURE 18.9 Concentration profile on the free surface, $[c^\Sigma] = \text{mol/m}^2$, during droplet formation process assuming Langmuir sorption at $t_1 = 0.0$ s, $t_2 = 0.03$ s, and $t_3 = 0.11$ s.

FIGURE 18.10 Surface tension profile, $[\sigma] = \text{N/m}$, during droplet formation process assuming Langmuir sorption at $t_1 = 0.0$ s, $t_2 = 0.03$ s, and $t_3 = 0.11$ s.

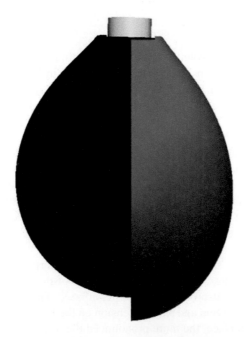

FIGURE 18.11 Comparison between a droplet with clean (left) and contaminated (right) free surfaces at $t = 0.11$ s.

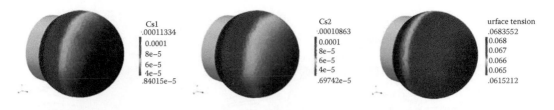

FIGURE 18.12 Concentration profiles and corresponding surface tension profile during droplet formation process assuming Langmuir sorption at $t_2 = 0.05$ s.

more slender due to lower surface tension values. This effect becomes even larger in case of higher surface coverages.

18.4.1.2 Surfactant Mixture

A mixture of two surfactant species is considered, one obeying to a slow adsorbing model (18.62) and the other to a fast one (18.42). Both surfactants are described through the Langmuir sorption processes and (Maxwell–Stefan) multicomponent surface transport according to Equation 18.97.

The concentrations of the two surfactants in the bulk phase are initialized to $c_1 = 1 \cdot 10^{-6}$ mol/L and $c_2 = 2 \cdot 10^{-6}$ mol/L. At the inlet, the bulk concentrations are set to $c_1^{in} = 1 \cdot 10^{-3}$ mol/L and $c_2^{in} = 2 \cdot 10^{-3}$ mol/L. Regarding the sorption process, the adsorption constants are $\kappa_1^{ads} = 3$ 1/s and $\kappa_2^{ads} = 40$ 1/s, while the desorption coefficients are $\kappa_1^{des} = 0.02$ mol/(m^{3}s) and $\kappa_2^{des} = 0.04$ mol/(m^{3}s). The bulk diffusivities are assumed as $D_1 = 1 \cdot 10^{-5}$ m^2/s and $D_2 = 2 \cdot 10^{-5}$ m^2/s. The surface Maxwell–Stefan diffusivities are assumed to be $D_1^\Sigma = 1 \cdot 10^{-7}$ m^2/s and $D_2^\Sigma = 1 \cdot 10^{-7}$ m^2/s.

Figure 18.12 shows the droplet formation process at a time step 0.05 s with the concentration profiles of both adsorbed surfactants and the corresponding surface tension profile. All transport effects are covered, in particular convection and diffusion in the bulk and on the interface as well as sorption processes. In fact, it is important to account for all transport mechanisms to be able to address droplet formation processes under the influence of surfactant mixtures, since the concentrations of adsorbed surfactants and the corresponding surface tension profile influence the hydrodynamics.

18.4.2 TWO-PHASE FLOW UNDER SURFACTANT INFLUENCE

As a test case, an air bubble rising in contaminated water is considered. The initial shape of the bubble is a sphere of $r_b = 1$ mm radius; the bubble is positioned in the center of a spherical domain with radius $R = 20r_b = 20$ mm. The Reynolds number Re expected for this kind of system is below 1000.

A soluble surfactant species is present in the system. For simplicity, we consider dilute concentrations of the surfactant both in the bulk and on the interface. At start, volumetric concentration of the surfactant in water is homogeneous and amounts to $c = 0.01$ mol/m^3, while the interface is clean $c^\Sigma = 0$ mol/m^2. The surfactant considered has the following properties: saturated concentration $c_\infty^\Sigma = 5 \cdot 10^{-6}$ mol/m^2, adsorption coefficient $\kappa^{ads} = 40$ m^3/smol, desorption coefficient $\kappa^{des} = 0.0893$ mol/m^3, bulk diffusivity $D = 1 \cdot 10^{-9}$ m^2/s, and surface diffusivity $D^\Sigma = 1 \cdot 10^{-7}$ m^2/s. The absolute temperature of the system is $T = 293$ K.

18.4.2.1 Moving Reference Frame

The calculation is performed in a moving reference frame that follows the bubble during its rise; in other words, the origin of the noninertial coordinate system is attached to the center of the bubble. The presence of a noninertial coordinate system implies a correction of the momentum equation and its boundary conditions. Thus, a frame acceleration term, $\rho\mathbf{a}_F$, must be added to the left-hand side of the momentum equation (18.13). Then, the velocity $\mathbf{v}$ becomes the fluid velocity with respect to the moving reference frame.

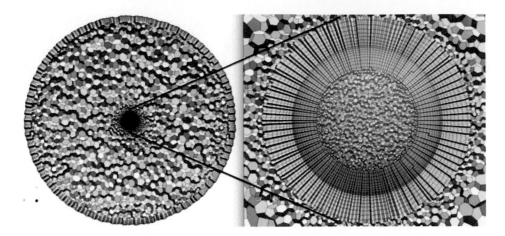

FIGURE 18.13   Polyhedral mesh for two-phase flow simulations.

Regarding the boundary condition for the velocity in this special case, if a bubble is rising in a stagnant liquid, the fluid velocity on the outer boundary of the spatial domain is equal to the opposite of the velocity of the moving reference frame, that is,

$$\mathbf{v}_{MRF} = -\mathbf{v}_{boundary}. \tag{18.102}$$

This velocity is specified only on the inlet part of the outer domain; on the outlet part, a zero normal derivative is imposed for both pressure and velocity.

18.4.2.2   Mesh Generation

OpenFOAM allows to use mixed cell types (polyhedral meshes) and moving meshes. The final mesh consists of a 3D unstructured mesh with two cell types: polyhedral cells in the most of the domain, that is, the bulk region, to enhance accuracy of the results and prisms around the free surface to well resolve the boundary layer. Moreover, another set of prism is added to the outer boundary of the spatial domain. The procedure to generate the mesh is briefly described as follows:

1. The initial mesh is generated by the mesh tool *Salome*: The procedure is completely automatized by the implementation of a *Python* script. The resulting mesh from this step consists of triangles for the surfaces, tetrahedral cells in the bulk region, and prisms with triangular base for the boundary layers. The mesh is saved in.*unv* (ideas) format.
2. The resulting mesh from Salome is imported in OpenFOAM and converted in the supported format through the utility *ideasToFoam*.
3. The polyhedral mesh can be obtained applying the OpenFOAM utility *polyDualMesh* and setting the feature angle to the recommended value of 60°. In Figure 18.13, the final mesh is shown.

18.4.2.3   Simulation Results

For two-phase flow results, we consider a case with one species of surfactant. A brief comparison between the clean case and the contaminated one is accomplished.

18.4.2.3.1   Pure Hydrodynamics

Consider an air bubble rising in quiescent water, Figure 18.14. As the bubble is rising, it is deforming; the final shape of the bubble resembles an ellipsoid. The rising velocity of the bubble, once it reaches the steady state, is about $v_y = 0.34$ m/s.

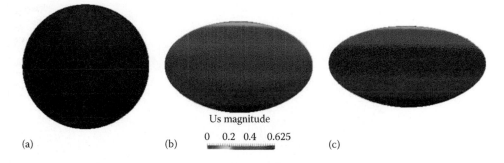

Us magnitude

0   0.2   0.4   0.625

(a)   (b)   (c)

FIGURE 18.14   Bubble shape and relative velocity field on the clean interface, $[v]$ = m/s. (a) t_1 = 0.0 s, (b) t_2 = 0.04 s, and (c) t_3 = 0.08 s.

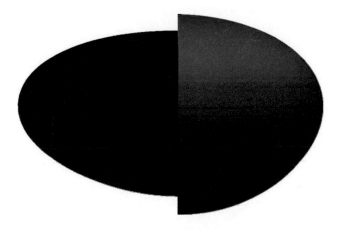

FIGURE 18.15   Comparison between the bubble with clean (left) and contaminated (right) interfaces at t = 0.08 s.

The DNS allows for detailed insights, for example, to study the inner circulation in the air bubble. The results are presented in Figure 18.16 in comparison with the ones for the contaminated case.

18.4.2.3.2   Single Surfactant

Consider the case with soluble surfactant present in the system. The steady-state rising velocity of the bubble is reduced by about 12%, and the final shape of the bubble is no more symmetric. Figure 18.15 shows the difference in the shape between the clean case and the contaminated one.

In Figure 18.16, the flow pattern inside the bubble and in its proximity is shown. The black line represents the current shape of the bubble. In both cases, two recirculation regions are present inside the bubble; nevertheless, in the contaminated case, see Figure 18.16b, these two regions are developing in the upper part of the bubble, and a stagnation cap is developing at the bottom. Regarding the flow field behind the bubble, in the pure water/air system, no recirculation is present, see Figure 18.16a, while in the contaminated system, two recirculation regions are visible, see Figure 18.16b.

In Figures 18.17 and 18.18, the relative velocity and the surfactant concentration fields are shown for different time instances. Surface velocity and shape of the bubble are significantly altered by the presence of the surfactant compared to the clean case.

Figure 18.19a shows the surfactant concentration field on the interface, highlighting the presence of a stagnant cap in the rear part of the bubble, where the surfactant concentration is the highest due to accumulation of surfactant. The streamlines corresponding to this case are shown in Figure 18.19b, colored by the relative velocity magnitude. Here, the recirculation region after the bubble is evident.

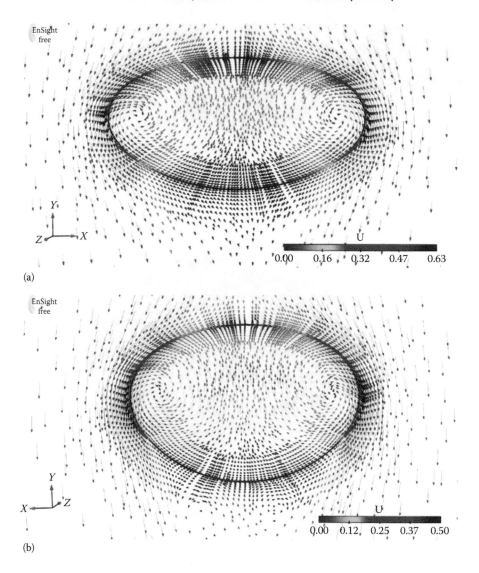

FIGURE 18.16 Circulation vector field at $t = 0.08$ s, vectors colored by relative velocity magnitude $[v] =$ m/s. No kink is present on the interface; optical effects due to the orientation of the velocity vectors could be misleading. (a) Pure system water/air and (b) contaminated system by a single surfactant.

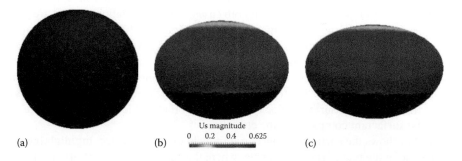

FIGURE 18.17 Bubble shape and relative velocity field on the contaminated interface, $[v] =$ m/s. (a) $t_1 = 0.0$ s, (b) $t_2 = 0.04$ s, and (c) $t_3 = 0.08$.

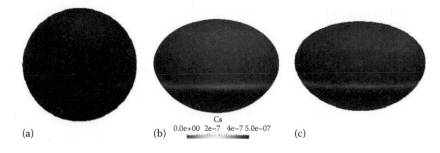

FIGURE 18.18 Surfactant concentration field on the contaminated interface, $[c^S]$ = mol/m². (a) t_1 = 0.0 s, (b) t_1 = 0.04s, and (c) t_1 = 0.08s.

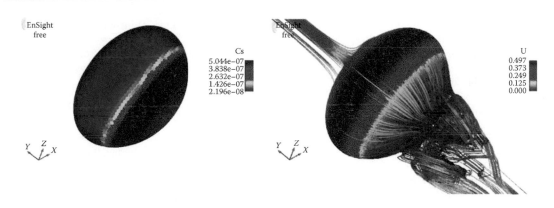

FIGURE 18.19 Contaminated system at t = 0.08 s. (a) Surfactant concentration field on the interface $[c^S]$ = mol/m² (b) Streamlines colored by relative velocity magnitude $[v]$ = m/s. Kinks in the lower part of the streamlines are due to a coarse mesh resolution in that region.

ACKNOWLEDGMENTS

The simulations are performed within OpenFOAM; pre- and postprocessing are executed by SALOME, ParaView, and EnSight (free).

The authors thank the German Research Foundation (DFG) for financial support within the Priority Program SPP1506 "Transport Processes at Fluidic Interfaces" [BO1879/9-2].

REFERENCES

1. C. W. Hirt and B. D. Nichols, Volume of fluid (VOF) method for the dynamics of free boundaries, *Journal of Computational Physics*, 39, 201–225, 1981.
2. S. Osher and J. Sethian, Fronts propagating with curvature dependent speed—Algorithms based on Hamilton-Jacobi formulations, *Journal of Computational Physics*, 79, 12–49, 1988.
3. M. Sussman and E. G. Puckett, A coupled level set and volume-of-fluid method for computing 3D incompressible two-phase flows, *Journal of Computational Physics*, 162, 301–337, 2000.
4. D. M. Anderson, G. McFadden, and A. A. Wheeler, Diffuse interface in fluid mechanics, *Annual Review of Fluid Mechanics*, 30, 139–165, 1998.
5. Y. Y. Renardy, M. Renardy, and V. Christini, A new volume-of-fluid formulation for surfactants and simulations of drop deformations under shear at a low viscosity ratio, *European Journal of Fluid Mechanics—B/Fluids*, 21, 49–59, 2002.
6. A. J. James and J. Lowengrub, A surfactant-conserving volume-of-fluid method for interfacial flows with insoluble surfactant, *Journal of Computational Physics*, 201, 685–722, 2004.
7. A. Alke and D. Bothe, 3D Numerical modelling of soluble surfactant at fluid interfaces based on the volume-of-fluid method, *Fluid Dynamics & Materials Processing*, 5(4), 345–372, 2009.

8. J. J. Xu and H. Zhao, An Eulerian formulation for solving partial differential equations along a moving interface, *Journal of Scientific Computing*, 19, 573–594, 2003.
9. C. Pozrikidis, *Boundary Integral and Singularity Methods for Linearized Viscous Flow*. Cambridge University Press, Cambridge, U.K., 1992.
10. S. Muzaferija and M. Perić, Computation of free-surface flows using the finite-volume method and moving grids, *Numerical Heat Transfer, Part B: Fundamentals*, 32(4), 369–384, 1997.
11. G. Tryggvason, B. Bunner, A. Esmaeli, D. Juric, N. Al-Rawahi, W. Tauber, J. Han, S. Nas, and Y. J. Jan, Front tracking method for the computation of multiphase flow, *Journal of Computational Physics*, 169, 708–759, 2001.
12. Z. Tuković and H. Jasak, A moving mesh finite volume interface tracking method for surface tension dominated interfacial fluid flow, *Computers & Fluids*, 55, 70–84, 2012.
13. H. Jasak and Z. Tuković, Automatic mesh motion for the unstructured finite volume method, *Transactions of FAMENA*, 30(2), 1–20, 2006.
14. D. Schmidt, M. Dai, H. Wang, and J. Blair Perot, Direct interface tracking of droplet deformation, *Atomization and Sprays*, 12(5–6), 721–735, 2002.
15. S. Quan and D. Schmidt, A moving mesh interface tracking method for 3D incompressible two-phase flows, *Journal of Computational Physics*, 221, 761–780, 2007.
16. K. Mooney, S. Menon, and D. Schmidt, A computational study of viscoelastic drop collisions, in *ILASS—Americas 22nd Annual Conference on Liquid Atomization and Spray Systems*, Cincinnati, OH, May 2010.
17. Z. Tuković and H. Jasak, Simulation of free-rising bubble with soluble surfactant using moving mesh finite volume/area method, in *Sixth International Conference on CFD in Oil & Gas, Metallurgical and Process Industries SINTEF/NTNU*, Trondheim, Norway, June 10–12, 2008.
18. D. Bothe, J. Prüss, and G. Simonett, Well-posedness of a two-phase flow with soluble surfactant, in *Nonlinear Elliptic and Parabolic Problems* (J. E. M. Chipot, ed.), pp. 37–61, Verlag, Basel, Switzerland, 2005.
19. D. A. Edwards, H. Brenner, and D. Wasan, *Interfacial Transport Processes and Rheology*. Boston, MA: Butterworth-Heinemann, 1991.
20. H. A. Stone, A simple derivation of the time-dependent convective-diffusion equation for surfactant transport along a deforming interface, *Physics of Fluids A: Fluid Dynamics*, 2(1), 111–112, 1990.
21. J. C. Slattery, L. Sagis, and E. Oh, *Interfacial Transport Phenomena*, 2nd edn. Springer, U.S., 2007.
22. L. Sagis, The Maxwell-Stefan equations for diffusion in multiphase systems with intersecting dividing surfaces, *Physica A*, 254(3), 365–376, 1998.
23. D. Bothe, On the multi-physics of mass transfer across fluid interfaces, in *Seventh International Berlin Workshop—IBW7 on Transport Phenomena with Moving Boundaries* (F. P. Schindler and M. Kraume, eds.), Berlin, Germany, accepted cf. arXiv:1501.05610 [*physics.u-dyn*], 2015.
24. D. Bothe, On the Maxwell-Stefan approach to multicomponent diffusion, in *Parabolic Problems. Progress in Nonlinear Differential Equations and Their Applications* (P. Guidotti and C. W. et al., eds.), vol. 80, pp. 81–93. Basel, Switzerland: Springer, 2011.
25. V. Giovangigli, Convergent iterative methods for multicomponent diffusion, *Impact of Computing in Science and Engineering*, 3(3), 244–276, 1991.
26. D. Bothe and W. Dreyer, Continuum thermodynamics of chemically reacting fluid mixtures, *Acta Mechanica*, 226(6), 1757–1805, 2015.
27. R. Krishna, Multicomponent surface diffusion of adsorbed species: A description based on the generalized Maxwell-Stefan equations, *Chemical Engineering Science*, 45(7), 1779–1791, 1990.
28. P. A. Kralchevsky, D. D. Krassimir, and N. D. Denkov, Chemical physics of colloid systems and interfaces, in *Handbook of Surface and Colloid Chemistry*, second expanded and updated edition. New York: CRC Press, 2002.
29. K. Dieter-Kissling, H. Marschall, and D. Bothe, An object oriented model library for surfactant sorption processes in the interface-tracking framework of OpenFOAM, in *Euromech Colloquium 555: Small-Scale Numerical Methods for Multi-Phase Flows* (Pessac, Bordeaux, France), I2M-ENSCBP, August 28–30, 2013.
30. K. Dieter-Kissling, H. Marschall, and D. Bothe, Direct numerical simulation of droplet formation processes under the influence of soluble surfactant mixtures. *Computers & Fluids* 113, 93–105, 2015.
31. R. Miller et al., Surfactant adsorption layers at liquid interfaces, in *Surfactant Science and Technology: Retrospects and Prospects* (L. Romsted, ed.). Florida: CRC Press, 2014.
32. V. Fainerman, R. Miller, and H. Mohwald, General relationship of the adsorption behaviour of surfactants at the water/air interface, *Journal of Physical Chemistry B*, 106, 809–819, 2002.

33. S. Menon and D. P. Schmidt, Conservative interpolation on unstructured polyhedral meshes: An extension of the Supermesh approach to cell-centered finite-volume variables. *Computer Methods in Applied Mechanics and Engineering*, 200, 2797–2804, 2011.

34. S. Patankar, *Numerical Heat Transfer and Fluid Flow: Computational Methods in Mechanics and Thermal Science*. Washington, DC: Hemisphere Publishing Corp., 1980.

35. M. Perić, R. Kessler, and G. Scheuerer, Comparison of finite-volume numerical methods with staggered and collocated grids, *Computers & Fluids*, 16(4), 389–403, 1988.

36. H. Weller, G. Tabor, H. Jasak, and C. Fureby, A tensorial approach to computational continuum mechanics using object-oriented techniques, *Computers in Physics*, 12, 620–631, 1998.

37. K. Dieter-Kissling, H. Marschall, and D. Bothe, Numerical method for coupled interfacial surfactant transport on dynamic surface meshes of general topology, *Computers & Fluids*, 109, 168–184, 2015.

38. H. Jasak, H. Weller, and A. Gosman, High resolution NVD differencing scheme for arbitrarily unstructured meshes, *International Journal for Numerical Methods in Fluids*, 31, 431–449, 1999.

39. J. Ferziger and M. Perić, *Computational Methods for Fluid Dynamics*. Berlin, Heidelberg, New York: Springer-Verlag, 1996.

40. H. Jasak, Error analysis and estimation for the finite volume method with applications to fluid flows. PhD thesis, Imperial College London, London, U.K., 1996.

41. R. Issa, Solution of the implicitly discretised fluid flow equations by operator-splitting, *Journal of Computational Physics*, 62(1), 40–65, 1986.

42. I. Demirdzić and M. Perić, Space conservation law in finite volume calculations of fluid flow, *International Journal for Numerical Methods in Fluids*, 8(9), 1037–1050, 1988.

43. K. Dieter-Kissling, M. Karbaschi, H. Marschall, A. Javadi, R. Miller, and D. Bothe, On the applicability of drop profile analysis Tensiometry at high flow rates using an interface tracking method, *Colloids and Surfaces A: Physicochemical and Engineering Aspects*, 441, 837–845, 2014.

Section IV

Specific Interfacial Physics Simulations

19 Approximate Analytical Solution via ADM and Numerical Simulation of Acoustic Cavitation
Bubble Dynamics

Mahmoud Najafi, Mohammad Taeibi Rahni,
Hamid Reza Massah, Zahra Mokhtari-Wernosfaderani,
and Mehdi Daemi

CONTENTS

19.1 INTRODUCTION

Cavitation is a phenomenon involving the formation of bubbles when a liquid locally vaporizes due to pressure variations. However, bubble formation refers to both the creation of a new bubble and expansion of preexisting bubble nuclei. Pressure variations in a liquid can be caused by acoustic waves, or by other techniques, which lead to acoustic cavitation [1]. For example, in a spherical container, one can propagate sound, such that standing waves form, trapping existing bubbles. Furthermore, when a bubble implodes under a certain condition, it experiences a brief light emission accompanied by a sound spike. This particular phenomenon is called *sonoluminescence*. In 1893, a boat that was meant to travel at 27 knots reached a lesser speed. Barnaby and Parsons [2] and Barnaby and Thornycroft [3] explained that, when pressure around the boat's blades decreased, a bubble cloud was formed around the propeller and its implosion downstream lowered the propeller performance. In 1894, Reynolds investigated cavitation and explained the reason for its occurrence [4]. Also in

1933, Marinesco and Trillat [5] applied ultrasonic waves to a photographic liquid and observed some light dots on a film. The following year, Frenzel and Schultes [6] explained that the reason for that event was due to sonoluminescence.

Normally, the physics explained previously is part of the multiphase flow problems containing large interfaces. To solve the governing equations of motion regarding such physics, there are a number of different approaches. If these equations are in their general form (i.e., the full Navier–Stokes), they do not have any closed form solutions, thus computational fluid dynamics needs to be used. To obtain closed form solutions, these equations must be simplified, leading to *exact* analytical solutions. In this work, however, an approximate analytical series solution for acoustic cavitation— bubble dynamics was derived from the Adomian decomposition method (ADM) and from a specific simplified form of the Navier–Stokes equations—Rayleigh–Plesset.

Concentrating on the bubble dynamics and kinetics in an infinite liquid with constant density, Besant, in 1895, presented the first mathematical model for bubble behavior [7]. He assumed that the pressure at infinity was constant and that the flow was ideal (i.e., incompressible and inviscid). Besant's derivation employed spherically symmetric equations of conservation of mass and momentum. Furthering Besant's work, in 1917, Rayleigh presented the most fundamental analysis of bubble behavior. He assumed a bubble full of gas and ignored surface tension, viscosity, and compressibility. In addition, he assumed that gravity was negligible, the gas mass inside the bubble was constant, and the internal vapor pressure was the vapor pressure at bulk temperature of the liquid [8]. Using Rayleigh's equation, Plesset derived the Rayleigh–Plesset equation [9] in 1949. Moreover, other investigators, such as Noltingk, Neppiras, and Poritsky [1], further developed the Rayleigh–Plesset equation. Later in 1956, Keller and Kolodner [10] presented the so-called Rayleigh–Plesset–Keller (RPK) equation, which includes the compressibility of the gas inside the bubble.

Many engineers and scientists have tried to solve the RPK equation with numerical techniques [11,12]. Nevertheless, an analytical solution would be preferred. With an analytical solution of the RPK equation, we can, for instance, predict the behavior of the bubble system. Consequently, the system of nonlinear RPK equations could be verified. In control problems, if the desired trajectory exists, it can provide the required information for predicting the system response. The desired trajectory is the trajectory defined by an analytical solution (see Chapter 20).

The present work in this part is organized as follows. Having used the modified ADM (MADM), we obtain a piecewise approximate analytical series solution for the RPK equation described later. The derivation of the RPK equation is presented in Section 19.2, while Section 19.3 is devoted to ADM plus its modification and its related piecewise approach. Section 19.4 presents some useful examples to support this approach. Then, in Section 19.5, the related results are verified, using some available experimental as well as numerical data. Finally, concluding remarks are made and the future research directions are discussed in Section 19.6.

19.2   MATHEMATICAL MODEL

In most liquids there are many small gas bubbles (i.e., bubble nuclei) that can grow or collapse under acoustic pressure fields. Such bubbles keep growing, when the tensile pressure is larger than the Blake threshold (see Section 19.3.1). On the other hand, the study of bubble behavior in acoustic fields, assuming an incompressible and inviscid flow, leads to the Rayleigh–Plesset equation. If the bubble is empty, this equation leads to a much simpler form, called the Rayleigh equation. In addition, the compressibility effect has been considered in many recent related works (e.g., Ref. 13). One of the most complete equations including this effect is the RPK equation.

In this section, the Blake critical radius and the Rayleigh–Plesset equation, including its extension to RPK, are derived in details.

19.2.1 BLAKE CRITICAL RADIUS

Assume a single bubble in an infinite liquid containing both liquid and gas vapors. Then, the interior bubble pressure (p_i) is the sum of both gas and vapor pressures, as

$$p_i = p_g + p_v. \tag{19.1}$$

Due to surface tension, the pressure inside the bubble is greater than the liquid pressure (p_L), that is,

$$p_i = p_g + p_v = p_L + p_\gamma, \tag{19.2}$$

where p_γ is the pressure related to surface tension. In Ref. 14, the surface tension related pressure is defined as

$$p_\gamma = \frac{2\gamma}{R_0}, \tag{19.3}$$

where
  R_0 is the bubble initial radius
  γ is the surface tension coefficient

Let p_{i_0} and p_{g_0} be the inner and gas pressures at equilibrium, respectively. Equilibrium occurs when the liquid pressure is equal to the liquid pressure at infinity, p_∞. One can assume that mass transfer is rapid in a way that the vapor pressure is constant with respect to the bubble radius variation. Hence, according to Equations 19.1 and 19.3, we have

$$p_{i_0} = p_{g_0} + p_v = p_\infty + \frac{2\gamma}{R_0}, \tag{19.4}$$

and, in static equilibrium, we have

$$p_{g_0} = p_0 + \frac{2\gamma}{R_0} - p_v, \tag{19.5}$$

where $p_0 = p_\infty = p_L$. If the bubble radius experiences a quasi-+static variation, with quasi-static pressure variation inside the bubble, and if the gas pressure changes, one can claim (based on polytrophic law) that

$$p_g = p_{g_0} \left(\frac{R_0}{R} \right)^{3\kappa}, \tag{19.6}$$

where κ (gas constant) is the ratio of specific heats, c_p/c_v. Substituting Equation 19.5 into Equation 19.6, we get

$$p_g = \left(p_0 + \frac{2\gamma}{R_0} - p_v \right) \left(\frac{R_0}{R} \right)^{3\kappa}. \tag{19.7}$$

Using Equations 19.2 and 19.7, the liquid pressure is derived to be

$$p_L = \left(p_0 + \frac{2\gamma}{R_0} - p_v \right) \left(\frac{R_0}{R} \right)^{3\kappa} + p_v - \frac{2\gamma}{R}. \tag{19.8}$$

To generalize the amount of pressure required to preserve the static equilibrium of a bubble, it is assumed that $2\gamma/R_0 \gg p_0$. If p_L is smaller than $p_0 + 2\gamma/R_0$, the bubble can overcome the surface tension related pressure and thereby expands. In return, the decreasing of p_L leads to the expansion of the bubble radius. Thus, a minimum value should be considered for p_L [14]. If p_L is less than this value, the bubble would expand rapidly. In other words, the bubble radius should be less than its critical maximum value, R_{crit}, while p_L is equal to its critical minimum value. Consequently, the bubble is unstable for $R > R_{crit}$. Differentiating p_L with respect to R, one can derive the critical radius, R_{crit}, as follows:

$$\frac{\partial p_L}{\partial R} = -3\kappa\left(p_0 + \frac{2\gamma}{R_0} - p_v\right)R_0^{3\kappa}R^{-(3\kappa+1)} + \frac{2\gamma}{R^2}. \tag{19.9}$$

Setting Equation 19.9 to zero, the critical radius can be found as

$$(R_{crit})^{3\kappa-1} = \frac{3\kappa}{2\gamma}\left(p_0 + \frac{2\gamma}{R_0} - p_v\right)R_0^{3\kappa}, \tag{19.10}$$

and thus

$$R_{crit} = \sqrt[3\kappa-1]{\frac{3kR_0^{3\kappa}}{2\gamma}\left(p_0 + \frac{2\gamma}{R_0} - p_v\right)}. \tag{19.11}$$

Substituting Equation 19.11 into Equation 19.8 and assuming an isothermal process ($\kappa = 1$), the critical value for p_L, known as the Blake threshold pressure, is derived to be [14]

$$p_L = \left(p_0 + \frac{2\gamma}{R_0} - p_v\right)\left(\frac{R_0}{R_{crit}}\right)^3 + p_v - \frac{2\gamma}{R_{crit}} = p_v - \frac{4\gamma}{3R_{crit}}. \tag{19.12}$$

19.2.2 Rayleigh–Plesset Equation

The Rayleigh–Plesset equation is the governing equation of a pulsating single spherical bubble in which surface tension, viscosity, and gas pressure are considered.

Note that the kinetic energy (KE) of liquid is the summation of work due to both p_∞ and p_L, that is,

$$W_\infty = KE + W_L, \tag{19.13}$$

where W_L is the work due to p_L as

$$W_L = p_L \cdot \Delta V = \frac{4}{3}\pi p_L\left(R_0^3 - R^3\right), \tag{19.14}$$

and where, ΔV is the bubble volume variations. On the other hand, the KE of liquid is equal to the summation of KE of layers with different radii within the whole liquid domain, that is,

$$KE = \frac{1}{2}\rho\int_R^\infty \dot{r}^2 \cdot 4\pi r^2\,dr = 2\pi\rho\dot{R}^2R^3. \tag{19.15}$$

The kinetic energy of liquid is equal to the work done on the liquid by pressure at infinity, p_∞. The work done by p_∞ is

$$W = p_\infty \cdot \Delta V = \frac{4\pi p_\infty}{3}\left(R_0^3 - R^3\right).$$

(19.16)

Substituting Equations 19.14, 19.15, and 19.16 into Equation 19.13 leads to

$$\frac{4\pi p_\infty}{3}\left(R_0^3 - R^3\right) = 2\pi\rho\dot{R}^2 R^3 + \frac{4}{3}\pi p_L\left(R_0^3 - R^3\right).$$

(19.17)

On the other hand, one can find

$$\frac{\partial(\dot{R}^2)}{\partial R} = 2\dot{R}\ddot{R}.$$

(19.18)

Now, differentiating Equation 19.17 with respect to R and using Equation 19.18, the Rayleigh–Plesset equation is derived as [15]

$$R\ddot{R} + \frac{3}{2}\dot{R}^2 = \frac{p_L - p_\infty}{\rho}.$$

(19.19)

In particular, if we have an empty bubble (i.e., $p_L = 0$), Equation 19.19 leads to the Rayleigh equation, as

$$R\ddot{R} + \frac{3}{2}\dot{R}^2 = \frac{-p_\infty}{\rho}.$$

(19.20)

19.2.3 Extensions of Rayleigh–Plesset Equation

Since 1949, many scientists have struggled with the governing equation of a pulsating bubble and some of them developed the Rayleigh–Plesset equation by adding useful physical terms. By neglecting vapor pressure, Noltingk and Neppiras [16] explored the liquid pressure as

$$p_L = \left(p_0 + \frac{2\gamma}{R_0}\right)\left(\frac{R_0}{R}\right)^{3\kappa} - \frac{2\gamma}{R}.$$

(19.21)

Substituting the previous equation into Equation 19.19, they obtained the following equation [16]:

$$R\ddot{R} + \frac{3}{2}\dot{R}^2 = \frac{1}{\rho}\left[\left(p_0 + \frac{2\gamma}{R_0}\right)\left(\frac{R_0}{R}\right)^{3\kappa} - \frac{2\gamma}{R} - p_\infty\right].$$

(19.22)

Later, Poritsky [17] added the viscous effect into Equation 19.22 to get

$$R\ddot{R} + \frac{3}{2}\dot{R}^2 = \frac{1}{\rho}\left[\left(p_0 + \frac{2\sigma}{R_0}\right)\left(\frac{R_0}{R}\right)^{3\kappa} - \frac{2\gamma}{R} - \frac{4\mu\dot{R}}{R} - p_\infty\right],$$

(19.23)

where μ is the liquid viscosity coefficient. Note that in Equation 19.23, viscosity is only considered in the bubble boundary conditions, that is, at the bubble surface.

Let $p_\infty = p_0 - p(t)$, wherein $p(t)$ can be a sinusoidal pressure applied by some loudspeakers, for example, $p(t) = p_a \sin(2\pi\omega t)$. The Rayleigh–Plesset equation then becomes

$$R\ddot{R} + \frac{3}{2}\dot{R}^2 = \frac{1}{\rho}\left[\left(p_0 + \frac{2\gamma}{R_0}\right)\left(\frac{R_0}{R}\right)^3 - \frac{2\gamma}{R} - \frac{4\mu\dot{R}}{R} - (p_0 - p(t))\right].$$ (19.24)

Here

ω is the frequency of the sinusoidal pressure
p_a is the ambient pressure

Assuming that the vapor pressure is constant as the radius changes from R_0 to R, Equation 19.6 becomes

$$p_g = \left(p_0 + \frac{2\gamma}{R_0} - p_v\right)\left(\frac{R_0}{R}\right)^{3\kappa}.$$ (19.25)

Therefore, we obtain

$$p_L = \left(p_0 + \frac{2\gamma}{R_0} - p_v\right)\left(\frac{R_0}{R}\right)^{3\kappa} + p_v - \frac{2\gamma}{R} - \frac{4\mu\dot{R}}{R}.$$ (19.26)

Hence, the Rayleigh–Plesset equation will finally be derived in the following form:

$$R\ddot{R} + \frac{3}{2}\dot{R}^2 = \frac{1}{\rho}\left[\left(p_0 + \frac{2\gamma}{R_0} - p_v\right)\left(\frac{R_0}{R}\right)^{3\kappa} + p_v - \frac{2\gamma}{R} - \frac{4\mu\dot{R}}{R} - (p_0 - p(t))\right].$$ (19.27)

In order to consider compressibility effects, this equation needs to be modified in a way that the sound radiation by the bubble is considered. Keller and Kolodner [10] presented a sensible means to apply the compressibility effect into the Rayleigh–Plesset equation. Their effort led to the RPK equation as

$$\left(1 - \frac{\dot{R}}{c}\right)\rho R\ddot{R} + \frac{3}{2}\dot{R}^2\rho\left(1 - \frac{\dot{R}}{3c}\right)$$

$$= \left(1 + \frac{\dot{R}}{c}\right)[p_g - (p_0 - p(t))] + p_v - \frac{2\gamma}{R} - \frac{4\mu\dot{R}}{R} + \frac{R}{c}\dot{p}_g,$$ (19.28)

where c is the speed of sound. Equations 19.19 through 19.28 are all different forms of the Rayleigh–Plesset equation, each of which contains some particular properties of the pulsating bubble problem. Recall that all these equations assume the following [1]:

1. The bubble remains spherical; the spatial condition around the bubble is uniform.
2. Body forces are negligible.
3. In comparison with the gas, the liquid's density is high.
4. The gas inside the bubble has a constant property.

Recall in this work that the process is considered isothermal, that is, $\kappa = 1$. The gas pressure inside the bubble is [5]

$$p_g = \left(p_0 + \frac{2\gamma}{R_0} - p_v \right)\left(\frac{R_0}{R} \right)^3.$$

(19.29)

To find $\dot{p}_g$ in Equation 19.28, one can differentiate Equation 19.29 with respect to time as [18]

$$\dot{p}_g = 3\left(p_0 + \frac{2\gamma}{R_0} - p_v \right)\left(\frac{R_0}{R} \right)^3\left(-\frac{\dot{R}}{R} \right).$$

(19.30)

Thus, substituting Equations 19.29 and 19.30 into Equation 19.28, we obtain [18]

$$\ddot{R} = \frac{1}{(1-(\dot{R}/c))}\left\{ \begin{array}{l} -\dfrac{3}{2}\dfrac{\dot{R}^2}{R}\left(1 - \dfrac{\dot{R}}{3c} \right) + \dfrac{(1+(\dot{R}/c))}{\rho R}\left[\left(p_0 + \dfrac{2\gamma}{R_0} - p_v \right)(R_0)^3\left(\dfrac{1}{R^3} \right) + p_v \\ -(p_0 - p(t)) \right] \\ -\dfrac{2\gamma}{\rho}\dfrac{1}{R^2} - \dfrac{4\mu}{\rho}\dfrac{\dot{R}}{R^2} + \dfrac{3}{c}\dfrac{(p_0 + (2\gamma/R_0) - p_v)(R_0)^3}{\rho}\left(\dfrac{1}{R^3} \right)\left(-\dfrac{\dot{R}}{R} \right) \end{array} \right\},$$

(19.31)

where $R(0) = R_0$ and $\dot{R}(0) = \dot{R}_0$ are the related initial conditions. To make Equation 19.31 dimensionless, let $r = R/R_0$ and $t = t/t_p$, in which t_p is the oscillation period of p_∞ (i.e., $t_p = 1/\omega$). The dimensionless form of the RPK Equation 19.31 is then [18]

$$\ddot{r} = \frac{1}{(1-(R_0/Tc)\dot{r})}\left\{ -\frac{3}{2}\frac{\dot{r}^2}{r}\left(1 - \frac{R_0}{3t_pc}\dot{r} \right) \right.$$

$$+ \left(1 + \frac{R_0}{t_pc}\dot{r} \right)\left[\left(\frac{t_p}{R_0} \right)^2\frac{(p_0 + (2\gamma/R_0) - p_v)}{\rho}\left(\frac{1}{r^4} \right) + \left(\frac{t_p}{R_0} \right)^2\frac{p_v}{\rho}\frac{1}{r} \right.$$

$$\left. - \left(\frac{t_p}{R_0} \right)^2\frac{(p_0 - p(t))}{\rho r} - \left(\frac{t_p}{R_0} \right)^2\frac{2\gamma}{\rho R_0}\frac{1}{r^2} - \left(\frac{t_p}{R_0} \right)^2\frac{4\mu}{\rho t_p}\frac{\dot{r}}{r^2} \right.$$

$$\left. + \left(\frac{t_p}{R_0} \right)\frac{3}{c}\frac{(p_0 + (2\gamma/R_0) - p_v)}{\rho}\left(\frac{1}{r^3} \right)\left(-\frac{\dot{r}}{r} \right) \right\},$$

(19.32)

where

$r(0) = 1$

$\dot{r}(0) = (t_p/R_0)R$

the dimensionless time varies from zero to one

In this work, we have used both Equations 19.32 and 19.38 to obtain the results presented in Section 19.4. In addition, the constant parameters used here are listed in Table 19.1.

TABLE 19.1

Constant Parameters of RPK Equation 19.31

Parameter	Value	Unit
ρ	1000	kg/m^3
ω	26.5	kHz
γ	0.073	N/m
μ	10^{-3}	Pa s
p_v	4.24×10^3	Pa
p_0	1	atm
p_a	1.2	atm
$\dot{R}_0$	0	m/s
R_0	4.5	μm
c	1481	m/s

19.3   ADOMIAN DECOMPOSITION METHOD

Consider an ordinary differential equation of the general form $Fu = g(t)$, where F is a nonlinear operator (e.g., a nonlinear differential operator). This equation can be rewritten as

$$Lu + Ru + Nu = g(t), \tag{19.33}$$

where
  L is the linear invertible operator
  N is the nonlinear part
  R is the remaining part

If L^{-1} is the inverse of L, we can have

$$L^{-1}Lu = -L^{-1}Ru - L^{-1}Nu + L^{-1}g(t). \tag{19.34}$$

Note, that here, a linear operator L is a second order differential (i.e., $L = \mathrm{d}^2/\mathrm{d}t^2$) and hence L^{-1} is a double integral:$\iint(.)\mathrm{d}t\,\mathrm{d}t$. Accordingly, Equation 19.34 can be written as [19]

$$u = \underbrace{u(0) + tu'(0) + L^{-1}g(t)}_{u_0} - L^{-1}Ru - L^{-1}Nu. \tag{19.35}$$

For simplicity, let $u_0 = u(0) + tu'(0) + L^{-1}g(t)$ and propose the solution to be $u = \sum_{n=0}^{\infty} u_n$. Note that Nu can be decomposed into an infinite series of Adomian polynomials as [20]

$$Nu = \sum_{n=0}^{\infty} A_n(u_0, u_1, \ldots, u_n), \tag{19.36}$$

where A_n's are the Adomian polynomials depending only on $u_0, u_1, \ldots, u_n$ and are given as [21]

$$A_n(u_0, u_1, \ldots, u_n) = \frac{1}{n!}\frac{\mathrm{d}^n}{\mathrm{d}\lambda^n}\left(N\sum_{k=0}^{\infty} u_k \lambda^k\right)\Bigg|_{\lambda=0}. \tag{19.37}$$

Another approach to Adomian polynomials was proposed by Wazwaz in 2000 [22]. He used the series: $u = \sum_{n=0}^{\infty} u_n$ and assumed that (1) A_n depends only on $u_0, u_1, \ldots, u_n$ and (2) the summation of indices of u_n in A_n is equal to n [23].

19.3.1 Collapse of an Empty Spherical Bubble

An equivalent form of Equation 19.20 is

$$R\ddot{R} + \frac{3}{2}\dot{R}^2 = -\frac{p_0}{\rho},$$
$$\begin{cases} R(0) = R_0, \\ \dot{R}(0) = 0. \end{cases}$$
(19.38)

Assuming the bubble mentioned in the previous condition collapses at time t_c, normalizing the radius as $r \equiv R/R_0$, and time as $t \equiv t/t_c$, Equation 19.38 becomes

$$r\ddot{r} + \frac{3}{2}\dot{r}^2 = -\frac{t_c^2 p_0}{\rho R_0^2} = -\xi^2,$$

$$\xi = \frac{t_c}{R_0}\sqrt{\frac{p_0}{\rho}},$$
(19.39)

$$r(0) = 1, \quad \dot{r}(0) = 0,$$

where $\xi = 0.914681$ is a universal constant called the Rayleigh factor [12]. Equation 19.39 does not have a closed form solution, but it does have an approximate analytical solution as [12]

$$r_0(t) = (1 - t^2)^{2/5}.$$
(19.40)

On the other hand, substituting r with u, the solution of Equation 19.39 via ADM to the 30th term is

$$u_0 = 1.000000000000000,$$
$$u_1 = -0.418320665880500\,t^2,$$
$$u_2 = -0.116661453001803\,t^4,$$
$$u_3 = -0.0618157358229147\,t^6,$$
$$\vdots$$
$$u_{30} = -1.46377481906057 \times 10^{-12}\,t^{60}.$$
(19.41)

Finally, the analytical approximation of Equation 19.38 is found to be

$$r = \sum_{n=0}^{\infty} u_n,$$
(19.42)

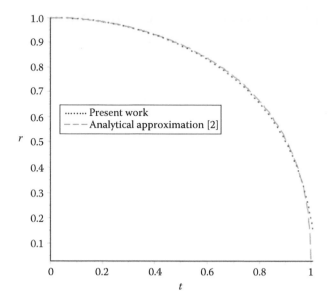

FIGURE 19.1 Dimensionless bubble radius versus dimensionless time.

where, $u_n = -\alpha t^{2n}$ and $\alpha > 0$. In Figure 19.1, the solutions using Equations 19.40 and 19.41 are compared to confirm the accuracy of our ADM results.

19.3.2 PIECEWISE ADOMIAN DECOMPOSITION METHOD

The conventional ADM and its modifications [24] normally solve the equations from time $t = 0$ to $t = \infty$. Due to the difficulties of nonlinearity, as well as the singularity in the RPK equation, the time interval can be divided into fine subdomains. This way, the final solution is a piecewise function, where the analytical solution in each subdomain is computed by the modified ADM [25]. Clearly, the radius and the velocity at the end of each subdomain are the initial conditions for the next one.

19.4 PROBLEM SOLUTION AND ANALYSIS

In this section, Equation 19.43 is considered and will be solved in subdomains $[t_k, t_{k+1}]$, where $t_{k+1} = t_k + \Delta t$ with initial conditions:

$$\begin{cases} r(t_0) = r_0 = 1, \\ \dot{r}(t_0) = \dot{r}_0 = \dfrac{t_p}{R_0} \dot{R}_0. \end{cases} \tag{19.43}$$

Remark 1: Since we are dealing with a nonlinear problem with multiple time scales [26], especially because the bubble behavior becomes very stiff at the imploding instant (at the singularity r→0), Δt needs to be changed with time. The RPK equation 19.32 is of the form

$$Lr = Nr. \tag{19.44}$$

In order to define linear and nonlinear parts in Equation 19.32, note that $L = \dfrac{d^2}{dt^2}(\cdot)$. Thus, we have

$$
\begin{aligned}
Nr = \frac{1}{(1-(R_0/t_pc)\dot{r})} &\left\{ -\frac{3}{2}\frac{\dot{r}^2}{r}\left(1-\frac{R_0}{3t_pc}\dot{r}\right) \right. \\
&+ \left(1+\frac{R_0}{t_pc}\dot{r}\right)\left[\left(\frac{t_p}{R_0}\right)^2\frac{(p_0+(2\gamma/R_0)-p_v)}{\rho}\left(\frac{1}{r^4}\right)+\left(\frac{t_p}{R_0}\right)^2\frac{p_v}{\rho}\frac{1}{r}\right. \\
&\left. -\left(\frac{t_p}{R_0}\right)^2\frac{(p_0-p(t))}{\rho r}-\left(\frac{t_p}{R_0}\right)^2\frac{2\gamma}{\rho R_0}\frac{1}{r^2}-\left(\frac{t_p}{R_0}\right)^2\frac{4\mu}{\rho T}\frac{\dot{r}}{r^2}\right] \\
&\left. +\left(\frac{t_p}{R_0}\right)\frac{3}{c}\frac{(p_0+(2\gamma/R_0)-p_v)}{\rho}\left(\frac{1}{r^3}\right)\left(-\frac{\dot{r}}{r}\right)\right\}.
\end{aligned}
\tag{19.45}
$$

Remark 2: Since we are using a piecewise method, $p(t)$ is assumed to be a step function with subdomains defined previously. As mentioned before, the values of the step function in each subdomain $[t_k, t_{k+1}]$ is considered to be $p(t) = p_a\sin(2\pi t_{k+1})$. According to the modified ADM, applying the inverse operator $L^{-1} = \iint(\cdot)\,dt\,dt$ to Equation 19.44, the series solution $r = \sum_{n=0}^{\infty}f_n$ will be derived as

$$
r = \sum_{n=0}^{\infty}c_{1n} + t\sum_{n=0}^{\infty}c_{2n} + L^{-1}\left(\sum_{n=0}^{\infty}A_n\right) = \sum_{n=0}^{\infty}f_n,
\tag{19.46}
$$

where

$$
f_0 = c_{1_0} + c_{2_0}t,
$$

and where

$$
f_{n+1} = c_{1_{n+1}} + c_{2_{n+1}}t + \iint A_n\,dt\,dt.
\tag{19.47}
$$

19.4.1 ADOMIAN POLYNOMIALS

Letting $f = r = \sum_{n=0}^{\infty}f_n$ and $f' = \dot{r}$, Equation 19.45 can be rewritten as

$$
\begin{aligned}
N(f) = \frac{1}{(1-(R_0/Tc)f')} &\left\{ -\frac{3}{2}\frac{f'^2}{f}\left(1-\frac{R_0}{3t_pc}f'\right) \right. \\
&+ \left(1+\frac{R_0}{t_pc}f'\right)\left[\left(\frac{t_p}{R_0}\right)^2\frac{(p_0+(2\gamma/R_0)-p_v)}{\rho}\left(\frac{1}{f^4}\right)+\left(\frac{t_p}{R_0}\right)^2\frac{p_v}{\rho}\frac{1}{f}\right. \\
&\left. -\left(\frac{t_p}{R_0}\right)^2\frac{(p_0-p(t))}{\rho f}-\left(\frac{t_p}{R_0}\right)^2\frac{2\gamma}{\rho R_0}\frac{1}{f^2}-\left(\frac{t_p}{R_0}\right)^2\frac{4\mu}{\rho t_p}\frac{f'}{f^2}\right] \\
&\left. -\left(\frac{t_p}{R_0}\right)\frac{3}{c}\frac{(p_0+(2\gamma/R_0)-p_v)}{\rho}\left(\frac{f'}{f^4}\right)\right\}.
\end{aligned}
\tag{19.48}
$$

The nonlinear part of Equation 19.48 contains many terms and computing the Adomian polynomials is extremely cumbersome. However, we can decompose it into smaller nonlinear parts and add them at the end. Assuming that the decomposed nonlinear parts are

$$N_{1_1}(f) = f'^2,$$

$$N_{1_2}(f) = \frac{1}{f},$$

$$N_{1_3}(f) = \left(1 - \frac{R_0}{3t_p c} f'\right),$$

$$N_1 = -\frac{3}{2} N_{1_1} N_{1_2} N_{1_3},$$

$$N_2 = \left(\frac{t_p}{R_0}\right)^2 \frac{(p_0 + (2\gamma/R_0) - p_v)}{\rho} \frac{1}{f^4} = \left(\frac{t_p}{R_0}\right)^2 \frac{(p_0 + (2\gamma/R_0) - p_v)}{\rho} (N_{1_2})^4,$$

$$N_3 = \left(\frac{t_p}{R_0}\right)^2 \frac{p_v}{\rho} \frac{1}{f} = \left(\frac{t_p}{R_0}\right)^2 \frac{p_v}{\rho} N_{1_2},$$

$$N_4(f) = \left(\frac{t_p}{R_0}\right)^2 \frac{(p_0 - p_a \sin(2\pi t))}{\rho} N_{1_2},$$

$$N_5 = \left(1 + \frac{R_0}{t_p c} f'\right),$$

$$N_6(f) = N_5 * (N_2 + N_3 + N_4),$$

$$N_7 = -\left(\frac{t_p}{R_0}\right)^2 \frac{2\gamma}{\rho R_0} \frac{1}{f^2} = -\left(\frac{t_p}{R_0}\right)^2 \frac{2\gamma}{\rho R_0} N_{1_2}^2,$$

$$N_{8_1}(f) = f',$$

$$N_{8_2}(f) = \frac{1}{f^2} = N_{1_2}^2,$$

$$N_8 = -\left(\frac{t_p}{R_0}\right)^2 \frac{4\mu}{\rho t_p} \frac{f'}{f^2} = -\left(\frac{t_p}{R_0}\right)^2 \frac{4\mu}{\rho t_p} N_{8_1} N_{8_2},$$

$$N_{9_1}(f) = f',$$

$$N_{9_2}(f) = \frac{1}{f^4} = (N_{1_2})^4,$$

$$N_9 = -\left(\frac{t_p}{R_0}\right)^3 \frac{(p_0 + (2\gamma/R_0) - p_v)}{\rho} N_{9_1} N_{9_2},$$

$$N_{10} = \frac{1}{(1 - (R_0/t_p c)f')}.$$

(19.49)

The total nonlinear equation is

$$N(f) = N_{10}(N_1 + N_6 + N_7 + N_8 + N_9). \tag{19.50}$$

To see the analytical trend, note that the Adomian polynomials for N_{1_1} are as follows:

$$A_{1_{10}} = f_0'^2; \quad A_{1_{11}} = 2f_0'f_1'; \quad A_{1_{12}} = 2f_0'f_2' + f_1'^2; \quad A_{1_{13}} = 2f_0'f_3' + 2f_1'f_2';\dots \tag{19.51}$$

While, for N_{1_2} we get

$$A_{1_{20}} = \frac{1}{f_0}; \quad A_{1_{21}} = -\frac{f_1}{f_0^2}; \quad A_{1_{22}} = -\frac{f_2}{f_0^2} + \frac{f_1^2}{f_0^3};$$

$$A_{1_{23}} = -\frac{f_3}{f_0^2} + 2\frac{f_1 f_2}{f_0^3} - \frac{f_1^3}{f_0^4};\dots \tag{19.52}$$

Similarly, for N_{1_3} we have

$$A_{1_{30}} = 1 - \frac{R_0}{3t_p c} f_0'; \quad A_{1_{31}} = f_1'\left(-\frac{R_0}{3t_p c}\right); \quad A_{1_{32}} = f_2'\left(-\frac{R_0}{3t_p c}\right);\dots \tag{19.53}$$

Hence, according to Equation 19.37 we get

$$\sum_{n=0}^{\infty} A_{1_n} = -\frac{3}{2}(A_{1_{10}} + A_{1_{11}} + A_{1_{12}} + A_{1_{13}} + \cdots)(A_{1_{20}} + A_{1_{21}} + A_{1_{22}} + A_{1_{23}} + \cdots)(A_{1_{30}} + A_{1_{31}} + A_{1_{32}} + A_{1_{33}} + \cdots),$$
$$\tag{19.54}$$

As a result, the Adomian polynomials for N_1 are of the form

$$A_{1_0} = -\frac{3}{2} A_{1_{10}} A_{1_{20}} A_{1_{30}},$$

$$A_{1_1} = -\frac{3}{2}(A_{1_{11}} A_{1_{20}} A_{1_{30}} + A_{1_{10}} A_{1_{21}} A_{1_{30}} + A_{1_{10}} A_{1_{20}} A_{1_{31}}),$$

$$A_{1_2} = -\frac{3}{2}(A_{1_{10}} A_{1_{20}} A_{1_{32}} + A_{1_{10}} A_{1_{21}} A_{1_{31}} + A_{1_{10}} A_{1_{22}} A_{1_{30}} + A_{1_{11}} A_{1_{20}} A_{1_{31}}$$

$$+ A_{1_{11}} A_{1_{21}} A_{1_{30}} + A_{1_{12}} A_{1_{20}} A_{1_{30}}),$$

$$A_{1_3} = -\frac{3}{2}(A_{1_{10}} A_{1_{20}} A_{1_{33}} + A_{1_{10}} A_{1_{21}} A_{1_{32}} + A_{1_{10}} A_{1_{22}} A_{1_{31}} + A_{1_{10}} A_{1_{23}} A_{1_{30}}$$

$$+ A_{1_{11}} A_{1_{20}} A_{1_{32}} + A_{1_{11}} A_{1_{21}} A_{1_{31}} + A_{1_{11}} A_{1_{22}} A_{1_{30}} + A_{1_{12}} A_{1_{20}} A_{1_{31}} + A_{1_{12}} A_{1_{21}} A_{1_{30}} + A_{1_{13}} A_{1_{20}} A_{1_{30}}). \tag{19.55}$$

Through tedious calculations, Adomian polynomials for other nonlinear parts can be obtained. Then, one can find the series $\sum_{n=0}^{\infty} A_{1_n}, \sum_{n=0}^{\infty} A_{2_n}, \dots, \sum_{n=0}^{\infty} A_{10_n}$. Consequently, using Equation 19.50, the Adomian polynomials for $N(f)$ can be calculated as

$$N(f) = \sum_{n=0}^{\infty} A_n = \left(\sum_{n=0}^{\infty} A_{10_n} \right)\left(\sum_{n=0}^{\infty} (A_{1_n} + A_{6_n} + A_{7_n} + A_{8_n} + A_{9_n}) \right). \tag{19.56}$$

Accordingly, the Adomian polynomials are obtained as follows:

$$A_n = A_{10_n}(A_{1_0} + A_{6_0} + A_{7_0} + A_{8_0} + A_{9_0}) + A_{10_{n-1}}(A_{1_1} + A_{6_1} + A_{7_1} + A_{8_1} + A_{9_1})$$
$$+ \cdots + A_{10_0}(A_{1_n} + A_{6_n} + A_{7_n} + A_{8_n} + A_{9_n}). \tag{19.57}$$

Having calculated A_n, one needs to substitute it into Equation 19.47 and find f_{n+1}. Then, considering $\phi(n+2) = \sum_{m=0}^{n+1} f_m$ and according to initial conditions (Equation 19.43), the system of equations that follow is a linear system of equations with two unknowns $c_{1_{n+1}}$ and $c_{2_{n+1}}$, which can be solved easily:

$$\begin{cases} \phi(n+2)\big|_{t=t_0} = r_0 \\ \dfrac{d\phi(n+2)}{dt}\bigg|_{t=t_0} = \dot{r}_0 \end{cases} \tag{19.58}$$

Remark 3. As mentioned before, in the piecewise method just discussed, considering the subdomain $[t_k, t_{k+1}]$, the initial condition should be defined with respect to the results of its previous subdomain. Hence, the system of Equation 19.58 can be rewritten as

$$\begin{cases} \phi(n+2)\big|_{t=t_k} = r_{t_k} \\ \dfrac{d\phi(n+2)}{dt}\bigg|_{t=t_k} = \dot{r}_{t_k} \end{cases} \tag{19.59}$$

The algorithm for the piecewise modified ADM is as follows:

Input: $p_0, \rho, \omega, \gamma, \mu, p_v, R_0, \dot{R}_0, p_a, \Delta t.$

Output: A piecewise series solution of the Rayleigh–Plesset–Keller Equation 19.32.

1) *Let $k:=1; t_k := 0$*
2) *for $k=1$ to p: (p is the number of piecewise functions, such that $t_p = 1.0$)*
   2.1) *Let $f_0 = c_{1_0} + c_{2_0}t$ and $\phi(1) = f_0$*
   2.2) *$c_{1_0} = r_{k_0}$ and $c_{2_0} = \dot{r}_{k_0}$*
   2.3) *for $n=1$ to q; (q is the number of f_ns which are considered to form*
      $\phi(q+1) = \sum_{n=0}^{q} f_n$)
      2.3.1) *compute A_{n-1}*

2.3.2) *calculate* $f_n = c_{1_n} + c_{2_n} t + \iint A_{n-1} dt\,dt$

2.3.3) *let* $\phi(n+1) = \sum_{m=0}^{n} f_m$

2.3.4) *solve the system* $\begin{cases} \phi(n+1)\big|_{t=t_k} = r_{t_k}, \\ \dfrac{d\phi(n+1)}{dt}\bigg|_{t=t_k} = \dot{r}_{t_k} \end{cases}$ *to find* c_{1_n} *and* c_{2_n},

2.4) $k = k + 1$ *and* $t_k = t_k + \Delta t$

3) *Print the piecewise function which is of the form* $\phi = \begin{cases} \phi(1) & t_0 \le t < t_1 \\ \phi(2) & t_1 \le t < t_2 \\ \vdots \\ \phi(p) & t_{p-1} \le t < t_p \end{cases}$

19.5 RESULTS AND DISCUSSION

Having solved the dimensionless RPK equation (Equation 19.32), using the piecewise modified ADM, a piecewise function of the following form is introduced:

$$\phi = \begin{cases} \phi(1), & t_0 \le t < t_1, \\ \phi(2), & t_1 \le t < t_2, \\ \vdots \\ \phi(p), & t_{p-1} \le t < t_p. \end{cases} \tag{19.60}$$

By computing the qth order of Adomian polynomials, the analytical solution derived using the modified ADM, namely $\phi(i)$, $t_{i-1} \le t < t_i$ in Equation 19.60, is of the form

$$\phi(i) = \left(a_{i_0} + a_{i_1} X + a_{i_2} X^2 + a_{i_3} X^3 + \cdots + a_{i_{2q+1}} X^{2q+1} \right)$$

$$\times \left(b_{i_0} + b_{i_1} (\ln X) + b_{i_2} (\ln X)^2 + b_{i_3} (\ln X)^3 + \cdots + b_{i_q} (\ln X)^q \right), \tag{19.61}$$

where $X = (e_{i_1} + e_{i_2} t)$. For each $i(1 \le i \le p)$, the constants K25613 are different, but the form of Equation 19.61 is the same. The piecewise function of Equation 19.60 is plotted in Figure 19.2, where the behavior of the bubble, its expansion, collapse, implosion, and rebound are shown.

Equation 19.32 is solved here by applying an eighth order Runge–Kutta–Fehlberg 78 (RKF78) method. Comparison of the modified ADM with RKF78 method is shown in Figure 19.3. According to this figure, these two solutions are considerably close (a difference of about 10^{-3}), which shows the trustworthiness of the piecewise modified ADM for such problems.

The dimensional version of Figure 19.2 is redrawn in Figure 19.4, illustrating that our results are similar.

Finally, Figure 19.5 shows the locus of all local maximum bubble radii with respect to time, showing an exponential decay. The reason for this is the fact that the first implosion leads to the loss of most of the system energy in different ways (e.g., sound, heat, and light energies).

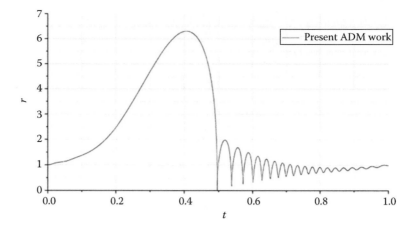

FIGURE 19.2   Dimensionless solution of RPK Equation 19.32, via piecewise modified ADM.

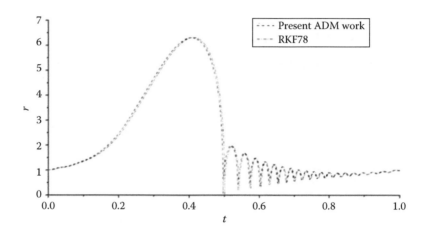

FIGURE 19.3   Comparison of the solution of RPK Equation 19.32 via piecewise modified ADM and RKF78.

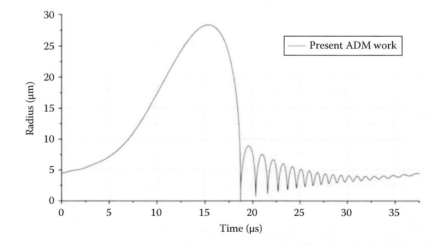

FIGURE 19.4   Dimensional solution of RPK Equation 19.31, via piecewise modified ADM.

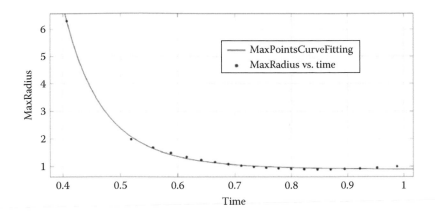

FIGURE 19.5   The exponential trend of maximum points of RPK equation.

19.6   CONCLUSION AND FUTURE WORKS

In this work, our major task was developing an approximate analytical solution of acoustic cavitations—bubble dynamics. The modified ADM (MADM) was utilized and the approximate analytical series solution was obtained. It was shown that the solution agrees considerably well with our numerical results, that is, RKF78 solution. In addition, our results showed that the locus of all local maximum bubble radii with respect to time has an exponential decay. Also that the first bubble implosion leads to the loss of most of the system energy as sound, heat, and light (i.e., a sonoluminescence process). This important observation needs to be worked on as one of the extensions of this research in the future. Also, since it was assumed that the entire process was isothermal (i.e., $\kappa = 1$), further investigations can be performed for an adiabatic process, in which expansion and modification of the proposed algorithm is necessary.

REFERENCES

1. F.R. Young, *Cavitation*, Imperial College Press, London, UK, 1999.
2. S.W. Barnaby and C.W. Parsons, *Trans. Inst. Naval Arch.*, 38, 232, 1897.
3. S.W. Barnaby and J. Thornycroft, *Proc. Inst. Civil Eng.*, 122, 57, 1895.
4. O. Reynolds, *Philos. Mag.*, 20, 578, 1894.
5. M. Marinesco and J.J. Trillat, *Comp. Trend*, 196, 858, 1933.
6. J. Frenzel and H. Schultes, *J. Phys. C*, 27, 421, 1934.
7. F.G. Hammitt, ONR Report No. UMICH 014456-7-1, 1977.
8. J.S. Rayleigh, *Philos. Mag.*, Series 6, 34, 94, 1917.
9. M.S. Plesset, *Appl. Mech.*, 16, 277, 1949.
10. J.B. Keller and I.I. Kolodner, *J. Appl. Phys.*, 27, 1152, 1956.
11. P.M. Bernner, S. Hilgenfeldt, and D. Lohse, *Rev. Modern Phys.*, 74, 425, 2002.
12. D. Obreschkow, M. Bruderer, and M. Farhat, *Phys. Rev. E*, 85, 066303, 2012.
13. Q.X. Wang and J.R. Blake, *J. Fluid Mech.*, 659, 191, 2010.
14. C.E. Brennen, *Cavitation and Bubble Dynamics*, Oxford University Press, London, U.K., 1995.
15. G. Vacca, Ultrafast optical studies of single-bubble sonoluminescence, PhD dissertation, Department of Applied Physics, Stanford University, Stanford, CA, 2001.
16. B.E. Noltingk and E.A. Neppiras, *Proc. Phys Soc. B*, 63B, 674, 1950.
17. H. Poritsky, *J. Appl. Mech.—Trans. ASME*, 18(3), 332, 1951.
18. Z. Mokhtari-Wernosfaderani, On lattice Boltzmann study for solving a nonlinear ODE, MSc thesis, Department of Applied Mathematics, Shahed University, Tehran, Iran, 2012.
19. M. Najafi, *Math. Comput. Model.*, 49, 46–54, 2009.
20. M. Najafi, *J. Vibrat. Contr.*, 15(2), 283, 2009.

21. G. Adomian, *Solving Frontier Problems of Physics: The Decomposition Method*, Kluwer Academic Press, New York, 1994.
22. A.M. Wazwaz, *Appl. Math. Comput.*, 111, 33, 2000.
23. F. Fekri, Investigation of the Navier-Stokes equations for an incompressible flow over a flat plate based on Adomian decomposition method, MSc thesis, Department of Applied Mathematics, Shahed University, Tehran, Iran, 2009.
24. A.M. Wazwaz and S.M. El-Sayed, *Appl. Math. Comput.*, 122, 393, 2001.
25. A.A. Mobasher, A.P. Biazar, and Z. Deng, *Int. J. Modern Eng.*, 13(1), 13, 2012.
26. B.P. Barber, P. Robert, A. Hiller, R. Lofstedt, S.J. Putterman, and K.R. Weninger, *Phys. Rep.*, 281(2), 65, 1997.

20 Numerical Solution, Stability, and Control of Acoustic Cavitation-Bubble Dynamics

Mohammad Taghi Hamidi Beheshti, Mahmoud Najafi,
Masoumeh Azadegan, and Mohammad Taeibi Rahni

CONTENTS

20.1 INTRODUCTION

In this chapter, we are interested in applying control techniques to a single spherical bubble modeled by Rayleigh–Plesset (RP) equation and represented by differential equations of the state-space form. By this representation, different control objectives, such as regulation, tracking, robustness, and optimization can be achieved via related control approaches. The results presented here seem to be the first to use control techniques for stability analysis and control of a single spherical bubble.

Cavitation phenomenon is the formation of bubbles when a liquid locally vaporizes due to pressure variations. Bubble formation however refers to both creation of a new bubble and expansion of preexisting bubble nuclei [1]. Of interest to a wide range of engineers, cavitation and bubbly flows are applicable to topics ranging from damages in hydroelectric equipment, ship propellers, and internal combustion engines to the performance of turbines and pumps [2]. The extremely high temperatures and pressures which can occur in noncondensable gases during collapse are believed to be responsible for the phenomenon known as luminescence, the emission of light observed during cavitation-bubble collapse. When it occurs in the context of acoustic cavitation, it is called sonoluminescence (SL) [2]. Luminescence seems to be of significant interest in utilizing the chemical-processing potential of high temperatures and pressures.

For example, it is possible to use cavitation to break up harmful molecules in water [3,4]. Harris et al. (2014) proposed the activation of photosensitizers (PSs) by SL. PS are molecules activated by light and can be preferentially taken up by cancer and microbial cells. They have become well-established therapeutic agents used in photodynamic therapy to kill cancer cells. During insonation, presence of bubbles can play an important role, creating strong microstreaming effects in solution and in more dramatic circumstances, leading to the formation of energetic microjets [5], plasmas [6], and the production of other highly reactive species [7].

A number of research efforts are focusing on the dynamics of two-phase flows, but few have concentrated on applying control techniques [8,9]. In Ref. [8], the governing two-phase two-component Darcy flow PDE system is described and the related optimal control problem has been formulated. In addition, a dynamical model for a vapor compression cycle has been established in Ref. [9], wherein the model is linearized at a certain operating point and the optimal control strategy is proposed.

Having considered the initial formation of a bubble and its important applications, in this chapter we have examined the behavior of a single bubble in an infinite liquid domain at rest (and with uniform temperature far from the bubble). Most previous works on bubble dynamics were based on experimental results [10–12]. Recently, some concepts of control have been implemented in this field [13]. For the first time, stochastic collocation approach was employed for analyzing uncertainties in shock-bubble interactions by simulations [13]. In that work, the influence of the uncertain initial conditions on the output quantities of interest is quantified and studied. Hegedus (2014) proved the existence of stable bubble motion, based on numerical techniques of modern nonlinear bifurcation theory and achieved stable bubble oscillations beyond Blake's critical threshold (see Chapter 19). Here, we concentrate on the regulation problem, which is a very important aspect of modern technology due to efficiency, quality control, safety, and reliability. Industrial processes require regulation in order to guarantee that the key variables (temperatures, pressures, velocity, etc.) are kept at nominal values. The problem of regulation in control theory is to devise strategies in order to keep the control variables at constant values against external disturbances acting on the plant under regulation and/or against parameter uncertainties [14]. Note, even though the bubble model considered in this chapter is the system of RP equation, the presented techniques can be easily extended to more complex bubble dynamics applications.

This chapter is organized as follows: in Section 20.1, bubble dynamics have been introduced and analyzed through simulations. Section 20.2 is devoted to the state-space representation of RP equation, while Section 20.3 deals with controllability and observability (classical concepts in control theory). Then, the study of the stability due to certain conditions is discussed in Section 20.4. We derive the linearized form of the system and give the classical stability conditions in terms of the eigenvalues of the system matrix. We also discuss the stability of the system via Lyapunov method. Then, the stabilizing controller is derived in terms of linear matrix inequality (LMI) and, consequently, the controller gain will be provided. Section 20.5 is concerned with control of the nonlinear system. Finally, two nonlinear feedback controls, including feedback linearization (FBL) and sliding mode control are designed and applied to the system to ensure stability and convergence of the desired performance. Finally, Section 20.6 concludes this chapter and points out future research directions.

20.2 NUMERICAL SOLUTION

In this section, bubble dynamics has been analyzed through simulations. We consider RP equation for describing a single spherical bubble dynamics, which is a second order and nonlinear ordinary differential equation given as [15–19]

$$R\ddot{R} + \frac{3}{2}\dot{R}^2 = \frac{1}{\rho}\left[\left(P_0 + \frac{2\gamma}{R_0} - P_v\right)\left(\frac{R_0}{R}\right)^3 + P_v - \frac{2\gamma}{R} - 4\mu\frac{\dot{R}}{R} - P_0 - P(t)\right], \qquad (20.1)$$

where

R is the radius of the bubble

ρ, γ, and μ are the density of liquid, surface tension coefficient, and liquid viscosity, respectively

As in Chapter 19, an isothermal process has been considered ($\kappa = 1$). The left-hand side of Equation 20.1 consists of dynamical pressure terms. Note, P_v is the vapor pressure, P_0 is the atmospheric pressure, and $P(t)$ can be an ultrasound driving pressure, modeled as a spatially homogeneous, standing sound wave, that is, $P(t) = -P_a\cos(\omega t)$, where P_a is the ambient pressure and ω is the fixed frequency. In addition, $P_\infty(t) = P_0 + P(t)$ is assumed to be a control input, which regulates the growth or the collapse of the bubble. The first term on the right-hand side of Equation 20.1 represents the gas pressure inside the bubble and the other terms containing γ and μ model the influence of surface tension. Also, R_0 is the radius of a static (unforced) bubble, neglecting the effect of gas diffusion.

The nonlinear model is simulated, using the parameters given in Table 19.1, which are common values in many experiments [18,19]. Figure 20.1 shows that the radius of the bubble, $R(t)$, increases with time in the presence of dynamical pressure terms and gas pressure, P_{gas}. The influence of P_{sur}, P_{vis}, P_v, P_0, and $P_\infty(t)$ on the bubble's behavior has been shown in Figures 20.2, 20.3, 20.4, 20.5, and 20.6, respectively. Considering the simplified form of the RP equation, that is, $R\ddot{R} + \dfrac{3}{2}\dot{R}^2 = \dfrac{1}{\rho}\left(P_0 + \dfrac{2\gamma}{R_0}\right)\left(\dfrac{R_0}{R}\right)^3$, one can observe the bubble spherical deformation with time in Figure 20.1.

Figure 20.2 shows the bubble dynamics after including surface pressure. This figure illustrates that surface tension has a stabilizing effect on the bubble surface. In the absence of dissipation (such as viscosity), the rebounds would continue indefinitely without attenuation [17]. Thus, liquid viscosity has an important contribution to the damping of bubble oscillations (as shown in Refs. [18,19]).

One can observe in Figure 20.3 that the amplitude of oscillations of the bubble decreases over time due to viscosity. The influence of vapor pressure in increasing the amplitude of oscillations and its steady state values are shown using dashed line in Figure 20.3.

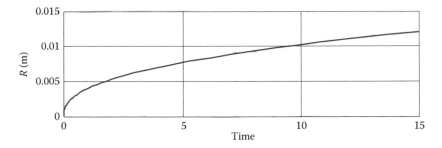

FIGURE 20.1 Bubble dynamics resulting from equation: $R\ddot{R} + \dfrac{3}{2}\dot{R}^2 = \dfrac{1}{\rho}\left(P_0 + \dfrac{2\gamma}{R_0}\right)\left(\dfrac{R_0}{R}\right)^3$.

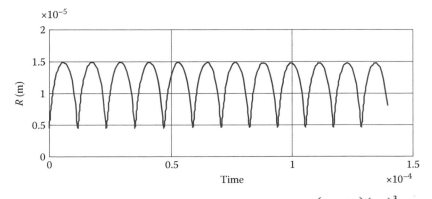

FIGURE 20.2 Bubble dynamics resulting from equation: $R\ddot{R} + \dfrac{3}{2}\dot{R}^2 = \dfrac{1}{\rho}\left(P_0 + \dfrac{2\gamma}{R_0}\right)\left(\dfrac{R_0}{R}\right)^3 - \dfrac{2\gamma}{R}$.

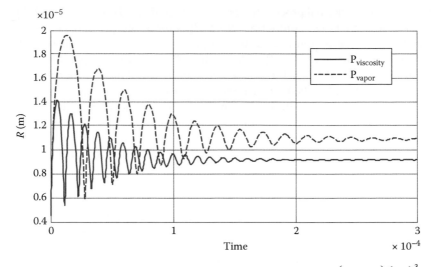

FIGURE 20.3 Bubble dynamics resulting from equation: $R\ddot{R} + \dfrac{3}{2}\dot{R}^2 = \dfrac{1}{\rho}\left(P_0 + \dfrac{2\gamma}{R_0}\right)\left(\dfrac{R_0}{R}\right)^3 - \dfrac{2\gamma}{R} - 4\mu\dfrac{\dot{R}}{R}$

(solid line) and $R\ddot{R} + \dfrac{3}{2}\dot{R}^2 = \dfrac{1}{\rho}\left(P_0 + \dfrac{2\gamma}{R_0} - P_v\right)\left(\dfrac{R_0}{R}\right)^3 - \dfrac{2\gamma}{R} + P_v - 4\mu\dfrac{\dot{R}}{R}$ (dashed line).

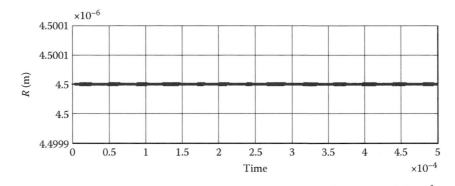

FIGURE 20.4 Bubble dynamics resulting from equation: $R\ddot{R} + \dfrac{3}{2}\dot{R}^2 = \dfrac{1}{\rho}\left(P_0 + \dfrac{2\gamma}{R_0} - P_v\right)\left(\dfrac{R_0}{R}\right)^3 +$ $P_v - \dfrac{2\gamma}{R} - 4\mu\dfrac{\dot{R}}{R} - P_0.$

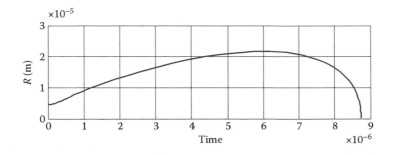

FIGURE 20.5 Bubble dynamics resulting from RP Equation 20.1.

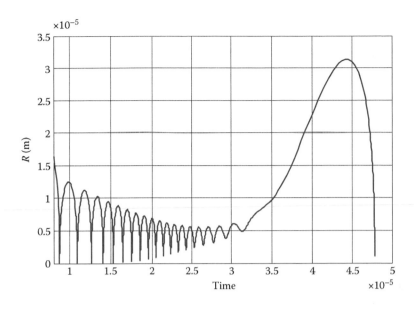

FIGURE 20.6 Bubble dynamics after the first collapse.

Next, the constant ambient pressure (P_0) has been included and the related result is shown in Figure 20.4. As expected, the radius of the bubble stays in its initial value, R_0, based on the definition of the equilibrium point (see Section 20.3).

Finally, the pressure produced by a sound wave and the bubble dynamics resulting from RP Equation 20.1 is shown in Figure 20.5. This figure demonstrates the slow expansion of the bubble's radius. As time approaches $8.73 * 10^{-6}$ s, the derivative of the radius (the bubble interface velocity) becomes undefined and consequently the first burst occurs strongly (note, the dimension of time is second throughout the paper, unless stated otherwise). The behavior of the bubble after this time is shown in Figure 20.6.

Two other cycles and external pressure $P_\infty(t)$ are shown in Figure 20.7. As shown in this figure, the maximum radius of the bubble occurs after the minimum pressure is reached. Then, after a rapid

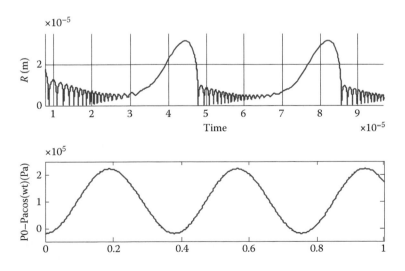

FIGURE 20.7 Two cycles of bubble dynamics and pressure $P_\infty(t)$.

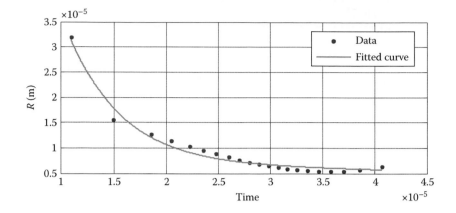

FIGURE 20.8 Curve fitting of the points in which $\dot{R} = 0$.

burst, a series of side lobes occur. We have indefinite cycles with the same behavior at the same period of external pressure, which is called stable SL [16]. The results obtained can be compared with those in Chapter 19 and also in Ref. [16].

The points in which bubble velocity approaches zero in a cycle have been shown in Figure 20.8. This figure reveals that as time gets larger, the maximum radius reduces exponentially due to the loss of energy after each burst. This is the most fundamental aspect of exponential stability in control problems due to the stabilizability of the system [20]. The exponential function fitted to R_{max} points is as follows:

$$R = ae^{bt} + ce^{dt}, \quad a = 0.0002183, \quad b = -1.997*10^5, \quad c = 7.733*10^{-6}, \quad d = -7667.$$

20.3 STATE-SPACE REPRESENTATION

To look at the dynamical behavior of the bubble, we need to represent the input-output model of the system (20.1) as a standard state-space model. As is given in any dynamical modeling text, state-variable form of the system of Equation 20.1 is in general described in the form

$$\dot{x}(t) = f\big(x(t), u(t), t\big), \quad x(t_0) = x_0,$$

$$y(t) = h\big(x(t), u(t), t\big),$$

(20.2)

where
  t represents time
  $x(t)$ is an n-dimensional vector, denoting the state of the system at time t
  $u(t)$ is the input or the control signal
  $y(t)$ is the output of the system, meaning the available information that can be measured or observed
  The functions $f(.)$ and $h(.)$ indicate how the system changes over time
  $x(t_0)$ denotes the value of $x(t)$ at the initial time $t = t_0 \geq 0$.

Having chosen the radius of the bubble R and the bubble interface velocity $\dot{R}$ as states of the system x_1 and x_2, respectively, and the bubble radius as the output of the system y, then the equivalent state-variable representation of Equation 20.1, is as follows:

$$\dot{x}_1 = x_2,$$

$$\dot{x}_2 = -\frac{3}{2}\frac{x_2^2}{x_1} + \frac{1}{\rho}\left(P_0 + \frac{2\gamma}{R_0} - P_v\right)\frac{R_0^3}{x_1^4} + \frac{P_v}{\rho x_1} - \frac{2\gamma}{\rho}\frac{1}{x_1^2} - \frac{4\mu}{\rho}\frac{x_2}{x_1^2} - \frac{u(t)}{\rho x_1}, \qquad (20.3)$$

$$y = x_1,$$

where $u(t) = P_0 + P(t)$ denotes an input pressure provided by a user and/or the environment that regulates the growth or attenuation of the bubble radius. Also, $y = R(t)$ is the output of the system, that is, the variable that one wants to control. If a nonlinear system operates around an equilibrium point, where the system is at rest, it is possible to study the behavior of the system in a neighborhood of such a point.

Definition 20.1 (Equilibrium Point): Consider the system in state-variable form—Equation 20.2. Suppose $u(t)$ is set to a constant value u^*. Then, x^* is said to be an equilibrium point for Equation 20.2, if $f(x^*, u^*) = 0$.

Let us set $u = u^* = P_0$. Using the definition above to find equilibrium points of Equation 20.3, we set $f(x^*, u^*) = [0\ 0]^T$. In other words, we have

$$x_2 = 0,$$

$$-\frac{3}{2}\frac{x_2^2}{x_1} + \frac{1}{\rho}\left(P_0 + \frac{2\gamma}{R_0} - P_v\right)\frac{R_0^3}{x_1^4} + \frac{P_v}{\rho x_1} - \frac{2\gamma}{\rho}\frac{1}{x_1^2} - \frac{4\mu}{\rho}\frac{x_2}{x_1^2} - \frac{P_0}{\rho x_1} = 0. \qquad (20.4)$$

Thus, the equilibrium point of the bubble with $u = P_0$ is given by $x^* = \begin{bmatrix} R_0 & 0 \end{bmatrix}^T$. Physically, this means that the bubble is at equilibrium whenever the radius R is equal to R_0 and the velocity $\dot{R}$ is zero. Qualitatively, the equilibrium $x^* = \begin{bmatrix} R_0 & 0 \end{bmatrix}^T$ is stable.

Now suppose a desired pressure $u = u^* \neq 0$ has been produced. The corresponding equilibrium must satisfy the equation $f(x^*, u^*) = [0\ 0]^T$, that is,

$$x_2 = 0,$$

$$-\frac{3}{2}\frac{x_2^2}{x_1} + \frac{1}{\rho}\left(P_0 + \frac{2\gamma}{R_0} - P_v\right)\frac{R_0^3}{x_1^4} + \frac{P_v}{\rho x_1} - \frac{2\gamma}{\rho}\frac{1}{x_1^2} - \frac{4\mu}{\rho}\frac{x_2}{x_1^2} - \frac{u^*}{\rho x_1} = 0. \qquad (20.5)$$

Setting

$$u = u^* = \left(P_0 + \frac{2\gamma}{R_0} - P_v\right)\left(\frac{R_0}{x_1^*}\right)^3 + P_v - \frac{2\gamma}{x_1^*}, \qquad (20.6)$$

the state $x^* = \begin{bmatrix} R^* & 0 \end{bmatrix}^T$ is an equilibrium point of the bubble. Physically, this means that by applying the constant pressure of Equation 20.6 to the bubble, one can make the bubble be at rest at any desired radius R^*.

20.4 CONTROLLABILITY AND OBSERVABILITY

Briefly, controllability and observability are dual aspects of the same problem. Roughly speaking, the concept of controllability (accessibility*) denotes the ability to move a system around its configuration space, using only certain admissible manipulations. Observability is a measure of how well internal states of a system can be inferred by knowledge of its external outputs (for more details, see Ref. [21]).

In addition, in studying controllability and observability of nonlinear systems, one needs to know the concepts of Lie bracket and Lie derivative. The reader is referred to Ref. [22] for the definition of these concepts.

Theorem 20.1 [22]: Assume we have an affine system as

$$\dot{x} = f(x) + \sum_{i=1}^{m} g_i(x) u_i. \tag{20.7}$$

This system is locally accessible about x_0, if the accessibility distribution (C) spans n space, where n is the rank of x and C is defined as

$$C = \left[g_1, g_2, \dots, g_m, [g_i, g_j], \dots, \left[ad_{g_i}^k, g_j \right], \dots, [f, g_i], \dots, \left[ad_{g_i}^k, g_j \right], \dots \right],$$

where $[X,Y]$ denotes the Lie bracket of X and Y. Now, we investigate the accessibility of the bubble system, which is of the form $\dot{x} = f(x) + g(x)u$, where

$$f(x) = \begin{bmatrix} x_2 \\ -\dfrac{3}{2}\dfrac{x_2^2}{x_1} + \dfrac{1}{\rho}\left(P_0 + \dfrac{2\gamma}{R_0} - P_v \right)\dfrac{R_0^3}{x_1^4} + \dfrac{P_v}{\rho x_1} - \dfrac{2\gamma}{\rho}\dfrac{1}{x_1^2} - \dfrac{4\mu}{\rho}\dfrac{x_2}{x_1^2} \end{bmatrix}, \quad g(x) = \begin{bmatrix} 0 \\ \dfrac{1}{\rho x_1} \end{bmatrix}.$$

One can compute the Lie bracket, as

$$[f, g] = \begin{bmatrix} -\dfrac{1}{\rho x_1} \\ -\dfrac{2 x_2}{\rho x_1^2} - \dfrac{4\mu}{\rho^2 x_1^3} \end{bmatrix}.$$

Therefore, dim $C(x) = 2$ for all $x \in R^2$ and the system is locally accessible based on Theorem 20.1.

Theorem 20.2 [22]: Consider the system defined as follows:

$$\dot{x} = f(x, u),$$
$$y = h(x) = \left[h_1(x), h_2(x), \dots, h_p(x) \right]. \tag{20.8}$$

---

* It is too difficult to investigate controllability in general for nonlinear systems; so, a weaker (local) form of controllability which is known as accessibility has been defined.

Let G denotes the set of all finite linear combinations of Lie derivatives of $h_1, h_2, \ldots, h_p$ with respect to f, which is represented by $\mathcal{L}_f\left(h_i\right)$ and dG denotes the set of all gradients of G

If we can find n linearly independent vectors within dG, then the system is locally observable. Now, let us investigate the observability of the bubble system which is of the form

$$\dot{x} = f\left(x, u\right), \quad y = x_1.$$

We have $\mathcal{L}_f\left(h\right) = \dot{x}_1 = x_2$ and hence dim $dG = 2$ and the system is locally observable.

20.5   STABILITY ANALYSIS: LINEARIZATION

This section is concerned with stabilization of the linearized bubble system. Since almost every physical system contains nonlinearities, oftentimes its behavior can be reasonably approximated by a linear model within a certain operating range of an equilibrium point. One reason for approximating the nonlinear system by a linear model is so that one may apply simple and systematic linear control design techniques. Note, a linearized model is valid only when the system operates at a sufficiently small neighborhood around an equilibrium point.

Given the nonlinear system of Equation 20.2 and an equilibrium point x^* obtained when $u = u^*$, we define new coordinates Δx, Δu, and Δy representing the deviations of x, u, and y at their equilibrium states, for example, $\Delta x = x - x^*$. The linearization of Equation 20.2 at x^* is given by

$$\begin{aligned}
\Delta \dot{x} &= A\Delta x + B\Delta u, \\
\Delta y &= C\Delta x + D\Delta u,
\end{aligned} \tag{20.9}$$

where

$$A = \left[\frac{\partial f}{\partial x}\right]_{x^*, u^*}, \quad B = \left[\frac{\partial f}{\partial u}\right]_{x^*, u^*}, \quad C = \left[\frac{\partial h}{\partial x}\right]_{x^*, u^*}, \quad D = \left[\frac{\partial h}{\partial u}\right]_{x^*, u^*}.$$

Remark 20.1: Note that matrices A, B, C, and D are the Jacobian matrices, evaluated at the nominal values $\left(x_1^*, \ldots, x_n^*, u^*\right)$. Recall that setting $u^* = P_0$ results in $R^* = R_0$, which is reasonable. Consider the equilibrium point $x^* = \left[R^* 0\right]^T$, obtained by setting $x_1^* = R^*$ in Equation 20.6. Following the procedure outlined above, we define

$$\Delta x = \left[x_1 - R^* x_2 - 0\right], \quad \Delta u = u - \left(\left(P_0 + \frac{2\gamma}{R_0} - P_v\right)\left(\frac{R_0}{R^*}\right)^3 + P_v - \frac{2\gamma}{R^*}\right), \quad \Delta y = y - R^*.$$

The Jacobian matrices are as follows:

$$\frac{\partial f}{\partial x} = \begin{bmatrix} 0 & 1 \\ \dfrac{3}{2}\dfrac{x_2^2}{x_1^2} - \dfrac{4R_0^3}{\rho}\left(P_0 + \dfrac{2\gamma}{R_0} - P_v\right)\dfrac{1}{x_1^5} - \dfrac{P_v}{\rho x_1^2} + \dfrac{4\gamma}{\rho x_1^3} + \dfrac{8\mu}{\rho}\dfrac{x_2}{x_1^3} + \dfrac{u}{\rho x_1^2} & -\dfrac{3x_2}{x_1} - \dfrac{4\mu}{\rho x_1^2} \end{bmatrix}, \quad \frac{\partial f}{\partial u} = \begin{bmatrix} 0 \\ -\dfrac{1}{\rho x_1} \end{bmatrix},$$

$$\frac{\partial h}{\partial x} = \begin{bmatrix} 1 & 0 \end{bmatrix}, \quad \frac{\partial h}{\partial u} = 0.$$

Next, evaluating the matrices above at $\left(x_1^*, x_2^*, u^*\right) = \left(R^*, 0, \left(P_0 + \dfrac{2\gamma}{R_0} - P_v\right)\left(\dfrac{R_0}{R^*}\right)^3 + P_v - \dfrac{2\gamma}{R^*}\right)$ we can write the linearized model as follows:

$$\Delta \dot{x} = A\Delta x + B\Delta u,$$
$$\Delta y = C\Delta x + D\Delta u, \tag{20.10}$$

where

$$A = \begin{bmatrix} 0 & 1 \\ -\dfrac{3R_0^3}{\rho}\left(P_0 + \dfrac{2\gamma}{R_0} - P_v\right)\dfrac{1}{R^{*5}} + \dfrac{2\gamma}{\rho R^{*3}} & -\dfrac{4\mu}{\rho R^{*2}} \end{bmatrix}, \quad B = \begin{bmatrix} 0 \\ -\dfrac{1}{\rho R^*} \end{bmatrix}, \quad C = \begin{bmatrix} 1 & 0 \end{bmatrix}, \quad D = 0.$$

The linearized model is only valid for a small neighborhood of the equilibrium $x^* = \begin{bmatrix} R^* & 0 \end{bmatrix}^T$.

20.5.1 STABILITY ANALYSIS

With the linearized model of the bubble derived previously, we now present a control theoretical stability analysis of the bubble dynamic behavior, using Theorem 20.3 as follows.

Theorem 20.3: The nonlinear system of Equation 20.1 is asymptotically stable near the equilibrium point $x^* = \begin{bmatrix} R^* & 0 \end{bmatrix}^T$, if the following condition is satisfied:

$$R^* < \sqrt{\dfrac{3R_0^3}{2\gamma}\left(P_0 - P_v + \dfrac{2\gamma}{R_0}\right)}. \tag{20.11}$$

Proof: The eigenvalues of matrix A in the linearized model of the bubble (20.10) are

$$\lambda = \dfrac{-\dfrac{4\mu}{\rho R^{*2}} \pm \sqrt{\left(\dfrac{4\mu}{\rho R^{*2}}\right)^2 - 4\overbrace{\left(\dfrac{3R_0^3}{\rho}\left(P_0 + \dfrac{2\gamma}{R_0} - P_v\right)\dfrac{1}{R^{*5}} - \dfrac{2\gamma}{\rho R^{*3}}\right)}^{M}}}{2},$$

which are obtained by setting the characteristic polynomial of A equal to zero, that is, $\det(\lambda I_2 - A) = 0$.

The linear system of Equation 20.10 is asymptotically stable, if and only if all eigenvalues of A satisfy $\mathrm{Re}(\lambda_i) < 0$. It can easily be seen that the equilibrium point $(R^*, 0)$ is a stable focus, if $\left(\dfrac{4\mu}{\rho R^{*2}}\right)^2 - M < 0$ and is a saddle point, if $-\dfrac{4\mu}{\rho R^{*2}} + \sqrt{\left(\dfrac{4\mu}{\rho R^{*2}}\right)^2 - M} > 0$, which leads to Equation 20.11. Note that if the origin of the linearized state equation is a stable focus or a saddle point, in a small neighborhood of the equilibrium point, the trajectories of the nonlinear state equation will behave like a stable focus or a saddle point [23]. This completes the proof.

Remark 20.2: The stability condition (20.11) depends on R_0, P_0, P_v, and γ. The liquid viscosity plays an important role on stability. Neglecting viscosity, both eigenvalues are on the imaginary axis and consequently we cannot determine the stability of the equilibrium point through the linearized model.

Remark 20.3: Condition (20.11) derived for the stability analysis of the bubble dynamics matches the Blake critical radius given in Ref. [24]. Choosing $R^* > R_c$ results in an infinite growth of the bubble, whereas choosing $R^* < R_c$, the bubble remains stable, meaning that its radius converges to the desired radius R^* within finite time.

The nonlinear model of Equation 20.1 is simulated under the above obtained controller (u^*) in Equation 20.6 with parameters given in Table 19.1, which are common values in many experiments [16]. The dynamics of $R(t)$ are shown in Figure 20.9. As seen in this figure, the system is asymptotically stable and converges to the desired value R^* within finite time. In Figure 20.10, R^* is set to

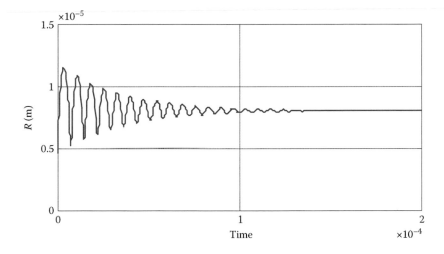

FIGURE 20.9 $R(t)$ for $R_0 = 4.5$, $R^* = 8$ μm.

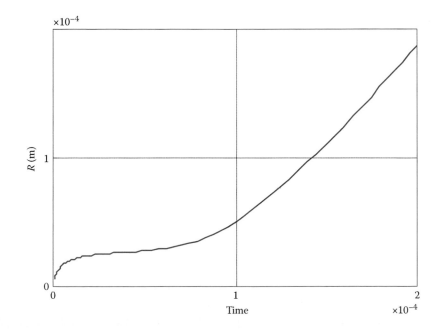

FIGURE 20.10 $R(t)$ for $R_0 = 4.5$, $R^* = 15$ μm.

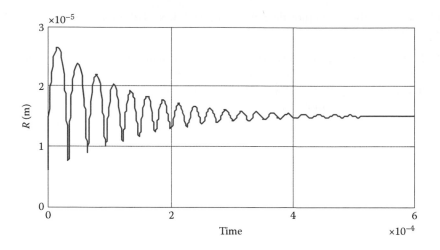

FIGURE 20.11 $R(t)$ for $R_0 = 6$, $R^* = 15$ μm.

15 μm (less than critical radius $R_C = 15.5$ μm), but the system becomes unstable. This is due to the fact that, the linearized model is valid only around the equilibrium point. Therefore, much larger bubbles can be stable if R_0 is increased (Figure 20.11).

In the coming subsection, Lyapunov stability analysis for the linear time invariant system is presented.

20.5.2 STABILITY ANALYSIS (LYAPUNOV METHOD)

Theorem 20.4 (Lyapunov stability for linear systems): Assume the autonomous linear system given by

$$\dot{x} = Ax, x(0) = x_0, \tag{20.12}$$

which may represent an open or a closed-loop system. System (20.12) is asymptotically stable about $x = 0$, if for any positive definite Q, there exists a positive definite P, such that

$$A^T P + PA = -Q. \tag{20.13}$$

This is known as the Lyapunov equation.

According to the Lyapunov theory, one can check the stability of a system by finding some Lyapunov function $V(x)$, which satisfies

$$V(x) > 0, \quad V(0) = 0, \tag{20.14}$$

$$\dot{V}(x) = \frac{dV}{dt} = \frac{\partial V}{\partial x}\frac{dx}{dt} \leq 0. \tag{20.15}$$

Choosing a Lyapunov function $V = x^T P x$, $P = P^T > 0$ and using Equation 20.15 leads to

$$\dot{V}(x) = x^T \left(A^T P + PA \right) x.$$

Thus, the system of Equation 20.12 is asymptotically stable, if Equation 20.13 is satisfied. This completes the proof.

In other words, to show the stability, we should find a positive definite matrix P, such that

$$A^T P + PA < 0, \quad P > 0. \tag{20.16}$$

This is the concept of LMI approach, in which matrix P is a variable. Efficient tools have been developed to quickly solve LMI such as LMI control toolbox in MATLAB® [25].

The stabilization condition for the linearized system of bubble equations with the state feedback controller in term of LMI is derived in the following subsection.

20.5.3 STABILIZATION

Consider the linearized model of bubble dynamics, that is, Equation 20.10. We look for a static stabilizing state feedback controller of the form:

$$u = -Kx, \tag{20.17}$$

to stabilize the closed-loop system

$$\Delta \dot{x} = A\Delta x - BK\Delta x, \tag{20.18}$$

where K is the constant controller gain.

Theorem 20.5: The system of Equation 20.10, with control gain $= YQ^{-1}$, is stable if there exist symmetric positive definite matrices Q and Y, such that the LMI shown in the following Equation 20.19, holds

$$QA^T - Y^T B^T + AQ - BY < 0. \tag{20.19}$$

Proof: Replacing A in Equation 20.16 with $A-BK$ pre and post multiplying the result by P^{-1} and defining $Q = P^{-1}$ and $Y = KQ$ leads to Equation 20.19. Then, we can find the state feedback controller gain by $K = YQ^{-1}$.

To illustrate the validation of our proposed method, consider the bubble system with parameters given in Table 19.1. In the following simulation, we assume $R^* = 16$ μm, which based on Equation 20.11 is larger than the critical radius and thus the open-loop system becomes unstable. Using LMI toolbox of the MATLAB®, the gain is calculated as $K = \begin{bmatrix} -2.9997 * 10^7 & 249.9850 \end{bmatrix}$. As shown in Figure 20.12, by applying the designed control signal to the system, the bubble radius converges to the desired value.

20.6 CONTROLLER DESIGN

The design of a controller that can modify the behavior of an unknown plant to reach a desired performance is a challenging problem in many control applications. Any process is characterized by a certain number of inputs u and outputs y, as shown in Figure 20.13. The control design task is to determine the input signal $u(t)$ so that the output response $y(t)$ satisfies given the performance requirements. The control design framework is often followed by most control engineers in choosing the input u shown in Figure 20.14.

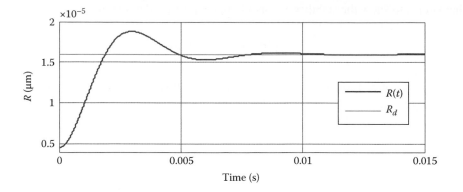

FIGURE 20.12   Dynamics of bubble with state feedback controller for $R_0 = 4.5$, $R^* = 16$ μm.

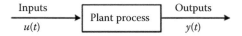

FIGURE 20.13   Representation of the plant.

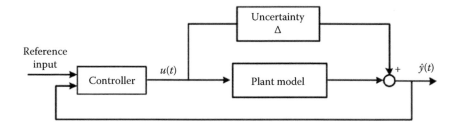

FIGURE 20.14   Block diagram of a feedback system with uncertainty.

To overcome the restrictions of control design of a linearized model obtained by conventional (Jacobian) linearization, which results in local stability, the following section illustrates a common approach used in control of nonlinear systems.

20.6.1   FEEDBACK LINEARIZATION

FBL is an approach for nonlinear control design and has attracted a lot of researches in recent years. The main idea is to transform nonlinear system dynamics into an equivalent linear system through change of variables and suitable control input, so that linear control techniques can be applied. This differs entirely from conventional linearization because FBL is achieved by exact state transformation and feedback (rather than by linear approximations of the dynamics). This linearization is exact and has no approximations.

The main idea of this method is to find state feedback $u = g(x,v)$, so that we have the following nonlinear system:

$$x^{(n)} = f(x) + b(x)u,$$
(20.20)

where n denotes the order of the system. Then it turns into the following linear system:

$$\dot{x} = Ax + Bv. \tag{20.21}$$

Then, linear control design methods can be carried out based on v. The control objective is to find a control law $u(t)$, so that $y(t)$ follows the desired trajectory. Equation 20.20 can be represented in the form of a controllable canonical model as

$$\frac{d}{dx}\begin{bmatrix} x_1 \\ \vdots \\ x_{n-1} \\ x_n \end{bmatrix} = \begin{bmatrix} x_2 \\ \vdots \\ x_n \\ f(x)+b(x)u \end{bmatrix}. \tag{20.22}$$

For nonzero $b(x)$, the control input is of the general form:

$$u = \frac{1}{b(x)}\left[v - f(x)\right], \tag{20.23}$$

where v is an appropriate function of the desired state values, tracking errors, and possibly some of their derivatives: $v = x_d^n - k_0 - k_1 e - \cdots - k_{n-1}e^{n-1}$ ($e = x - x_d$ is the tracking error). The gains $k_0,\ldots,k_{n-1}$ are determined so that the error converges to zero exponentially. The block diagram illustrating the FBL of a nonlinear system is shown in Figure 20.15.

Now, consider the bubble system which is of the form (20.22), where

$$f(x) = -\frac{3}{2}\frac{x_2^2}{x_1} + \frac{1}{\rho}\left(P_0 + \frac{2\gamma}{R_0} - P_v\right)\frac{R_0^3}{x_1^4} + \frac{P_v}{\rho x_1} - \frac{2\gamma}{\rho}\frac{1}{x_1^2} - \frac{4\mu}{\rho}\frac{x_2}{x_1^2}, \quad b(x) = -\frac{1}{\rho x_1}. \tag{20.24}$$

Substituting (20.24) into (20.23) yields

$$u = -\rho x_1 v - \frac{3}{2}\rho x_2^2 + \left(P_0 + \frac{2\gamma}{R_0} - P_v\right)\frac{R_0^3}{x_1^3} + P_v - \frac{2\gamma}{x_1} - 4\mu\frac{x_2}{x_1}, \tag{20.25}$$

Applying Equation 20.25 to the bubble system of Equation 20.3 results in $\dot{x}_2 = v$, which is a linear dynamic. Choosing the equivalent input $v = \ddot{R}_d - 2\lambda\dot{e} - \lambda^2 e$, where $e = R(t) - R_d$ (tracking error), results in the following equation, which is a second order dynamic equation of the tracking error:

$$\ddot{e} + 2\lambda\dot{e} + \lambda^2 e = 0.$$

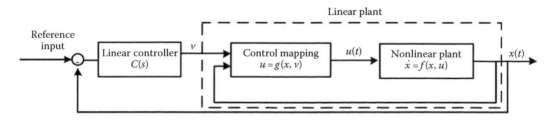

FIGURE 20.15   Block diagram of the feedback linearization approach.

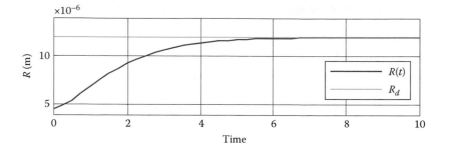

FIGURE 20.16   Desired radius, bubble radius, and control input of FBL method with $\lambda = 1$, $R_d = 12$ μm.

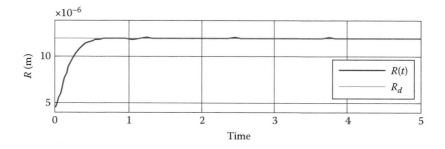

FIGURE 20.17   Desired radius, bubble radius, and control input of FBL method with $\lambda = 10$, $R_d = 12$ μm.

Any positive constant λ guarantees that the tracking error converges to zero exponentially. So, the bubble system of Equation 20.3 with the control input of Equation 20.25 is stable and the tracking performance is achieved.

The performance of the proposed controller is investigated via simulations with parameters given in Table 19.1, under different desired values. Figure 20.16 shows that applying the proposed controller with $\lambda = 1$ leads to the desired radius in a few seconds. Increasing $\lambda = 10$, we see from Figure 20.17 that the radius of the bubble approaches the desired value in shorter time, compared to Figure 20.16.

The bubble's response to *a step function input* with sudden change of the desired value of radius at times 1 and 2 s is shown in Figure 20.18. As shown in this figure, the bubble radius retracts the desired trajectory quickly with the proposed controller (pressure). The simulation results confirm the validity of the proposed control approach.

The FBL approach has two shortcomings and limitations as follows:

1. This approach is based on an exact model of the system and requires a perfect model (leading to a robust control problem).
2. If the relative degree, r (number of differentiating y to find explicit relation with u) is less than the degree of the system, n, that is, $(r < n)$, we have internal dynamics. In this case, whether or not the controller is applicable depends on the stability of the internal dynamics.

In our design, r is the same as the order of system, n, and no internal dynamics would be presented. However, this method may lead to instabilities because complete knowledge of the state equations and exact mathematical cancelation of terms are almost impossible.

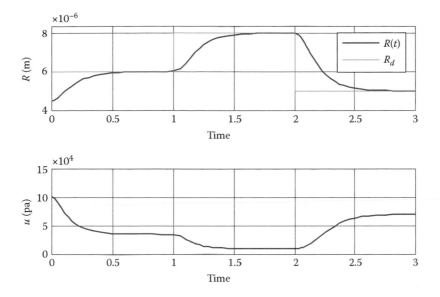

FIGURE 20.18 Desired radius, bubble radius, and control input of FBL with reference input: $R_d = 6u(t) + 2u(t-1) = 3u(t-2)$.

We can consider every parameter of the system, which is undetermined as parametric uncertainty. Reference [26] has introduced three components of damping in the bubble oscillations, which are due to liquid viscosity, liquid compressibility through acoustic radiation, and thermal conductivity. Let us consider a 10% perturbation in the magnitude of the liquid viscosity to investigate its effect on the bubble radius in the presence of an uncertainty. The bubble radius and the control input are shown in Figure 20.19. Clearly, the control objectives are not achieved when uncertainty exists.

In the next subsection, an alternative approach is presented to tackle this problem.

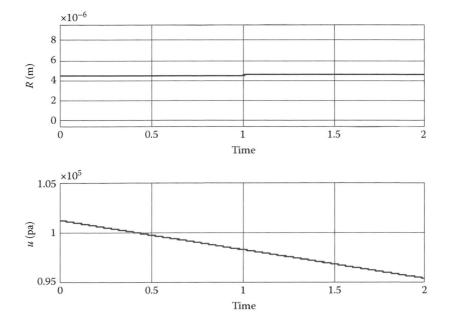

FIGURE 20.19 Bubble dynamics and control input of FBL method with $\lambda = 1$, $R_d = 12$, $\mu = \mu + 0.1\,\mu$.

20.6.2   SLIDING MODE CONTROLLER

Sliding mode controller (SMC) is a robust nonlinear feedback control approach against the uncertainties, including uncertainty in parameters. Consider the nonlinear system of Equation 20.20.

A time-varying sliding surface, $S(t)$, is defined as

$$S(t) = \left(\frac{\mathrm{d}}{\mathrm{d}t}\right)^{n-1} e(t), \tag{20.26}$$

where $e(t) = x(t) - x_d(t)$ is the error in the output state, in which $x_d(t)$ is the desired output. The following switching condition

$$\frac{1}{2}\frac{\mathrm{d}}{\mathrm{d}t}\left(S(t)^2\right) \leq -\eta\left|S(t)\right|, \quad \eta > 0 \tag{20.27}$$

makes the surface $S(t)$ an invariant set. All trajectories outside $S(t)$ point toward the surface and the trajectories on the surface remain there. It takes a finite time to reach the surface $S(t)$ from the outside. Moreover, Equation 20.26 implies that once the surface is reached, the convergence to zero error is exponential. Chattering is caused by nonideal switching around the switching surface. Delay in digital implementation causes $S(t)$ to pass to the other side of the surface, which in turn, produces chattering [27]. Design details presented in the following theorem provide a control that guarantees satisfaction of Equation 20.27.

Theorem 20.6: For the nonlinear system of bubbles given by Equation 20.3, if $k > \rho x_1 \eta$, the following control law ensures exponential convergence of radius tracking error to zero:

$$u = -\frac{3}{2}\rho x_2^2 + \left(P_0 + \frac{2\gamma}{R_0} - P_v\right)\frac{R_0^3}{x_1^3} + P_v - \frac{2\gamma}{x_1} - 4\mu\frac{x_2}{x_1} - \rho x_1\left(\ddot{x}_{1d} - \lambda\dot{e}\right) + k\,\mathrm{sign}(S), \tag{20.28}$$

where
  $k > 0$ is time dependent
  sign(.) is the sign function
  $S(t)$ is the sliding surface defined by Equation 20.26 and is

$$S(t) = \dot{e}(t) + \lambda e(t), \lambda > 0, \tag{20.29}$$

where
  λ is a strictly positive constant
  $e(t)$ is defined as the tracking error between the bubble radius and the desired radius as

$$e(t) = R(t) - R_d(t). \tag{20.30}$$

Proof: Consider the following Lyapunov function:

$$V = \frac{1}{2}S^2. \tag{20.31}$$

It is clear that V is a positive definite function. Taking its derivative with respect to time and substituting $\dot{S}$ yields

$$\dot{V} = S\dot{S} = S(\ddot{x}_1 - \ddot{x}_{1d} + \lambda\dot{e}).$$ (20.32)

Substituting Equation 20.3 into 20.32 yields

$$\dot{V} = S\left(-\frac{3}{2}\frac{x_2^2}{x_1} + \frac{1}{\rho}\left(P_0 + \frac{2\gamma}{R_0} - P_v\right)\frac{R_0^3}{x_1^4} + \frac{P_v}{\rho x_1} - \frac{2\gamma}{\rho}\frac{1}{x_1^2} - \frac{4\mu}{\rho}\frac{x_2}{x_1^2} - \frac{u(t)}{\rho x_1} - \ddot{x}_{1d} + \lambda\dot{e}\right).$$ (20.33)

Control input $u(t)$ is composed of two parts, one for removing the nonlinear part and the other for making $\dot{V}$ a negative definite function. Substituting Equation 20.28 into 20.33 yields

$$\dot{V} = S\left(-\frac{1}{\rho x_1}(k\,\mathrm{sign}(S))\right).$$

Choosing the gain as $k > \rho x_1 \eta$, we have

$$\dot{V} < -\eta S\big(\mathrm{sign}(S)\big) = -\eta|S| < 0.$$

Thus, Equation 20.31 is a Lyapunov function and by choosing a proper value for gain, radius tracking is guaranteed.

Using the control law given by Equation 20.28 with $\eta = 1$, $k = 2\rho x_1 \eta$, and $R_d = 12$ μm, the closed-loop response is shown in Figure 20.20. Due to the switching delay caused by $\mathrm{sign}(S)$, we see unwanted oscillations in the system, as shown in Figure 20.20. A common method for chattering reduction is to replace the term $\mathrm{sign}(S)$ with the saturation function $sat(S/\phi)$, where ϕ is the boundary layer thickness. Applying Equation 20.28 with $sat(.)$ function, the chattering is eliminated in the solution, but the bubble radius approaches the desired value at a slower rate (see Figure 20.21).

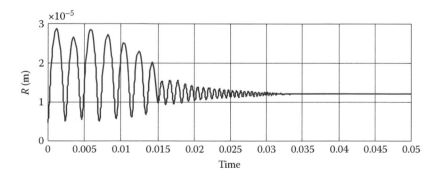

FIGURE 20.20 Bubble dynamics under SMC with $\eta = 1$, $k = 2\rho x_1 \eta$, $R_d = 12$ μm.

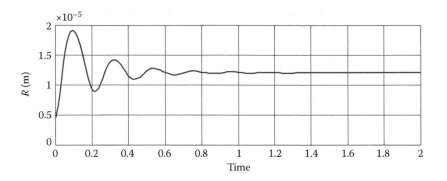

FIGURE 20.21   Bubble dynamics under SMC; replacing sign(.) with sat(.) function with $\phi = 1$ (chattering removed).

20.6.2.1   Modified SM Controller

If $f(x)$ is not exactly known, its estimate $\hat{f}(x)$ should be used in Equation 20.28. Substituting $\hat{f}(x)$ for $f(x)$ in Equation 20.28 and repeating the same procedure as before, the control gain (k) becomes

$$k > \rho x_1 \eta + \left| \hat{f}(x) - f(x) \right|, \tag{20.34}$$

where $\hat{f}(x)$ is the mathematical model estimation of the unknown function $f(x)$ that is the actual plant. Assuming the estimation error on $f(x)$ is bounded by some known function; that is, $\left| \hat{f}(x) - f(x) \right| \le F(x,t)$, a lower bound on the gain k is achieved as follows:

$$k > \rho x_1 \eta + F(x,t). \tag{20.35}$$

The closed-loop system can be stabilized with a suitable choice of the gain (k) as in Equation 20.35, which is large enough to overcome the maximum bound of the perturbation term $F(x,t)$. Let us consider liquid viscosity and vapor pressure inside the uncertain bubble and assume the range of the uncertainty to be $|\Delta \mu| < 0.0073$ and $|\Delta P_v| < 500$, respectively. The simulation results using the control law developed in this section are shown in Figures 20.22 and 20.23. Figure 20.22 shows the tracking performance with uncertain viscosity. Both viscosity and vapor pressure are assumed uncertain in Figure 20.23. As shown in this figure, the bubble radius is driven to the desired value even with parameter uncertainties and our control objectives are achieved.

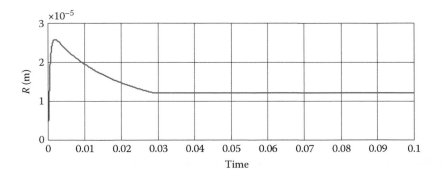

FIGURE 20.22   Bubble dynamics under SMC with uncertain μ.

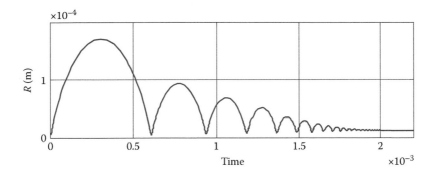

FIGURE 20.23   Bubble dynamics under SMC with uncertain μ and *Pv*.

20.7   CONCLUSION AND FUTURE WORKS

In this chapter, we have considered the challenging problem of regulation for a class of nonlinear systems, namely, acoustic cavitation-bubble dynamics. Interestingly, after developing a linear model for the RP equation, we analyzed the stability of the model and presented the tools for testing its controllability and observability. Next, nonlinear strategies for achieving certain performance objectives were developed and simulated, using nominal parameter values. It was shown that, under parameter uncertainties, the regulation objectives fail to be achieved. This was the incentive for developing the formulation of another advanced nonlinear algorithm, namely, SMC to tackle this problem. Further research to be conducted in this subject may include "Fuzzy control of bubble dynamics," "adaptive control of bubble system," and "delayed RP equation of bubble system control." More complex models of bubble behavior, such as RPK equation (see Chapter 19), also warrant further research.

REFERENCES

1. F.R. Young, *Cavitation*, Imperial College Press, London, U.K., 1999.
2. C.E. Brennen, *Cavitation and Bubble Dynamics*, Cambridge University Press, London, U.K., 2013.
3. E. Dahi, *In Ozonization Manual for Water and Wastewater Treatment*, John Wiley & Sons, New York, 1982.
4. F. Harris, S.R. Dennison, and D.A. Phoenix, *Trends in Molecular Medicine*, 20(7), 363–367, 2014.
5. P. Prentice, A. Cuschieri, K. Dholakia, M.R. Prausnitz, and P.A. Campbell, *Nature Physics*, 1, 107–110, 2005.
6. D.J. Flannigan and K. Suslick, *Nature Physics*, 6, 598–601, 2010.
7. D. Suhr, F. Brummer, and D.F. Hulser, *Ultrasound Medical Biology*, 17(8), 761–768, 1991.
8. M. Simon and M. Ulbrich, *Advanced Computing, Lecturer Notes in Computational Science and Engineering*, 93, 81–98, 2013.
9. T. Patel, J. Shah, and M. Satria, *International Journal of Current Engineering and Technology*, 3(5), 2047–2052, 2013.
10. G. Nagashima, E.V. Levine, D.P. Hoogerheide, M.M. Burns, and J.A. Golovchenko, *Physical Review Letters*, 113(2), 024506.1–024506.5, 2014.
11. B.B. Li, H.C. Zhang, and J. Lu, *Optics & Laser Technology*, 43(8), 1499–1503, 2011.
12. F. Hegedűs, S. Koch, W. Garen, Z. Pandula, G. Paal, L. Kullmann, and U. Teubner, *International Journal of Heat and Fluid Flow*, 42, 200–208, 2013.
13. F. Hegedus, *Ultrasonics*, 54, 1113–1121, 2014.
14. J.W. Polderman and J. Willems, *Introduction to the Mathematical Theory of Systems and Control*, Springer, New York, 1998.
15. R. Lofstedt, B.P. Barber, and S.J. Putterman, *Physics Fluids A*, 5(11), 2911–2928, 1993.
16. S. Hilgenfeldt, M.P. Brenner, S. Grossmann, and D. Lohse, *Journal of Fluid Mechanics*, 365, 171–204, 1998.

17. C.E. Brennen, *Fundamentals of Multiphase Flow*, Cambridge University Press, London, U.K., 2005.

18. S. Hilgenfeldt, D. Lohse, and M.P. Brenner, *Physics Fluids*, 8, 2808–2826, 1996.

19. S. Grossmann, S. Hilgenfeldt, D. Lohse, and M. Zomack, *The Journal of Acoustical Society of America*, 102, 1223–1230, 1997.

20. M. Najafi, G.R. Sarhangi, and H. Wang, *IEEE Transactions on Automatic Control*, 42(9), 1308–1312, 1997.

21. H. Nijmeijer and A.J. van der Schaft, *Nonlinear Dynamical Control Systems*, Springer-Verlag, New York, 1990

22. J.K. Hedrick and A. Girard, *Control of Nonlinear Dynamic Systems: Theory and Applications*, Berkeley press, Berkeley, 2005.

23. H.K. Khalil and J.W. Grizzle, *Nonlinear Systems*, Vol. 3, Prentice Hall, Upper Saddle River, NJ, 2002.

24. F.G. Blake, I. Acoustics Research Laboratory, Technical Memorandum, Harvard University, Cambridge, MA, No. 12, 1949.

25. S. Boyd, L. El Ghaoui, E. Feron, and V. Balakrishnan, 15, Society for industrial and applied mathematics, Philadelphia, 1994.

26. R.B. Chapman and M.S. Plesset, *ASME Journal of Basic Engineering*, 93, 373–376, 1971.

27. P. Kachroo and M. Tomizuka, *IEEE Transactions on Automatic Control*, 41, 1063–1068, 1996.

21 Liquid Bridge and Drop Transfer between Surfaces

Huanchen Chen, Tian Tang, and Alidad Amirfazli

CONTENTS

21.1 INTRODUCTION

The study of liquid drop transfer from one surface to another through elongating a liquid bridge formed between surfaces has undergone an intense development in recent years. Such a transfer process can be widely seen in our daily lives and has a number of applications, for example, printing industry, drop deposition, packaging industry, and microgripping [1–11]. For example, during the printing, ink is transferred from the substrate to the target surface through the liquid bridge. Recently, several studies have been performed to fabricate high resolution microscale electrical circuits and semiconductors with such a transfer technique [7–9].

A typical transfer process can be represented by the procedure shown in Figure 21.1 and consists of five steps. Firstly, a liquid drop is placed on the donor surface (Figure 21.1a). The acceptor surface is brought to the donor surface to form the liquid bridge (Figure 21.1b). Then the liquid bridge will be further compressed until the separation reaches a certain value (H_{min}; Figure 21.1c). There may be a short pause, and then the acceptor surface will retreat with some speed (U) (Figure 21.1d) until the liquid bridge ruptures (Figure 21.1e). Part of the liquid can be transferred from the donor surface to the acceptor surface after the breakage of liquid bridge. The transfer ratio (α), which is usually defined as the volume of the liquid transferred onto the acceptor surface divided by the total liquid volume, is of central importance for practical matters in the applications. In most applications, there exists a desired value or range of values for the transfer ratio, for example, in offset printing, the ink transfer ratio should be 1 to avoid image quality defects. Therefore, in order to be able to control the transfer ratio, it is important and necessary to know what physical parameters affect the transfer ratio, and their quantitative connections to the value of α.

This chapter aims to provide an overall description of the liquid transfer process and a quantitative characterization using the transfer ratio. First, we categorize the liquid transfer process into three regimes according to (1) different dominating forces and (2) the dependence of transfer ratio on the stretching speed, U. Next, a discussion is provided on what physical parameters may affect α in each regime. Their quantitative relation to α is then addressed, from which an empirical equation

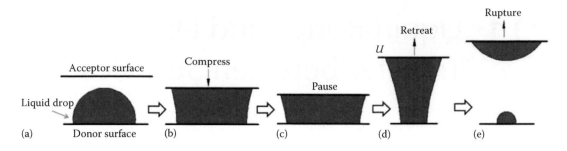

FIGURE 21.1   A schematic illustration of a typical liquid drop transfer process. (a) Placement of liquid drop on the donor surface, (b) formation of liquid bridge, (c) compression of liquid bridge, (d) stretching of liquid bridge, and (e) rupture.

is given to predict the transfer ratio. It should be noted that in most of the applications, droplets of very small volumes are used. Therefore, the liquid transfer systems usually have a very small Stokes number ($St = \rho g L^2 / \mu U$, the ratio of gravity to viscous forces, where ρ and μ are the liquid density and dynamic viscosity, g is the gravitational acceleration, and L is the characteristic length), and Bond number ($Bo = \rho g L^2 / \gamma$, the ratio of gravity to interfacial tension, γ), hence the effect of gravity is negligible. Therefore, there is no urgent need to include the discussion on gravity in this chapter.

21.2   LIQUID TRANSFER CATEGORIES

The transfer ratio as a function of stretching speed U ($\alpha = F(U)$) for a typical liquid transfer is shown in Figure 21.2. Here, a 2.0 µL glycerol drop is transferred from an octadecyltrichlorosilane (OTS) to a poly(ethyl methacrylate) (PEMA) surface with $H_{min} = 0.45$ mm. The experiments were performed at 20.5°C. Both surfaces have contact angle hysteresis (*CAH*): the OTS surface has an advancing contact angle of $\theta_a = 96.8°$ and receding contact angle of $\theta_r = 83.9°$. The corresponding values for the PEMA surface are θ_a: 72.1°, θ_r: 61.7°. The images of the liquid bridge at the breakup (breakup shape) for various transfer cases are also shown in Figure 21.2. It can be seen that when U is small ($U = 0.05$, 0.1, and 0.2 mm/s), most of the liquid tends to be transferred to the acceptor surface. In addition, the breakup shapes are almost the same and not affected by the change in U. Consequently, the transfer ratio stays at a constant value (0.986) when $U < 0.2$ mm/s. When U is larger than 0.5 mm/s, the breakup shape starts to become more and more symmetric with the increase of U. This change results in the liquid being partitioned more evenly between the donor and acceptor surfaces; hence the transfer ratio converges to 0.5 with the increase of U. For sufficiently large U (>16.5 mm/s), the breakup shape becomes almost symmetric, and the liquid is equally split between the two surfaces, leading to a constant transfer ratio of 0.5.

Based on the typical observation, the liquid transfer can be categorized into three regimes each associated with a different range of stretching speed U. At very small stretching speeds, the transfer ratio does not change with the value of U, but it is strongly dependent on the surface wettability. In Figure 21.2 for instance, the acceptor surface has a much smaller value of contact angle than that of the donor surface, and the transfer ratio is close to one. In this regime, which will be referred to as the *quasistatic regime* the transfer process is dominated by the surface forces, whereas viscous and inertial forces are negligible. At very large stretching speeds, the transfer ratio is again insensitive to the value of U. In addition, the droplet is split equally between the two surfaces and the transfer ratio stays at 0.5 irrespective of the wettability of the surfaces. This is the regime where surface forces are negligible whereas viscous and inertial forces dominate, that is, the *dynamic regime*. For the stretching speed in between, where the transfer ratio exhibits dependence on U and changes from the low-speed plateau to the high-speed plateau (0.5), surfaces, viscous, and inertial forces all play a role, and this will be referred to as the *transition regime*.

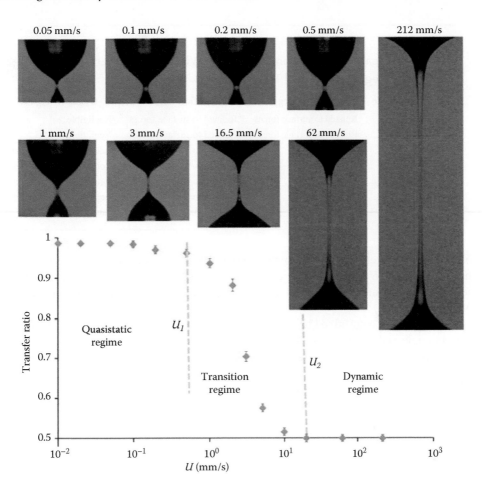

FIGURE 21.2 Relation between transfer ratio and U of glycerol transferred from OTS to PEMA surfaces. Shapes of the liquid bridge, at breakup, are also shown for different stretching speeds. U_1 and U_2 are respectively the start and end of the transition regime.

21.3 PARAMETER SPACE

In each of the regimes defined earlier, the process of liquid transfer is governed by certain forces which are determined by particular physical parameters [1,2,12–31]. The parameters that can affect the transfer process in each regime are summarized in Table 21.1, where γ, SCA, μ, ρ, and U are, respectively, the liquid surface tension, surface contact angle (including θ_a, θ_r, and CAH), liquid dynamic viscosity, liquid density, and stretching speed. For liquid drop transfer in the quasistatic regime, the process is only controlled by surface forces and hence the only governing parameters are the liquid surface tension γ and surface contact angles SCA [2,15–25]. When either the effects of viscous or inertial forces become much larger than the effects of surface forces, the transfer process belongs to the dynamic regime. And the controlling parameters in this regime become liquid viscosity μ (affecting viscous force), density ρ (affecting inertial forces), volume V (affecting inertial forces), and stretching speed U (affecting both inertial and viscous forces) [1,15,16,26–32]. Since in the transition regime, all three forces play a role, the transfer process is controlled by all six physical parameters listed in Table 21.1.

Given the parameter space identified earlier, it is desirable to be able to predict the transfer ratio α as a function of these parameters. Shown in Figure 21.2, with increasing U, α always changes from one plateau value (α_0, transfer ratio in quasistatic regime) to another (0.5, transfer ratio in

TABLE 21.1

Effect of Physical Parameters on the Surface, Viscous, and Inertial Forces in the Three Regimes of Liquid Transfer

Physical Parameter	Quasistatic Regime	Transition Regime	Dynamic Regime
γ	Related to surface forces	Related to surface forces	Negligible
SCA	Related to surface forces	Related to surface forces	Negligible
μ	Negligible	Related to viscous forces	Related to viscous forces
ρ	Negligible	Related to inertial forces	Related to inertial forces
V	Negligible	Related to inertial forces	Related to inertial forces
U	Negligible	Related to both viscous and inertial forces	Related to both viscous and inertial forces

dynamic regime). Therefore, the shape of the curve is actually dependent on the start value (α_0), and two threshold speeds (U_1 and U_2 in Figure 21.2) that correspond to the start and end of the transition regime, that is, the boundaries between quasistatic and transition regimes, and between transition and dynamic regimes.

It is clear that α_0 is determined by the two governing parameters in the quasistatic regime: surface contact angle and liquid surface tension. The location of U_1 and U_2 can have a complex dependence on the physical parameters listed in Table 21.1. In principle, one can use the ratio of viscous to surface forces, and the ratio of inertial to surface forces to identify the boundaries. These ratios correspond to two dimensionless numbers of Capillary ($Ca = U\mu/\gamma$, ratio of viscous to surface forces) and Weber ($We = \rho U^2 V^{1/3}/\gamma$, ratio of inertial to surface forces, where the characteristic length is taken to be the cubic root of liquid volume). Besides Ca and We, the two boundary speeds can also be affected by the surface contact angle. This is because although contact angles do not affect transfer in the dynamic regime, they still play an important role in affecting the profile of the liquid bridge in the transition regime, which in turn influences the spread of momentum from the acceptor to the donor surfaces. In the majority of practical situations, for example, printing, small volumes of viscous liquid are normally used, where the effects of the inertial forces are typically small, compared with the surface and viscous effects. Therefore, the remainder of this chapter will focus on addressing how α_0 and the two boundary speeds depend on surface tension, contact angle, and Ca, whereas the effect of We will not be discussed.

21.4 QUASISTATIC REGIME

21.4.1 GENERAL BEHAVIOR

An example of a quasistatic liquid transfer is provided in Figure 21.3, which shows the evolutions of contact angles and contact radii measured in an experiment. Here a 2 µL water drop is transferred from a PEMA surface (θ_a: 77.6°, θ_r: 68.2°) to a poly(methyl methacrylate) (PMMA) (θ_a: 72.5°, θ_r: 60.3°) with $U = 0.005$ mm/s. The liquid bridge is formed at point A, where the contact angles on both surfaces were between their θ_a and θ_r values. When the liquid bridge is compressed, on each surface the contact angle first increases to the advancing contact angle and then stays at this value until the compression stops (point B). Considering the contact line, it only starts to expand after the contact angle increases to θ_a. Afterward, the contact radii on both surfaces simply expand until the end of the compression stage.

When the acceptor surface starts to separate, due to CAH, the contact lines on both surfaces are first pinned while the contact angles decrease to θ_r, after which the contact radii on the two surfaces begin to shrink. It can be seen that due to CAH, at a given separation H the profile of liquid bridge is not unique. Instead, the shape of the liquid bridge is strongly affected by the history of the contact line motion. For example, the snapshots 1 and 2 in Figure 21.3 are two profiles of the liquid bridge

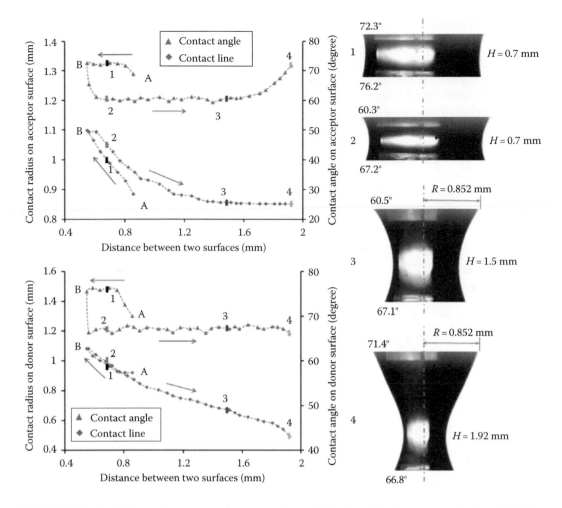

FIGURE 21.3 Evolutions of contact angles and contact radii during a 2 µL water transfer from PEMA to PMMA surfaces. The snapshots 1, 2, 3, and 4 are taken at $H = 0.7$ mm (compression stage), 0.7 mm (stretching stage), 1.5 and 1.92 mm, respectively.

at the same $H = 0.7$ mm, but during the compression and stretching stages, respectively. The contact angles on both surfaces in snapshot 1 are near θ_a, whereas those in snapshots 2 are close to θ_r. With the different values of the contact angles, the liquid profiles in snapshots 1 and 2 are different.

At the end of the stretching stage, the contact lines on the two surfaces exhibit substantially different behaviors. On the PEMA (donor) surface, the contact angle stays at θ_r and the contact radius shrinks until the breakage. However, on the PMMA (acceptor) surface, after the liquid bridge is stretched to some extent, instead of continuing shrinking, the contact radius becomes pinned again at 0.852 mm (this pinned value of contact radius at the end of stretching stage is denoted as R_{min} in this chapter), and the contact angle starts to increase until the breakage. Snapshots 3 and 4 in Figure 21.3 show the profile of liquid bridge at $H = 1.5$ mm and $H = 1.92$ mm, respectively. Clearly on the PMMA surface the contact radius stays the same, but the contact angle increases from 60.5° to 71.4°.

Through this example, it can be seen that due to *CAH*, the shape of the liquid bridge is strongly dependent on the history of the contact line motion. The contact angles mainly stay at θ_a during the compression stage and at θ_r during the stretching stage. The occurrence of contact line pinning in the late stage of stretching, caused by *CAH*, can significantly change the breakage shape and hence the transfer ratio. Studies have shown [18,20] that the value of the contact angle can also significantly affect transfer ratio in this regime in addition to *CAH* and history of the process.

21.4.2 Modeling

A number of numerical models that calculate equilibrium liquid bridge shapes can be found in the literature [15–19]. Given the importance of *CAH* shown in Figure 21.3, any analytical model that attempts to capture the real physical phenomenon must consider *CAH*. One such model was developed in [19] based on the Young–Laplace EQ.

Considering an axisymmetric liquid bridge in equilibrium as shown in Figure 21.4, for a given *H*, the shape of the liquid bridge can be obtained by solving a set of differential equations that describes the geometric relations and equilibrium (Young–Laplace EQ) [33,34]. This set of equations can be written in nondimensionalized form as follows:

$$\frac{dr^*}{ds^*} = \cos\theta, \tag{21.1}$$

$$\frac{dz^*}{ds^*} = \sin\theta, \tag{21.2}$$

$$\frac{d\theta}{ds^*} = \Delta P^* - \frac{\sin\theta}{r^*}. \tag{21.3}$$

Here the dimensionless quantities are defined as

$$r^* = \frac{r}{V^{1/3}}, \quad z^* = \frac{z}{V^{1/3}}, \quad s^* = \frac{s}{V^{1/3}}, \quad \Delta P^* = \frac{\Delta P V^{1/3}}{\gamma}, \tag{21.4}$$

with

    r and z being the radial and vertical coordinates
    s being the arc length measured from the contact point of the liquid with the lower surface
    θ being the angle between the local tangent of the liquid interface and the horizontal axis
    ΔP being the Laplace pressure (pressure difference across the meniscus)

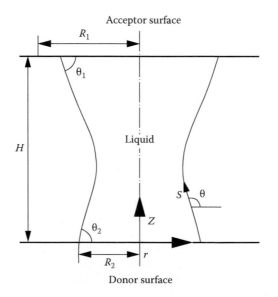

FIGURE 21.4 Geometry and coordinate systems for an axisymmetric liquid bridge in equilibrium between two solid surfaces.

It should be noted that ΔP^* is not known *a priori* and is part of the solution. In order to incorporate *CAH* and solve the Equations 21.1 through 21.3, appropriate boundary conditions at the upper and lower surface need to be prescribed. Two of the boundary conditions are

$$z^*\left(s^* = s^*_{max}\right) = H^* \equiv \frac{H}{V^{1/3}}, \tag{21.5}$$

$$\int_0^{H^*} \pi(x^*)^2 \, dz^* = 1, \tag{21.6}$$

where (21.6) is a constraint equation that states that the volume of the liquid is a constant; s^*_{max} is the normalized arc length between the lower and upper surfaces. Like ΔP^*, s^*_{max} is solved simultaneously with the differential equations. Additional boundary conditions also exist for the contact between the liquid and the surfaces. For each surface, when the contact angle is between θ_r and θ_a, the contact lines are pinned, in which case the contact radius on this surface is used as a boundary condition other than the contact angle. When the contact angle reaches θ_r or θ_a, the contact line is allowed to move and the value of the contact angle is used as a boundary condition. With this strategy, the effect of *CAH* can be captured (more details can be found in [18]).

It is clear from the previous formulation that the nondimensionalized problem is independent of the surface tension, γ, and the liquid volume, V. This implies that all the dimensions of the liquid bridge should simply scale with $V^{1/3}$ and that the liquid profile should be independent of the surface tension. This seemingly surprising result is very useful, as it suggests the transfer ratio, being the ratio between the volume of transferred liquid and the total volume, should be independent of γ and V. Therefore the only other physical parameters that can affect the solution of Equations 21.1 through 21.3 are the contact angles, and near the end of the transfer, the advancing contact angles, are not expected to play significant roles. As such, one can propose that the transfer ratio in the quasistatic regime should be primarily determined by the receding contact angles of the two surfaces.

21.4.3 Prediction of α_0

Based on the previous analysis, the dependence of the transfer ratio α_0 on the receding contact angles of the acceptor surface, $(\theta_r)_{acc}$, and donor surface, $(\theta_r)_{don}$, is explored. Figure 21.5a shows α_0 of a 2 µL water transferred from 10 different donor surfaces to the one and the same acceptor surface (named as cases 1– 10) as a function of the receding contact angle difference $(\Delta\theta_r = (\theta_r)_{don} - (\theta_r)_{acc})$, where $(\theta_r)_{acc} = 60.3°$. A clear monotonic relation between α_0 and $\Delta\theta_r$ can be seen from Figure 21.5a. When $\Delta\theta_r$ is small (negative), for example, cases 1 and 2, the transfer ratio is small and close to zero. When $\Delta\theta_r$ is close to zero, a medium transfer ratio is found. When $\Delta\theta_r$ is positive and large, the transfer ratio approaches one, that is, complete transfer. Such an observation is related to the contact line pinning phenomenon, discussed in Section 21.4.1, at the end of the transfer process. A detailed examination of contact line pinning for the 10 cases here reveals the existence of three domains: Domain I where $\Delta\theta_r$ is small (negative) and contact line pinning can be observed only on the donor surface, Domain II where $\Delta\theta_r$ is near zero and contact line is pinned on both surfaces, and Domain III where $\Delta\theta_r$ is positive and large, and contact line pining can be found only on the acceptor surfaces.

Using the model introduced in Section 21.4.2, a systematic study can be performed for large ranges of $(\theta_r)_{acc}$ and $(\theta_r)_{don}$ to obtain a comprehensive picture of contact line pinning [20]. The result is a two-dimensional map with $(\theta_r)_{acc}$ and $(\theta_r)_{don}$ being the two axes (Figure 21.5b), on which the three domains are clearly identified. Reading the map from left to right along a horizontal line, that is, keeping $(\theta_r)_{acc}$ fixed while increasing $(\theta_r)_{don}$, and hence $\Delta\theta_r$, the contact line pinning changes from Domain I to Domain II and then to Domain III. In addition, in Domains II and III, the pinned

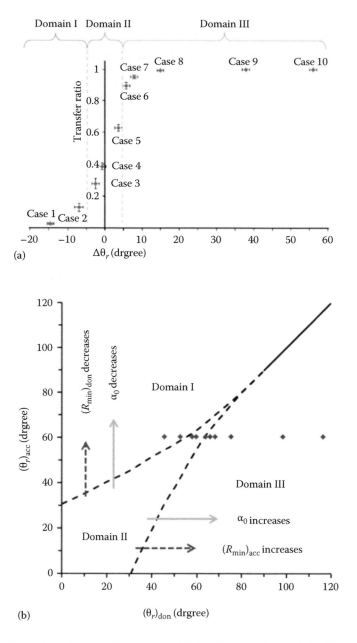

FIGURE 21.5 (a) Transfer ratio as a function of $\Delta\theta_r$ for the experimental liquid transfer cases 1–10. (Reproduced from Chen, H. et al., *Soft Matter*, 10, 2503, 2014.) (b) Map of three domains for contact line pinning near the end of liquid transfer, in terms of $(\theta_r)_{acc}$ and $(\theta_r)_{don}$. The filled diamonds (♦) show $(\theta_r)_{don}$ and $(\theta_r)_{acc}$ of the experimental cases (cases 1–10 from left to right).

contact radius $(R_{min})_{acc}$ on the acceptor surface increases, which is associated with increase in α_0. Reading the map from bottom to top along a vertical line, that is, keeping $(\theta_r)_{don}$ fixed, while increasing $(\theta_r)_{acc}$ and hence decreasing $\Delta\theta_r$, the contact line pinning changes from Domain III to Domain II and then to Domain I. Correspondingly $(R_{min})_{don}$ in Domains II and I increases and α_0 decreases.

It is also interesting to see from Figure 21.5b that the occurrence of contact line pinning is controlled not only by $\Delta\theta_r$, but also by the absolute values of $(\theta_r)_{acc}$ and $(\theta_r)_{don}$. Firstly, when $(\theta_r)_{acc}$ or $(\theta_r)_{don}$ is below 30.7°, only two domains are available. Secondly, the width of Domain II decreases with the increase of $(\theta_r)_{acc}$ and $(\theta_r)_{don}$. If either $(\theta_r)_{acc}$ or $(\theta_r)_{don}$ is larger than $90°$, Domain II vanishes (two dashed lines converge to the solid line), which implies that contact line pinning on both surfaces can occur only when $(\theta_r)_{acc} = (\theta_r)_{don}$. It also suggests that when $(\theta_r)_{acc}$ and $(\theta_r)_{don}$ are greater than $90°$, even if they are only slightly different, the transfer ratio will either be close to zero (when $(\theta_r)_{acc} > (\theta_r)_{don}$) or close to one ($(\theta_r)_{acc} < (\theta_r)_{don}$).

The previous domain map is universal: it only depends on the two receding contact angles and is independent of other liquid properties including surface tension and volume, as long as the transfer is quasistatic and gravity can be neglected. Based on these observations, an empirical model is proposed to predict α_0 in terms of $(\theta_r)_{acc}$ and $(\theta_r)_{don}$ as follows [22]:

$$\alpha_0 = \frac{1}{1 + e^{-m\left((\theta r)acc + (\theta r)don\right)^n * ((\theta r)don - (\theta r)acc)}}, \qquad (21.7)$$

where

$(\theta_r)_{acc}$ and $(\theta_r)_{don}$ are in radians
m and n are positive coefficients to be determined through calibration of experimental results

It can be seen from this equation that when $\Delta\theta_r = (\theta_r)_{don} - (\theta_r)_{acc}$ is very negative, the exponential in the denominator is very large; resulting in α_0 being close to zero. For very large and positive $\Delta\theta_r = (\theta_r)_{don} - (\theta_r)_{acc}$, the exponential function is vanishingly small, and hence α_0 converges to one. The term $(\theta_r)_{don} + (\theta_r)_{acc}$ is added to describe how fast the transfer ratio changes from 0% to 100% because the width of Domain II decreases with the increases of $(\theta_r)_{acc}$ and $(\theta_r)_{don}$. Therefore, Equation 21.7 captures the essential characteristics of the quasistatic transfer ratio observed earlier. The value of the two unknown coefficients m and n were found to be 3.14 and 2.53, respectively, through a regression analysis based on experimental results of four types of liquid (water, glycerol, ethylene glycol, and silicon oil) transfer between 21 pairs of surfaces with very different $(\theta_r)_{acc}$ and $(\theta_r)_{don}$ [22]. Using Equation 21.7, the transfer ratio in the quasistatic regime can be estimated by only knowing $(\theta_r)_{acc}$ and $(\theta_r)_{don}$.

21.5 PREDICTING α IN ALL THREE REGIMES

At higher stretching speeds, the liquid transfer enters the transition and dynamic regimes and the transfer ratio deviates from α_0. As discussed in Section 21.2, the transfer ratio in most applications will be affected by surface tension, contact angle (mainly receding contact angles), and Ca. The effect of Ca can be represented in Figure 21.6a, where four types of silicon oil with very similar surface tension, density, but very different viscosity were transferred from a Teflon AF surface to a PEMA surface with $H_{min} = 0.45$ mm. Due to a similar surface tension, the value of θ_r between the four types of silicon oil and the surfaces are also very similar ($(\theta_r)_{acc} \backsim 0°$ and $(\theta_r)_{don} \backsim 47°$); hence the effect of θ_r is minimal. The transfer ratio as a function of U shows almost the same shape for the four types of liquid, but the curves are shifted to the right with the decrease in liquid viscosity. Because $Ca = U\mu/\gamma$, this suggests that when U is fixed, the ratio of viscous over surface forces of each system is linearly related to the liquid viscosity. Therefore, the viscosity μ serves as a velocity shift in affecting the transfer ratio. This is confirmed by Figure 21.6b where the transfer ratio is plotted as function of Ca and all the four sets of data collapse onto a master curve. The transfer ratio is a constant value ($\backsim 0.92$) when $Ca < 10^{-4}$ (quasistatic regime) and stays at 0.5 when $Ca > 0.05$ (dynamic regime). That is, for the system with $(\theta_r)_{acc} \backsim 0°$ and $(\theta_r)_{don} \backsim 47°$, $Ca = 10^{-4}$ and $Ca = 0.05$ are the two boundary capillary numbers that separate the three regimes of liquid transfer based on the magnitude of slope value of this master curve (0.05 is chosen as the threshold value).

As the values of $(\theta_r)_{acc}$ and $(\theta_r)_{don}$ change, the dependence of α on Ca changes as well. Figure 21.7 shows the transfer ratio as a function of Ca for glycerol transfer from five different donor surfaces to one acceptor surface with $H_{min} = 0.45$ mm at $20.5°C$. All the five groups of data have different

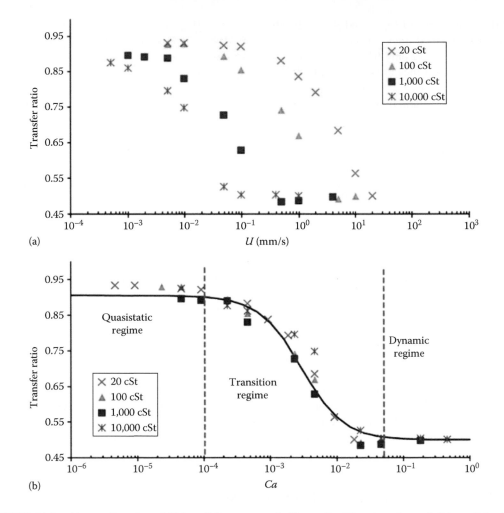

FIGURE 21.6 (a) α as function of U for all four types of silicon oils. The viscosity and $(\theta_r)_{don}$ of the four silicon oils are: 20 cSt, 100 cSt, 1,000 cSt, 10,000 cSt, and 46.1°, 47.2°, 46.5°, 48.3°, respectively. $(\theta_r)_{acc}$ for all four types of silicon oils are ~0°. (b) α as function of Ca for the four types of silicon oils. The solid line is the curve $\alpha = F(Ca)$ fitted using Equation 21.8 with the experimental data.

start values (α_0), due to different $(\theta_r)_{don}$. It can also be noticed that, although all the groups of data converge to 0.5 with the increase of Ca, the boundaries at which α begins to deviate from α_0 and 0.5 are very different. For example, for the transfer from silicon surface, the transfer ratio starts to deviate from α_0 when $Ca \sim 4*10^{-3}$, while for the transfer from Teflon, the transfer ratio remains at α_0 till $Ca \sim 4*10^{-2}$. Since only one type of liquid (glycerol) was used here, the very different relations between α and Ca are caused by the different values of $(\theta_r)_{don}$.

Based on the previous discussion, an empirical equation to predict the transfer ratio $\alpha = F(Ca)$ is proposed as follows:

$$\alpha = 0.5 + \frac{(\alpha_0 - 0.5)}{1 + pCa^q}, \tag{21.8}$$

where

　　α_0 is the transfer ratio in quasistatic regime, and can be calculated using Equation 21.7 knowing
　　　　$(\theta_r)_{acc}$ and $(\theta_r)_{don}$
　　p and q are two positive coefficients

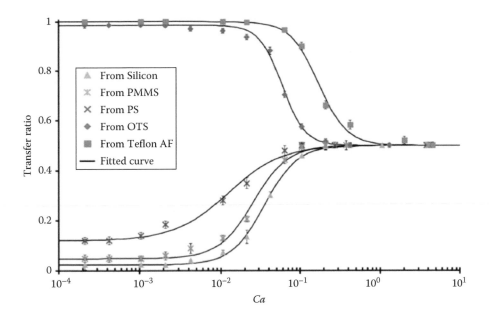

FIGURE 21.7 Transfer ratio as a function of Ca for glycerol transfer from five different donor surfaces to a PEMA surfaces. θ_r between glycerol and PEMA, silicon, PMMA, polystyrene (PS), OTS and Teflon AF surfaces are 61.7°, 37.8°, 53.2°, 56.5°, 83.9°, and 106.7°, respectively. The solid lines are the curves $\alpha = F(Ca)$ fitted using Equation 21.8 with the experimental data. For the glycerol transfer from silicon, PMMA, PS, OTS, and Teflon AF, the values of coefficient p are respectively 1081.53, 1471.03, 301.31, 5313.21, and 82.62; the values of coefficient q are respectively 2.09, 1.99, 1.31, 3.04, and 2.50.

When Ca is very small, the term pCa^q is almost zero. Equation 21.8 predicts the transfer ratio to be α_0, which is in the quasistatic regime. When Ca is very large, the term pCa^q is very large so that the second term on the right side of Equation 21.8 is negligible. The transfer ratio equals to 0.5, which is in the dynamic regime. When the system has a medium value of Ca, the transfer ratio gradually changes from α_0 to 0.5 with the increase of Ca, in the same manner as seen in the experiments. In fact, the solid line in Figure 21.6b is the curve $\alpha = F(Ca)$ fitted with the experimental data. The value of p and q for this specific case are 5408.72 and 1.52, respectively. Based on Table 21.1 in Section 21.2, it can be known that γ, SCA, and μ are the three parameters that could affect the curve $\alpha = F(Ca)$ when the inertial effects are negligible; hence these three parameters govern the value of p and q for a transfer system. As shown in Figure 21.6, μ only serves as a velocity shift in affecting the transfer ratio, and cannot affect the curve $\alpha = F(Ca)$. Therefore, γ and SCA are the two only remaining parameters that could govern the coefficients p and q.

The empirical model Equation 21.8 can be used to estimate the transfer ratio for any specific system. First, by knowing $(\theta_r)_{acc}$ and $(\theta_r)_{don}$ of the system, α_0 can be completely determined using Equation 21.7. Then with only two experimentally measured transfer ratios in the transition regime, the two coefficients p and q in Equation 21.8 can be calculated; hence the expression of $\alpha = F(Ca)$ is completely determined so as to predict the transfer ratio at any Ca.

21.6 SUMMARY AND FUTURE PERSPECTIVES

In this chapter, liquid transfer between two surfaces is categorized into quasistatic, transition, and dynamic regimes. In the quasistatic regime, contact line pinning due to CAH at the end of the transfer process plays an important role; as a result, the transfer ratio is controlled by the receding contact angles $(\theta_r)_{acc}$ and $(\theta_r)_{don}$. Viscosity of the liquid is shown to serve as a velocity shift so that the transfer ratio depends on the stretching speed through the capillary number Ca. An empirical relation $\alpha = F(Ca)$ is proposed, which involves α_0 (as an explicit function of $(\theta_r)_{acc}$ and $(\theta_r)_{don}$), along with two unknown coefficients. With two experiments in the transition regime, the transfer ratio of the system can be obtained for any values of Ca.

In closing, some recommendations for the future study on this topic are provided. First of all, the coefficient p and q in Equation 21.8 are shown to be governed by γ and SCA. However, the relations between p, q, and γ, $(\theta_r)_{acc}$, $(\theta_r)_{don}$ still need to be explored. All of the discussions in this chapter are based on the assumption that the two surfaces are parallel, the liquid is Newtonian fluid, the stretching speed is a constant value and always normal to the surfaces. One or more of these assumptions can be violated in practice. The liquid transfer process in those situations can become more complex than the one discussed in this chapter.

REFERENCES

1. Darhuber, A. A., Troian, S. M. Physical mechanisms governing pattern fidelity in microscale offset printing. *J. Appl. Phys.* 2001, 90, 3602–3609.
2. Kang, H. W., Sung, H. J., Lee, T.-M., Kim, D.-S., Kim, C.-J. Liquid transfer between two separating plates for microgravure-offset printing, *J. Micromech. Microeng.* 2009, 19, 015025.
3. Prakash, M., Quere, D., Bush, J. W. M. Surface tension transport of prey by feeding shorebirds: The capillary ratchet. *Science* 2008, 320, 931–934.
4. Gorb, S. The design of the fly adhesive pad: Distal tenent setae are adapted to the delivery of an adhesive secretion. *Proc. R. Soc. Lond. B* 1998, 265, 747–52.
5. Eisner, T., Aneshansley, D. J. Defense by foot adhesion in a beetle (hemisphaerota cyanea). *Proc. Natl. Acad. Sci.* 2000, 97, 6568–6573.
6. Dixon, A. F. G., Croghan, P. C., Gowing, R. P. The mechanism by which aphids adhere to smooth surfaces. *J. Expl. Biol.* 1990, 152, 243–253.
7. Pudas, M., Hagberg, J., Leppavuori, S. Printing parameters and ink components affecting ultra-fine-line gravure-offset printing for electronics applications. *J. Eur. Ceram. Soc.* 2004, 24, 2943–2950.
8. Hong, C. M., Wagner, S. Inkjet printed copper source/drain metallization for amorphous silicon thin-film transistors. *IEEE Electron. Device Lett.* 2000, 21, 384–386.
9. Kopola, P., Aernouts, T., Guillerez, S., Jin, H., Tuomikoski, M., Maaninen, A., Hast, J. High efficient plastic solar cells fabricated with a high-through put gravure printing method. *Solar Energy Mater. Solar Cells* 2010, 94, 1673–1680.
10. Reis, P. M., Jung, S., Aristoff, J. M., Stocker, R. How cats lap: Water uptake by felis catus. *Science* 2010, 330, 1231–1234.
11. Barbulovic-Nad, I., Lucente, M., Sun, Y., Zhang, M., Wheeler, A. R., Bussmann, M. Biomicroarray fabrication techniques—A review. *Crit. Rev. Biotechnol.* 2006, 26, 237–259.
12. Chadov, A. V., Yakhnin, E. D. Investigation of the transfer of a liquid from one solid surface to another. 1. slow transfer method of approximate calculation. *Kolloidn. Zh.* 1979, 41, 817.
13. Yakhnin, E. D., Chadov, A. V. Investigation of the transfer of a liquid from one solid surface to another. 2. Dynamic transfer. *Kolloidn. Zh.* 1983, 45, 1183.
14. Fortes, M. A. Axisymmetric liquid bridges between parallel plates. *J. Colloid Interface Sci.* 1982, 88, 338–352.
15. Samuel, B., Zhao, H., Law, K.-Y. Study of wetting and adhesion interactions between water and various polymer and superhydrophobic surfaces. *J. Phys. Chem. C* 2011, 115, 14852–14861.
16. Qian, B., Loureiro, M., Gagnon, D., Tripathi, A., Breuer, K. S. Micron-scale droplet deposition on a hydrophobic surface using a retreating syringe. *Phys. Rev. Lett.* 2009, 102(16), 164502.
17. Qian, B., Breuer, K. S. The motion, stability and breakup of a stretching liquid bridge with a receding contact line. *J. Fluid Mech.* 2011, 666, 554–572.
18. Qian, J., Gao, H. Scaling effects of wet adhesion in biological attachment systems. *Acta Biomater.* 2006, 2, 51–58.
19. Chen, H., Amirfazli, A., Tang, T. Modeling liquid bridge between surfaces with contact angle hysteresis. *Langmuir* 2013, 29, 3310–3319.

20. Lambert, P., Chau, A., Delchambre, A. Comparison between two capillary forces models. *Langmuir* 2008, 24, 3157–3163.
21. Chen, H., Tang, T., Amirfazli, A. Liquid transfer mechanism between two surfaces and the role of contact angles. *Soft Matter*, 2014, 10, 2503–2507.
22. Chen, H., Tang, T., Amirfazli, A. Transfer by pinning: how contact angle hysteresis governs quasi-static liquid transfer. *Soft Matter,* 2015, submitted.
23. Cheung, E., Sitti, M. Adhesion of biologically in spired oil-coated polymer micropillars. *J. Adhesion Sci. Tech.* 2008, 22, 569–589.
24. Hong, S.-J., Chou, T.-H., Chan, S. H., Sheng, Y.-J., Tsao, H.-K. Droplet compression and relaxation by a superhydrophobic surface: Contact angle hysteresis. *Langmuir* 2012, 28, 5606–5613.
25. Souza, E. J. D., Brinkmann, M., Mohrdieck, C., Crosby, A., Arzt, E. Capillary forces between chemically different substrates. *Langmuir* 2008, 24, 10161–10168.
26. Meseguer, J., Sanz, A. Numerical and experimental study of the dynamics of axisymmetric slender liquid bridges. *J. Fluid Mech.* 1985, 153, 83–101.
27. Zhang, X., Padgett R. S., Basaran, A. Nonlinear deformation and breakup of stretching liquid bridges. *J. Fluid Mech.* 1996, 329, 207–245.
28. Yildirim, O. E., Basaran, O. A. Deformation and breakup of stretching bridges of Newtonian and shear-thinning liquids: Comparison of one- and two-dimensional models. *Chem. Eng. Sci.* 2001, 56, 211–233.
29. Bai, S.-E., Shim, J.-S., Lee, C.-H., Bai, C.-H., Shin, K.-Y. Dynamic effect of surface contact angle on liquid transfer in a low speed printing process. *Jpn. J. Appl. Phys.* 2014, 53, 05HC05.
30. Ahmed, D. H., Sung, H. J., Kim, D.-S. Simulation of non-Newtonian ink transfer between two separating plates for gravure-offset printing. *Inter. J. Heat Fluid Flow* 2011, 32, 298–307.
31. Dodds, S., Carvalho, M., Kumar S. Stretching and slipping of liquid bridges near plates and cavities. *Phys. Fluids* 2009, 21, 092103.
32. Dodds, S., Carvalho, M., Kumar, S. Stretching liquid bridges with moving contact lines: The role of inertia. *Phys. Fluids* 2011, 23, 092101.
33. Lam, C. N. C., Wu, R., Li, D., Hair, M. L., Neumann, A. W. Study of the advancing and receding contact angles: Liquid sorption as a cause of contact angle hysteresis. *Adv. Colloid Interface Sci.* 2002, 96, 169–191.
34. Decker, E. L., Frank, B., Suo, Y., Garoff, S. Physics of contact angle measurement. *Colloids Surf. A* 1999, 156, 177–189.

22 Solutal Marangoni Convection: Challenges in Fluid Dynamics with Mass Transfer

Mohsen Karbaschi, Nina M. Kovalchuk, Aliyar Javadi, Dieter Vollhardt, and Reinhard Miller

CONTENTS

22.1 INTRODUCTION

Marangoni instabilities at liquid–fluid interfaces reveal itself as a spontaneous convection caused by interfacial tension gradients supported over time by feedback mechanisms. The development of convective motion in otherwise diffusive systems intensifies considerably mass transfer and can result in local interfacial deformations. That is why studies of such instabilities are of great interest for many fundamental and applied areas, such as foams and emulsions, food and cosmetic products, bio- and nanotechnology, and pharmacy.

The phenomenon of convection due to surface tension gradients was first qualitatively described by Carlo Marangoni [1]. However, earlier in 1855, Thomson observed remarkable movements at the surface of wines and other alcoholic beverages [2]. The easiest way to observe Marangoni convection is to touch a liquid interface with a hot needle or to deposit a droplet of a surfactant solution. In both cases convection will terminate very quickly, as it promotes fast equilibration of temperature or concentration distribution over the interface. To support convection during a long time a mechanism should be provided to maintain the surface tension gradient, for example, by continuous heating one side of an interface while cooling the opposite side. A Marangoni instability enables convection to be established in the system without any imposed tangential gradient of surface tension. It amplifies up to a macroscopic scale any microscopic gradients of surface tension caused by fluctuations of concentration or temperature. The development of any instability is only possible if there is a positive feedback in the system. For Marangoni instabilities, a feedback usually appears due to the imposed temperature or concentration gradient in the direction perpendicular to the interface. The easiest mechanism of Marangoni instability will be considered in more detail later on. Here it should be only stressed that the presence of normal, but not a tangential concentration gradient, is a necessary condition for the onset of a Marangoni instability.

Marangoni instabilities are caused by temperature [3,4] or concentration gradients of surface active agents [5]. For the last type of instabilities, the diffusion–convection mass transfer of surfactant molecules at the surface and in the adjacent bulk, and also the kinetics of surfactant adsorption at the fluid surface influence the dynamics of this interfacial instability [6].

Pearson was the first who, in 1958 [7], considered the problem theoretically of thermal Marangoni instabilities, followed by Sternling and Scriven [8], who considered solutal instabilities. Since then much work was done on Marangoni instabilities at fluid interfaces [9], however, there are still many unsolved problems in the understanding of interfacial instabilities [10–13].

The problem is more complicated when considering the dynamics of momentum, heat, and mass transfer, which are influenced by any type of interfacial instabilities. The understanding of this complex mutual interaction in the dynamics of interfacial layers requires accurate experimental studies combined with numerical investigations [14]. In literature, there are various attempts to study experimentally the complexities of systems with Marangoni instabilities. For example, the importance of geometrical parameters as well as convective flow patterns for Marangoni instabilities in growing drop experiments was recently reported in [15]. For more examples, one can also see [16–18]. The use of computational techniques to study interfacial instabilities was also proposed by many authors. For example, the effect of Marangoni instabilities on mass transfer was numerically studied by Mao and Chen [19]. In another study, Grahn [20] presented experimental investigations and numerical simulations of density-driven instabilities, together with surface tension-driven Marangoni convection. More examples can be found in [18,21,22].

Although there are many works published on numerical studies of interfacial instabilities, the use of dynamic boundary conditions for the accumulation and transport of surfactant molecules across and along the interface is a challenge in particular with respect to the conservation of mass and momentum for solving the Navier–Stokes equations (see Chapter 9). Because of these complexities in the numeric handling of dynamic interfacial properties, computations are often restricted to problems with special assumptions that are sometimes far from reality. Therefore, not many numerical studies performed on interfacial instabilities are accompanied by corresponding experimental investigations. These challenges together with numerical difficulties in the accurate description of surface deformations can lead to results that can be only qualitatively verified by experiments [18,22]. A literature overview on numerical studies of Marangoni instabilities was recently published in [10]. On the other hand, due to the mentioned complexities, any analytical solutions are restricted to only the simplest cases. Thus, to have a clear understanding of the onset and development of interfacial instabilities, comprehensive computations are required and must be well combined with accurate experimental measurements. In this chapter, examples are considered wherein mathematical models match well with the real experimental systems with planar and curved surfaces. Therefore, experimental results on Marangoni instabilities are discussed in detail on the basis of numerical simulations and in turn numerical results are validated by tailored experiments. This chapter is arranged as follows. First, the criteria of Marangoni instabilities generated at fluid interfaces due to surfactant transfer are discussed in detail. Thereafter, two distinct cases are discussed in detail based on interconnected experimental and numerical studies. In the first case, surfactant is transferred to and through a planar surface generating quasistationary convections or oscillations depending on the physicochemical parameters and geometry. In the second example, experimental studies are reported on convective instabilities caused by forced convective mass transport of surfactant molecules into a drop during the process of in situ drop-bulk exchange. These experiments are modeled by using computational fluid dynamic to define the effect of the convective flow field on the properties of Marangoni instabilities developed at the mobile fluid drop interface. In addition, an overview is given on the state of the art computational fluid dynamics (CFD) for solutal Marangoni instabilities. In this chapter, we only report on the details of applied computational techniques, while the basis of computations in interfacial science is reported in Section II of the book.

22.2 CRITERIA OF INSTABILITY

A theoretical consideration of solutal Marangoni instabilities was first motivated by developments in theory and praxis of liquid–liquid extractions. It was noticed that, in many cases, the mass transfer coefficients were much larger than predicted by the diffusion theory [23]. In many of these cases,

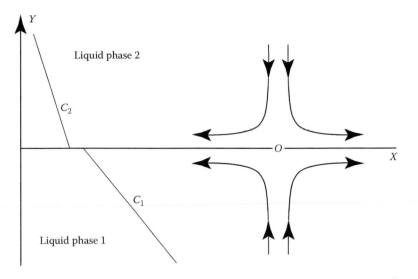

FIGURE 22.1   Mechanism of solutal Marangoni instability.

the extraction was accompanied by a well visible interfacial turbulence, that is, the mass transfer resulted in the development of a spontaneous interfacial motion. That is why the Marangoni instability was assumed to be the reason of the intensified mass transfer [23]. Therefore, to find conditions at which a Marangoni instability can develop was not only of scientific, but also of industrial interest, because such criteria would enable an essential improvement in extractor performance.

To derive necessary criteria, Sternling and Scriven [8] considered a simple model presented in Figure 22.1. Two liquid phases, 1 and 2, containing a surfactant are brought into contact to form planar interface. The concentration in phase 1 is higher than in phase 2 and therefore mass transfer occurs through the interface. The model presented in [8] assumes that both phases are semi-infinite and the surfactant concentration changes linearly in each phase.

Following Sternling and Scriven [8], let us consider the following stationary initial state: the liquid in both phases is motionless, the concentration gradients do not change with time, and the flux to the interface from phase 1 is equal to the flux from the interface into phase 2. It is assumed that due to natural fluctuations, which always appear in real systems, the surfactant concentration in point O (Figure 22.1) increases and therefore the interfacial tension in this point decreases. The fluctuation of interfacial tension makes the liquid move at the interface from point O. Due to the flow continuity condition, the liquid in the bulk of both phases will move in the direction perpendicular (normal) to the interface in point O as shown in Figure 22.1. Note, in this stage the liquid velocity remains so small that the contribution of convection to the mass transfer is negligible.

The further scenario depends on the ratio of convective fluxes of surfactant from phase 1 and phase 2 to the interface. The flux from the phase 1 is directed from the region of higher concentration and therefore transports the solution rich in surfactant to the interface, whereas the flux from phase 2 brings diluted (in comparison with that near the interface) solution. If the convective flux from phase 2 is larger, then the concentration in point O and the interfacial velocity decrease, that is, the fluctuation decays and an instability does not develop. Therefore, it can be concluded that in this case the system is stable with respect to possible fluctuations and only a diffusion mass transfer occurs. If, however, the flux from phase 1 is larger, then the concentration in point O increases further, resulting in an increase of the velocity and subsequently of the surfactant flux. This feedback leads to a rapid increase in the convective velocity, which becomes noticeable on a macroscopic level and contributes considerably to the mass (surfactant) transfer. Hence, the system is unstable and the Marangoni instability leads to the development of spontaneous convection in the initially motionless system.

The ratio of fluxes depends on the ratio of the diffusion coefficients and viscosities in the two liquid phases. If the diffusion coefficient D_1 in phase 1 is smaller than D_2, then the concentration gradient in this phase is larger (due to the equality of diffusion fluxes). Hence, the convective flux should be larger in this phase at the same flow velocity. There is also an additional effect working in the same direction. For a smaller diffusion coefficient, the loss of surfactant from the center to the interface is smaller due to side diffusion. This facilitates an increase of the surface concentration gradient, that is, the smaller diffusion coefficient in phase 1 favors instabilities. The increase in kinematic viscosity, ν, results in an increase of the bulk velocity, provided that the interfacial velocity is the same, and therefore in an increase of the normal flux. Therefore, a higher viscosity also favors instabilities.

Sternling and Scriven [8] performed a quantitative analysis of the stability of the system shown in Figure 22.1, assuming that the velocities and the changes in concentrations are so small that the nonlinear terms in the Navier–Stokes and convective diffusion equations can be neglected. That is why this approach is called linear stability analysis. The results obtained in [8] are presented in Table 22.1 and are in agreement with the qualitative analysis discussed previously: the system is stable if the transfer happens from the phase with the larger diffusion coefficient and smaller kinematic viscosity. In all other cases, an instability is possible and its regime (stationary convection or oscillation) depends on the ratios of D_1/D_2 and ν_1/ν_2.

The model discussed in [8] has been further improved by accounting for deformations of the interface, interfacial adsorption, and mass transfer within the interface [24–26]. According to these studies the effect of an interfacial deformation can be neglected in many cases, but the system could be unstable at any direction of surfactant transfer. Only the ratios of diffusion coefficients and viscosities determine character of instability. If $D_1\nu_2/D_2\nu_1 < 1$, instability develops as a steady convection, whereas for $D_1\nu_2/D_2\nu_1 > 1$ an oscillatory regime can be expected.

It has been also shown that the depth of a liquid layer [27,28] and the adsorption kinetics [29] influence considerably the onset of instabilities. Criteria for spherical interfaces were derived in [30].

It should be stressed that the criteria discussed previously are only necessary, but not sufficient conditions for the generation of an instability. An instability can develop only if the concentration gradient exceeds a certain threshold value. This critical value depends on other system parameters, and it is included in the characteristic Marangoni number. Instabilities evolve to a macroscopic level if the Marangoni number exceeds a certain threshold value [25,27].

The criteria discussed previously are derived in the framework of a linear approximation. However, when an instability develops, the velocities of spontaneous convection become very large. Then the question appears whether the instability criteria are still true for macroscopic instabilities or if nonlinear effects can change the evolution of an instability. So far, there is no analytical answer to this question. Therefore, numerical studies are of interest. Two examples of such studies are given as follows.

TABLE 22.1
Criteria of Instability Caused by Surfactant Transfer from Phase 1 to Phase 2

D_1/D_2	ν_1/ν_2	System Stability
≥ 1	<1	Stable
>1	≤ 1	Stable
≤ 1	>1	Unstable (oscillatory regime or stationary convective cells)
<1	<1	Unstable (stationary convective cells)
>1	>1	Unstable ($D_1/D_2 > \nu_1/\nu_2$—oscillatory regime; $D_1/D_2 < \nu_1/\nu_2$—oscillatory regime or stationary convective cells)

Source: According to Sternling, C.V. and Scriven, L.E., *AIChE J.*, 5, 514, 1959.

22.3 AUTO-OSCILLATION BEHAVIOR OF FLUID INTERFACES: DIFFUSIONAL INITIAL STATE

The material presented here was motivated by experimental data being in contradiction to the instability criteria considered in the previous section. Three systems presented in Figure 22.2 have been considered: The system A resembles the one shown in Figure 22.1: two immiscible liquids, one of them containing a surfactant out of partition equilibrium, are brought into contact giving rise to surfactant transfer [6,31–34]. Both liquid phases can contain other solutes, also out of partition equilibrium, so that mass transfer can occur in both directions. In the system B, a surfactant is transferred from a source positioned in one of the liquid bulks. The source can be a droplet of sparingly soluble surfactant formed at the tip of a capillary, as shown in Figure 22.2b [35–38], or a surfactant solution slowly supplied by a syringe [39]. In this configuration diffusional mass transfer can be accompanied by buoyancy convection due to the concentration dependent density of the solutions and/or forced convection due to the supply from the syringe. The liquid membrane in system C is composed of three liquid phases [40]. Each phase can contain various solutes, but it necessarily contains a surfactant in phase 1, which is transferred through phase 2 (liquid membrane) to phase 3. Very often another configuration, such as a U-shaped tube, is used to study liquid membrane oscillators [41,42]. At closer consideration it is clear that system C is a combination of systems A and B, with conditions at the interface "a" similar to case A, whereas conditions at the interface "b" are similar to case B. In all three systems, the instability criteria predict a stationary convection, but experimentally oscillations were observed.

The measured parameter for the analysis of the oscillations is the interfacial tension or, for ionic surfactants, the electric potential difference across the interface. It was shown in [39,43] that the oscillations of interfacial tension are completely synchronized with the oscillations of the electrical potential. The oscillations have an asymmetric shape with an abrupt decrease of interfacial tension followed by a gradual increase, as shown in Figure 22.3. As one can see, the abrupt decrease of the interfacial tension is accompanied by the appearance of a very fast interfacial flow [35,41]. Therefore, it was assumed that the oscillations are the result of Marangoni instabilities.

It should be noted that the characteristics and even the development of an oscillatory regime depended essentially on the system geometry. For example, in system B the oscillation developed only if the ratio of source distance from the interface to the beaker diameter was larger than a certain critical value [36,44], otherwise a quasisteady convection was observed in agreement to the instability criteria.

To understand why there is a contradiction between the linear stability theory and experimental observations, and why in the same system just a change of geometrical parameters causes a change of the instability regime, numerical simulations have been performed based on a nonlinear model,

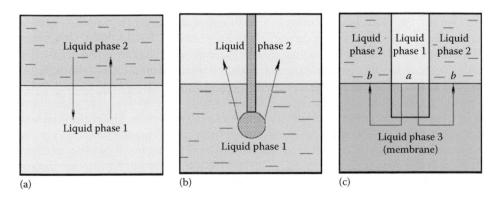

FIGURE 22.2 Sketch of three systems displaying oscillatory regimes of Marangoni instability in contradiction to the predictions of linear stability analysis. (a) The two phase system with a surfactant out of partition equilibrium, (b) the system with a dissolving surfactant droplet, and (c) the liquid membrane system.

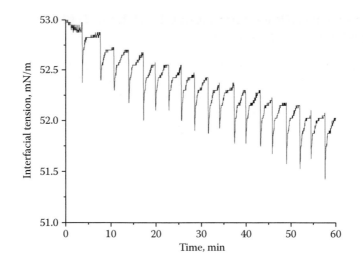

FIGURE 22.3 Oscillations of interfacial tension at the water/dodecane interface caused by the dissolution of a heptanol droplet in water (experimental data). (Redrawn from Kovalchuk, N.M. and Vollhardt, D., *Adv. Colloid Interface Sci.*, 120, 1, 2006.)

including the full nonsteady Navier–Stokes, convective diffusion, and continuity equations. The system geometry was directly incorporated into the model. The further features of the mathematical model were as follows: A cylindrical symmetry was imposed in accordance to the experimental observations on flow visualization, allowing to reduce the 3D case to a 2D one by introducing a respective stream function and vorticity. Only longitudinal deformations of the interface have been considered, that is, transverse deformations of the interface were neglected. A diffusion limited adsorption kinetics was assumed with an instantaneous local adsorption equilibrium obeying the Langmuir adsorption model. Langmuir–Szyszkowski equation was used for the interfacial tension isotherm (see Chapter 1). The equation for the interfacial mass transfer took the compressibility/expansibility of the interfacial layer into account. The buoyancy force was taken into account where appropriate. The details of the mathematical models and main results of simulations can be found in [33] for the system A, in [45] for the system B (liquid-air interface), in [37] for the system B (liquid-liquid interface) and in [46] for the system C, surface "b".

An explicit finite difference method with a regular grid has been used for these simulations. The equation for the stream function was solved by the Gauss–Seidel iterative method. To solve the equation for the vorticity and convective diffusion equation, a two-point forward difference approximation was used for the time derivatives, three-point centered differences were used for the diffusion terms, and modified upwind differences were used for the convective terms. The use of upwind differences for the convective terms introduced an artificial diffusivity into the numerical scheme, but it was proved in [45] that the contribution of this diffusivity to the total mass transfer is negligible. The used numerical scheme is very simple; nevertheless, it produced meaningful results in terms of surfactant and velocity distribution over the system. The performed simulations allowed to develop a consistent theory of auto-oscillations and to predict protocols for the control of the appearance of oscillations and their characteristics. The last results were confirmed by precisely tailored experiments. Taking into account that the oscillation mechanism has the same main features in all three systems, we are allowed to consider the simplest case, namely, the transfer of a surfactant to the liquid–air interface from a small droplet placed at the tip of a capillary positioned in the liquid bulk (see Figure 22.2b).

If buoyancy is negligible, then the initial surfactant transfer in the system presented in Figure 22.2b occurs due to diffusion and the surfactant distribution has a spherical symmetry around the droplet. When the surfactant molecules reach the interface this symmetry breaks, because the surfactant transfer begins to occur not only through the bulk but also along the interface. Moreover, the surfactant concentration directly above the droplet, in the vicinity of the capillary, is always higher than at any more distant place of the interface because of the difference in the path's length. As a result, a convection develops in the system. It is important to note, that similar to the case of Figure 22.1, the convection brings more surfactant from the droplet to the interface in the vicinity of the capillary, that is, the same feedback as discussed before exists in this system and results in the development of a Marangoni instability. The numerical simulations show a very quick increase in the convection velocity (up to values of the order of mm/s) accompanied by the adsorption of a large amount of surfactant in the vicinity of the capillary and spreading over the interface. As a result, the interfacial tension decreases abruptly. The abrupt changes in the surface tension are of the order of several mN/m depending on the surfactant properties.

Until this point, the results of the numerical simulations are in good agreement with the instability criteria. If the beaker diameter is large or the droplet immersion depth is small enough, the numerical simulations predict that an instability develops in the regime of quasisteady convection, in full accordance with the experimental results and theoretical criteria. However, in agreement with experimental data, numerical simulations predict oscillatory instabilities (Figure 22.4a) at larger droplet immersion depths and/or smaller beaker diameters.

The results of the numerical simulations show the reason for the oscillatory behavior. During the development of the instability the surface velocity increases considerably as does the convective surfactant transfer over the major part of the interface. At the same time the velocity drops down to zero at the beaker wall. This velocity gradient results in a local surface contraction near the wall and an essential increase of the surfactant concentration in the wall region. A reverse concentration gradient near the wall causes the termination of the instability. Then the system does not only return to a motionless state, but even demonstrates for a short time a very slow reverse convective motion at the surface directed from the wall to the capillary. At this time the supply of surfactant from the droplet to the interface is interrupted and only desorption from the interface takes place. This is the moment of a gradual decrease of surface tension. After a certain time the surfactant supply to the interface is renewed and the instability develops again giving, in such a way, rise to repeated oscillations. The value and duration of the reverse concentration gradient at the interface depends of the surfactant properties and the geometry of the system. That is why the increase in the beaker diameter or decrease in the droplet immersion depth results in transition from an oscillatory regime to a quasisteady convection. The effect of the system's geometry is discussed in detail in [47].

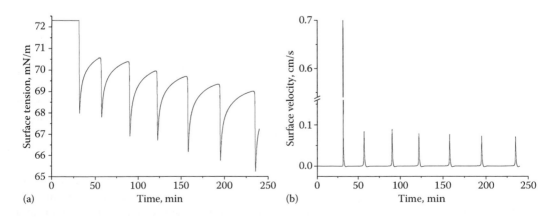

FIGURE 22.4 Oscillation of surface tension (a) and surface velocity (b) at water/air interface produced by dissolution of octanol droplet (numerical results, model without buoyancy). (Redrawn from Kovalchuk, N.M. and Vollhardt, D., *J. Phys. Chem. B*, 107, 8439, 2003.)

The comparison of graphs in Figure 22.4a and b displays clearly a correlation between the changes in velocity and surface tension. One can clearly see that the fast stage, when the instability develops, terminates very quickly while the slow diffusional stage lasts longer. Note, the velocity increases to much higher values at the first oscillation as compared to the subsequent ones, which is also confirmed by experimental observations.

When comparing Figure 22.4 with Figure 22.3 one can see that the oscillation amplitude and especially the oscillation period are rather different in these two cases. The reason for the difference in amplitudes is the dependence on the surfactant properties [48] and especially on the partition coefficients of the solution components [37]. In the water–air system the whole surfactant remains inside the aqueous phase and at the interface, whereas in the water–dodecane system the partition of heptanol is in favor of dodecane. Numerical simulations have shown that the oscillation amplitude decreases considerably with the increasing partition coefficient, as confirmed experimentally in [37]. The difference in periods is mainly due to the buoyancy force [37].

The results of numerical simulations provide guidance to how to control the dynamic regimes during the mass transfer in the considered systems. Transition from quasisteady convection to an oscillatory regime can be achieved by adding initially some amount of surfactant to the interface [49,50]. The oscillation amplitude can be changed by changes of the partition coefficient or pH, as also confirmed by experiments [51].

To summarize, numerical simulations have shown that, in the three considered systems, Marangoni instabilities initially develop as a quasisteady convection in full accordance with the instability criteria derived in a linear approximation. However, when the flow velocity increases significantly, the nonlinear effects due to the local contraction of the interface near the beaker walls terminate the instability. Periodically arising and terminating Marangoni instabilities result in the experimentally observed oscillations.

22.4   CONVECTIVE MARANGONI INSTABILITIES AT A DROP SURFACE

This section is dedicated to experiments of Marangoni instabilities in a pendent drop by using a coaxial double capillary system. The inflow pattern of the surface active dye injected into a droplet in a pulse-like manner is visualized by images of the droplet. In addition, the dynamic surface tension is measured by fitting the Gauss–Laplace equation to the drop profiles. From the results, the dynamics of Marangoni instabilities due to the local arrival of surface active molecules at the drop surface is studied for different injection rates and experimental conditions. The experiments are supported by CFD simulations for a quantitative understanding of interfacial phenomena occurred in the system. The numerical results can well describe the development of Marangoni instabilities at the drop surface. However, there is a delay in the onset of instabilities that is not recognized by numeric simulations. This delay in the presence of a Marangoni instability is most probably due to the mobility of the drop surface that prevents the formation of a quasistatic adsorbed layer as discussed in [10,52]. The dynamic flow values in the bulk and at the drop surface are used to characterize the mobility of the drop surface during the process of drop-bulk exchange.

The experiments were performed by using the profile analysis tensiometry (PAT) equipped with a so-called coaxial double capillary system (SINTERFACE Technologies, Berlin, Germany). Figure 22.5. shows a schematic picture of the setup.

In brief, a drop of water is formed at the circular tip of a capillary by using syringe 1. After a while, the drop is quiescent and the solution of surface active dye is injected into the drop through syringe 2 (picture c). The injection is performed in a pulse-like manner with a controlled dosing rate and time interval between the subsequent pulses. The drop volume is kept at a constant value through a feedback control system performed by the software of the PAT via syringe 1. During the exchange process, the concentration profile of the dye solution inside the drop can be recorded by a CCD camera. The surface tension values are also measured simultaneously from the drop profile. More details of the setup can be found in [52].

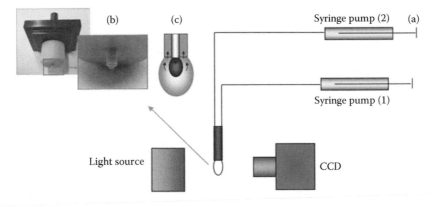

FIGURE 22.5 (a) Syringes for injection and withdrawal of liquid to and from the droplet interior, (b) coaxial double capillary (SINTERFACE Technologies, Berlin, Germany), and (c) schematic picture of the double capillary during the bulk exchange process.

The process of drop-bulk exchange was visualized experimentally and also modeled via CFD simulations. In these experiments, a solution of the low molecular weight dye brilliant green ($C = 3.3$ mg/mL, $D \approx 5 \times 10^{-6}$ cm^2/s) is used for injection into the drop. The density of the dye solution is slightly larger than that of water as it flows toward the drop bottom. The pulse-like imposed inflow is accurately controlled at time intervals of 0.1 s between two subsequent pulses. The experiments are performed for different dosing rates of 0.033 and 0.1 mm^3/pulse with a 20 mm^3/s pumping rate. Figure 22.6. presents some experimental snapshots and CFD simulations of the concentration contours for different dosing rates and times from the beginning of injections. As one can see, there is a good agreement between the experimental and simulation results. This presents well the capability of the applied numerical approach to describe flow properties in the drop-bulk exchange process. However, the simulation results show a perfect axisymmetric propagation due to the considered ideal boundary conditions. Such conditions do not really exist in the experiments due to small perturbations or a slightly asymmetric geometry of the capillary. The effect of different experimental and geometrical parameters on the flow field and exchange process was discussed in [52].

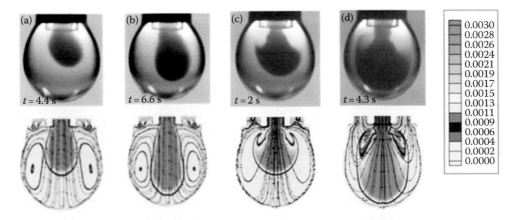

FIGURE 22.6 Experimental snapshots and numerical results of the dye solution injection into the drop (values in g/mL): (a) $dV = 0.033$ mm^3/pulse and $t = 4.4$ s, (b) $dV = 0.033$ mm^3/pulse and $t = 6.6$ s, (c) $dV = 0.1$ mm^3/pulse and $t = 2$ s, and (d) $dV = 0.1$ mm^3/pulse and $t = 4.3$ s.

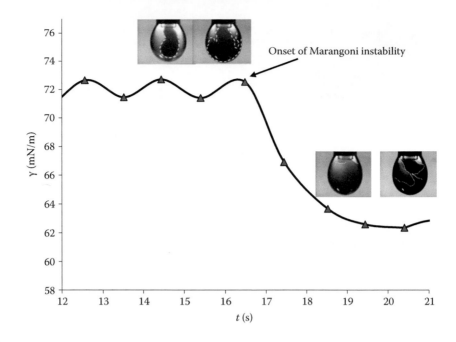

FIGURE 22.7 Dynamic surface tensions measured close to the onset of the Marangoni convection for $dt = 0.1$ s and $dV = 0.033$ mm³/pulse.

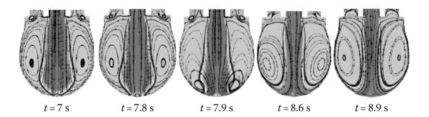

FIGURE 22.8 Simulated evolution of the Marangoni convection at the drop apex for $dt = 0.1$ s and $dV = 0.033$ mm³/pulse.

After a certain time of the drop-bulk exchange process the surfactant solution reached the drop apex. The adsorption of the surfactant molecules at some places of the drop surface leads to a surface tension gradient and thus a Marangoni instability occurs. The dynamic surface tension measurements show the onset and evolution of the Marangoni instability. According to Figure 22.7, the surface instability is developed in about 2 s. The development of the Marangoni instability was also numerically investigated via CFD simulations, the results of which are displayed in Figure 22.8. It starts with the moments when the surfactant solution reaches the drop apex and shows also the subsequent onset of the surface instability. The local variation of surface tension during the evolution of the Marangoni instability is also presented in Figure 22.9. The CFD results are in good agreement with the experimental data. However, in the computations, the Marangoni instability starts right after the surfactant molecules have reached the drop's surface. On the other hand, experiments show a delay of around 10 s for the onset of the Marangoni instability. There might be different reasons for such a delay in interfacial instability, however, the most probable reason can be the mobility of the drop's surface, which resists the formation of a quasistatic adsorbed layer at the drop surface, as discussed in [10].

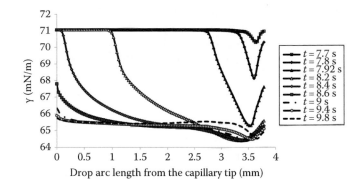

FIGURE 22.9 Surface tension distribution versus drop arc length from the capillary tip during the Marangoni instabilities for $dt = 0.1$ s and $dv = 0.033$ mm³/pulse.

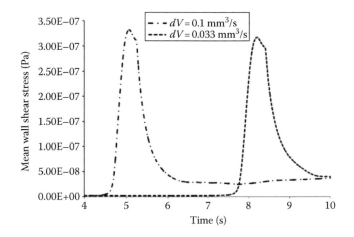

FIGURE 22.10 Mean wall shear stress and the development of Marangoni convection at the drop surface.

Marangoni convection is described by means of a shear stress tangential to the surface. This shear stress is calculated from the tangential surface tension gradient and is balanced by a viscose stress resulting from a velocity gradient in the bulk. The variation of surface shear stress values at the drop surface correlates with the strength of surface instability. The computationally obtained mean wall surface shear stress values are presented in Figure 22.10. A comparison between the results of surface instability for two different dosing rates show the independence of the strength of instability from the flow patterns in the surface adjacent bulk. However, the dynamics of the flow field affected by this interfacial instability is directly defined by the characteristics of surfactant adsorption at the drop surface, as discussed in [14].

22.5 CONCLUSIONS AND REMARKS

In this chapter, we discussed the solutal interfacial instabilities and Marangoni flows in different systems. The criteria of instability are presented for different system parameters or geometrical conditions. Two different individual cases of interfacial Marangoni convections are discussed here. In the first case, surfactant molecules are transferred mostly by diffusion to and through a planner surface, generating quasistationary interfacial convection or oscillations depending on the physicochemical properties of the system and its geometry. In the second case, surfactant molecules are transferred to a curved free surface of a droplet using a coaxial double capillary system. For both cases, computational simulations match very closely the obtained experimental results.

The target of this chapter was to emphasize different critical conditions where Marangoni convection may result in a surface deformation or, on the contrary, in a quasistationary interfacial convection. In addition, the combination of Marangoni convection with mass transfer is discussed in detail. This discussion includes interfacial deformations and oscillations, different mechanisms of surfactant transport to and through the interface, subsequent interfacial convections, and also conservation of mass in the bulk and at the surface coupled with the adsorption kinetics of surfactants at the mobile fluid interfaces. However, there are more complexities connected with Marangoni interfacial convection and instabilities, which are not well understood yet, such as interface restrictions against instabilities and delayed Marangoni convections reported in Sections 22.3 and 22.4. Therefore, for a better understanding of interfacial phenomena, more data from accurate experiments and a data analysis supported by quantitative computational simulations are required.

REFERENCES

1. C. Marangoni, Über die Ausbreitung der Tropfen einer Flüssigkeit auf der Oberfläche einer anderen, *Ann. Phys. Leipzig*, 143 (1871) 337–354.
2. J. Thomson, On certain curious motions observable at the surface of wine and other alcoholic liquors, *Phil. Mag.*, 10 (1855) 330–333.
3. B.S. Dandapat, B. Santra, H.I. Andersson, Thermocapillarity in a liquid film on an unsteady stretching surface, *Int. J. Heat Mass Transf.*, 46 (2003) 3009–3015.
4. R.V. Birikh, V.A. Briskman, M.G. Velarde, J.-C. Legros, *Liquid Interfacial Systems: Oscillations and Instability*, Marcel Dekker, New York (2003).
5. N.M. Kovalchuk, D. Vollhardt, Marangoni instability and spontaneous nonlinear oscillations produced at liquid interfaces by surfactant transfer, *Adv. Colloid Interface Sci.*, 120 (2006) 1–31.
6. R. Tadmouri, N.M. Kovalchuk, V. Pimienta, D. Vollhardt, J.-C. Micheau, Transfer of oxyethylated alcohols through water/heptane interface: Transition from nonoscillatory to oscillatory behavior, *Colloids Surf. A*, 354 (2010) 134–142.
7. J.R.A. Pearson, On convection cells induced by surface tension, *J. Fluid Mech.*, 4 (1958) 489–500.
8. C.V. Sternling, L.E. Scriven, Interfacial turbulence: Hydrodynamic instability and the Marangoni effect, *AIChE J.*, 5 (1959) 514–523.
9. P. Colinet, J.C. Legros, M.G. Velarde, *Nonlinear Dynamics of Surface-Tension-Driven Instabilities*, Wiley-VCH, Berlin, Germany (2001).
10. A. Javadi, M. Karbaschi, D. Bastani, J.K. Ferri, V.I. Kovalchuk, N.M. Kovalchuk, K. Javadi, R. Miller, Marangoni instabilities for convective mobile interfaces during drop exchange: Experimental study and CFD simulation, *Colloids Surf. A*, 441 (2014) 846–854.
11. R.F. Engberg, M. Wegener, E.Y. Kenig, The impact of Marangoni convection on fluid dynamics and mass transfer at deformable single rising droplets—A numerical study, *Chem. Eng. Sci.*, 116 (2014) 208–222.
12. N.M. Kovalchuk, Spontaneous oscillations due to solutal Marangoni instability: Air/water interface, *Cent. Eur. J. Chem.*, 10 (2012) 1423–1441.
13. N.M. Kovalchuk, Spontaneous nonlinear oscillations of interfacial tension at oil/water interface, *Open Chem.*, 13 (2015) 1–16.
14. M. Lotfi, M. Karbaschi, A. Javadi, N. Mucic, J. Krägel, V.I. Kovalchuk, R.G. Rubio, V.B. Fainerman, R. Miller, Dynamics of liquid interfaces under various types of external perturbations, *Curr. Opin. Colloid Interface Sci.*, 19 (2014) 309–319.
15. A. Javadi, D. Bastani, J. Krägel, R. Miller, Interfacial instability of growing drop: Experimental study and conceptual analysis, *Colloids Surf. A*, 347 (2009) 167–174.
16. N.M. Kovalchuk, D. Vollhardt, Comparison of surface tension auto-oscillations in fatty acid–water and aliphatic alcohol–water systems, *Mater. Sci. Eng. C*, 22 (2002) 147–153.
17. K. Schwarzenberger, T. Köllner, H. Linde, T. Boeck, S. Odenbach, K. Eckert, Pattern formation and mass transfer under stationary solutal Marangoni instability, *Adv. Colloid Interface Sci.*, 206 (2014) 344–371.

18. M. Wegener, A.R. Paschedag, The effect of soluble anionic surfactants on rise velocity and mass transfer at single droplets in systems with Marangoni instabilities, *Int. J. Heat Mass Transf.*, 55 (2012) 1561–1573.

19. Z.-S. Mao, J. Chen, Numerical simulation of the Marangoni effect on mass transfer to single slowly moving drops in the liquid–liquid system, *Chem. Eng. Sci.*, 59 (2004) 1815–1828.

20. A. Grahn, Two-dimensional numerical simulations of Marangoni–Bénard instabilities during liquid–liquid mass transfer in a vertical gap, *Chem. Eng. Sci.*, 61 (2006) 3586–3592.

21. M. Wegener, T. Eppinger, K. Bäumler, M. Kraume, A.R. Paschedag, E. Bänsch, Transient rise velocity and mass transfer of a single drop with interfacial instabilities—Numerical investigations, *Chem. Eng. Sci.*, 64 (2009) 4835–4845.

22. J. Wang, Z. Wang, P. Lu, C. Yang, Z.S. Mao, Numerical simulation of the Marangoni effect on transient mass transfer from single moving deformable drops, *AIChE J.*, 57 (2011) 2670–2683.

23. H. Sawistowski, Interfacial phenomena, in: C. Hanson (Ed.) *Recent Advances in Liquid–Liquid Extraction*, Pergamon Press, Oxford, U.K. (1971).

24. T.S. Sørensen, M. Hennenberg, A. Sanfeld, Deformational instability of a plane interface with perpendicular linear and exponential concentration gradients, *J. Colloid Interface Sci.*, 61 (1977) 62–76.

25. M. Hennenberg, T.S. Sørensen, A. Sanfeld, Deformational instability of a plane interface with transfer of matter. Part 1.—Nonoscillatory critical states with a linear concentration profile, *J. Chem. Soc., Faraday Trans.* 2, 73 (1977) 48–66.

26. M. Hennenberg, P. Bisch, M. Vignes-Adler, A. Sanfeld, Mass transfer, Marangoni effect, and instability of interfacial longitudinal waves: I. Diffusional exchanges, *J. Colloid Interface Sci.*, 69 (1979) 128–137.

27. J. Reichenbach, H. Linde, Linear perturbation analysis of surface-tension-driven convection at a plane interface (Marangoni instability), *J. Colloid Interface Sci.*, 84 (1981) 433–443.

28. S. Slavtchev, M. Hennenberg, J.-C. Legros, G. Lebon, Stationary solutal Marangoni instability in a two-layer system, *J. Colloid Interface Sci.*, 203 (1998) 354–368.

29. M. Hennenberg, P. Bisch, M. Vignes-Adler, A. Sanfeld, Mass transfer, Marangoni effect, and instability of interfacial longitudinal waves. II. Diffusional exchanges and adsorption–desorption processes, *J. Colloid Interface Sci.*, 74 (1980) 495–508.

30. T.S. Sorensen, Marangoni instability at a spherical interface. Breakdown of fluid drops at low surface tension and cytokinetic phenomena in the living cell, *J. Chem. Soc., Faraday Trans.* 2, 76 (1980) 1170–1195.

31. E. Nakache, M. Dupeyrat, M. Vignes-Adler, Experimental and theoretical study of an interfacial instability at some oil–water interfaces involving a surface-active agent: I. Physicochemical description and outlines for a theoretical approach, *J. Colloid Interface Sci.*, 94 (1983) 187–200.

32. D. Lavabre, V. Pradines, J.-C. Micheau, V. Pimienta, Periodic Marangoni instability in surfactant (CTAB) liquid/liquid mass transfer, *J. Phys. Chem. B*, 109 (2005) 7582–7586.

33. N.M. Kovalchuk, D. Vollhardt, Oscillation of interfacial tension produced by transfer of nonionic surfactant through the liquid/liquid interface, *J. Phys. Chem. C*, 112 (2008) 9016–9022.

34. V. Pradines, R. Tadmouri, D. Lavabre, J.-C. Micheau, V. Pimienta, Association, partition, and surface activity in biphasic systems displaying relaxation oscillations, *Langmuir*, 23 (2007) 11664–11672.

35. V.I. Kovalchuk, H. Kamusewitz, D. Vollhardt, N.M. Kovalchuk, Auto-oscillation of surface tension, *Phys. Rev. E*, 60 (1999) 2029.

36. N.M. Kovalchuk, D. Vollhardt, Auto-oscillations of surface tension in water-alcohol systems, *J. Phys. Chem. B*, 104 (2000) 7987–7992.

37. N.M. Kovalchuk, D. Vollhardt, Nonlinear spontaneous oscillations at the liquid/liquid interface produced by surfactant dissolution in the bulk phase, *J. Phys. Chem. B*, 109 (2005) 22868–22875.

38. N.M. Kovalchuk, D. Vollhardt, Spontaneous nonlinear oscillation produced by alcohol transfer through water/alkane interface: An experimental study, *Colloids Surf. A*, 291 (2006) 101–109.

39. Y. Ikezoe, S. Ishizaki, T. Takahashi, H. Yui, M. Fujinami, T. Sawada, Hydrodynamically induced chemical oscillation at a water/nitrobenzene interface, *J. Colloid Interface Sci.*, 275 (2004) 298–304.

40. J. Srividhya, M.S. Gopinathan, Modeling experimental oscillations in liquid membranes with delay equations, *J. Phys. Chem. B*, 107 (2003) 1438–1443.

41. K. Arai, F. Kusu, Electrical potential oscillations across a water-oil-water liquid membrane in the presence of drugs, in: A.G. Volkov (Ed.) *Liquid Interfaces in Chemical, Biological and Pharmaceutical Applications*, Marcel Dekker, New York (2001).

42. M. Szpakowska, I. Czaplicka, E. Płocharska-Jankowska, O. Nagy, Contribution to the mechanism of liquid membrane oscillators involving cationic surfactant, *J. Colloid Interface Sci.*, 261 (2003) 451–455.

43. V. Pimienta, D. Lavabre, T. Buhse, J.-C. Micheau, Correlation between electric potential and interfacial tension oscillations in a water-oil-water system, *J. Phys. Chem. B*, 108 (2004) 7331–7336.

44. O.V. Grigorieva, N.M. Kovalchuk, D.O. Grigoriev, D. Vollhardt, Experimental studies on the geometrical characteristics determining the system behavior of surface tension auto-oscillations, *J. Colloid Interface Sci.*, 261 (2003) 490–497.

45. N.M. Kovalchuk, D. Vollhardt, Theoretical description of repeated surface-tension auto-oscillations, *Phys. Rev. E*, 66 (2002) 026302.

46. N.M. Kovalchuk, D. Vollhardt, Instability and spontaneous oscillations by surfactant transfer through a liquid membrane, *Colloids Surf. A*, 309 (2007) 231–239.

47. N.M. Kovalchuk, D. Vollhardt, Influence of the system geometry on the characteristics of the surface tension auto-oscillations: A numerical study, *J. Phys. Chem. B*, 107 (2003) 8439–8447.

48. N.M. Kovalchuk, D. Vollhardt, Effect of substance properties on the appearance and characteristics of repeated surface tension auto-oscillation driven by Marangoni force, *Phys. Rev. E*, 69 (2004) 016307.

49. O.V. Grigorieva, N.M. Kovalchuk, D.O. Grigoriev, D. Vollhardt, Spontaneous nonlinear surface tension oscillations in the presence of a spread surfactant monolayer at the air/water interface, *Colloids Surf. A*, 250 (2004) 141–151.

50. O. Grigorieva, D. Grigoriev, N. Kovalchuk, D. Vollhardt, Auto-oscillation of surface tension: Heptanol in water and water/ethanol systems, *Colloids Surf. A*, 256 (2005) 61–68.

51. N.M. Kovalchuk, V. Pimienta, R. Tadmouri, R. Miller, D. Vollhardt, Ionic strength and pH as control parameters for spontaneous surface oscillations, *Langmuir*, 28 (2012) 6893–6901.

52. A. Javadi, J. Ferri, T.D. Karapantsios, R. Miller, Interface and bulk exchange: Single drops experiments and CFD simulations, *Colloids Surf. A*, 365 (2010) 145–153.

23 Hierarchical Marangoni Roll Cells

Experiments and Direct Numerical Simulations in Three and Two Dimensions

Karin Schwarzenberger, Thomas Köllner,* Thomas Boeck,
Stefan Odenbach, and Kerstin Eckert

CONTENTS

23.1 INTRODUCTION

Many technological processes, such as extraction, evaporation, or absorption are accompanied by solutal Marangoni instability. This type of convection, which is driven by gradients in interfacial tension, can significantly influence process performance in industrial production. The patterns originating from this convection are of a complex and unsteady nature. One particularly remarkable feature is the formation of hierarchical structures, that is, larger patterns emerging on a background

---

* Karin Schwarzenberger and Thomas Köllner contributed equally to this work.

of smaller flow patterns. For the example of liquid–liquid extraction, hierarchical roll cells were already observed in the early experiments on solutal Marangoni convection [1–5].

Despite the continuing research [6–13], the details of the multiscale patterns and the mechanisms underlying the hierarchy formation remained unresolved for a long time. Only recently, a thorough characterization of hierarchical Marangoni roll cells could be achieved by a combination of highly resolved three-dimensional (3D) simulations and specifically designed validation experiments [14,15]. However, the hierarchical nature requires the use of a well adapted numerical technique and of large parallel computers to simultaneously resolve both the very fine and the large-scale features of solutal Marangoni convection.

To reduce computational cost compared to full 3D simulations, two-dimensional (2D) models are frequently employed [9,16]. Experimentally, such a reduction can be realized to a certain degree in the Hele-Shaw (HS) cell. Here, the liquids are placed between two parallel plates that are sufficiently close together such that the fluid motion becomes mainly 2D [17]. This can be represented by gap-averaged equations [6,7] which—contrary to pure 2D models—take into account the influence of wall friction. Despite the benefits of the HS cell, differences may arise by reducing 3D dynamics to a 2D situation.

In this chapter, we present mathematical models, which are able to reproduce the hierarchical patterns observed in the experiments. Therefore, particular focus is laid on providing sufficiently comparable situations in experiments and simulations. The methods are applied to an exemplary two-layer system where hierarchical Marangoni roll cells develop due to the mass transfer of a weakly surface-active solute. In combination with the validation experiments, the results of the simulations provide new insights into the hierarchical nature of the patterns. On this basis, we discuss the applicability of the simplified theoretical models and point out limitations when comparing experimental and numerical results.

23.2 STATIONARY SOLUTAL MARANGONI INSTABILITY

The foundation for a theoretical understanding of the structures arising from solutal Marangoni instability was laid by Sternling and Scriven [18]. According to their linear stability analysis, solutal Marangoni convection can occur either via an oscillatory or a stationary mode. This chapter focuses on the stationary mode which can set in if an interfacial tension lowering solute is transferred out of the phase with the higher kinematic viscosity and the lower solute diffusivity. The mechanism that drives the flow is as follows: Consider a location at the interface with a positive variation of solute. Inevitably, interfacial forces by lowered interfacial tension will create a flow that spreads the solute tangentially along the interface, away from this initial variation. Fluid from the bulk is carried to the interface by the ensuing flow. This transport increases the interfacial concentration due to lower diffusivity in the solute-rich layer. Hence, interfacial tension gradients are amplified and Marangoni instability occurs in the form of roll cell convection as depicted in Figure 23.1. In a 2D situation, such as that in the HS cell, the roll cells appear as a double vortex structure, see Figure 23.1a. At the extended interface (with insignificant curvature and remote solid boundaries), convection in form of a torus is driven, the boundaries of which form the cell border. Adjoining convection cells form a spatially fixed and dense network of polygonal cells as shown in Figure 23.1b.

These cells belong to the lowest level in the hierarchy, since they are free of any substructure. According to the terminology of Linde et al. [19], we refer to them as Marangoni roll cells of the first order, hereafter abbreviated as RC-I. Depending on the characteristic length scales across which significant differences in interfacial tension occur, second-order roll cells (RC-II) may emerge in the temporal evolution of a system. This complex, hierarchical flow pattern cannot be explained on the basis of the linear theory. Figure 23.2 depicts an example of RC-IIs which host a substructure of smaller RC-I in their interior.

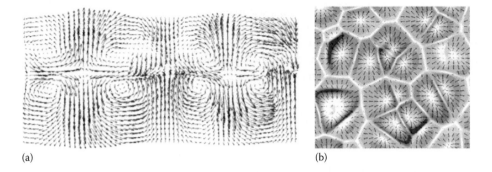

(a) (b)

FIGURE 23.1 (a) Flow structure of first-order Marangoni roll cells (RC-I) in a Hele-Shaw geometry measured by particle image velocimetry. (With permission from Eckert, K., Acker, M., Tadmouri, R., and Pimienta, V., Chemo-Marangoni convection driven by an interfacial reaction: Pattern formation and kinetics, *Chaos: Interdisciplinary J. Nonlinear Sci.*, 22(3), 037112. Copyright 2012, American Institute of Physics.) (b) Network of Marangoni roll cells (RC-I) at an extended interface (numerical results). (From *Adv. Colloid Interface Sci.*, 206, Schwarzenberger, K., Köllner, T., Linde, H., Boeck, T., Odenbach, S., and Eckert, K., Pattern formation and mass transfer under stationary solutal Marangoni instability, 344–371. Copyright 2014, with permission from Elsevier.) Velocity magnitude (gray scale) with velocity vectors, dark gray value represents high velocity.

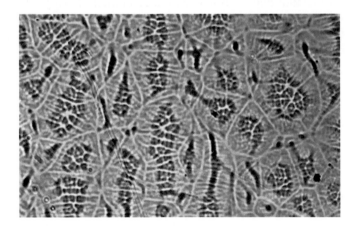

FIGURE 23.2 Second-order Marangoni cells (RC-II) with internal substructure of smaller Marangoni cells of first order (RC-I); shadowgraph image from mass transfer experiment described in Section 23.3.1.

To study the nonlinear evolution toward RC-II, a two-layer system similar to the analysis of Sternling and Scriven [18] is considered by using following simplifications:

- Interface is assumed as plane and undeformable.
- Partition of the solute is governed by a Henry condition.
- Interfacial tension depends linearly on solute concentration.
- Density of the phases changes linearly with solute concentration as well.
- All other material properties are supposed as constant.
- Mass flux is continuous between bulk phases.
- Liquids are assumed isothermal and incompressible.

As shown, for example, in Chapter 1, extended models for complex interfacial physics are available, for example, for adsorption kinetics of surfactants. However, to unravel the basic cause for the formation of hierarchic structures, it seems pertinent to start with this simple model as it contains the main physical ingredients common to most mass transfer processes. In fact, as can be seen in Refs. 13–15 and in Section 23.4, this model is able of successfully reproducing the RC-II observed in the experiments. The next sections detail the equations resulting from these assumptions, as well as the numerical and experimental methods.

23.3 METHODS TO STUDY SOLUTAL MARANGONI INSTABILITY

23.3.1 Two-Layer Mass Transfer System

To characterize the evolution of hierarchical Marangoni roll cells, a simple model system was chosen which is sensitive to the stationary mode of the primary Marangoni instability: cyclohexanol with dissolved butanol (7.5 vol%) superposed on an aqueous layer. The binary phases cyclohexanol and water are in thermodynamic equilibrium due to mutal saturation to reduce effects from phase changes and multicomponent diffusion. The surface tension and density lowering butanol is transferred to the aqueous layer. Since cyclohexanol has higher kinematic viscosity and lower solute diffusivity, this configuration is unstable with respect to stationary Marangoni instability as described in Section 23.2. Numerical simulations of interfacial convection in this system require the knowledge of all relevant material properties. They are taken from the literature [20,21] or estimated by suitable relationships as briefly described in the following. The applied parameters are summarized in Table 23.A.1. A comprehensive description can be found in Ref. 14.

Interfacial tension (σ) and density ($\rho^{(i)}$) in layer (i) are assumed to depend linearly on dimensional solute (i.e., butanol) concentration ($\tilde{c}^{(i)}$), that is,

$$\rho^{(i)} = \rho_{ref}^{(i)} + \rho_{ref}^{(i)}\beta_c^{(i)}\tilde{c}^{(i)}, \tag{23.1}$$

$$\sigma = \sigma_{ref} + \sigma_{ref}\alpha_c\tilde{c}^{(1)} \quad \text{for } z = 0. \tag{23.2}$$

Note that a tilde is added for the dimensional concentration to differentiate from the dimensionless counterpart, for example, as used in the mathematical models in Sections 23.3.2.2 and 23.3.3.2. The linear dependence is quantified by the solutal expansion coefficients ($\beta_c^{(i)}$) and the coefficient of interfacial tension change (α_c). All other material properties are constant in each phase.

The concentration at the interface is supposed to be in thermodynamic equilibrium with the local excess concentration of the interface Γ. Furthermore, we apply Henry's model, stating that the excess concentration depends linearly on the concentrations adjacent to the interface $\Gamma = K_1\tilde{c}^{(1)}$ and $\Gamma = K_2\tilde{c}^{(2)}$. Both relations yield the concentration partition coefficient or Henry's constant H,

$$\tilde{c}^{(1)} = \frac{K_2}{K_1\tilde{c}^{(2)}} = \frac{\tilde{c}^{(2)}}{H} \quad \text{for } z = 0. \tag{23.3}$$

H is approximated by means of a correlation method [22], which estimates the equilibrium concentration of butanol to be 31 times higher in the organic phase compared to the aqueous phase, that is, $H = 31$. Thus, the absolute concentration of butanol in both layers changes only slightly for our configuration, making the applied linearizations and the further assumption of constant material properties more robust.

23.3.2 THREE-DIMENSIONAL GEOMETRY

23.3.2.1 Experiment

As the conditions in the validation experiments have to agree with those of the simulations as far as possible, the preparation of the two-phase system demands an appropriate setup. On the one hand, disturbances of the Marangoni flow by the superposition of the phases should be minimized, on the other hand, the procedure of superposition should be completed quickly so that the desired step-like initial concentration profile can be maintained.

These requirements are mostly realized by the setup employed in Refs. 13 and 14 where two cubical cuvettes, each filled with one phase, are superposed. The cuvette heights correspond to the parameters $d^{(1)}$, $d^{(2)}$ that are used in the mathematical model discussed in the next section. The evolution of the flow patterns and their length scales is recorded by a shadowgraph optics. This measurement method visualizes variations in the concentration gradients as the solute concentration changes the refractive index of the liquids.

23.3.2.2 Mathematical Model [14]

To study the nonlinear evolution toward hierarchical flow structures theoretically, direct numerical simulations are required. In accordance with the experimental system, the mathematical model comprises two superposed immiscible, isothermal liquid phases in a cubical computational domain. In both layers, the momentum transport is modeled by the incompressible Navier-Stokes–Boussinesq equations and the solute transport by an advection–diffusion equation. The dimensionless equations are derived by introducing the following scales. The mass is measured in multiples of $\tilde{M} = \rho_{ref}^{(1)}(d^{(1)})^3$, time in viscous units $\tilde{T} = (d^{(1)})^2/\nu^{(1)}$, length in multiples of the lower layer height $\tilde{L} = d^{(1)}$ and molar concentration relative to the initial concentration c_0. For the experimental system with a layer height of $d^{(1)} = 20$ mm and initially 7.5 vol% butanol dissolved in the organic phase, the nondimensional parameters are defined and calculated in Table 23.A.1 (see Appendix 23.A) along with the relevant material properties.

Figure 23.3 shows a sketch of the computational domain. The horizontal size is $l_x \times l_y$. The vertical dimensions are $-1 \le z \le 0$ for the lower water-rich phase and $0 \le z \le d$ for the upper cyclohexanol-rich phase. Both phases are in contact along a plane interface at $z = 0$. Initially, the system is at rest and the solute is solely present in the upper organic phase with a homogeneous concentration of $c^{(1)} = 1$.

The nondimensionalization yields the following equations [23], which are the basis of the numerical investigations:

$$\partial_t \mathbf{u}^{(1)} = -\mathbf{u}^{(1)} \cdot \nabla \mathbf{u}^{(1)} - \nabla p_d^{(1)} + \Delta \mathbf{u}^{(1)} - c^{(1)} G \mathbf{e}_z, \tag{23.4}$$

$$\nabla \cdot \mathbf{u}^{(1)} = 0, \tag{23.5}$$

$$\partial_t \mathbf{u}^{(2)} = -\mathbf{u}^{(2)} \cdot \nabla \mathbf{u}^{(2)} - \frac{1}{\rho} \nabla p_d^{(2)} + \nu \Delta \mathbf{u}^{(2)} - c^{(2)} G \beta \mathbf{e}_z, \tag{23.6}$$

$$\nabla \cdot \mathbf{u}^{(2)} = 0, \tag{23.7}$$

$$\partial_t c^{(1)} = -\mathbf{u}^{(1)} \cdot \nabla c^{(1)} + \frac{1}{Sc^{(1)}} \Delta c^{(1)}, \tag{23.8}$$

$$\partial_t c^{(2)} = -\mathbf{u}^{(2)} \cdot \nabla c^{(2)} + \frac{D}{Sc^{(1)}} \Delta c^{(2)}. \tag{23.9}$$

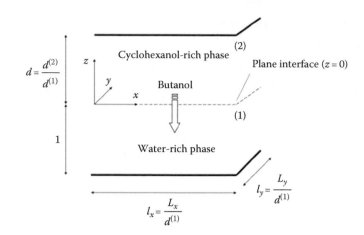

FIGURE 23.3 Sketch of the computational domain, which corresponds to the experimental setup. (From *Phys. Fluids*, 25, Köllner, T., Schwarzenberger, K., Eckert, K., and Boeck, T., Multiscale structures in solutal Marangoni convection: Three-dimensional simulations and supporting experiments, 092109. Copyright 2013, with permission from Elsevier).

The dimensional molar concentration $\tilde{c}^{(i)}$ of butanol and the velocity $\mathbf{u}^{(i)}$ in phase i follow from $\tilde{c}^{(i)} = c^{(i)} \cdot c_0$ and $\mathbf{u}^{(i)} \cdot \tilde{L}/\tilde{T}$, respectively. The matching conditions at the plane interface ($z = 0$) are

$$u_x^{(1)} = u_x^{(2)}, \quad u_y^{(1)} = u_y^{(2)}, \quad u_z^{(1)} = u_z^{(2)} = 0, \quad \partial_z c^{(1)} = D\partial_z c^{(2)}, \quad c^{(1)}H = c^{(2)}, \tag{23.10}$$

$$\frac{Ma}{Sc^{(1)}} \partial_x c^{(1)} = -\mu \partial_z u_x^{(2)} + \partial_z u_x^{(1)}, \quad \frac{Ma}{Sc^{(1)}} \partial_y c^{(1)} = -\mu \partial_z u_y^{(2)} + \partial_z u_y^{(1)}. \tag{23.11}$$

These relations arise from the continuity of velocity, the conservation of solute in the bulk, Henry's model, and the shear stress balance at the interface. The boundary conditions are periodic in the x–y dimension; no-slip and impermeable boundaries are supposed for the solid walls at the bottom ($z = -1$) and top ($z = d$). The various nondimensional parameters are defined in Table 23.A.1.

The shadowgraph technique used in the experiments is mimicked by averaging the horizontal Laplacian of the concentration distribution over both layers [24]:

$$s(x,y) = \int_{[-1, d]} (\partial_x^2 + \partial_y^2)c(\mathbf{x},t)dz. \tag{23.12}$$

The quantity $s(x, y)$ permits a visual comparison of the emerging structures with the experiments. Locations with high values of $s(x, y)$ correspond to a gain of surfactant by horizontal diffusion. These points are usually areas of low concentration. Thus $s(x, y)$ primarily displays the horizontal surfactant distribution in close proximity to the interface, since surfactant gradients are mainly produced there by interfacial convection. In the later stages of mass transfer, it also includes information from locations further away from the interface. However, note that the *experimental* shadowgraph records are affected by slight deflections of the interface and nonlinearities of the refractive index as well as of the optical devices.

23.3.2.3 Numerical Method [14]

The presented simple geometry is suitable for the application of pseudospectral methods [25,26]. A significant advantage of the pseudospectral discretization is that the solution of large linear systems of equations, which appear as a result of most other discretization strategies, is avoided. As a result, a very efficient numerical scheme is obtained. However, this advantage comes at the cost of low flexibility in terms of geometry and boundary conditions, since the basis functions (in which the field quantities are expanded globally) have to respect those conditions. Although efforts have been undertaken to relax those restrictions, for example, by immersed boundary methods [27] (boundary condition) or mapping methods [28] (curved interfaces), the main application of pseudospectral discretizations is on cubical domains with at least one periodic direction.

The particular pseudospectral scheme that is used was comprehensively presented in Ref. 29, where it was applied to thermal convection in a two-layer system. A wide variety of physical problems have been solved with this code (of course, slightly adapted and extended), for example, magnetohydrodynamic channel flow [30] and chemical-reaction-driven buoyant convection [31]. To study mass transfer in the two-layer system, temperature in Ref. 29 is replaced by the concentration field, and the respective boundary conditions for this field are modified [14].

Both planar layers are treated as separate computational domains that are coupled at the interface. In each layer, the fields are expanded in truncated Fourier series in the two periodic horizontal directions x, y. The vertical direction z is expanded in Chebyshev polynomials T_p of order p. The smallest wavenumbers for the x, y directions are $k_{x0} = 2\pi/l_x$, $k_{y0} = 2\pi/l_y$. For example, the expansion of the vertical velocity in the upper layer, where $(x,y,z) \in [0,l_x] \times [0,l_y] \times [0,d]$, reads

$$u_z^{(2)}(x,y,z,t) = \sum_{m=-N_x/2}^{N_x/2-1} \sum_{n=-N_y/2}^{N_y/2-1} \sum_{p=0}^{N_z} e^{imk_{x0}x+ink_{y0}y} T_p\left(\frac{2z}{d}-1\right)\hat{u}_z^{(2),m,n,p}. \tag{23.13}$$

The data output in physical space and the calculation of the nonlinear terms is done on the collocation points, for example, for layer (2) they are

$$(x_i,y_j,z_k) = \left(\frac{i \cdot l_x}{N_x}, \frac{j \cdot l_y}{N_y}, d \cdot 0.5\left(1+\cos\left(\frac{k\pi}{N_z^{(2)}}\right)\right)\right), \tag{23.14}$$

with $i \in \{0,1,\dots,N_x\}$, $j \in \{0,1,\dots,N_y\}$, $k \in \{0,1,\dots,N_z^{(2)}\}$. The velocity field is represented through the poloidal–toroidal decomposition, whereby incompressibility is automatically satisfied. Parallelization with message passing interface (MPI) is based on a domain decomposition in one horizontal direction. The number of grid points in each direction is a power of two because only base two fast Fourier transformations are used. The numerical resolution requirements turned out to be quite severe due to the high Schmidt number Sc of the system. A very high vertical resolution near the interface is required to properly capture the vertical structures. The horizontal resolutions (N_x, N_y) are set to resolve the similar fine solute structures in the horizontal directions.

The time step is adjusted according to the current grid CFL number:

$$C_g = \max\left\{\frac{u_x\delta t}{\Delta x}, \frac{u_y\delta t}{\Delta y}, \frac{u_z\delta t}{\Delta z}\right\}. \tag{23.15}$$

This number is calculated every time step by dividing the local displacement $u_\alpha \delta t$ by the local values Δx, Δy, Δz of the spacing between collocation points. We force C_g to be smaller than a constant C_b and to be larger than $C_b/2$, that is,

$$\frac{C_b}{2} \leq C_g \leq C_b. \tag{23.16}$$

This is done by adapting the new time step δt if this latter condition is violated. It turned out that a value of $C_b = 0.3$ is appropriate for a stable time-stepping scheme, so, we use that value.

The standard initial conditions are set as follows. The velocity field is initialized (at $t = 0$) with pseudorandom numbers. Specifically, the gridpoint values of the vertical velocity component $u_z(x_i, y_i, z_k)$ and the vertical vorticity $\nabla \times \mathbf{u}(x_i,y_j,z_k) \cdot \mathbf{e}_z$ are uniformly distributed between $[0, 10^{-3}]$ in space. The solute is initialized with a homogeneous concentration of unity for layer 2, that is, $c^{(2)} = 1$, and zero for layer 1, that is, $c^{(1)} = 0$.

To provide an order of magnitude, the simulation presented in Section 23.4.1 consumed approximately 650 GB main memory for a resolution of about three billion collocation points. On 512 processors it took 1.24×10^5 core hours for the simulation to be advanced until $t = 2.07$ with 17,800 time steps. They are governed by the Courant-Friedrichs-Lewy (CFL) restriction given in Equation 23.15 except before the onset of convection.

23.3.3 Two-Dimensional Geometry

23.3.3.1 Experiment

To obtain a quasi 2D situation in experiments, the liquid–liquid system is placed in a narrow gap between two glass plates, the HS cell. This setup provides simplified access to the vertical structure of quantities recorded by optical methods such as shadowgraphy, interferometry, and particle image velocimetry. By varying the orientation of the HS cell, it is possible to reveal information about the influence of gravity [11,17,32–34].

Figure 23.4 sketches the experimental HS setup (a) and the computational domain (b), which is drawn as a gray inset in (a) for comparison. A spacer made of polytetrafluoroethylene (PTFE) foil, whose inner contour is shown as a dashed line in Figure 23.4a, acts as a container for the liquids. The shape of the PTFE foil was optimized to provide a robust filling procedure for different two-phase systems [35]. The gap width 2ε, see Figure 23.4b, is set by the thickness of the foil.

23.3.3.2 Hele-Shaw model

According to the experimental setup, the nondimensional plate distance is $2\varepsilon/d^{(1)} = 2\gamma^{-1/2}$ where $\gamma = (d^{(1)})^2/\varepsilon^2$ is the ratio of squared layer height to half plate distance, see Figure 23.4b. In x- and z-direction, the geometry is defined according to the 3D domain in Section 23.3.2.2.

The flow in this HS gap is simulated by a common gap-averaged hydrodynamic model [6,7,36]. It assumes a parabolic flow profile and no solute variation along the y-direction, thereby simplifying the 3D governing equations of Section 23.3.2.2. The modeling [32] follows the formulation of Ref. 6 in applying the aforementioned assumptions: First, the nondimensional velocity field $\mathbf{u}^{(i)}(x,y,z,t)$ in layer (i) has only two nonzero components ($u_y^{(i)} = 0$) with a parabolic dependence on y, that is,

$$\mathbf{u}^{(i)}(x,y,z) = \frac{3}{2}(1 - y^2\gamma)(v_x^{(i)}(x,z)\mathbf{e}_x + v_z^{(i)}(x,z)\mathbf{e}_z). \tag{23.17}$$

In what follows, we are concerned with velocity fields that are gap-averaged over y and denoted by $\mathbf{v}^{(i)}(x,z,t) = \langle \mathbf{u}^{(i)} \rangle_y$. The second assumption is that solute concentration is constant across the gap, that is, it does not depend on the y-coordinate: $c^{(i)} = c^{(i)}(x,z,t)$.

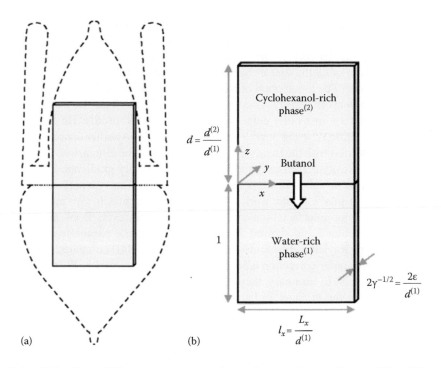

FIGURE 23.4 Hele-Shaw (HS) geometry as experimental setup (a) according to Shi and Eckert (From Shi, Y. and Eckert, K., *Chem. Eng. Sci.*, 63, 3560, 2008) with gray inset of numerical domain as defined in Kollner et al. [32]; detailed numerical domain (b) in dimensionless lengths according to Kollner et al. (From Kollner, T. et al., *Eur. Phys. J. Special Top.*, 2015, in press) (scaled by lower layer height $d^{(1)}$).

With this, the 3D Navier-Stokes–Boussinesq and advection–diffusion equations for the solute yield the 2D HS model:

$$\partial_t \mathbf{v}^{(1)} = -\frac{6}{5}\mathbf{v}^{(1)} \cdot \nabla \mathbf{v}^{(1)} - \nabla p_d^{(1)} + \Delta \mathbf{v}^{(1)} - c^{(1)}G\mathbf{e}_z - 3\gamma\mathbf{v}^{(1)}, \qquad (23.18)$$

$$\partial_t \mathbf{v}^{(2)} = -\frac{6}{5}\mathbf{v}^{(2)} \cdot \nabla \mathbf{v}^{(2)} - \frac{1}{\rho}\nabla p_d^{(2)} + \nu\Delta \mathbf{v}^{(2)} - c^{(2)}G\beta\mathbf{e}_z - 3\gamma\nu\mathbf{v}^{(2)}, \qquad (23.19)$$

$$\nabla \cdot \mathbf{v}^{(1)} = 0, \quad \nabla \cdot \mathbf{v}^{(2)} = 0, \qquad (23.20)$$

$$\partial_t c^{(1)} = -\mathbf{v}^{(1)} \cdot \nabla c^{(1)} + \frac{1}{Sc^{(1)}}\Delta c^{(1)}, \quad \partial_t c^{(2)} = -\mathbf{v}^{(2)} \cdot \nabla c^{(2)} + \frac{D}{Sc^{(1)}}\Delta c^{(2)}. \qquad (23.21)$$

The additional damping with prefactor 3γ in Equation 23.18 is well known from Darcy's law.

The matching conditions at the interface as well as the boundary conditions are adapted from the 3D case. The set of equations underlying the HS model again is solved by the pseudo-spectral algorithm described in Section 23.3.2.3.

23.3.3.3 Discussion of the Hele-Shaw Model

After the introduction of the gap-averaged model, we give a short review on the modeling of flows in a HS cell and show the problems which are inherent in the underlying assumptions. Generally, the averaging over the gap is clearly restricted to a physical situation where viscous and diffusive perturbations equilibrate sufficiently fast across the gap on a comparable time scale. Former works have proposed certain model corrections for this problem. However, they have typically focused either on the momentum or on the species balance but not on both.

Momentum transport: The assumption of a parabolic profile in the Navier-Stokes equation may be violated due to gradients in density, viscosity, or geometrical effects. Ruyer-Quil [37] derived a gap-averaged equation as a first-order correction to the parabolic profile. He found inertial terms with other prefactors: $(6/5)\partial_t \mathbf{v}^{(1)} + (54/35)\mathbf{v}^{(1)} \cdot \nabla \mathbf{v}^{(1)} = -\nabla p - 3\gamma \mathbf{v}^{(1)}$. Another approach was pursued by Zeng et al. [38], who started with the steady Stokes equation in three dimensions and solved different examples including a density-driven case and a case with viscosity gradients. It is concluded there that the viscous term (the so-called Brinkmann correction) should carry a prefactor $\beta = 12/\pi^2$, which yields $0 = -\nabla p - 3\gamma \mathbf{v}^{(1)} + \beta\Delta\mathbf{v}^{(1)}$. A later work [39] adopted this approach with modifications of β.

In the Stokes limit for interfacial-tension driven flows, however, Boos and Thess [36] and Gallaire et al. [40] discussed and showed the good quality of the parabolic assumption, that is, $\beta = 1$. By examining the Rayleigh–Taylor instability, Martin et al. [41] compared Ruyer-Quil's model (with additional second-order correction $6/5\Delta\mathbf{v}^{(1)}$) to the one introduced in Section 23.3.3.2, that is, Equations 23.18 and 23.19. In summary, they recommended the model of Section 23.3.3.2.

As a final remark on the momentum balance, let us note that when inertia is small compared to viscous forces (Reynolds number $Re \ll 1$), the Brinkman equation [36] can be used instead of Equations 23.18 and 23.19, that is, the terms $\partial_t \mathbf{v}^{(i)} + (6/5)\mathbf{v}^{(i)} \cdot \nabla \mathbf{v}^{(i)}$ are omitted in these equations.

Species transport: The second assumption is the quasi-instantaneous equilibration of the concentration field across the gap. It demands that the diffusion time $\tau_d = \varepsilon^2/D^{(i)}$ is much lower than the characteristic time for advection $\tau_a = \varepsilon/\tilde{U}$, that is, for the *cross-gap* Péclet number $Pe = \tau_d/\tau_a = \tilde{U}\varepsilon/D^{(i)} \ll 1$ should hold. Otherwise, this leads to the effect of Taylor dispersion [42], that is, solute in the middle of the gap is transported faster than in the vicinity of the plates. Taylor and later Aris [43] showed, for a simpler geometrical configuration, that this process could be accounted for by a transport with the mean velocity and an additional dispersion term. Recently, the dispersion effect for *unidirectional* flows has been incorporated in density-driven flows [39,44].

For the *nonunidirectional* flows that we are concerned within this study, Zimmerman and Homsy [45] proposed a model based on the analysis of Horne and Rodriguez [46], which Zimmerman et al. and Petitjeans et al. [47] used to study miscible displacement in the HS setup. To the best of our knowledge this is the most reasonable 2D model that includes the effect of Taylor dispersion. Zimmerman et al. cast the effect of Taylor dispersion into an anisotropic and velocity-dependent diffusivity tensor $\mathbf{D}$. The new transport equation for solute in layer (i) then reads

$$\partial_t \tilde{c}^{(i)} + \tilde{\mathbf{v}}^{(i)} \cdot \tilde{\nabla} \tilde{c}^{(i)} = \tilde{\nabla} \cdot (\mathbf{D}^{(i)} \cdot \tilde{\nabla} \tilde{c}^{(i)}), \tag{23.22}$$

where a tilde was added for the nondimensional counterpart of the corresponding dimensional quantities.

The dispersion tensor is given as follows:

$$\mathbf{D}^{(i)} = D^{(i)}\mathbf{I} + \frac{2\varepsilon^2}{105D^{(i)}} \tilde{\mathbf{v}}^{(i)} \otimes \tilde{\mathbf{v}}^{(i)}, \tag{23.23}$$

where $\mathbf{I}$ is the unit tensor. This formulation leads to an enhanced diffusion in flow direction while there is only molecular diffusion orthogonal to the flow. Since this model is just a heuristic extension of the analysis of Horne and Rodriguez [46], the validity of this approach is not assured, especially near the interface where the dispersion effect should increase the transport of solute in the center of the HS cell.

Nevertheless, to roughly estimate the influence of Taylor dispersion, this model is applied to the cyclohexanol system (see Section 23.3.1) placed in a 0.5 mm HS gap. The simulations with the dispersion model only *marginally* depart from the standard model without dispersion [32]. After an initial phase ($t < 1$) for which the dispersion model evolves faster, the difference in the vertical cell size as comparative parameter amounts to the typical variance between different simulation runs. In view of these small differences and the weak rigorous foundation of the dispersion correction, the formulation described in Section 23.3.3.2 is used to study Marangoni pattern formation in the HS cell.

23.4 EXEMPLARY RESULTS ON HIERARCHICAL MARANGONI ROLL CELLS

The methods introduced in the preceding sections were recently applied to our standard system cyclohexanol with dissolved butanol (7.5 vol%) superposed to an aqueous layer [14,32]. For this system, the reference unit for dimensionless times is 333.3 s (viscous time) and lengths are scaled in 20 mm (lower layer height). In the following, selected patterns are shown to illustrate features of hierarchical Marangoni convection. Furthermore, we critically assess numerical results with the help of observations from the experiments and point out difficulties in this comparison.

For the 3D case in Section 23.4.1, the optical flow method is considered, which is used in experiments to estimate the fluid velocity at the interface [13,14]. The point in question for this procedure is the relation between the optical flow field and the fluid velocity. It will be addressed by comparing a known velocity (from simulation) to its related optical flow (from numerical shadowgraph image). Conclusions from this comparison can then be naturally extended to the interpretation of the experimental data. Next, in Section 23.4.2, the applicability of the 2D HS model (Section 23.3.3.2) to the experimental situation in a HS cell is discussed regarding the influence of 3D flow effects and interfacial deformations. Finally, a direct comparison of 2D to 3D simulations in Section 23.4.3 evaluates the capability of the simpler 2D model.

23.4.1 THREE-DIMENSIONAL GEOMETRY: OPTICAL FLOW

For the mass transfer of butanol from the cyclohexanol-rich to the water-rich layer, hierarchical Marangoni roll cells emerge with time [14]. The starting point for the hierarchical evolution is the network of simple, small RC-I depicted in Figures 23.1 and 23.7. Successively, the size of the patterns increases to compensate the growing zone of equilibrated fluid adjacent to the interface. However, at the same time small scales are produced according to the instability mechanism described in Ref. 14 (see Section 23.4.3). This leads to the hierarchical pattern of RC-II shown in Figures 23.2, 23.5, and 23.6.

Basic information about the hierarchical Marangoni convection can already be obtained from the experiments where the evolution toward RC-II is recorded by shadowgraphy. From the resulting images, relevant length scales can be readily extracted for a direct comparison with the numerical results [14]. On the basis of shadowgraph records, it is also possible to obtain some insight into the flow field of the RC-II (Figure 23.5). This is done by calculating the optical flow field $\mathbf{u}_{of}$ [48] from the cross-correlation function of two consecutive shadowgraph images. The optical flow procedure detects the movement of structures in the concentration field, that is, it mainly shows the advection of substructure RC-I by the secondary flow of the RC-II. As visible from the experimental data in Figure 23.5, the radial convection of the RC-II is directed from the center toward the periphery in analogy with surface tension-driven Bénard convection [49–51]. At the inflow points, where fluid from the bulk impinges on the interface, small length scales (substructure RC-I) are produced as a result of the high vertical concentration gradients there. The substructure is then advected to the periphery and disappears in the outflow region, where the fluid is diverted to the bulk.

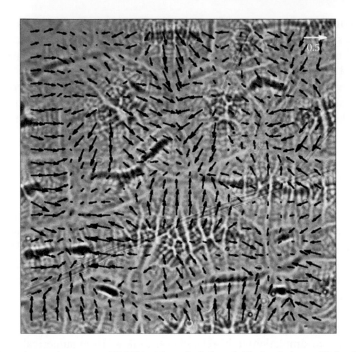

FIGURE 23.5 Experimental pattern of RC-II with optical flow field. A window of 0.5×0.5 (corresponding to 1 cm × 1 cm) is shown.

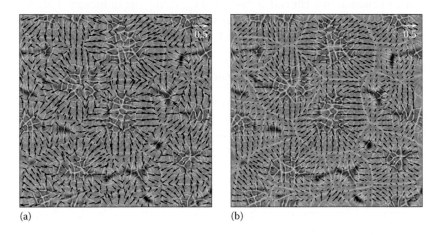

(a)   (b)

FIGURE 23.6 Interfacial velocity (a) and optical flow (b) for a pattern of RC-II at $t = 1.5$. For comparison with (b), the interfacial velocity in (a) is plotted on a coarse grid, that is, averaged over 64×64 points. The domain size is 0.5×0.5.

From the numerical results, the full concentration and velocity fields during the pattern evolution are accessible. Figure 23.6a displays the synthetic shadowgraph distribution along with the velocity field at the interface $\mathbf{u}(z = 0)$ (plotted on a coarse grid).

Before discussing optical flow, let us compare the shadowgraph pattern between simulated data (Figure 23.6a) and experimental data (Figure 23.5). It can be seen that the simulations reproduce the hierarchical pattern observed in the experiments remarkably well. The simulations therefore show that the emergence of a higher hierarchy level is not primarily the result of interactions with interfacial deformations or complex surfactant sorption kinetics, which are absent from our model with a plane interface and Henry condition. In addition to the agreement in the qualitative features of the RC-II, quantitative agreement could also be found by comparing the length scales of the RC-I and RC-II to our own validation experiments and to experiments in the literature on the same mass transfer system [20].

In line with this agreement in length scales, the experimental (Figure 23.5) and theoretical flow fields (Figure 23.6a) display qualitatively identical features and are of similar magnitude as well. The comparison of the optical flow from experiments (Figure 23.5) with the interfacial velocity (Figure 23.6a) is motivated by the fact that the sharpest solute fronts (schlieren) are located near the interface, where they are produced by the Marangoni convection. However, as the quantities $(\mathbf{u}(z=0) \leftrightarrow \mathbf{u}_{of})$ for this comparison are not identical, the interpretation of the optical flow requires special care. This will be demonstrated in the following. An optical flow field $\mathbf{u}_{of}$ is calculated from the synthetic shadowgraph distribution by applying the same procedure as for the experimental data and plotted in Figure 23.6b for comparison with $\mathbf{u}(z = 0)$ in Figure 23.6a. In fact, both flow fields agree to a large extent for the hierarchical RC-II. A quantitative comparison, however, reveals that the optical flow underestimates the actual velocity at the interface: The mean magnitude of the interfacial velocity is $\langle|\mathbf{u}|\rangle = 0.2303$ and the mean magnitude of the optical flow is $\langle|\mathbf{u}'_{of}|\rangle = 0.1228$.

This deviation is caused by three effects: The first one is the spatial averaging inherent in the calculation of the optical flow field, since it determines the cross-correlation for discrete image subregions of finite width. The extent of the spatial averaging can be estimated by downsampling the interfacial velocity field to the coarse grid shown in Figure 23.6a. When the mean interfacial velocity is computed from those data, its value reduces to $\langle|\mathbf{u}_{coarse}|\rangle = 0.1878$. Second, the optical flow underestimates the interfacial flow to a certain degree since the velocity in the bulk decreases with the distance from the interface. This affects the movement of the solute fronts which are sharpest at $z = 0$ but also include information on structures further away from the interface. The third cause for the deviation is due to the fact that the moving solute fronts belong to the RC-I which are advected by the secondary flow of the RC-II. Therefore, the flow induced by the smallest structures cannot be detected. This is visualized in Figure 23.7 for the early state of mass transfer, where the pronounced hierarchical patterns have not developed yet and mainly RC-I exist. Whereas the radial convection of the RC-I can be clearly recognized in the interfacial velocity field in (a), the optical flow in (b) disregards the stationary pattern of RC-I. Flow is only detected in regions where RC-II begin to form. Hence, the mean magnitude of the optical flow is very small $\langle|\mathbf{u}'_{of}|\rangle = 0.0355$ by comparison with the mean magnitude of the interfacial velocity $\langle|\mathbf{u}|\rangle = 0.3338$. We conclude that only in the case of the hierarchical RC-II, the optical flow field can be used to qualitatively describe the velocity field and to provide a sound estimate for the interfacial velocity.

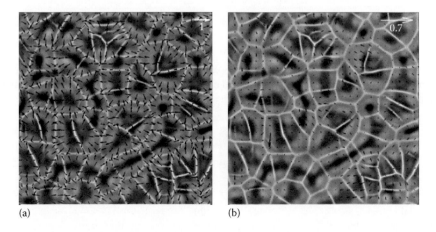

(a)   (b)

FIGURE 23.7   Interfacial velocity (a) and optical flow (b) for a pattern of mainly RC-I at $t = 0.2$. For comparison with (b), the interfacial velocity in (a) is plotted on a coarse grid, that is, averaged over 16×16 collocation points. The size of the shown domain is 0.125×0.125.

23.4.2 HELE-SHAW GEOMETRY: DEPENDENCE ON GAP WIDTH

This section reports on the quality of representing the experimental observations by the HS model. For this purpose, two configurations differing in plate distance (=gap size) 2ε are examined [32]. Accordingly, a different factor $\gamma = (d^{(1)})^2/\varepsilon^2$ in front of the Darcy friction term (Equation 23.18) is obtained. The corresponding values for both configurations are

- *Thin* HS cell: $2\varepsilon = 0.5$ mm, $\gamma = (d^{(1)})^2/\varepsilon^2 = 6400$
- *Thick* HS cell: $2\varepsilon = 1$ mm, $\gamma = (d^{(1)})^2/\varepsilon^2 = 1600$

The shadowgraph distribution from experiment and its theoretical counterpart are used for a qualitative comparison here, while a quantitative evaluation can be found in Ref. 32. Figure 23.8 shows the pattern in a 0.5 mm gap while the pattern for the doubled plate distance of 1 mm is depicted in Figure 23.9. With increasing plate distance, the influence of wall friction in the HS model decreases due to the lower friction factor γ. Besides, 3D effects (e.g., edge convection [52]) are expected to be more pronounced in the experiments with large gap.

The general structure of the Marangoni roll cells in the HS configuration can be seen nicely in Figure 23.8. In the upper organic phase, the region of mixed fluid (poor in solute, i.e., butanol) is bordered by a dark-gray rim to fluid rich in solute at the top. A second boundary at the bottom is given by the interface which manifests in the experimental image (a) as a solid black rim due to the meniscus formed between both liquid phases (detailed later). The dark-gray rim of the mixed fluid is bent toward the interface at the inflow regions, where butanol-rich fluid from the bulk flows to the interface. Apparently, the horizontal size of the roll cells (distance from inflow to inflow) is larger in the experiments (a) than in the simulation (b). Qualitatively, however, the experimental flow structures are well represented by the simulation [32]. In both cases, Marangoni roll cells without internal substructure develop (RC-I).

For the doubled plate distance in Figure 23.9, the hierarchical pattern of small and large roll cells (RC-II) appears well visible in the numerical shadowgraph image (b). This can be explained by the decreasing influence of wall friction that gives rise to an intensified flow [32]. In Figure 23.9, the

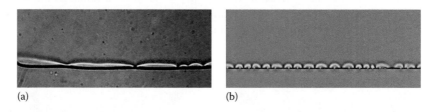

(a)   (b)

FIGURE 23.8 Pattern of mainly RC-I in a *thin* HS cell at $t = 1.0$. The experimental and numerical shadowgraph images show the same domain with horizontal extent of one length unit $l_x = 1$ (20 mm). (a) Experiment, $2\varepsilon = 0.5$ mm and (b) simulation, $\gamma = 6400$.

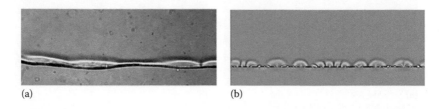

(a)   (b)

FIGURE 23.9 Pattern of RC-II in a *thick* HS cell at $t = 1.0$. The experimental and numerical shadowgraph images show the same domain with horizontal extent of one length unit $l_x = 1$ (20 mm). (a) Experiment, $2\varepsilon = 1$ mm and (b) simulation, $\gamma = 1600$.

patterns reach further into the bulk and the inflow region has a broader horizontal extent compared to Figure 23.8. This enhanced advection of solute again leads to high gradients at the interface so that small scales can be produced according to the theory [18]. In the experimental image (a), the small substructure of RC-I is only faintly visible by a dim horizontal modulation near the interface: Contrary to the supposed plane interface in the numerical model, a concave meniscus (viewed from the aqueous layer) across the plates is formed in the experiments. As a result, the small substructures are shadowed in the experimental images due to the deflection of light at this meniscus.

Furthermore, we expect that the curved interface is a main cause for the observed differences in pattern size between simulation and experiment [32]. Apart from the concave meniscus, the position of the interface also changes in lateral direction as a result of the experimental filling procedure, see Figure 23.9a. This imposes a higher concentration at parts of the interface which reach deeper into the delivering phase. The resulting Marangoni convection accelerates the experimental evolution compared to the simulations. Besides, the interfacial area increases by 10%–15% due to interfacial deformations.

3D flow effects, that is, a variation of concentration across the gap and deviations from the parabolic velocity profile, are another source of discrepancies between experimental and numerical results [32]. In addition to the 3D flow originating from the meniscus convection explained earlier, Marangoni cells emerge, which are smaller than the gap size. Therefore, it is expected that Marangoni cell are also amplified across the gap ($\partial_y c \neq 0$). This effect is particularly important at a large plate distance leading to the pronounced deviations between experiment and simulation in Figure 23.9. On this basis, we can conclude that the HS model is expected to be more precise for systems where large Marangoni cells without substructure develop, for example, in Ref. 12, since 3D flow effects will be less pronounced in that case. These limitations similarly concern 2D simulations without the influence of wall friction. Hence, this comparison likewise provides some guidance for their validity. Nevertheless, the qualitative agreement of structures encourages further studies based on the HS configuration. For example, an extension of the model to include interfacial deformations has the potential to approach the experimental situation more closely and to estimate the contribution of interface deformations to the overall deviation discussed before.

23.4.3 Two-Dimensional versus Three-Dimensional Simulation

In the preceding section, limitations were pointed out when representing an experimental HS cell by a 2D model. This section is devoted to the purely theoretical comparison of the full 3D to the 2D case. To provide comparability, the 2D model without influence of wall friction ($\gamma = 0$) is used.

Figure 23.10 shows two space-time plots of the concentration variation at the interface $c^{(1)}(x,t) - \langle c^{(1)}(x,t) \rangle_x$: one for the 2D simulation in (a) and one for the 3D simulation in (b). In the 3D case, we consider the concentration variation along a fixed value of the y-coordinate ($y = 0.25$). Dark gray values mainly correspond to inflow regions (low interfacial tension), bright gray values mark outflow regions (high interfacial tension). Small-scale modulations indicate substructure RC-I, which are advected by the RC-II. For both models, the coarsening and successive substructuring in the temporal evolution from small RC-I to a pattern of few large-scale RC-II at $t = 2$ takes place. However, some differences can be observed for the 2D and 3D case. First, the instability sets in earlier for the pure 2D simulation in Figure 23.10a. Moreover, the additional degree of freedom in the 3D simulation affects the appearance of substructure cells. At locations with considerable flow in y-direction (i.e., out-of-plane), the substructures manifest as closed cellular structures in Figure 23.10b. Since this figure only depicts a slice of the dense network of patterns, cells which are cut off-center give the impression that much finer structures develop for the 3D case.

In fact, Figure 23.11a shows that the deviation between 2D and 3D in the size of the RC-I is rather small. This characteristic size λ of the RC-I is derived as follows. The number of convection cells N_c is counted as the minimal number of continuous (connected) locations C_j with a flow toward the interface, that is, inflow zones. In the 3D case, λ is set to the edge length of supposed square cells that fill the complete area $l_x \times l_y$:

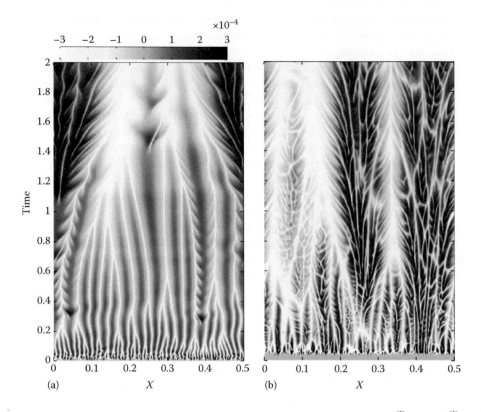

FIGURE 23.10 Temporal evolution of concentration variation at the interface: (a) $c^{(1)}(x,t) - \langle c^{(1)}(x,t) \rangle_x$ for the 2D simulation without influence of wall friction ($\gamma = 0$), (b) $c^{(1)}(x,t) - \langle c^{(1)}(x,t) \rangle_x$ for a horizontal line $y = 0.25$ of the 3D simulation shown in Section 23.4.1.

$$\lambda = \sqrt{\frac{l_x \times l_y}{N_c}}, \qquad (23.24)$$

whereas in the 2D case the mean interval length is used:

$$\lambda = \frac{l_x}{N_c}. \qquad (23.25)$$

Due to the higher variance in time for the 2D system, the ensemble average over three independent runs is plotted in Figure 23.11a. For both cases, λ grows until it reaches a fairly constant value. The end of the RC-I growth phase coincides with the occurrence of the RC-II according to the following mechanism. In the inflow zones of the higher-order structures, solute-rich fluid from the bulk is constantly brought to the interface. Therefore, high vertical concentration gradients can be maintained, which leads to the steady production of small scales as demonstrated in Ref. 14. Apart from the higher variance even in the averaged 2D curve, the progression of λ is quite similar in Figure 23.11a but with slightly smaller values in 3D.

Nevertheless, the decision for a 2D or 3D model also affects the estimation of technologically relevant parameters. Mass transfer efficiency is frequently described by the enhancement factor R (see e.g. [15]), which relates the averaged concentration of solute in the accepting phase (layer 1) for the case of Marangoni convection ($Ma \neq 0$) to mass transfer by pure diffusion ($Ma = 0$).

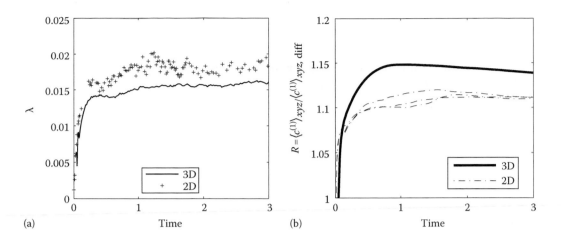

FIGURE 23.11   Comparison between the 2D simulation without influence of wall friction ($\gamma = 0$) and the 3D simulation shown in Section 23.4.1: temporal evolution of (a) the cell size λ and (b) the enhancement factor R. In (a), we averaged over three simulations for the 2D case.

Figure 23.11b depicts the temporal evolution of the enhancement factor for the 2D and 3D case. To visualize the stronger variance of the 2D simulations mentioned before, R is plotted for the three different runs (dashed curves). In 3D (full curve), deviations between independent runs are negligible due to the larger effective system size (the interfacial area amounts to an effective ensemble of one-dimensional distributions). After the initial phase, the 3D simulation yields higher values in R. Hence, the additional lateral mixing of the hierarchical roll cells in 3D more effectively accelerates mass transfer compared to the reduced dynamics in 2D.

23.5   CONCLUSION

In this chapter, theoretical models to describe pattern formation during mass transfer across two liquid layers subject to Marangoni instability were presented for 3D and 2D HS geometry. A pseudospectral method was applied for the direct numerical simulation of this problem [14,32]. Thereby, the complex hierarchical flow patterns observed in experiments could be reproduced. However, the comparison of the results obtained by the 3D [14] and the 2D HS model [32] in Sections 23.4.1 and 23.4.2 confirms the expectation that experimental situations are described more accurately by taking into account the third dimension. Even if the experimental geometry is approximated to a 2D situation in the HS cell, the 3D flow effects prevent a quantitative agreement between the experimental and numerical results. Furthermore, the experiments in the cubical cuvettes (see Section 23.3.2.1) more closely correspond to the assumption of a plane and undeformable interface in the numerical model. In contrast, pronounced influence of contact angle and interfacial deformations due to the filling procedure have to be considered in the HS cell. Nevertheless, the qualitative features of the temporal evolution toward hierarchical RC-II can be reproduced by the 2D model, for example, the progression of the RC-I size (see Figure 23.11a) and the occurrence of RC-II depending on wall friction (see Section 23.4.2). Therefore, computationally less expensive 2D simulations can provide first insights into pattern formation of a system subject to solutal Marangoni instability.

23.A   APPENDIX

This appendix serves to summarize in Table 23.A.1 all important material properties of the system (cyclohexanol + butanol)/water as used in the numerical simulation as well as to explain the nomenclature.

TABLE 23.A.1
Basic Parameters for Simulation in the System Cyclohexanol +7.5 vol% Butanol over Water

Description	Symbol	Value
Mass density aqueous phase (1)	$\rho_{ref}^{(1)}$	997 kg m^{-3}
Mass density organic phase (2)	$\rho_{ref}^{(2)}$	955 kg m^{-3}
Kinematic viscosity (1)	$v^{(1)}$	1.2×10^{-6} m^2 s^{-1}
Kinematic viscosity (2)	$v^{(2)}$	20×10^{-6} m^2 s^{-1}
Diffusivity butanol (1)	$D^{(1)}$	5×10^{-10} m^2 s^{-1}
Diffusivity butanol (2)	$D^{(2)}$	7×10^{-11} m^2 s^{-1}
Interfacial tension of the binary system	σ_{ref}	3.4×10^{-3} N m^{-1}
Change in interfacial tension via $\tilde{c}^{(1)}$	$\sigma_{ref}\alpha_c$	-8.77×10^{-3} Nm^{-1}/(mol L^{-1})
Solutal expansion coefficient (1)	$\beta_c^{(1)}$	-0.0172 mol^{-1} L
Solutal expansion coefficient (2)	$\beta_c^{(2)}$	-0.0128 mol^{-1} L
Schmidt number (1)	$Sc^{(1)} = \dfrac{v^{(1)}}{D^{(1)}}$	2400
Schmidt number (2)	$Sc^{(2)} = \dfrac{v^{(2)}}{D^{(2)}}$	2.86×10^5
Marangoni number	$Ma = \dfrac{c_0\alpha_c\sigma_{ref}d^{(1)}}{\rho^{(1)}v^{(1)}D^{(1)}}$	-2.4×10^8
Grashof number	$G = \dfrac{c_0\beta_c^{(1)}g(d^{(1)})^3}{(v^{(1)})^2}$	-7.67×10^5
Partition coefficient	$H = c_{eq}^{(2)}/c_{eq}^{(1)}$	31
Density ratio	$\rho = \dfrac{\rho_{ref}^{(2)}}{\rho_{ref}^{(1)}}$	0.96
Kinematic viscosity ratio	$v = \dfrac{v^{(2)}}{v^{(1)}}$	16.7
Diffusivity ratio	$D = \dfrac{D^{(2)}}{D^{(1)}}$	0.14
Description	**Symbol**	**Value**
Ratio of expansion coefficients	$\beta = \dfrac{\beta_c^{(2)}}{\beta_c^{(1)}}$	0.75
Layer height ratio	$d = \dfrac{d^{(2)}}{d^{(1)}}$	1
Dynamic viscosity ratio	$\mu = v\rho$	16.03

Source: The data sources are detailed in, with permission from, Köllner, T., Schwarzenberger, K., Eckert, K., and Boeck, T., Multiscale structures in solutal Marangoni convection: Three-dimensional simulations and supporting experiments, *Phys. Fluids*, 25, 092109, Copyright 2013, American Institute of Physics.

Phase 1 is the water-rich phase marked with upper index (1) and phase 2 the organic cyclohexanol-rich phase marked with (2). The acceleration due to gravity is $g = 9.81$ m s^{-2}. Both layer heights are set to $d^{(1)} = d^{(2)}$ mm.

ACKNOWLEDGMENTS

Financial support by the Deutsche Forschungsgemeinschaft in the form of the Priority Program 1506 and Grants Nos. Bo1668/6 (T.B.) and Ec201/2 (K.E.) is gratefully acknowledged. Furthermore, we thank the computing center (UniRZ) of TU Ilmenau and the computing center of FZ Jülich (NIC) for access to its parallel computing resources. Prof. H. Linde is thanked for numerous fruitful discussions.

REFERENCES

1. A. Orell and J. W. Westwater, Spontaneous interfacial cellular convection accompanying mass transfer: Ethylene glycol-acetic acid-ethyl acetate, *AIChE Journal*, 8, 350–356, 1962.
2. H. Linde and E. Schwarz, Untersuchungen zur Charakteristik der freien Grenzflächenkonvektion beim Stoffübergang an fluiden Grenzen, *Zeitschrift für physikalische Chemie*, 224, 331–352, 1963.
3. C. Bakker, F. Fentener van Vlissingen, and W. Beek, The influence of the driving force in liquid–liquid extraction—A study of mass transfer with and without interfacial turbulence under well-defined conditions, *Chemical Engineering Science*, 22(10), 1349–1355, 1967.
4. E. Schwarz, Zum Auftreten von Marangoni-Instabilität, *Wärme- und Stoffübertragung*, 3, 131–133, 1970.
5. H. Linde, P. Schwartz, and H. Wilke, Dissipative structures and nonlinear kinetics of the Marangoni-instability, in *Dynamics and Instability of Fluid Interfaces*, T. S. Sørensen (ed), Springer, Berlin/Heidelberg, 1979, pp 75–119.
6. D. Bratsun and A. De Wit, On Marangoni convective patterns driven by an exothermic chemical reaction in two-layer systems, *Physics of Fluids*, 16(4), 1082–1096, 2004.
7. A. Grahn, Two-dimensional numerical simulations of Marangoni-Bénard instabilities during liquid-liquid mass transfer in a vertical gap, *Chemical Engineering Science*, 61, 3586–3592, 2006.
8. Z. Mao, P. Lu, G. Zhang, and C. Yang, Numerical simulation of the Marangoni effect with interphase mass transfer between two planar liquid layers, *Chinese Journal of Chemical Engineering*, 16(2), 161–170, 2008.
9. Y. Sha, H. Chen, Y. Yin, S. Tu, L. Ye, and Y. Zheng, Characteristics of the Marangoni convection induced in initial quiescent water, *Industrial & Engineering Chemistry Research*, 49, 8770–8777, 2010.
10. S. Chen, B. Fu, X. Yuan, H. Zhang, W. Chen, and K. Yu, Lattice Boltzmann method for simulation of solutal interfacial convection in gas–liquid system, *Industrial & Engineering Chemistry Research*, 51, 10955–10967, 2012.
11. K. Schwarzenberger, K. Eckert, and S. Odenbach, Relaxation oscillations between Marangoni cells and double diffusive fingers in a reactive liquid–liquid system, *Chemical Engineering Science*, 68, 530–540, 2012.
12. K. Eckert, M. Acker, R. Tadmouri, and V. Pimienta, Chemo-Marangoni convection driven by an interfacial reaction: Pattern formation and kinetics, *Chaos: An Interdisciplinary Journal of Nonlinear Science*, 22(3), 037112, 2012.
13. K. Schwarzenberger, T. Köllner, H. Linde, S. Odenbach, T. Boeck, and K. Eckert, On the transition from cellular to wave-like patterns during solutal Marangoni convection, *The European Physical Journal Special Topics*, 219, 121–130, 2013.
14. T. Köllner, K. Schwarzenberger, K. Eckert, and T. Boeck, Multiscale structures in solutal Marangoni convection: Three-dimensional simulations and supporting experiments, *Physics of Fluids*, 25, 092109, 2013.
15. K. Schwarzenberger, T. Köllner, H. Linde, T. Boeck, S. Odenbach, and K. Eckert, Pattern formation and mass transfer under stationary solutal Marangoni instability, *Advances in Colloid and Interface Science*, 206, 344–371, 2014.
16. R. Birikh, A. Zuev, K. Kostarev, and R. Rudakov, Convective self-oscillations near an air-bubble surface in a horizontal rectangular channel, *Fluid Dynamics*, 41(4), 514–520, 2006.
17. Y. Shi and K. Eckert, Acceleration of reaction fronts by hydrodynamic instabilities in immiscible systems, *Chemical Engineering Science*, 61, 5523–5533, 2006.
18. C. V. Sternling and L. E. Scriven, Interfacial turbulence: Hydrodynamic instability and the Marangoni effect, *AIChE Journal*, 5, 514–523, 1959.

19. H. Linde, K. Schwarzenberger, and K. Eckert, Pattern formation emerging from stationary solutal Marangoni instability: A roadmap through the underlying hierarchic structures, in *Without Bounds: A Scientific Canvas of Nonlinearity and Complex Dynamics*, R. Rubio, Y. Ryazantsev, V. Starov, G.-X. Huang, A. Chetverikov, P. Arena, A. Nepomnyashchy, A. Ferrus, and E. Morozov (eds.). Springer, Berlin/ Heidelberg, 2013.

20. E. Schwarz, *Hydrodynamische Regime der Marangoni-Instabilität beim Stoffübergang über eine fluide Phasengrenze*. PhD thesis, HU Berlin, Berlin, Germany, 1967.

21. D. Lide, *CRC Handbook of Chemistry and Physics: A Ready-Reference Book of Chemical and Physical Data*. CRC Press, Boca Raton, 2004.

22. A. Leo and C. Hansch, Linear free energy relations between partitioning solvent systems, *The Journal of Organic Chemistry*, 36(11), 1539–1544, 1971.

23. A. Nepomnyashchy, I. Simanovskii, and J. Legros, *Interfacial Convection in Multilayer Systems*. Springer, New York, 2006.

24. W. Merzkirch, *Flow Visualization*. Academic Press, London, U.K., 1987.

25. C. Canuto, M. Hussaini, A. Quarteroni, and T. Zang, *Spectral Methods in Fluid Dynamics*. Springer-Verlag, Berlin/Heidelberg, 1988.

26. R. Peyret, *Spectral Methods for Incompressible Viscous Flow*. Springer, New York, 2002.

27. F. Mariano, L. Moreira, A. Silveira-Neto, C. da Silva, and J. Pereira, A new incompressible Navier–Stokes solver combining Fourier pseudo-spectral and immersed boundary methods, *Computer Modeling in Engineering & Sciences*, 59(2), 181–216, 2010.

28. M. Fulgosi, D. Lakehal, S. Banerjee, and V. De Angelis, Direct numerical simulation of turbulence in a sheared air–water flow with a deformable interface, *Journal of Fluid Mechanics*, 482, 319–345, 2003.

29. T. Boeck, A. Nepomnyashchy, I. Simanovskii, A. Golovin, L. Braverman, and A. Thess, Three-dimensional convection in a two-layer system with anomalous thermocapillary effect, *Physics of Fluids*, 14, 3899, 2002.

30. T. Boeck, D. Krasnov, and E. Zienicke, Numerical study of turbulent magnetohydrodynamic channel flow, *Journal of Fluid Mechanics*, 572, 179–188, 2007.

31. T. Köllner, M. Rossi, F. Broer, and T. Boeck, Chemical convection in the methylene-blue–glucose system: Optimal perturbations and three-dimensional simulations, *Physical Review E*, 90(5), 053004, 2014.

32. T. Köllner, K. Schwarzenberger, K. Eckert, and T. Boeck, Solutal Marangoni convection in a Hele-Shaw geometry: Impact of orientation and gap width, *The European Physical Journal Special Topics*, 224, 261–276, 2015.

33. K. Eckert, M. Acker, and Y. Shi, Chemical pattern formation driven by a neutralization reaction. I. Mechanism and basic features, *Physics of Fluids*, 16(2), 385–399, 2004.

34. H. Linde, S. Pfaff, and C. Zirkel, Strömungsuntersuchungen zur hydrodynamischen Instabilität flüssig-gasförmiger Phasengrenzen mit Hilfe der Kapillarspaltmethode, *Zeitschrift für physikalische Chemie*, 225, 72–100, 1964.

35. Y. Shi and K. Eckert, A novel Hele-Shaw cell design for the analysis of hydrodynamic instabilities in liquid–liquid systems, *Chemical Engineering Science*, 63, 3560–3563, 2008.

36. W. Boos and A. Thess, Thermocapillary flow in a Hele-Shaw cell, *Journal of Fluid Mechanics*, 352, 305–330, 1997.

37. C. Ruyer-Quil, Inertial corrections to the Darcy law in a Hele–Shaw cell, *Comptes Rendus de l' des Sciences-Series IIB-Mechanics*, 329(5), 337–342, 2001.

38. J. Zeng, Y. C. Yortsos, and D. Salin, On the Brinkman correction in unidirectional Hele-Shaw flows, *Physics of Fluids*, 15(12), 3829–3836, 2003.

39. N. Jarrige, I. B. Malham, J. Martin, N. Rakotomalala, D. Salin, and L. Talon, Numerical simulations of a buoyant autocatalytic reaction front in tilted Hele-Shaw cells, *Physical Review E*, 81(6), 066311, 2010.

40. F. Gallaire, P. Meliga, P. Laure, and C. N. Baroud, Marangoni induced force on a drop in a Hele Shaw cell, *Physics of Fluids*, 26(6), 062105, 2014.

41. J. Martin, N. Rakotomalala, and D. Salin, Gravitational instability of miscible fluids in a Hele-Shaw cell, *Physics of Fluids*, 14(2), 902–905, 2002.

42. G. Taylor, Dispersion of soluble matter in solvent flowing slowly through a tube, *Proceedings of the Royal Society of London A: Mathematical, Physical and Engineering Sciences*, 219(1137), 186–203, 1953.

43. R. Aris, On the dispersion of a solute in a fluid flowing through a tube, *Proceedings of the Royal Society of London. Series A. Mathematical and Physical Sciences*, 235(1200), 67–77, 1956.

44. M. Leconte, J. Martin, N. Rakotomalala, D. Salin, and Y. Yortsos, Mixing and reaction fronts in laminar flows, *The Journal of Chemical Physics*, 120(16), 7314–7321, 2004.

45. W. Zimmerman and G. Homsy, Nonlinear viscous fingering in miscible displacement with anisotropic dispersion, *Physics of Fluids A: Fluid Dynamics*, 3(8), 1859–1872, 1991.

46. R. N. Horne and F. Rodriguez, Dispersion in tracer flow in fractured geothermal systems, *Geophysical Research Letters*, 10(4), 289–292, 1983.

47. P. Petitjeans, C.-Y. Chen, E. Meiburg, and T. Maxworthy, Miscible quarter five-spot displacements in a Hele-Shaw cell and the role of flow-induced dispersion, *Physics of Fluids*, 11(7), 1705–1716, 1999.

48. J. Barron, D. Fleet, and S. Beauchemin, Performance of optical flow techniques, *International Journal of Computer Vision*, 12(1), 43–77, 1994.

49. E. Koschmieder and M. Biggerstaff, Onset of surface-tension-driven Bénard convection, *Journal of Fluid Mechanics*, 167, 49–64, 1986.

50. J. Bragard and M. Velarde, Bénard–Marangoni convection: Planforms and related theoretical predictions, *Journal of Fluid Mechanics*, 368(1), 165–194, 1998.

51. K. Eckert, M. Bestehorn, and A. Thess, Square cells in surface-tension-driven Bénard convection: Experiment and theory, *Journal of Fluid Mechanics*, 356, 155–197, 1998.

52. J. C. Berg and C. R. Morig, Density effects in interfacial convection, *Chemical Engineering Science*, 24, 937–946, 1969.

24 Modeling Foam Stability

Marcel B.J. Meinders and Ruud G.M. Van der Sman

CONTENTS

24.1 INTRODUCTION

Foams are dispersions of gas bubbles in a continuous matrix that find many applications in industry and daily life. Examples include aerated foods, packaging, and isolation materials. A big challenge is to control foam formation and deterioration via creaming, drainage, coalescence, and disproportionation. Many foams, for example, food foams, consist of various interacting components, like proteins, small molecular surfactants, and carbohydrate polymers, that exhibit different physical and chemical properties. The role of these components on the various phenomena involved in foam formation and stability are not yet sufficiently well understood. In order to get better insights, it is crucial to develop generic quantitative relations between the properties of the ingredients and that of the multicomponent foam. Therefore we model and experimentally study ingredient interactions, air/water interfaces, thin liquid films, single air bubbles, and foams of model food systems containing proteins and low molecular weight surfactants.

In the following, we distinguish between wet foams and dry foams. Wet foams consist of spherical gas bubbles and have a liquid volume fraction ϵ that is about larger than that of the random closed packing configuration (ϵ_c is larger then about 0.36 for a monodisperse foam). Wet foams can be realized by an appropriate choice of stabilizers. In this case, coalescence between neighboring bubbles can be neglected and disproportionation is the main deterioration mechanism. Disproportionation, or Ostwald ripening, is a coarsening process where gas molecules diffuse from small to large bubbles via the continuous phase as result of the difference in Laplace pressure between bubbles. It is (almost) impossible to stop and is of significant importance for the long-term stability of foams and emulsions. Various studies have been published [46] that indicate that it may be retarded or stopped when the dispersed phase is poorly soluble or insoluble or by bulk or interfacial viscosity or elasticity [6,14,23–25,34,35]. Also a so-called microbubble dispersion can be considered as a wet foam. Such a dispersion consist of micron-sized bubbles with a thick protein layer. Protein-stabilized microbubbles can be manufactured using ultrasound. Such bubbles cream hardly under normal gravitation and can be stable against disproportionation for months [1,8,41,47,49,50].

A gas bubble can also be stabilized against disproportionation by so-called Pickering stabilization [39], where colloidal particles are adsorbed at the interface that are partially wetted by the continuous phase [5]. These colloidal particles have an adsorption energy $E_a = \pi r_p^2 \sigma (1 \pm \cos\theta_c)$, with r_p being the particle radius and θ_c the contact angle. The adsorption energy of a particle with $r_p = 10$ nm and $\theta_c = \pi/2$ is already in the order of 1000 times larger than $k_B T$, with k_B is Boltzmann constant. These colloidal particles do not desorb due to the build-up stress in the interface and can withstand the Laplace pressure. This results in a stable bubble with a monolayer of particles at the interface that are closed packed [7].

In Section 24.2, we provide an overview of a numerical model that describes the effect of interfacial rheology on the dissolution of single gas bubbles in an aqueous environment as well as a wet foam [25,34,35]. Some examples are given.

In Section 24.3, we will focus on the formation and stability (in particular drainage and bubble coalescence) of dry polyhedral foams.

24.2 FOAM STABILITY OF WET FOAMS AND THE EFFECT OF INTERFACIAL RHEOLOGY

24.2.1 SINGLE BUBBLE DISSOLUTION

The bubble is considered as a spherical gas bubble with radius R inside a liquid phase. For simplicity's sake, let us consider that the gas consists only of one component, for example, nitrogen. Due to a difference in the concentration of the gas in the liquid phase and just at the interface of the bubble, there will be transport of gas molecules across the bubble surface. Using Fick's law of diffusion and considering the steady-state solution, this transport can be written as

$$\frac{\partial n}{\partial t} = 4\pi R D (C_m - C_s) \tag{24.1}$$

with
  n is the amount of gas molecules
  t is the time
  D is the diffusion coefficient of the gas in the liquid phase
  C_m and C_s are the gas concentrations in the liquid phase and at the interface, described by
    Equations 24.2 and 24.3, respectively

Here, only diffusion is considered, and the velocity of the bubble interface as well as possible mass transfer effects due to the interfacial layer are neglected. Often the concentration is related to the solubility of the gas C_∞ (at the temperature T and pressure P_0). Then

$$C_m = f C_\infty \tag{24.2}$$

and according to Henry's law

$$C_s = C_\infty \frac{P_R}{P_\infty} = C_\infty \left(1 + \frac{2\sigma}{P_\infty R}\right)$$ (24.3)

with
  f is the saturation factor
  P_R is the gas pressure inside the bubble

This equation is directly related to the Kelvin equation describing the vapor pressure of small droplets. The gas pressure inside the gas bubble is given by

$$P_R = P_\infty + P_L = P_\infty + \frac{2\sigma}{R}$$ (24.4)

with
  P_∞ is the environmental pressure
  σ is the interfacial tension
  $P_L = 2\sigma/R$ is the Laplace pressure.

The diffusion equation then reads

$$\frac{\partial n}{\partial t} = 4\pi RDC_\infty \left(f - \left(1 + \frac{2\sigma}{P_\infty R}\right)\right)$$ (24.5)

The relation between the number of gas molecules n and the bubble radius R is given by the universal gas law $P_R V = nR_g T$ with $V = \frac{4}{3}\pi R^3$ is the volume of the bubble, R_g is the universal gas constant, and T is the temperature.

24.2.2   INTERFACIAL RHEOLOGY: STABILIZATION BY VISCOELASTIC SURFACE COATINGS

Air bubbles are normally stabilized by surface active molecules like surfactants and proteins, which lower the interfacial tension σ upon adsorption. Furthermore, they can provide a repulsion between two bubble surfaces approaching each other, thereby slowing down or preventing coalescence. Adsorbed interfacial layers show viscoelastic behavior, where the interfacial tension depends on the amount and speed of a change in the interfacial area. For example, the interfacial tension is closely related to the amount of adsorbed molecules. When the interfacial area is increased and no additional molecules adsorbed, the interfacial tension will also increase. In time surface, active molecules can adsorb, thereby decreasing the interfacial tension until a new equilibrium is obtained. In the case of adsorbed proteins, reorientations and possible conformational changes play a role in the viscoelastic behavior of the interfacial layer. Figure 24.1 shows an example of the viscoelastic behavior of an adsorbed protein layer measured with automated drop tensiometry [2].

One of the most simple models that shows viscoelastic behavior is the standard linear solid model. As shown in Figure 24.2, it consists of a spring with an elastic spring constant K_1 that is connected in parallel with a spring (elasticity constant K_2) and dashpot (viscosity η) that are connected in a series. It is a combination of a Maxwell and Kelvin–Voigt model. The interfacial tension of such a system can be expressed in the following differential equation:

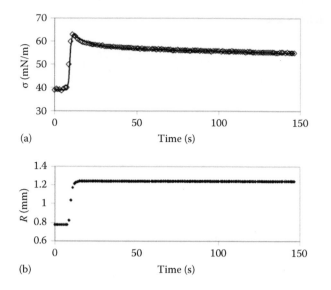

(a)

(b)

FIGURE 24.1   Measured interfacial tension σ ($\diamond$) as a function of time for an air bubble in a 0.1 mg/mL glycinin solution at pH = 3 (a). The bubble radius (b) was suddenly increased.

FIGURE 24.2   Schematic representation of the standard linear solid.

$$\frac{\partial \sigma}{\partial t} = \frac{K_1 K_2}{\eta} \ln\left(\frac{A}{A_0}\right) - \frac{K_2}{\eta}\sigma + (K_1 + K_2)\frac{\partial \ln A}{\partial t} \tag{24.6}$$

where A is the area of the interface and $A_0 = A(t = 0)$. The interfacial tension σ can then be calculated given a certain initial interfacial tension $\sigma_0 = \sigma(t = 0)$. After a change in interfacial area, the system relaxes toward an equilibrium surface tension with characteristic relaxation time of

$$\tau = \frac{\eta}{K_2} \tag{24.7}$$

When $K_1 = 0$ and τ approaches 0, the system behaves purely viscous. When only $\tau = 0$, the system behaves purely elastic with an interfacial dilatation or Youngs modulus $E = K_1$. When $\tau = \infty$, the system also behaves purely elastic with $E = K_1 + K_2$. Otherwise, the system behavior is viscoelastic. The standard linear solid can be easily extended to include multiple relaxation times by addition of one or more spring-damper systems.

When the system behaves purely elastic (assume $\tau = 0$) than the interfacial tension as a function of the bubble area is given by

$$d\sigma_E = K_1 d \ln A \tag{24.8}$$

or expressed in the bubble radius

$$\sigma_E = \sigma_0 - K_1 \ln\left(\frac{A}{A_0}\right) = \sigma_0 - 2K_1 \ln\left(\frac{R}{R_0}\right) \qquad (24.9)$$

A bubble will stabilize when the Laplace pressure equals zero, or when $\sigma_E = 0$. In that case the radius of the bubble equals

$$R = R_0 \exp\left(-\frac{\sigma_0}{2K_1}\right) \qquad (24.10)$$

A bubble is only stable when there is no relaxation of the stresses build up in the elastic layer due to desorption and/or rearrangements of the adsorbed species.

24.2.3 DISPROPORTIONATION OF A FOAM, AN ENSEMBLE OF BUBBLES

For the simulation of a wet foam, an ensemble of N bubbles is considered with radii R_i. In case the bubble interface behaves purely viscous or elastic, the evolution in time of the radii can be calculated by solving N coupled ordinary differential equations (ODEs), for each bubble i given by the Equation 24.5, assuming a certain initial bubble size distribution $\{R_{i0}\}$, similar to that introduced by De Smet et al. [12]. When the system behaves viscoelastic, $2N$ coupled ODEs for the radii (Equation 24.5) and interfacial tension (Equation 24.6) needs to be solved, assuming a certain initial bubble size distribution $\{R_{i0}\}$ and interfacial tension $\{\sigma_{i0}\}$.

The gas saturation f is estimated assuming the total number of gas molecules in the bubbles to be constant. Then from $\sum_i \partial n_i/\partial t = 0$ it follows that

$$f = \frac{C_m}{C_\infty} = 1 + \sum_i \frac{2\sigma_i}{P_\infty} \Big/ \sum_i R_i \qquad (24.11)$$

24.2.4 MODEL EQUATIONS AND DIMENSIONLESS PARAMETERS

24.2.4.1 Diffusion

For the calculations, it is convenient to rewrite the set of $2N$ ODEs in a dimensionless form. In the following, for clarity, we omit the subscript i indicating a bubble. The derived differential equations are valid for each bubble in the ensemble. Introducing $v = n/n_0$, $\varepsilon = R/R_0$, $\varsigma = \sigma/\sigma_0$, and $\theta = Dt/R_m^2$ with n_0, R_0, R_m, and σ_0, the initial amount of gas molecules, bubble radius, mean radius, and interfacial tension, respectively, one finds for the diffusion equation 24.5, after some straightforward algebra,

$$\frac{\partial v}{\partial \theta} = \frac{3V_0 C_\infty}{n_0}\left(\frac{R_m}{R_0}\right)^2 \varepsilon\left(f - \left(1 + \frac{2\sigma_0}{P_\infty R_0}\frac{\varsigma}{\varepsilon}\right)\right) \qquad (24.12)$$

One might introduce the dimensionless variables $\alpha_1 = 3V_0 C_\infty/n_0$, $\alpha_2 = 2\sigma_0/P_\infty R_0$, and $\alpha_3 = R_m/R_0$, so that Equation 24.5 can be written as

$$\frac{\partial v}{\partial \theta} = \alpha_1 \alpha_3^2 \varepsilon\left(f - \left(1 + \alpha_2 \frac{\varsigma}{\varepsilon}\right)\right) \qquad (24.13)$$

Similar on finds for the universal gas law

$$\varepsilon^3 + \alpha_2 \varsigma \varepsilon^2 = \left(1 + \alpha_2\right) v \tag{24.14}$$

or differentiating to θ and rearranging yields

$$\frac{\partial v}{\partial \theta} = \left(\frac{3\varepsilon^2 + 2\alpha_2 \varsigma \varepsilon}{1 + \alpha_2}\right)\frac{\partial \varepsilon}{\partial \theta} + \left(\frac{\alpha_2 \varepsilon^2}{1 + \alpha_2}\right)\frac{\partial \varsigma}{\partial \theta} \tag{24.15}$$

Combining with the dimensional diffusion equation 24.13 gives

$$A_1 \frac{\partial \varepsilon}{\partial \theta} + A_2 \frac{\partial \varsigma}{\partial \theta} = A_0 \tag{24.16}$$

with

$$A_1 = \left(\frac{3\varepsilon^2 + 2\alpha_2 \varsigma \varepsilon}{1 + \alpha_2}\right) \tag{24.17}$$

$$A_2 = \left(\frac{\alpha_2 \varepsilon^2}{1 + \alpha_2}\right) \tag{24.18}$$

$$A_0 = \alpha_1 \alpha_3^2 \varepsilon \left(f - \left(1 + \alpha_2 \frac{\varsigma}{\varepsilon}\right)\right) \tag{24.19}$$

24.2.4.2 Interfacial Rheology

The relation between ε and σ is determined by the rheological properties of the interfacial layer. For the standard linear solid, Equation 24.6, it reads, using the dimensionless parameters defined before,

$$\frac{\partial \varsigma}{\partial \theta} = 2\left(\frac{K_1}{\sigma_0}\right)\left(\frac{R_m^2}{D}\frac{K_2}{\eta}\right)\ln \varepsilon - \left(\frac{R_m^2}{D}\frac{K_2}{\eta}\right)\varsigma + 2\left(\frac{K_1}{\sigma_0} + \frac{K_2}{\sigma_0}\right)\frac{\partial \ln \varepsilon}{\partial \theta} \tag{24.20}$$

or

$$\frac{\partial \varsigma}{\partial \theta} = 2\kappa_1 \lambda \ln \varepsilon - \lambda \varsigma + 2(\kappa_1 + \kappa_2)\frac{\partial \ln \varepsilon}{\partial \theta} \tag{24.21}$$

with $\kappa_1 = K_1/\sigma_0$, $\kappa_2 = \dfrac{K_2}{\sigma_0}$, and $\lambda = (R_m^2/D)(K_2/\eta) = \tau_{diffusion}/\tau_{relaxation}$. The rheology equation can then be written as

$$B_1 \frac{\partial \varepsilon}{\partial \theta} + B_2 \frac{\partial \varsigma}{\partial \theta} = B_0 \tag{24.22}$$

with

$$B_1 = -\frac{2(\kappa_1 + \kappa_2)}{\varepsilon} \tag{24.23}$$

$$B_2 = 1 \tag{24.24}$$

$$B_0 = 2\kappa_1 \lambda \ln \varepsilon - \lambda \varsigma \tag{24.25}$$

Combining Equations 24.16 and 24.22 yields $2N$ coupled differential equations for an ensemble of N bubbles:

$$\frac{\partial \varepsilon}{\partial \theta} = \frac{A_2 B_0 - B_2 A_0}{B_1 A_2 - B_2 A_1} \tag{24.26}$$

$$\frac{\partial \varsigma}{\partial \theta} = \frac{A_0 B_1 - B_0 A_1}{B_1 A_2 - B_2 A_1} \tag{24.27}$$

The saturation factor f can also be expressed in the dimensionless variables. According to Equation 24.11, this gives

$$f = 1 + \frac{\sum \alpha_2 \varsigma R_{i0}}{\sum_i \varepsilon R_{i0}} \tag{24.28}$$

The $2N$ coupled differential equations (Equations 24.26 and 24.27) can be solved numerically together with Equation 24.28 and initial conditions for the bubble radii R_{i0} and interfacial tension σ_{i0}, using standard ODE-solvers. We used the fortran routine DLSODI from the ODEPACK library.

24.2.5   SIMULATION RESULTS

24.2.5.1   Constant Surface Tension

Figure 24.3a shows the calculated bubble size distributions of a wet foam initially consisting of $N = 500$ bubbles at different times. The initial bubble size distribution $\{R_{i0}\}$ is log-normal with a mean of $\langle R_0 \rangle = 200$ μm. The surface tension for all bubbles is constant in time and taken to be $\sigma = \sigma_0 = 40$ mN/m. Furthermore, $T = 293$ K, gas (N_2) solubility $C_\infty = 0.688$ mol/m³, diffusion constant $D = 2 \times 10^{-9}$ m²/s, and environmental pressure $P_\infty = 1 \times 10^5$ Pa. Evolved distributions are obtained by solving N coupled ODEs (Equation 24.26) in combination with Equation 24.28.

The initial distribution evolves toward the so-called LSW distribution. This time invariant distribution is derived by Lifshitz and Slyozov [31] and Wagner [52], independently. This is shown in the middle panel where the bubble size distributions are depicted, but now scaled with the mean value. The right panel shows the cube of the averaged radii, scaled with respect to the initial mean $R_{kl} = \langle R_0 \rangle^{l-k} \langle R^k \rangle / \langle R^l \rangle$. For θ about larger than 2×10^{-4}, the stationary regime is reached where a linear increase in R_{10}^3, R_{21}^3, and R_{32}^3 as a function of time is observed with a slope of about 1.00, 1.05, and 1.08, respectively, corresponding to the LSW prediction.

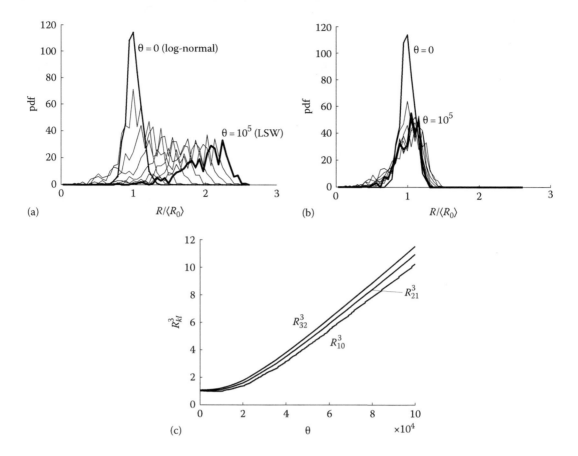

FIGURE 24.3 Calculated evolution of the bubbles distribution of wet foam initially consisting of 500 bubbles having an initial log-normal size distribution ($\langle R \rangle = 200$ μm). The surface tension is constant in time $\sigma = \sigma_0 = 40$ mN/m. The initial distribution evolves toward the LSW distribution. (a) shows the size distributions at different times θ, scaled with the initial mean $\langle R_0 \rangle$. (b) shows the normalized size distributions but scaled with the mean value, indicating that the initial log-normal distribution evolves to the stationary LSW distribution. (c) shows the averaged radii $R_{32} = \langle R^3 \rangle / \langle R^2 \rangle / \langle R_0 \rangle$ and $R_{10} = \langle R \rangle / \langle R_0 \rangle$ as a function of time.

24.2.5.2 Purely Elastic Interface

Figure 24.4a shows the calculated bubble size distributions of a wet foam initially consisting of $N = 500$ bubbles at different times, stabilized by a fully elastic interface. The interfacial rheology is described by the standard linear solid (see Figure 24.2) with an interfacial dilatation or Youngs modulus $E = K_1 = 10$ mN/m and $\tau = \eta / K_2 = 0$. The initial interfacial surface tension for all bubbles is taken to be $\sigma_0 = 40$ mN/m. The initial bubble size distribution $\{R_{i0}\}$ is log-normal with a mean of $\langle R_0 \rangle = 200$ μm. Furthermore, $T = 293$ K, gas (N$_2$) solubility $C_\infty = 0.688$ mol/m^3, diffusion constant $D = 2 \times 10^{-9}$ m^2/s, and environmental pressure $P_\infty = 1 \times 10^5$ Pa. Evolved distributions are obtained by solving N-coupled ODEs (Equation 24.26) in combination with Equation 24.9 for the interfacial tension and Equation 24.28 for the gas saturation factor. Figure 24.4b shows examples of the evolution of the radii of individual bubbles.

The initial size distribution evolves toward a stable biomodal distribution consisting of N—bubbles that decreased in size and have about zero interfacial tension due to decrease in interfacial area (see Equation 24.10) and one large bubble. The interfacial tension of the small bubbles is not exactly equal to 0 due to the fact that the gas saturation factor f stays somewhat larger than 1 ($f(\theta = 10^6) = 1.002$, data not shown). The right panel displays the cube of the scaled averaged radii $R_{kl} = \langle R^k \rangle / \langle R^l \rangle / \langle R_0 \rangle$.

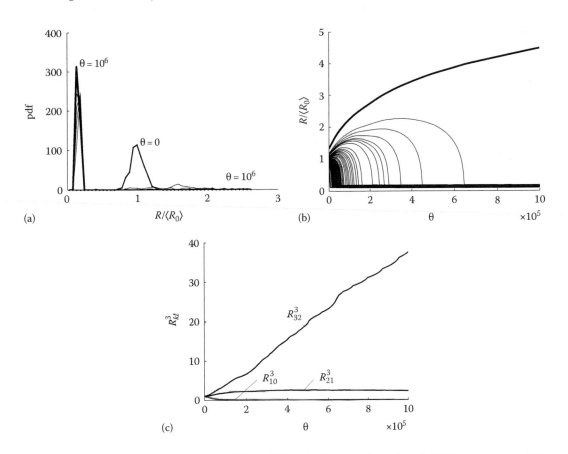

FIGURE 24.4 (a) Calculated evolution of the bubbles distribution of wet foam initially consisting of 500 bubbles with a initial log-normal size distribution ($\langle R \rangle = 200$ μm). The interface behave pure elastic with an elastic modulus $K_1 = E = 10$ mN/m. The initial interfacial tension of all bubbles equals $\sigma_0 = 40$ mN/m. cial. (b) Evolution of the radii of individual bubbles as a function of time during the disproportionation process. (c) Scaled averaged radii $R_{32} = \langle R^3 \rangle / \langle R^2 \rangle / \langle R_0 \rangle$ and $R_{10} = \langle R \rangle / \langle R_0 \rangle$ as a function of time.

This behavior of an (ensemble of) bubbles stabilized by a purely elastic interface that evolves to a stable situation can occur in the case of Pickering stabilization and can also occur for the microbubbles mentioned previously that consist of a thick protein shell that behaves elastic in the time frame of months. Besides being elastic, the interfacial layer needs also to be stable against buckling and subsequent collapse. The elastic buckling pressure for a spherical vessel with a thin wall is given by the so-called called Zoelly–van der Neut equation [55]:

$$P_c = \frac{2 E_b d^2}{R^2 \sqrt{(3(1 - v_p^2))}} \tag{24.29}$$

where
 E_b is the bulk elastic modulus
 d is the thickness of the wall
 R is the radius of the vessel
 v_p is the Poisson ratio

This formula is valid when about $d < R/10$. For a typical $v_p = 0.5$ this gives $d^2 = 2\sigma R/3 E_b$. For a micron-sized bubble and an E_b of about 10 MPa [17], a thickness of a layer of about 40 nm is needed, close to what is measured experimentally [41].

24.2.5.3   Viscoelastic Interface

In general, the interfacial layer will also show viscous behavior due to desorption and rearrangements of the molecules at the interface. Due to this relaxation, behavior of the disproportionation process is continued and cannot be stopped. This is seen in Figure 24.5 showing an example of the calculated bubble size distributions of a wet foam initially consisting of $N = 500$ bubbles at different times. The interface is stabilized by a viscoelastic interface, whose behavior is described by the standard linear solid (see Figure 24.2) with $K_1 = E = 0$ mN/m, $K_2 = E = 10$ mN/m, $\eta = 5 \times 10^5$ Pa·s. The initial interfacial surface tension for all bubbles is taken to be $\sigma_0 = 40$ mN/m. The initial bubble size distribution $\{R_{i0}\}$ is log-normal with a mean of $\langle R_0 \rangle = 200$ μm. Furthermore, $T = 293$ K, gas (N_2) solubility $C_\infty = 0.688$ mol/m^3, diffusion constant $D = 2 \times 10^{-9}$ m^2/s, and environmental pressure $P_\infty = 1 \times 10^5$ Pa. Evolved distributions are obtained by solving $2N$ coupled ODEs (Equations 24.26 and 24.27) in combination with Equation 24.28. Figure 24.4b shows examples of the evolution of the radii of individual bubbles and Figure 24.4c shows the cube of the scaled averaged radii $R_{kl} = \langle R^k \rangle / \langle R^l \rangle / \langle R_0 \rangle$. The disappearance of the small bubbles is delayed but eventually all dissolve until only one large bubble is left.

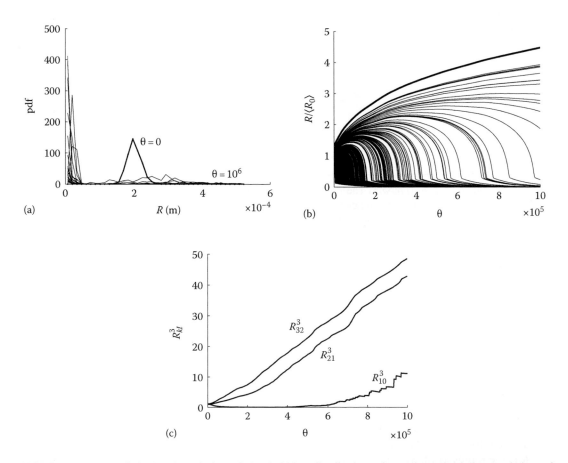

FIGURE 24.5 (a) Calculated evolution of the bubbles distribution of wet foam initially consisting of 500 bubbles with a initial log-normal size distribution ($\langle R \rangle = 200$ μm). The interface behaves viscoelastic with $K_1 = E = 0$ mN/m, $K_2 = E = 10$ mN/m, $\eta = 5 \times 10^5$ Pa·s. The initial interfacial tension of all bubbles equals $\sigma_0 = 40$ mN/m. (b) Evolution of the radii of individual bubbles as a function of time during the disproportionation process. (c) Scaled averaged radii $R_{32} = \langle R^3 \rangle / \langle R^2 \rangle / \langle R_0 \rangle$ and $R_{10} = \langle R \rangle / \langle R_0 \rangle$ as a function of time.

24.3 FORMATION AND STABILITY OF DRY FOAMS

In order to get insight in the role of ingredients on formation and stability of foams with a liquid volume fraction smaller than about the random closed packing of 0.36, the approach of Ruckenstein et al. is followed [3,4,36,37], which is in a later stage simplified by Hilgenfeldt, Koehler, Stone, and coworkers [19,20,26].

In short, the foam is regarded as a three-dimensional porous system with a polyhedral structure described by Kelvin or Weaire–Phelan cells. Variables of interest are described on a macroscopic scale that is larger than the typical cell size, in a coarse-grained model. An important variable of interest is the liquid volume fraction $\epsilon = V_l/V$, with V_l is the volume of the liquid in macroscopic unit volume V. Flow or drainage of the liquid is through the Plateau borders. It is driven by gravity and Laplace pressure gradients, due to differences in curvature of the Plateau borders.

To describe coalescence, also the liquid in the thin liquid films between the foam bubbles, the lamellae, is taken into account. Liquid can drain from the thin film to the Plateau border due to a difference between Laplace pressure in the Plateau borders and the disjoining pressure between the opposite interfaces of the film. The disjoining pressure is modeled using extended Derjaguin–Landau–Verwey–Overbeek (DLVO) theory, incorporating van der Waals, electrostatic, and hydrophobic forces, which are estimated from the molecular properties and surface coverage of the ingredients. Rupture of the thin film or bubble coalescence is estimated using the theory developed by Sheludko and Vrij [33,11,44,51].

The model is described in more detail later and some results for protein foams are presented.

24.3.1 FOAM STRUCTURE

Here we follow the approach of Ruckenstein et al. [4,3,36] who considered a foam structure consisting of regular pentagonal dodecahedrons, following Leonard and Lemlich [56] and Desai and Kumar [13]. From the exact geometrical model of the foam, relations between the foam structure and liquid volume fraction can be deduced. The regular dodecahedron consists of $n_F = 12$ faces and $n_E = 30$ edges, and thus the number of thin films and the number of Plateau border per cell or bubble equals $n_f = 6$ and $n_p = 10$, respectively. The volume of the cell $V_c = \delta_V L_c^3$ with $\delta_V = (15 + 7\sqrt{(5)})/4 \approx 7.663$ and L_c is the length of the edge.

Furthermore, the area of the plateau border A_p is related to its radius of curvature R_p and the thickness of the thin film h according to $A_p = \delta_1 R_p^2 + \delta_2 R_p h$, with $\delta_1 = 9\sqrt{3} - 9\pi/2$ and $\delta_2 = 3\sqrt{3}$. The radius of the thin film $R_F \approx 0.99L$.

The liquid volume fraction $\epsilon = (V_c - V_B)/V_c$, with V_B the volume gas bubble, in the dodecahedral cell can be written as

$$\epsilon = \epsilon_p + \epsilon_f \tag{24.30}$$

where $\epsilon_p = n_p A_p L/V_c$ corresponds to the liquid volume fraction in the Plateau border and $\epsilon_f = n_f A_f h/V_c$ to that in the thin liquid films.

24.3.2 PLATEAU BORDER DRAINAGE

Various models are proposed to describe the liquid flow through the Plateau borders, thin films, and nodes of a foam (for an overview, see, e.g., [10]). As a first approximation, it is assumed that the channels dominate the flow and thus nodes are neglected.

A foam column is considered and a coordinate system with the z-axis in the vertical direction, increasing downwards. The foam is assumed to be homogeneous in the xy plane. The fluid flow is described on a macroscopic scale that is larger than the typical cell size. Conservation of mass gives

$$\frac{\partial \epsilon}{\partial t} = -\frac{\partial}{\partial z} J_\epsilon \tag{24.31}$$

where J_ϵ is the liquid flux in the z direction. It is assumed that the flux is due to the flow of liquid in the Plateau borders and that liquid flow between the thin films and the Plateau borders only occurs in the horizontal plane. In that case one can write

$$J_\epsilon = (1 - \epsilon) J_{PB} \tag{24.32}$$

with J_{PB} the liquid flux (volume per unit area per unit time) through the Plateau borders. The term $1 - \epsilon$ follows from mass conservation.

J_{PB} can be estimated from the expression for the average liquid velocity v_p in a vertical triangular Plateau border as derived by Desai and Kumar [13]

$$v_p = \frac{c_v A_p}{20\sqrt{3}\mu}\left(\rho g + \frac{\partial}{\partial z}\frac{\sigma}{R_p}\right) \tag{24.33}$$

where
    μ is the liquid viscosity
    ρ is the density of the liquid
    g is the acceleration due to gravity

The constant c_v accounts for the mobility of the interface. Here a large surface viscosity is assumed for which c_v is about unity. This corresponds to a rigid immobile interface that might, for example, be formed by adsorbed proteins. The last term in Equation 24.33 corresponds to possible gradients in Laplace pressure in the Plateau borders, which depends on bubble size, liquid volume fraction, and film thickness, as described earlier.

The total liquid flux across a horizontal plane can now be estimated by adding the liquid flows through all Plateau borders that crosses the plane per unit area. Assuming a random orientation of the Plateau borders and considering that $n_p/5$ Plateau borders per bubble and $2N_B R_B$ bubbles per unit area intersect the plane, with N_B the number of bubbles per unit volume and $R_B = (3V_B/(4\pi))^{1/3}$, the radius of the equivalent sphere, one finds that

$$J_{PB} = N_B R_B n_p A_p v_p / 5 \tag{24.34}$$

Combining now this Equation 24.34 with that of the average liquid velocity v_p in the Plateau border, Equation 24.33, and the conservation of mass, Equation 24.31, one finds

$$\frac{\partial \epsilon}{\partial t} = -\frac{c_v n_p}{100\sqrt{3}\mu}\frac{\partial}{\partial z}\left\{N_B R_B A_p^2 (1-\epsilon)\left(\rho g + \frac{\partial}{\partial z}\frac{\sigma}{R_p}\right)\right\} \tag{24.35}$$

Using the geometrical relation derived here, $N_B = (1-\epsilon)/V_B$, R_B, A_p, and r_p can all be expressed in ϵ. Using appropriate boundary conditions, as will be discussed later, the partial differential equation 24.35 can be solved numerically, using standard solvers for coupled ODEs. We used the fortran routine LSODI form, the freely available Odepack collection.

24.3.3 Film Drainage and Stability

The difference in curvature of the thin film and the Plateau border results in a pressure difference that causes liquid to flow from the film to adjacent Plateau borders. At small film thickness below about 100 nm, the two interfaces of the thin film interact, which is expressed in the disjoining pressure Π. The thin foam film is considered as a plane parallel plate for which the film thinning velocity v_f can be approximated by the Reynolds equation for film thinning [11,22,45]:

$$\frac{\partial h}{\partial t} = -v_f = -\frac{2c_f h^3}{3\mu R_f^2}\left(\frac{\sigma}{R_p} - \Pi\right) \tag{24.36}$$

where
  h is the thickness of the film
  R_f is the radius of the film, and
  the constant c_f accounts for the mobility of the interfaces, similar as before for the Plateau border drainage

For immobile surfaces, as considered here, c_f is about unity. The driving force for the film thinning is the difference in Laplace pressure of the adjacent Plateau borders σ/R_p and the film disjoining pressure Π. The interfacial tension σ, disjoining pressure Π, and the stability of the film depend on the details of the interface, like the amount of the adsorbed species and their properties like size and charge.

24.3.3.1 Disjoining Pressure

The stability of thin aqueous films stabilized by small molecular surfactants is well understood and can be described by the DLVO theory developed by Derjaguin, Landau, Verwey, and Overbeek [27,33,48]. Attractive interactions, enhancing film thinning and destabilizing the film, include long range dispersion forces summarized as van der Waals forces. Repulsive interactions, working against film thinning thereby stabilizing the film, include electrostatic interactions between two similar charged interfaces. The van der Waals and electrostatic interactions are described quantitatively by the DLVO theory. Besides the DLVO forces, other interactions are recognized to play a role in film stability, like repulsive steric (excluded volume), hydration, and hydrophobic forces, as well as attractive hydrophobic and oscillatory solvation forces [16,21,27,29,54]. The sum of these surface forces determine the disjoining pressure Π, which depends on the film thickness h. Here we will summarize the DLVO contributions.

a. *Van der Waals interactions*: For a macroscopic plane parallel film, the van der Waals force per unit area can be written as

$$\Pi_{vw} = -\frac{A_H}{6\pi h^3} \tag{24.37}$$

with A_H is the compound Hamaker constant. When opposing interfaces consist of equal components, $A_H > 0$, thus they attract each other. For a medium 1 that interacts with medium 2 across medium 3, the compound Hamaker constant can be approximated by

$$A_H = A_{132} \approx A_{33} + A_{12} - A_{13} - A_{23} \tag{24.38}$$

with $A_{ij} \approx \sqrt{A_{ii}A_{jj}}$ is the Hamaker constant for medium i interacting with medium j across vacuum.

  The Hamaker constant for water is about $A_{33} \approx 4.38 \times 10^{-20}$ J. The Hamaker constant for proteins is about $1 \times 10^{-20} \lesssim A_{11} \lesssim 2 \times 10^{-20}$ J [32]. Often it is taken to be $A_{11} = 23.4 k_B T$ [40].

b. *Electrostatic interactions*: In the case of ionic adsorbed species, the charged interfaces and the induced concentration profile between the two (equally) charged interfaces of the present ions, a counter ion causes a repulsive interaction that counteracts the thinning of the film. This is described by the so-called double layer model. It consists of a dense layer of counter ions firmly attached to the interface, the Stern-layer, and a diffusive layer where the concentration of counter ions is gradually decreasing, and that of co-ions increasing, with distance from the interface.

The electrostatic part of the disjoining pressure can be expressed in the concentration of the ions at the plane in the middle between the two charged interface, with respect to the concentrations in the bulk (at infinite separation):

$$\Pi_{el}(h) = k_B T \left(\sum_i \rho_{mi}(h) - \sum_i \rho_{mi}(\infty) \right) \tag{24.39}$$

where
  the sum i runs over all ions in solution
  $\rho_{mi}(h)$ is the number density of ion i (number per unit volume) at the midplane of the two inter-
    faces separated at a distance h
  $\rho_{mi}(\infty)$ is the concentrations at the midplane when the interface are infinitely far separated
  The latter thus corresponds to the electrolyte concentration in the bulk $\rho_{mi}(\infty) = \rho_{i\infty}$.

To obtain $\rho_{mi}(h)$, one need to solve the Poisson–Boltzmann equation, which for the one-dimensional case reads

$$\frac{\partial^2 \Psi}{\partial x^2} = -\frac{e}{\varepsilon_0 \varepsilon_r} \sum_i z_i \rho_{i\infty} e^{-z_i e \Psi / k_B T} \tag{24.40}$$

where
  Ψ is the electrostatic potential that depends on the spatial coordinate x between the two interfaces
  e is the electron charge
  ε_0 is the electric permittivity of free space
  ε_r is the relative permittivity of solution
  z_i is the valency of electrolyte i
  $\rho_{i\infty}$ is its number density at the bulk reservoir (here the Plateau borders) where the electrostatic
    potential is taken to be zero, $\Psi_\infty = 0$

Because $\rho_{i\infty}$ is known, the earlier differential equation 24.40 can be solved given that the electrical field at the midplane is equal to zero, $E_{x=h/2} = \partial \Psi / \partial x |_{h/2} = 0$, while at the interface at $x = 0$ it is given by

$$E_s = \frac{\partial \Psi}{\partial x}\bigg|_{x=0} = -\frac{\sigma_c}{\varepsilon_0 \varepsilon_r} \tag{24.41}$$

with σ_c is the surface charge density, which is directly related to the adsorbed amount or interfacial coverage Γ_i of species i (number of adsorbed molecules per unit area). The surface charge density can be written as

$$\sigma_c = e \sum_i \Gamma_i \sum_j \alpha_{ij0} \tag{24.42}$$

where j runs over all ionizable groups of species i and α_{ij0} is the fraction of ionizable groups ij at the surface that is dissociated. The latter depends on the dissociation or equilibrium constant. For example, for the acid reaction $AH \leftrightarrow A^- + H^+$ with equilibrium constant $K_a = [A^-][H^+]/[AH]$ one finds for the fraction of dissociated groups at the interface $\alpha_{a0} = [A^-]([AH]+[H^+]) = K_a/(K_a +[H^+]_0)$, with the ion concentration at the interface equal to $[H^+]_0 = [H^+]_\infty \exp(-e\psi_0/k_BT)$. Thus in the case of so-called charge regulation, the surface charge density depends on the surface electrostatic potential ψ_0. This causes that the Poisson–Boltzmann equation 24.40 must be solved with the boundary condition, Equation 24.42, determined self-consistently (see, e.g., [9,15,18,38]).

For film thicknesses h larger than about the Debeye screening length λ_D,

$$\lambda_D = 1/\kappa_D = \sqrt{\sum_i \frac{\epsilon_0 \epsilon_r k_B T}{2e^2 z_i^2 \rho_{i\infty}}} \tag{24.43}$$

the electrostatic contribution to the disjoining pressure can be rather well approximated by the so-called weak overlap approximation for monovalent electrolytes

$$\Pi_{el} \approx 64 k_B T \rho_\infty \tan h^2 \frac{e\psi_0}{4k_BT} e^{-\kappa_D h} \tag{24.44}$$

with ρ_∞ is the bulk concentration of electrolytes. The surface potential can be estimated from the Gouy–Chapman theory that gives the general relation between the surface potential and surface charge density of an isolated surface

$$\sum_i \rho_{i\infty} \left(e^{-z_i e \Psi_0 / k_B T} - 1 \right) = \frac{\sigma_c^2}{2\epsilon_0 \epsilon_r k_B T} \tag{24.45}$$

which for a $z{:}\,z$ electrolyte reduces to the Grahame equation

$$\sigma_c = \sqrt{8\rho_\infty \epsilon_0 \epsilon_r k_B T} \sin h\left(\frac{-ze\Psi_0}{2k_BT} \right) \tag{24.46}$$

Neglecting charge regulation, and thus estimating the surface charge density from Equation 24.42 and assuming $\alpha_{ij0} = \alpha_{ij\infty}$, the electrostatical contribution to the disjoining pressure $\Pi_{el}(h)$ can be estimated from Equations 24.46 and 24.44.

The total surface pressure is then estimated by

$$\Pi = \Pi_{vw} + \Pi_{el} \tag{24.47}$$

24.3.3.2 Film Rupture and Coalescence

Film rupture has successfully been described by the theories developed by Sheludko and Vrij [11,33,44,51,29]. They identified instability regions in the disjoining pressure isotherm for certain thicknesses. In these instability regions, small spontaneous surface perturbations grow in time until both interfaces touch each other and the film ruptures. The unstable regions are given by

$$\frac{\partial \Pi}{\partial h} < -\frac{\pi\sigma}{R_f^2} \tag{24.48}$$

In these regions there are fluctuations with a certain wavelength that can grow exponentially in time, while in other regions all fluctuations will disappear. Spontaneous rupture of the film occurs when these exponentially growing fluctuations from the opposite interfaces touch each other. A critical thickness h_c at which the film breaks can be estimated from the characteristic growing time of the fluctuation that grows fastest and the draining velocity. A quantitative prediction of the lifetime of a draining thin film, rupture probability, and critical film thickness at which the film breaks is despite the many theoretical and experimental studies that have been devoted to it is still challenging. In general the characteristic growing time of the fluctuation is much smaller than the draining time. Therefore, as an approximation, the onset of the instability region, corresponding to the largest thickness that satisfies Equation 24.48, is taken as the critical film thickness.

The surface tension in Equation 24.48 is related to the amount of energy it costs to increase the size of the interface due to a fluctuation. When the interface is also elastic, extra energy is needed to increase the interfacial area. This effect could be approximated by adding the interfacial elasticity modulus E to the surface tension σ giving as a cryterium for the unstable regions

$$\frac{\partial \Pi}{\partial h} < -\frac{\pi(\sigma + E)}{R_f^2} \qquad (24.49)$$

The thickness of the foam films in a certain volume element is described by the film thinning Equation 24.36. Starting initially with a certain film thickness, the film thickness will decrease until it reaches the critical thickness h_c, after it will rupture. In order to avoid a sudden unrealistic spontaneous rupture of all films in the volume element, a certain distribution ϕ of film radii R_f around its average value $\langle R_{f0} \rangle$ is assumed at the beginning of film thinning. Because the film draining velocity depends on R_f, only a certain fraction of the films will reach the critical film thickness at a certain time. A Gaussian distribution is assumed at $t = 0$, thus

$$\phi_0 = \frac{1}{\varpi\sqrt{2\pi}} \exp\left[\frac{\left(R_f - \langle R_{f0} \rangle\right)^2}{2\varpi_0^2}\right] \qquad (24.50)$$

with ϖ_0^2 the initial variance of R_f. Thus each film in a volume element is characterized by a film radius R_f and film thickness h that will evolve during film and Plateau border drainage. The number of films per unit volume between R_f and $R_f + \delta R_f$ is given by $n_f N_{B0} \phi \delta R_f$. Films with R_f that reach the critical film thickness h_c will rupture. Per unit time, this number is proportional to the thinning velocity v_f of these films at the critical thickness h_c and given by $v_f \phi$. The change in the relative number of films between R_f and $R_f + \delta R_f$ can then be approximated by

$$\frac{\partial \phi}{\partial t} \delta R_f = -\frac{\left(v_f \phi\right)_{R_f, h_c}}{1 + \exp\left[(h - h_c)/\lambda_c\right]} \qquad (24.51)$$

where the Fermi function $1/(1 + \exp[(h - h_c)/\lambda_c])$, with λ_c controlling the steepness, is used to control numerical stability and to smooth the abrupt change in the probability of rupture going from 0 for $h > h_c$ to 1 for $h \le h_c$.

24.3.4 FOAM STRUCTURE CHANGE

Due to the rupture of thin films and coalescence of bubbles, the average bubble size in a volume element will increase. Realizing that the number of bubbles in a volume element is proportional to the number of thin films in that volume element, and that the total gas volume in the volume element does not change when a film ruptures, the mean bubble volume V_B at time t can be calculated from the integral of the thin film radius distribution at time t, giving

$$V_B = \frac{V_{B0}}{\int_0^\infty \phi dR_f} \qquad (24.52)$$

Similarly, mean values for the film radius $\langle R_f \rangle = \int R_f \phi dR_f / \int \phi dR_f$ and film thickness $\langle h \rangle = \int h \phi dR_f / \int \phi dR_f$ can be calculated. From the mean bubble size and liquid volume fraction (state variable ε) follows the average volume of the dodecahedral cell $V_c = V_B/(1-\varepsilon)$, and so the average length of the Plateau border $L = (V_c/\delta_V)^{1/3}$. Further, the liquid volume fraction in the films is calculated from the average film thickness, size, distribution, and average cell size, $\epsilon_f = n_f \pi R_f^2 h / V_c$. Then the average Plateau border area can be estimated from the liquid volume fraction in the Plateau borders, $\varepsilon_p = \varepsilon - \varepsilon_f$, and Plateau border length, $A_p = V_c \varepsilon_p / L$. Finally, the average Plateau border radius R_p can be estimated from $A_p = \delta_1 R_p^2 + \delta_2 R_p h$. This allows calculation of the differential equations for plateau border drainage (Equation 24.35) and film thinning (Equation 24.36).

24.3.5 Change in Foam Height

Summarizing, the development of the liquid volume fraction (Equation 24.35) and thin film properties (Equation 24.36) are described in a set of coupled differential equations that can be solved numerically using appropriate boundary and initial conditions. The state variables to be solved are the total liquid volume fraction ε, which is a function of the spatial coordinate z and time t, and the film thickness h, which is a function of the film radius R_f, z, and t. So the initial conditions have to be set for ε at each z and h at each R_f and z. Because only liquid flow between the thin films and Plateau borders is considered at a certain z and no liquid is exchanged between films at different z, only boundary conditions for the liquid volume fraction have to be defined.

Suppose that the foam is produced with initial volume fraction ε_0, which is smaller than the volume fraction corresponding to a closed packed system of spherical bubbles $\varepsilon_c \approx 0.36$. When at the bottom $\varepsilon(z=z_2) < \varepsilon_c$, no liquid will drain out of the foam due to the capillary pressure or Plateau border suction. This implies $J_\varepsilon = 0$. When after a certain time the liquid fraction at the bottom becomes that of the closed packed systems, liquid can drain out and the liquid volume fraction at the bottom will remain about the critical value of the closed packed system with spherical bubbles. This gives for the boundary condition at the bottom

$$J\varepsilon(z_2,t) = 0, \quad \text{when} \quad \epsilon < \epsilon_c \qquad (24.53)$$

$$\epsilon(z_2,t) = \epsilon_c, \quad \text{when} \quad \epsilon \geq \epsilon_c \qquad (24.54)$$

The amount of drained liquid at time t will be equal to the integral from time 0 to t of the drainage flux at the bottom of the foam. The change of z_2 with time due to liquid drainage will be equal to J_{PB}. During foam generation by gas sparging there will also be additional gas flux J_G, which will move the bottom of the foam upwards. Adding the foam generation and drainage flux, gives for the movement of the bottom of the foam

$$\frac{dz_2}{dt} = -\frac{J_\epsilon - J_G}{1-\epsilon}\bigg|_{z_2} \qquad (24.55)$$

At the top of the foam, $z = z_1$, the liquid flux into the foam will be zero if there is no coalescence at the top. However, when a thin film at the top of the foam between a bubble and the outer environment ruptures, the foam height will decrease (z_1 will increase). The change in height corresponds to the volume of the bubble that coalesced with the outer environment, and the liquid associated with this bubble will flow into the foam. If N_B is the number of bubbles per unit volume, then $N_{BA} = N_B^{2/3}$ is the number of bubbles per unit area. Then for $z = z_1$ one finds that the number of bubbles that break at the top per unit area is approximately equal to

$$\frac{\partial N_{BA}}{\partial t} = \frac{2}{3} N_B^{-1/3} N_{B0} \int \frac{\partial \phi}{\partial t} dR_f \qquad (24.56)$$

Multiplying this with the bubble volume $V_B = (1-\epsilon)/N_B$ gives the change in height per time

$$\frac{dz_1}{dt} = -\frac{2}{3} N_{B0}^{-1/3} (1-\epsilon) \left(\int \phi dR_f \right)^{-4/3} \int \frac{\partial \phi}{\partial t} dR_f \Bigg|_{z=z_1} \qquad (24.57)$$

Multiplying this with ϵ gives the liquid flux at the top at $z = z_1$ into the foam. The top boundary condition thus becomes

$$J_\epsilon(z_1,t) = \epsilon \frac{dz_1}{dt} = -\frac{2}{3} N_{B0}^{-1/3} \epsilon (1-\epsilon) \left(\int \phi dR_f \right)^{-4/3} \int \frac{\partial \phi}{\partial t} dR_f \Bigg|_{z=z_1} \qquad (24.58)$$

24.3.5.1　Coordinate Transformation and Dimensionless Model Equations

During sparging, drainage, and coalescence, the position of the bottom z_2 and top z_1 of the foam will change with time. The calculation are much simplified by transforming this moving boundary problem to static boundary conditions, using

$$\xi = \frac{z - z_1}{z_2 - z_1} \qquad (24.59)$$

To transform the derivatives to ξ-space one needs to apply the chain rule

$$\frac{\partial}{\partial t}\Big|_z = \frac{\partial}{\partial t}\Big|_\xi + \frac{\partial \xi}{\partial t}\Big|_z \frac{\partial}{\partial \xi} \qquad (24.60)$$

$$\frac{\partial}{\partial z} = \frac{\partial \xi}{\partial z} \frac{\partial}{\partial \xi} \qquad (24.61)$$

Furthermore, it is convenient to transform to dimensionless parameters, for example, to scale the variables of interest with their initial values. Introducing the following dimensionless parameters $\zeta = z/H_0$, $\zeta_1 = z_1/H_0$, $\zeta_2 = z_2/H_0$, $R_{pr} = R_p/R_{p0}$, $R_{fr} = R_f/R_{f0}$, $V_{Br} = V_B/V_{B0}$, $h_r = h/h_0$, and $\Pi_r = \Pi R_{p0}/\sigma$, gives the following coupled partial differential equations

$$\frac{\partial \epsilon}{\partial \vartheta} = -A_1 \frac{\partial \xi}{\partial \zeta} \frac{\partial}{\partial \xi} \left\{ \frac{A_{pr}^2 (1-\epsilon)}{V_{Br}^{2/3}} \left(1 + A_2 \frac{\partial \xi}{\partial \zeta} \frac{\partial}{\partial z} \frac{1}{R_{pr}} \right) \right\} = -\frac{\partial \xi}{\partial \zeta} \frac{\partial J_{\epsilon r}}{\partial \xi} \qquad (24.62)$$

$$\frac{\partial h_r}{\partial \vartheta} = -\mathcal{B}_1 \frac{h_r^3}{R_{fr}^2}\left(\frac{1}{R_{pr}} - \Pi_r\right) \tag{24.63}$$

$$\frac{\partial \phi_r}{\partial \vartheta} = -\frac{\left(\frac{\partial h_r}{\partial \vartheta}\right)_{h_{rc}} \phi_r/\delta h_r}{1+\exp\left[(h_r - h_{rc})/\lambda_{rc}\right]} \tag{24.64}$$

$$\frac{\partial \zeta_1}{\partial \vartheta} = \frac{\mathcal{D}_1(1-\epsilon)\left(\int \phi_r \delta R_{fr}\right)^{\frac{4}{3}}\left(\int \frac{\partial \phi_r}{\partial \vartheta}\delta R_{fr}\right)}{1+\mathcal{D}_1(1-\epsilon)\left(\int \phi_r \delta R_{fr}\right)^{\frac{4}{3}}\left(\int \frac{\partial \phi_r}{\partial \vartheta}\delta R_{fr}\right)\frac{\partial \phi_r}{\partial \xi}\frac{\partial \xi}{\partial \zeta}}\Bigg|_{\zeta_1} \tag{24.65}$$

$$\frac{\partial \zeta_2}{\partial \vartheta} = -\frac{J_{\epsilon r} - J_{Gr}}{1-\epsilon}\Bigg|_{\zeta_2} \tag{24.66}$$

where $\mathcal{A}_1 = (\tau_x c_v n_p \rho g A_{p0}^2)/(100\sqrt{3}(4\pi/3)^{1/3}\mu H_0 V_{B0}^{2/3})$, $\mathcal{A}_2 = \sigma/(\rho g R_{p0} H_0)$, $\mathcal{B}_1 = (\tau_x 2 c_f h_0^2 \sigma)$. $(3\mu R_{f0}^2 R_{p0})$, $\mathcal{C}_1 = \tau_x p_0$, $\mathcal{D}_1 = 2\tau_x N_{B0}^{-1/3}/3H_0$, $\partial \xi/\partial \zeta = 1/(\zeta_2 - \zeta_1)$, $J_{Gr} = \tau_x J_G/H_0$, and $\vartheta = t/\tau_x$, with τ_x is an arbitrary constant controlling time. H_0 is the initial foam height, R_{p0} is the initial radius of the Plateau border that can be derived from the initial set liquid volume fraction ϵ_0, initial set film thickness h_0, initial set bubble size $V_{B0} = (4/3)\pi R_{B0}^3$, initial Plateau border length L_0 and area A_{p0}, using the structural relations as described before. Similarly is the mean initial film radius R_{f0} estimated. The earlier set of coupled differential equations can be solved numerically by discretization of the ξ and R_{fr}-space. The total system consists of $n_\xi + 2n_\xi n_{R_{fr}} + 2$ coupled ODEs where n_ξ and $n_{R_{fr}}$ are the number of grid points in ξ and R_{fr}-space, respectively; n_ξ ODEs correspond to the evolution of the liquid volume fraction $\epsilon(\xi, \vartheta)$ (Equation 24.62), $n_\xi n_{R_{fr}}$ correspond to the evolution of the film thickness $h_r(\xi, R_{fr}, \vartheta)$ (Equation 24.63), $n_\xi n_{R_{fr}}$ correspond to the evolution of the number of thin liquid films at a certain film radius $\phi_r(\xi, R_{fr}, \vartheta)$ (Equation 24.64), two correspond to the evolution of the position of the top and bottom of the foam (Equations 24.65 and 24.66, respectively).

The boundary conditions are given by

$$J_{\epsilon r}(\xi = 0, \vartheta) = \epsilon \frac{\partial \zeta_1}{\partial \vartheta}\Bigg|_{\zeta_1} \tag{24.67}$$

$$J_{\epsilon r}(\xi = 1, \vartheta) = 0, \quad \text{when} \quad \epsilon < \epsilon_c \tag{24.68}$$

$$\epsilon(\xi = 1, \vartheta) = \epsilon_c, \quad \text{when} \quad \epsilon < \epsilon_c \tag{24.69}$$

$$J_{Gr} = J_{Gr0}, \quad \text{when} \quad \vartheta < \vartheta_G \tag{24.70}$$

where ϑ_G is the time when the foam has reached the initial set height, thus when $\zeta_2 - \zeta_1 = 1$. The initial values are given by

$$\epsilon(\xi, \vartheta = 0) = \epsilon_0 \tag{24.71}$$

$$h_r(\xi, R_{fr}, \vartheta = 0) = 1 \tag{24.72}$$

$$\zeta_1(\vartheta = 0) = 0 \tag{24.73}$$

$$\zeta_2(\vartheta = 0) = 0.01 \tag{24.74}$$

24.3.6 RESULTS

An example of the model calculations for an 0.55 mM solution of sodium dodecyl sulfate (SDS) in 10 mM sodium phosphate buffer solution at pH = 7 is shown in Figure 24.6. Figure 24.6a shows the estimated disjoining pressure (Equation 24.47), assuming a constant coverage of 3 μM/m² and Hamaker constant $A_H \approx 3.7 \times 10^{-20}$ J [53] and total ionic strength of 20 mM. Figure 24.6a shows the stable regions calculated positions using Equation 24.49 and surface tension σ = 50 mN/m and surface elasticity E = 10 mN/m. From this a critical film thickness of about 170 nm is obtained. The middle panel shows the calculated top z_1 and bottom z_2 of the foam as well as the foam height $z_2 - z_1$

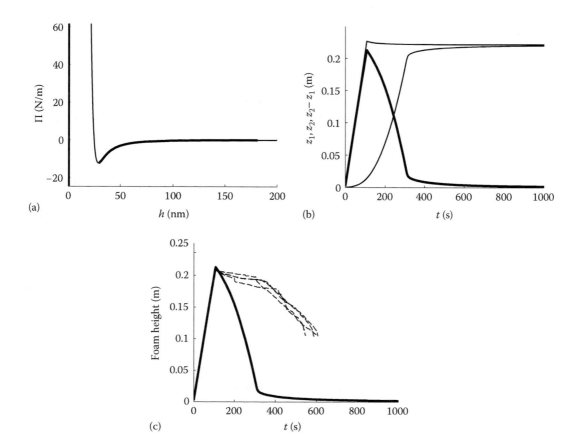

FIGURE 24.6 (a) Calculated disjoining pressure Π as a function of the film thickness h of an 0.55 mM SDS solution. The instable regions (Equation 24.49) are indicated by the thicker lines. (b) Calculated position of the top z_1 and bottom z_2 of the foam as well as the foam height $z_2 - z_1$ as a function of time t, assuming rupture to occur at the border of the instable region giving $h_c \approx 170$ nm. (c) Calculated (continuous line) and experimental (dashed lines) foam heights as a function of time t. See the text for more information on the model parameters.

using Equations 24.66, with $J_G = 1.4$ m/s, $H_0 = 0.22$ m, $\epsilon_0 = \epsilon_c = 0.36$, $R_{B0} = 0.4$ mm, $h_{f0} = 200$ nm, $\varpi_{r0} = 0.01$, and $\sigma = 50$ mN/m. Values of other parameters are mentioned in the text or can be derived from the equations mentioned. The ξ and R_{fr} coordinate was discretized in $n_\xi = 30$ and $n_{R_{fr}} = 3$ grid points, respectively, giving a total of 122 coupled ODEs. Figure 24.6c shows comparison of experimentally derived foam height as measured using an automated foaming device (Foamscan, Teclis IT-Concept, Longessaigne, France) [30]. It is seen that the foam production is predicted well but that the calculated foam collapse due to coalescence does not correspond to the measured one. This is due to the fact that the critical film thickness is not at the border of the instable region. The estimation of the thickness at which the film ruptures needs a more elaborated analysis where the breaking time of a film has to be compared to the time the film is in an unstable region (see, e.g., [11,33,51,29]. A prediction of the rupture time on basis of such an analysis, which is very sensitive for the exact shape of the disjoining pressure and interfacial energy, and thus needs precise knowledge surfactant coverage, surface rheology, Hamaker constants, and possible hydrophobic interactions, is outside the scope of this chapter.

Instead of estimation of the critical rupture thickness from the molecular properties, we estimated it from the experimental foam data. Figure 24.7 shows the calculated foam heights as a function of time t for the foam made by sparging of the 0.55 mM SDS solution, but now assuming a critical rupture thickness of $h_c = 90$ nm. It is seen that for $h_c = 90$ nm, a good comparison of the

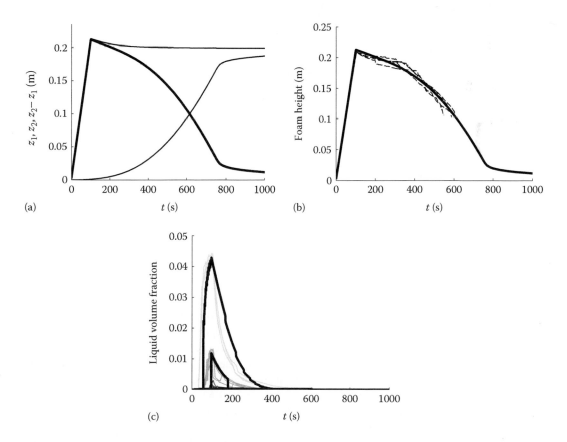

FIGURE 24.7 (a) Calculated top z_1 and bottom z_2 and foam height $z_2 - z_1$ of a foam made by sparging of an 0.55 mM SDS solution with a critical rupture thickness of $h_c = 90$ nm. Corresponding disjoining pressure is shown in the left part of Figure 24.6. (b) Calculated (continuous line) and experimental (dashed lines) foam heights as a function of time t. (c) Calculated (thick black continuous lines) and experimental (thin gray lines) liquid volume fractions as a function of time t. For more information on the model parameters, see text.

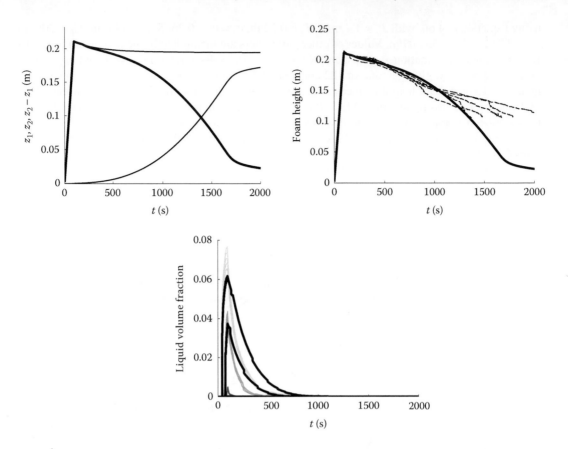

FIGURE 24.8 Similar as Figure 24.7, but now for a foam generated by sparging of an 1.1 mM SDS solution. A critical rupture thickness of $h_c = 52$ nm is assumed.

calculated foam height with the experimental foam heights is obtained. This also counts for the calculated liquid volume fractions ε compared to the experimental ε measured using the Teclis Foamscan by means of conductivity sensors at specific heights in the foaming column.

Figure 24.8 shows the calculated and experimental foam heights and liquid volume fractions of a foam generated by sparging of a 1.1 mM SDS solution and critical rupture time of $h_c = 52$ nm. A surface tension of 45 mN/m and surface coverage of 3.5 μM/m^2 was taken. However, these relatively small differences with respect to the corresponding values of the 0.55 mM SDS foam have no significant effect on the draining velocity and disjoining pressure and borders of the instable regions. This also indicates that a more elaborated analysis is needed to predict rupture times from molecular properties. However, the good correspondence between experimental and calculated foam properties allows estimation of the critical rupture time from the experiment.

Similar conclusions can be drawn when simulating the foaming behavior of protein foams. Figure 24.9a shows the calculated disjoining pressure Π as a function of the film thickness h of an aqueous solution of 0.1 mg/mL β-lactoglobulin (BLG) in 10 mM sodium phosphate buffer at pH = 7. For the calculation an interfacial coverage of 1.5 mg/m^2 was assumed, a surface tension σ of about 57 mN/m, a surface dilatational modulus E of about 60 mN/m [57], and a protein Hamaker constant of 1.3×10^{-20}J [40]. According to the stability criterion equation 24.49, the critical rupture thickness would be about 140 nm. However, calculating foam heights on basis of this h_c did not correspond to experiments. From experiments a critical rupture thickness of about 50 nm was estimated, as can be seen in Figure 24.9b where we compare the calculated and measured foam heights as a function of time. In Figure 24.9c, we show a comparison of calculated and measured evolution in time of

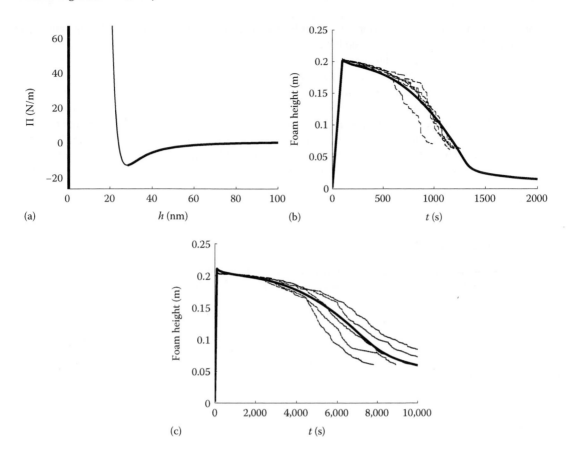

FIGURE 24.9 (a) Calculated disjoining pressure Π as a function of the film thickness h of a 0.1 mM BLG solution (pH = 7, 10 mM phosphate buffer). The instable regions (Equation 24.49) are indicated by the thicker lines. (b) Calculated (continuous line) and experimental (dashed lines) foam heights as a function of time t of the foam generated by sparging the 0.1 mM BLG solution. (c) Similar but for a 0.25 mg/mL BLG solution. For more information on the model parameters, see text.

the foam heights of the foam generated by sparging a 0.25 mg/mL BLG solution (10 mM sodium phosphate buffer, pH = 7). Also in the case of the protein foam, the disjoining pressure and stability regions are rather insensitive to the relative small deviations in molecular and interfacial properties like tension and dilational modulus between a 0.1 and 0.25 mg/mL BLG solution. Furthermore, it is far from clear that the rupture of protein films can be described by the rupture criteria as formulated by Sheludko and Vrij, used here. It might also be expected that at least for protein films other stabilization mechanisms related to the protein interfacial gel forming properties and presence of bulk aggregates are important [28,42,43,48]. These mechanisms and quantification thereof are far from understood and need further investigation.

24.4   SUMMARY AND CONCLUSIONS

In order to increase the insights in the key parameters that control the formation and stability of multicomponent foams, it is essential to develop generic quantitative relations between the properties of the ingredients and that of the foam. We presented here two different quantitative models. One model concerns the role of interfacial rheology in the disproportionation of wet foams and microbubble dispersions. It is shown that for a gas that can dissolve into the continuous liquid,

Ostwald ripening or disproportionation is only retarded and cannot be stopped by an interfacial layer that shows any viscous or creep behavior.

The second presented model concerns the formation and drainage and coalescence of dry foams. Bubble coalescence is estimated using the theory developed by Sheludko and Vrij, where the onset of the instable region is used at the critical thickness at which the foam film will rupture. Model examples are given for surfactant as well as protein foams at ionic strengths relevant for foods (about 20 mM). Foams are generated by sparging. Calculated foam heights and height-dependent liquid volume fractions are compared with measured values. It is shown that calculated foam properties can be described well. However, the calculation of the critical rupture thickness needs a more elaborate treatment than the one used here. The stability and rupture mechanisms and the description in physical models of foam films, including protein interfacial gel forming, non-DLVO forces, and presence of bulk aggregates, are still not well understood and needs further study.

ACKNOWLEDGMENT

We thank Frederik Lech, Tijs Rovers, Guido Sala, and Peter Wierenga for valuable discussions and collaboration. Special thanks goes to Frederik Lech for letting us use some of his experimental data.

REFERENCES

1. E. S. Basheva, P. A. Kralchevsky, N. C. Christov, K. D. Danov, S. D. Stoyanov, T. B. J. Blijdenstein, H. J. Kim, E. G. Pelan, and A. Lips. Unique properties of bubbles and foam films stabilized by hfbii hydrophobin. *Langmuir*, 27(6):2382–2392, 2011.
2. J. Benjamins, A. Cagna, and E. H. Lucassen-Reynders. Viscoelastic properties of triacylglycerol/water interfaces covered by proteins. *Colloids and Surfaces A: Physicochemical and Engineering Aspects*, 114:245–254, 1996.
3. A. Bhakta and E. Ruckenstein. Decay of standing foams: Drainage, coalescence and collapse. *Advances in Colloid and Interface Science*, 70:1–124, 1997.
4. A. Bhakta and E. Ruckenstein. Drainage and coalescence in standing foams. *Journal of Colloid and Interface Science*, 191(1):184–201, 1997.
5. B. P. Binks. Particles as surfactants—Similarities and differences. *Current Opinion in Colloid & Interface Science*, 7:21–41, 2002.
6. M. A. Borden and M. L. Longo. Dissolution behavior of lipid monolayer-coated, air-filled microbubbles: Effect of lipid hydrophobic chain length. *Langmuir*, 18(24):9225–9233, 2002.
7. C. Buchcic, R. H. Tromp, M. B. J. Meinders, and M. A. Cohen Stuart. Assembly of jammed colloidal shells onto micron-sized bubbles by ultrasound. *Soft Matter*, 11(7):1326–1334, 2015.
8. F. Grieser, F. Caruso, F. Cavalieri, and M. Ashokkumar. Ultrasonic synthesis of stable, functional lysozyme microbubbles. *Langmuir*, 24(18):10078–10083, 2008.
9. Y. C. D. Chan, M. Richard Pashley, and R. Lee White. A simple algorithm for the calculation of the electrostatic repulsion between identical charged surfaces in electrolyte. *Journal of Colloid and Interface Science*, 77(1):283–285, 1980.
10. S. Cohen-Addad, R. Höhler, and O. Pitois. Flow in foams and flowing foams. *Annual Review of Fluid Mechanics*, 45(1):241, 2013.
11. J. E. Coons, P. J. Halley, S. A. McGlashan, and T. Tran-Cong. A review of drainage and spontaneous rupture in free standing thin films with tangentially immobile interfaces. *Advances in Colloid and Interface Science*, 105(1):3–62, 2003.
12. Y. De Smet, L. Deriemaeker, and R. Finsy. A simple computer simulation of ostwald ripening. *Langmuir*, 13(26):6884–6888, 1997.
13. D. Desai and R. Kumar. Liquid holdup in semi-batch cellular foams. *Chemical Engineering Science*, 38(9):1525–1534, 1983.
14. P. Brent Duncan and D. Needham. Test of the epstein-plesset model for gas microparticle dissolution in aqueous media: Effect of surface tension and gas undersaturation in solution. *Langmuir*, 20(7):2567–2578, 2004. PMID: 15835125.
15. R. Ettelaie and R. Buscall. Electrical double layer interactions for spherical charge regulating colloidal particles. *Advances in Colloid and Interface Science*, 61:131–160, 1995.

16. D. Grasso, K. Subramaniam, M. Butkus, K. Strevett, and J. Bergendahl. A review of non-dlvo inter-actions in environmental colloidal systems. *Reviews in Environmental Science and Biotechnology*, 1(1):17–38, 2002.

17. T. Hagiwara, H. Kumagai, and T. Matsunaga. Fractal analysis of the elasticity of bsa and β-lactoglobulin gels. *Journal of Agricultural and Food Chemistry*, 45(10):3807–3812, 1997.

18. T. W. Healy, D. Chan, and L. R. White. Colloidal behaviour of materials with ionizable group surfaces. *Pure and Applied Chemistry*, 52(5):1207–1219, 1980.

19. S. Hilgenfeldt, S. A. Koehler, and H. A. Stone. Dynamics of coarsening foams: Accelerated and self-limiting drainage. *Physical Review Letters*, 86(20):4704–4707, 2001.

20. S. Hilgenfeldt, A. M. Kraynik, S. A. Koehler, and H. A. Stone. An accurate von Neumann's law for three-dimensional foams. *Physical Review Letters*, 86(12):2685–2688, 2001.

21. J. N. Israelachvili. *Intermolecular and Surface Forces: Revised Third Edition*. Academic Press, Amsterdam, 2011.

22. I. B. Ivanov and D. St. Dimitrov. Hydrodynamics of thin liquid films. *Colloid and Polymer Science*, 252(11):982–990, 1974.

23. A. S. Kabalnov. Ostwald ripening and related phenomena. *Journal of Dispersions Science and Technology*, 22(1):1–12, 2001.

24. A. Katiyar, K. Sarkar, and P. Jain. Effects of encapsulation elasticity on the stability of an encapsulated microbubble. *Journal of Colloid and Interface Science*, 336(2):519–525, 2009.

25. W. Kloek, T. van Vliet, and M. B. J. Meinders. Effect of bulk and interfacial rheological properties on bubble dissolution. *Journal of Colloid and Interface Science*, 237(2):158–166, May 15, 2001.

26. S. A. Koehler, S. Hilgenfeldt, and H. A. Stone. A generalized view of foam drainage: Experiment and theory. *Langmuir*, 16(15):6327–6341, 2000.

27. P. A. Kralchevsky, K. D. Danov, and N. D. Denkov. Chemical physics of colloid systems and interfaces. In K. Birdi and S, eds., *Handbook of Surface and Colloid Chemistry*, 3rd edition, Vol. 7, pp. 197–377. CRC Press, 2008.

28. D. Langevin. Aqueous foams: A field of investigation at the frontier between chemistry and physics. *ChemPhysChem*, 9(4):510–522, 2008.

29. F. J. Lech, P. W. Wierenga, H. Gruppen, and M. B. J. Meinders. Stability properties of surfactant free thin films at different ionic strengths: Measurements and modeling. *Langmuir*, 31(9):2777–2782, 2015.

30. F. J. Lech, P. Steltenpool, M. B. J. Meinders, S. Sforza, H. Gruppen, P. A. Wierenga. Identifying changes in chemical, interfacial and foam properties of beta-lactoglobulin-sodium dodecyl sulphate mixtures. In *Colloids and Surfaces A: Physicochemical and Engineering Aspects*, Vol. 462, pp. 34–44, Elsevier, 2014.

31. I. M. Lifshitz and V. V. Slyozov. The kinetics of precipitation from supersaturated solid solutions. *Journal of Physics and Chemistry of Solids*, 19(1–2):35–50, 1961.

32. M. Lund and B. Jönsson. A mesoscopic model for protein-protein interactions in solution. *Biophysical Journal*, 85:2940–2947, 2003.

33. E. D. Manev and J. K. Angarska. Critical thickness of thin liquid films: Comparison of theory and experiment. *Colloids and Surfaces A: Physicochemical and Engineering Aspects*, 263(1):250–257, 2005.

34. M. B. J. Meinders, W. Kloek, and T. van Vliet. Effect of surface elasticity on ostwald ripening in emul-sions. *Langmuir*, 17(13):3923–3929, June 26, 2001.

35. M. B. J. Meinders and T. van Vliet. The role interfacial rheological properties on ostwald ripening in emulsions. *Advances in Colloid and Interface Science*, 108–109:119–126, 2004.

36. G. Narsimhan and E. Ruckenstein. Structure, drainage, and coalescence of foams and concentrated emulsions. *Surfactant Science Series*, 57:99–188, 1996.

37. G. Narsimhan and F. Uraizee. Kinetics of adsorption of globular proteins at an air-water interface. *Biotechnology Progress*, 8(3):187–196, 1992.

38. B. W. Ninham and V. A. Parsegian. Electrostatic potential between surfaces bearing ionizable groups in ionic equilibrium with physiologic saline solution. *Journal of Theoretical Biology*, 31(3):405–428, 1971.

39. S. Pickering. Pickering: Emulsions. *Journal of the Chemical Society, Transaction*, 91:2001–2021, 1907.

40. C. M. Roth, B. L. Neal, and A. M. Lenhoff. Van der Waals interactions involving proteins. *Biophysical Journal*, 70(2):977–987, 1996.

41. T. A. M. Rovers, G. Sala, E. van der Linden, and M. B. J. Meinders. Temperature is key to yield and stability of BSA stabilized microbubbles. *Food Hydrocolloids*, 52:106–115, 2015.

42. B. Rullier, M. A. V. Axelos, D. Langevin, and B. Novales. [Beta]-Lactoglobulin aggregates in foam films: Correlation between foam films and foaming properties. *Journal of Colloid and Interface Science*, 336(2):750–755, 2009.

43. A. Saint-Jalmes, M. L. Peugeot, H. Ferraz, and D. Langevin. Differences between protein and surfactant foams: Microscopic properties, stability and coarsening. *Colloids and Surfaces A: Physicochemical and Engineering Aspects*, 263(1–3):219–225, 2005.

44. A. Sheludko. Certain peculiarities of foam lamellas, parts i-iii. *Proceedings of the Koninkl. Ned. Akad. Wetenschap. B*, 65:76–108, 1962.

45. A. Sheludko. Thin liquid films. *Advances in Colloid and Interface Science*, 1(4):391–464, 1967.

46. P. Stevenson. Inter-bubble gas diffusion in liquid foam. *Current Opinion in Colloid & Interface Science*, 15(5):374–381, 2010.

47. K. S. Suslick, M. W. Grinstaff, K. J. Kolbeck, and M. Wong. Characterization of sonochemically prepared proteinaceous microspheres. *Ultrasonics Sonochemistry*, 1(1):S65–S68, 1994.

48. S. Tcholakova, N. D. Denkov, and A. Lips. Comparison of solid particles, globular proteins and surfactants as emulsifiers. *Physical Chemistry Chemical Physics*, 10(12):1608–1627, 2008.

49. F. L. Tchuenbou-Magaia, N. Al-Rifai, N. E. M. Ishak, I. T. Norton, and P. W. Cox. Suspensions of air cells with cysteine-rich protein coats: Air-filled emulsions. *Journal of Cellular Plastics*, 47(3):217, 2011.

50. F. L. Tchuenbou-Magaia, I. T. Norton, and P. W. Cox. Hydrophobins stabilised air-filled emulsions for the food industry. *Food Hydrocolloids*, 23(7):1877–1885, 2009.

51. A. Vrij. Possible mechanism for the spontaneous rupture of thin, free liquid films. *Discussion of the Faraday Society*, 42:23–33, 1966.

52. C. Z. Wagner. Theorie der Alterung von Niederschl\"agen durch Uml\"osen, *Zeitschrift f\"ur Elektrochemie,* 65:581–594, 1961.

53. L. Wang. Surface forces in foam films. PhD thesis, Virginia Polytechnic Institute and State University, Blacksburg, VA, 2006.

54. L. Wang and R.-H. Yoon. Effect of ph and nacl concentration on the stability of surfactant-free foam films. *Langmuir*, 25(1):294–297, 2008.

55. W. C. Young Creighton. *Roark's Formulas for Stress and Strain*. McCraw-Hill, 1989.

56. R. A. Leonard and R. Lemlich. A study of interstitial liquid flow in foam. part i. Theoretical model and application to foam fractionation. *AIChE Journal*, 11(1):18–25, 1965.

57. P. A. Wierenga, M. R. Egmond, A. G. J. Voragen, and H. H. J. de Jongh. The adsorption and unfolding kinetics determines the folding state of proteins at the air-water interface and thereby the equation of state, Journal of colloid and interface science, 299(2): 850–857, 2006.

Index